ALLE · ZEIT · WACH
1842

Karl G. Grell

Protozoology

With 437 Figures and 15 Tables

Springer-Verlag
Berlin · Heidelberg · New York 1973

Professor Dr. Karl G. GRELL
Direktor des Zoologischen Instituts der Universität, D-7400 Tübingen,
Hölderlinstraße 12

ISBN 3-540-06239-4 Springer-Verlag Berlin · Heidelberg · New York
ISBN 0-387-06239-4 Springer-Verlag New York · Heidelberg · Berlin

Typesetting, printing and binding: Brühlsche Universitätsdruckerei, Gießen, Germany.

Preface

When I prepared the first German edition of this book in 1955, it was my intention to acquaint biologists in my country with the new and exciting results being obtained on the other side of the Atlantic Ocean (incl. the English Channel). In the meantime, especially after publication of the second German edition in 1968, Dr. Konrad F. Springer and many colleagues, too, suggested that I should prepare an English version. Though this was the exact opposite of my original intention, I finally agreed despite the risks involved.

Since 1968 our knowledge in Protozoology increased considerably. Though I tried to concentrate the text as much as possible, an enlargement of up to pages 554 was unavoidable. Many figures have been changed, replaced and added. Altogether their number increased from 422 to 437.

In my opinion, it is only a matter of time before the "true" protozoologists disappear. There will be cell biologists, biochemists, geneticists and others working with certain Protozoa, but very few who are interested in the group as a whole, their morphological and physiological diversity, their various types of reproduction and their relationships to other groups of organisms. Even at the present time, the Society of Protozoologists, comprising more than thousand members, consists for the most part of specialists who concentrate their efforts specifically upon *Chlamydomonas*, *Amoeba*, *Plasmodium*, *Tetrahymena* or some other protozoans.

Consequently, it is becoming more and more difficult for a "true" protozoologist to review the work all these specialists are doing. This, indeed, is the crucial problem in writing a book such as this. There is no doubt that chlamydomonadologists, amoebologists, plasmodiologists, tetrahymenologists and others will find some mistakes and maybe problematical formulations.

Nevertheless, I hope the book will help to convince more and more biologists that many Protozoa are suitable tools for solving problems of general biological interest.

Finally, I have to thank many people. First of all JOANNE RUTHMANN for translating the text of the second German edition, Dr. GERTRUD BENWITZ for preparing the book for the press and all those who studied parts of the manuscript and made valuable suggestions for improvements. Dr. HANS MACHEMER's help in updating the chapter on motility may be mentioned as an example. All collaborators and colleagues who gave permission to publish drawings, light or electron micrographs are mentioned in the legends. The new drawings were carried out by HEINER BAUSCHERT.

Last, but not least, I whish to thank Dr. Konrad F. Springer and his staff who put the book into an attractive form.

KARL G. GRELL

Contents

A. Introduction

Within the animal kingdom the *Protozoa* are considered to represent one sub-kingdom, while all multicellular animals (Metazoa) belong to the second sub-kingdom. In spite of this high taxonomic evaluation, Protozoa do not belong to those organisms which are generally known to the layman. Because of their *unicellularity*, they are generally very *small organisms* visible only with the aid of the microscope. Few Protozoa, like the extinct nummulites, ever reached a size of several centimeters.

In spite of their minute size, Protozoa play an important role *in the household of nature*. Flagellates which are capable of photosynthesis represent the basic link in the food chain of organisms. The calcareous shells and the silicon skeletons of planktonic Foraminifera and Radiolaria sink in a continual rain down to the bottom of the ocean. Whole geological deposits like limestone, green sandstone and fusuline chalk have been formed in this way and have contributed, in the course of geological time, to high mountain ranges. The shells of many foraminiferan species are also found in layers containing oil and are therefore used as indicators for the characterization of these layers.

Many Protozoa, as *parasites of man*, are of immediate importance to medicine. Although most of the 25 odd species found in man are not pathogenic, some are the causes of dangerous diseases such as amoebic dysentery, sleeping sickness and malaria.

Aside from these practical viewpoints, Protozoa play an equally important role in furthering our *insight into natural science* as such. Their value for scientific research is primarily due to the fact that many problems of scientific biology can be more easily studied with unicellular organisms than with multicellular systems. Protozoa offer a unique opportunity for biologists to study single cells easily and directly. Many species can be cultured with relatively little effort in special nutritive media under controlled conditions.

In many respects, they are therefore ideal objects for scientific research.

It must be stressed, however, that the number of protozoan species which have been used in scientific laboratories represents only a minute fraction of the *total number of species* described by taxonomists thus far. In a compilation from 1958 [*1029*], the number of known species is estimated at 44250. Of these, 20182 are fossils (mostly Foraminifera). Of the remaining species, 17293 are free-living and 6775 are parasitic. Actually, only a small fraction of existing species is catalogued in this estimate. In a 1948 monograph of peritrichous ciliates of the vicinity of Erlangen, Germany [*1224*], 62 new species were described. In the course of a 1954 field trip to the Grand Canyon, 13 new species of coccidians were discovered in 25 species of rodents. While the protozoans in fresh water and those parasitic in terrestrial species are relatively well known, the protozoan fauna of the seas is likely to be known only to a small extent.

The following pages can thus give only an incomplete impression of the variety of existing Protozoa and their biological relationships.

B. The Subject Nature

The year 1675 when ANTHONY VAN LEEUWENHOEK first discovered numerous "animalcules" in a drop of rain water can be considered the birth date of research in protozoology. After HUYGENS made similar observations in 1678, a period of lively discovery began which continued throughout the 18th century and resulted in the description of many species, some of which can still be recognized today. In 1765, WRISBERG coined the term "*Infusoria*" for all organisms found in a variety of infusions of vegetable matter.

While LEEUWENHOEK and HUYGENS assumed that Infusoria arose from "aerial germs", the opinions of researchers of the 18th century differed widely regarding their origins. Some thought them to be larval stages of insects. Most investigators, however, subscribed to BUFFON and NEEDHAM's (1750) theory of spontaneous generation, which held that Infusoria could arise at any time from decomposing matter. This theory was disproved as early as 1776 by SPALLANZANI, who showed that Infusoria would arise in previously heated flasks only if the latter had been exposed to the air. In those flasks which were closed right away by melting the glass necks, Infusoria would not develop, no matter how long they were standing. But SPALLANZANI's results could not quiet the defenders of the theory of spontaneous generation who were inspired by the ideas of natural philosophy prevalent at the turn of the 19th century. It was only as the reproduction of Infusoria was studied that SPALLANZANI was fully confirmed.

In 1817, GOLDFUSS called all those animals which he thought most primitive "*Protozoa*". They included some multicellular groups. A new impetus to research on microscopic organisms was given by EHRENBERG, who, although an adversary of the theory of spontaneous generation and of the speculations of natural philosophy, held to the prejudice that all animals must, in principle, have identical body plans. Thus, he thought he recognized an alimentary tract, gonads and various other organs in Infusoria. The title of his main work (1838) was „Die Infusionsthierchen als vollkommene Organismen" (The Infusoria as Complete Organisms). DUJARDIN (1841), on the other hand, maintained that Infusoria have a very primitive organization, consisting, in principle, of a "simple substance capable of movement" which he called "sarcode", later termed "protoplasm" by other scientists (PURKINJE, VON MOHL).

After SCHLEIDEN and SCHWANN had propagated the *cell theory* in 1838/39, a clarification of the concept Protozoa began which eventually replaced the purely methodological concept of Infusoria. In 1845 VON SIEBOLD proposed the first scientifically useful definition of Protozoa as "animals without clear-cut separation of the different organ systems, whose variable form and simple organization can be reduced to a single cell".

In a corresponding manner the unicellular plants were designated as "*Protophyta*". Since it was soon found that no sharp line could be drawn between unicellular animals and plants, HAECKEL (1866) proposed in his "Generelle Morphologie" to unite the Protophyta and Protozoa as "*Protista*" to represent a third and "neutral" kingdom between the plant and animal kingdoms referred to as Metaphyta and Metazoa, respectively. It was only in a later edition (1873) of his work, however, that he considered the ciliates to belong to the Protozoa.

Although HAECKEL's proposal met a widely recognized necessity, it could not prevail in the end, as further research showed that the gradation from unicellular phototrophic organisms to Algae is so continuous that the concept "Protophyta" became taxonomically useless. Instead, it became customary to classify all flagellates regardless of their "animal-like" or "plant-like" nutrition as Protozoa in zoological systematics and as primitive Algae in botanical taxonomy. Since we are probably still far from the ultimate goal of representing the natural relationships of all organisms in one system, it seems reasonable to maintain the concept of Protozoa which meets at least the requirements of zoological taxonomy.

In the century which has elapsed since HAECKEL's proposal, our knowledge of other and related forms of life has grown considerably. It is therefore no longer possible to simply define Protozoa as primitive unicellular organisms. Recourse must be taken to new concepts in order to maintain the usefulness of the concept of Protozoa.

We know today that the appearance of the first cells must have been preceded by a long period of evolution characterized by the presence of macromolecular organizations which were not yet cells. In the primeval seas, organic substances whose synthesis was once thought possible only in organisms must have been plentiful. Model experiments have shown that such substances could arise by electrical discharge, high temperature and ultraviolet irradiation in the primeval atmosphere. Since the latter contained no free oxygen as yet, they could not become decomposed by oxidation. It is not known if these precursors of living cells contained only protein as the essential macromolecular constituent at first or if they were from the beginning nucleoprotein complexes capable of autoreduplication and of incorporation of other organic compounds into their molecular framework.

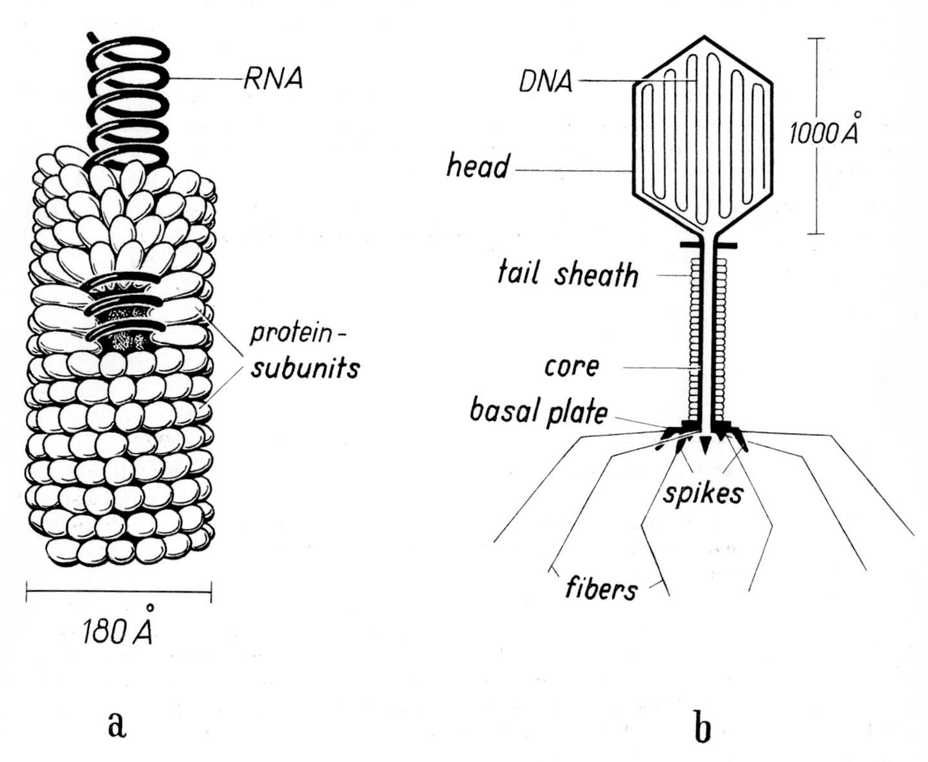

Fig. 1. Viruses. Organizational diagram: a of a tobacco mosaic virus particle (only partly depicted), b of a phage particle. After various authors

The so-called *viruses* are also nucleoprotein complexes. Since all viruses which have been studied thus far are only capable of multiplication as parasites within cells, the viruses cannot themselves be precursors of cellular organization. It is well known today that many diseases of man, his domestic animals and his cultivated plants are caused by viruses. Research on the causative agents of these diseases has long been hampered, however, by the fact that they cannot be made visible in the ordinary microscope. Only the electron microscope showed that they are corpuscular particles of definite size and form.

Two especially thoroughly studied virus particles are shown diagrammatically in Fig. 1. The *tobacco mosaic virus* (a), the causative agent of the mosaic disease of tobacco plants, is a hollow cylinder of about 2800 Å length and a diameter of about 180 Å. Attached to a helix of ribonucleic acid (RNA) like the grains of an ear of corn are more than 2000 identical protein-subunits. Each of these subunits consists of 158 amino acids whose sequence within the polypeptide chain is known. Much more complicated are *bacterial viruses* or *phages* (b) which can lead to a lysis of bacterial cells. Many are characterized by a head and a tail portion, both about 1000 Å in length. The phage attaches itself with the tail portion to the bacterial cell wall. Special structures (basal plate, spikes and fibers) which consist of many different proteins are involved in the establishment of contact. The outer sheath of the tail is contractile. The protein coating of the head portion surrounds a cavity within which a deoxyribonucleic acid (DNA) double helix of nearly 50 μm length is packed. Following attachment, this thread of DNA is "injected" through the central canal of the tail as through a syringe into the bacterium, while all other structures of the phage particle remain outside.

The RNA of the tobacco mosaic virus as well as the DNA of phage is capable of identical autoreduplication within the host cell and of building up complete virus particles from substances found in this cell. The nucleic acids must therefore contain all the genetic information which is required for this synthesis of virus material.

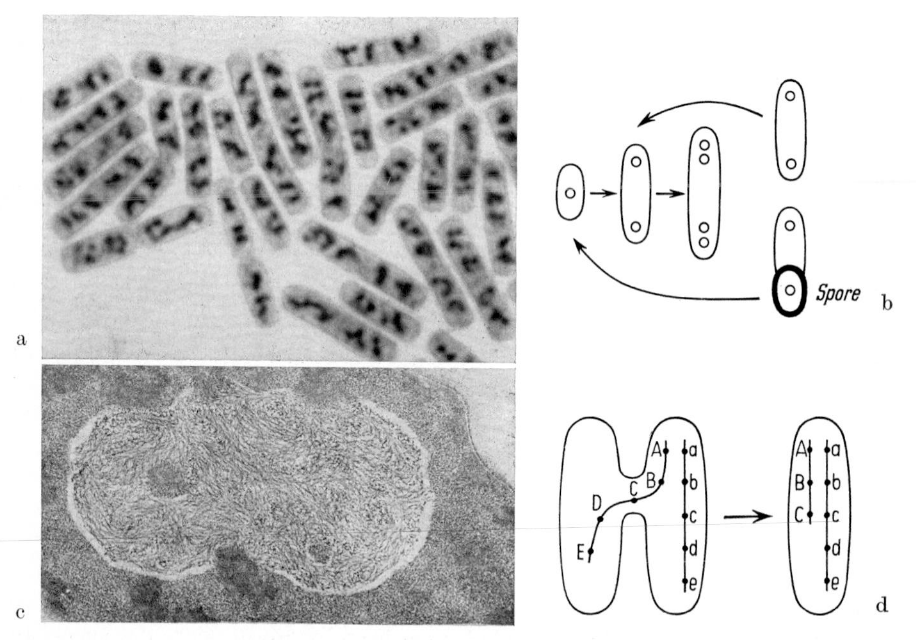

Fig. 2. Bacteria. a Cells of *Bacillus cereus* with nuclear equivalents (nucleoids). Osmium tetroxide, HCl-Giemsa. × 3000. Micrograph by C. ROBINOW. b Behavior of the nucleoid during growth, division and spore formation. After a diagram from PIEKARSKI, 1949. c *Staphylococcus aureus*. Thin section through the area of the nucleoid. × 53000. After KLEINSCHMIDT from D. JACHERTS and B. JACHERTS, 1964 [*833*]. d Diagram of "conjugation"

In contrast to viruses which have no metabolism of their own, but are wholly dependent upon the host cell, the *Bacteria* represent true cells whose components are subject to constant metabolic renewal. They are bounded by an external, complex cell wall whose inner component is a unit membrane (p. 14). Compared to Protozoa and Metazoa, however, the organization of the bacterial cell is much more simple. The ribosomes which are essential for protein synthesis are free within the cytoplasm. Mitochondria and Golgi complexes are not recognizable although structures may be present which fulfill their functions. Whenever flagella are present, they consist of a few filaments twisted about each other and not of the 9 + 2 pattern found in cross sections of the flagella of all other organisms (p. 290). Protoplasmic streaming does not seem to be present in Bacteria.

For a long time Bacteria were throught to lack a nucleus. Later on, it was possible to show that even they contain centers which can be compared with the nuclei of Protozoa and Metazoa (Fig. 2a). These structures can be stained by the Feulgen-reaction indicating the presence of deoxyribonucleic acid. Light microscope studies showed that they are only formed from each other by division and that the process of their duplication and distribution is clearly related to cell growth and spore formation (b). In most cases, Bacteria possess two or four such structures which have been called *nuclear equivalents* or *nucleoids*.

The nature of the nucleoid was better understood when it became possible to carry out *genetic experiments* with Bacteria [*734*, *834*]. The species *Escherichia coli* proved to be especially suitable for such studies. These experiments showed that all genes belong to a single linkage group which must be ring-shaped. The nucleoid of *Escherichia coli* is therefore not equivalent to a whole nucleus but to a single chromosome. It seems to consist essentially of a DNA double helix wound up to form a compact structure in a manner which has not been fully clarified as yet. In sections, the nucleoid shows a fibrillar composition. It is not separated from the cytoplasm by a special envelope (c).

The studies which led to these insights are based essentially on an analysis of genetic recombination. Gene recombination, which occurs in all higher organisms (see p. 78) at meiosis, takes place in Bacteria after a process called "*conjugation*" (d). In "*conjugation*", a protoplasmic bridge is formed between two cells. Whereas the union of two sexual cells of higher organisms leads to the formation of a zygote whose nucleus contains equal amounts of genetic material from both partners, the "conjugating" cells of Bacteria contribute very unequal portions of their genetic make-up to the formation of the common product. One of the cells functions as the donor, the other as the recipient. Once the ring chromosome of the donor cell has opened at one point, it glides via the protoplasmic bridge into the recipient. Only rarely is a whole chromosome transferred in this way. Usually, it is only a more or less short chromosome segment which gets into the recipient cell. This piece can now exchange genes with the intact chromosome of the recipient. "Conjugation" in Bacteria, therefore, does not lead to the formation of a zygote in the usual sense because the alleles to a large number of genes in the chromosome of the recipient cell are missing (p. 232). It is for this reason that the term "*merozygote*" has been used. The process which leads to the formation of a merozygote has been called "*meromyxis*".

A cell nucleus is also absent in the photosynthesizing *Cyanophycea*, formerly classified as blue-green algae. Their genetic information is stored in a few, possibly two, identical DNA-strands. Though recombination indicating sexuality has been reported [*104*], genetic analysis is still in its beginning [*71*]. Since the Cyanophycea also show similarities with the Bacteria in the organization and chemical composition of their cell walls, both groups of unicellular organisms are thought to be related and are classed together as *prokaryotes*.

All other organisms have true nuclei and are therefore called *eukaryotes*. We do not know whether they descended from prokaryotes or whether both groups originated independently from a stock of primordial unicellular organisms. It is certain, however, that the Protozoa represent the most primitive eukaryotes and that they, in turn, must have given rise to all multicellular plants and animals in the course of evolution.

On the other hand, it is difficult to draw the line between unicellular and multicellular forms. Although unicellularity is certainly the dominating feature of Protozoa, multicellular colonies are so frequently encountered that unicellularity as such is not sufficient to define the group.

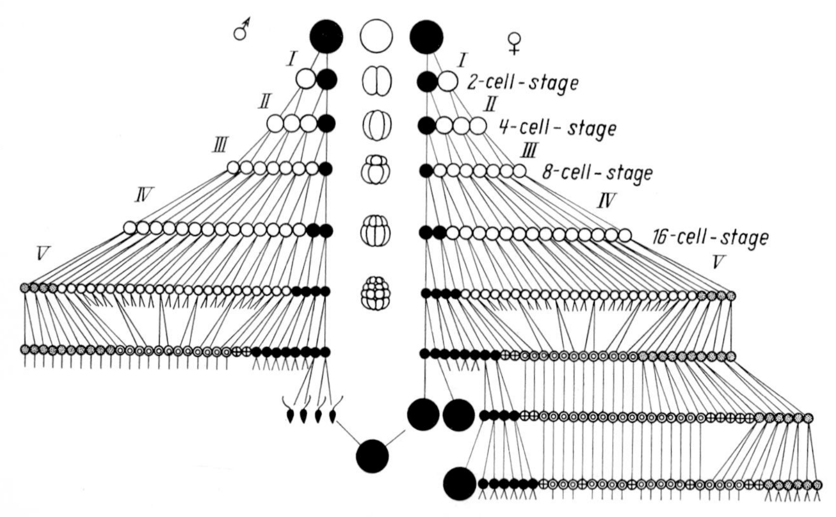

Fig. 3. Diagram of cell lineage in the development of a multicellular animal from the fertilized egg cell. At the right, the female, at the left, the male individual. Germ-line black, somatic cells light. Among these, various differentiations are marked: a cross indicates cells which die at a certain stage; a circle, cells which, without further divisions, go through several consecutive stages; points indicate cells which, although still capable of division, supply only certain tissue cells. In accordance with BELAR from KÜHN, 1965 [978]

In order to appreciate the differences between *Metazoa* and protozoan colonies, it is best to start with a review of metazoan development (Fig. 3). Characteristic for the latter is the fact that the capacity for unlimited reproduction is restricted in most cases to special cells connected in the time sequence of cell divisions during development by the so-called *germ line*. These *generative cells* are distinct from the multitude of body cells which, though ultimately descended from the germ line, are sooner or later differentiated for special functions. In their totality, they form the bulk of the metazoan body, or, to use the Greek word, the *soma*, and are therefore called *somatic cells*.

While the somatic cells are destroyed with the death of the individual at the natural termination of the life cycle *(death due to age)*, the generative cells can form the starting point of a new generation.

Like the generative cells of Metazoa, most Protozoa have an unlimited capacity for reproduction under suitable conditions. Therefore, the phrase "*potential immortality of Protozoa*" has been used. An amoeba, which divides into two, ceases to

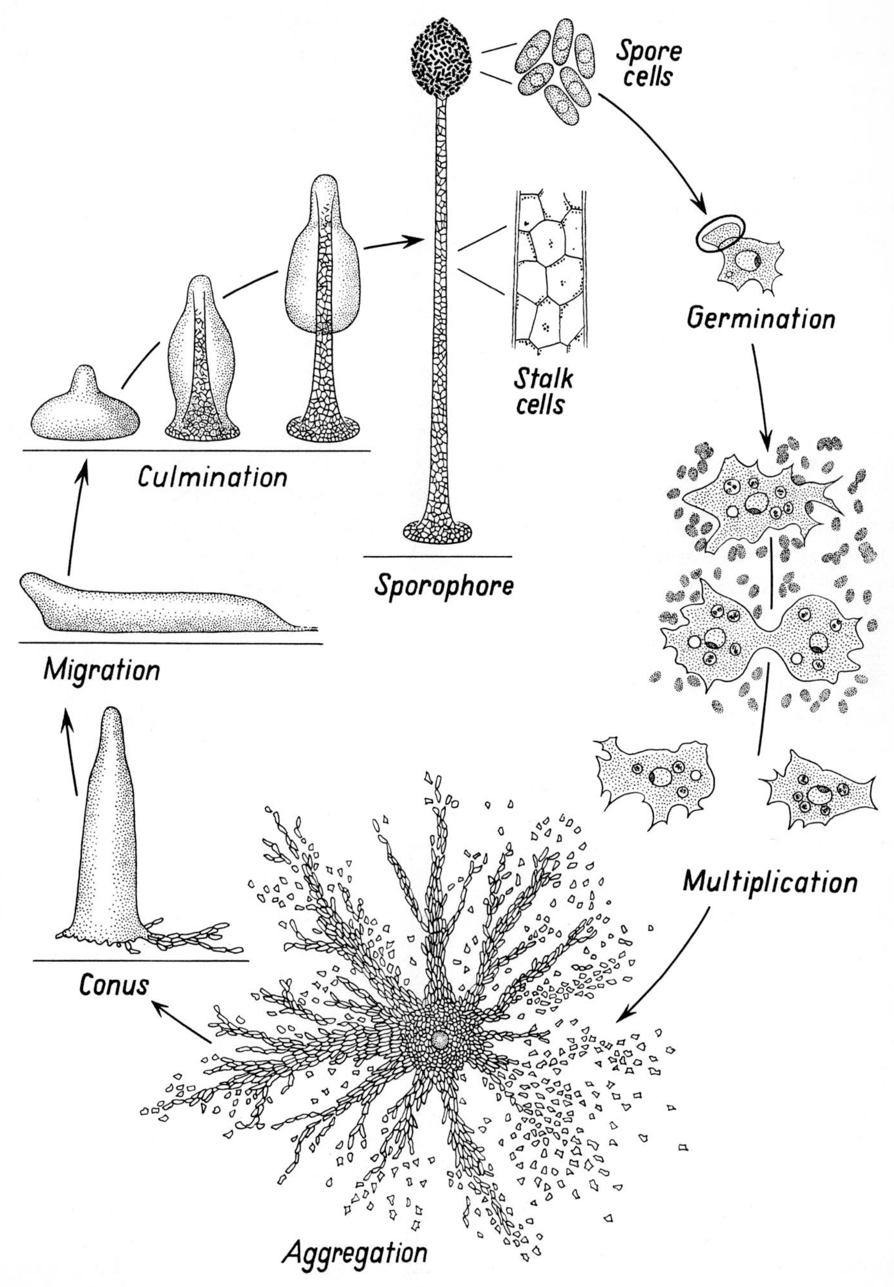

Fig. 4. *Dictyostelium discoideum* (collective amoeba). Diagram of the developmental cycle. In the right half of the picture the cells are depicted at significantly higher magnification than at the left and in the middle. After GERISCH, 1964 [*574*]

exist as an individual, but the whole of its living substance is handed on to its daughter cells. Unlimited multiplication is only checked by continuous destruction of individuals through external influences *(death due to catastrophy)*.

On the other hand, it is just among the amoebae that we find an example of differentiation between germ line and soma in Protozoa. Some species, which have been combined as "*collective amoebae*" *(Acrasina)*, are able to form multicellular sporophores of definite size and form. Within the spores, which correspond to the cysts of other amoebae, they are able to endure periods of starvation and drought. One of the species studied most is *Dictyostelium discoideum* (Fig. 4). When the spores encounter a suitable substrate, e.g. an area covered by bacteria, they germinate. Once emerged, the amoebae can multiply indefinitely. Their solitary phase will only come to an end when the food supply is exhausted. Then a process called *aggregation* sets in. Without fusing, the amoebae form aggregations and move in radially directed strands toward a common center to form a vertical *conus*. This complex, which may contain several thousands of amoebae, is also called *pseudoplasmodium*. It tips over eventually and can crawl about for some time as a single unit *(migratory phase)*. At rest again, it rises up *(culmination)* and transforms into a *sporophore*. This consists of a stalk with a broad base and the terminal mass of spores. Only some of the amoebae, the so-called *spore cells*, are transformed into spores. They become enclosed by a firm envelope and are characterized by an especially dense cytoplasm. The remaining amoebae become *stalk cells*. They are highly vacuolized and keep the stalk upright by their turgidity. Since only the spore cells assure the continuity of the species, they can be called generative cells. The stalk cells have lost their capacity for reproduction. They die after dissemination of the spores and are the somatic cells.

Whereas in "collective amoebae" the individuals form a cell aggregate after a multiplication phase, we find in other Protozoa that the daughter cells formed by fission, instead of separating, remain united to form a *cell colony*. In the most simple case, this is nothing but an accumulation of loosely connected individual cells or a chain of cells. However, due to the fixation of a definite number of fissions, colonies with a certain size and regular arrangement of cells are formed, surrounded by a common coating and of a characteristic shape.

In the flagellate order of *Phytomonadina*, colony formation is especially extensive in the family *Volvocidae*. In most cases, the individual cells lie within a common gelatine and are arranged in a way characteristic for the various species. The spherical cell colonies are clearly polarized. They always move with the same pole pointing forward and rotate clockwise about a definite axis as seen from the anterior pole. There is also a directed orientation toward the light (p. 315).

In the genus *Eudorina* (syn. *Pleodorina*) the species can be arranged in a series with increasing polarity of the cell colonies (Fig. 5). Although all cells of *E. elegans* are capable of reproduction, the four cells at the anterior pole frequently carry out one fission less than the other cells. In *E. illinoisensis* the cells of the anterior tier are smaller than the others and have usually lost the ability to reproduce. As purely somatic cells, they are different from the generative cells of the colony whose capacity for reproduction by fission is not limited. In *E. californica*, whose colonies may contain 128 cells, the ratio between somatic and generative cells is

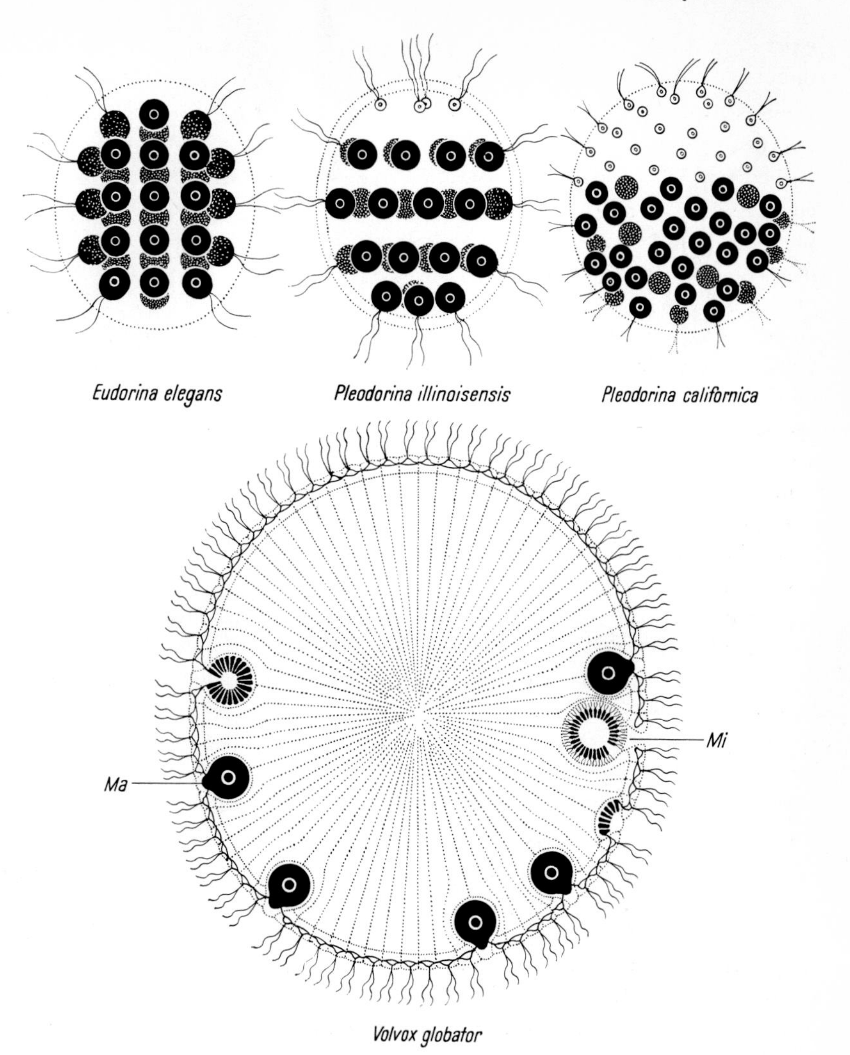

Fig. 5. Steps in the differentiation of generative (dark) and somatic cells (light) of colonial phytomonads (Volvocidae). *Ma* Macrogamete. *Mi* Microgametes. After various authors

about 3:5. Since most somatic cells are further apart than the generative ones, they occupy the entire anterior half of the colony. It could be shown in this species that only the somatic cells have an eyespot (stigma) and are capable of reacting to light (p. 315). The climax of "somatization" is reached in *Volvox globator:* almost all of its about 10000 cells are somatic. Only in the posterior half of the colony are some scattered generative cells which give rise, during asexual reproduction, to daughter colonies and to the sex cells in sexual reproduction.

In ciliates, colony formation is most frequently found among the *Peritricha*. Usually, all cells of a colony are identical. All can separate from the colony and, by repeated fission, give rise to a new colony. It is only in species of *Zoothamnium* that two different cell types are found, numerous small *microzooides* and large

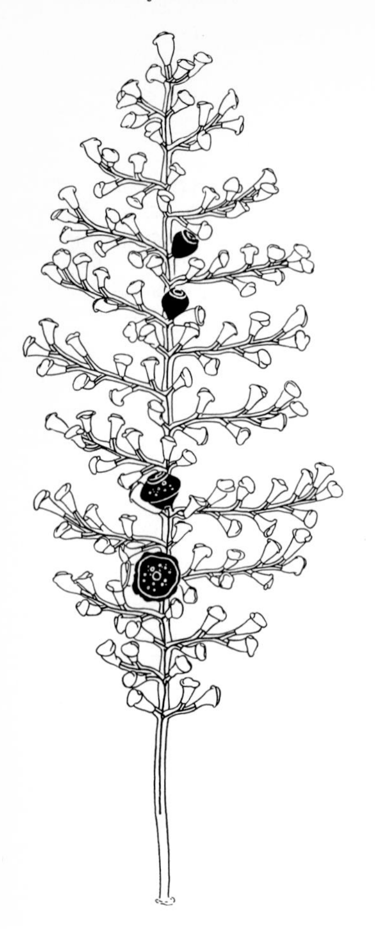

Fig. 6. *Zoothamnium alternans* (Ciliata). Differentiation into microzooides (light) and macrozooides (black). Only those macrozooides are emphasized which because of their size can easily be distinguished from the microzooides. Modified after SUMMERS, 1938 [*1685*]

macrozooides arranged singly at the points of ramification (Fig. 6). Although the microzooides are still capable of division, they cannot detach themselves and form a new colony. Colony formation is due to the macrozooides only.

All species discussed above show differentiation between somatic and generative elements at the cellular level. In other Protozoa, a corresponding differentiation is found between the *cell nuclei*. Heterokaryotic Foraminifera and the majority of ciliates (p. 96ff.) have generative nuclei capable of unlimited reproduction as well as somatic nuclei which perish sooner or later to be re-formed by descendants of the generative nuclei.

A differentiation of germ line and soma can therefore also occur in Protozoa, among the cells of a colony as well as between the nuclei of a single, multinuclear cell. It is not a distinguishing characteristic of the metazoan body plan.

However, it is characteristic of Metazoa in general that there is not only a differentiation of germ line and soma, but also a further *differentiation among the cells of the somatic lines* (Fig. 3). The cells of the various tissues and organs become differentiated and specialized for different functions. Their very structure is frequently indicative of the special function they fulfill within the organism.

If we try to define the Protozoa as eukaryotic unicellular organisms or cell colonies not endorsed with somatic differentiation, we are heading for another difficulty. The spores of *Myxosporidia*, a group of parasites usually combined with Actinomyxidia and Microsporidia in a special class of Protozoa ("Cnidosporidia"), are the result of a divisional process which is – at least to a certain degree – connected with somatic differentiation (Fig. 433). It might be that the Myxosporidia derive from metazoan ancestors, their relatively simple organization being due to parasitism. But, since there are no tissues, it is equally impossible to regard them as Metazoa.

On the other hand, it should not be overlooked that the concept of a "germ line" in the sense of a fixed lineage between the germ cells of successive generations cannot be maintained in many cases. Generative cells may arise from somatic cells by transformation or from primordial cells which – under other conditions – would differentiate into somatic cells.

Hence, it is difficult to draw a clear borderline between Protozoa and Metazoa if one approaches the problem from the protozoological viewpoint only. As already pointed out by HAECKEL in his studies on the Gastraea-theory (1874), there is, however, one criterion which unites all Metazoa with the exception of the parasitic "Mesozoa": the possession of two different epithelia called ectoderm and endoderm by embryologists. Even *Trichoplax adhaerens*, probably the most primitive metazoon of all [*676*, *681*], has a dorsal epithelium which may correspond to the ectoderm and a ventral epithelium comparable to the endoderm. Each cell of both epithelia has one flagellum, an indication that *Trichoplax adhaerens* is derived from a colonial heterotrophic flagellate.

C. Morphology

Aside from the special cases mentioned in the preceding chapter, the individual protozoan cell represents a whole organism which must perform all the functions necessary for the survival of the species. Certain functions are often carried out by special cell areas which form a structural entity and are called *organelles* in analogy to the organs of the Metazoa. Thus, the individual cells of the Protozoa generally reach a *higher degree of differentiation* than metazoan tissue cells which are only required to fulfill a part of the functions in the entire mechanism of the organism.

Like all animal and plant cells, the Protozoa also consist of *cytoplasm*, separated from the surrounding medium by a special *cell envelope*, and the *cell nucleus*.

I. The Cytoplasm

The fine structure of the cytoplasm in Protozoa and multicellular organisms is identical in its principle features (Fig. 7). Embedded within the ground cytoplasm are various structures which, due to their arrangement and differentiation, bestow upon the cell its characteristic appearance.

1. The Ground Cytoplasm

In the electron microscope, the ground cytoplasm (hyaloplasm) appears to be more or less homogeneous. The fact that it sometimes shows a granular, reticular or fibrillar texture depends to a certain extent upon the method and quality of fixation. Although the ground cytoplasm is defined as the structureless matrix of the cytoplasm, it cannot be thought of as only a mixture of organic and inorganic substances in a watery phase. Certain properties suggest that it has a *submicroscopic structure* [*547, 959*]. This is possible because the protein molecules, which play the major role in its make-up, are linked to each other through special binding forces (so-called "weak bonds") at certain attachment points („Haftpunkte") thereby forming a three-dimensional *molecular framework* whose interstices store the remaining materials (especially water). The protein molecules may themselves be fibrous, or fibers may be formed through the linear aggregation of globular molecules. However, the cohesion of this molecular framework is only a loose one. The linkages can easily be disrupted and are just as easily re-formed. In addition to regions in which the submicroscopic structure imparts a certain amount of rigidity, there are others of a nearly liquid consistency.

It appears that protein molecules can easily be directed parallel to each other to form submicroscopic fibrils. Contractility, as far as it is a property of the ground cytoplasm, is probably always based on submicroscopic fibrils. In some amoebae such fibrils can be seen in the electron microscope and must therefore be counted among the "structures" of the cytoplasm. However, although there is some indirect evidence of their contractility (p. 287), there is no direct proof of it.

In many Protozoa a distinction between the outer *ectoplasm* and the inner *endoplasm* can be made. The ectoplasm of amoebae is more gel-like, the endoplasm

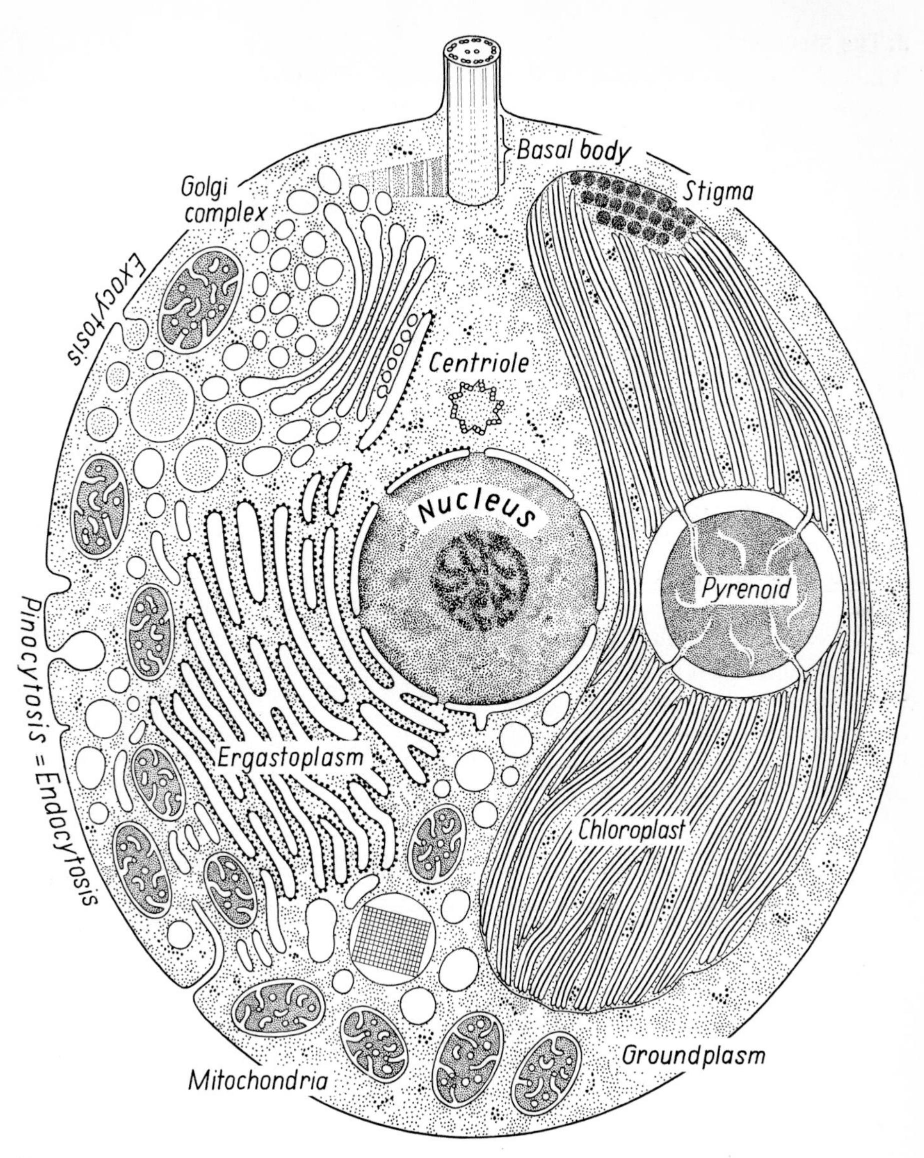

Fig. 7. Diagram of the fine structure of a cell

more sol-like in its consistency. In the electron microscope, the change from one layer to another appears to be gradual. Structures are, however, predominantly found in the endoplasm. How this change of consistency of the ground cytoplasm is related to the localization of the structures is still an open question.

There are also amoebae without differentiation of ecto- and endoplasm. Due to their homogeneous cytoplasmic organization, even cell nuclei lie directly below the cellular envelope.

2. The Structures

Some of the structures seen in electron micrographs of cells seem to originate directly from the ground cytoplasm. Others are formed from pre-existing structures or arise by division from them. Although not much is known as yet about this *structural turn-over* within the cell, it should be kept in mind that an electron micrograph can only display the momentary state. Only the comparison of many such momentary states permits insight into the typical architecture of this particular type of cell.

Aside from structures which are clearly basic to the life process as such, the cytoplasm contains *reserve substances* such as lipid droplets, starch and paramylon granules which can be utilized at a later date, and *end products of metabolism* which are voided to the outside by defecation.

Membranes

An essential component of many cytoplasmic structures are *membranes*. Biochemical and electron microscopic studies led to the concept of the *unit membrane* as the basic element of many biological membranes [*959*, *1441*, *1426*]. A hypothetical diagram of its molecular composition is shown in a simplified form in Fig. 8. The inner layer is a bimolecular leaflet of lipid molecules (L) pointing toward each other with their hydrophobic ends. This is bounded at both surfaces by a layer of protein molecules (P) which are connected to the hydrophilic poles of the lipid molecules. As a diagram of electron microscopic thin sections shows (b), the ultrastructural appearance agrees with this concept in many respects: there are two outer layers of higher contrast and an inner layer, which may correspond to the lipid component.

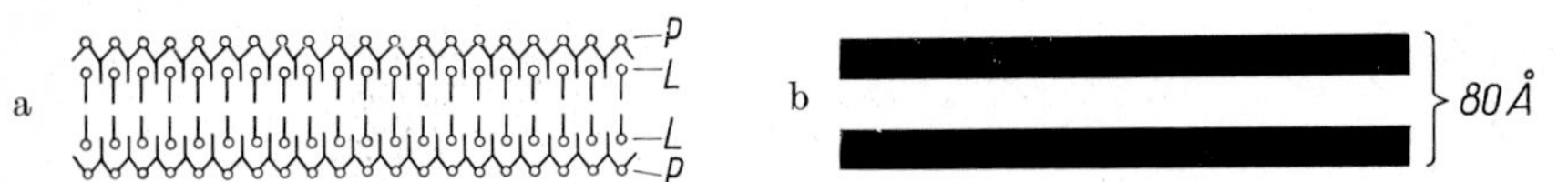

Fig. 8. Diagram of a unit membrane. a Molecular arrangement (hypothetical). P Protein layer, L Lipid layer. b Electron micrograph

It should be mentioned, however, that certain membrane phenomena cannot be adequately explained by the "unit membrane" concept. Therefore, several alternatives have been proposed. Most of these alternatives maintain that the molecular components are arranged in subunits which allow the possibility of different patterns of association, even within the same functionally differentiated membrane. Hence, the term "unit membrane" is used in this context in the sense of its ultrastructural appearance only.

Membranes are found within the cytoplasm as the outer borders of *vesicles* (Fig. 9). In some Protozoa, especially in amoebae, vesicles are so abundant that they predominate in the overall appearance of the cytoplasm. Many of these membranes are smooth; the outer surfaces of others are covered with granules. Occasionally, the granular membranes are concentrated in a certain region of the cytoplasm, the *ergastoplasm*. Together with their enclosed cavities, the membranes form sac-like structures which are often arranged parallel to each other. These structures are referred to as *endoplasmatic cisternae*. They can be connected to each other and

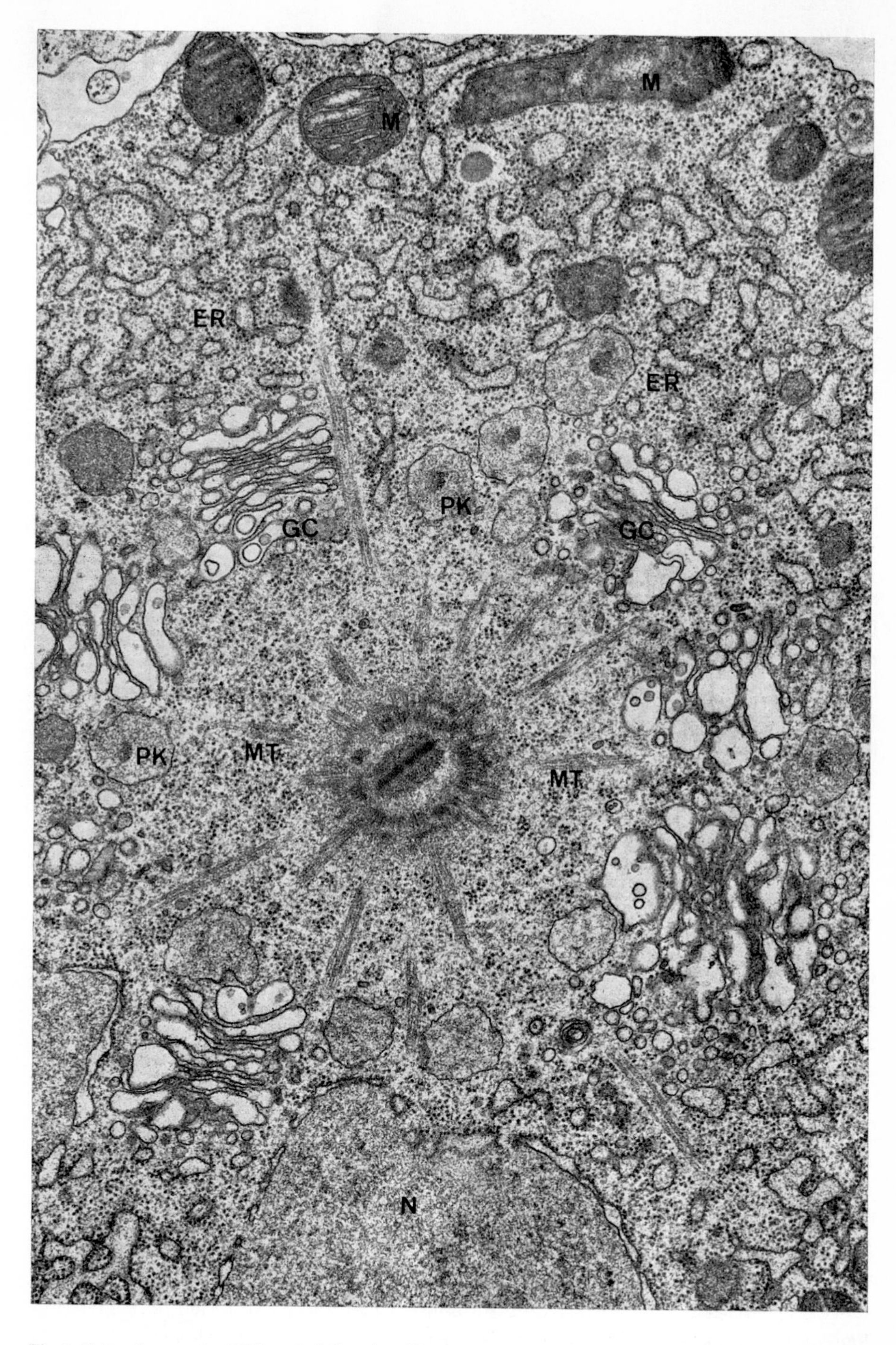

Fig. 9. *Heterophrys marina* (Heliozoa). Cell center with central granule (centroplast = nucleating site for the microtubules of axopodia: *MT*), surrounded by a sphere of Golgi complexes (*GC*) with prekinetocysts (*PK*) in their neighborhood. *ER* Endoplasmic reticulum with ribosomes. M Mitochondria. N Nucleus. × 26000. Courtesy of C. F. Bardele

thus form a more or less continuous network of canals and lacunae, the so-called *endoplasmic reticulum*. The latter probably plays a part in the intracellular transport of substances.

Ribosomes

The granules on the outside of the membranes have a diameter of 120–150 Å and consist of protein and ribonucleic acid. They are therefore called *ribosomes*. The high affinity of the ergastoplasm for basic dye stuffs ("basophilia") depends to a large extent upon the RNA content of the ribosomes. The precise function of ribosomal RNA has not been clarified as yet. The role of the ribosome as such, however, is well known today. Experiments with cell-free extracts have shown that *proteins are synthesized at the ribosomes*. Messenger RNA, which transmits the genetic information from the nucleus into the cytoplasm becomes attached to the ribosomes. With the aid of molecules of transfer RNA which are specific for the various amino acids, the polypeptide chains of the protein are then synthesized at the ribosomes according to the genetic information.

Ribosomes can also occur free within the ground cytoplasm. Frequently they are arranged in clusters ("polysomes"). This arrangement is due to the connection of the individual ribosomes by the same strand of messenger RNA. At every ribosome of the chain, molecules of the same protein are formed in succession.

As far as is known, conspicuous accumulations of granular membranes to form an ergastoplasm are exceptional among Protozoa. Usually, the granular membranes are irregularly scattered in the cytoplasm.

Golgi complexes

Golgi complexes (dictyosomes) are piles of membranous sacs whose surfaces are always free of ribosomes. At the outer margins of the Golgi complexes the membranous sacs are usually inflated. These regions are pinched off as vesicles. They are replaced by growth of the membranes (Fig. 7).

A definite *polarity* is frequently recognizable: The Golgi complex has a "proximal" surface where the flattened sacs are stacked very close together and a "distal" surface where the widened sacs are lost as vesicles. In some Protozoa, e.g. in phytomonads (Fig. 15), the "proximal" surface is always directed toward the nuclear envelope [*872*, *873*].

Occasionally an endoplasmatic cisterna is observed in close apposition to the Golgi complex. Though both membrane systems are not directly connected, there are numerous vesicles between them which might serve in the transport of materials from the cisterna to the membranous sacs of the Golgi complex [*82*, *506*, *1775*].

With the exception of some ciliates, Golgi complexes have been found in all Protozoa whose fine structure has been studied in detail. Their number and size, however, varies considerably. Some amoebae are good examples of this: while *Stereomyxa angulosa* (Fig. 10) contains many small Golgi complexes, *Paramoeba eilhardi* (Fig. 307) has only two, though of a large size. These are usually located close to the nucleus and can be recognized in vivo by phase contrast [*670*, *678*].

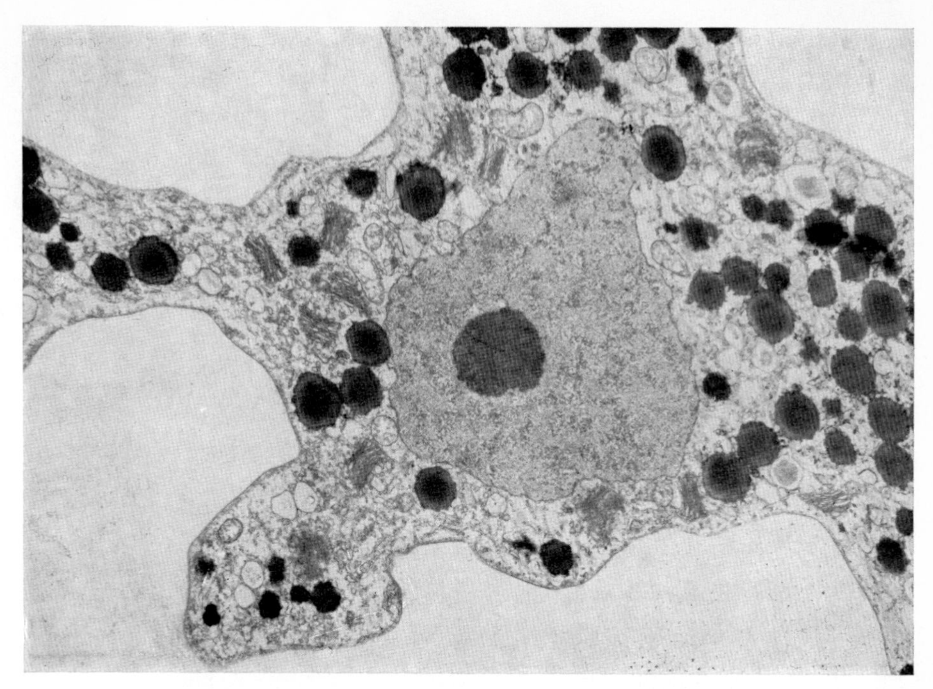

Fig. 10. *Stereomyxa angulosa* (amoeba). Section through the central area of the cell with the cell nucleus and numerous Golgi complexes. × 10000. Courtesy of G. BENWITZ

The *origin* of Golgi complexes is not completely clarified. Some observations indicate that they can divide by simply breaking in two halves near the middle. Evidently, their formation is under nuclear control. After enucleation of *Amoeba proteus*, the Golgi complexes disappear. When supplied with a transplanted nucleus from another amoeba, Golgi complexes regenerate [*523*, *524*]. Cells of *Chlamydomonas* whose nuclei were polyploidized by colchicine have more Golgi complexes than normal [*1426*].

Also, little is known concerning their *function*. It has been shown in some cases that they can take part in the formation of secretions, especially of polysaccharides which are transported within the vesicles to the surface of the cell. The latter can be mucoid substances as well as formed secretions which are either incorporated into the cell envelope or added to its outside (p. 41).

In addition to this function, the Golgi complexes seem to be concerned with production of, or perhaps packaging of hydrolytic enzymes into special vesicles, the so-called *lysosomes* (p. 344).

Parabasal bodies

The characteristic *parabasal bodies* of the Polymastigina seem to correspond to Golgi complexes. However, their structure is more complicated and they always occupy a definite position within the cell. They are rod- or sausage-shaped bodies which originate at the front end of the cell (Fig. 11 *pb*). Trichomonads have only a single, though occasionally branched, parabasal body which is anchored at the

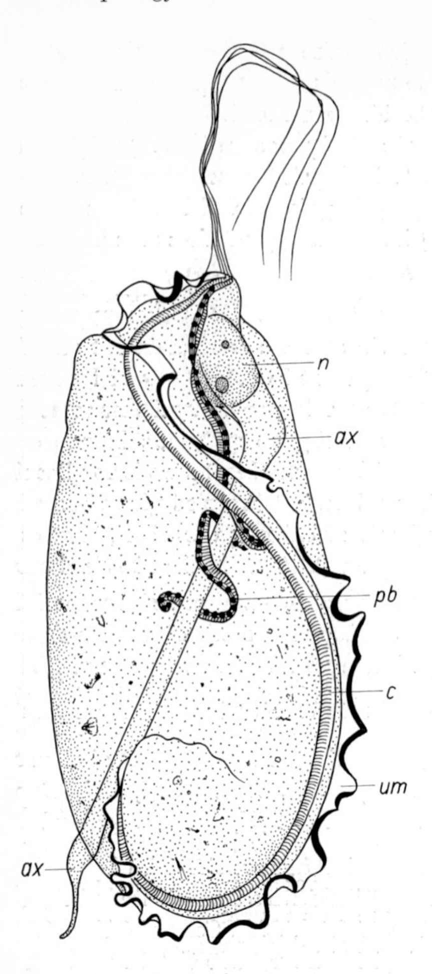

Fig. 11. *Trichomonas termopsidis* (Polymastigina). Diagram of cellular organization. *n* Nucleus. *ax* Axostyle. *pb* Parabasal body. *c* Costa. *um* Undulating membrane. After HOLLANDE and VALENTIN, 1968 [*798*]

base of the flagella and can either wind about an axostyle or follow a straight course to the posterior end (Fig. 21 a, b). Only *Mixotricha paradoxa*, an especially large trichomonad, has no parabasal body but several Golgi complexes scattered in its cytoplasm instead. Hypermastigids, on the other hand, have numerous parabasal bodies grouped in a circle around the nucleus (Fig. 21 c—e).

A cross-striated protein fiber, the parabasal filament, represents a continuous structural element of each parabasal body. Cross sections through the parabasal bodies of *Trichonympha* (Fig. 12) show that the filaments (arrows) are always directed towards the nuclear envelope. Each filament is connected with a stack of stretched out membranous sacs which look like cross sections of Golgi complexes. At the surface opposite the filament, vesicles are pinched off and distributed throughout the cytoplasm.

▶

Fig. 12. *Trichonympha* (Polymastigina). Bottom left: survey picture. The electron micrograph shows a cross section at the level of the cell nucleus and the parabasal bodies (*pb*), whose filaments (arrows) are oriented toward the nuclear envelope. A zone of granulated membranes (*gm*), which also occur sporadically in the cytoplasm, is found next to the nuclear envelope. Magnification of the electron micrograph × 20700. After GRIMSTONE, 1959 [*695*]

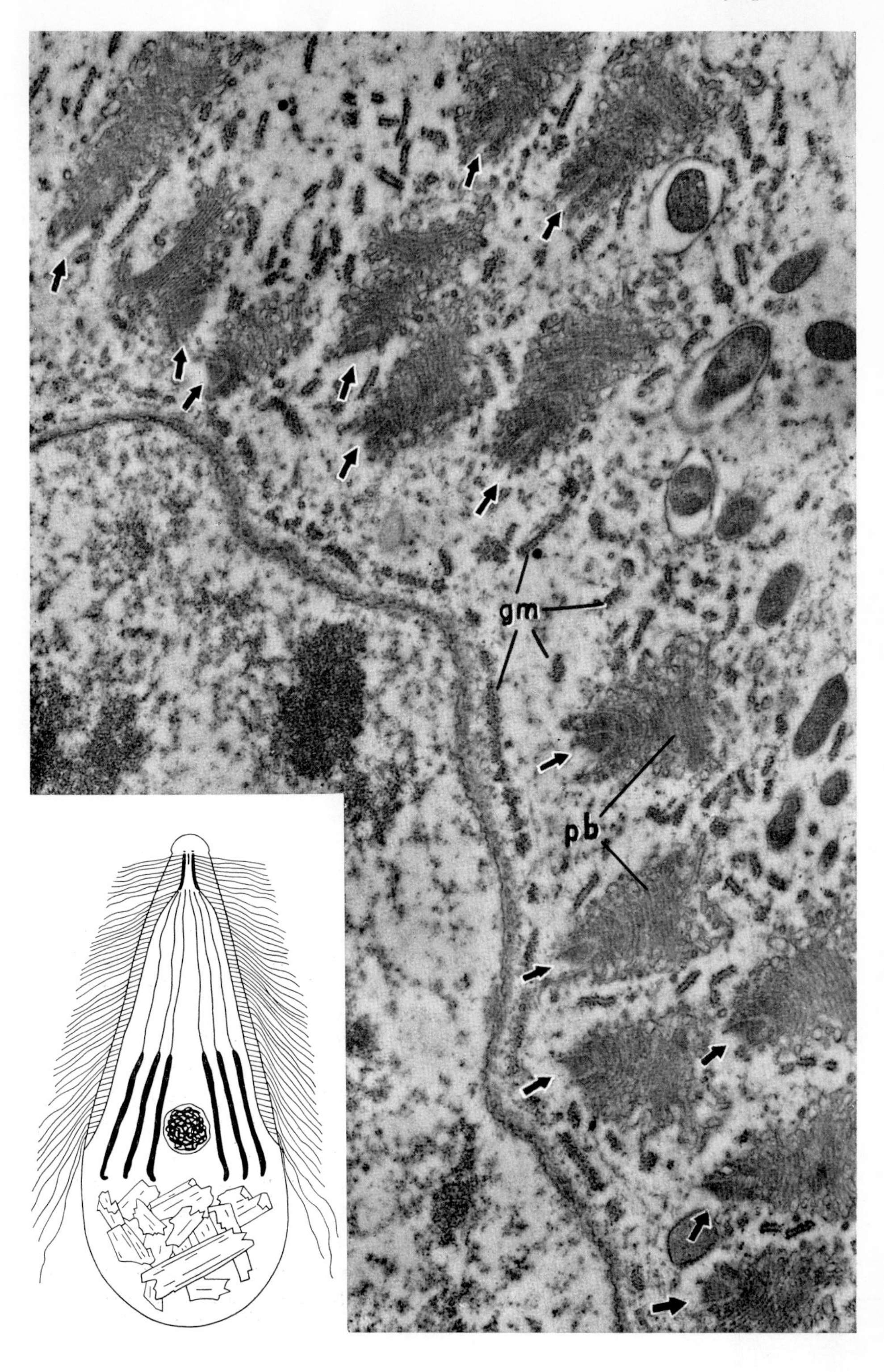
gm
pb

Cytochemical studies have shown that the vesicles contain polysaccharides. They seem to be transported to the posterior end of the cell and to be utilized in the renewal of the pellicle. As symbionts living in the gut of termites, the hypermastigids (p. 388) take up pieces of wood at the posterior end and must therefore constantly regenerate their pellicle. This interpretation is supported by starvation experiments: with decreasing phagocytosis, the parabasal bodies show at first only widened sacs and finally none at all [*695*].

Mitochondria

While the membrane systems discussed above are in a steady state of loss and renewal in the course of their physiological activity, the *mitochondria* are of a greater stability. As granular or filamentous structures of variable size they can be carried about by cytoplasmic cyclosis. Occasionally, they are accumulated locally or show a fixed relationship to other structures. While mitochondria could formerly only be characterized by their special affinity to certain dyes (e.g. Janus green B), it is now possible to clearly distinguish them from other organelles on the basis of their ultrastructure.

Mitochondria are bounded by two membranes. The inner membrane forms *invaginations* into the interior of the mitochondrion, which is filled with a ground substance termed *matrix*, or *chondrioplasm*. The shape and arrangement of the invaginations and their relationship to the matrix is highly variable in different cells. In the majority of Protozoa, the infoldings are *tubules* which form a dense tangle in the mitochondrial interior (Figs. 7, 23, 25). In others, sheet- or sac-like extensions *(cristae, sacculi)* are present which occupy less space and leave a rather voluminous matrix (Fig. 13). Apart from individual differences in size and form, the various structures are specific for the particular species. Mitochondria with cristae are characteristic for euglenoids but are also found among other flagellates *(Pedinomonas, Cercomonas)*. In amoebae of the genus *Pelomyxa*, the tubuli can take a zig-zag course and are somewhat swollen at the angles [*167*, *1270*]. Occasionally, the matrix is traversed by a bundle of dense fibers [*506*].

That mitochondria multiply by fission is shown not only by observations with the light microscope but also by labeling with radioactive substances which are incorporated by the mitochondria [*1291*]. The presence of a special self-replicating mitochondrial DNA, distinct from nuclear DNA [*1679*, *1690*] and all other components for protein synthesis (messenger-RNA, transfer-RNA, ribosomes) does not mean, however, that the autoreduplication of mitochondria is completely autonomous. As will be discussed later (p. 265), synthesis of most proteins is under nuclear control. It is reasonable, therefore, to speak of a "semi-autonomous" autoreduplication.

The large increase in surface area expressed in the fine structure of mitochondria becomes intelligible from their *function*. Studies with cell-free extracts show that they contain many enzymes, especially those of the citric acid cycle and of the respiratory chain. The energy gained by oxidation is stored in adenosine triphosphate (ATP) which is synthesized in the mitochondria and utilized for energy-consuming reactions. They have therefore been called the "power stations of the cell". Many enzymes, notably those of the respiratory chain and the enzymatic

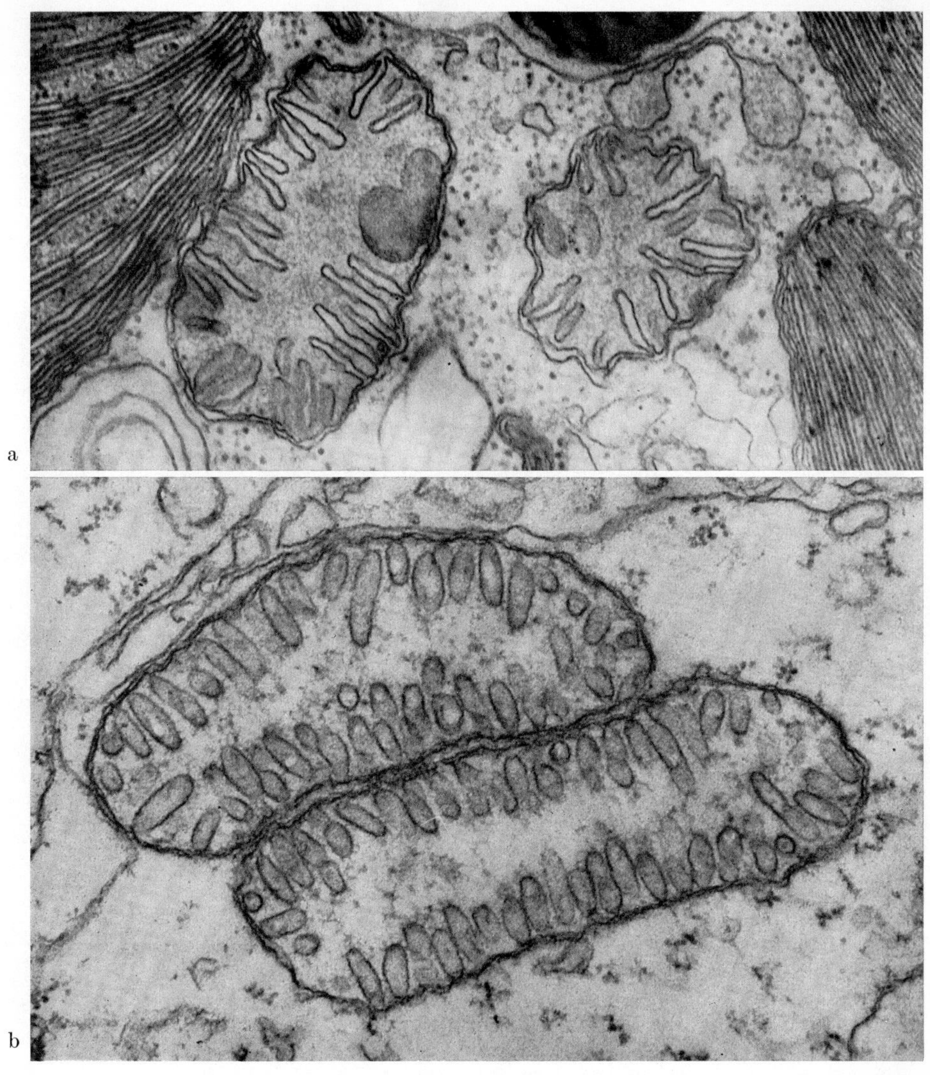

Fig. 13. Mitochondria. a *Euglena spirogyra* (cristae type). On the left and right margins of the picture are sectioned chloroplasts. × 47000. After LEEDALE, MEEUSE, and PRINGSHEIM, 1965 [*1016*]. b *Oxyrrhis marina* (sacculi type). × 60000. Courtesy of G. SCHWALBACH

machinery for ATP-synthesis, are localized at the mitochondrial membranes, probably sequentially arranged corresponding to the successive reactions catalyzed by them.

Anaerobic Protozoa whose ATP store is not due to oxidative reactions (flagellates of termites, species of *Entamoeba*, *Pelomyxa palustris* and rumen ciliates) may lack mitochondria. On the other hand, they are concentrated in cellular regions of high energy consumption, e.g. in the vicinity of basal bodies and of pulsating vacuoles.

Kinetoplast

The characteristic *kinetoplasts* (blepharoplasts) of the flagellate families Bodonidae and Trypanosomidae may be regarded as a special type of mitochondria [*1154*, *1602*, *1738*]. They are, however, larger than most mitochondria. There is only one kinetoplast per cell, always located right behind the basal body of the flagellum. Undoubtedly, they are autoreduplicating bodies. Their division precedes that of the cell nucleus (Fig. 115). That they contain and synthesize DNA has been known for a long time and was proved by various means [*459*, *1664*, *1669*].

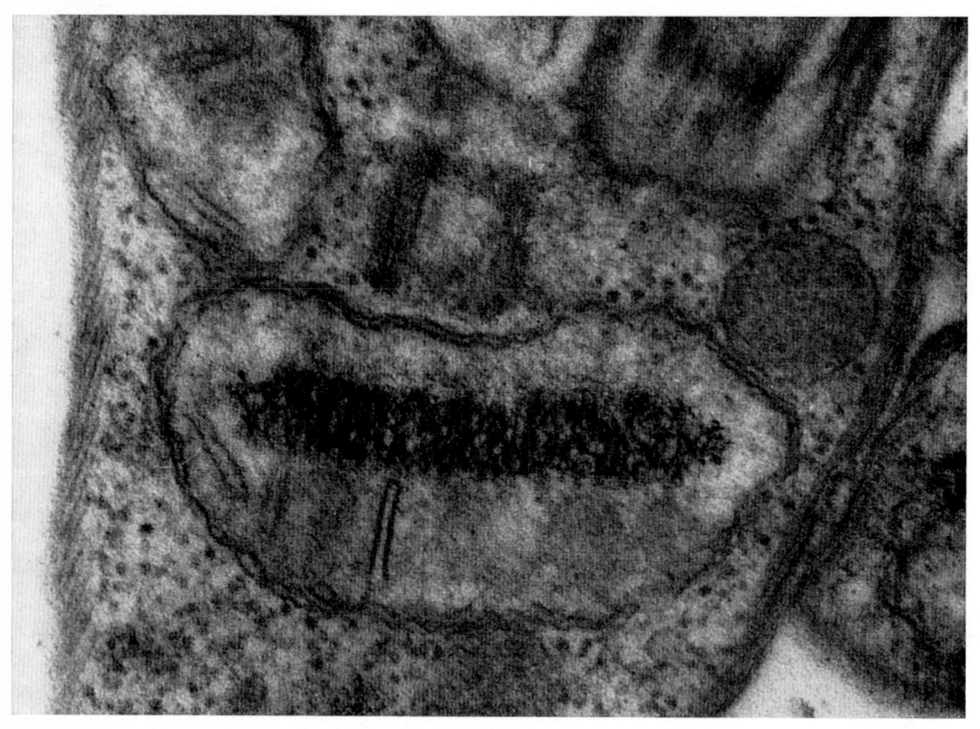

Fig. 14. *Trypanosoma lewisi*. Longitudinal section through the region of the kinetoplast. The kinetoplast is bordered by a double membrane from which cristae originate. A fibrillar material (DNA) is found within. Above the kinetoplast: a longitudinally sectioned basal body. × 46000. After ANDERSON and ELLIS, 1965 [*56*]

Electron microscope studies [*56*, *268*, *1154*, *1369*, *1666*] have shown that their fine structure is largely identical with that of mitochondria (Fig. 14). Their outer envelope is formed by two membranes. Infoldings of the inner membrane correspond to the cristae. In the interior, a fibrillar material is found, representing the DNA component. It has also been shown that the enzyme content of kinetoplasts is very similiar to that of mitochondria [*342*].

Recent investigations have shown that the kinetoplast-DNA has a peculiar configuration: Apart from longer molecules it consists of a multitude of nearly equal minicircles [*1602*].

In the variously differentiated cell types of the trypanosomids (p. 382), the size and structure of the kinetoplast differs accordingly [*56*, *1135*, *1462*]. The extracellular *Leptomonas*-form of *Leishmania donovani*, for example, has a large kinetoplast with many cristae while the intracellular *Leishmania*-form has a small one with few cristae. Though the rate of oxygen uptake is different in both forms (5–7 fold more in the *Leptomonas*-form), it is not probable that this change in size and structure is connected with a fundamental difference in the electron transport chain as has been assumed previously [*969, 1600, 1602*].

Enlargement of the kinetoplast may be accompanied by a subdivision: a relatively small part contains only fibrillar material (DNA) while the remainder extending far into the cytoplasm, forms the cristae.

After treatment with certain dyes (pyronin, acriflavine, ethidium bromide) a variable percentage of trypanosomes, living in the vertebrate bloodstream, become "*dyskinetoplastic*": The kinetoplasts alter and reduce their DNA-structure so that they are no longer stainable with basic dyes, but still retain their membranes. Though such "dyskinetoplastic" trypanosomes can live and multiply in the bloodstream they cannot undergo transformation to the insect vector forms which require normal mitochondrial respiration [*1683*].

On the other hand, it is interesting to note that only those species that lack a developmental phase in an insect vector (*T. equiperdum, T. equinum*) are normally "dyskinetoplastic".

It is probable, therefore, that kinetoplast-DNA is functional mostly in forms which do not live in vertebrate bloodstreams. But what these functions might be, is completely unknown.

Plastids

In flagellates which are capable of photosynthesis like the green plants ("phytoflagellates"), a large part of the cell is taken up by the *plastids*. Because of their chlorophyll content, they are usually green and are then called *chloroplasts*. Due to the admixture of different carotinoids, the plastids can also have a yellow, brown or red color.

The number and size of the plastids are variable. A single, frequently cup-shaped chloroplast is characteristic for the phytomonads. It surrounds the nucleus and the major portion of the cytoplasm (Fig. 15). The remaining flagellates generally have several, sometimes even many plastids, which are then smaller and more or less evenly distributed in the cytoplasm (euglenoids, dinoflagellates).

The plastids are separated from the remaining cytoplasm by a double membrane. In some chrysomonads (for example *Ochromonas dancia*, Fig. 16) and cryptomonads, they are surrounded by another double membrane which is connected with the nuclear envelope [*582, 583*].

The ground substance of the plastids is called *stroma*. In the flagellates, the stroma is traversed by the so-called *lamellae* which are more or less densely packed and are predominantly parallel to each other. However, each lamella represents an aggregate of subunits, the so-called *membrane sacs* or *thylakoids*.The two membranes of each thylakoid merge at the edges and leave only a narrow space between themselves. The stacking of the thylakoids is generally so tight that the adjoining membranes appear as one in the electron microscope and are then twice as thick as the membranes which form the outer limits of the lamella. When the lamellae

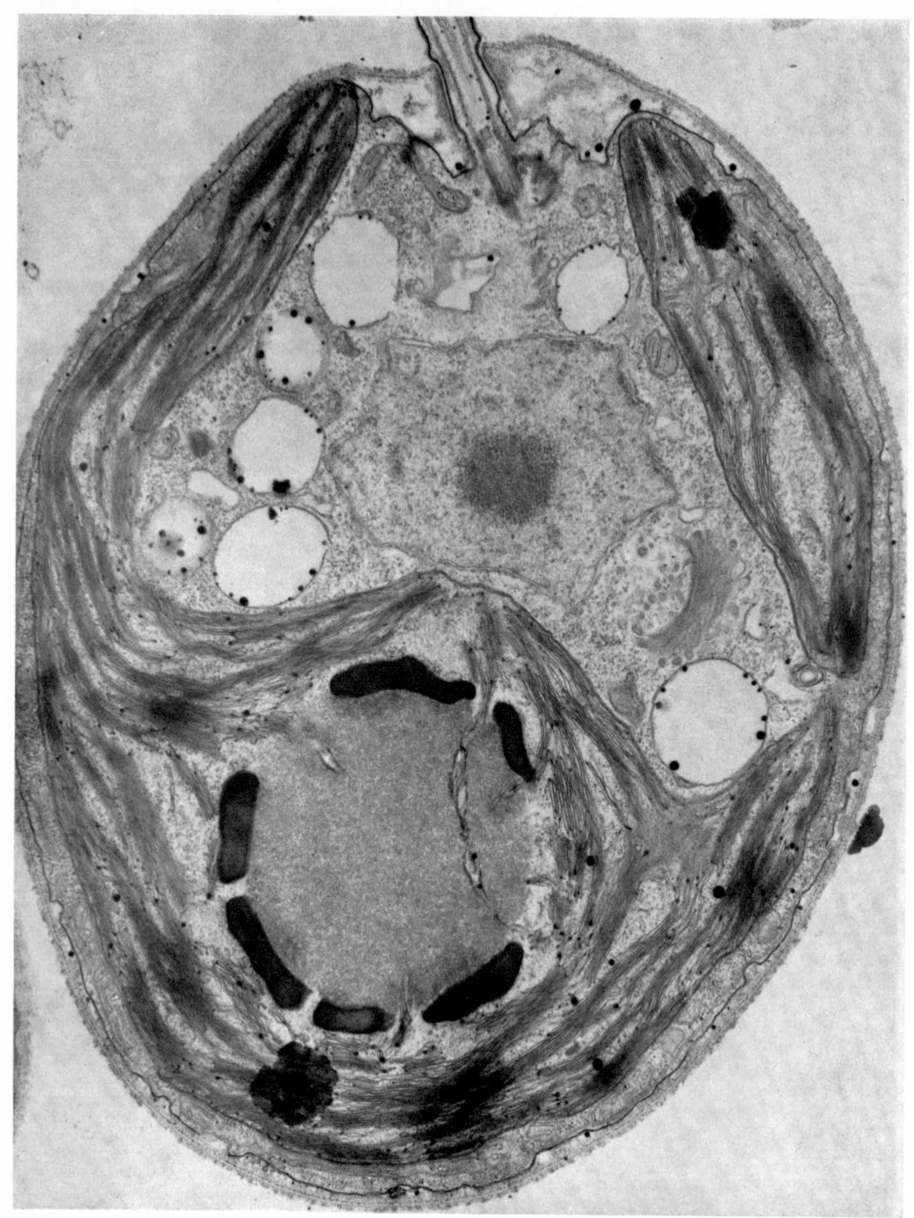

Fig. 15. *Chlamydomonas reinhardi*. Longitudinal section. At the anterior end (top) one of the two flagella originates. The cup-shaped chloroplast (with lamellae and pyrenoid) surrounds the cell nucleus (with nucleolus) and the cytoplasm (vacuoles, Golgi complex). The cell membrane is ensheathed by a cellulose layer. × 17000. Micrograph by G. PALADE from SAGER, 1965 [*1478*]

consist of only two thylakoids (cryptomonads), one gets the impression of three membranes, two thin outer ones and a thicker one at the inside. In most cases the lamellae consist of three thylakoids so that they show two thin outer and two thick inner membranes (Fig. 13a). Occasionally the thylakoids change over from one lamella to another or they form stacks of more than three (up to 12) thylakoids.

In higher plants the thylakoids do not form continuous lamellae. Besides "stroma thylakoids", we find "grana thylakoids", stacked like piles of coins. They owe their appearance to the local superposition of projections of the stroma thylakoids. These "grana" are also recognizable in the light microscope. Among the Protozoa, they have thus far only been observed in the phytomonad, *Carteria acidicola* [*875*].

The membranes of the thylakoids consist of proteins, lipids and chlorophyll. The molecular arrangement of these components has not been clarified as yet. That chlorophyll is part of the membrane structure itself is evident in a mutant of *Chlamydomonas reinhardi* which, unlike the wild type, has lost the ability to form chlorophyll in the dark. In continuous dark culture, the lamellae of the chloroplast disappear. When the cells are exposed to light again, first the synthesis of chlorophyll and then the formation of lamellae take place. A faintly green mutant which forms some chlorophyll (about 5% of that found in the wild type) but no carotinoids also has some lamellae in the chloroplasts [*1474, 1488*]. This finding cannot, however, be generalized since chlorophyll-deficient mutants may have apparently normal lamellae in other cases.

Although, in higher plants, chloroplasts can multiply by division, they are always formed from small *proplastids* without thylakoids during sexual reproduction. Recent studies have shown that such proplastids are also common among the flagellates. They are *rudiments of true plastids*, capable of division, and thus enable the cells to return to photosynthesis even after a prolonged period of culture in the dark.

In many cases the proplastids are difficult to identify because they are not much different from other vesicles in the cytoplasm. The chrysomonad *Ochromonas danica*, however, has a plastid which is enclosed by a fold of the nuclear envelope (Fig. 16a). During dark culture it reduces to a proplastid, easily recognizable because of its connection with the nucleus (b). When the cells of a dark culture are exposed to light, the proplastid grows to a plastid within two days, even

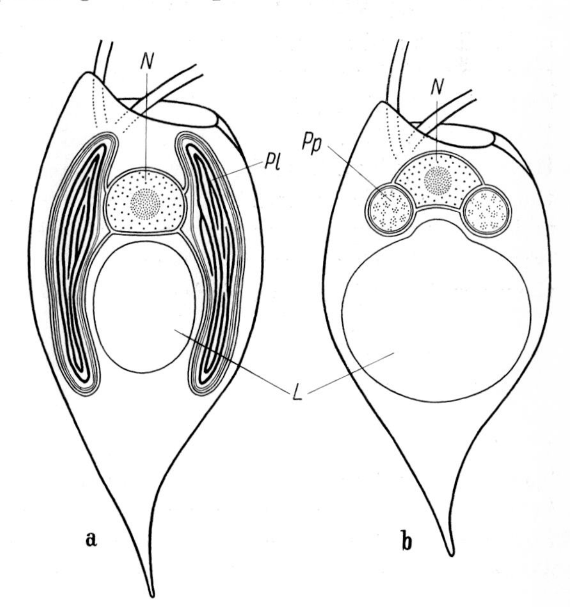

Fig. 16. *Ochromonas danica* (Chrysomonadina). Diagram of structural changes of the plastid. a Cell cultured in light, b in the dark. *N* Nucleus. *Pl* Plastid. *Pp* Proplastid. *L* Leucosin vacuole. After investigations by GIBBS, 1962 [*583*]

though the resynthesis of chlorophyll and of carotinoids as well as the formation of lamellae is complete only three days later [*582, 583*]. If the cells during that period are treated with chloramphenicol, which inhibits protein synthesis on chloroplastic but not on cytoplasmic ribosomes, chlorophyll synthesis and the formation of thylakoids is reduced almost completely [*1615*].

If *Euglena gracilis*, which contains about 10 chloroplasts, is either cultured in the dark or at elevated temperature, or if it is irradiated with ultraviolet light or treated with certain chemicals such as streptomycin or antihistamine, chlorophyll synthesis stops after some time and the chloroplasts seem to disappear. In reality, however, they fall apart to form about 30 proplastids. The latter are small vesicles of 1–2 μm diameter without lamellae and without chlorophyll although they show a reddish fluorescence due to porphyrin. However, only the proplastids of cells bleached by dark treatment can again differentiate to form chloroplasts, while those bleached by other means are irreversibly damaged [*584, 585*]. In those cells which become green again, three proplastids always seem to merge to form one chloroplast.

Colorless, exclusively heterotrophic forms are found among the phytomonads and the euglenoids. They are morphologically very similar to the green forms capable of photosynthesis. Thus, the colorless *Polytoma* corresponds to the green *Chlamydomonas* and the colorless *Astasia* to the green *Euglena*. The demonstration of vesicles, which can be interpreted as degenerated plastids or as genetically chlorophyll-deficient proplastids, can be cited in support of the point of view that the colorless forms originated from the green ones. In *Polytoma* these vesicles contain starch granules [*993*].

Also, the colorless cryptomonad *Chilomonas paramecium* contains a "leucoplast" with starch granules. In this case even the organelle-specific DNA and ribosomes could be demonstrated. Evidently, the "leucoplast" derives from a chloroplast of a colored *Cryptomonas*–ancestor [1572a].

In contrast to the typical cells of higher plants, *pyrenoids* are found in the plastids of flagellates and algae. They are characterized by the prevalence of a densely granular or fibrillar ground substance. In the most simple cases, e.g. in the chrysomonad *Olisthodiscus*, there is only a wider spacing of the lamellae in the region of the pyrenoid. Usually, however, the ground substance is so extensive that it is traversed only by a few lamellae whose course is no longer parallel. In addition, the lamellae of the pyrenoid are usually reduced to one or two thylakoids only, which, though continuous with those of the plastid, are more or less inflated and transformed into tube-like structures within the pyrenoid [*416, 579, 580*]. Sometimes the pyrenoids project like buds from the plastid as in some dinoflagellates and in the phytomonad, *Carteria acidicola* (Fig. 17), where they consist of several parts, each with a different lamellar course. The material between the lamellae has a fibrillar texture in this case, oriented at right angles to the lamellae [*875*]. In *Prasinocladus marinus*, an extension of the nucleus projects deep into the pyrenoid [*1086, 1289*].

The fact that polysaccharides are condensed either as starch or paramylon granules in the vicinity of the pyrenoid serves as an indication of its function. The granules can appear outside the envelope of the plastid or, as in phytomonads, inside the chloroplast (Fig. 17). Figure 18 shows an example of different functional states.

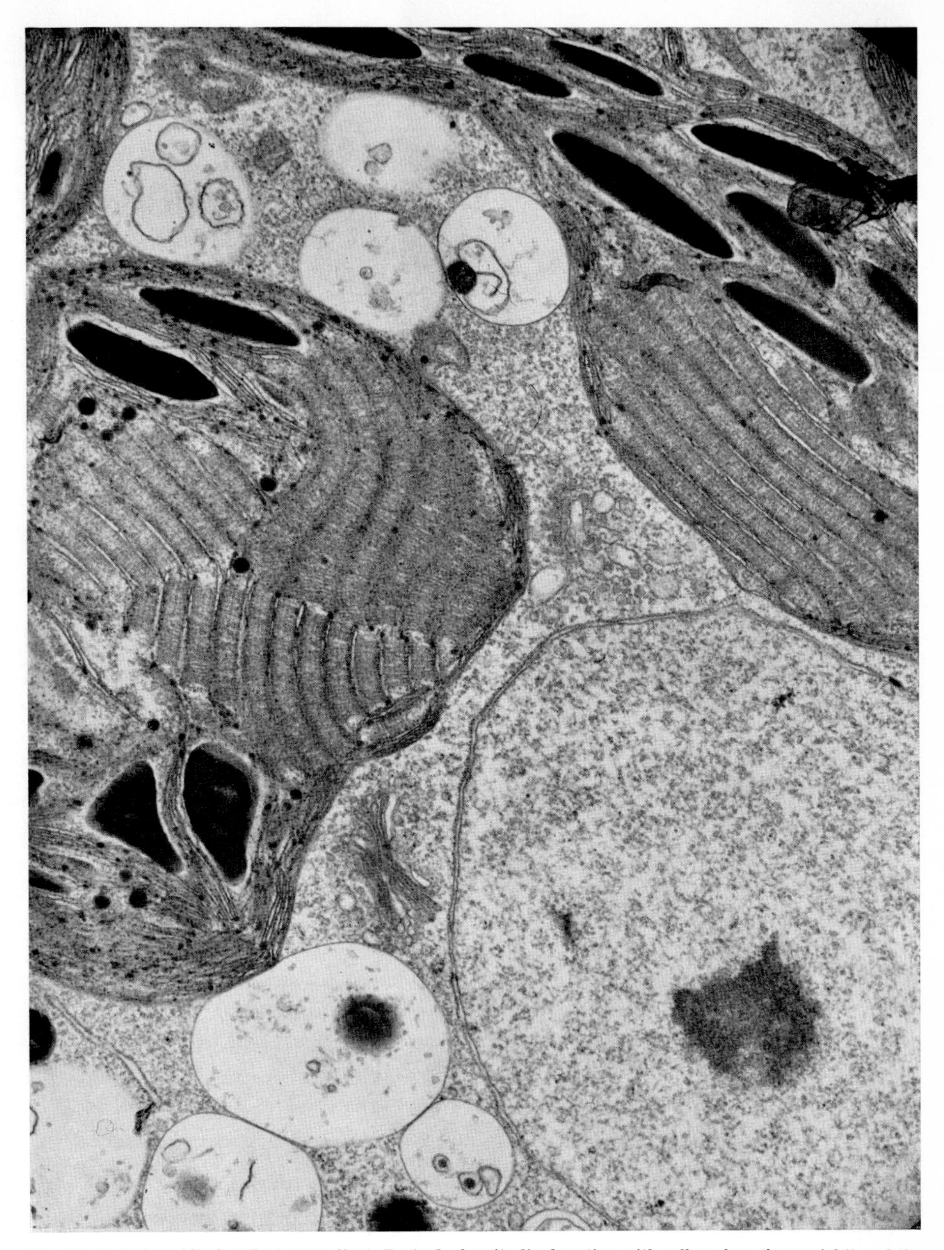

Fig. 17. *Carteria acidicola* (Phytomonadina). Part of a longitudinal section with cell nucleus (lower right) and the chloroplast. The latter forms two projections containing the pyrenoids. The black inclusions are starch granules. Two Golgi complexes are seen in the vicinity of the nuclear envelope. × 16400. After JOYON and FOTT, 1964 [*875*]

In many cases, the *eyespot (stigma)* is also a differentiation of the plastids (p. 314).

That the plastids are autoreduplicating bodies is no longer contested. In flagellates with only one or a few plastids, division can be observed directly. In addition, it was shown that the plastids contain DNA which is different from nuclear DNA

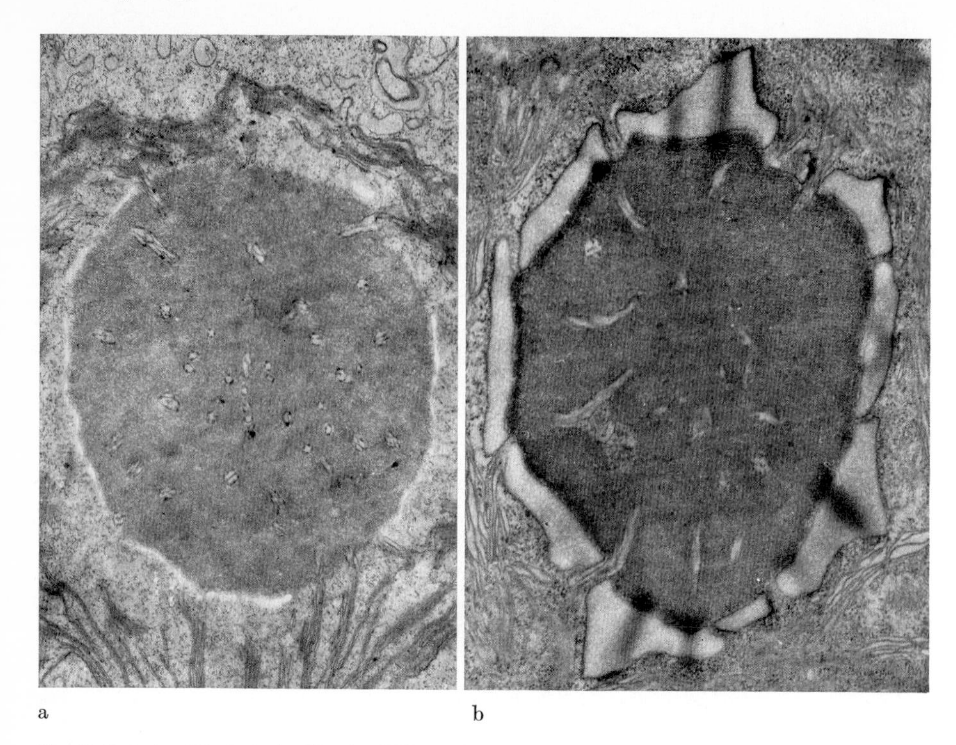

a b

Fig. 18. *Eudorina (Pleodorina) californica.* Different functional states of the pyrenoid. a Sparsely formed starch envelope (generative cell). × 18000. b Thick envelope of starch granules (somatic cell). × 24000. Courtesy of G. SCHWALBACH

[*171, 475, 476, 1417, 1487*]. DNA fibers are concentrated in certain regions. They disappear after treatment with DNA-ase. In *Euglena gracilis*, the plastid DNA amounts to 1–5% of the entire DNA content of the cell. Statistical evaluation of irradiation experiments (UV) led to the conclusion that there are approximately 30 units in *Euglena gracilis* which can be inactivited and 10 units which are segregated during cell division. These numbers correspond to those of the proplastids and chloroplasts [*494, 1068*]. The possibility that "cytoplasmic" genes discovered in *Chlamydomonas reinhardi* are bound to the DNA of the plastids will be considered later (p. 241).

Since it has been proved that ribosomes and several enzymes necessary for RNA- and protein synthesis are found in the plastids, it must be taken into account that the plastids control a part of their structural and functional properties through their "own" genes. However, they are not "a cell within a cell". Like mitochondria they depend on nuclear genes and have to be regarded as "semi-autonomous" (p. 20).

Even though the functions of the various previously discussed membrane systems are still not clarified in detail, it is certain that they play a role in the intermediate metabolism of the cell. The meaning of the structures discussed in the following section is, however, to be sought in another direction.

Fibrils

Fibrillar structures are found in the cytoplasm of many Protozoa. Their diversity is so great that only examples can be discussed here [*1312*]. Fibrillar systems which have a topographical relation to the cellular envelope (p. 39) and to the basal bodies of flagella and cilia (p. 288) are discussed in a later chapter.

Whether the fibrils have a static importance or whether they have the ability to contract is usually hard to decide. In many cases they fulfill both functions by maintaining a definite cell form as well as permitting changes of cell shape.

Many ciliates of ruminants have fibrils which form a kind of intracellular framework. Sometimes it is located more superficially (Fig. 19) and in other cases it also traverses the endoplasm even including the cell nuclei (Fig. 20).

A number of structures have been shown in recent years to be built of *microtubules*. A microtubule is an apparently hollow tube of 210–250 Å in diameter. Its wall is composed of globular *subunits* each about 40 Å in diameter. These subunits align

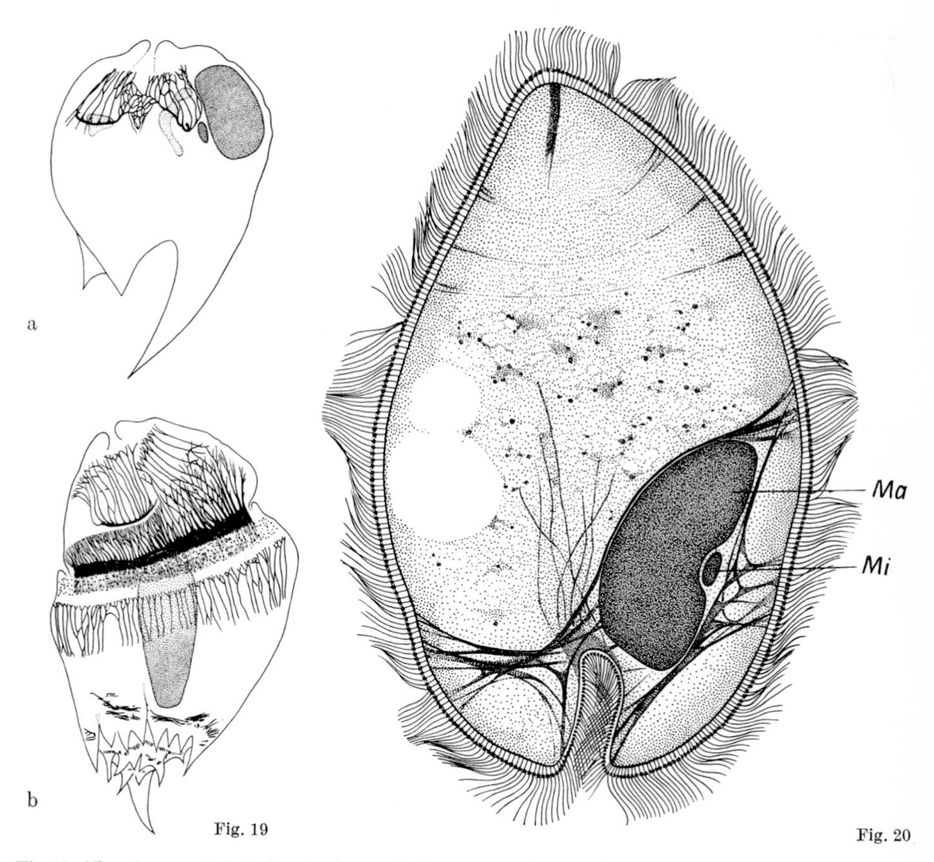

Fig. 19 Fig. 20

Fig. 19. Silver impregnated fibrils of ophryoscolecids. a *Entodinium caudatum*. × 730. b *Ophryoscolex caudatus*. × 350. After NOIROT-TIMOTHÉE, 1960 [*1237*]

Fig. 20. *Isotricha prostoma*. Ciliate from the ruminant paunch with fibrillar supporting apparatus at the cytostome and around the nuclei: *Mi* micronucleus, *Ma* macronucleus. × 950. After BELAR from HARTMANN, 1953 [*720*]

into 11–13 longitudinal strands. Probably, there is a pool of such "monomeric" subunits in the cytoplasm which under certain conditions permits them to aggregate in a very short time into the "polymeric" structure of the microtubule.

Microtubules can be joined by thin, threadlike *bridges*. One of their functions may be to stabilize the structure to be composed of microtubules. In certain cases they seem to play a role in motility. Microtubules to be linked by bridges may actively slide against each other, according to the "sliding filament concept" to be developed for explaining the contraction of striated muscles.

Arms, which do not join microtubules but terminate free in the cytoplasm, may serve to link particles or membranes which move alongside them. This has first

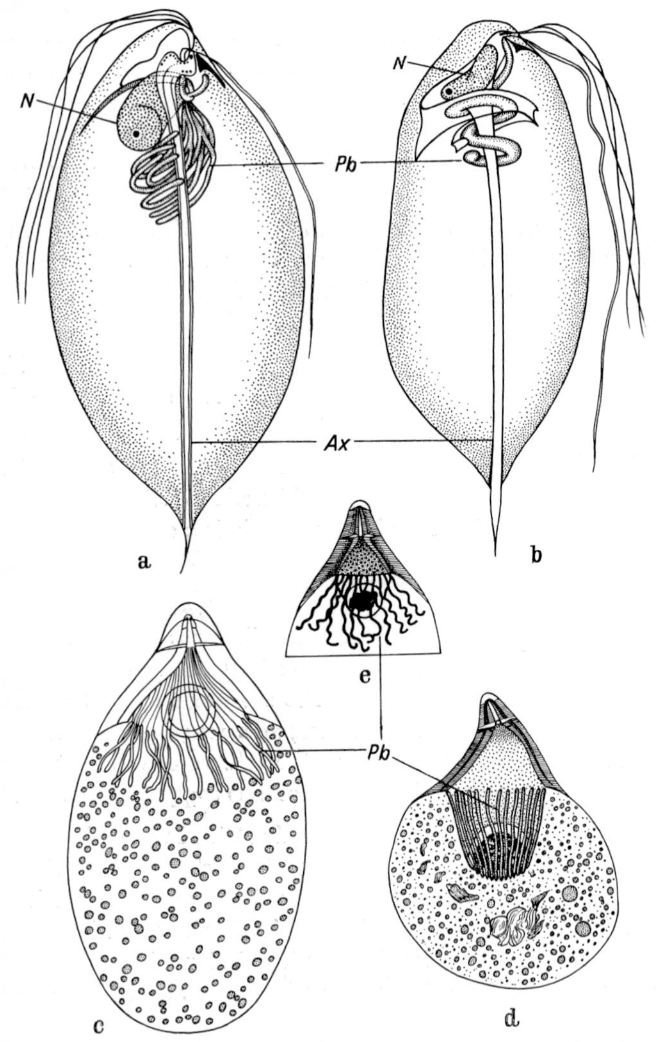

Fig. 21. Cell differentiations in various species of Polymastigina. a *Pseudodevescovina ramosa*, b *Metadevescovina magna*, c *Trichonympha chula*, d *Trichonympha teres*, e *Trichonympha chattoni*. *N* Nucleus, *Ax* axostyle, *Pb* parabasal body. After KIRBY, 1944 and 1949 [*937*]

been demonstrated for the "neurotubules" which join synaptic vesicles during their motion within the axon (axoplasmic transport). It may also be the structural basis for many cases in Protozoa where the cytoplasm streams in a fixed direction or particles are ingested by means of leading (e.g. "cytopharyngeal") structures [*90, 1461, 1751*].

The organelles composed of microtubules are formed at "*nucleating sites*" where the assembly of subunits starts. Nucleating sites can be centrioles and basal bodies, but also other structures of various appearances (Fig. 9), or even cytoplasmic regions without any visible structures at all (p. 65).

Characteristic for many Polymastigina are the *axostyles* which can either be present only once, then traversing the whole cell (Figs. 11, 21), or in large numbers (Fig. 343). They can be rod-shaped, tape-like or thread-like. Electron microscopic pictures show that they consist of numerous microtubules which are disposed in rows parallel to each other (Fig. 22a, b). The number of microtubules in each row as well as the number of rows differs from species to species.

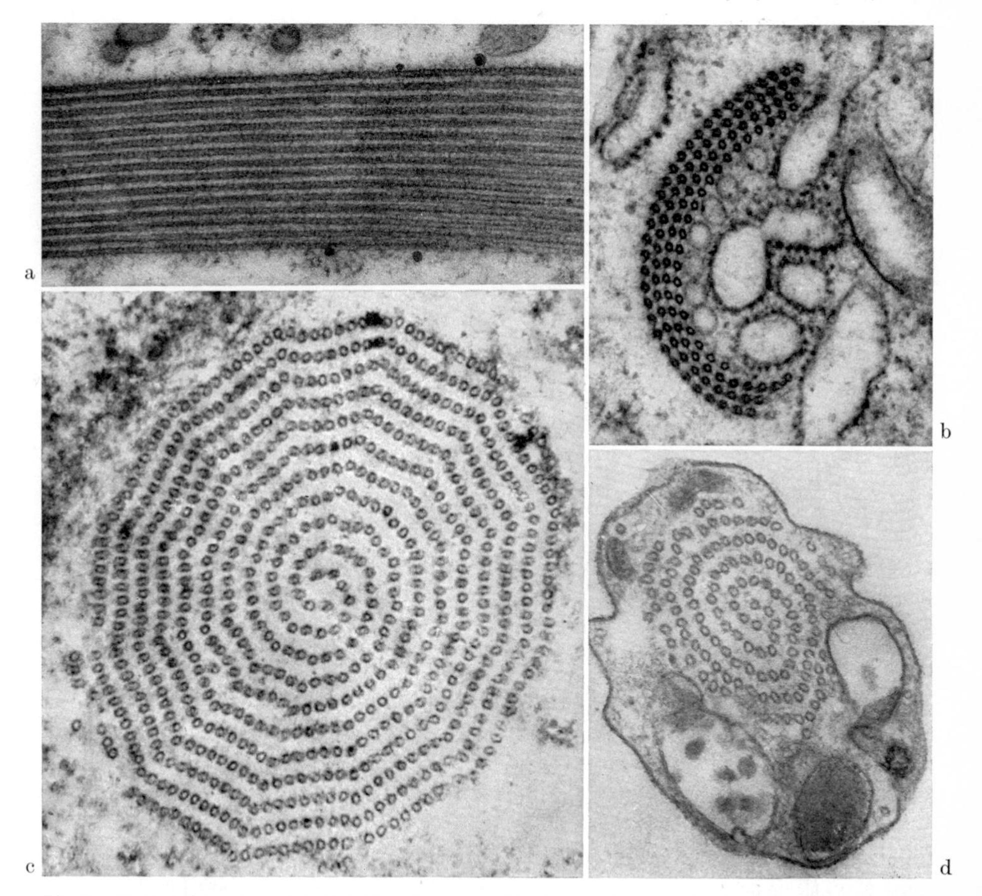

Fig. 22. Microtubules. a Longitudinal section, b cross section through the axostyle of a polymastigote flagellate (probably *Saccinobaculus*). × 54000. After GRIMSTONE and CLEVELAND, 1965 [*698*]; c, d Cross section through the axoneme of the heliozoan *Echinosphaerium nucleofilum* (c in the vicinity of the nucleus, d at the level of the axopodium). × 44000. After KITCHING and CRAGGS, 1965 [*944*]

Active bending of the axostyles has been observed in *Trichomonas*. In *Pyrsonympha, Oxymonas, Saccinobaculus* and *Notila* [*698*] wave-like motions of the axostyles are found. However, to what extent these undulations permit locomotion and how they are coordinated with the action of the flagella is not well known as yet. In many cases, for instance in *Mixotricha paradoxa* (p. 297), the axostyle certainly has no locomotor function [*322*].

Although the *axonemes* which support the axopodia of many Heliozoa and Radiolaria cannot be homologized with the axostyles of Polymastigina, they are, like axostyles, composed of microtubules (Fig. 22c, d). Their arrangement is described on page 280.

Even among the flagellates (for instance, *Peranema*, Fig. 322a) organelles can be formed which facilitate the entrance of food into the cell and dissect it into smaller portions. In some ciliates, especially in the cyrtophorine gymnostomes, this function is fulfilled by the so-called *cytopharyngeal basket* which consists of a circle of rods. These are connected with other structures, which cannot be discussed here in detail, to form a functional unit (p. 335). The rods also consist of densely packed microtubules connected by cross bridges.

While the microtubules have an exclusively static importance in this case, they seem to be involved in food ingestion as well as in contraction of the suctorian tentacles (p. 343). In the prehensile tentacles used for capturing the prey, they form complexes which have been interpreted as "myonemes" on the basis of light microscopic observations (Fig. 293b).

Also the many fibrils which run below the cell surface (p. 310) or originate from the basal bodies (p. 298) as well as the spindle fibers and the sub-fibrils of flagella and cilia (p. 288) are microtubules.

Although the extent to which the microtubules are functionally identical cannot yet be decided, it can be said that they are widely distributed elementary structures which may have a similar chemical composition.

Extrusomes

Many flagellates and ciliates form complicated structures in their cytoplasm which can wholly or partially be extruded in response to certain stimuli and which are always located below the cell membrane. Although their homology has not been demonstrated, it seems reasonable to treat them under a common heading. Therefore, they are here called extrusive bodies or *extrusomes*. According to their structure, different types can be distinguished.

The *trichocysts* of ciliates have been known for the longest period of time. Among the Holotricha (Trichostomata, Hymenostomata) which feed on bacteria, the trichocysts can be distributed as rod- or thread-like elements below the whole cell surface. With their distal ends they are anchored to the cell envelope in whose pattern of differentiation they are incorporated in a definite way (p. 45). Longitudinal sections through the trichocysts of *Paramecium* and *Frontonia* (Fig. 23) show that they consist of an elongated shaft and a tip. At the distal end a cap is formed which encloses only the tip in *Paramecium*, in *Frontonia* also part of the shaft.

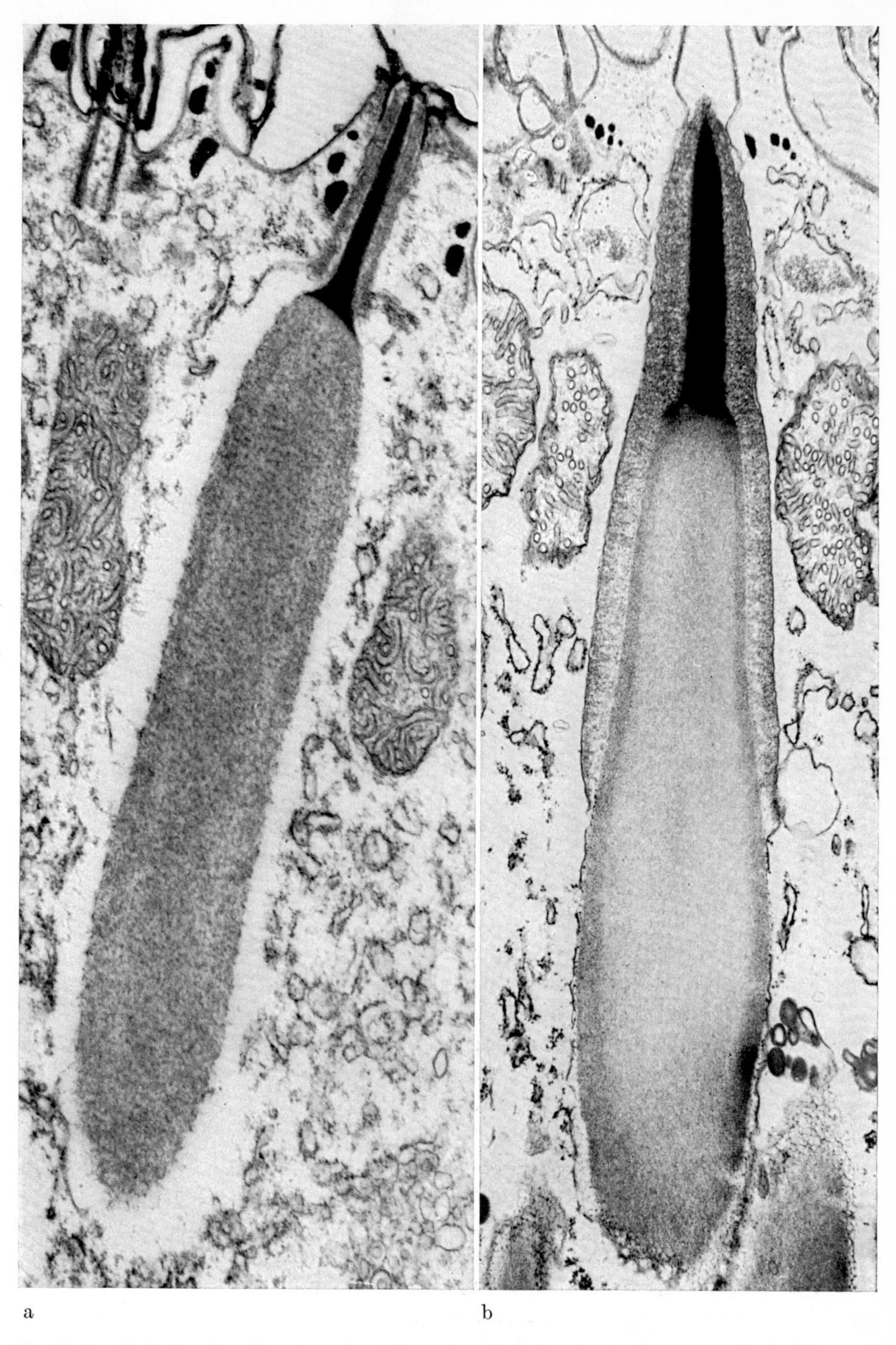

Fig. 23. Longitudinal sections through resting trichocysts: a of *Paramecium caudatum*, × 31000, b of *Frontonia vesiculosa*, × 29000. After Yusa, 1963/65 [*1835/1836*]

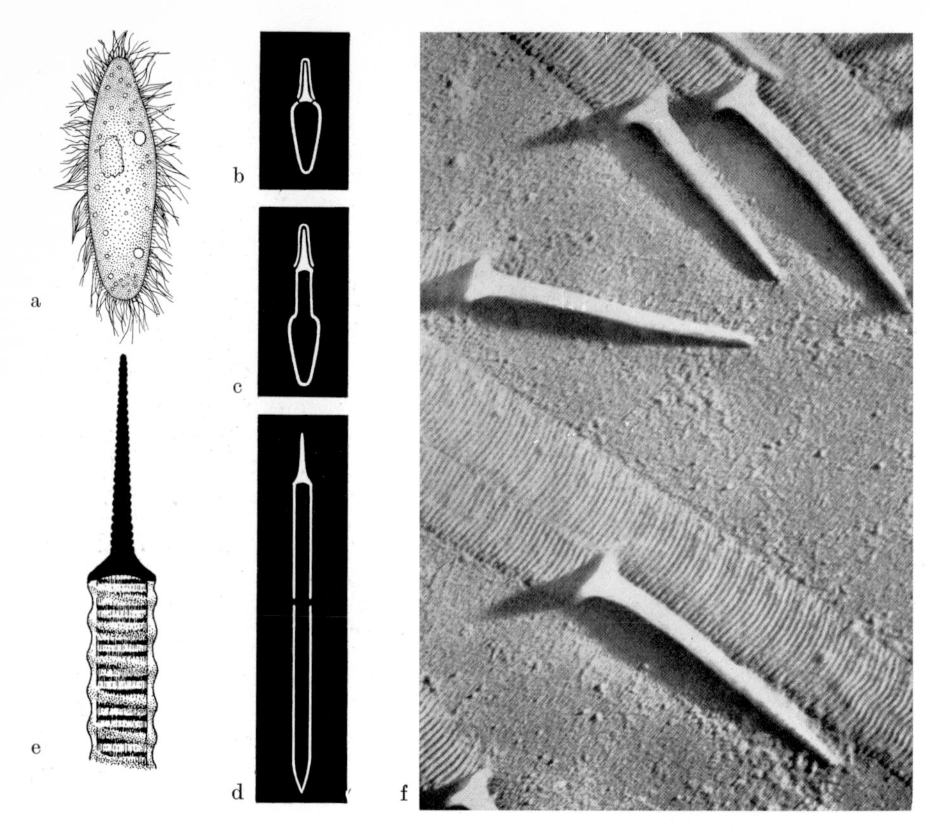

Fig. 24. Trichocysts of *Paramecium*. a *Paramecium* with exploded trichocysts (after treatment with picric acid). After JENNINGS, 1931; b—d Single trichocysts in dark field, b resting, c, d during elongation. Combined after KRÜGER, 1930 [*970*] and JAKUS, 1945 [*850*]. e Tip of the exploded trichocyst based upon electron microscopic investigations (partly hypothetical). After KRÜGER and WOHLFARTH-BOTTERMANN, 1952 [*971*], modified. f Trichocyst tips of *Paramecium*, electron micrograph (shadowed with chromium). × 14000. After JAKUS and HALL, 1946 [*851*]

The explosion of the trichocysts can be triggered by mechanical, chemical or electrical stimulation. It is difficult to analyze because it occurs in a span of a few milliseconds. In its exploded state the trichocyst reaches about 10 times its original length (20—30μm). The shaft becomes a long, periodically cross-striated rod which collapses when dried and still carries the unchanged tip at the anterior end (Fig. 24). Electron microscopic investigations have revealed — mostly with the negative staining method — many ultrastructural details touching the molecular level. The striae are due to a regular three-dimensional arrangement of a fibrous protein called "trychinin". Since the resting trichocyst already shows a periodical though denser cross-striation, it seems possible that the discharge process consists in a sudden extension of a pre-existing "paracristalline" lattice. The mechanism of this molecular "unfolding" is still unknown, however [*80*, *731*, *732*].

Whereas the trichocysts were formerly thought to originate from the basal bodies of cilia [*1066*], recent investigations [*1552*, *1835*, *1836*] have shown that they are derived from vesicles of the cytoplasm. After they have been exploded due to an electrical shock, their reformation can be studied (Fig. 25). The vesicles are at

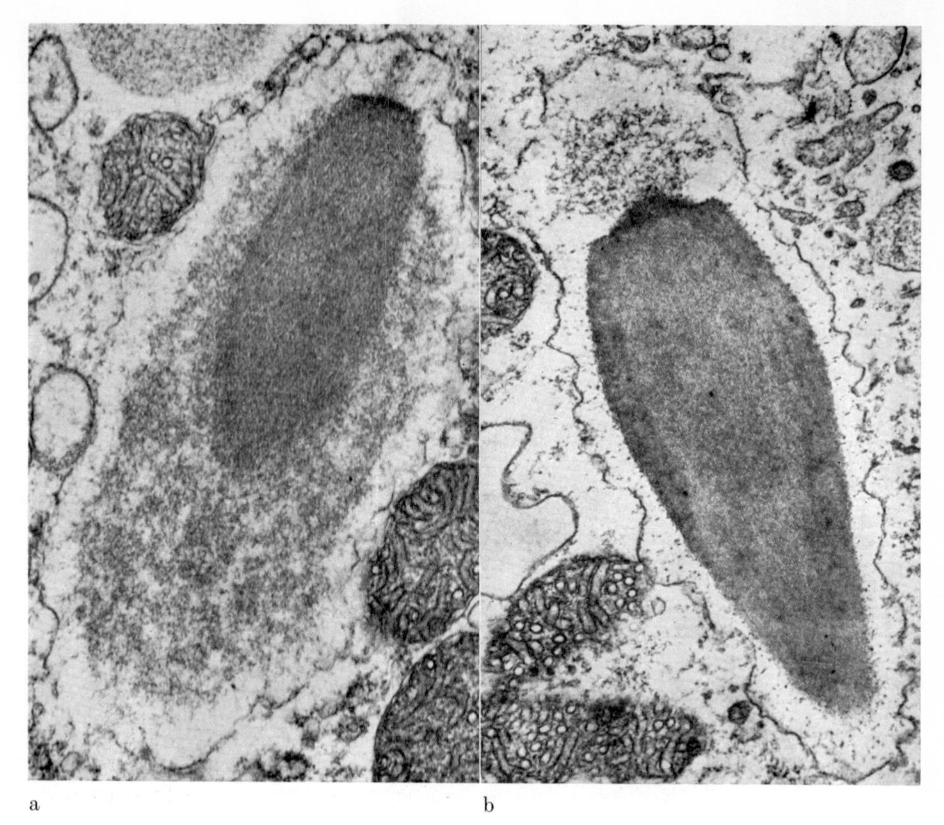

a b

Fig. 25. *Paramecium caudatum.* Developmental stages of trichocysts in vesicles of the cytoplasm. a × 30000, b × 27000. After Yusa, 1963 [*1835*]

first only filled by a granular mass. Then, a dense condensation appears (a) which enlarges at the expense of the ground substance and elongates (b). Finally the tip is formed which shows at first the same structure of parallel fibers as the shaft. Just how the trichocysts are anchored at the "correct" position in the pellicle is still a riddle.

The function of trichocysts is not satisfactorily known. At first glance one may suppose that they are organelles of defense: Through their massed and sudden "explosion" a kind of barrage is formed which may afford a certain amount of protection against enemies. When *Paramecium* is attacked by *Didinium* it always reacts with the expulsion of trichocysts (p. 337). However, the trichocysts are completely ineffective in this case.

In the carnivorous holotrichs (Gymnostomata) the extrusomes are usually called *toxicysts*. They are found in the vicinity of the cytostome and contain a toxin which paralyzes the prey.

Toxicysts are also formed in vesicles of the cytoplasm, but they have a structure which is completely different from that of trichocysts [*448*]. In a tubular capsule, which is frequently bent like a banana, there is a thread which can be expelled. In *Dileptus* the thread seems to be pushed together like a telescope within the capsule. An "occlusion" at its distal end might contain the toxin which becomes distributed

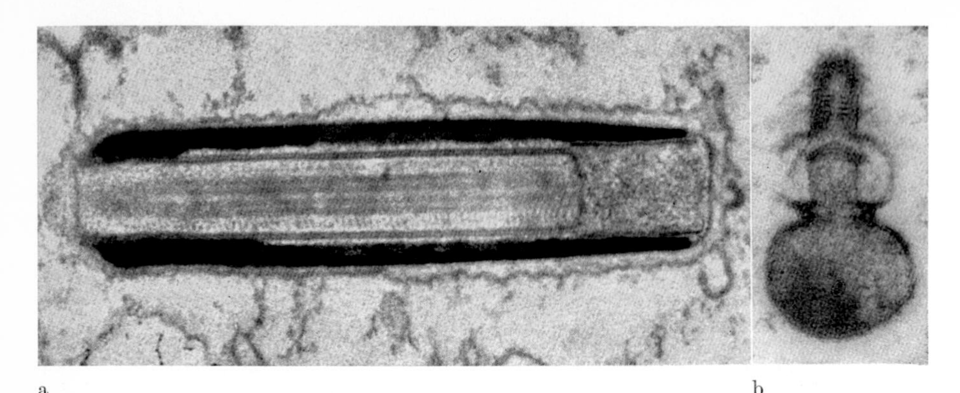

a b

Fig. 26. Toxicysts and haptocysts. a Toxicyst of *Dileptus anser*. × 59000. After DRAGESCO, AUDERSET, and BAUMANN, 1965 [*448*]. b Haptocyst of *Podophrya parameciorum*. × 73000. After JURAND and BOMFORD, 1965 [*882*]

along the surface as the thread is expelled (Fig. 26a). In other cases the thread is everted in a manner similar to that of the nematocysts of Cnidaria. In addition to toxicysts, *Didinium* has short attachment rods (Fig. 286) which lie below the surface of the oral cone and function in establishing contact with the prey. They have been termed *pexicysts* [*1793, 1794*].

Functionally similar are the so-called *haptocysts* which have recently been demonstrated in the tentacular swellings of Suctoria [*91, 727, 882, 1054, 1458*]. However, these formations are much smaller than the structures discussed thus far. Their length amounts to only 0.3–0.4 μm whereas even the smallest trichocysts are about 2 μm long. In spite of their small size, the haptocysts have a complicated structure. As shown in Fig. 26b, in one example, three portions can be distinguished: a narrow tube closed at the tip whose wall is cross-striated, a delicate middle piece, and a part swollen like a balloon from which a stamp-like projection extends into the middle piece. The tubular section is anchored in the envelope of the tentacular swelling and penetrates the pellicle of the prey as soon as it is touched by the suctorian (p. 341).

Some ciliates *(Tetrahymena, Colpidium, Ophryoglena, Holophrya, Balantidium, Didinium)* have sac-like extensions below their pellicle whose mucoid content is expelled upon stimulation and forms a kind of protective covering. Mucoid sacs, recently called *mucocysts*, originate in the endoplasm and are attached underneath the pellicle [*238, 630, 1731*].

In certain flagellates there are structures which bear a certain resemblance to the trichocysts of ciliates and are therefore usually called by the same name. In recent years they have been demonstrated in many dinoflagellates [*153, 449*]. As shown in Fig. 27 in one example, a shaft and a tip can be distinguished. The shaft is quadratic or rhombic in cross section and has a paracrystalline fine structure. The tip, which contains an axial rodlet in *Oxyrrhis marina*, seems to consist of a bundle of slightly twisted filaments in most dinoflagellates.

Species with a thick pellicle *(Prorocentrum, Gonyaulax)* expel their trichocysts through pre-formed pores. After explosion, the trichocysts appear as long threads which are cross-striated like those of ciliates.

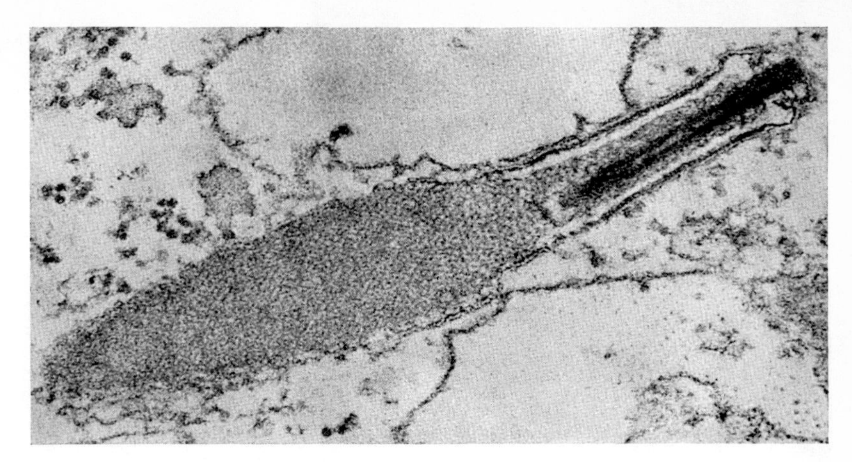

Fig. 27. *Oxyrrhis marina*. Trichocyst. × 90000. Courtesy of G. SCHWALBACH

In addition to trichocysts some genera of dinoflagellates *(Nematodinium, Polykrikos)* possess large and highly complicated organelles called *nematocysts* or *cnidocysts*. Since their function is completely unknown, no details will be given on their ultrastructure [*689, 1148*].

The extrusomes of cryptomonads, the so-called *ejectisomes*, are enigmatic organelles, too [*50, 818, 820, 1132, 1546, 1547, 1549*]. They are highly refractile bodies lying singly below the outer body surface, especially below the pellicle which surrounds the characteristic pharyngeal invagination (vestibulum) of cryptomonads (Fig. 312).

In electron micrographs (Fig. 28a), each ejectisome is seen to be surrounded by a membrane. It consists of two parts of the same shape but of different size. Each is a cylinder, scooped out to a cone from two sides, in the larger part more at the side directed toward the vestibulum, in the smaller one more at the side directed away from it.

The cross sections (b) give the impression of concentric layers in the wall of the cylinder. However, this appearance is obviously due to a continuous, coiled tape as demonstrated for the R-bodies of *kappa*-symbionts (Fig. 306).

At discharge, the ejectisome is transformed into a thread consisting of a long and a short section. It is assumed that the two sections correspond to the two parts of the quiescent ejectisome which become attached to each other in the process of explosion. Apparently, explosion is similiar to pulling apart a role of paper: The compact cylinder is transformed into a long, pointed tube with the outer coils of the tape at its base and the inner ones at its tip [*821*].

The trichocysts of dinoflagellates as well as the ejectisomes of cryptomonads are formed within vesicles pinched off from the Golgi complexes.

Organelles which may be functionally related to the haptocysts of Suctoria (p. 36), have recently been found in the axopods of some Heliozoa *(Raphidiophrys, Clathrulina* et al.). Since they travel up and down, they have been termed *kinetocysts* [*85, 87, 89*].

As all organelles described before, the kinetocysts show a heteropolar structure lying with their long axis perpendicular to the axopod. One of their poles is always

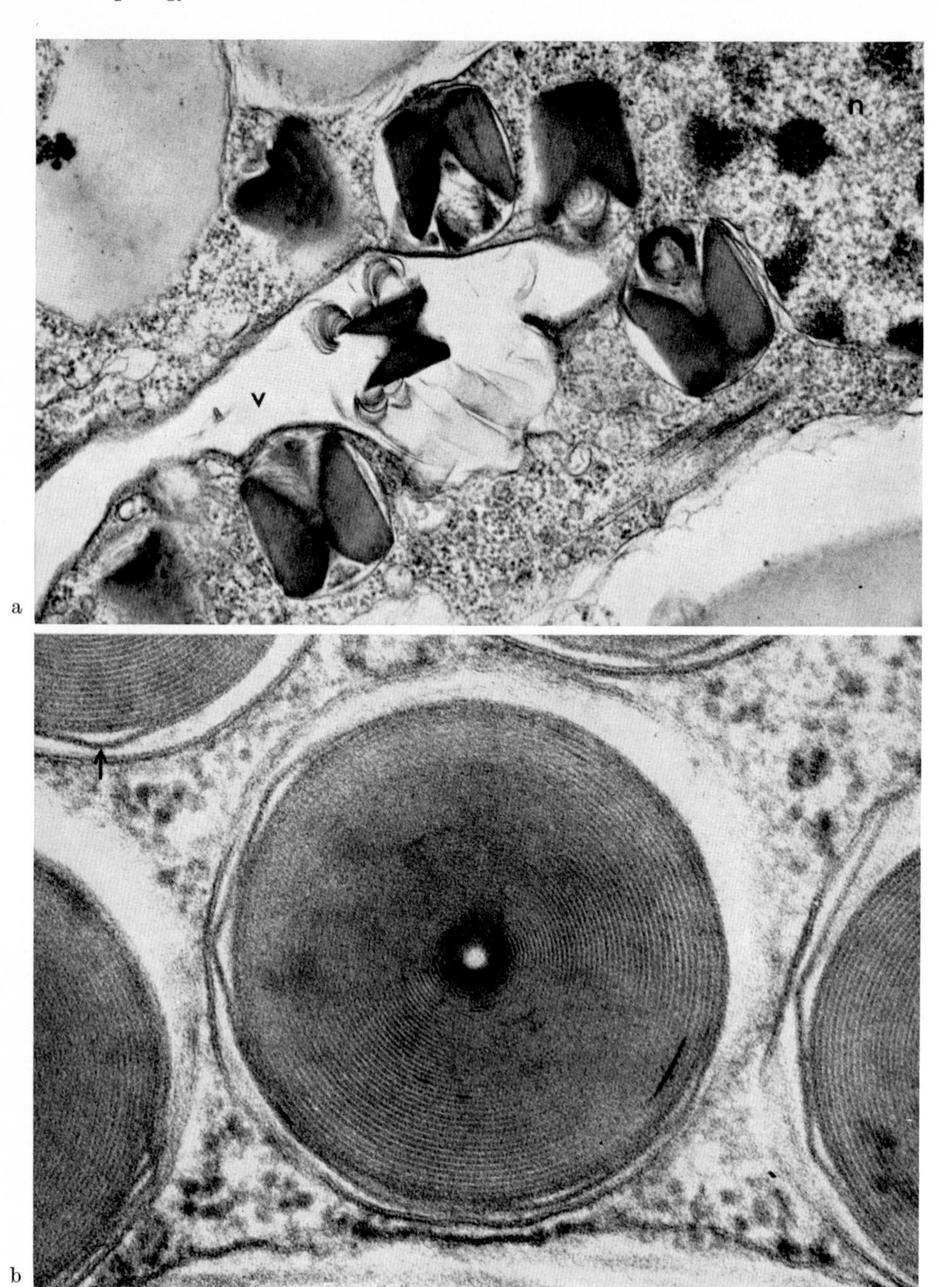

Fig. 28. *Chilomonas paramecium*. (Cryptomonadina). Ejectisomes. a Section through the region of the vestibulum (*v*) and the cell nucleus (*n*). An ejectisome in the lumen. × 30000. b Single ejectisome in cross section. The arrow points to an area where the "tape" has become loose. × 108000. Courtesy of L. JOYON (CLERMONT-FERRAND)

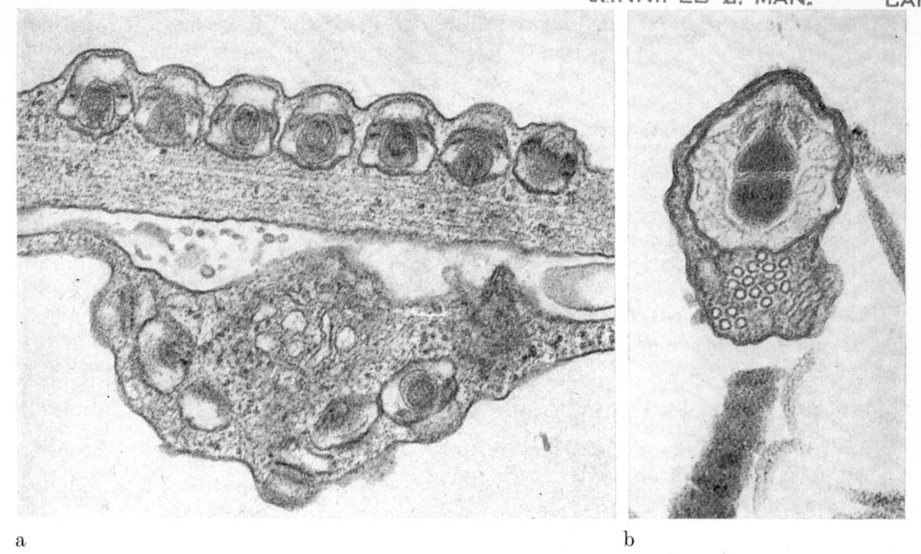

Fig. 29. Kinetocysts in the axopods of the Heliozoa *Clathrulina elegans* (*a*) and *Acanthocystis myriospina* (*b*). a × 50000, b × 60000. After BARDELE, 1971 [*88*]

oriented towards the plasma membrane while the other pole is directed towards the microtubules. Though already known to light microscopists, who called them „Körnchen“ (granules), the kinetocysts are the first specific organelles to be discovered in rhizopods (Fig. 29).

II. The Pellicle

The cytoplasm is always separated from the external environment by a cell envelope (pellicle). This is of special importance in Protozoa not only because it provides protection from harmful influences, but also because it permits a controlled exchange of substances coordinated with their life activities, the perception of mechanical and chemical stimuli and the establishment of contact with other cells.

The diversity of function indicates that, rather than being a static envelope, the pellicle must be involved in a constant exchange with the cytoplasm. Such structural dynamics, which also hold for the membrane systems of the cytoplasm, are, among other things, expressed by the fact that certain gene-dependent materials (sex and mating type specific substances, antigens) can temporarily become incorporated into the pellicle [*109*].

In the simplest case the pellicle consists of a single membrane, designated as the *cell membrane* or the *plasmalemma*.

The cytoplasm of amoebae has apparently at all times and everywhere the capacity of membrane formation. If an amoeba is cut in half, a new membrane is formed immediately at the cut surface, even in the anucleate fragment. Protoplasmic droplets taken from the plasmodium of a myxomycete surround themselves at once with a membrane, whose presence can be demonstrated in the electron microscope. On the other hand, the plasmalemma can just as easily be broken down, for instance when two pseudopodia flow together.

In pinocytosis (p. 329) as well as in phagocytosis (p. 331), i.e. during the normal activities of the amoeba, a constant membrane flux must take place. Membranous portions taken into the cytoplasm can be demonstrated as vesicles by electron microscopy. They must be renewed by a different pathway. The constant renewal of the cell membrane of an amoeba can be demonstrated directly by labeling with fluorescent antibodies. A creeping *Amoeba proteus*, whose pinocytotic activity has not been accelerated by inductor substances, replaces within four to five hours about one half of its cell membrane [*1823*]. On the other hand, it is unlikely that movement as such is correlated with a continuous breakdown and renewal of the cell membrane, because renewal occurs much more slowly than expected if both processes were correlated and because other findings also argue against such an assumption [*349*].

On the basis of its fine structure, the pellicle of amoebae is considered a unit membrane. Usually, no further differentiations are associated with it. In some species (*Amoeba proteus, Hyalodiscus simplex, Pelomyxa carolinensis*), however, it is covered by a diffuse layer of little contrast from which a fine fringe of extensions originates at regular intervals. This layer consists of mucopolysaccharides and may play a role in the adhesion of the amoeba at the substrate or in the adsorption of substances before they are taken into the cytoplasm by pinocytosis (p. 330).

In *Paramoeba eilhardi* the cell membrane is covered by densely packed little box-like structures whose shape could be reconstructed from thin sections (Fig. 30). Each box has a spindle-shaped outline and is 330–370 nm long. It has a bottom,

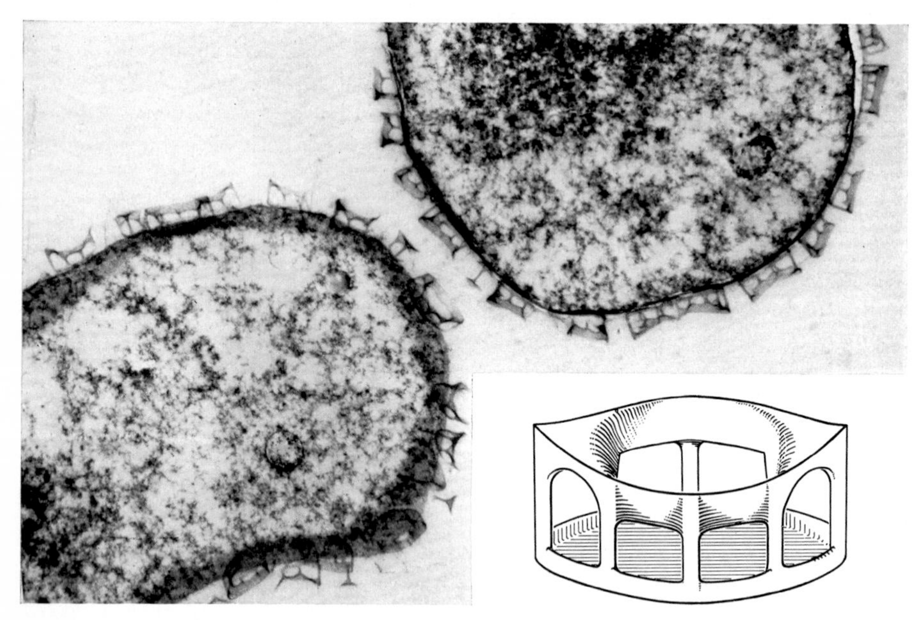

Fig. 30. *Paramoeba eilhardi*. Sections through the outer cell region. The cell membrane is covered by "little boxes". Below right, a reconstruction. × 27000. After GRELL and BENWITZ, 1966 [*677*]

and its side walls are perforated between eight pillars. The upper surface has a cup-like depression leading to a hexagonal opening. Since the boxes are elastic, their function was thought to be that of "suction cups" which are compressed when touched and thus facilitate adhesion of the amoeba [*677*].

While the pellicle of Heliozoa is represented by a simple membrane, the *Radiolaria* show a special condition. A central portion of the cell, the *central capsule* containing one or several nuclei, is separated from the so-called extracapsular cytoplasm by a special envelope of organic material. Pores, whose number and arrangement is characteristic for the various radiolarian suborders, provide the connections between both regions of the cell.

Little is known about the pellicle of the other rhizopods because they secrete firm shells (Testacea, Foraminifera) whose formation and structure cannot as yet be investigated by the preparative techniques of electron microscopy.

In *flagellates* the pellicle is of an astounding diversity. High rigidity is achieved by a variety of means, especially in free-living species which are exposed to a constantly changing environment. Only a few examples can be discussed in the following sections.

Chrysomonads have a simple cell membrane. In species which do not form pseudopodia, the cell membrane is frequently covered with a layer of minute scales or platelets of silicon with a species-specific structure. Species of the genus *Chrysochromulina* have two sorts of scales of different size, form and sculpture [*1091, 1092, 1096, 1097*]. Occasionally, a fine spur arises from the center of the scale (Fig. 311).

In some genera formerly classified as phytomonads but now considered to represent a special order (Prasinophyceae)*, a layer of minute scales is found outside the cell membrane which even covers the flagella [*1089, 1093, 1101*]. Especially complicated structures are found in *Mesostigma viride* [*1093*]: outside the small scales which also cover the flagella there is a layer of considerably larger scales which look like diatoms of the genus *Navicula* when seen from above. This is followed by a layer of delicate "baskets" on top of the basal plates. All these structures arise in vesicles of the Golgi complexes. They are probably transported within these vesicles to the depression at the anterior pole of the cell where the flagella originate and are shifted there through the cell membrane to the outside.

Phytomonads have a relatively simple pellicle. The cell membrane (unit membrane) is surrounded by a homogeneous layer containing cellulose and pectin (Fig. 15). It is usually too rigid to permit changes of cell shape. Fissions take place within the cellulose wall, the daughter cells emerging through this outer covering (p. 125). The gelatinous layer present in some genera (*Haematococcus*, *Stephanosphaera* et al.) can probably be thought of as a swollen cellulose layer. On its outside it is covered by an additional thin and finely striated layer, probably of pectin [*872, 873*].

The excessive rigidity due to additional layers can be compensated by elastic regions between the stiff plates.

This is the case in the *euglenoids* which are capable of changes of shape in spite of a fairly rigid pellicle. This is most clearly expressed in species capable of metaboly (p. 310).

* *Micromonas, Nephroselmis, Pyramimonas, Halosphaera, Heteromastix, Mesostigma.*

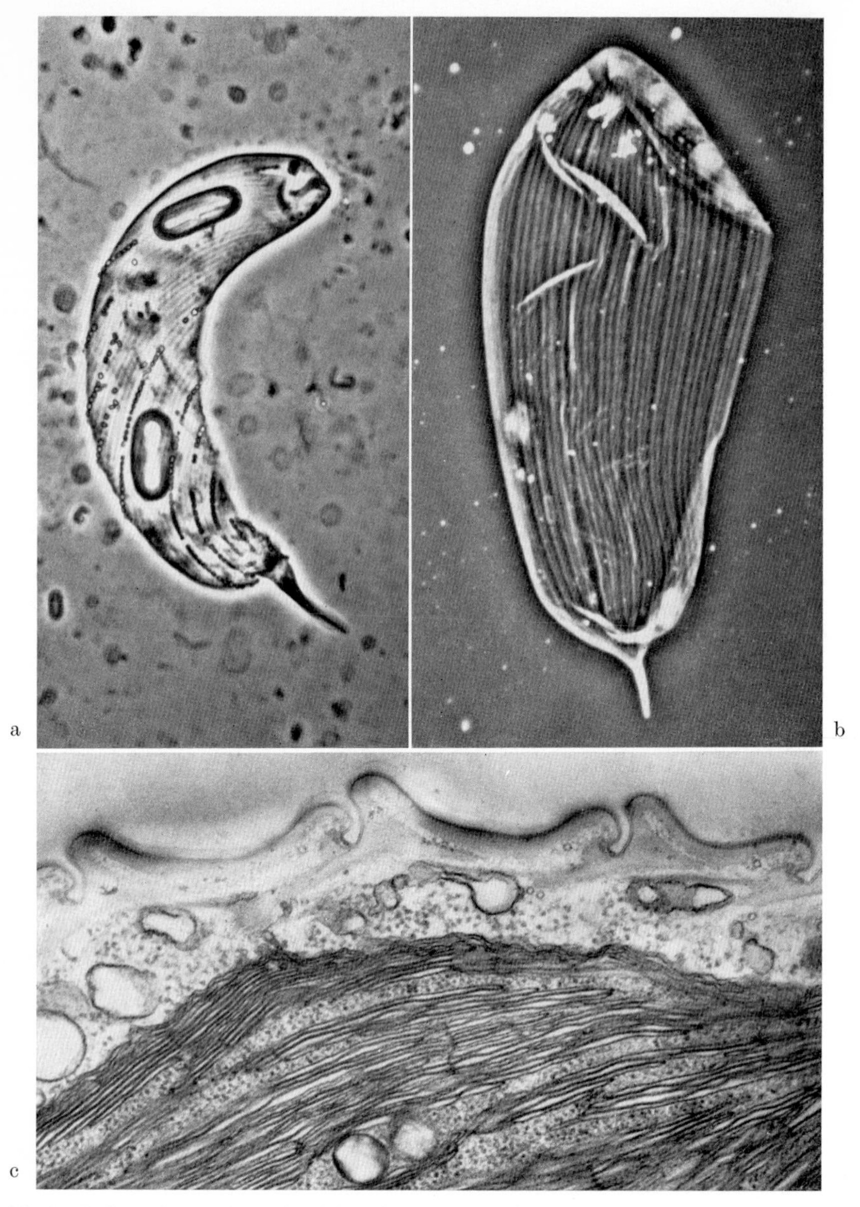

Fig. 31. *Euglena spirogyra.* Structure of the cell envelope. a Cell whose content, except for the two paramylon granules, has flowed out. b Isolated pellicle, view from the inner surface. c Part of a cross section. Underneath the pellicle (ridges, grooves, tooth-like projections) a chloroplast (lamellar structure) has been sectioned. a, b × 1000, c × 40000. After LEEDALE, MEEUSE, and PRINGSHEIM, 1965 [*1016*]

Fig. 31 shows spiral stripes in the pellicle of *Euglena spirogyra*; these originate at the reservoir and impart an external asymmetry to the cell. The stripes frequently show a series of small knobs whose number is variable (a). Along a stripe the pellicle can be ripped open and separated from the cell content. This shows that the stripes are really ridges separated by grooves (b). Cross sections demonstrate that the pellicle is bounded by a membrane. The layer below it, consisting predominantly of proteins, is of uniform thickness at the ridges but widened below the grooves to form tooth-like projections (c). Mucous sacs open in the grooves, providing perhaps a sort of lubrication for the pellicle [*1009, 1016*].

The protein layer of other euglenoids is thinner and without tooth-like projections below the grooves. Even the number of microtubules which run lengthwise in the cell and which have a regular position relative to the ridges of the pellicle may vary from species to species [*1126–1128*]. *Euglena spirogyra* has only one microtubule below each ridge (Fig. 31c).

In the order *Opalinina*, the pellicle consists of one cell membrane only (Fig. 32). This forms parallel longitudinal folds between the rows of cilia which continue without interruption from the anterior to the posterior end. Within each fold there is a row of 20–25 longitudinal fibrils whose function is unknown [*1236*].

The considerable surface enlargement of the pellicle might be connected with the mode of nutrition of the Opalinina. Since they have no cytostome, they may take up nutriments only by selective permeation (p. 329) directly through the pellicle. On the other hand, it has recently been shown that in *Cepedea dimidiata* pinocytotic vesicles are continuously pinched off at the base of the longitudinal folds. At least a part of the nutriments must get into the cell by this pathway [*1238*].

Among the *Sporozoa*, the gregarines have a similarly folded pellicle [*979, 1539, 1765, 1772, 1778*]. It is probable that also in Sporozoa food uptake is not solely due to permeation. One indication of this is the finding of so-called "micropores" in the pellicle of certain developmental stages of Coccidia (p. 333). Although these are not true pores, they might well be spots where pinocytotic vesicles are formed.

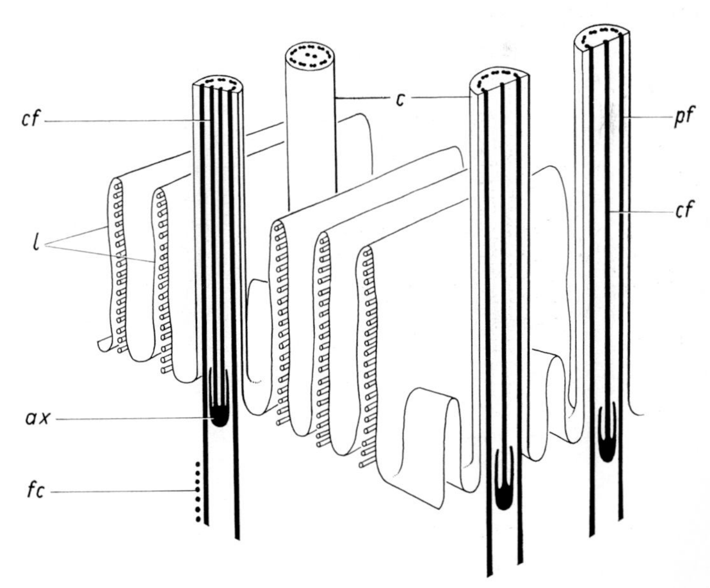

Fig. 32. *Opalina ranarum*. Diagram of the pellicle, based on electron micrographs. *l* Longitudinal folds of the cell membrane with fibrils. *c* Cilia with peripheral (*pf*) and central fibrils (*cf*), ending at the axosome (*ax*). *fc* Sectioned fibrils at the basal body. After Noirot-Timothée, 1959 [*1236*]

The erythrocytic stages of species of *Plasmodium* always have one such "micropore" which might even be considered a cytostome (Fig. 284).

Special structures connected with the pellicle are characteristic for the anterior end of sporozoites and merozoites, as shown diagrammatically in Fig. 33. The *conoid* (C), a conical differentiation composed of a certain number of spiral fibers, has been found in most cases, except *Plasmodium*. Inside the conoid, paired formations called *rhoptries* (R) arise which probably have the function of secreting a lytic enzyme, when the infectious stage penetrates a host cell. In their neighborhood, sections of the micronemes (Mn) can be seen which may have some topographical connection with the rhoptries [*82, 236, 242, 559–564, 746, 750, 1530–1532*].

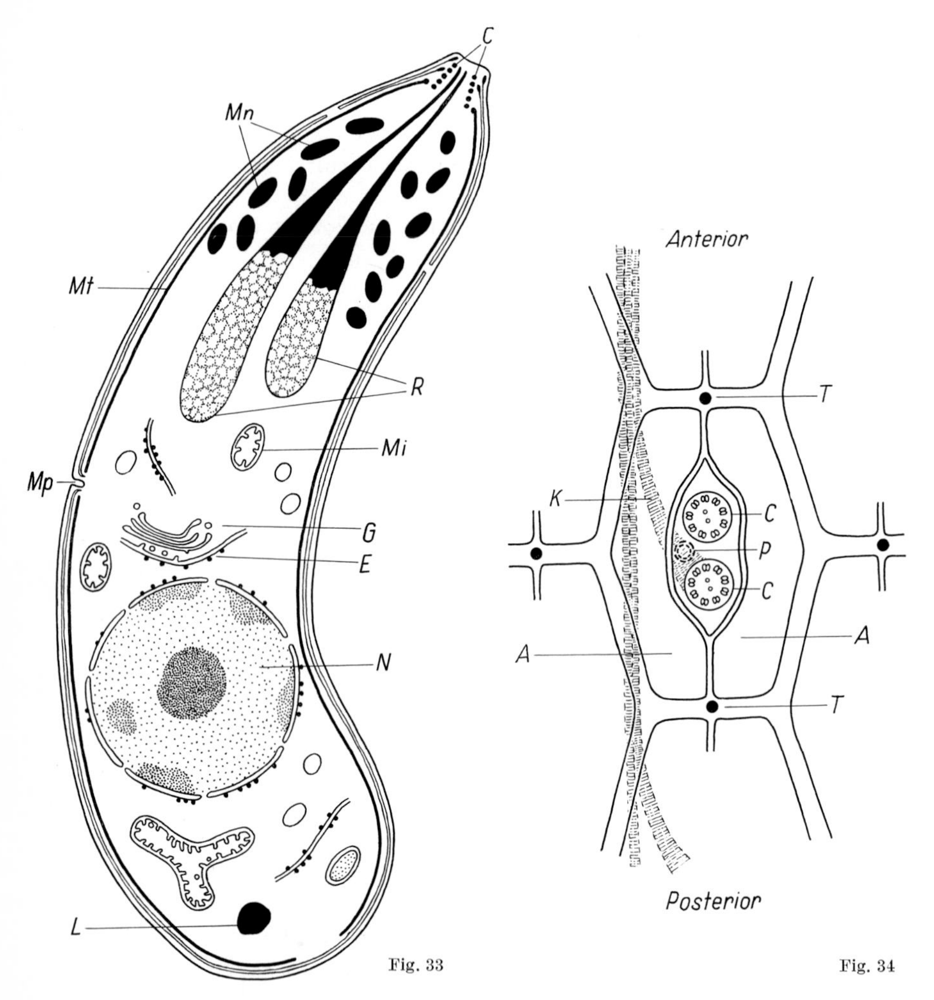

Fig. 33. Diagram of an infectious stage of a sporozoon. *C* Conoid. *R* Roptries. *Mi* Mitochondrion. *G* Golgi complex. *E* Cisterna of endoplasmic reticulum. *N* Nucleus. *Mn* Micronemes. *Mt* subpellicular microtubule. *Mp* Micropore. *L* Lipoid. After PORCHET-HENNERÉ and VIVIER, 1971 [*1332*]

Fig. 34. *Paramecium aurelia*. Diagram of a ciliary field, showing the relative position of cilia (*C*), the parasomal sac (*P*), trichocyst tips (*T*), alveoli (*A*) and kinetodesmal fibrils (*K*). After JURAND and SELMAN, 1969 [*885*]

The pellicle of *ciliates*, forming together with its associated structures the so-called *cortex*, shows such a variety of differentiations that the following discussion must be restricted to *Paramecium*.
As in most holotrichs, the cilia are arranged in *longitudinal rows*, the *kineties*. At the oral surface where the peristome and the cytostome are located, the kineties of both body halves meet at an acute angle along an anterior (preoral) and a posterior (postoral) suture, whose formation is due to the differential growth of the pellicle during fission (p. 134). At the aboral surface, all kineties run parallel from the anterior to the posterior end (Fig. 35).
Corresponding to the course of the rows of cilia, the pellicle is divided into small morphological subunits called *ciliary fields*. Each of the fields is limited by a ridge-like tetragonal or hexagonal elevation. Its center bears a single cilium or a pair of *cilia* (Fig. 34C). At the midpoint, where two ciliary fields touch longitudinally, a *trichocyst* (T) is anchored in the pellicle. Cilia and trichocysts are thus arranged in regular alternation.
A so-called *kinetodesmal fibril* (K) originates from the basal body of each cilium, if only one is present, and from the posterior basal body, if two are present. The kinetodesmal fibrils of a longitudinal row of successive cilia run anteriad and to the right (as seen from inside the pellicle). They join to form a bundle ("kinetodesma"). In addition, there is a small pellicular depression to the right of the basal bodies, the *parasomal sac* (P). Each ciliary field has thus an asymmetrical structure.

Electron microscopic studies have shown that the pellicle is bounded by a unit membrane which is continuous with the sheath of the cilia. Below each ciliary field are two membranous sacs which touch meridionally. Every cilium or ciliary pair is thus enclosed by two kidney-shaped *pellicular alveoli* (A) which leave the spots free where the trichocysts are anchored. Thus there are usually three consecutive membranes in the pellicle, the continuous outer cell membrane and both membranes of the pellicular alveoli [*1311*].
An impression of the cortical pattern is obtained by silver impregnation of *Paramecium* (Fig. 35). It must be pointed out, however, that silver is sometimes taken up more by the basal bodies, and the surrounding alveolar walls (a), and at other times more by the ridges of the ciliary fields (b).
Apparently the ciliary fields are also, in morphogenetic respects, units which grow independently, reduplicate their structural elements and divide to form new units. The localization of new structural elements is precisely fixed in the course of these events. New basal bodies, for example, always arise in front of the old ones [*408*, *409*].
In addition, the cell bodies of Protozoa are frequently enclosed by a special cyst, lorica, test or shell. Such formations may consist of very different materials. Usually they have an organic matrix which may be reinforced by incrustation or addition of inorganic substances (calcium carbonate, silica) or of foreign bodies.

The *cysts* are frequently only temporary sheaths. Many Protozoa can encyst and in this way protect themselves from harmful changes in their environment. Encystment is particularly widespread in species living in ephemeral fresh-water puddles. Exhaustion of the food supply, drying out and putrefaction are factors which favor encystment. In other cases, reproduction is regularly connected with

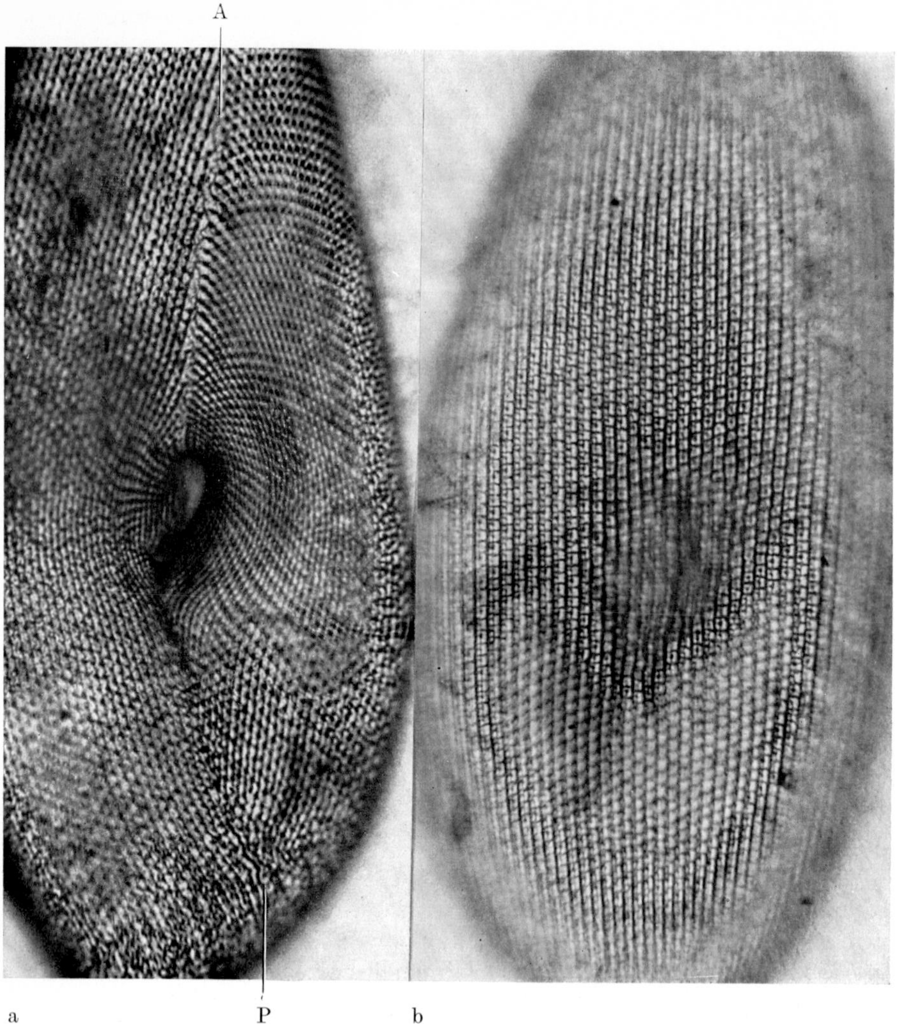

A

a P b

Fig. 35. *Paramecium multimicronucleatum.* Silver impregnation. a Oral side. It is mainly the basal bodies and the alveolar walls surrounding them which are impregnated. *A* Anterior suture, *P* Posterior suture. × 670. b Aboral side. Impregnation of the ridges bordering the ciliary fields. × 850. After SCHWARTZ, 1963 [*1564*]

the formation of a cyst. Some flagellates and some ciliates *(Colpoda, Ichthyophthirius)* divide only within the cyst. The developmental stages of parasitic species which are transmitted to another host are usually ensheathed by a resistant cyst (parasitic amoebae, Sporozoa).

Occasionally, the cyst has several layers (ectocyst, endocyst) instead of a single one and is provided with a special pore for hatching (Fig. 36). It may also consist of several valves which gap under certain conditions, e.g. the spore of Sporozoa under the influence of digestive enzymes of the host.

In a *lorica, test* or *shell* the cell body is either totally separated from the outer envelope, or it is only connected with it at certain points. Special openings provide

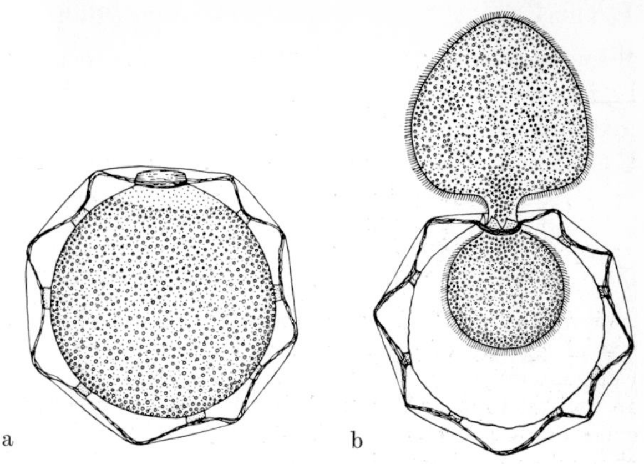

Fig. 36. *Bursaria truncatella* (Ciliata). Cyst with an intricately structured wall and an opening for hatching (hatching is depicted in b). × 190. After BEERS, 1948 [*127*]

the connection with the surrounding medium. Envelopes of this kind are found in almost all protozoan groups and are discussed in the taxonomic survey.

III. The Nucleus

Every protozoan cell has at least one *nucleus*. Many Protozoa are always multinucleated (Diplomonadina, Calonymphida, Opalinina, some Amoebina, Testacea and Heliozoa, all ciliates) or form multinuclear developmental stages (Radiolaria, Foraminifera, Sporozoa).

The consequences of enucleation demonstrate the importance of the nucleus for the cell. The results differ depending upon the cell type.

When the nucleus of *Amoeba proteus* is sucked into a pipette or destroyed by puncture, the enucleated amoeba will continue to live for an average span of about one week. Although its movements become uncoordinated, it can still react to stimuli like a normal amoeba with a nucleus. Its respiration is considerably decreased. Food, which is still taken up, can no longer be digested. It will finally die without having divided. In multinuclear individuals, occasionally found in *Amoeba proteus*, removal of a single nucleus has no such consequences [*263—266*]. An enucleated amoeba can be fully reactivated by transplanting the nucleus of another amoeba before changes have progressed too far [*356*, *357*, *1059*, *1060*, *1252*].

High oxygen tension can lead to the destruction of the nuclear contents in some of the Polymastigina living in the roach, *Cryptocercus punctulatus*, although the cytoplasm shows no visible damage. The nuclear envelope is usually retained. Gamonts of *Trichonympha* (p. 167) with such "empty" nuclei are not only capable of encystment but also of normal division. The centrioles form a typical spindle which elongates and pulls the nuclear envelope apart. Fission results in two gametes which will, however, die soon after division [*291*].

In the alga *Acetabularia*, which has only a single nucleus in its rhizoid, enucleated portions will not only survive for up to three months, but they are even capable of photosynthesis and growth. If they contain certain substances obtained from the nucleus, they are even able to form a new hat [*710*].

These examples show that an enucleated cell, although able to fulfill certain life functions and even to complete fission once it has set in, is nevertheless destined to die sooner or later. The cell nucleus is therefore absolutely necessary for normal reproduction.

The situation in heterokaryotic Foraminifera and in ciliates is discussed in a later chapter (p. 96).

1. The Resting Nucleus and the Chromosomes

Between divisions, the nucleus of the cell is called the *resting nucleus*. Physiologically, it is not in a resting phase during this period, which is one of lively exchange of materials with the cytoplasm.

Comparative studies have shown that four structural elements take part in the construction of cell nucleus, namely:

1. the chromosomes,
2. the nucleolar substance,
3. the karyoplasm,
4. the nuclear envelope.

As it is usually difficult to recognize the structure of nuclei in life, recourse must be taken to special methods of *fixation* and *staining* in order to distinguish between these structural elements.

In contrast to other structural elements of the nucleus, the chromosomes and the nucleolar substance can be stained with *basic dyestuffs*. This basophilia is due to the *nucleic acids* of chromosomes and nucleolar substance, combined with certain proteins to form nucleoproteins. Of the two types of nucleic acids, *desoxyribonucleic acid* is found mainly in the chromosomes. Its nucleotide sequence stores the genetic information. The *ribonucleic acids* are in part also associated with the chromosomes as the so-called messenger RNA, which transmits genetic information into the cytoplasm. The major portion of the RNA is accumulated in the nucleolar substance and is responsible for its intense staining with basic dyes. Substances containing DNA, i.e. the chromosomes and structures derived from them, can be specifically stained by the *Feulgen reaction*. Other methods permit the cytochemical localization of RNA, e.g. methyl green/pyronin staining. By incorporation of radioactive precursors labeled with tritium (^{3}H-thymidine for DNA, ^{3}H-uridine for RNA) the synthesis of both nucleic acids can be localized *(autoradiography)*.

Protozoan nuclei are of very different appearances. This *diversity* is due, on the one hand, to differential coiling of the chromosomes in the resting nucleus and, on the other hand, to different proportions of the various structural elements discussed above.

In only a few cases can all four structural components of the nucleus readily be distinguished. An example is provided by the large nucleus of the foraminiferan,

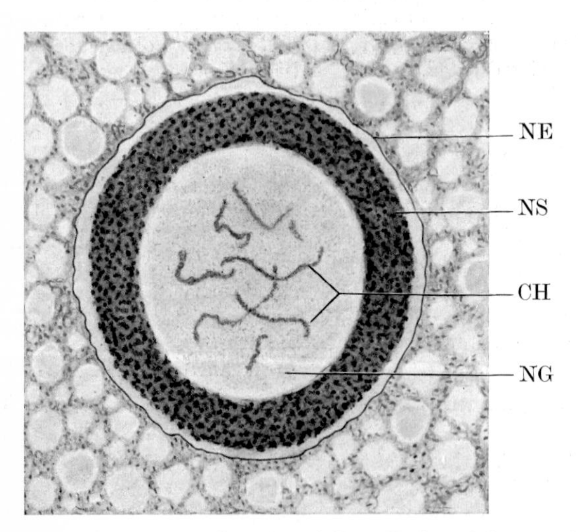

Fig. 37. *Myxotheca arenilega* (Foraminifera). Nucleus of the gamont. The cytoplasm is more basophilic than the picture indicates. *NE* Nuclear envelope. *NS* Nucleolar substance. *CH* Chromosomes. *NG* Nuclear ground substance. Bouin, Iron hematoxylin; × 850

Myxotheca arenilega (Fig. 37). Below the nuclear envelope, a dense layer of nucleolar substance in the form of small, irregular lumps is seen, while the chromosomes are present in the interior as fine threads. The remainder of the nucleus is filled with karyoplasm.

The resting nuclei of dinoflagellates (Fig. 38 and 50) and of euglenoids also show the chromosomes clearly. Their number is usually very large. They occupy the whole nucleus and are usually highly condensed as in stages of nuclear division in other organisms. They can frequently be recognized in the living cell. A similar situation is observed in some Polymastigina (Fig. 39).

As a rule, the chromosomes are not as condensed in the resting nucleus as they are during nuclear division. Usually, they are largely uncoiled and can then no longer

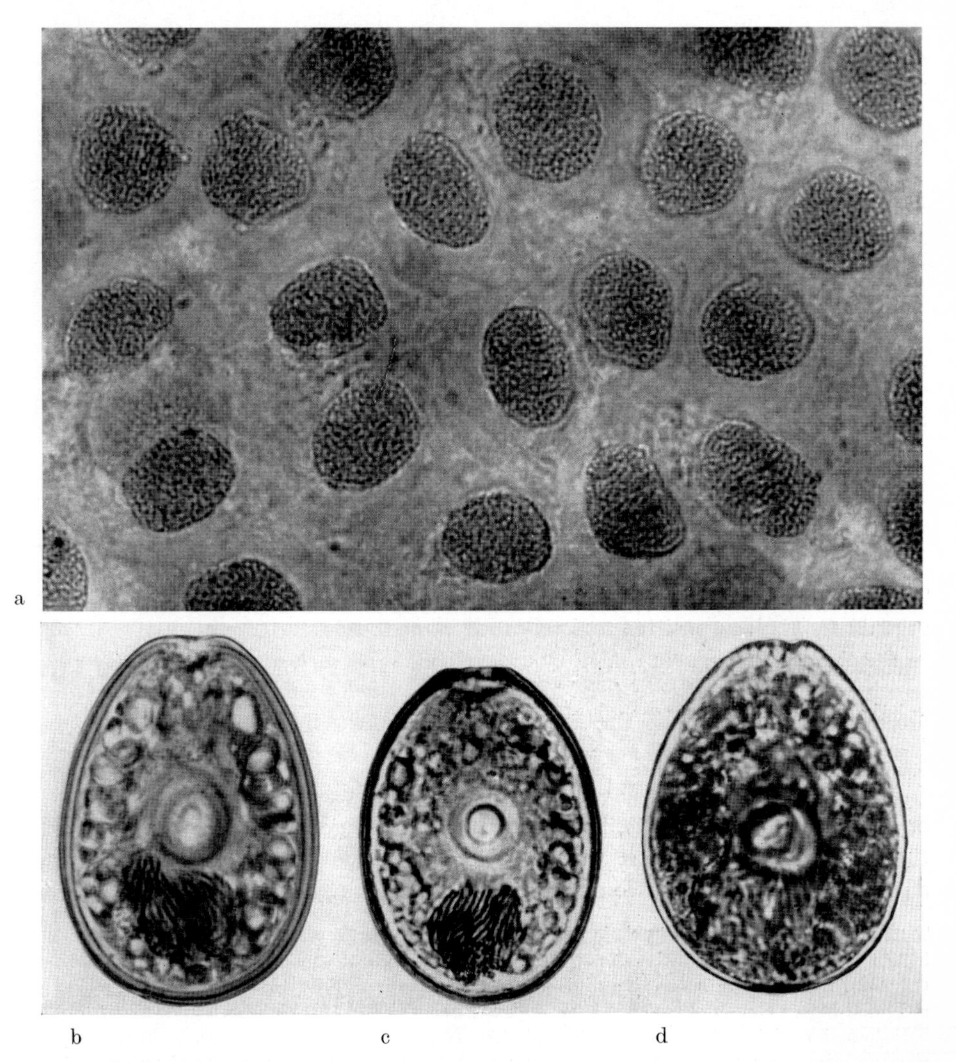

Fig. 38. Resting nuclei of dinoflagellates. a Part of a germinative body of *Blastodinium*. b—d *Exuviaella marina*, entire cells, nucleus in the lower half. a—c Alcohol-acetic acid, aceto-carmine; d live; b—d: × 750

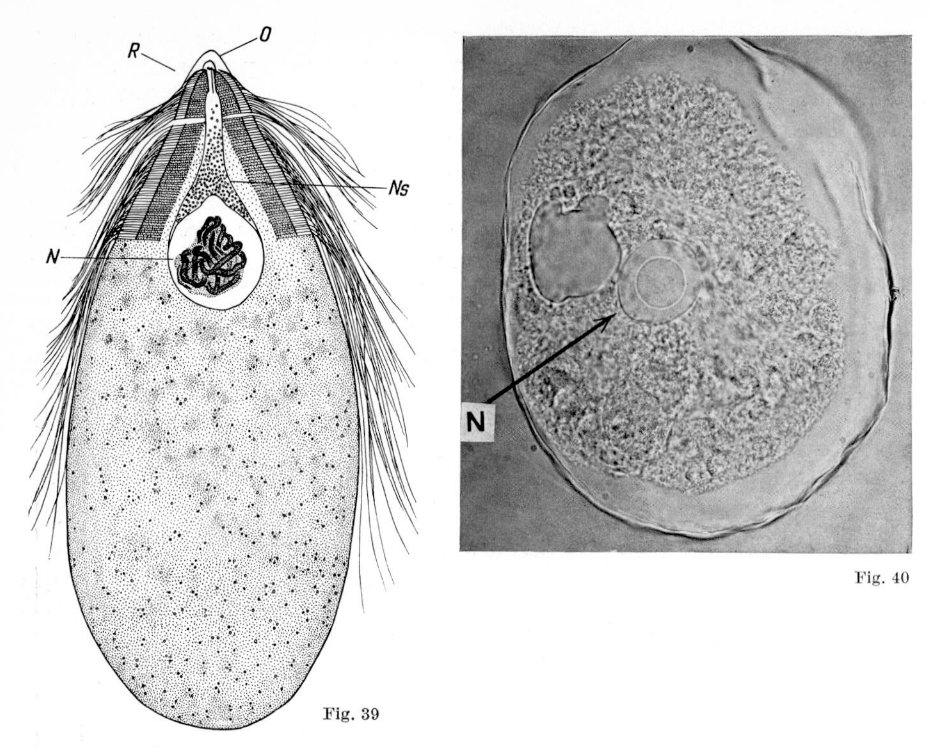

Fig. 39

Fig. 40

Fig. 39. *Trichonympha* (Polymastigina). Cell nucleus (*N*) with chromosomes in a "nuclear sac" (*Ns*), attached to the anterior end (rostrum, *R*) of the cell. Iron hematoxylin. × 400. After CLEVELAND, 1949 [*276*]

Fig. 40. *Amoeba sphaeronucleolosus* in life. Nucleus (*N*) with nucleolus. × 450. After BELAR, 1928 [*135*]

be demonstrated with basic dyes or by the Feulgen reaction. In the living cell, such a nucleus appears as a light vesicle within which the nucleolus is clearly visible (Fig. 40).

The resting nuclei of many Protozoa are of this "interphasic" type. In higher organisms, Feulgen-positive regions called *chromocenters* are frequently observed in them. It can be shown that they are derived from certain chromosome regions which are heterochromatic (Fig. 52 and p. 62). Nuclei with such chromocenters are extremely rare among Protozoa. Occasionally, the nucleolus is attached to a bud-like projection which may perhaps represent a chromocenter (Fig. 41).

In discussing the various structural components of protozoan nuclei, we shall begin with the *chromosomes*, because they carry the genetic information and are thus the most important structures found in the nucleus.They are the only nuclear structures which are capable of *identical reduplication*.

That the chromosomes retain their individuality, although they are not recognizable as such in the resting nucleus, can be shown in sporogony stages of the coccidian *Aggregata eberthi* (Fig. 42). In this case, successive divisions may follow so rapidly that the resting nucleus is reformed before the chromosomes have separated completely at telophase. This leads to dumb-bell-shaped nuclei which were formerly interpreted as stages of "amitosis". When such a nucleus, actually a double nucleus, divides again, the configuration of the chromosomes is identical with that of the previous telophase.

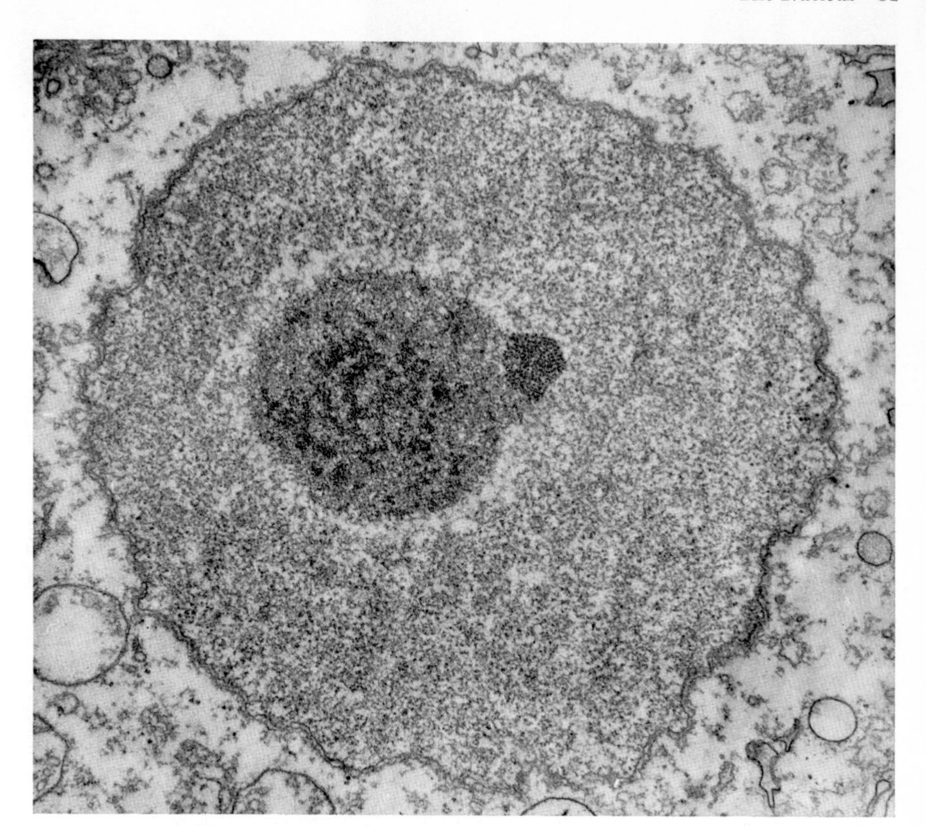

Fig. 41. *Corallomyxa mutabilis* (Amoeba). Cell nucleus. The nucleolus is located close to a "chromocenter". × 20000. Courtesy of G. BENWITZ

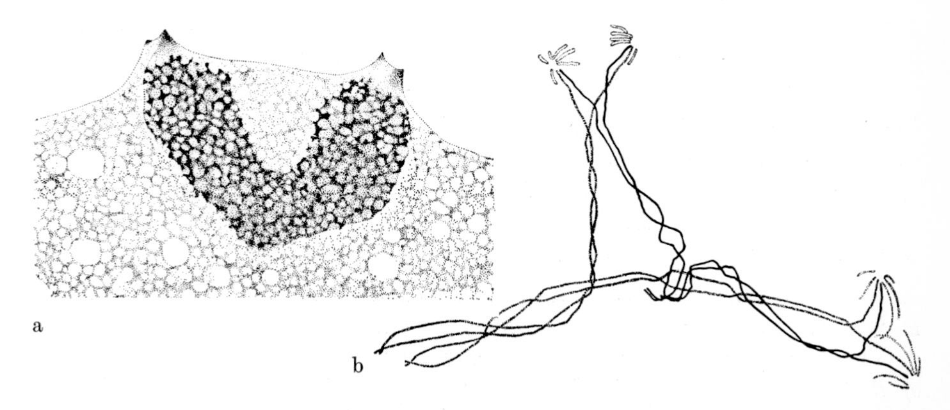

Fig. 42. *Aggregata eberthi* (Coccidian). Abnormal sporogony division. a Interphase stage of a "double nucleus" after incomplete separation of the chromatids. b The next anaphase of such a nucleus. Iron hematoxylin. × 2600. After BELAR, 1926 [*134*]

Table 1. Chromosome numbers in Protozoa

Species	Order	Haploid number	Authors
Spirotrichonympha polygyra	Polymastigina	2	CLEVELAND (1938)
Holomastigotoides tusitala	Polymastigina	2	CLEVELAND (1949)
Trypanosoma lewisi	Protomonadina	3	WOLCOTT (1952)
Diplocystis schneideri	Gregarinida	3	JAMESON (1920)
Gregarina blattarum	Gregarinida	3	SPRAGUE (1941)
Actinocephalus parvus	Gregarinida	4	WESCHENFELDER (1938)
Stylocephalus longicollis	Gregarinida	4	GRELL (1940)
Klossia loossi	Coccidia	4	NABIH (1938)
Volvox globator	Phytomonadina	5	CAVE and POCOCK (1951)
Echinomera hispida	Gregarinida	5	SCHELLACK (1907)
Tetrahymena pyriformis	Holotricha	5	RAY (1954)
Entamoeba histolytica	Amoebina	6	KOFOID and SWEZY (1925)
Discorbis vilardeboanus	Foraminifera	6	Le CALVEZ (1951)
Zygosoma globosum	Gregarinida	6	NOBLE (1938)
Aggregata eberthi	Coccidia	6	DOBELL and JAMESON (1915)
Volvulina steinii	Phytomonadina	7	STEIN (1958)
Urinympha talea	Polymastigina	8	CLEVELAND (1951)
Rotaliella roscoffensis	Foraminifera	9	GRELL (1958)
Leptospironympha wachula	Polymastigina	10	CLEVELAND (1951)
Allogromia laticollaris	Foraminifera	10	ARNOLD (1955)
Pandorina morum	Phytomonadina	12	COLEMAN (1959)
Zelleriella intermedia	Opalinina	12	CHEN (1948)
Spirotrichosoma normum	Polymastigina	12	CLEVELAND and DAY (1958)
Notila proteus	Polymastigina	14	CLEVELAND (1950)
Chlamydomonas reinhardi	Phytomonadina	16	Mc VITTIE and DAVIES (1971)
Gonium pectorale	Phytomonadina	17	CAVE and POCOCK (1951)
Tracheloraphis phoenicopterus	Holotricha	17	RAIKOV (1958)
Rotaliella heterocaryotica	Foraminifera	18	GRELL (1957)
Actinophrys sol	Heliozoa	22	BELAR (1922)
Patellina corrugata	Foraminifera	24	LE CALVEZ (1950)
Trichonympha okolona	Polymastigina	24	CLEVELAND (1949)
Spirotrichosoma promagnum	Polymastigina	24	CLEVELAND and DAY (1958)
Barbulanympha ufalula	Polymastigina	26	CLEVELAND (1953)
Trachelonema sulcata	Holotricha	28	KOVALEVA (1972)
Spirotrichosoma paramagnum	Polymastigina	48	CLEVELAND and DAY (1958)
Spirotrichosoma magnum	Polymastigina	60	CLEVELAND and DAY (1958)

In all species studied in sufficient detail, it has been shown that in Protozoa, too, the *number of chromosomes* is constant. Special cases are found only in the polygenomic nuclei (p. 96). This constancy of number indicates that daughter chromosomes are regularly distributed to the two daughter nuclei at mitosis. Table 1 is a compilation of chromosome numbers of some species of Protozoa.

A relationship between the size of the nucleus and the number of chromosomes is most clearly expressed in species where the chromosomes are indeed the predominant component of the nucleus. One example is furnished by the *micronuclei of ciliates* which, like sperm heads of Metazoa, are highly condensed nuclei. Fig. 43 of *Paramecium bursaria* illustrates the differences in the size of micronuclei of various races. In meiotic prophase when the nuclei swell up considerably, it is

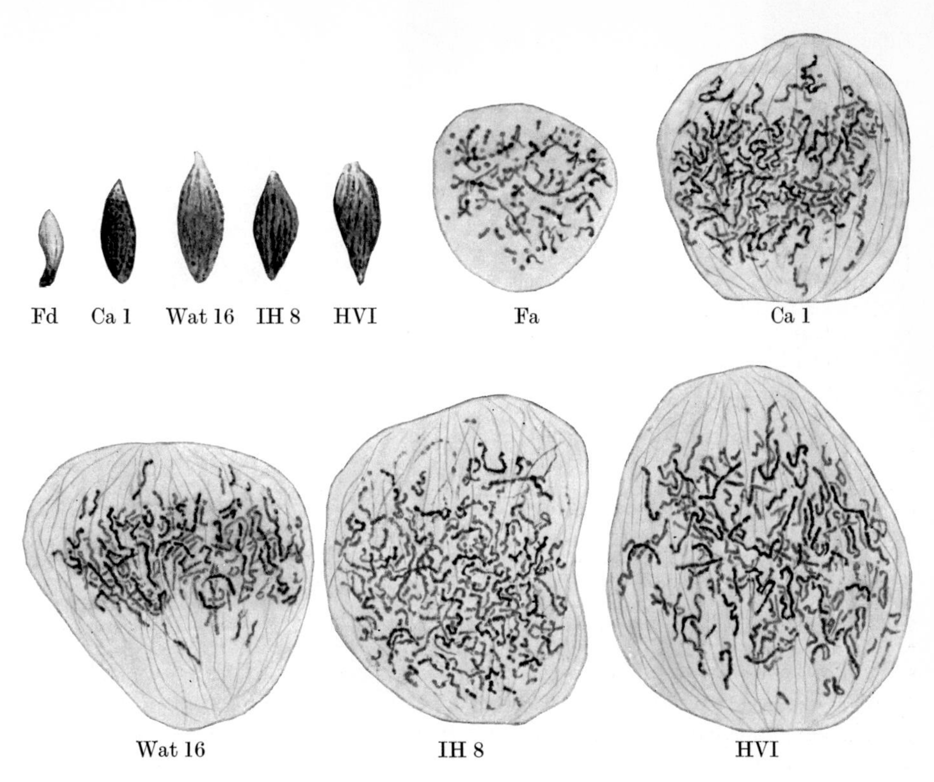

Fig. 43. *Paramecium bursaria.* Micronuclei from five races of the same mating type. Above left the vegetative micronuclei. The remaining figures show the micronuclei of the same races during prophase of the first progamic division. Iron hematoxylin. × 1390. After CHEN, 1940 [*244*]

easily recognizable that the differences in chromosome number are considerable. While only about 70 chromosomes are found in race Fd, there are several hundred in the other races.

We speak of *polyploidy* whenever there are several chromosome sets in a single nucleus. Although it was not possible, in *Paramecium bursaria*, to obtain a precise chromosome count, we may assume that the high number of chromosomes is due to polyploidy. This may also be true for *Amoeba proteus* in whose equatorial plate more than 500 chromosomes have been identified (Fig. 44).

Polyploidy is, of course, only proved when it is possible to demonstrate that the chromosome numbers of closely related species or races are *multiples* of a certain *basic number*. As an example, the species of the genus *Spirotrichosoma* (Hypermastigida) of the New Zealand termite *Stolotermes ruficeps* may be mentioned [*321*]. Even though the smallest species has the lowest chromosome number ($n = 12$) and the largest one the highest ($n = 60$), there is no consistent correlation of body size and degree of polyploidy. The foraminiferan, *Rotaliella roscoffensis* ($n = 9$), has half as many chromosomes as the closely related *Rotaliella heterocaryotica* ($n = 18$) but its chromosomes are about twice as large (Fig. 80). In the hypermastigids, *Holomastigotoides tusitala*, and *H. diversa*, whose basic number is $n = 2$ (Fig. 45), there are *euploid* races whose chromosomes are present twice, or in triplicate, as well as *heteroploid* races where only single chromosomes of the set are

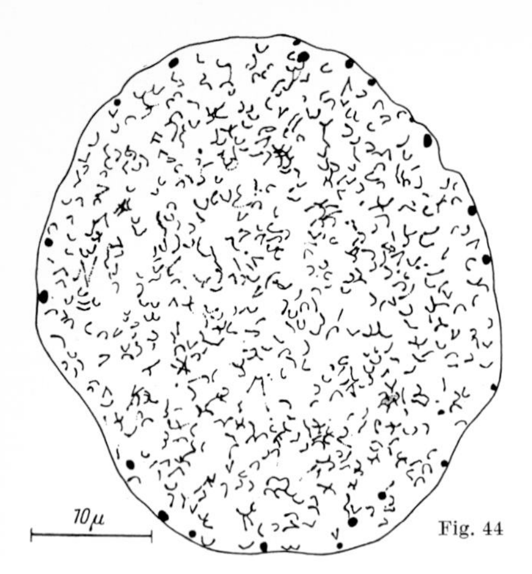

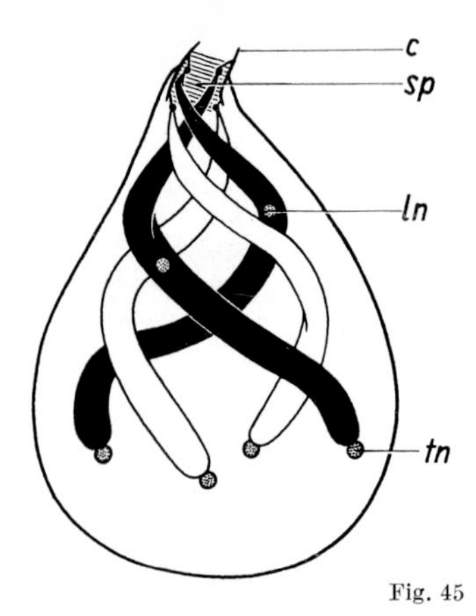

Fig. 44. *Amoeba proteus.* Late prophase (combined from three consecutive sections). Sections of the spindle fibers are not included. BOUIN, Iron hematoxylin. After LIESCHE, 1938 [*1043*]

Fig. 45. *Holomastigotoides tusitala* (Polymastigina). Diagram of the "resting nucleus" (late prophase stage). Chromatids of the same chromosome are shown black or white respectively. *c* Centriole. *sp* Spindle. *ln* Lateral nucleolus. *tn* Terminal nucleolus. After CLEVELAND, 1953 [*285*]

found in multiple [275]. It is difficult to decide whether the large variations of chromosome number found in *Holomastigotiodes psammotermitides* are due to differences of race. In this species only about 0.5% of the individuals within the same population had the basic number of $n = 2$. Most had tetra-or octoploid chromosome sets and many showed heteroploidy with 3, 5 or 7 chromosomes [*627*].

The *size of the chromosomes* varies among the Protozoa within rather wide limits. Very small chromosomes are, for instance, found in the cryptomonads and the amoebae, very large ones in the hypermastigids (Fig. 45) and in the radiolarian, *Aulacantha scolymantha* (Fig. 105–108).

If their number is small, as in the hypermastigids mentioned above, differences of size and form become recognizable among the chromosomes of a set. These differences are an expression of their individuality.

Our knowledge of chromosomal morphology is based upon the study of favorable objects. As a rule each chromosome is provided with a special point where it is attached to the spindle at mitosis (Fig. 52), the *kinetochore* or *centromere*. The position of the kinetochore is characteristic for each chromosome. In most chromosomes it is intercalary, i.e. it divides the chromosome into two arms of different or equal length. Chromosomes of this type are called *metakinetic*. In many cases the point of spindle fiber attachment is characterized by a special granule, the spindle granule or *kinosome* (Fig. 45) which may be connected to the kinetochore by a fine thread. In metakinetic chromosomes the point of spindle fiber attachment is frequently recognizable as a constriction. It has previously been denied that the kinetochore can be at the very end of a chromosome. However, such *telokinetic* chromosomes are indeed found in some Protozoa (Fig. 45). Aside from the primary constriction at the site of the kinetochore there may also be secondary ones. The secondary constriction separates a usually short segment, the *satellite*, from the main chromosome body. At these secondary constrictions the nucleolar substance may be formed.

Basically, a chromosome consists of a fine thread, the *chromonema*, which is helically coiled to varying degrees. The cycle of chromosomal *condensation* and *decondensation* during nuclear division depends upon the coiling and uncoiling of the chromonema. In nuclei which are incapable of division, e.g. the salivary gland nuclei of Diptera, *polytenic* chromosomes consisting of many threads may be formed. Similar polytenic chromosomes have been found recently in the newly forming macronucleus of some ciliates. This is evidently only a temporary state of chromosomal organization (p. 108). The existence of a special *matrix* surrounding the coiled chromonema is still under debate.

A special situation is found in species of the genus *Holomastigotoides* where no typical interphase nucleus is formed. Instead, nuclear division is arrested for an extended period in a late prophase stage (Fig. 45). Each of the two chromosomes has already reduplicated at this time to form two chromatids which are then distributed to the two daughter nuclei in the further course of mitosis. In *Holomastigotoides tusitala* both chromosomes are telokinetic. They are attached by fine threads from the kinosomes to the rod-shaped centrioles (c). The spindle fibers (sp) are stretched between the centrioles. The two chromosomes can be distinguished not only by their different lengths but also by the position of the nucleoli. Only one of the two chromosomes has a lateral nucleolus (ln) while both have a terminal one (tn). The terminal nucleoli are frequently fused with each other.

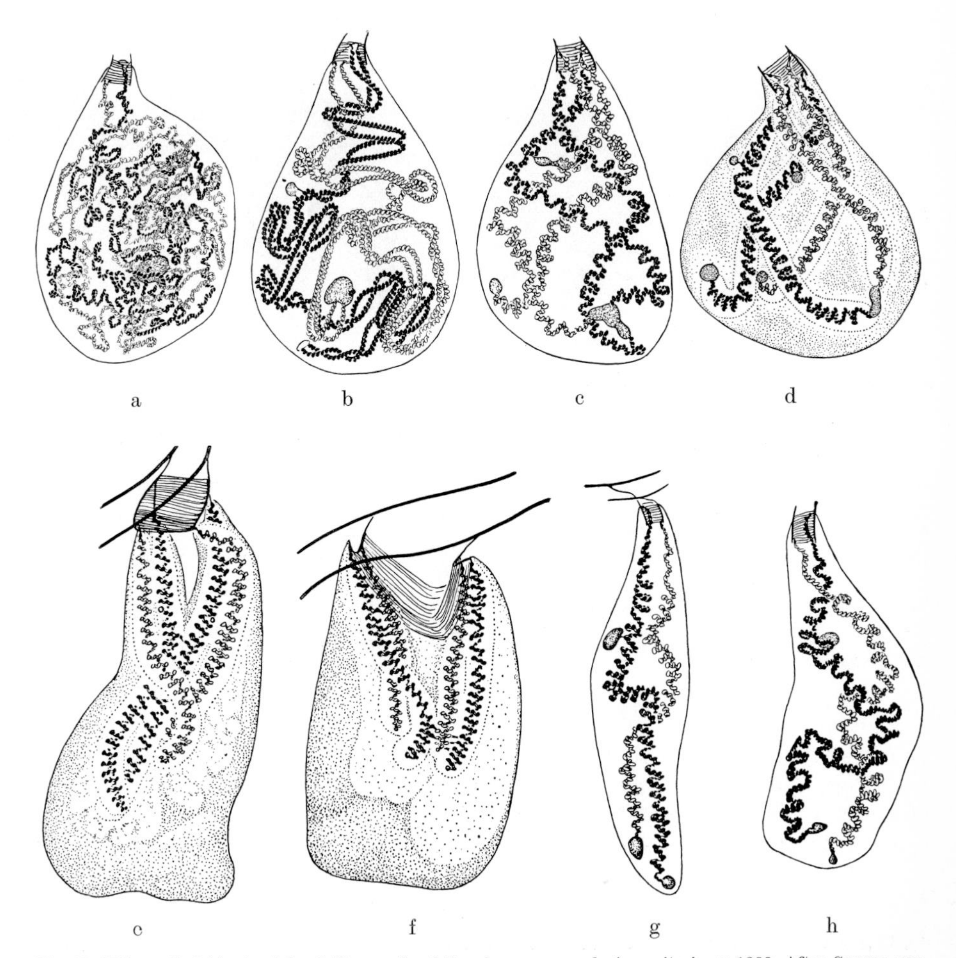

Fig. 46. *Holomastigotoides tusitala*. Coiling cycle of the chromosomes during mitosis. × 1200. After CLEVELAND, 1949 [275]

Because of their large size, the chromosomes of *Holomastigotoides* species are especially suited for a study of the coiling cycle. Since nuclear division in *Holomastigotoides* is of a highly specialized type, it is not discussed in detail at this point. As shown in Fig. 46, the chromosomes are very long and thin at the onset of nuclear division (a). They do, however, show *minor coils* which are retained during the whole period of nuclear division. Reduplication of the chromosomes must therefore take place in the presence of minor coiling. As soon as the daughter chromosomes have separated (b), *major coils* make their appearance (c). These lead to a shortening and condensation of the chromosomes (d), i.e. to the stage at which the nucleus remains for some time (see above). When division commences, the spindle elongates (e) and the daughter chromosomes move apart (f). In the daughter nuclei the major coils unwind (g, h). This is followed by the next reduplication of the chromosomes.
Fig. 47 shows the chromosomes of *Holomastigotoides psammotermitidis* in different stages of their coiling cycle. In this species the chromosomes are metakinetic [*627*].

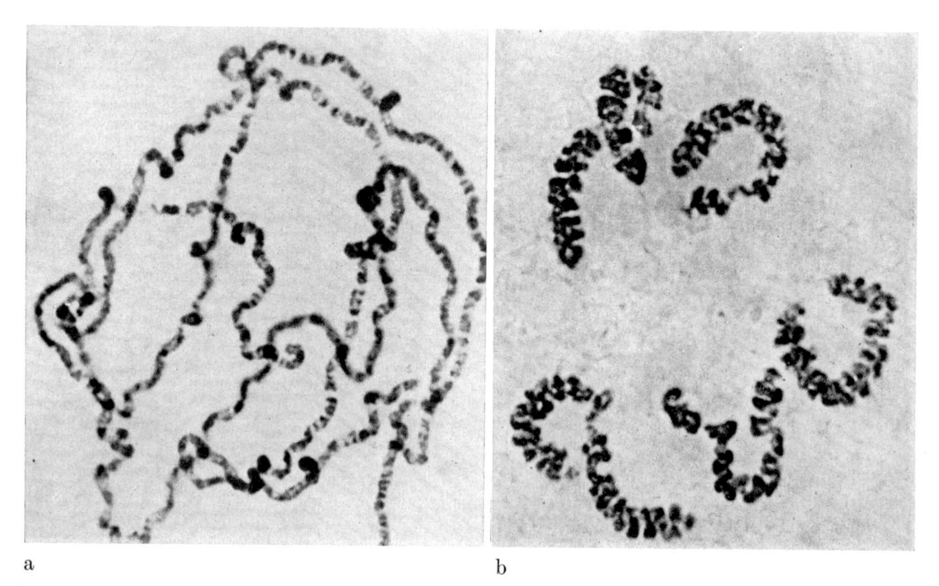

Fig. 47. *Holomastigotoides psammotermitidis*. Chromosomes of early (a) and late (b) prophase. In a the minor coils can be distinguished, in b the major coils. The chromosomes have reduplicated to form closely adjacent chromatids. Aceto-carmine squash preparation. × 1500. After Grassé and Hollande, 1963 [*627*]

The presence of localized kinetochores and the organization of nucleolar substance along definite chromosome regions is also recognizable in Opalinina of the genus *Zelleriella*. These are diplonts with 24 chromosomes. Fig. 48 shows the haploid set of *Zelleriella louisianensis*. The chromosomes are drawn singly, together with their spindle fibers from metaphase stages. All are metakinetic and show the point of spindle fiber attachment clearly. The latter divides each chromosome into two arms of unequal length. In chromosomes 1, 4 and 5 there is, on one of the arms, a segment associated with nucleolar substance.
The number of these "*nucleolus-chromosomes*" can be different in the individual

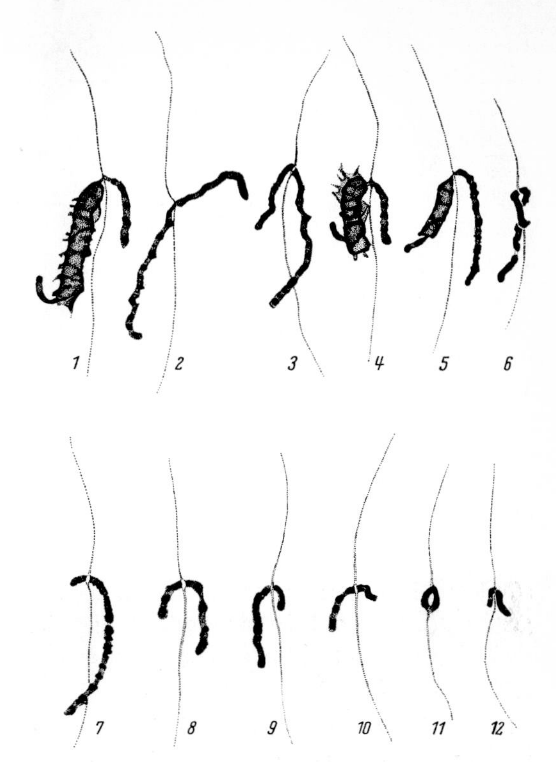

Fig. 48. *Zelleriella louisianensis* (Opalinina). Haploid chromosome set (different chromosomes are indicated by numbers). *1*, *4* and *5* are nucleolus chromosomes. Sublimate alcohol, iron hematoxylin. × 2640. After CHEN, 1948 [*250*]

species. Whereas *Zelleriella elliptica* has four, *Z. louisianensis* and *Z. intermedia* have six. In the resting nucleus, which is a typical interphase nucleus, the nucleolar substance forms elongated, often indistinctly delimited areas which can fuse with each other (Fig. 49).

Many Protozoa have so-called *karyosome nuclei* with a globular, usually central *nucleolus*, formerly called nuclear body or karyosome. When several nucleoli are present, they are most often found below the nuclear envelope. In *Amoeba proteus* the nucleoli also form a peripheral layer.

In some cases a tangle of threads ("nucleolonema") can be demonstrated in the nucleoli or a heterochromatic "knob" may be attached to it (Fig. 41). Very likely, this is also an indication of the connection to a "nucleolus-chromosome".

Electron micrographs show that the nucleolus is not separated from the karyoplasm by a membrane. The nucleolus is composed chiefly of small granules which are similar to the ribosomes of the cytoplasm. Whether the ribosomes are indeed formed within the nucleolus cannot be decided at the present time. So far it can only be considered probable that ribosomal RNA originates in the nucleolus.

Though the function of the nucleolar substance has not been satisfactorily clarified, there is no doubt that in one way or another it plays a role in cell metabolism. This is indicated by the observation that in the nuclei of growing cells, the nucleolar

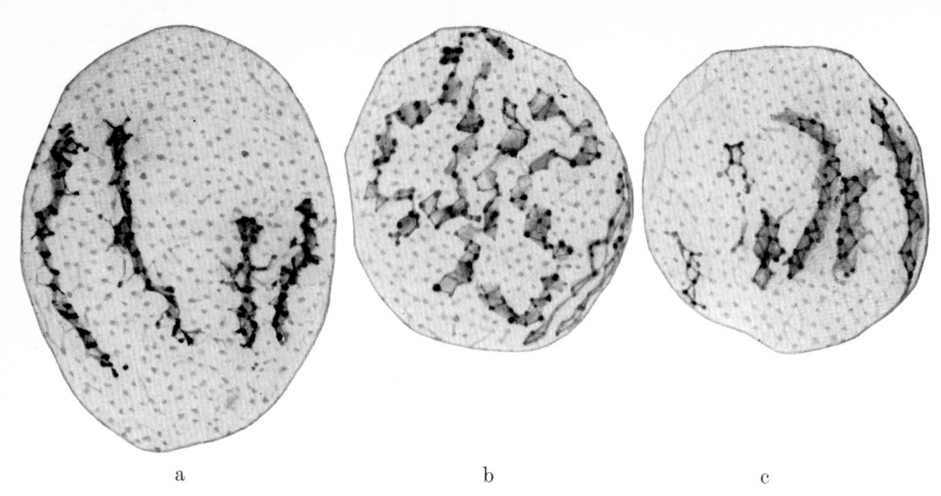

Fig. 49. Resting nuclei of various species of *Zelleriella*. a *Z. elliptica*, b *Z. lousianensis*, c *Z. intermedia*. Sublimate alcohol, iron hematoxylin. × 2640. After CHEN, 1948 [*250*]

substance is always clearly developed, whereas in the nuclei of cells which are not growing (for example in gametes or sporozoites) little or nothing is demonstrable. Sometimes structural changes in the nucleoli can be observed which could be interpreted as different functional states (e.g. gamonts of Sporozoa).

Thecamoeba verrucosa (Fig. 346a) has two bodies of different structure in its cell nucleus. One of these has been designated as nucleolus and the other as „Binnenkörper". Both contain RNA and are not stainable with the Feulgen reagent. During mitosis the nucleolus stays at one side of the spindle, whereas the material of the „Binnenkörper" is distributed to the spindle poles (see film C 943).

When *Amoeba proteus* is centrifuged long enough, the various components of the cell nucleus become stratified in different layers. The nucleoli flow together to form an amorphous mass at the centrifugal pole which is composed of two clearly different layers [*361*].

In these cases the nucleolar substance does not appear to be uniform but rather built up of different components which also could have a different function.

Electron microscopic investigations have as yet brought little information concerning the *fine structure of chromosomes*. In typical interphase nuclei they are usually so uncoiled that they can only be distinguished from the karyoplasm by special methods which increase the contrast.

The chromosomes of the *dinoflagellates* (Fig. 38), appearing even in the resting nucleus as clearly distinguishable threads, show regular horizontal banding and a fibrillar structure (Fig. 50). Since observations with the light microscope have proved that the chromosomes of the resting nucleus can become uncoiled, the obvious explanation is that the horizontal banding is due to the coiling of a "chromonema" [*411, 683*]. However, other model conceptions have also been set forth [*155, 588, 589, 628, 629*]. Above all it has not been clarified whether we are dealing with a single thread with complicated coiling or with numerous fibrils. If the statement that the chromosomes of the dinoflagellates contain no histone but only DNA is correct [*413, 1437*], they are then similar to the bacterial chromosomes in one essential respect (p. 5).

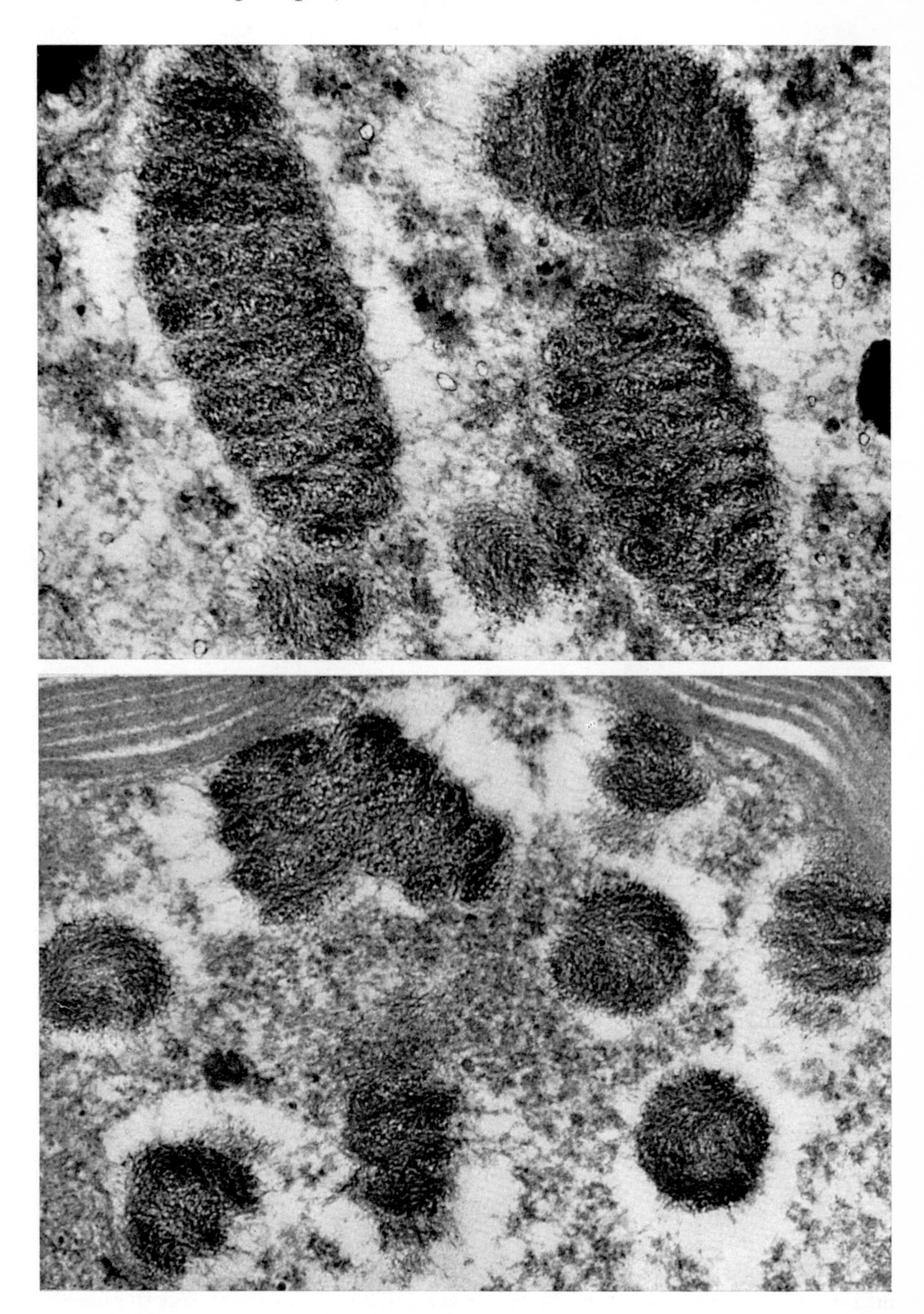

Fig. 50. *Amphidinium massarti* (Dinoflagellata). Chromosomes in the resting nucleus. × 60000. After GRELL and SCHWALBACH, 1965 [*683*]

In *Amoeba proteus* and in *Pelomyxa carolinensis*, groups of "*helices*" were found in the resting nuclei frequently originating from a common axis, which is not yet precisely defined. Occasionally, they have been found singly below the nuclear pores and even in the cytoplasm adjacent to the nucleus. By autoradiographic methods, RNA but not DNA could be demonstrated in the "helices". The original assumption that the "helices" represent a structural component of interphase chromosomes is therefore no longer defensible. There is the possibility that they represent aggregations of messenger RNA or RNA-protein complexes [*391, 1266—1268, 1450, 1672, 1811a, 1824*].

The *nuclear envelope* is the border between the nucleus and the cytoplasm. Although it may structurally be related to the cytoplasm, it can be looked upon as a part of the nucleus, as it remains connected with it during intracellular movements. Electron micrographs show that the nuclear envelope is composed of *two unit membranes* separated by a variable *perinuclear space* of 100–300 Å width. In all closely studied cases "*pores*" could be demonstrated in the nuclear envelope. They are more or less regularly distributed. Depending upon the species investigated, their diameter ranges from 500–1000 Å. At the margins of each pore, the two unit membranes merge with each other.

If a gold sol is injected into the cytoplasm of *Amoeba proteus*, the gold particles are found after 24 hours not only in the cytoplasm but also within the nucleus. In ultrathin sections, the gold particles are frequently found within the pores of the nuclear envelope [*513, 514*]. Occasionally the margins of the pores are surrounded by electron dense material. Tangential sections of the nuclear envelope show, therefore, ring-shaped annuli.

In *Tetrahymena pyriformis* where the macronuclear envelope contains 68 ± 14 pores per square-micron, a positive correlation between the number of pores containing a central granule and the nucleocytoplasmic RNA-efflux at the different physiological states could be demonstrated. Evidently, the central granules are ribonucleoproteins fixed at their passage into the cytoplasm [*1827*].

In *Amoeba proteus*, a honeycomb-like structure is found below the inner unit membrane of the nuclear envelope. Each tube of the honeycomb layer is concentric around a pore of smaller diameter [*1450*]. In *Entamoeba blattae* the arrangement of the honeycomb layer and the pores is reversed with the honeycomb pattern outside and the porous nuclear envelope below it [*124*].

We speak of *nuclear differentiation* when different types of nuclei are found in the same species. Nuclear differentiation can be *successive* or *simultaneous*. In the first case, different nuclear types are formed successively. They may differ in chromosomal condensation as well as in the amount of nucleolar substance and karyoplasm.

When the gamont of the monothalamic foraminifer, *Myxotheca arenilega*, grows, its nucleus reaches a considerable size (Fig. 51a). This increase in volume is exclusively due to the nucleolar substance and the karyoplasm. As soon as the gamont has reached a certain size, gamogony sets in, leading to the formation of mobile gametes (comp. Fig. 359). The nucleoli which form a thick layer below the nuclear envelope fuse to form large spheres which are irregularly distributed in the karyoplasm. The chromosomes, once located in the nuclear interior, migrate jointly to the periphery where a comparatively small spindle is formed (b). A complete disintegration of the nucleus follows. While the nucleolar substance and the karyoplasm are largely resorbed in the cytoplasm of the gamont, two small daughter

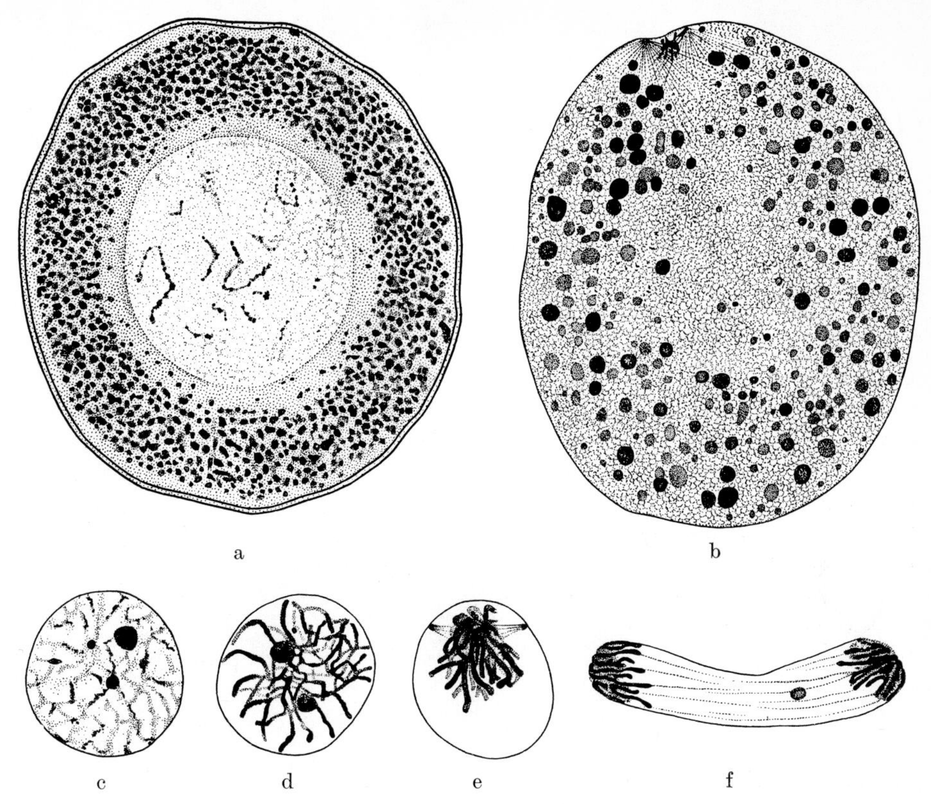

Fig. 51. *Myxotheca arenilega* (Foraminifera). Nucleus of the gamont. a Resting stage. b First gamogonic division. —f Later gamogonic divisions. a and b × 1100; c—f × 3000. After FÖYN, 1936 [*531*]

nuclei arise which continue to multiply and will eventually give rise to the gamete nuclei (c—f).

It will hardly be necessary to stress that both types of nuclei have the same haploid set of chromosomes. The first mitosis of gamogony must therefore be triggered by a specific state of the cytoplasm. This occurs only when the gamont has reached a certain critical size.

Simultaneous nuclear differentiation, whereby different nuclear types appear simultaneously in the same cell, will be discussed after the description of nuclear division (p. 96).

2. Nuclear Division

Mitosis is the usual mode of nuclear division in Protozoa. *Meiosis* which leads to a reduction of the chromosome number is derived from mitosis. The aberrant mode of division of *polygenomic* nuclei will be discussed in a separate chapter (p. 96).

a) Mitosis

The essential feature of mitosis is the regular distribution of previously reduplicated chromosomes, the *daughter chromosomes*, or *chromatids*, to the two daughter nuclei. Distribution is accomplished with the aid of a special *spindle apparatus*.

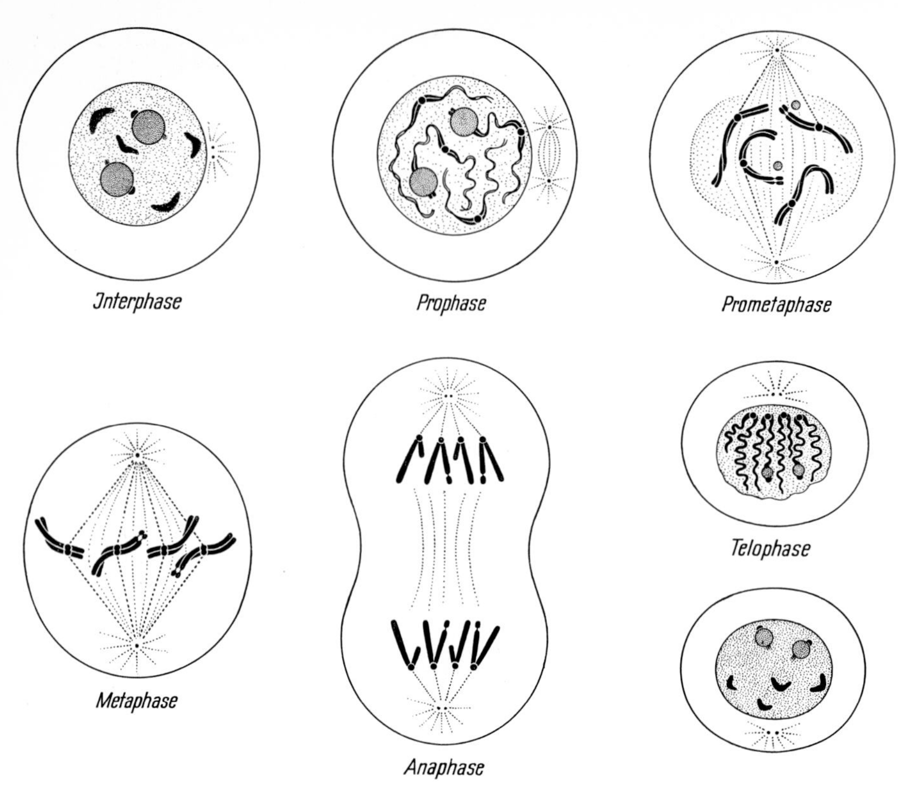

Fig. 52. Diagram of mitosis in a multicellular animal

The *course of mitosis* in multicellular animals follows essentially the scheme shown in Fig. 52. In this diagram, the nucleus is taken to be diploid and to contain two pairs of homologous chromosomes, of which one is able to form a nucleolus. The chromosomes consist of *euchromatic* segments which are uncoiled in the resting nucleus and of *heterochromatic* regions which are recognizable in the interphase nucleus as distinct, Feulgen-positive *chromocenters*. In our diagram, heterochromatic segments are shown chiefly at the sites of spindle fiber attachment as is usually the case.

Spindle formation proceeds from the *centriole*, a small granule at the periphery of the nucleus, which is reduplicated already during interphase. In *prophase* the chromosomes are recognizable as elongated threads which are reduplicated already at this stage into two chromatids. Simultaneously, the centrioles move apart, still connected by a spindlelike fibrous structure.

As soon as the centrioles have moved to opposite poles, the nuclear envelope breaks down. From the centrioles which are surrounded by a fine *polar radiation*, fibers grow toward the nuclear space which, in part, unite with each other to form a fibrous body, the *spindle*, while other fibers become attached to the kinetochores of the chromosomes (prometaphase).

During *metaphase*, the chromosomes arrange themselves with their points of spindle fiber attachment in a plane midway between the two poles, the so-called *equatorial plane*.

Following the division of the kinetochores, the chromatids move apart during the *anaphase* stage. The shortening and the condensation of the chromosomes, due to progressive coiling from early prophase on, now reaches its maximum.

It is only during *telophase* that the chromosomes begin once again to unwind and to elongate. As they approach their interphasic state, the formation of new nucleoli sets in which are again dissolved in the late prophase stage of the next mitosis.

In Protozoa, the mitotic pattern can be modified in various ways. This is largely due to differences of the spindle apparatus and, to a lesser extent, to peculiarities of the chromosomes themselves.

The nuclear divisions which lead to the formation of gametes in *gregarines* follow the mitotic pattern of Metazoa to a large extent.

In *Monocystis magna* (Fig. 53) [*134*] the centrioles are surrounded by a lightly staining region during nuclear division, the so-called *centrosome*. At prophase (a),

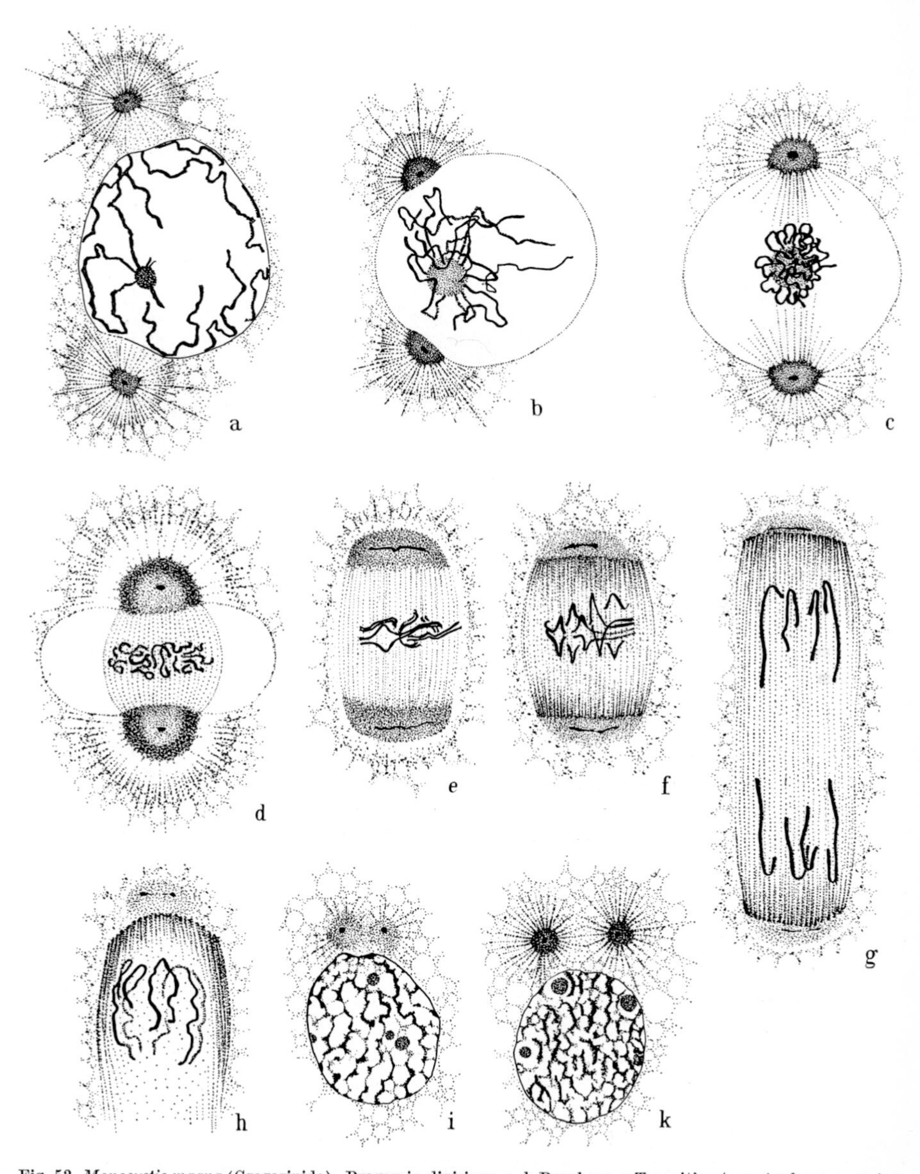

Fig. 53. *Monocystis magna* (Gregarinida). Progamic divisions. a, b Prophase. c Transition to metaphase. e—g Anaphase. h—k Telophase. × 1850. After BELAR, 1926 [*134*] from CLAUS-GROBBEN-KÜHN, 1932 [*269*]

the polar radiation is formed by the centrosomes and the two centers become directly attached to the nuclear envelope (b). The latter is locally dissolved, the polar rays grow into the nuclear space (c) and unite to form a barrel-shaped spindle (d). The metaphase stage is now reached and the chromosomes are arranged to form the equatorial plate. As the chromatids move apart at anaphase, the centrioles elongate at right angles to the axis of the mitotic spindle (e–g). After their reduplication, a new centrosome with polar radiation is formed around each daughter centriole during the telophase stage (h–k).

In another gregarine, *Stylocephalus longicollis* (Fig. 54) [*648*], the midportion of the spindle which separates the daughter plates at anaphase has a tube-like appearance. At telophase it stretches considerably while it is constricted and finally bent in the middle.

The spindle apparatus is also very prominent in the mitosis of *Polymastigina*. A more detailed account is called for in this group since the structures responsible for division are modified in various ways.

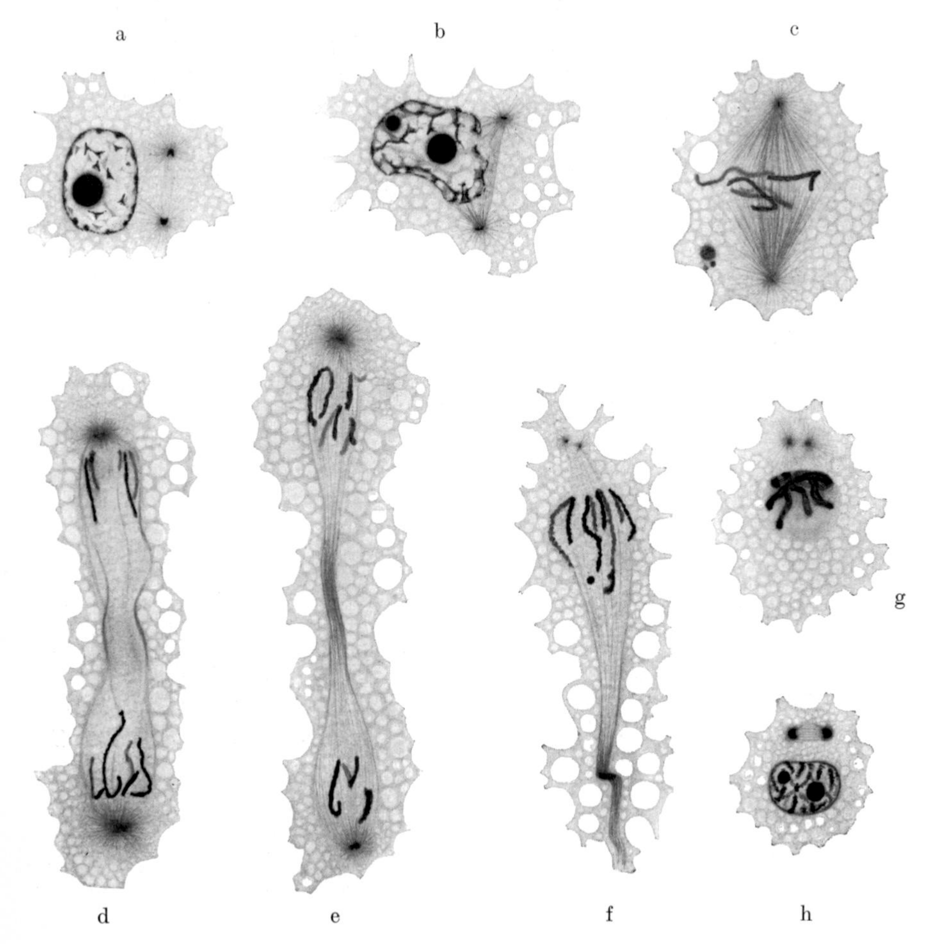

Fig. 54. *Stylocephalus longicollis* (Gregarinida). Progamic divisions. a, b Prophase. c Metaphase. d, e Anaphase. f—h Telophase. Carnoy-iron hematoxylin. × 2700. After GRELL, 1940 [*648*]

Electron microscopic investigations of some Polymastigina [*784—786, 792—799*] have shown that the "nucleating sites" (p. 31) at which the spindle fibers (microtubules) are formed do not have the ultrastructure of "typical" centrioles (p. 292). For this reason another term ("attractophores") has been proposed for them. On the other hand, it should be remembered that the term "centriole" (central granule of the centrosome) was introduced in cytology (BOVERI, 1895) before its ultrastructure was known. In addition, the "atypical" centrioles of Polymastigina vary in their ultrastructure. Other types of "nucleating sites" for spindle fiber formation exist in other groups of Protozoa (e.g. Amoebina, Foraminifera). Since it would be very confusing to introduce different terms for all such structures which coincide as "centers of spindle formation", the author prefers to retain the name "centriole" in this context*.

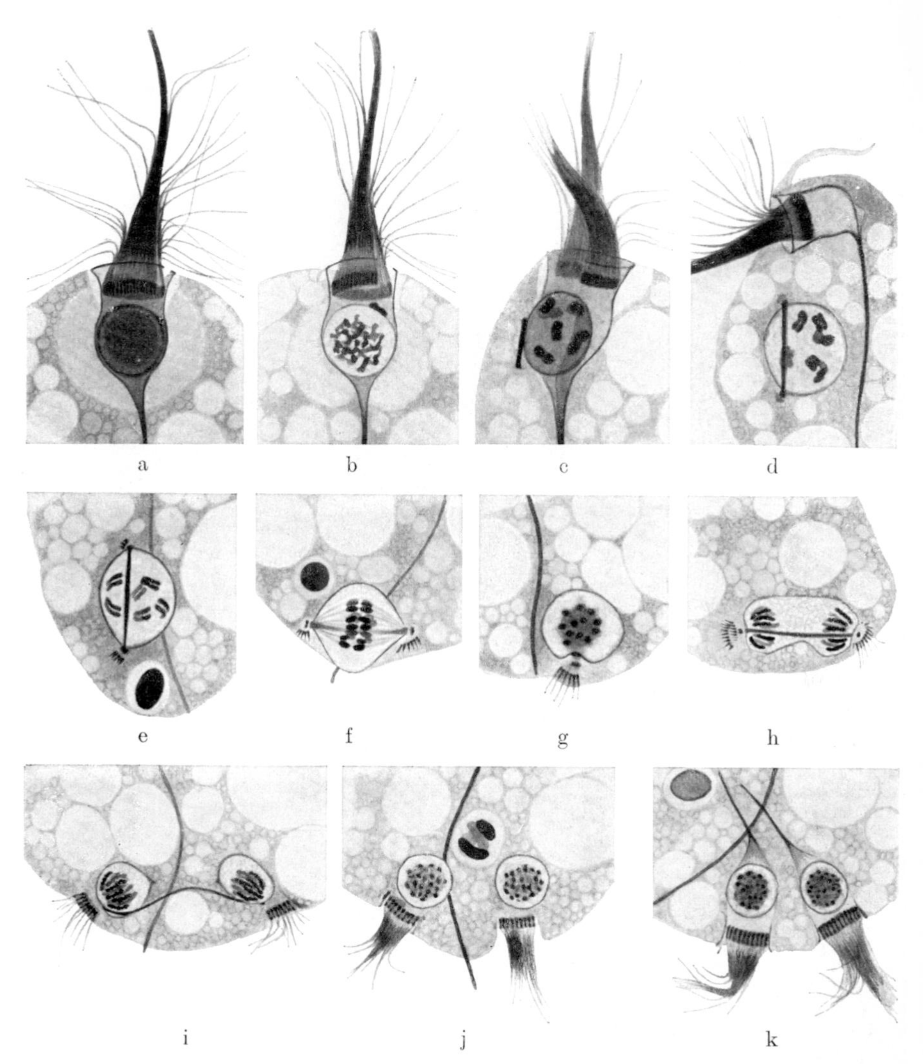

Fig. 55. *Lophomonas blattarum* (Polymastigina). Mitosis. a—d Shows a portion of the anterior, e—k of the posterior end of the cell. a Resting nucleus in the fibrillar calyx, which tapers toward the posterior end to the axostyle; over the nucleus, the wreath of basal bodies; b—d Prophase. e Metaphase. f Anaphase. g The same in polar view. h Late anaphase. i—k Telophase (formation of the new fibrillar calyces). Sublimate alcohol, iron hematoxylin. × 1900. After BELAR, 1926 [*134*]

* It might be advisable in the future to call "centers of spindle formation" which have the ultrastructure of basal bodies "eucentrioles" and all others "heterocentrioles".

In the Pyrsonymphida, these centers are located within the nucleus. An *intranuclear* spindle is therefore formed (Fig. 71 and 191).

It is, however, characteristic for most Polymastigina that the nuclear interior has no part in spindle formation, which is wholly *extranuclear*.

In *Lophomonas* (Fig. 55), which belongs to the most primitive Hypermastigida, the nucleus is surrounded by a goblet-shaped calyx which fixes its position in the anterior part of the cell. Toward the posterior end, the calyx tapers to form an axostyle. At the interior of the calyx, there is a circle of basal bodies from which a pointed tuft of flagella arises (a). At the beginning of mitosis, the nucleus slides out of the calyx (b, c). Two centrioles become recognizable at the nuclear envelope which move to opposite sides of the nucleus and form the spindle between them-

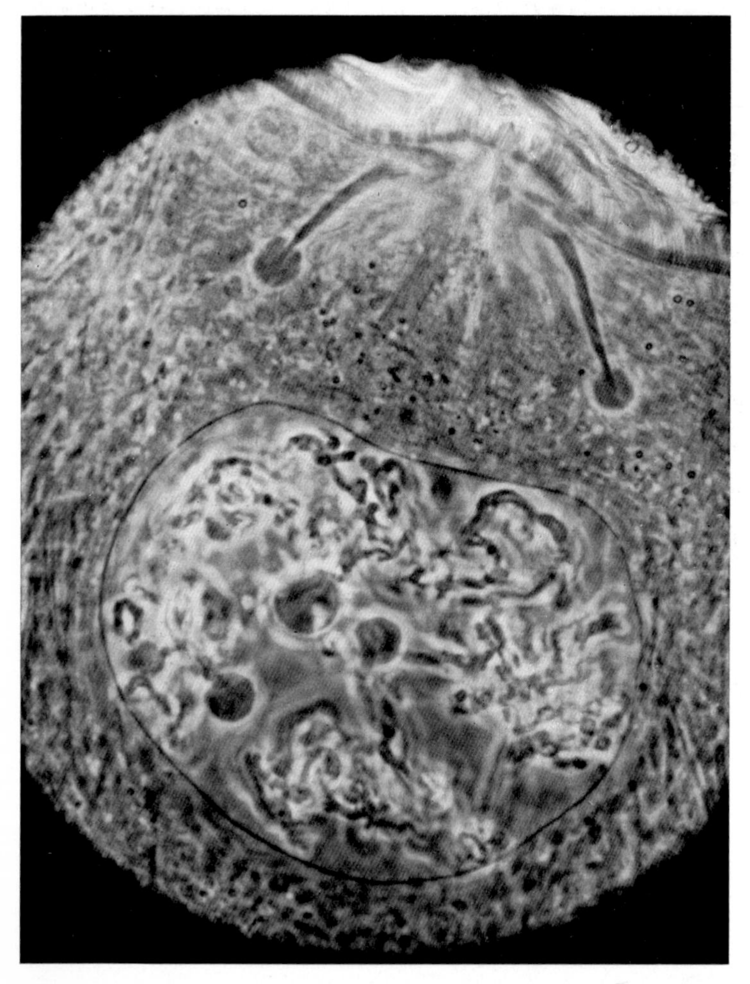

Fig. 56. *Barbulanympha*. Prophase of the first meiotic division. Centrosomes of the centrioles are still without polar radiation. In the nucleus the chromosomes have already reduplicated to form chromatids, which show relational coiling. Some nucleoli are visible within the nucleus. × 1500. Photographed in life with phase contrast, after CLEVELAND, 1953 [*285*]

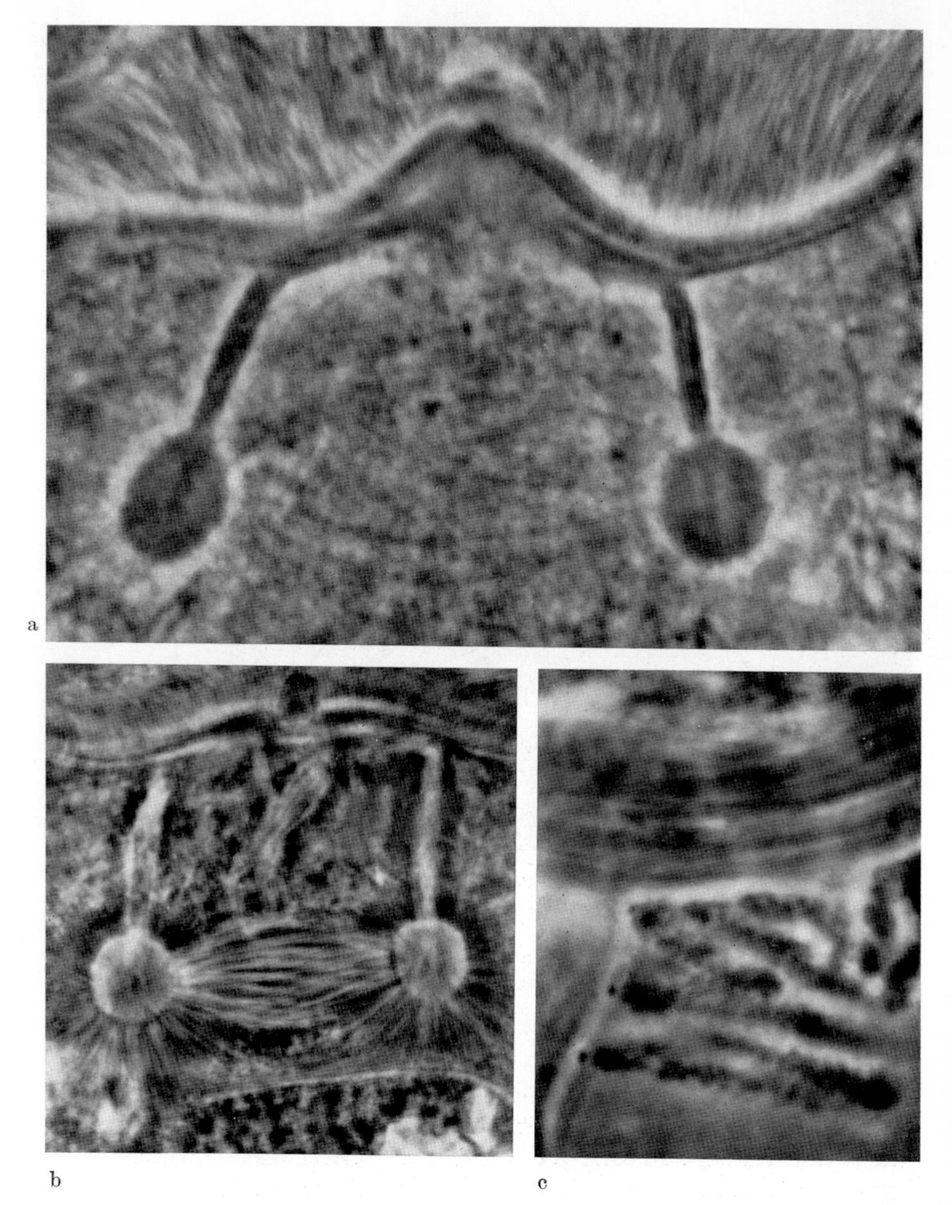

Fig. 57. *Barbulanympha*. Structures related to nuclear division. a Spindle fibers begin to form between the centrosomes. b Formation of the spindle during late prophase. c Three chromosomes (indications of coiling faintly discernible!) with spindle granules inserted into the nuclear envelope and the spindle above. Photographed in life with phase contrast. a Original by L. R. CLEVELAND, b after CLEVELAND, 1953 [*285*], c after CLEVELAND, 1954 [*288*]

selves (d, e). In anaphase (f) a circle of basal bodies appears in the vicinity of the centrioles.

In the highly differentiated Hypermastigida the centrioles can be permanent structures, attached in a definite fashion at the front end of the cell, and clearly recognizable in certain species even at times when no nuclear division takes place.

In *Barbulanympha* (Fig. 343) there are always two centrioles below the flagellar tuft amidst the parabasal bodies and axostyles. They are rod-like, contorted structures and consist of an elongated piece and the pointed posterior end which is surrounded by the spherical centrosome (Fig. 56) [*273, 296, 297*]. At prophase, asters are formed only at the centrosomes (Fig. 57a). Most fibers are connected to each other to form the spindle while others are attached to the spindle granules of the chromatids (b). The spindle granules are inserted into the nuclear envelope (c). That portion of the nuclear envelope which is between the spindle granules of daughter chromatids is therefore extraordinarily stretched during anaphase (Fig. 58). In spite of this the nuclear envelope, which must be rearranged during this process, remains whole until the daughter nuclei are reconstructed.

The nuclear division of *Barbulanympha* deserves special attention; not only the centrioles and the chromosomes with their spindle granules but also the spindle fibers, formerly often regarded as fixation artefacts, can be observed with incomparable clarity in life [*285*].

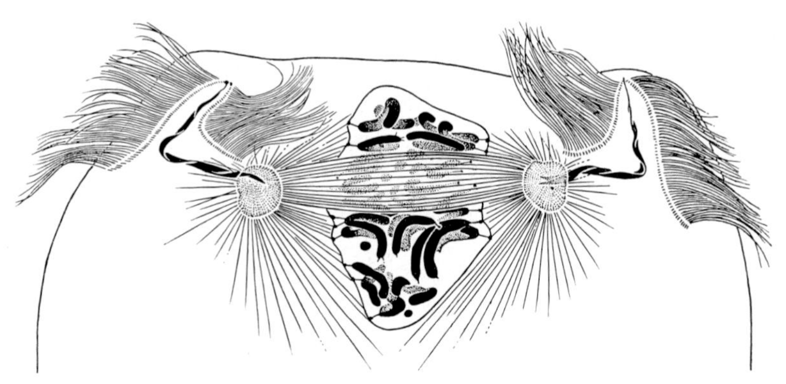

Fig. 58. *Barbulanympha* (Polymastigina). Mitosis. Anterior part of the cell (anaphase). After CLEVELAND, 1938 [*273*]

Electron microscopic pictures show (Fig. 59) that the spindle granules (kinosomes), being connected with the kinetochores by a fibrous material, are of high density and intimately integrated into the nuclear envelope.

In *Pseudotrichonympha* (Fig. 60) [*271, 625*], the centrioles are broken down after each nuclear division. In early prophase (a) they develop again, becoming rod-shaped structures whose ends are surrounded by centrosomes. Even during spindle formation, they continue to elongate (b). The centrosomes attach themselves directly to the nuclear envelope (c). After division (d), the centrioles become dissolved in the cytoplasm (e, f).

The behavior of the centrioles is even more unusual in *Macrospironympha* (Fig. 61) [*290*]. They are relatively small and difficult to recognize in this case, but they are

▶

Fig. 59. Attachment structures of chromosomes within the nuclear envelope of Polymastigina. a *Trichonympha agilis*. Early prophase of mitosis. Each attachment structure is connected with several microtubules (chromosomal fibers) × 14000. After HOLLANDE and CARRUETTE-VALENTIN, 1971 [*785*]. b *Barbulanympha ufalula*. Single attachment structure. × 120000. After HOLLANDE and VALENTIN, 1968 [*796*]

a

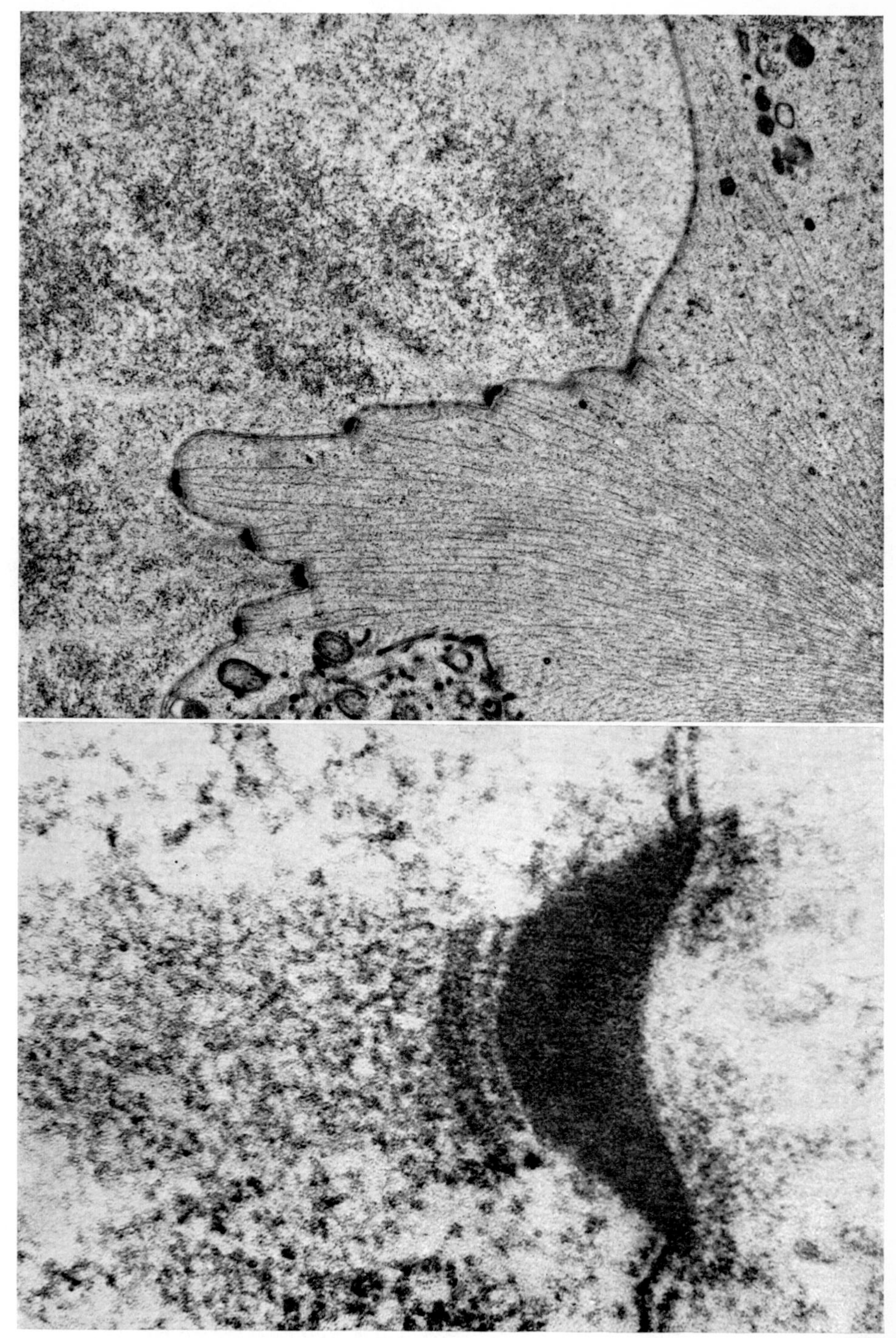

b

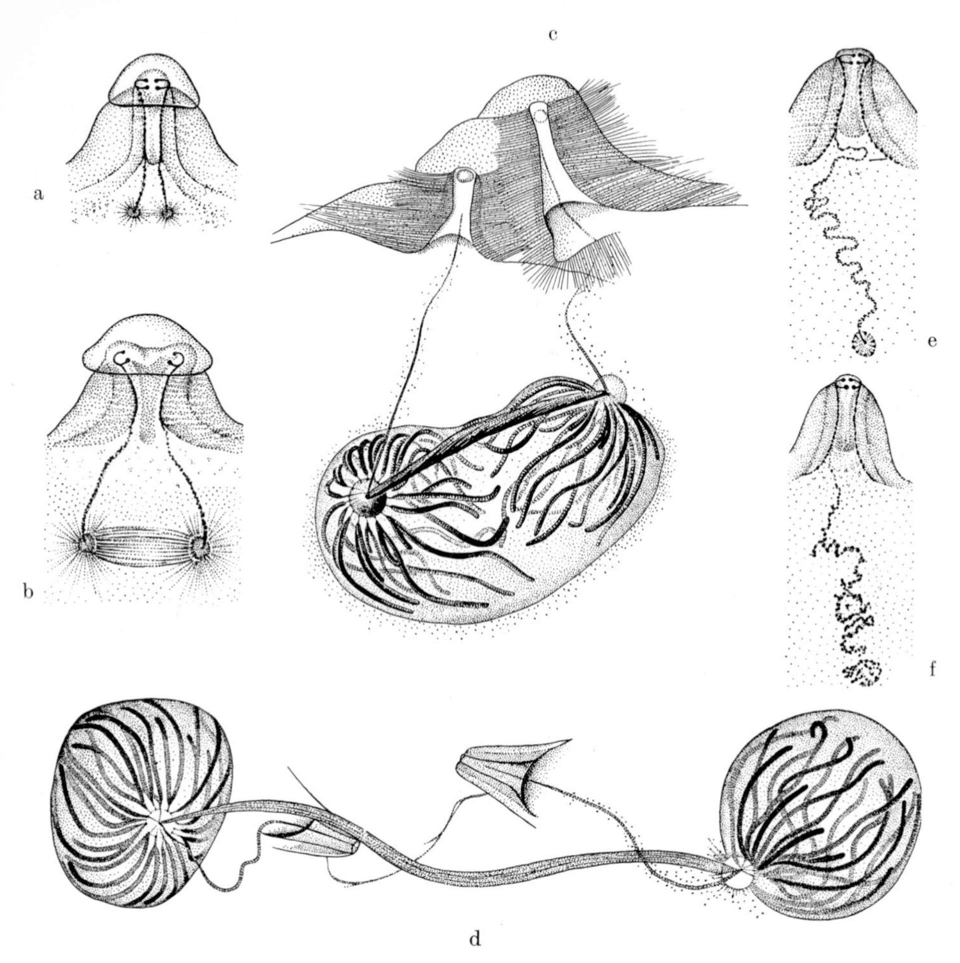

Fig. 60. *Pseudotrichonympha* (Polymastigina). Mitosis. a Early, b late prophase. c Early, d late telophase. e, f Degeneration of the middle piece and centrosome of the centriole after nuclear division. Sublimate alcohol, iron hematoxylin. a, b × 1100, c—f × 900. a, b, e, f after CLEVELAND, 1953 [*271*], somewhat schematized; c, d after GRASSÉ and HOLLANDE, 1951 [*625*]

also likely to be rod-shaped. At the beginning of division, the nucleus loses its connection with the rostrum and migrates toward the interior of the cell. It is followed by the so-called "rostral body" whose normal position is above the nucleus in the rostrum. The spindle-forming ends of the centrioles stick to the rostral body. They separate therefore from the other parts of the centriole and are transported by the rostral body to the vicinity of the nucleus (a). The rostral body is now dissolved while the centriolar ends form the spindle (b). After the nucleus has divided and the rostrum has reduplicated, new rostral bodies (c) are formed at the daughter rostra. While the nuclei regain their connection with the rostra, the spindle-forming ends are dissolved in the cytoplasm (d). Later on they are regenerated by the remaining portions of the centrioles.

In those Hypermastigida which inhabit the gut of the American wood roach, *Cryptocercus punctulatus*, sexual reproduction is initiated by a mitosis which has

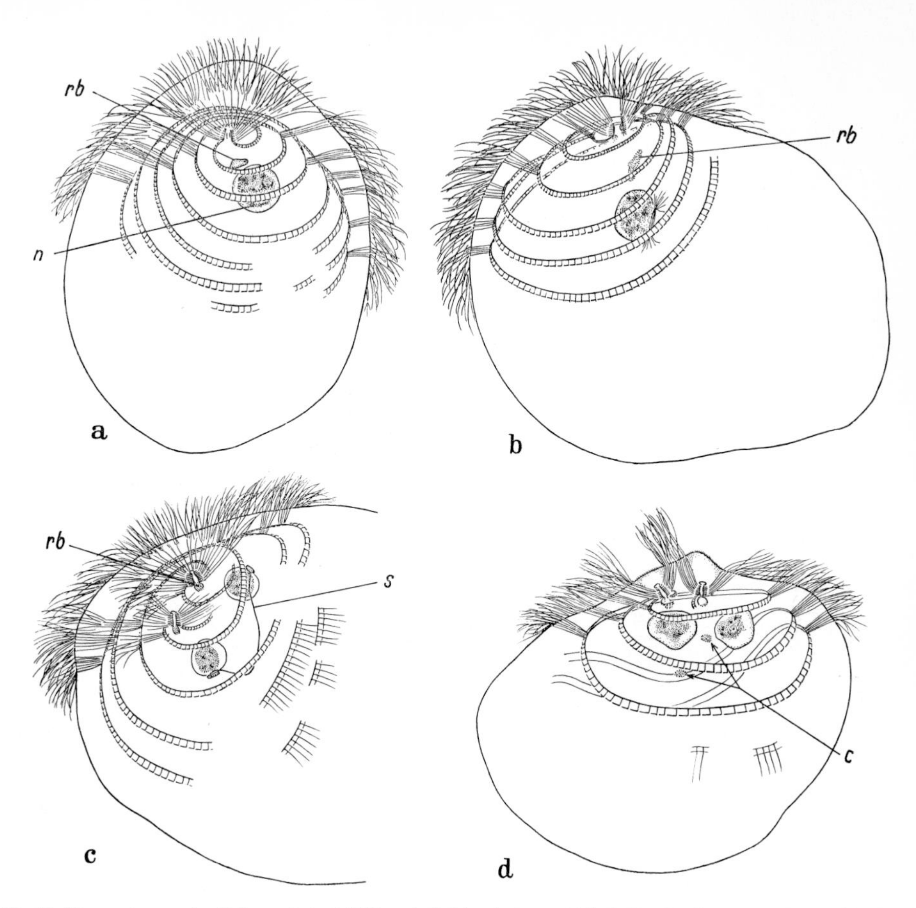

Fig. 61. *Macrospironympha* (Polymastigina). Different divisional stages (Meiosis I). a Nucleus (*n*) and rostral body (*rb*) have separated from the rostrum. The rostral body carries the spindle-forming centriolar ends with it. b Formation of the spindle and dissolution of the rostral body. c The centrosomes are still connected by the spindle (*S*). New rostral bodies are formed at the daughter rostra. d After the spindle dissolves, the centrosomes (*c*) are free in the cytoplasm and then degenerate too. The daughter nuclei approach the daughter rostra. × 330. After CLEVELAND, 1956 [*290*]

been curiously modified in some of the species. This mitosis is a differential division which leads to the formation of a male and a female gamete (see p. 168). In *Trichonympha* (Fig. 62), [*276*], encystment of the cells (gamonts) takes place first (a). In each cyst, a male and a female gamete is then formed. In prophase (b), the 24 chromosomes undergo reduplication to form the two chromatids. Since the sites of spindle fiber attachment of the chromosomes also separate completely in prophase, two sets of 24 chromatids are thus found which can, in many cases, be clearly distinguished because of differences in staining. Those chromatids which are destined for the future male gamete stain more deeply than those of the future female gamete. The differential character of the division is thus already expressed at the time of chromosomal reduplication. A peculiar process of attachment is responsible for the fact that the chromatids are not distributed at random to the two daughter cells: all deeply staining chromatids become incorporated into one daughter nucleus, all weakly staining chromatids get into the other one. The

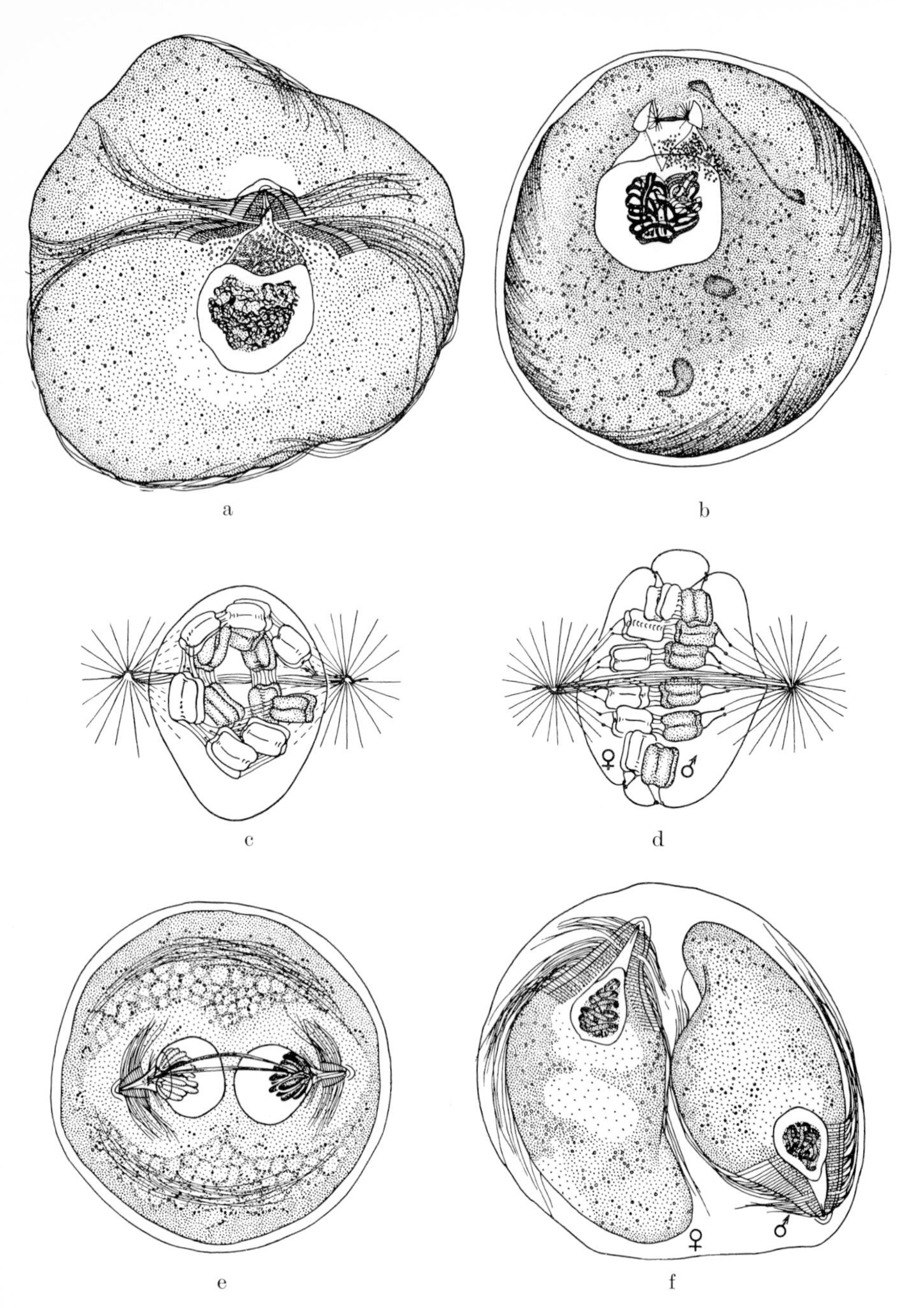

Fig. 62. *Trichonympha* (Polymastigina). Gamogonic division. a Gamont (encysted). b The same, nucleus in prophase. c, d Concatenation of the "pseudotetrads" in prophase and metaphase. e Telophase. f Gametes. Iron hematoxylin. a, b, e, f × 490; c, d × 680. After CLEVELAND, 1949 [*276*]

segregational process is preceded by the formation of "pseudotetrads" of four chromatids each which are then attached to each other to form a chain (c, d). In this manner two "composite chromosomes" are formed in the nucleus, each comprising a whole genome. Even at telophase (e) and in the nuclei of the fully formed gametes (f) different staining intensities are frequently recognizable. It is not before the onset of meiotic prophase in the synkaryon (Fig. 72) that the chromosomes appear once again as separate units.

A similar chain formation has also been observed in *Barbulanympha* [*284*], and in *Leptospironympha* [*281*]. In the latter, the single chromatids attach directly without prior formation of pseudotetrads (Fig. 163).

As usual in the mitoses of higher plants, centrioles as distinct, stainable bodies can also be absent in Protozoa. Even in such cases there are, as a rule, structures which play the role of divisional centers and thus correspond to centrioles in this respect. In the heliozoan, *Actinophrys sol*, the axonemes of the axopodia touch the resting nucleus (Fig. 63) [*132*]. At prophase, dense plasma regions develop as cap-like structures at opposite poles of the nucleus which push the polar axonemes away from the nuclear surface (b—d). In the further course of mitosis these polar caps increase in volume while the karyoplasm is transformed into a barrel-shaped spindle (e—g). Following anaphase (h), and the reconstruction of the daughter nuclei (i—l), the mass of the polar cap material seems to become evenly distributed as a superficial layer around the nucleus. The axonemes can therefore move right up to the nuclear envelope again (m).

While the monothalamous foraminifer *Myxotheca arenilega* shows typical centrioles from the ultrastructural point of view (gamogony) [*1551*], the spindle poles of the related *Allogromia laticollaris* are represented by centrosome-like globular bodies surrounded by a special folding of the nuclear envelope (agamogony) [*1554*]. In both cases, the spindle fibers, evidently induced by these structures, are separated from them by the nuclear envelope which persists during division.*

In addition to the cases to be discussed on the following pages, many Protozoa seem to have a type of mitosis in which the nuclear envelope persists, but no structures at the divisional poles are recognizable at all. Examples are *Chlamydomonas* [*868*] and *Trypanosoma* [*1768*]. It is an open question, why centrioles or comparable structures are necessary in other cases [*1305*].

As a further modification of mitosis in Protozoa, the chromosomal condensation cycle may be changed, the chromosomes being long at metaphase and short at telophase. One example of this is furnished by the sporogony divisions in the coccidian, *Aggregata eberthi* (Fig. 64) [*134*]. These divisions are further characterized by a reduction of the spindle apparatus. Between the centrioles which are located at the tips of small, conical elevations of the cell surface, a delicate, extranuclear spindle is formed tangential to the nuclear surface. A typical equatorial plate does not appear. The chromosomes are fastened to the spindle by their kinetochores. This is followed by the separation of the chromatids whose free arms may even at telophase extend far into the middle of the divisional figure (see p. 50).

In many instances the spindle apparatus is more or less masked by the presence of nucleolar material. In some amoebae, masking of the spindle is so complete that

* The occurence of typical centrioles in the first case may be correlated with the formation of flagellated free-swimming gametes while in the latter case, where typical centrioles are missing, non-flagellated agametes are formed.

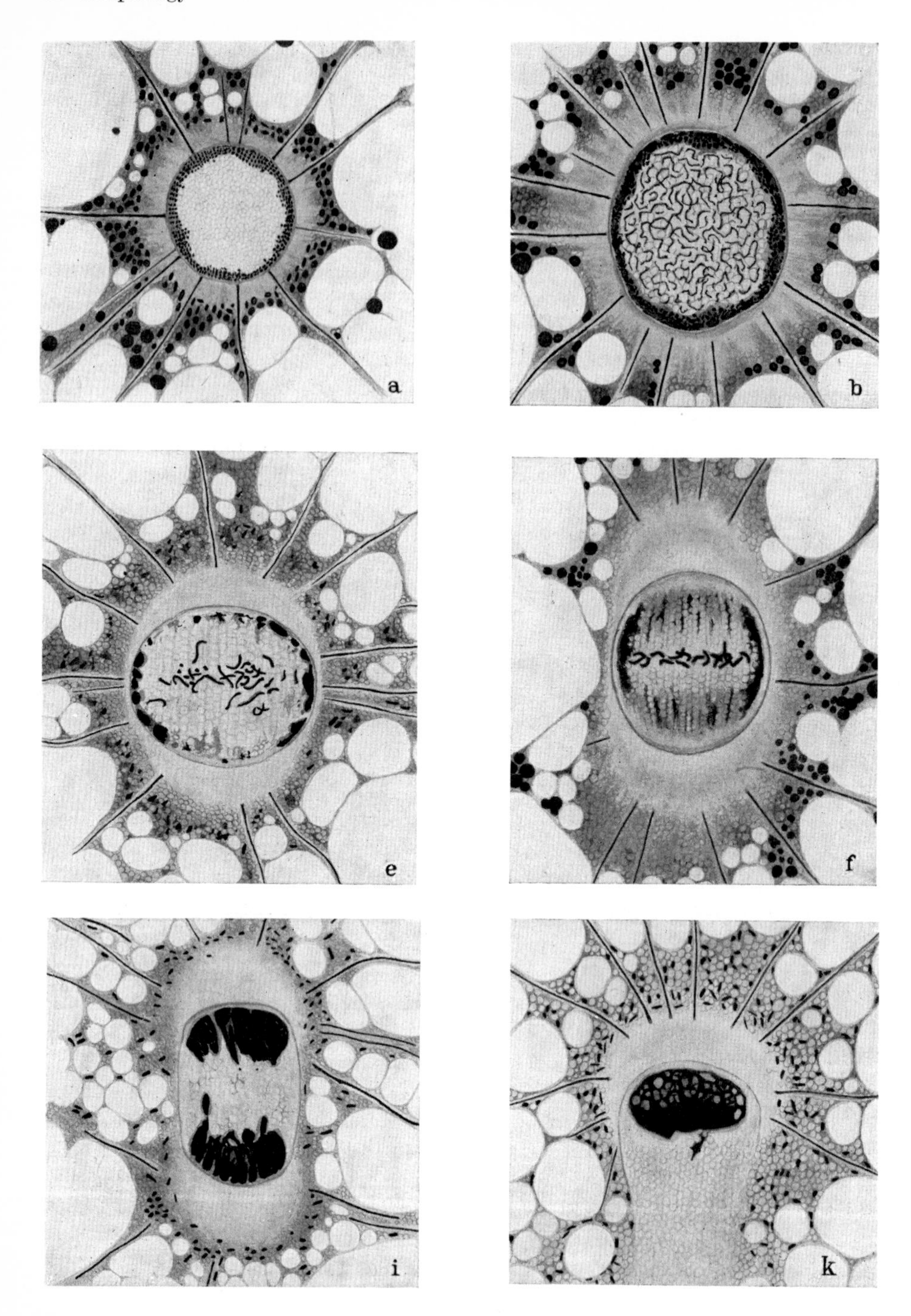

Fig. 63. *Actinophrys sol* (Heliozoa). Mitosis. Only the nucleus and the surrounding cytoplasm is shown. a Resting nucleus. b Early, c and d late prophase (development of polar caps). e Transition to metaphase. f Metaphase (disappearance of the nucleolar substance). g Begin of anaphase. h Midanaphase. i, k Telophase. l and m Reconstruction of the daughter nuclei. Iron hematoxylin. × 1900. After BELAR, 1922 [*132*]

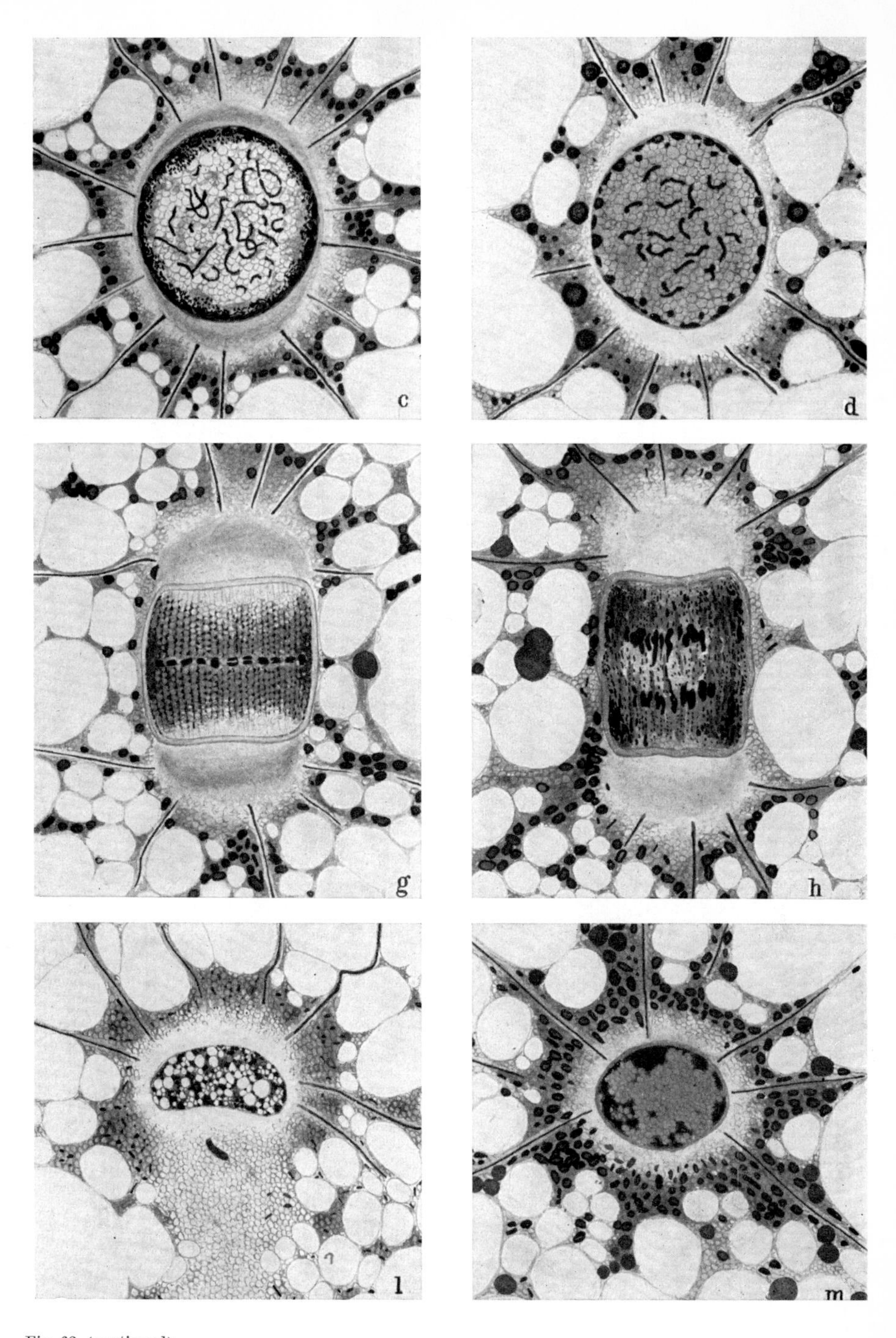

Fig. 63. (continued)

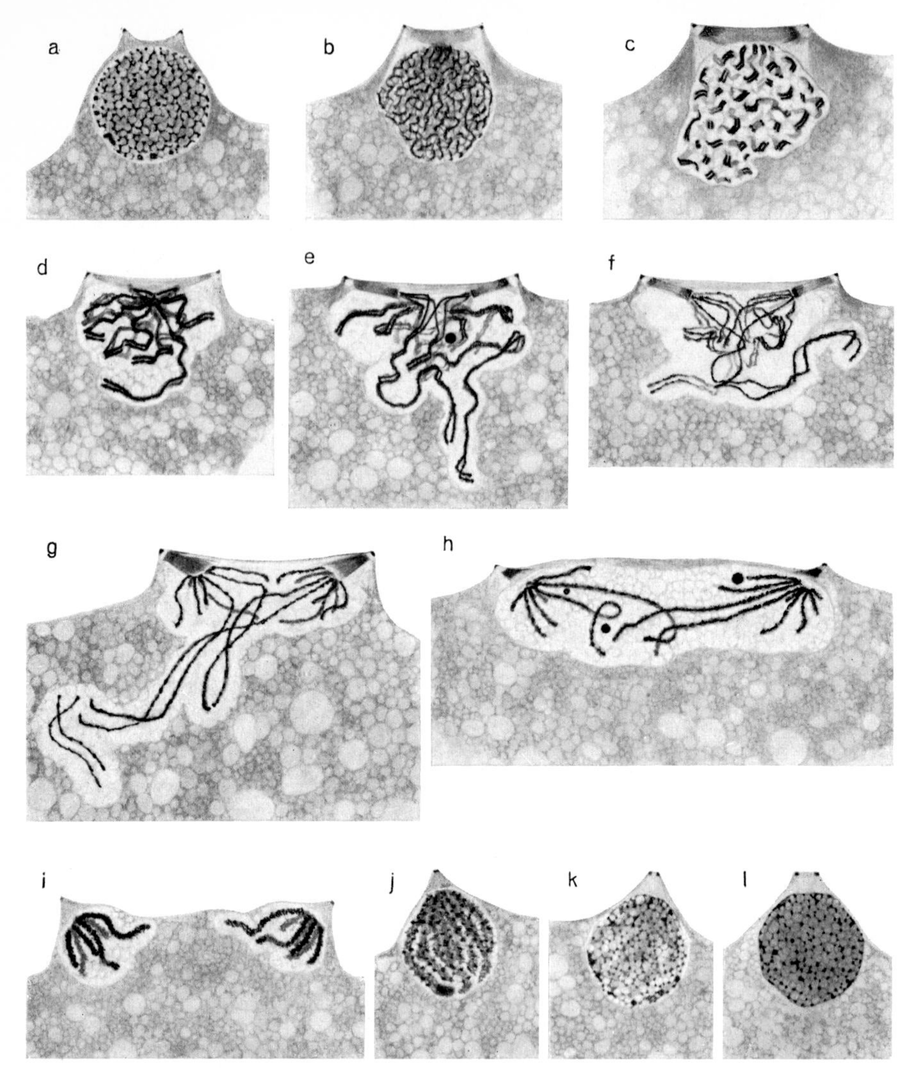

Fig. 64. *Aggregata eberthi* (Coccidia). Mitotic stages from sporogony. These mitoses take place on the surface of the oocyst, only small portions of the periphery are therefore shown. a Resting nucleus. b Prophase. c Early metaphase. d—g Anaphase. h and i Start of telophase. j Division of the centriole. k and l Reconstruction of the daughter nuclei. Iron hematoxylin. × 2100. After BELAR, 1926 [*134*]

division was formerly thought of as an "amitosis". Its mitotic character may then only be recognizable by the equatorial arrangement of the chromosomes.

In some cases it has not been possible to sufficiently clarify the course of mitosis. While the *micronucleus of ciliates* shows typical mitotic stages in some species (Fig. 65), the chromosomes of many other ciliates seem to be masked in a very peculiar way [*390*]. Masking is not due to the presence of nucleolar substance but to Feulgen-positive material whose relationship to the chromosomes is not clarified. During mitosis, longitudinally directed structures appear which resemble chromosomes. However, their number does not appear to be constant. It is certainly smal-

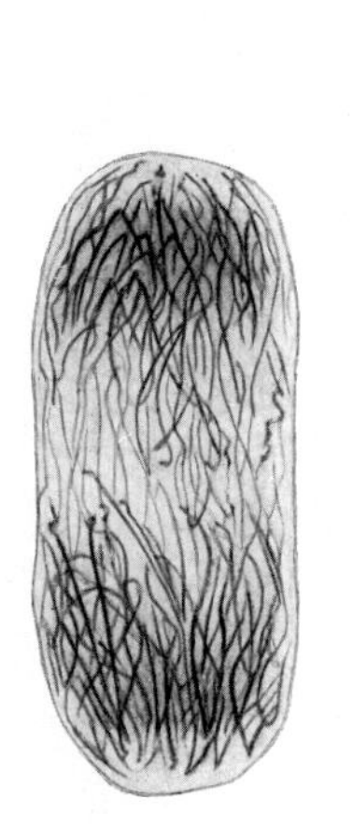

Fig. 65. *Paramecium caudatum.* Mitosis of the micronucleus, anaphase. × 1900. After CHEN, 1940 [*243*]

ler than the number of meiotic chromosomes. In addition, these structures are arranged not in an equatorial plate but parallel to the axis of the achromatic figure. At the beginning of anaphase they are divided transversely.

Divisional stages of micronuclei contain bundles of microtubules which show parallel arrangement during the phase of stretching. However, the microtubules which as a whole represent a spindle apparatus, are not polarized by typical centrioles [*728, 1744*].

The mode of nuclear division in *dinoflagellates* and *euglenoids*, whose chromosome numbers are usually very high, is also not known in all details. There is no cycle of chromosomal condensation and decondensation during nuclear division (Fig. 66). In most cases no centriolar structures have been found. The nuclear envelope remains fully intact throughout division. Though microtubules participate in the division process, they are not connected with the chromosomes.

While in dinoflagellates microtubular bundles run along the axis of the dividing nucleus in cytoplasmic channels, being exclusively extranuclear [*972*], in euglenoids microtubular bundles run from pole to pole within the elongating nucleus, closely associated with the surface of the dividing nucleolus [*1011, 1013*].

How the separation of the chromatids occurs is still unknown. Since no morphological differences between the chromosomes of a set have been demonstrated as yet, one hypothesis assumes that all chromosomes are homologous: If this is true, no regular separation of chromatids is needed, since all chromosomes have the same genetic information [*656*]. It may be mentioned in this connection, that nuclear fragmentation has been observed in *Euglena,* occasionally: The progeny of cells with either one half-nucleus or three half-nuclei turned out to be fully viable [*1013*].

On the other hand, some parasitic dinoflagellates have only few chromosomes and the course of their nuclear division seems to correspond more to the mitotic pat-

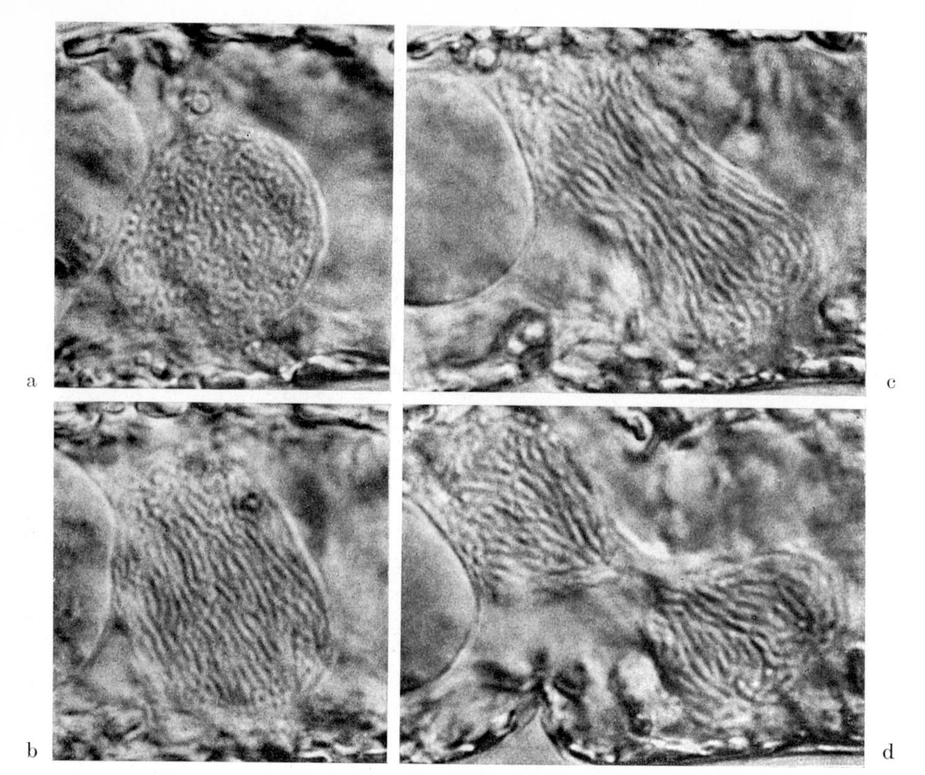

Fig. 66. *Dissodinium lunula* (Dinoflagellata). Nuclear division within the sickle-shaped cell (see Fig. 131). From the film E1634 (DREBES)

tern. *Syndinium globiforme* for instance has four chromosomes which are inserted by special structures into the nuclear envelope. These structures (comp. Fig. 59) are connected to the base of two typical centrioles by several microtubules (Fig. 67).

Even if we disregard the insufficiently analyzed cases, the examples discussed so far show that mitotic patterns are much more diverse in Protozoa than in higher organisms. Their study may therefore contribute to an understanding of the complicated physiological processes which are basic to the apparently simple act of nuclear division.

b) Meiosis

In all organisms with sexual reproduction, the fusion of the gametic nuclei must lead to a doubling of the chromosome number. The number of chromosomes can, of course, only remain constant if there is a reduction at some time during the developmental cycle. This *reduction of chromosome number* is intimately connected with a special type of nuclear division called m*eiosis*.

Comparative studies show that meiosis does not occur in all organisms at the same point of the developmental cycle (Fig. 68). In the most simple case it takes place immediately at the first division of the zygote (*zygotic meiosis*, a). The zygote itself is the only diploid stage in the life cycle; all other stages are haploid. This is

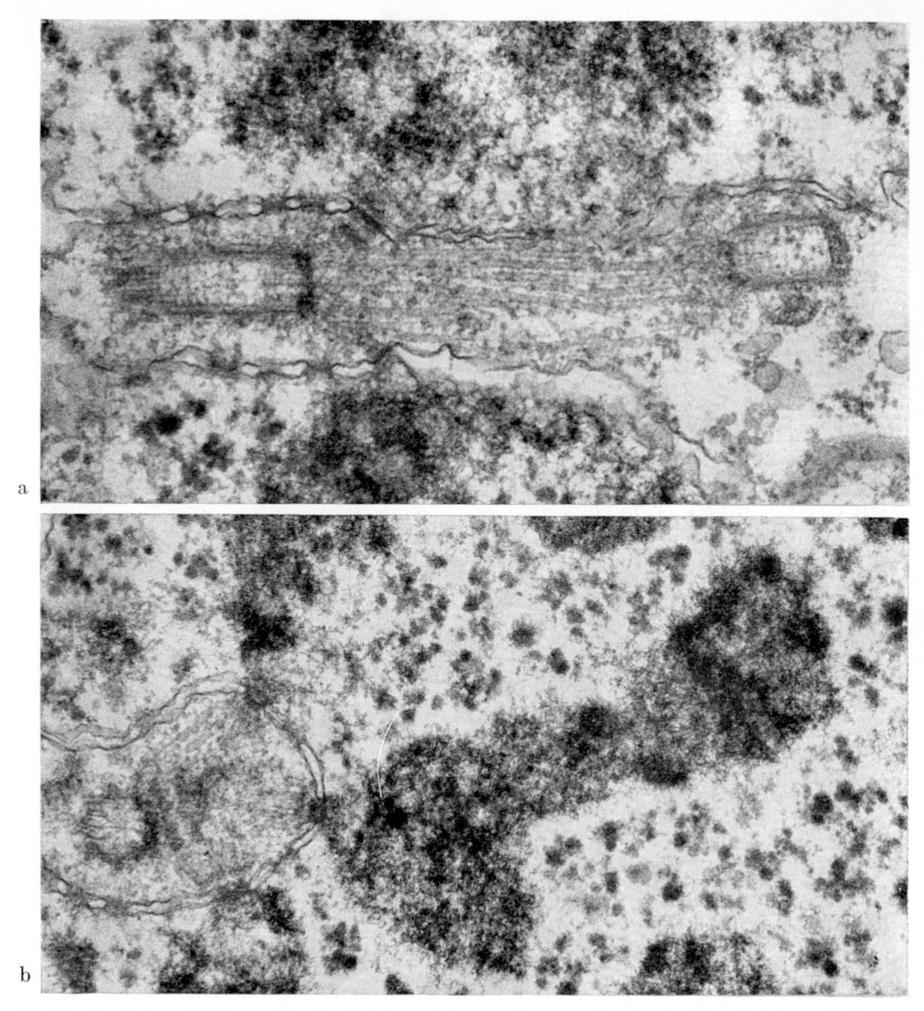

Fig. 67. *Syndinium globiforme* (Dinoflagellata), an intracapsular parasite of colonial radiolarians *(Collozoum, Sphaerozoum)*. Stages of nuclear division. × 42000. a Two centrioles connected by microtubules within a nuclear channel; b Cross section of a nuclear channel (left) with centriole. Attachment structures of chromosomes inserted into the nuclear envelope. Courtesy of H. Ris

found among Protozoa in Phytomonadina, some Polymastigina *(Trichonympha, Eucomonympha, Barbulanympha, Oxymonas, Leptospironympha, Saccinobaculus)* and in all Sporozoa.

If reduction occurs at the time of gamete formation (*gametic meiosis*, b), the major part of the developmental cycle takes place in the diploid phase. Only the gametes themselves are haploid. This type of meiosis, which occurs in all multicellular animals, is found among Protozoa in the Heliozoa *Actinosphaerium* and *Actinophrys*, in some Polymastigina (*Notila, Urinympha, Rhynchonympha* and *Macrospironympha*), in the Opalinina and in the ciliates.

In other cases meiosis follows asexual reproduction which leads to the formation of agametes (*intermediary meiosis*, c). The haploid agametes grow to form a sexually reproducing, i.e. gamete-forming generation, the gamonts. The gametes copulate

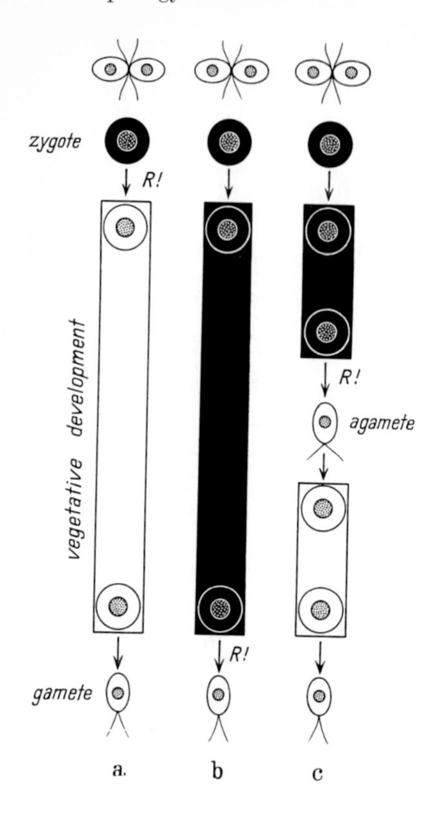

Fig. 68. Diagram of the alternation between haploidy (stages white) and diploidy (stages black) due to different positions of meiosis (*R*!) in the developmental cycle of the organisms. a Haplont with zygotic meiosis. b Diplont with gametic meiosis. c Heterophasic alternation of generations with intermediary meiosis

and the zygote again gives rise to an asexual generation (agamont) which forms the agametes. This position of meiosis is, of course, connected with a regular alternation of a diploid and a haploid generation (heterophasic alternation of generations, see p. 230). While intermediary meiosis is of widespread occurrence in the plant kingdom (many algae, mosses, ferns and flowering plants), it is found among the Protozoa only in the Foraminifera.

Meiosis itself is cytologically a modified mitosis which is characterized by two special features:

1. Doubling of the number of chromosomes, otherwise characteristic of every nuclear division, does not take place.
2. A pairing of homologous paternal and maternal chromosomes occurs.

Both features are intimately connected during meiosis and can therefore only be discussed within the framework of the whole process.

In most organisms meiosis always involves *two divisions* (*two-step meiosis*, Fig. 69). This is due to the different behavior of spindle fiber attachments and chromosome arms during division, a feature which is also expressed to a limited degree in mitosis. In the first meiotic division, the division of the kinetochores is suppressed while the chromosome arms are reduplicated. In the second division, the chromosome arms fail to reduplicate while the kinetochores are divided. Since the homologues pair in the prophase of the first division (c), *groups of four* are formed (four-

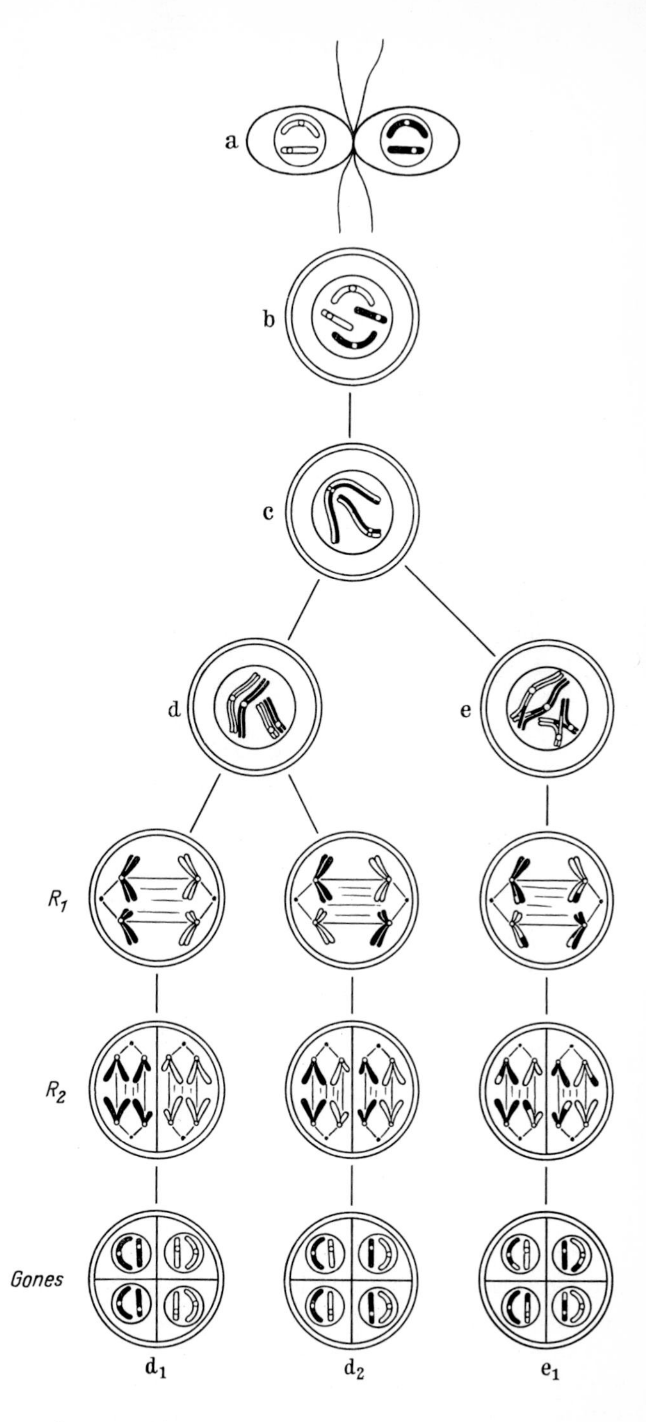

Fig. 69. Diagram of the external course of two-step meiosis. a Fertilization of the gametes. b Zygote. c Pairing of homologous chromosomes. d—d_1 and d_2 Various possibilities of the combination of paternal and maternal chromosomes due to different orientations of the kinetochores in the first division (R_1). e—e_1 Exchange of segments between the chromatids

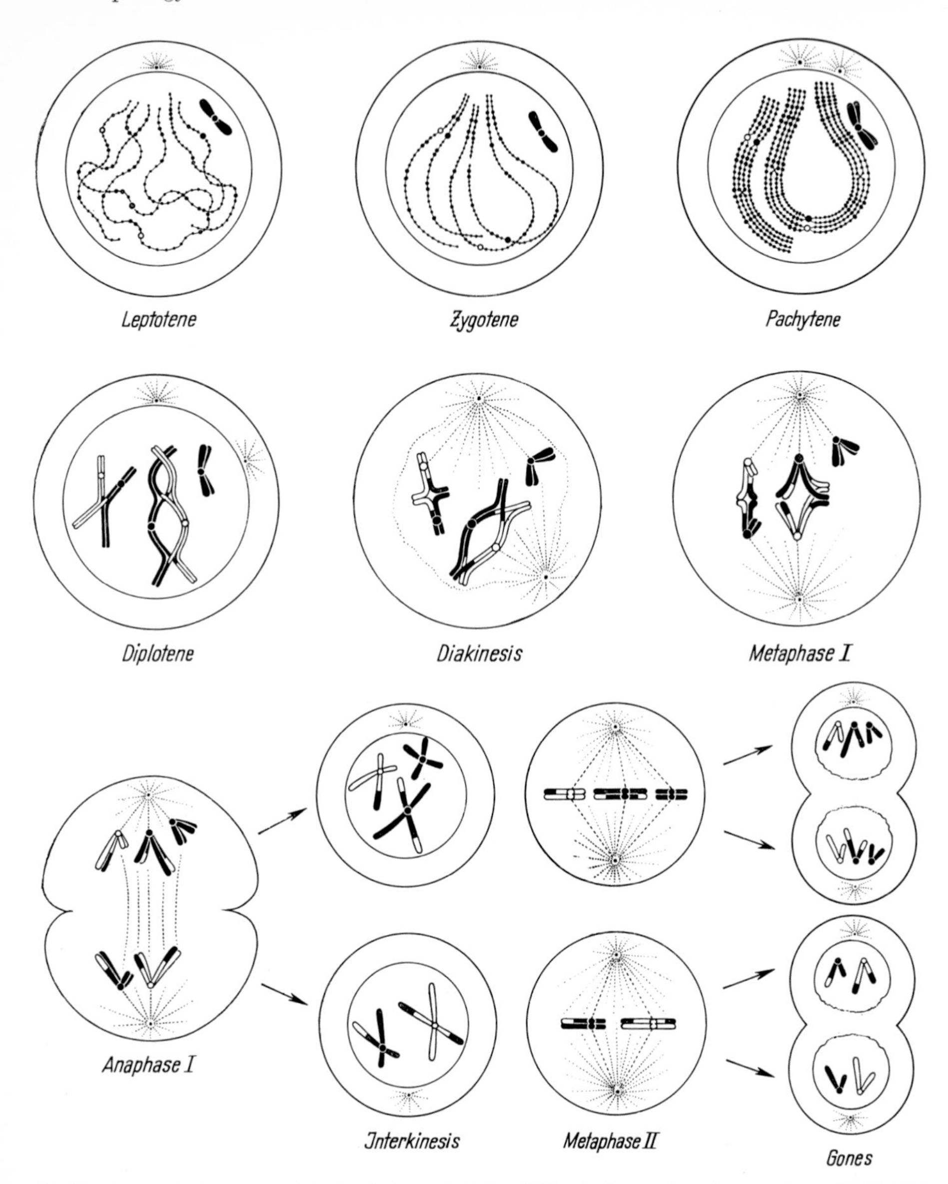

Fig. 70. Diagram of chromosome behavior during meiosis. In addition to the two homologous pairs, a chromosome without a partner (sex chromosome) is shown which is condensed during prophase

strand stage, d). Each of these groups of four is thus composed of two homologous chromosomes which are split into two chromatids and connected by the undivided kinetochore. When these groups of four have reached a state of maximal condensation, they are called *tetrads*. At metaphase, the two kinetochores of a tetrad can of course not become arranged in the equatorial plate like the kinetochores of mitotic chromosomes. Instead, they are arranged parallel to the spindle axis at a certain distance from the equatorial plane. This arrangement is referred to as *co-orientation*. In anaphase, each tetrad is divided into two dyads (R_1). Since the orientation of the two kinetochores of a tetrad with respect to the spindle poles is in no way predetermined, the combination of the different dyads in the two daughter nuclei of the first division is also a matter of chance. With two tetrads only, as shown in Fig. 69, there are only two possible arrangements: either both paternal and both maternal dyads pass to the same pole (R_1, left) or each daughter nucleus receives one paternal and one maternal dyad (R_1, center). Four different combinations of dyads are therefore possible.

The chromosomal processes are further complicated by the possibility of an *exchange of segments* between the chromatids of a tetrad at the four-strand stage. That such an exchange of segments does indeed take place during the prophase of meiosis is indicated by the occurrence of genetic crossing over. At mid-prophase, the homologues frequently show a tendency to separate from each other. It can be observed, however, that they remain connected at certain points. At these points, two of the chromatids are seen to cross each other giving the impression of a change of pairing. However, the generally accepted interpretation is that these *chiasmata* are – at least in most cases – the morphological expression of previous segmental exchange between chromatids rather than a mere change of pairing association. It must be mentioned, however, that homologous chromosomes are frequently wound about each other without being connected by chiasmata.

While the random orientation of the kinetochores at metaphase leads to different combinations of maternal and paternal chromosomes, the exchange of segments leads to chromosomes which are no longer uniform after meiosis but consist of different maternal and paternal parts.

Details of the course of meiosis are shown diagrammatically in Fig. 70. In early prophase the chromosomes are long and thin threads (*leptotene* stage). In favorable material, a longitudinal differentiation is recognizable. Little nodules (chromomeres), probably locally coiled segments, are distributed regularly over the whole length of the chromosomes. After some time the pairing of homologues commences (*zygotene* stage), in most cases from one end, proceeding zipper-like along the chromosomes. Frequently, a polar orientation of the chromosomes is recognizable at this stage (*bouquet* stage). Finally the paired chromosomes shorten and the splitting of each homologue into two chromatids becomes recognizable (*pachytene* stage). While the homologues show a tendency to move apart (*diplotene* stage), the chiasmata remain as localized connections between them. Eventually, the polar rays connect with the kinetochores *(diakinesis)* and the tetrads are arranged at the equatorial plate with their spindle fiber attachments co-orientated *(metaphase I)*. When the dyads have moved apart at *anaphase I*, the first division is completed. Frequently, no actual interphase nucleus is formed between the two divisions (interkinesis), the chromosomes remaining distinctly visible as separate elements. Their appearance is often that of a cross whose two arms are represented by the chromatids connected by the still undivided kinetochore. In the following division the chromosomes are arranged as in a normal mitosis *(metaphase II)*. The kinetochores divide and the two chromatids are separated in the usual manner.

Cells which arise by meiotic divisions are termed *gones*.

Among the *Protozoa*, meiosis is modified in a variety of ways. The most variable relationships are found among *Polymastigina*. So far, meiosis is only known from symbionts of *Cryptocercus punctulatus*.

In some species, it has been shown that chromosomal reduction is carried out by a single division. The term *one-step meiosis* has been used in this connection. Reduplication of the chromosome arms and of the kinetochores is suppressed in

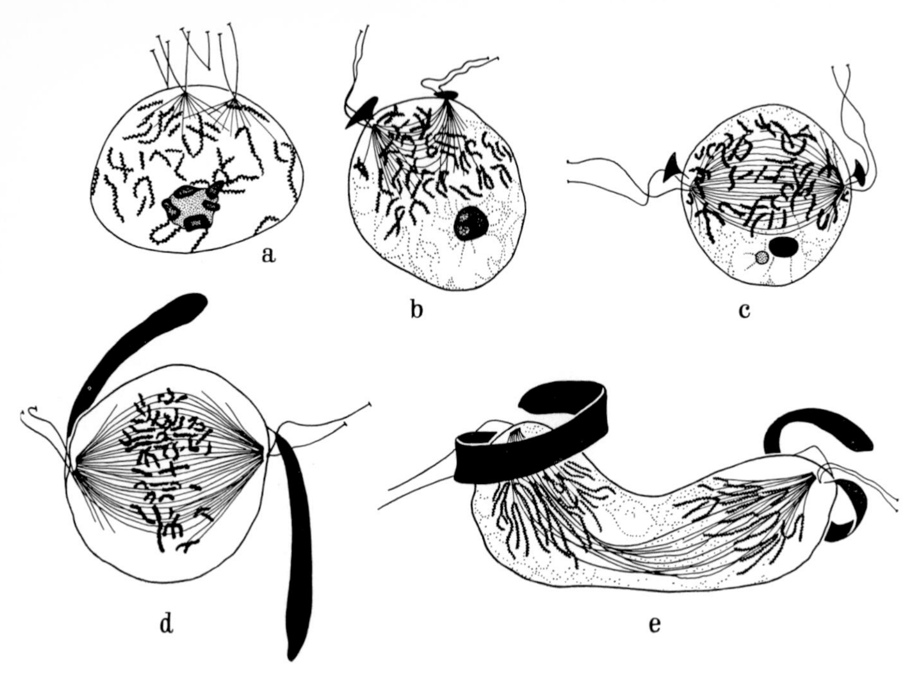

Fig. 71. *Oxymonas doroaxostylus* (Polymastigina). One-step meiosis. a Early, b and c late prophase. d Metaphase. e Early telophase. New axostyles (black) are formed from the centrioles, located intranuclearly in this case. × 1050. Iron hematoxylin. After CLEVELAND, 1950 [*277*]

the same division. Thus, neither tetrads nor chiasmata are formed. Recombination of paternal and maternal chromosomes is therefore restricted to the random co-orientation at metaphase.

In *Leptospironympha* [*281*], *Oxymonas* [*277*] and *Saccinobaculus* [*278*] one-step meiosis is of the zygotic type, in *Notila* [*279*] and in *Urinympha* [*282*] it is gametic. In *Notila*, meiosis takes place after the pairing of gamonts (Fig. 191). Each gamont nucleus divides into two gamete nuclei of the same sex. In *Urinympha*, on the other hand, meiosis takes place within one and the same gamont whose nucleus divides into two gamete nuclei of opposite sexes which fuse with each other (autogamy). This cannot, of course, lead to any recombination of the chromosomal material because homologous chromosomes are only separated temporarily and then reunited.

In the course of the one-step meiosis of *Oxymonas* (Fig. 71) the homologous chromosomes pair only very briefly. The pairing seems to be restricted to co-orientation at metaphase. The centrioles are located within the nucleus and an intranuclear spindle is formed. Already during division, new axostyles are formed in association with them.

All other genera found in the gut of *Cryptocercus punctulatus* carry out a *two-step meiosis*. In *Trichonympha*, *Barbulanympha* and *Eucomonympha* it takes place within the zygote. As shown in Fig. 72 [*276*] from *Trichonympha* and in Fig. 73 [*288*] from *Barbulanympha*, the formation of tetrads does not follow the general scheme (Fig. 70). The reduplication of the homologous chromosomes, which be-

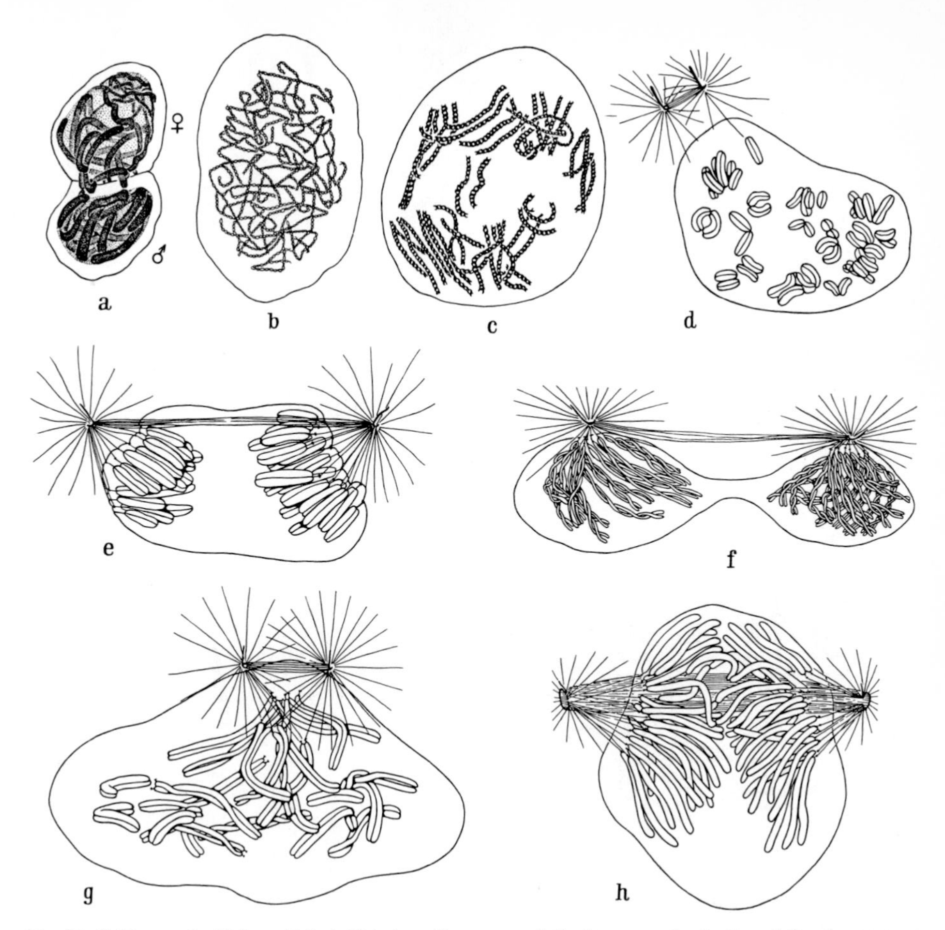

Fig. 72. *Trichonympha* (Polymastigina). Meiosis. a Karyogamy. b Synkaryon; reduplication of the chromosomes (the chromatids show relational coiling). c Begin of pairing of homologues. d Formation of the tetrads. e Anaphase I. f Telophase I. g Prophase II. h Anaphase II. Iron hematoxylin. × 900. After CLEVELAND, 1949 [*276*]

comes recognizable usually only after pairing, is visible very early. In prophase, the sister chromatids are at first wound about each other (relational coiling). Each shortens and condenses independently without losing connection. In this more or less condensed state the chromatid pairs connect to form the tetrads. Thus, although a stage comparable to zygotene is completely absent, typical tetrads are formed which are disassembled in two divisional steps.

In *Barbulanympha* (Fig. 74) each chromatid has its own point of spindle fiber attachment (a) as early as metaphase of the first division. The two chromatids of the dyad are thus not connected by their common kinetochore but through pairing forces alone. Their final separation takes place at the next division (b).

In genera with gametic two-step meiosis *(Rhynchonympha, Macrospironympha)* the first step of division is always connected with cytokinesis. While the second step in *Rhynchonympha* leads to an autogamic fusion of the daughter nuclei, it occurs in *Macrospironympha* within a cyst leading to the formation of a male and a female gamete [*290*].

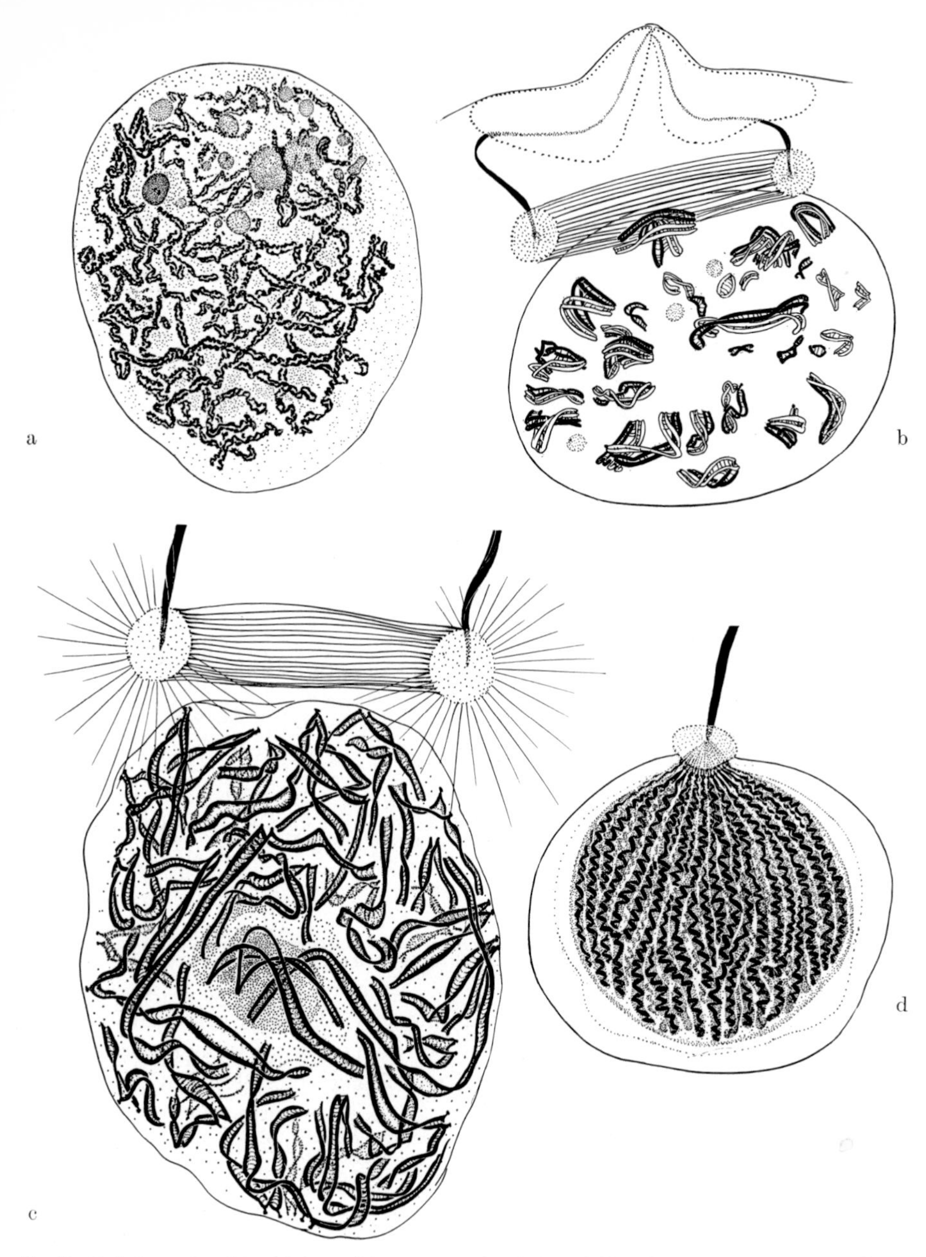

Fig. 73. *Barbulanympha ufalula* (Polymastigina). Stages of meiosis. a Midprophase. Chromosomes reduplicated (chromatids with relational coiling). b Late prophase. Formation of the tetrads (homologous pairs shown in black and white). c Prophase II. Each chromatid of a dyad has its own kinetochore. d Telophase. a × 720; b × 950; c and d × 1400. After CLEVELAND, 1954 [*288*]

In *Heliozoa*, meiosis consists also of two divisional steps which precede gamete formation. Cell division is not associated with this, since one of the two daughter nuclei of each nuclear division becomes pycnotic and is dissolved in the cytoplasm. Therefore, each cell gives rise to one gamete only (compare Fig. 172). In meiotic

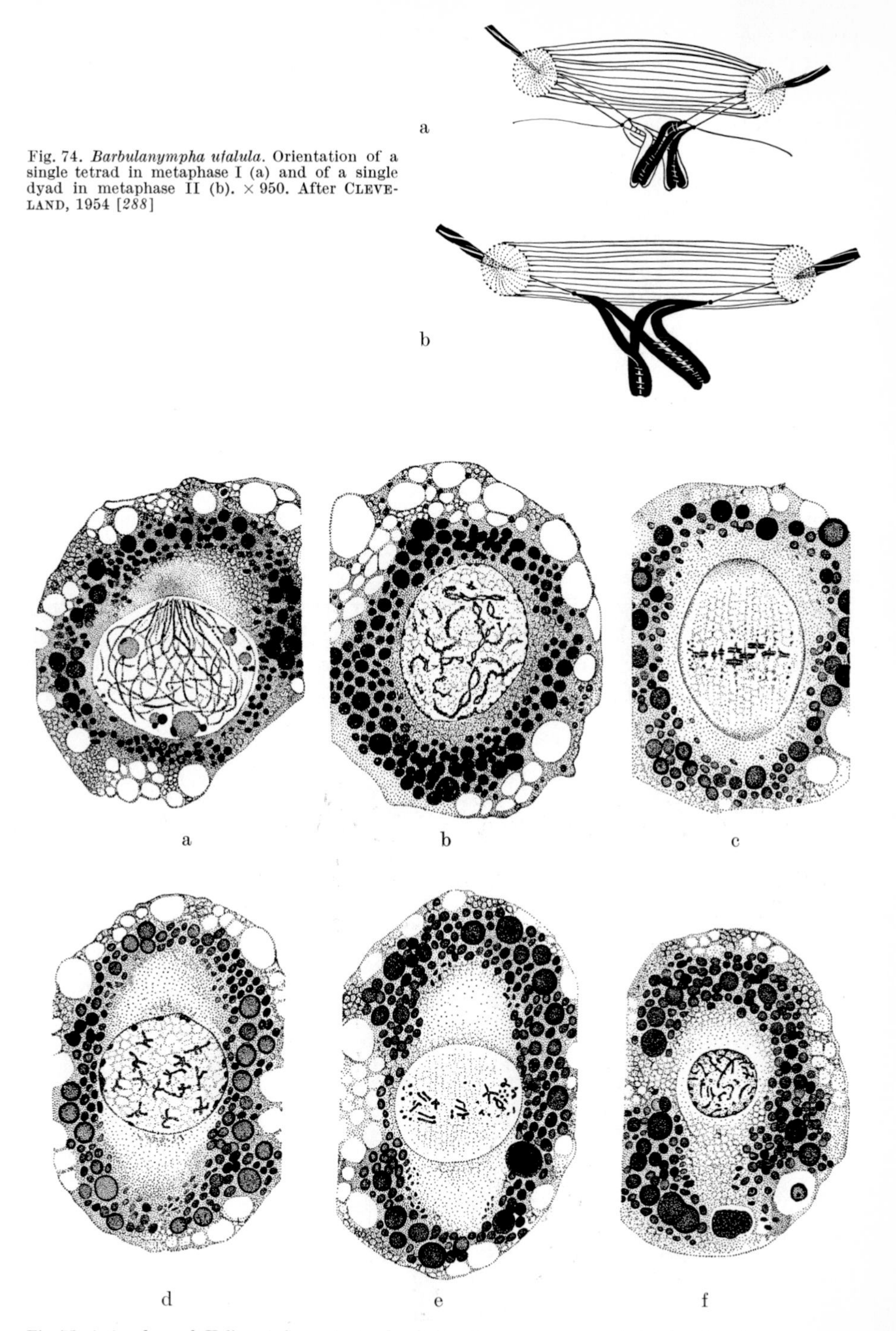

Fig. 74. *Barbulanympha ufalula*. Orientation of a single tetrad in metaphase I (a) and of a single dyad in metaphase II (b). × 950. After CLEVELAND, 1954 [*288*]

Fig. 75. *Actinophrys sol* (Heliozoa). Some stages of meiosis. a Bouquet stage. b Relational coiling of the homologues ("strepsitene"). c Metaphase I. d Interkinesis. e Metaphase II. f Gamete nucleus. On the lower margin of the figure the pycnotic second daughter nucleus, to the right in a vacuole the pycnotic first daughter nucleus. Iron hematoxylin. × 1950. After BELAR, 1922 [*132*]

prophase the chromosomes form a typical bouquet stage with their ends oriented toward the polar cap corresponding to the centriole (Fig. 75a). While they shorten, the chromosomes lose their polar orientation more and more. With increasing condensation, the doubleness of the threads becomes apparent (b). The homologues show again a tendency to separate. However, they cross each other at a number of points. It is not recognizable whether these are true chiasmata. At metaphase, double rods are clearly recognizable (c). That each dyad consists of two chromatids becomes apparent only at anaphase. At interkinesis the dyads are seen as typical cross configurations (d). In the second division, the two chromatids are separated in the usual manner (e, f).

In *Foraminifera*, meiosis is intermediary and apparently always involves two divisions.

The events in *homokaryotic* species where all nuclei of the agamont take part in meiosis shall be discussed first. When the agamonts have reached a certain size, all of their nuclei enter meiosis. The monothalamic genera *Myxotheca* and *Allogromia* have especially large nuclei whose structure has already been discussed (p. 49). As shown in Fig. 76 the homologues of *Myxotheca arenilega* arrange themselves at the beginning of prophase into a distinct "bouquet" (a). At the spot where the chromosome bundle becomes narrow, the layer of nucleolar substance is interrupted. Later on, the homologues lose their polar orientation and become wound around each other (b). It must remain open whether these are true connections of the chiasma type. After this stage the chromosome pairs condense considerably and gain the appearance of short double threads (c). During late prophase centrioles

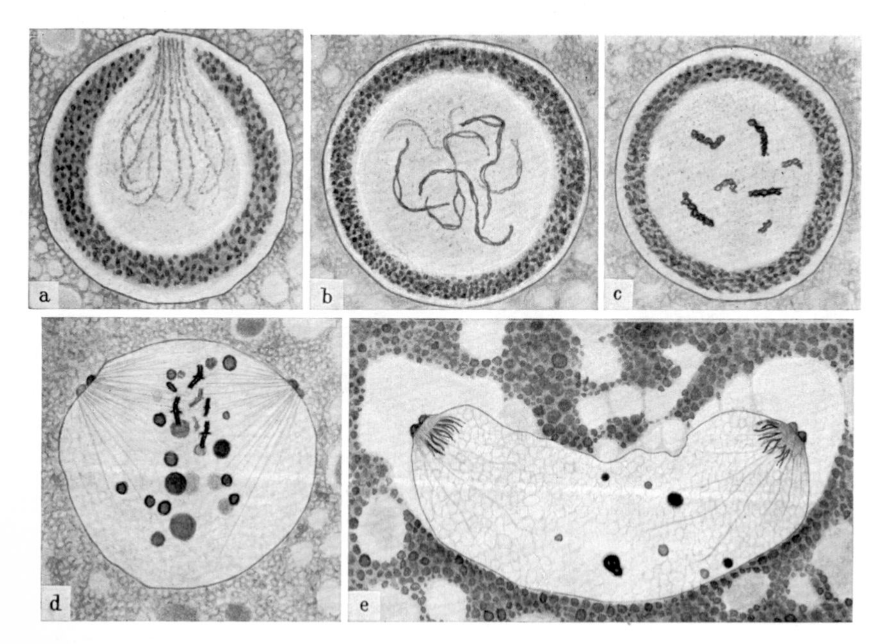

Fig. 76. *Myxotheca arenilega* (Foraminifera). Some stages of the first meiotic division. a—c Prophase. d Metaphase (combined from two sections). e Telophase (nucleus shrunk somewhat during fixation). Bouin (a—d) and Flemming (e), iron hematoxylin. × 850

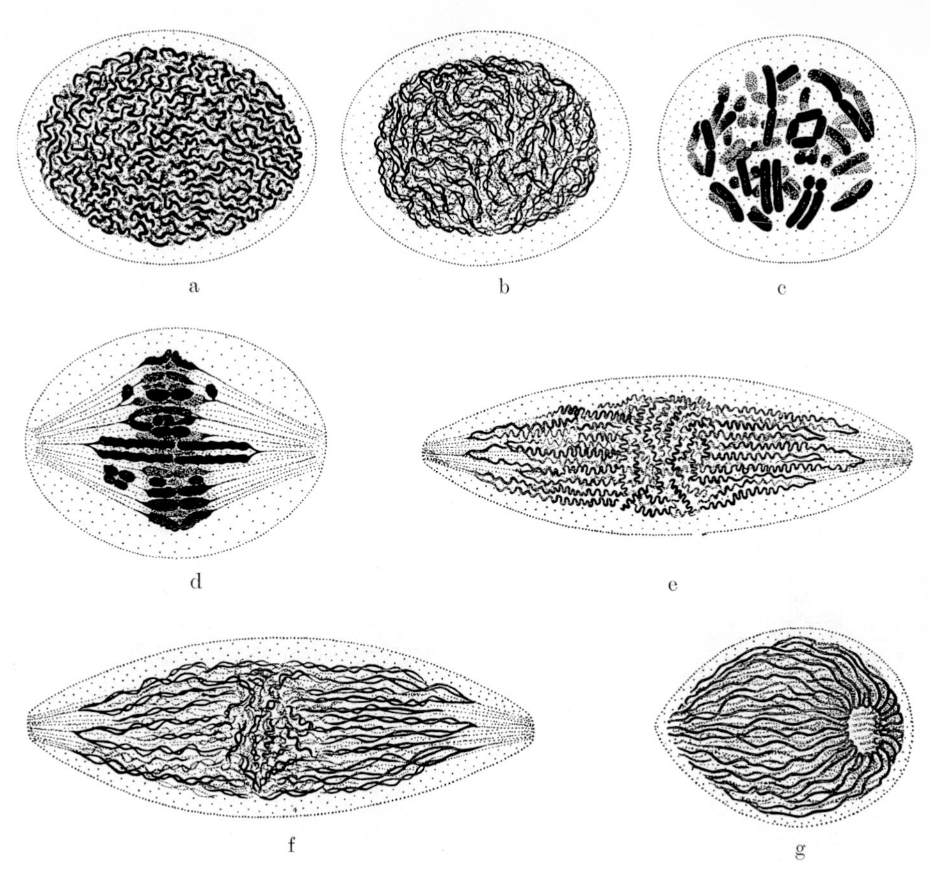

Fig. 77. *Patellina corrugata* (Foraminifera). Stages of the first meiotic division (somewhat schematized). The nucleoli are not depicted. a Early, b mid-, and c late prophase (diakinesis). d Metaphase. e Early, f late anaphase. g Telophase. Bouin-Duboscq, Feulgen staining. After GRELL, 1959 [*667*]

appear at the nuclear envelope. They are reduplicated very early and form polar fibers which radiate into the nuclear interior. The centrioles glide apart along the nuclear envelope but do not move to opposite poles. The metaphase spindle is therefore characterized by a clearly eccentric position (d). At the same time pieces of nucleolar substance which form the peripheral layer fuse to form larger nucleoli which will finally, during anaphase and telophase, dissolve completely in the karyoplasm. The nucleus is therefore free of nucleolar material after the first meiotic division. It appears again when the young agametes are formed.

Patellina corrugata (Fig. 77) is the first foraminiferan for which the intermediary position of meiosis has been proved [*1004*]. The onset of pairing is difficult to observe since the chromosomes are not arranged in a bouquet but form a dense ball of threads (a). In a stage evidently corresponding to diplotene, they are wound around each other in pairs (b). In late prophase (diakinesis) they are condensed to form bivalents without recognizable chiasmata (c). Size differences between the chromosomes are especially apparent at metaphase when co-orientation takes place. The points of spindle fiber attachment which do not take part in prophase coiling are now clearly distinguishable from the chromosome arms (d). Anaphase is peculiar

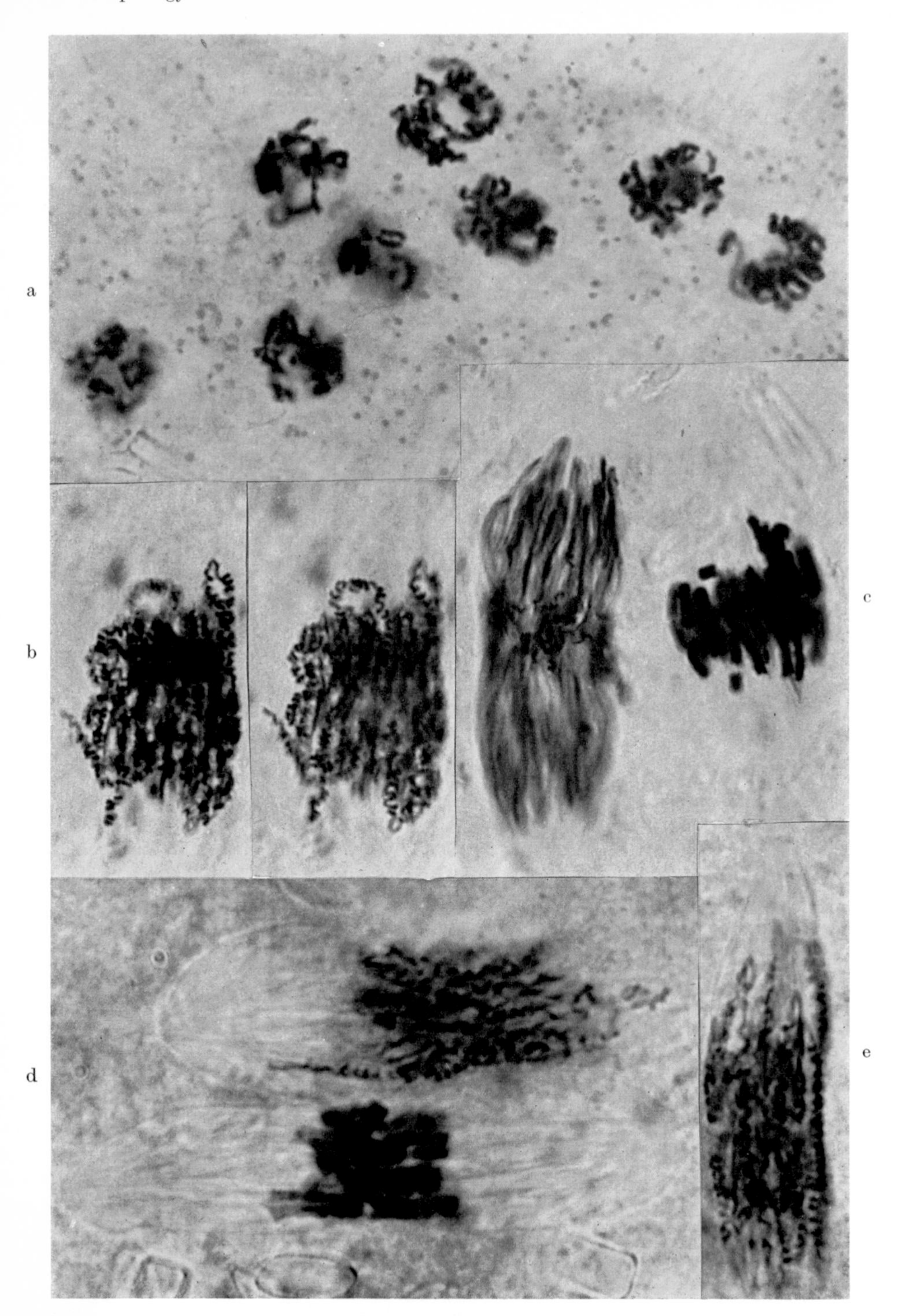

Fig. 78. *Patellina corrugata*. Stages of the first meiotic division. a Late prophase. b—e Metaphase and anaphase. Bouin-Duboscq, Feulgen staining. After GRELL, 1959 [*667*]

because the dyads do not remain condensed as they move apart. Instead, they uncoil more and more in the course of anaphase movement (e). In later stages, it is clearly recognizable that each dyad consists of two chromatids whose arms are relationally coiled (f).

Some stages of the first meiotic division are shown in Fig. 78.

In *Patellina corrugata* the number of nuclei in the agamonts may vary. A statistical study showed that the cell size at which the nuclei start meiosis increases with increasing numbers of nuclei. It seems that there must be a certain minimal amount of cytoplasm per nucleus before a cell reaches a state which permits the onset of meiosis. That this is due to a cytoplasmic rather than a nuclear condition is indicated by the behavior of left over gamete nuclei which were ingested by the agamonts. These nuclei, rather than being digested, take up the appearance of the other nuclei. As soon as the latter start meiosis, the chromosomes in the left over gamete nuclei also condense. Since gamete nuclei are haploid, this condensation cannot be preceded by pairing, i.e. the chromosomes remain as so-called univalents, which cannot be regularly distributed by the spindle to the daughter nuclei.

In *heterokaryotic* Foraminifera whose nuclear differentiation will be discussed in a later chapter (p. 96) it is only the generative nuclei which participate in meiosis.

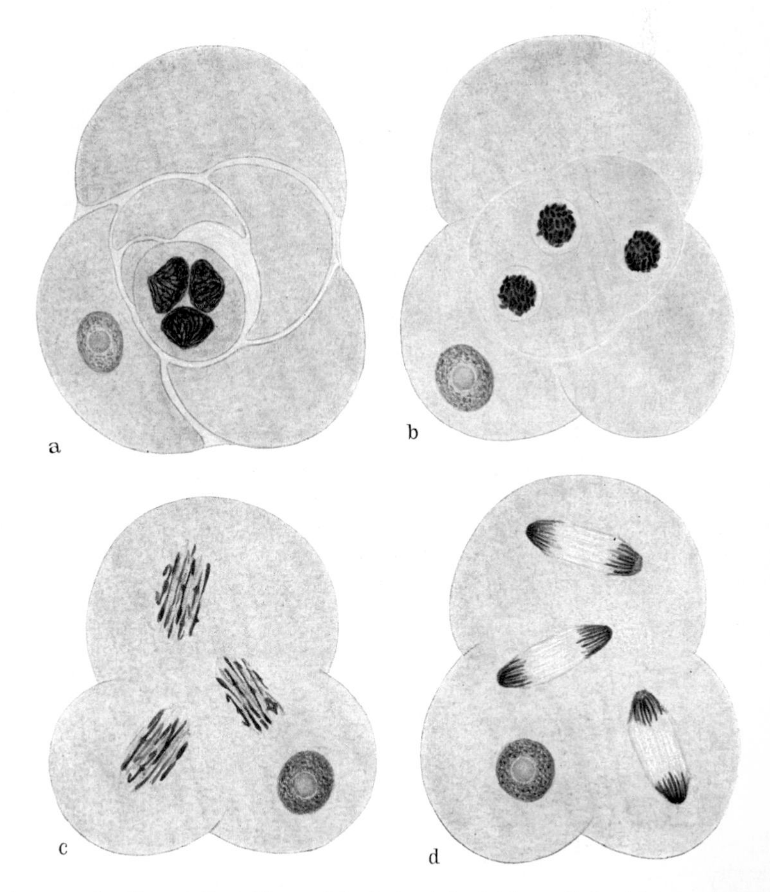

Fig. 79. *Rotaliella heterocaryotica* (Foraminifera). Stages of the first meiotic division. a Early, b late prophase. c Prometaphase. d Anaphase. Bouin-Duboscq, Feulgen staining. × 900. After GRELL, 1954 [*659*]

The somatic nucleus can take part in some phases of meiosis, although it is ultimately destined to die. In *Rotaliella heterocaryotica* (Fig. 79) it soon swells up temporarily to become pycnotic and then disintegrates into lumps which are resorbed in the cytoplasm. In *Rotaliella roscoffensis* (Fig. 80) and in the other heterokaryotic species the somatic nucleus will not only swell up but its chromosomes condense and even an intranuclear spindle is formed. However, pairing of homologues fails to take place in all cases. Counts of 18 univalents can be made in *Rotaliella roscoffensis* whose generative nuclei have 9 bivalents. When the somatic nucleus is stretched due to the formation of the spindle, the univalents become randomly distributed. They are frequently separated into two groups, but only rarely will both groups have the same number of chromosomes. The nuclear envelope finally breaks down and the chromosomes get into the cytoplasm where they become pycnotic and finally disappear.

The same cytoplasmic condition which releases the onset of meiosis of the generative nuclei is thus also responsible for the start of partial meiotic processes in the somatic nucleus. Since the pairing of homologous chromosomes fails to occur, an "asynaptic pseudomeiosis" is all that can take place.

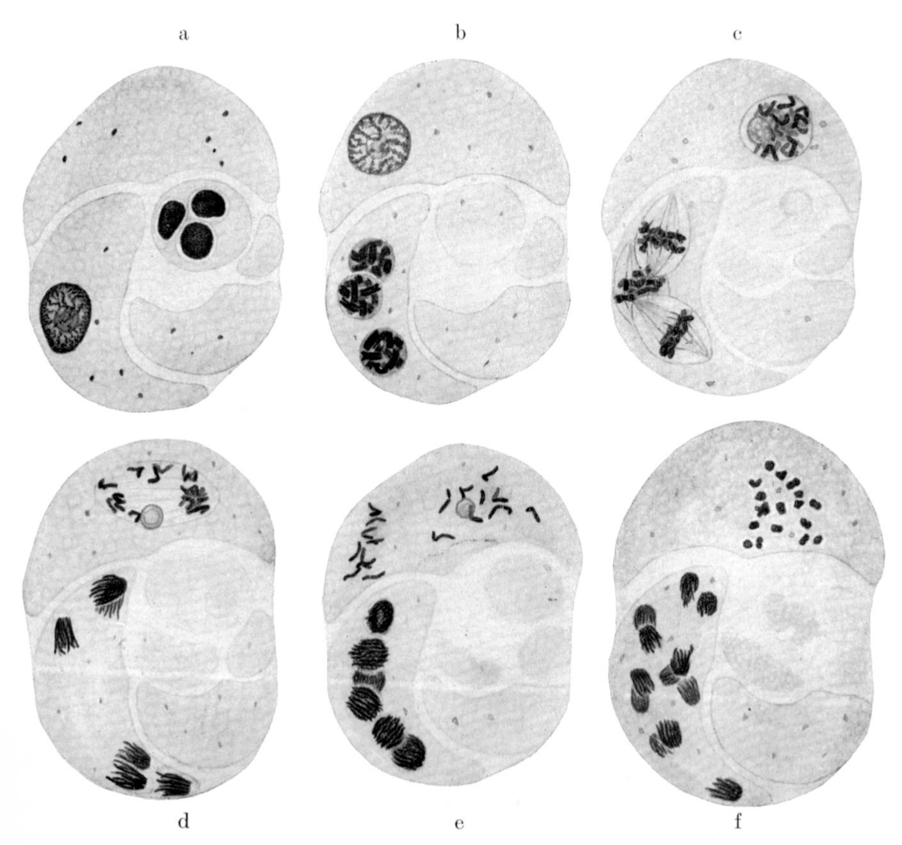

Fig. 80. *Rotaliella roscoffensis* (Foraminifera). Stages of meiosis. a, b Prophase, c metaphase. d anaphase of the first meiotic division. e Prophase, f anaphase of the second meiotic division. Bouin-Duboscq, Feulgen staining. × 830. After GRELL, 1957 [*663*]

In the *Sporozoa* it is the first nuclear division of sporogony which leads to a reduction of the chromosome number. This takes place, in *Gregarinida*, within the spore and is usually difficult to analyze because of the small size of the spores. In *Stylocephalus longicollis* (Fig. 81 and 390b) [*648*], a species with relatively large, purse-

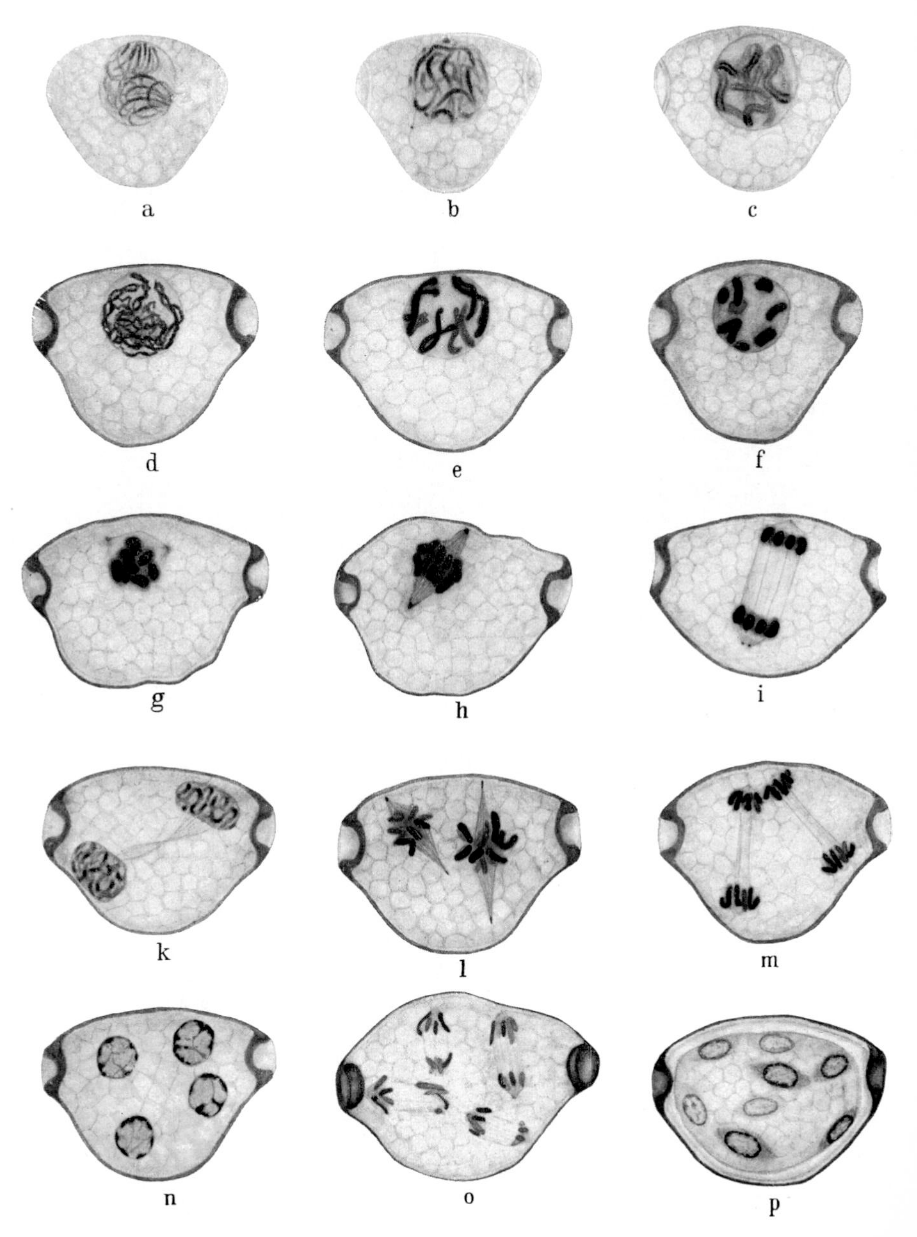

Fig. 81. *Stylocephalus longicollis* (Gregarinida). Stages of sporogony. a—f Prophase. g Transition to metaphase. h Metaphase. i Anaphase of the first sporogonial division (meiosis). k Binuclear stage. l Metaphase. m Anaphase of the second sporogonial division. n Tetranuclear stage. o Anaphase of the third sporogonial division. p Stage shortly before the formation of the sporozoites. Carnoy, iron hematoxylin. × 2700. After Grell, 1940 [*648*]

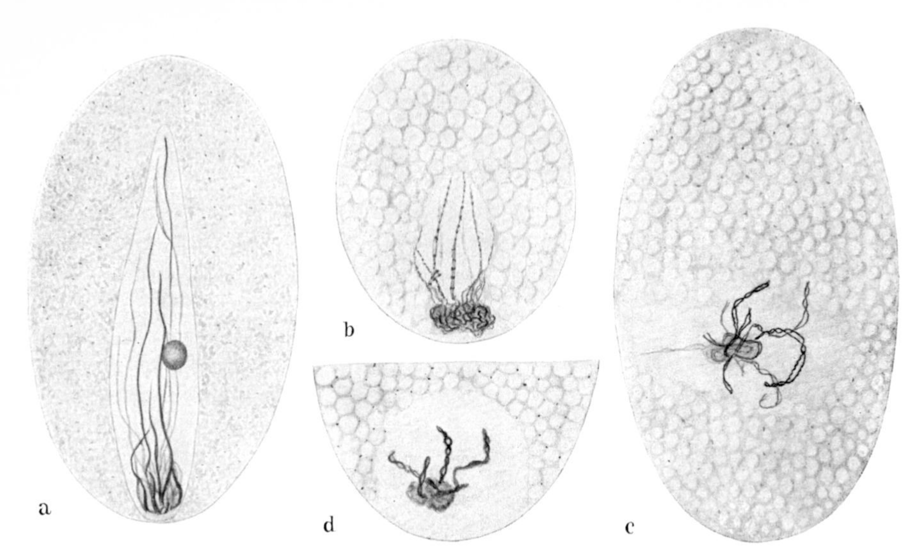

Fig. 82. *Eucoccidium dinophili* (Coccidia). Some stages of meiotic prophase. a—c Entire oocysts. d Part of an oocyst. a Bouin-Allen, Feulgen staining. b—d Aceto-carmine. × 1100. After GRELL, 1953 [*657*]

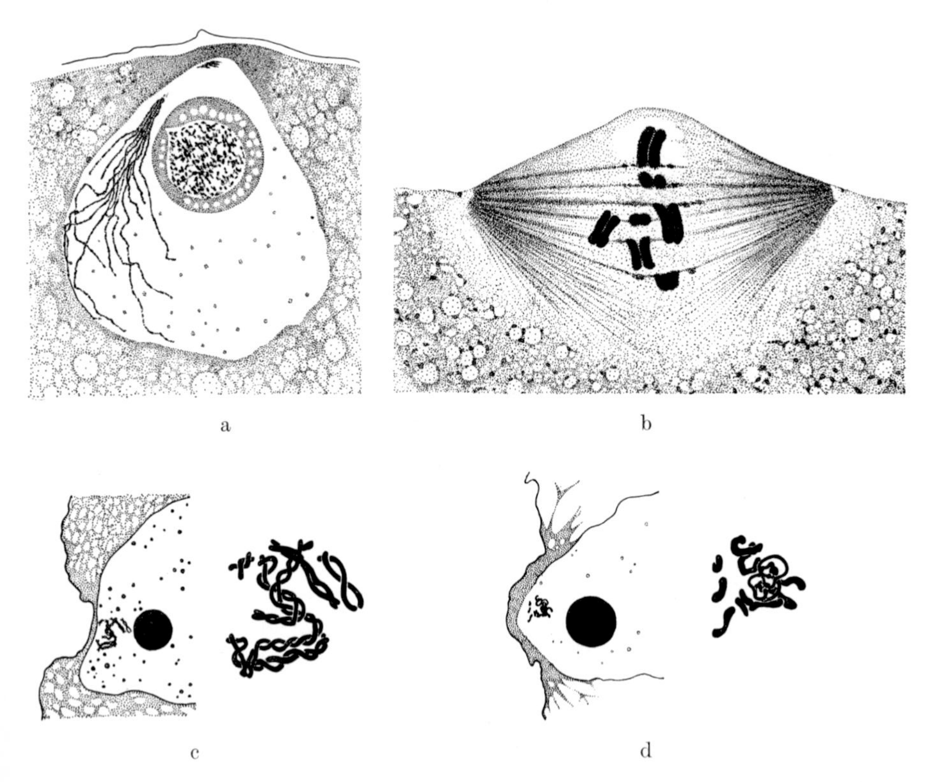

Fig. 83. *Aggregata eberthi* (Coccidia). Some stages of meiosis. a Bouquet stage ("fertilization spindle"); the bundle of chromosomes is bent near the upper pole of the nucleus, the actual bending point is in the following section. b Metaphase. c Coiling of the homologues in early prophase. d "Diakinesis stage" with the homologues more or less separated. In c and d the right half of the figure shows the same chromosome group enlarged. Iron hematoxylin. a × 1000. b × 2300. a, b after BELAR, 1926 [*134*]; c, d after NAVILLE, 1925 [*1220*]

shaped spores, the characteristic stages of pairing and mutual coiling of the homologues are nevertheless clearly recognizable. It is remarkable that the homologues separate completely before metaphase and condense independently. This chromosomal behavior which has also been reported from other Sporozoa indicates that no chiasmata are formed. Nevertheless, there is the usual co-orientation of the homologues at metaphase.

In contrast to the gregarines, the zygotes of *Coccidia* are very large cells (Figs. 82 and 83). During meiotic prophase, the chromosomes stretch enormously within the intact nuclear envelope, leading to a bundle of chromosomes which has formerly been termed "fertilization spindle". In some Coccidia this bundle can traverse the whole length of the zygote and may even be visible in life as a pointed, conical clear space in the cytoplasm. After the pairing of homologues, the two partners are wound around each other many times only to separate completely again at the end

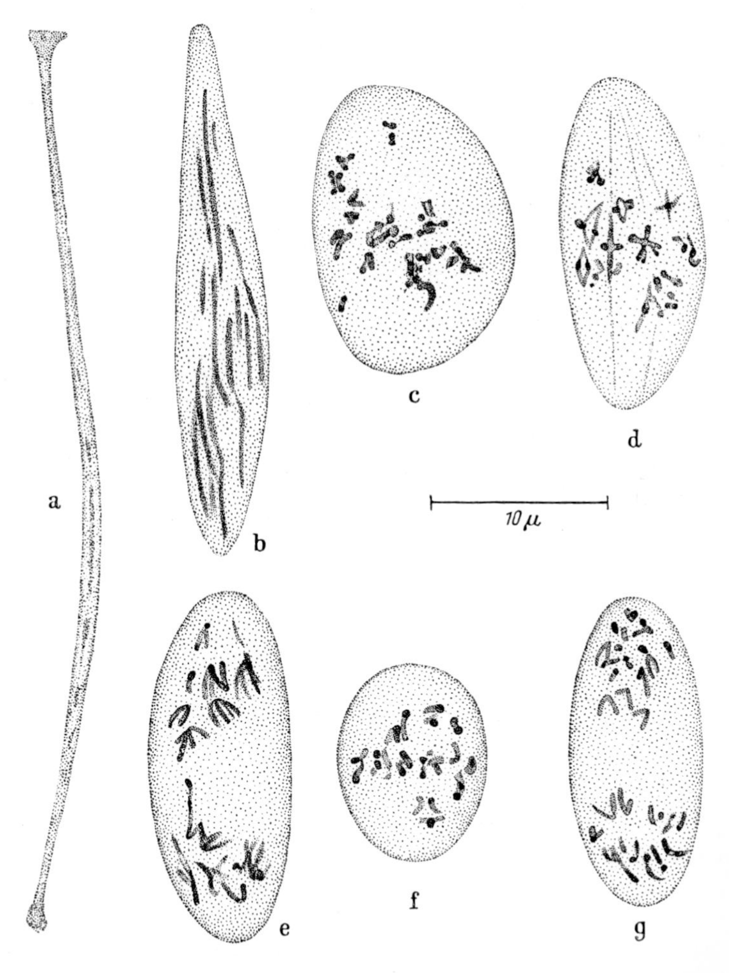

Fig. 84. *Colpidium campylum* (Ciliata). Meiosis. a Midprophase ("sickle stage"). b Transition to metaphase I. c Prometaphase. d Metaphase I. e Anaphase I. f Prometaphase II. g Anaphase II. Alcohol-acetic acid, acetocarmine. After DEVIDÉ and GEITLER, 1947 [*390*]

of prophase. There are absolutely no indications of chiasmata and tetrads in Sporozoa. Meiosis seems to involve only a single division.

In *ciliates*, meiosis takes place before the formation of gamete nuclei. It is carried out only by the generative micronucleus while the somatic macronucleus disintegrates. At prophase, the micronucleus often becomes extremely elongated (Fig. 84). As in the "fertilization spindle" of Coccidia, this is probably the leptotene stage. In other cases the micronucleus is only observed to swell considerably (Fig. 43). As the chromosomes of ciliates are very small, the details of the meiotic processes are difficult to recognize. Only in a few instances have structures been observed which could be interpreted as tetrads [*41, 390, 967, 1385, 1416*].

3. Nuclear Dimorphism and the Polygenomic State

The realization of different nuclear types within the same cell is termed *simultaneous nuclear differentiation* (p. 60). In Protozoa, this type is found only as *nuclear dimorphism:* there are *generative* nuclei which are capable of reproduction and *somatic* nuclei which are either incapable of reproduction or able to reproduce only for a limited period of time but which are active in other ways.

So far, nuclear dimorphism has been shown to occur only in two protozoan groups, the heterokaryotic Foraminifera and the ciliates.

Most ciliates have a *polygenomic* macronucleus which is thus different from other nuclear types. Since the nuclei of some Radiolaria also seem to be polygenomic, it seems suitable to discuss their nuclei after those of the ciliates.

a) Foraminifera

The nuclear dimorphism of Foraminifera can only be understood on the basis of their heterophasic life cycle (p. 230) with its regular alternation of haploid and diploid generations. The haploid generation (gamont) reproduces sexually, the diploid generation (agamont) asexually. The gamont which has a single nucleus only forms the gametes which fuse in pairs to form the zygotes. From the zygotes the agamonts originate. Through the so-called metagamic divisions the agamonts become multinuclear. The period of nuclear multiplication usually takes place before hatching, the agamonts having a fixed number of nuclei already at the onset of the growth period. As soon as the agamonts have reached a certain limit size, meiosis sets in. This is followed by a multiple division which leads to the formation of gamonts.

In *homokaryotic* Foraminifera (monothalamic species, as well as *Spirillina, Patellina* and others) the nuclei of the agamonts are all alike. All participate in meiosis and all have the same structure. Studies on *Patellina corrugata* (Fig. 360) showed that they are identical in their content of DNA, RNA and protein [*1840*].

In *heterokaryotic* Foraminifera the nuclei of agamonts are differentiated into two different types: only the generative nuclei carry out meiosis while the somatic ones are incapable of division and will ultimately disintegrate. That they can participate in meiosis to different degrees has already been discussed (p. 91). The first species found to have nuclear dimorphism is *Rotaliella heterocaryotica* (Fig. 85). Sexual reproduction (gamogony) leads to the formation of amoeboid gametes

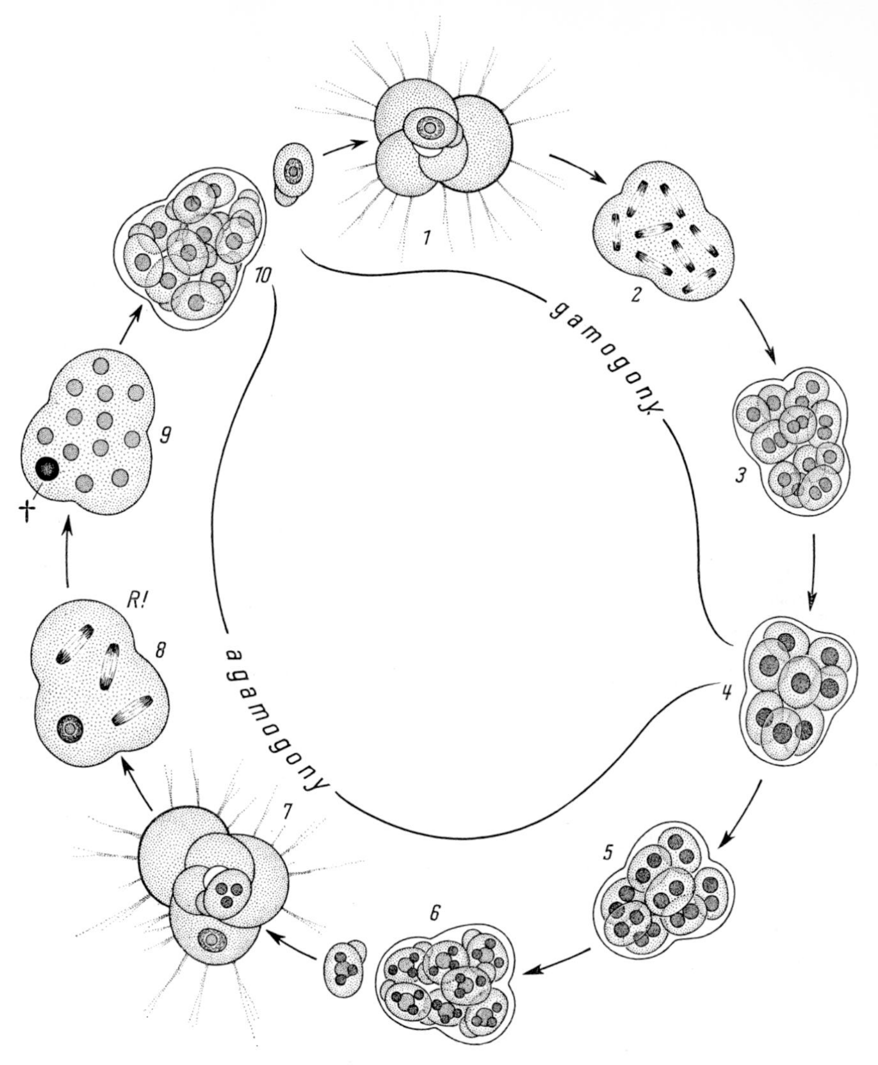

Fig. 85. *Rotaliella heterocaryotica*. Developmental cycle. *1* Adult gamont. *2* Last mitosis of gamogony. *3* Autogamic fusion of the gametes. *4* Zygotes. *5* Binucleated agamonts. *6* Agamonts with four nuclei. *7* Adult agamont. *8* First meiotic division. *9* End of the second meiotic division. *10* Agametes (= young gamonts).† Degenerating somatic nucleus. After GRELL, 1954 [*659*]

which unite in pairs within the shell of the gamont to form the zygotes (autogamy). Even the first phase of asexual reproduction (agamogony) takes place within the shell of the gamont. It consists of two metagamic divisions. The emerging agamonts have thus four nuclei which differentiate immediately after the second division into three generative and one somatic nucleus. While the generative nuclei remain small and condensed, the somatic nucleus increases in size and forms a nucleolus. As the agamont grows, the three generative nuclei remain in the initial chamber while the somatic nucleus moves into a younger chamber.

Cytochemical measurements on isolated nuclei have shown that the dry mass as well as the RNA content of the somatic nucleus is about three times as high as that of a generative nucleus. Both nuclear types have the same content of DNA which is doubled at the same time before meiosis [*1840*]. Although the somatic nucleus of *Rotaliella heterocaryotica* swells up only temporarily but shows no other signs of participating in meiosis, it reduplicates its DNA content just as the generative nuclei. Perhaps DNA synthesis is generally the first reaction to that condition of the cytoplasm which induces the onset of meiosis.

Rotaliella roscoffensis (Fig. 362) and *Metarotaliella parva* (Fig. 363), like *Rotaliella heterocaryotica*, have agamonts with normally three generative nuclei and one somatic nucleus. In contrast to the other species named, the generative nuclei of *Metarotaliella parva* can temporarily form nucleoli [*1789*]. Another species, *Metarotaliella simplex*, has agamonts with either one somatic nucleus and three generative nuclei, or one somatic and one generative nucleus only. Evidently, this depends on the number of metagamic divisions. In the first, most frequent case, two divisions occur, in the latter only one. The condition of these variations is not known. The agamonts of *Rubratella intermedia* contain five generative nuclei and one somatic nucleus (Fig. 365). There are three metagamic nuclear divisions. However, after the second division one of the four sister nuclei degenerates. Only three nuclei participate therefore in the third division.

In *Glabratella sulcata* (Fig. 367), whose gamont (Fig. 86a) attains a larger size than the agamont, the number of metagamic nuclear divisions is variable, but to a higher extent than in *Metarotaliella simplex*. The emerging agamonts may thus have a

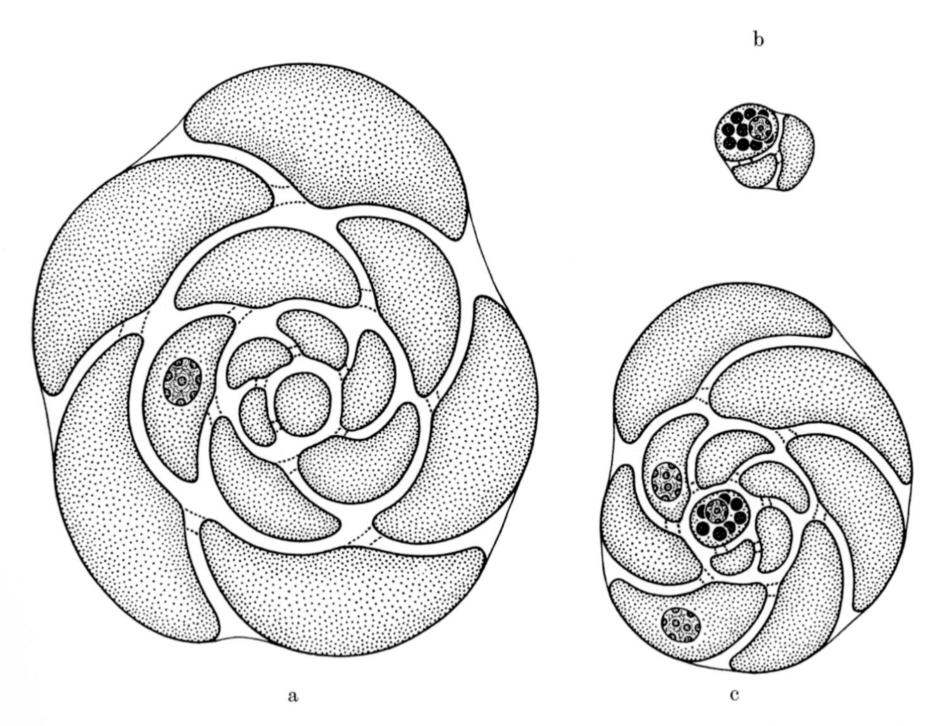

Fig. 86. *Glabratella sulcata*. Nuclei, a in gamont, b in a young, c in an older agamont. Somewhat diagrammatic. After GRELL, 1956 [*661*]

variable number of nuclei. In spite of this variability, only one of these nuclei differentiates into a somatic nucleus (b). As the agamont grows, the somatic nucleus moves into a younger chamber. Upon this, one of the nuclei in the initial chamber differentiates immediately into a new somatic nucleus. This process may be repeated, occasionally even two or three times. Fully grown agamonts may therefore contain 3–5 somatic nuclei and a varying number of generative ones (c).

Even though the physiological meaning of nuclear dimorphism as such is still a riddle, there are sufficient indications that the structural differentiation of the nuclei is accompanied by a functional differentiation.

In *Rotaliella roscoffensis* (Fig. 362) it happens frequently, that shortly after the second metagamic division one or two nuclei degenerate. Hence, the young agamonts contain three or two nuclei only, one of which always differentiates in the somatic direction.

In some cases agamonts have been found which had no generative nuclei at all but only a somatic nucleus (Fig. 87a). They can be distinguished from gamonts by the fact that the somatic nucleus is not located in the initial chamber but in one of the younger chambers. When such agamonts have reached the size at which the generative nuclei would begin meiosis, the somatic nucleus carries out an "asynaptic pseudomeiosis" (p. 92) and degenerates (b). This will also conclude the life of the cell.

This ,,Naturexperiment" (Boveri) permits the conclusion that the agamont of *Rotaliella roscoffensis* can grow without generative nuclei. Since an agamont without a somatic nucleus has never been found, it can be assumed that it is indispensable for cell growth ("metabolic nucleus") while the function of the generative nuclei is largely restricted to handing on the genetic information ("reproductive nuclei").

The observation that always *only one* nucleus is differentiated in the somatic direction in spite of variable numbers of nuclei (although, in *Glabratella sulcata*, only within the initial chamber) suggests that all generative nuclei are potentially

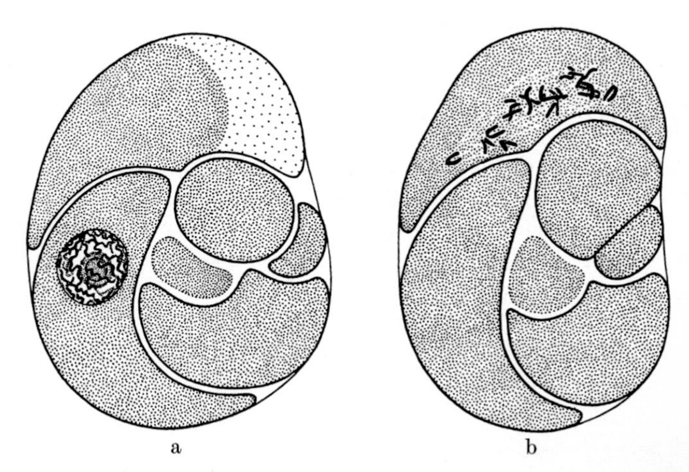

Fig. 87. *Rotaliella roscoffensis*. Agamonts without generative nuclei. a Somatic nucleus in the chamber next to the last one. b Elongated somatic nucleus (prior to its elimination). Bouin-Duboscq, Feulgen staining. × 960. After Grell, 1957 [*663*]

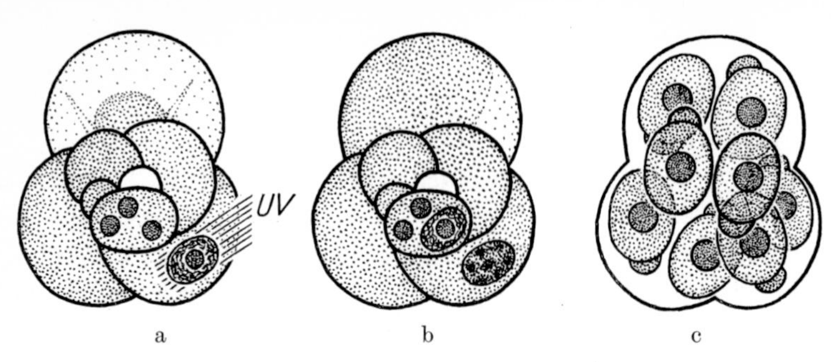

Fig. 88. *Rotaliella heterocaryotica.* Diagram of an experiment in which the somatic nucleus of a living agamont was inactivated by ultraviolet radiation (a). After inactivation one of the three generative nuclei differentiates in the initial chamber to a somatic nucleus (b). After meiosis only 8 gamonts originate from such an agamont (c). After CZIHAK and GRELL, 1960 [*350*]

somatic and that the determination of one nucleus prevents the determination of others.

This assumption has been verified experimentally (Fig. 88). In *Rotaliella heterocaryotica,* whose nuclear number is not variable, the somatic nucleus can be seen in life and be inactivated with a microbeam of ultraviolet light (a). In some cases one of the generative nuclei in the initial chamber will then swell and form a nucleolus (b). Since the inactivated nucleus as well as the newly determined one degenerates, eight instead of twelve gamonts are formed as a result of meiosis (c).

Cytochemical measurements showed that the secondary somatic nucleus has the same content of DNA, RNA and protein as the primary one. In some cases it has even been possible to eliminate the secondary somatic nucleus and thereby induce the formation of a tertiary one [*1840*].

The pathway of nuclear determination is not known as yet. It seems possible that determination may be due to a general cytoplasmic condition, perhaps a determinative substance whose concentration increases within the initial chamber. The "sensitivity" toward this substance might vary, the initially determined nucleus being the "most sensitive" one. One might theorize that either the synthesis of an enzyme which destroys the determinative substance or of a repressor which prevents production, is controlled by the determined nucleus. The fact that only one somatic nucleus is determined within the initial chamber might be due to a relatively simple chemical feedback cycle similar to the models provided by bacterial genetics.

b) Ciliates

The existence of heterokaryotic Foraminifera shows that nuclear dimorphism is not a "privilege" of ciliates which might set them apart from all other Protozoa. There are even indications that not all ciliates have nuclear dimorphism. Species found in the interstitial system of marine sands (mesopsammon) of the genus *Stephanopogon (mesnili, colpoda)* have been shown to be clearly *homokaryotic* [*1065*]. Although multinucleated they have only one type of nucleus which can neither be interpreted as macro- nor as micronucleus since it possesses a central nucleolus and divides mitotically.

The overwhelming majority of ciliates, however, is characteristically *heterokaryotic.* Their generative nuclei are called *micronuclei,* their somatic ones *macronuclei.* Studies of recent years have shown that *two types of nuclear dimorphism* can be distinguished as primary and secondary types respectively [*673, 1396*].

α) Primary Type

In a number of exclusively holotrichous ciliates, all with the exception of *Loxodes* of the marine sand interstitial fauna, a type of nuclear dimorphism has been found which is very much like that of heterokaryotic Foraminifera [*1385–1387*, *1390*, *1391*].

The always multiple macronuclei are *diploid*. They differ from micronuclei in size, less compact structure and the presence of one or several nucleoli. Furthermore, they are *incapable of division* and are always formed anew from descendants of the micronuclei. Autoradiographic studies showed that they synthesize RNA but are incapable of DNA synthesis [*1733*].

In species with a few nuclei only, the nuclear number is constant. This constancy is due to the synchrony of the nuclear divisions (Fig. 89).

Loxodes rostrum (a) has two macronuclei and one micronucleus. Every fission is connected with two divisions of the micronucleus. Each daughter cell receives one of the macronuclei and two micronuclei, one of which differentiates again into a macronucleus. In species with the same number of nuclei, division and reorganization take place in a like manner *(Geleia nigriceps, G. orbis, Remanella rugosa, R. granulosa)*.

Loxodes striatus (b) has two macronuclei and two micronuclei. At the time of fission, the two micronuclei are reduplicated first. One of the two daughter nuclei divides again and gives rise to the two micronuclei while the other remains undivided and differentiates into a macronucleus.

In species with numerous macro- and micronuclei, divisions take place more or less asynchronously. One example of this is provided by *Loxodes magnus* (Fig. 90).

The macronuclei are evidently not capable of unlimited function. In *Loxodes magnus* it has been observed that 2–10% of the macronuclei of a cell are pycnotic.

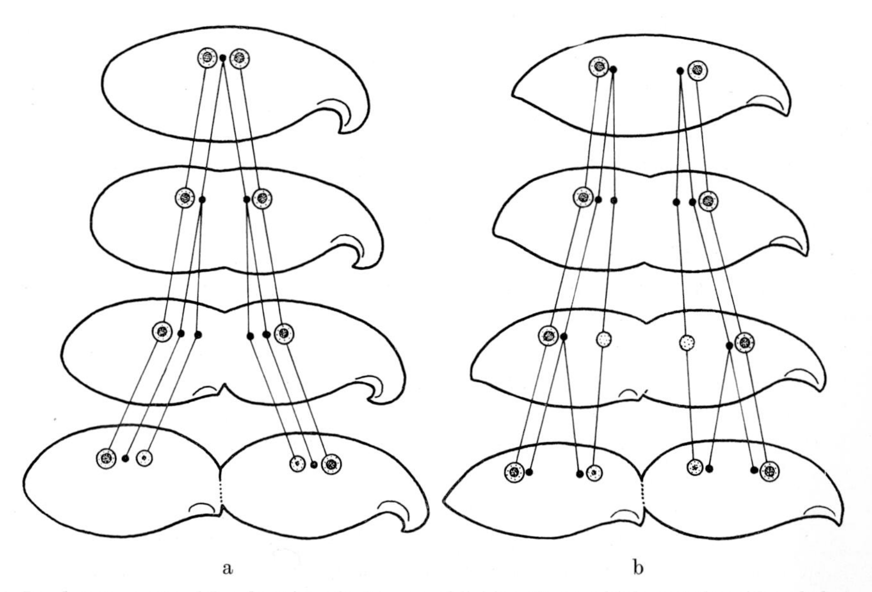

Fig. 89. *Loxodes rostrum* (a) and *Loxodes striatus* (b). Diagram of division. Macronuclei: large circles (with nucleolus) Micronuclei: small black circles. After RAIKOV, 1957

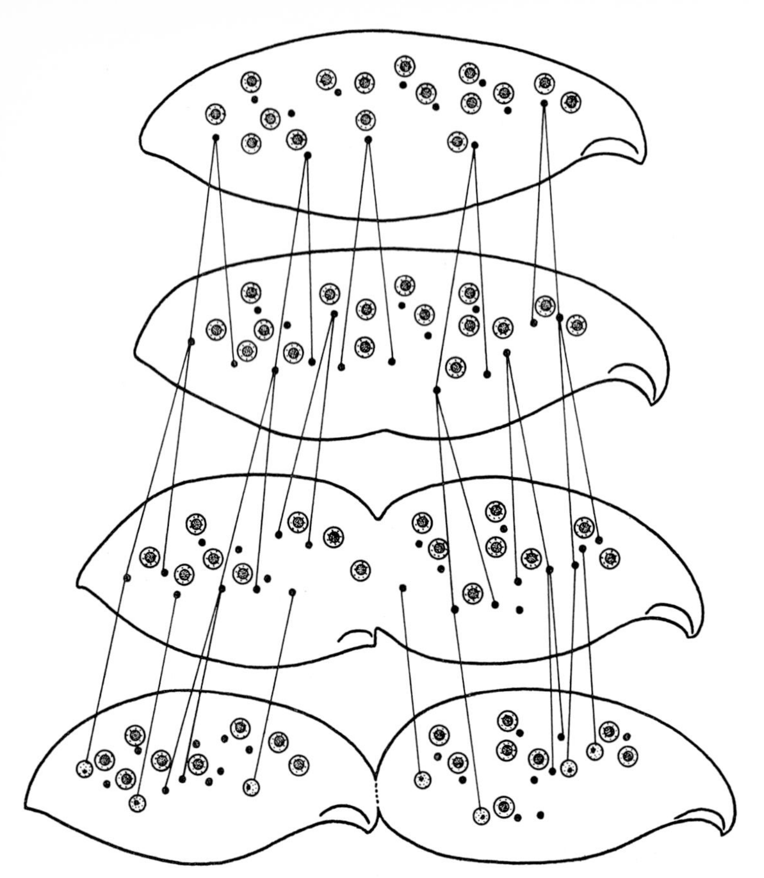

Fig. 90. *Loxodes magnus*. Diagram of division. After Raikov, 1957

Single macronuclei have therefore probably a life span of 4—7 divisional cycles only.

A most peculiar modification of the primary type is found in some *Trachelocercidae*. Macro- and micronuclei are fused to form a compound nucleus. In *Tracheloraphis phoenicopterus* (Fig. 91) it consists of six micronuclei surrounded by a "macronucleus" formed from six macronuclear anlagen (a). During division, the micronuclei first carry out a mitosis, leading to a total of twelve micronuclei (b). The compound nucleus then tears into two portions containing six micronuclei each (c, d). These divide again (e). Of the twelve daughter nuclei (f), six become micronuclei and six macronuclear anlagen (g). After the residues of the old macronucleus have been resorbed in the cytoplasm, the newly formed macronuclear anlagen fuse to give rise again to a compound nucleus which encloses the micronuclei (h).

Details of nuclear structure vary a great deal in the various Trachelocercidae. *Trachelocerca coluber*, for example, has only one compound nucleus with two micronuclei. *Tracheloraphis dicaryon* has two compound nuclei with four micronuclei each. Some of the species, like *Trachelonema poljanskyi*, have no compound nucleus at all but many macro- and micronuclei.

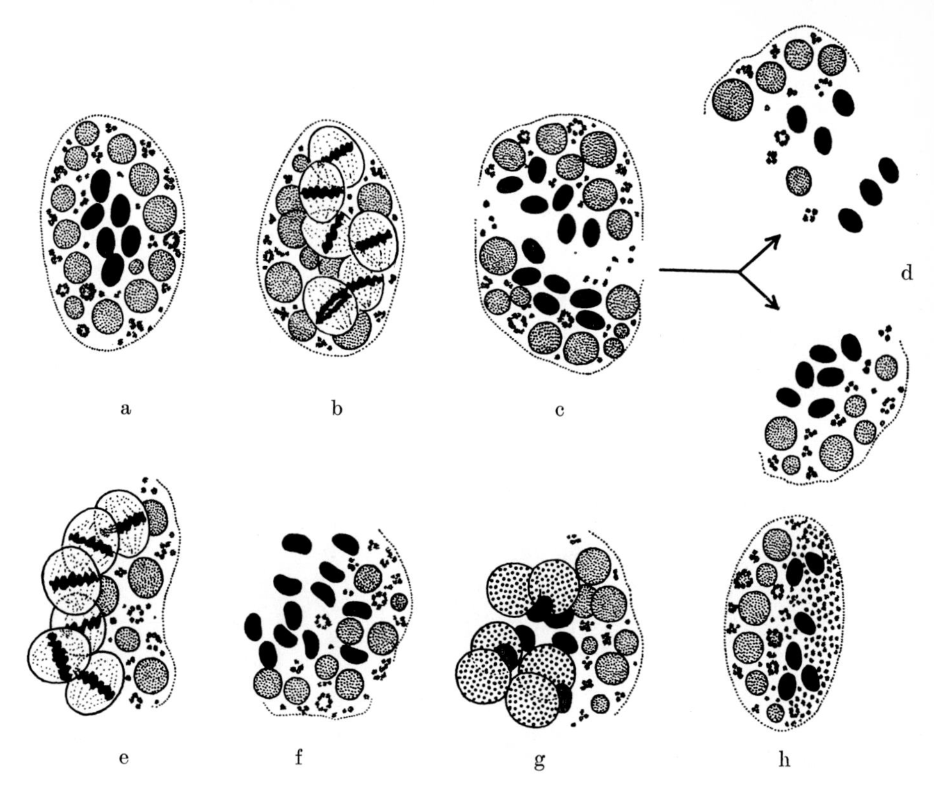

Fig. 91. *Tracheloraphis phoenicopterus*. Behavior of the "compound" nucleus during division. a "Compound" nucleus at rest. Within are six (black) micronuclei. b First mitosis of the micronuclei. c, d Division of the "compound" nucleus. e—h Reconstruction of the daughter nuclei. e Second mitosis of the micronuclei. f Twelve micronuclei. g Six micronuclei, six macronuclear anlagen. h Reconstructed "compound" nucleus. Combined after RAIKOV, 1958 [*1385*]

It has frequently been observed that nucleolar substance passes from the macronuclei into the cytoplasm. In *Geleia nigriceps* the whole nucleolus is periodically discharged into the cytoplasm [*1387*]. Like the somatic nuclei of Foraminifera the diploid macronuclei of ciliates thus seem to represent the metabolically active nuclei. Their irreversible determination is also in accord with foraminiferan somatic nuclei. The function of the micronuclei, on the other hand, is restricted to reproduction and the provision of a persisting reservoir for macronuclei.

β) Secondary Type

The secondary type is characteristic for the majority of ciliates. The macronucleus, which is usually present singly, is *polyploid* and *capable of division*. It is thus not destroyed in the course of asexual reproduction. It disintegrates only when sexual processes take place (conjugation, autogamy) and is then formed anew from a descendant of the synkaryon. In spite of its ability to divide it is therefore justified to consider it a somatic nucleus.

The fact that the primary type has only been found in some holotrichs suggests that the secondary type might have developed from it. Curiously enough, those ciliates which represent the primary type do not belong to one taxonomic category within the Holotricha. They belong to different suborders and families which include also genera of the secondary type with polyploid macronuclei. If the taxonomic relationships in our system do indeed have a phylogenetic basis, the conclusion that the secondary type must have evolved polyphyletically from the primary type is unavoidable. Ciliates with a diploid macronucleus incapable of division would in every case seem to represent "karyological relicts".

In contrast to macronuclei of the primary type those of the secondary type include a large *variety of forms* (Fig. 92). In many species the macronucleus is simply oval, roundish or droplet-shaped (a). Occasionally, it is divided into two parts (b), bent in a horseshoe shape (c) or it is long and cylindrical (d). In some, it is branched in a definite manner (f) or it forms a continuous net below the pellicle (g). Many ciliates have a beaded macronucleus (h) which brings to mind a rosary. Even highly irregular forms are found (i, k).

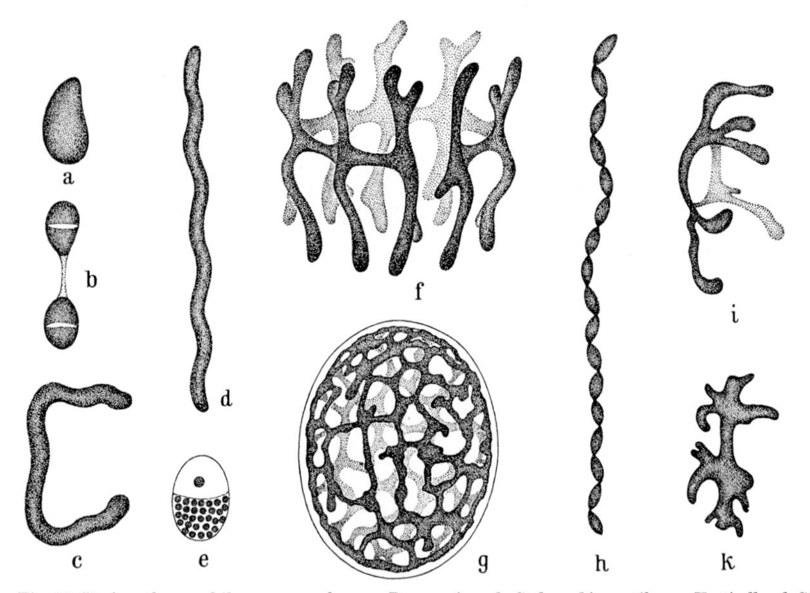

Fig. 92. Various forms of the macronucleus. a *Paramecium*. b *Stylonychia mytilus*. c *Vorticella*. d *Stentor roeseli*. e *Spirochona gemmipara*. f *Ephelota gemmipara*. g *Metaphrya sagittae* (pellicle indicated by line). h *Spirostomum ambiguum*. i *Ophryodendron porcellanum*. k *Conchophthirius caryoclada*. i after Collin, 1912; k after Kidder, 1933, the remaining figures are originals

A special type is the so-called "heteromerous" macronuclei which consist of two parts: an intensely staining (Feulgen-positive) mostly granular "orthomere" and a faintly staining "paramere", frequently with a Feulgen-positive granule ("endosome") (Fig. 92e) [*503*].

Macronuclei are easily distinguished from micronuclei by their *size*. They are intensely Feulgen-positive and contain much more DNA than the micronuclei. Their high DNA content supports the contention that the macronuclei are *polyploid*.

Spectrophotometric measurements have shown that the DNA content varies considerably among the various species. Taking the micronucleus as diploid, the macronucleus of *Stylonychia mytilus*, e.g., should be about 64-ploid, [*41*], that of *Nassula ornata* about 230-ploid [*1399*], in *Paramecium aurelia* about 860-ploid [*1825*] and in *Bursaria truncatella* it should be about 5000-ploid [*1469*].

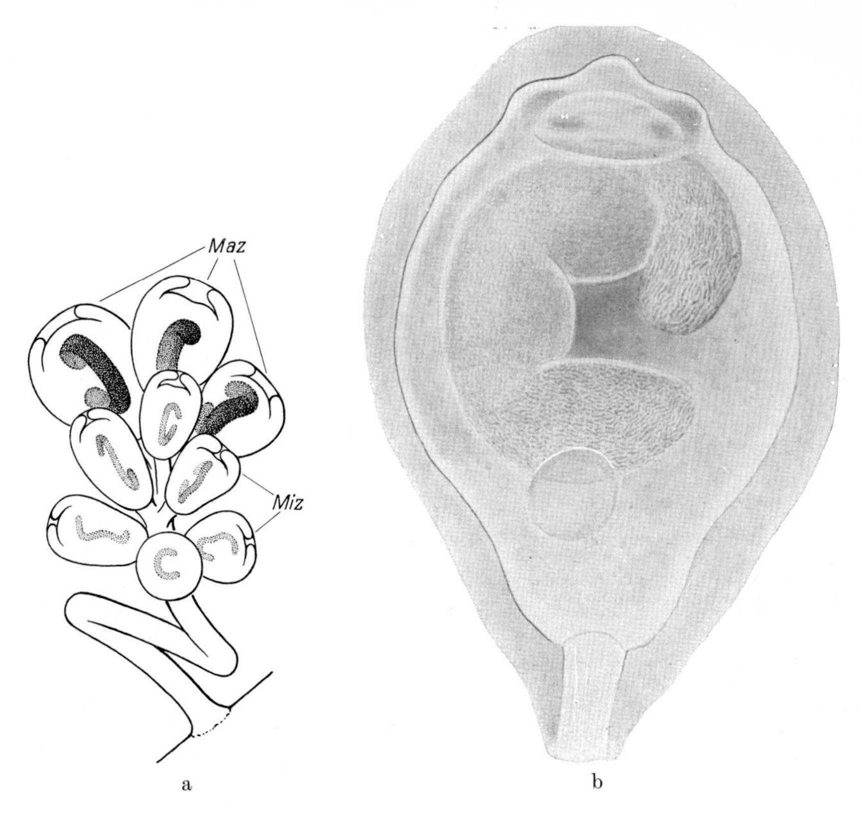

Fig. 93. *Zoothamnium*. a Colony with macrozooides (*Maz*) and microzooides (*Miz*). Picric acid, Feulgen. b Single macrozooide in life (phase contrast). After GRELL, 1950 [*651*]

Macronuclei of different cell types in the same species can differ in size and DNA content. Thus, the microzooides of *Zoothamnium* have smaller macronuclei of less DNA content than the macrozooides (Fig. 93a).

Even different developmental stages may differ in size and DNA content of their macronuclei, as shown by comparison of a suctorian which has just finished metamorphosis with a fully grown one (Fig. 94a, b).

The *structure* of the macronucleus has not been completely clarified. *Nucleolar substance* is easily demonstrated, usually as numerous nucleoli distributed throughout the whole macronucleus. The Feulgen-positive material, on the other hand, is difficult to analyze as it appears in fixed and stained preparations more or less homogeneous or, frequently, granular. It can nevertheless no longer be doubted that DNA is bound to the same structures in the macronucleus as in the nuclei of all eukaryotes, i.e. to *chromosomes*. In many cases chromosomes are clearly recognizable in life, using phase contrast equipment. One example is provided by the macrozooides of *Zoothamnium* (Fig. 93b). In *Tokophrya* a tangle of threads is recognizable in the macronucleus (Fig. 94a, b). It should be stressed that the threads have the same dimensions in the juvenile stage with its small macronucleus as they do in the large macronucleus of the fully grown suctorian. In the

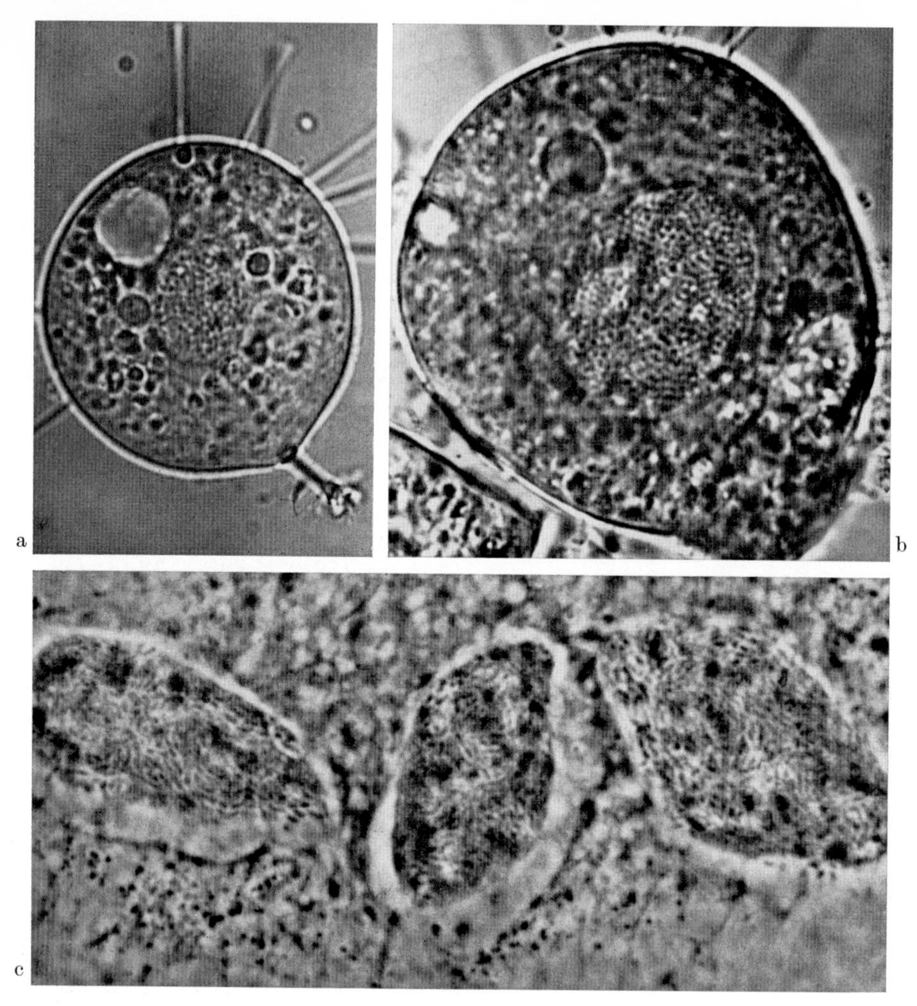

Fig. 94. Chromosomal structures in the macronucleus. a, b *Tokophrya sp.* Young and old growth stage. Photographed in life. × 500. After GRELL, 1953 [*654*]. c Three nodes of the macronucleus of *Loxophyllum meleagris*. Alcohol-acetic acid. × 1900. After RUTHMANN, 1963 [*1468*]

macronuclear beads of *Loxophyllum meleagris* (c) there are stretches with paired threads, the threads seemingly consisting of single segments.

In *Nassula ornata* (Fig. 95) with its relatively less compact macronucleus the arrangement of the chromosomes before nuclear division suggests stages of endomitosis (see below). There seems to be a certain synchrony: as the daughter threads move apart within the macronucleus, the chromatids of the micronucleus progress toward the spindle poles.

Since the macronucleus grows after conjugation or autogamy from a diploid nucleus (p. 199), its polyploidy should be due to autonomous chromosome divisions or endomitoses [*565, 1306*] as in many somatic nuclei of animals and plants. Indications of *endomitotic polyploidization* have indeed been found occasionally in the macronuclear anlagen, i.e. in those developmental stages which are passed as

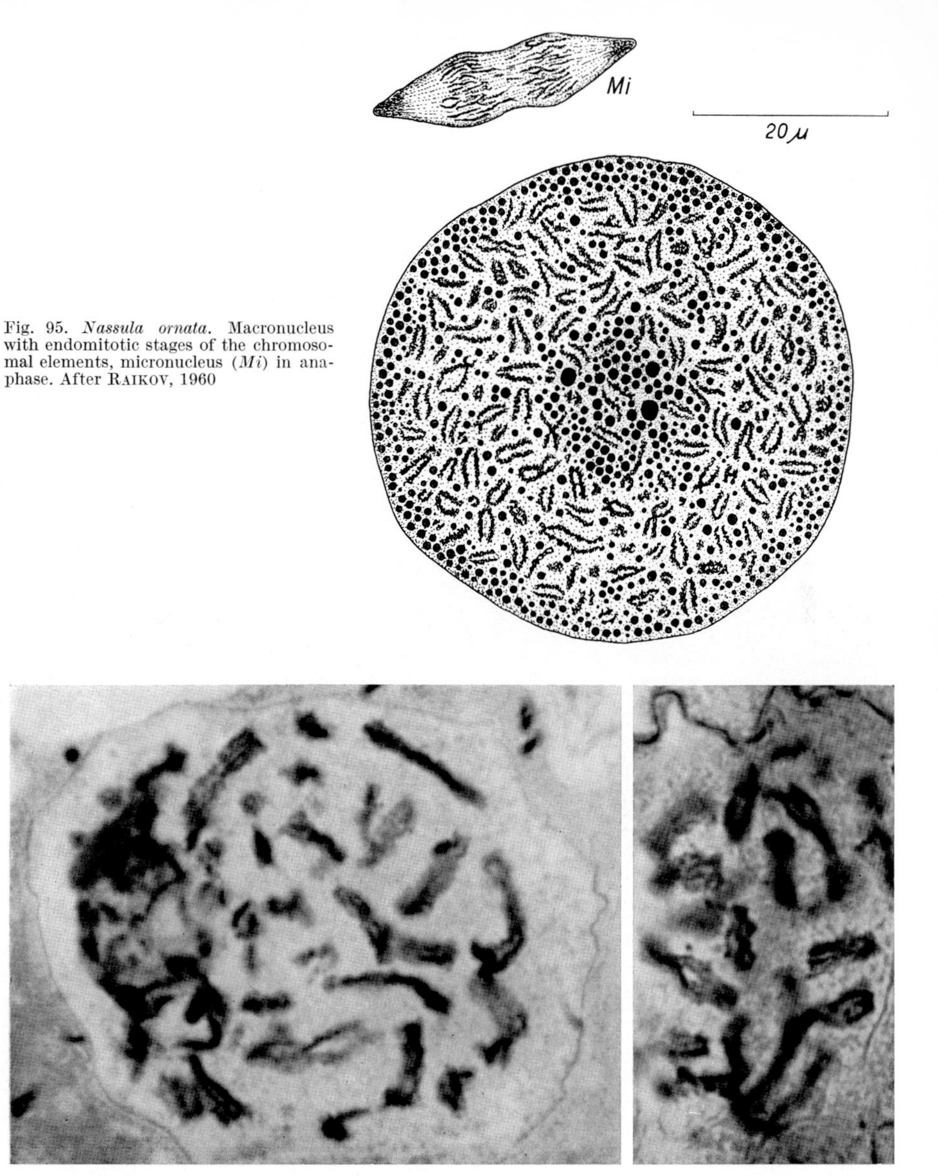

Fig. 95. *Nassula ornata*. Macronucleus with endomitotic stages of the chromosomal elements, micronucleus (*Mi*) in anaphase. After Raikov, 1960

Fig. 96. *Ephelota gemmipara* (Suctoria). Macronuclear anlage. Endomitotic stages of the chromosomes. Sections, sublimate alcohol, iron hematoxylin. × 1550. After Grell, 1953 [*653*]

the macronucleus grows to its definitive size [*649, 651, 1317*]. In the suctorian, *Ephelota gemmipara*, whose macronuclear development takes several days, chromosome bundles of four or eight strands have been observed (Fig. 96).

In other cases, especially in *Stylonychia* and *Euplotes*, highly condensed rodlets are first seen in the macronuclear anlagen. Their number corresponds approximately to that of the meiotic chromosomes (2n). These rodlets uncoil subsequently to thin threads which form a dense tangle within the nucleus. After some time the threads

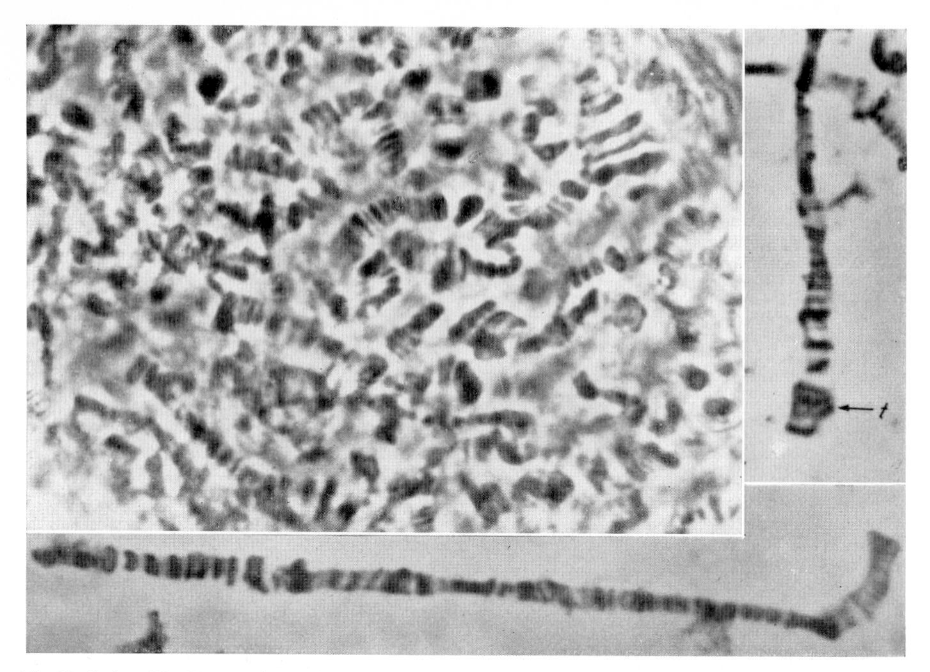

Fig. 97. *Stylonychia*. Macronuclear anlage with polytene chromosomes. The survey view is a section of a hemikaryotic macronuclear anlage of *St. mytilus*. Photographed in life. × 2200. After AMMERMANN, 1965 [*41*]. The individual chromosomes (right and below) are from a synkaryotic macronuclear anlage of *St. muscorum*. Aceto-orcein, squash preparation. × 2300. After PÉREZ-SILVA and ALONSO, 1966 [*1296*]

become broad tapes with a clearly recognizable pattern of crossbands as in the *polytenic chromosomes* of Diptera (Fig. 97). There is good evidence that the pattern is specific. However, in contrast to the giant chromosomes of Diptera, the homologues are not paired in *Stylonychia*. Probably, each homologue represents one "composite chromosome" consisting of the end-to-end connected single chromosomes of the haploid set [*41, 610, 1296, 1413, 1500*].

Although this seems to fit in the so-called "composite chromosomes" hypothesis to be discussed later (p. 113), light- and electron microscopic studies indicate that the polytene chromosomes disintegrate. As the chromosomes reach their maximum state of polyteny (Fig. 98a), sheets of fibrous material appear between the individual bands which transect the chromosomes in separate compartments. Subsequently, these compartments swell, resulting in a macronuclear anlage packed with thousands of separate spherical chambers (b). At the same time the DNA-content of the anlage decreases to a low (in *Stylonychia mytilus:* diploid) level. It is only after this DNA-poor stage, in which chromosomes are not longer discernible, that DNA synthesis takes place again, leading to the DNA content of the definitive macronucleus. This second phase of polyploidization is connected with the appearance of "replication bands" (p. 114) comparable to the bands which double the DNA in the vegetative macronucleus.

It is not known whether the linear integrity of the DNA comprising the polytene chromosomes is actually interrupted by the compartmentalization process, or — as is more acceptable from the genetic point of view — DNA strands are left to be

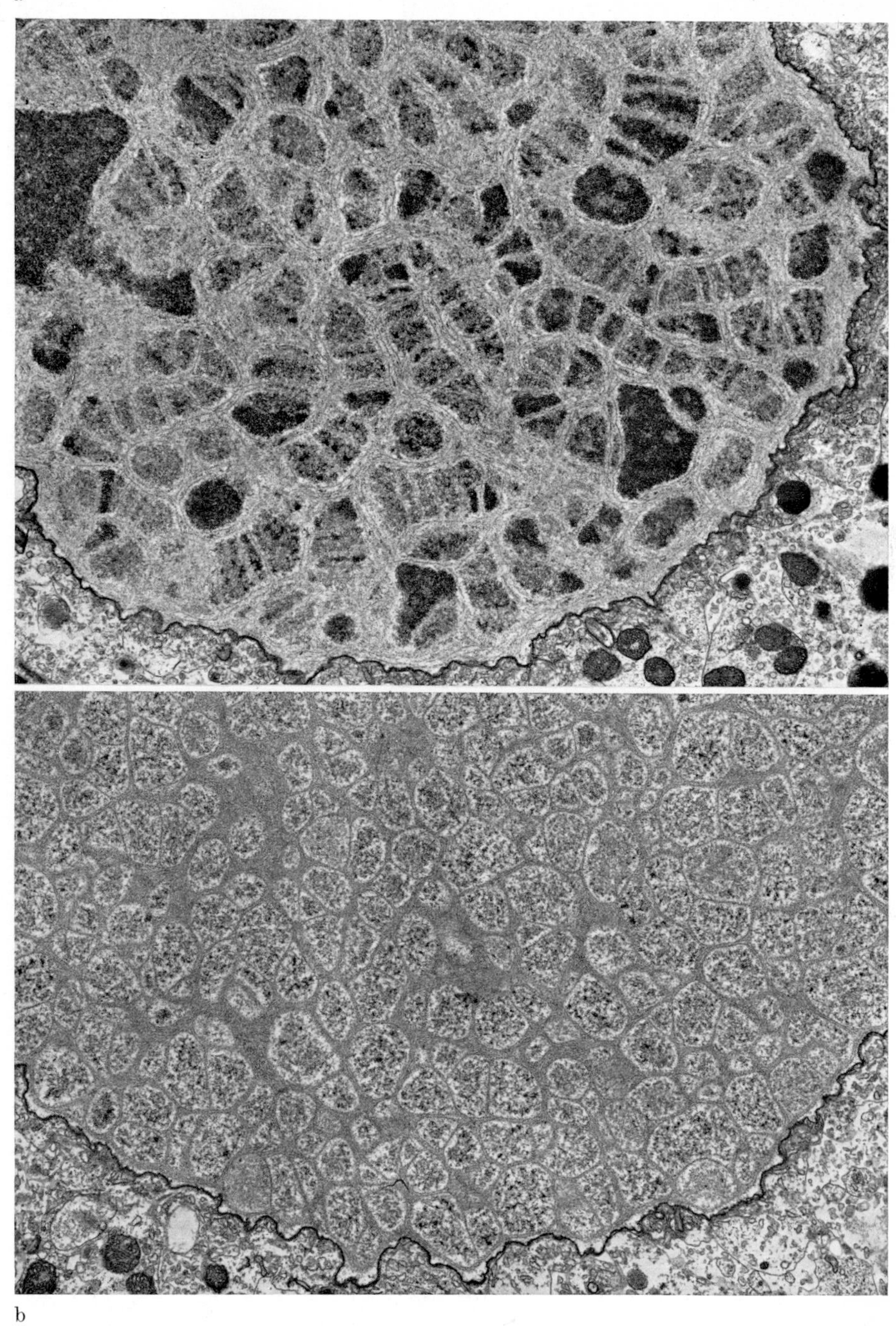

Fig. 98. *Stylonychia mytilus.* Macronuclear anlage. a Banded polytene chromosomes at their largest. At this stage lamellar partitions can be seen among the chromosomes passing between the bands. b Compartmentalization of the polytene chromosomes during the DNA-poor stage. Each compartment is interpreted to contain material from an individual band. ×7000. After KLOETZEL, 1970 [*945*]

the bases for later rounds of replication. On the other hand, the possibility cannot be excluded that a "chromatin diminution" occurs, restricting the genetic information to the amount which is necessary for the macronucleus to fulfill its functions.

That the latter possibility cannot be generalized, however, is proved by records of continuous DNA synthesis during macronuclear development of other ciliates not belonging to the hypotrichs.

In spite of the difficulties encountered in interpreting the different stages of macronuclear anlagen development in special cases, there is no doubt that the definitive macronucleus is filled with chromosome-like threads, which replicate by an endomitosis-like process. From the cytological point of view, the macronucleus must be regarded as an *endopolyploid* nucleus, regardless of whether these chromosomes correspond totally i.e. with respect to their full genetic information to the chromosomes at the beginning of anlagen development*.

As the following discussion will show, the macronucleus differs from endopolyploid nuclei of plants and animals principally by its *mode of division*. Hence, endopolyploidy is not a sufficient criterion to express its characteristic state with respect to other types of nuclei.

Since it has long been known that the division of the macronucleus differs from the well-known scheme of mitosis, it has frequently been referred to as an *amitosis*. This concept can, however, only be used with reservations as it refers to the direct constriction of an interphase nucleus into two without regular reduplication of the chromosomes and distribution of chromatids. However, as far as the division of the macronucleus is concerned, it has been shown to be preceded by a DNA synthesis. The products of division contain the full complement of genes. The transmission of genetic information must therefore be just as regular as in mitosis. This fact tends to be obscured by the term "amitosis".

In most ciliates the division of the macronucleus is *equal*, i.e. it stretches to some length and is then constricted in the middle. If it consists of several beads, these will at first fuse into a common mass. After this condensation, the nucleus stretches and is divided into two daughter nuclei of equal size which are then segmented into beads again.

Some ciliates of the secondary type contain many macronuclei. However, these represent an "operational unit" in the sense that they all fuse to form a single nucleus before cell division commences. This nucleus is then divided into two daughter nuclei which again divide and so forth, until many small macronuclei are present (Fig. 99).

In other ciliates, *multiple* and *unequal* divisions are found, the daughter nuclei being much smaller than the mother nucleus. This type of division is especially characteristic for *Suctoria* whose daughter cells are motile swarmers. The highly branched macronucleus of *Ephelota gemmipara* (Fig. 92f) is condensed to a uniform mass before division, followed by renewed branching. Each cytoplasmic bud is supplied by a projection of the macronucleus which is separated from the mother nucleus as the swarmer differentiates (Fig. 100a). It is only in the course of the so-called reactive budding (p. 149) that the whole mass of the macronucleus is subdivided. While division is *simultaneous* in this case, it is *successive* in the *Dactylo-*

* Even if the chromosomes become reduced to "chromomeres" as has been assumed for *Stylonychia* [*47*], the "chromomeres" must contain many genes packed in a chromosome-like structure.

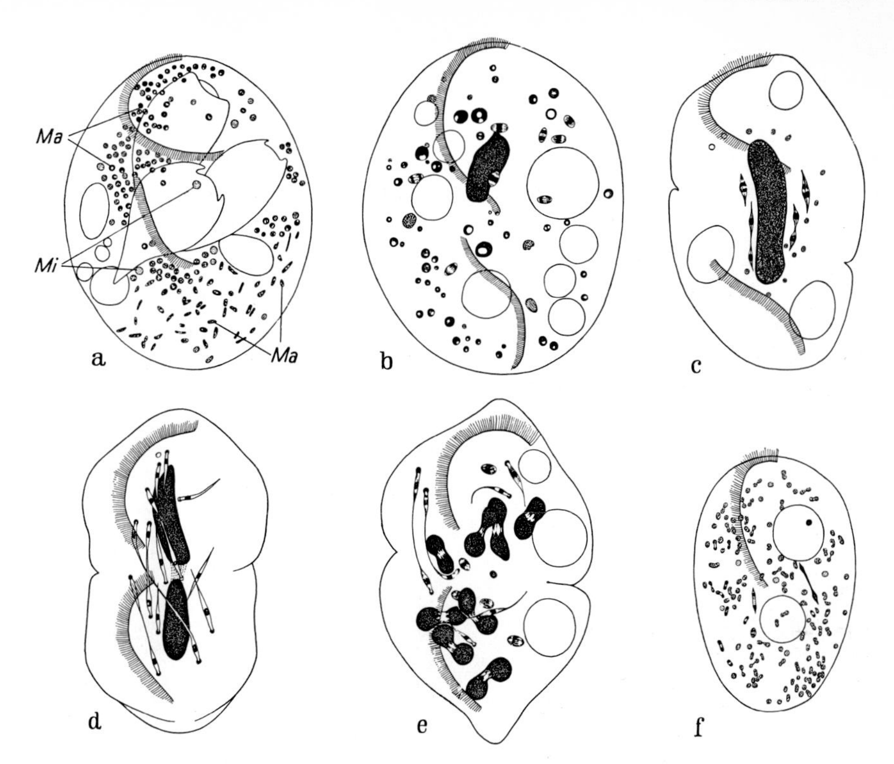

Fig. 99. *Urostyla grandis*. Nuclear and cell division. a Cell at the beginning of division. Numerous, already rounded macronuclei (*Ma*) and a few micronuclei (*Mi*) which are in prophase. b Beginning fusion of the macronuclei, micronuclei in meta- and anaphase. c Fusion of the small macronuclei leads to a large macronucleus which begins to stretch; micronuclei are partly in the resting stage, partly in ana- and telophase. d Division of the large macronucleus, micronuclei in telophase. e Third divisional step of the macronucleus in each daughter individual. f Late division stage of the macronuclei in a daughter individual. × 130. After RAABE, 1946 [*1374, 1375*]

phrya-stages of *Tachyblaston ephelotensis* (b). Through continued budding the macronucleus gives off one daughter nucleus after another without growing to its original size in between. On the whole, there are about 16 daughter nuclei formed in this way (compare Figs. 138 and 431). In most Suctoria, unequal division is restricted to the formation of a single bud, the macronucleus growing to its original size after budding (Fig. 140). The peculiar movements of the chromosome-like threads shall only be mentioned in this context (p. 149). Multiple and unequal divisions are also occasionally found in other ciliates. Many Astomata form buds by constriction at their posterior ends. They are not set free at once but remain connected with the mother cell or a bud which is formed later. Each bud has a portion of the mother nucleus which does, however, grow between successive acts of budding. Even the individual members of a chain may begin to grow (Fig. 100c).

It need not be stressed that the daughter nuclei of a multiple or an unequal division contain less DNA than the mother nucleus. This type of division is thus connected with a "*depolyploidization*". The ciliate *Ichthyophthirius multifiliis*, a parasite within the epidermis of fresh-water fishes, encysts after a prolonged phase of growth and forms, within this "multiplication cyst", numerous small swarmers

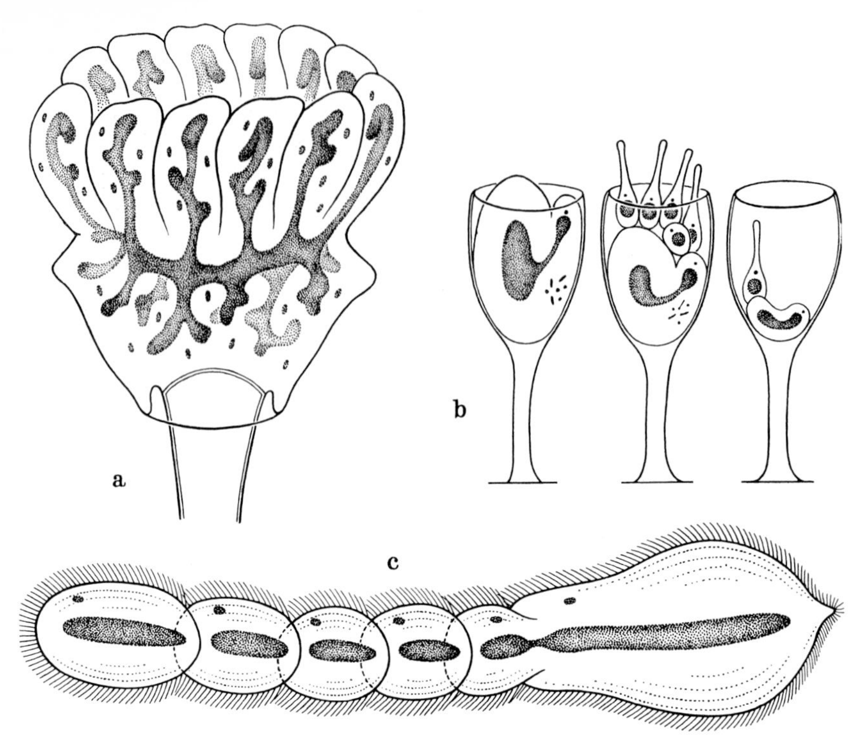

Fig. 100. Unequal and multiple division of the macronucleus. a Simultaneous multiple division in *Ephelota gemmipara*. b Successive multiple division in *Dactylophrya*-stages of the parasitic suctorian *Tachyblaston ephelotensis*. c Unequal division in *Anoplophrya* (Astomata). After GRELL, 1953 [*656*]

which again infect other hosts. While the large mother nucleus is about 12600-ploid, the small macronuclei of the swarmers are only about 48-ploid [*1759*].

The macronucleus is the only type of nucleus with the *ability to regenerate*. This can even be shown in those cases where it divides normally by binary fission. If *Stentor* is cut into several pieces (Fig. 101), each of these can become a new individual if it contains a fragment of the old macronucleus (b′, c′, d′). This fragment will then grow to the size of the old macronucleus. Pieces without a portion of the macronucleus will die (a′).

Even the fragments into which the macronucleus disintegrates at conjugation can under certain conditions grow to normal macronuclei. In *Paramecium aurelia* there are about 30 to 40 such fragments. The development of a new macronuclear anlage after conjugation can be suppressed by a rise in temperature. In this case all fragments grow to normal macronuclei which are distributed in subsequent fissions to the descendants of the temperature-treated *Paramecium* until every cell has only one regenerated macronucleus. Since regenerated macronuclei do not differ genetically from normal ones, each fragment must have contained at least one complete set of necessary genes [*1632*]. This type of macronuclear regeneration also takes place in untreated strains of *Paramecium aurelia* which are homozygous for the gene *am* and undergo autogamy.

In its ability to carry out unequal and multiple division and its power of regenera-

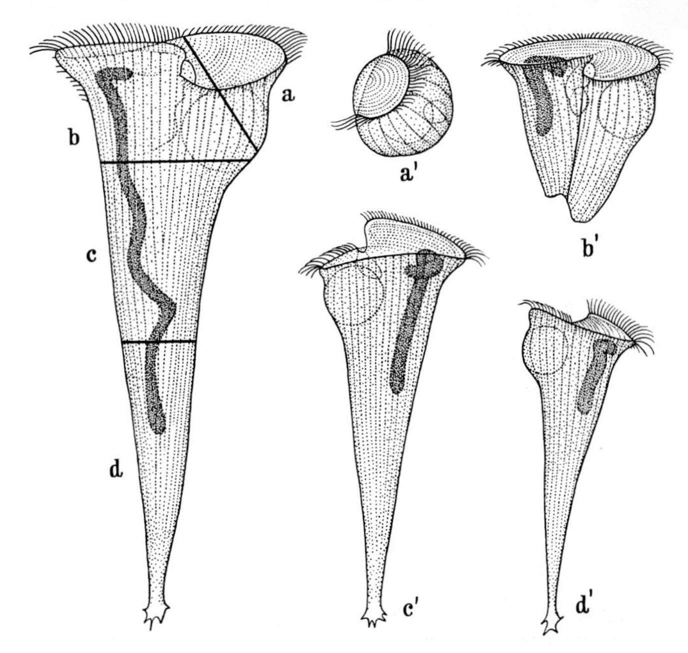

Fig. 101. *Stentor roeseli*. Transection experiment. The individual was cut into four pieces (a, b, c, d). Three (b, c, d) of them, which received fragments of the macronucleus, develop into completely normal individuals (b′, c′, d′), whereas the part without a fragment (a) rounds off and degenerates (a′). After BELAR from HARTMANN, 1953 [*720*]

tion the macronucleus is distinct not only from all diploid nuclei but also from all polyploid ones. Its polyploidy is therefore of a special type and should be referred to as "*polygenomic*".

Since the division of the macronucleus leads to a regular transmission of genetic information, even if it is unequal or multiple over an unlimited number of cell generations in many cases, the units of distribution must be whole "genomes", at least in the sense of macronuclear "genomes" i.e. structural units which comprise the genes to be expressed by the macronucleus. In other words: If the chromosome-like threads (p. 105) are qualitatively different, those belonging to the same "genome" must stay together at division. Macronuclear division, no matter how it occurs, must involve a "*segregation of genomes*".

How this is effected is not clarified as yet. According to the "*subnuclei*"-hypothesis [*1632*], the whole macronucleus consists of diploid "subnuclei" which multiply mitotically and are in some way separated from each other. The "*composite chromosomes*"-hypothesis [*651*] assumes that the chromosome-like threads belonging to the same "genome" are physically connected to form "composite chromosomes" multiplying by endomitosis. That chain formations of chromosomes are possible is indicated by the cytological findings in some Polymastigina (p. 71). An additional assumption would be that homologous "composite chromosomes" are paired as is usually the case with the dipteran polytene chromosomes.

Both hypotheses agree that the macronucleus consists of *subunits* which segregate at random during division.

A recent hypothesis, the so-called "master-slave"-hypothesis, has been established especially to explain certain genetical results in *Tetrahymena pyriformis* and cannot be discussed in this context [*32*].

The mechanism of nuclear division is still obscure. After polarization microscopy had led to the recognition of a birefringent achromatic substance formed at division [*1562*], bundles of spindle fibers (microtubules) could also be shown by light and electron microscopy within the

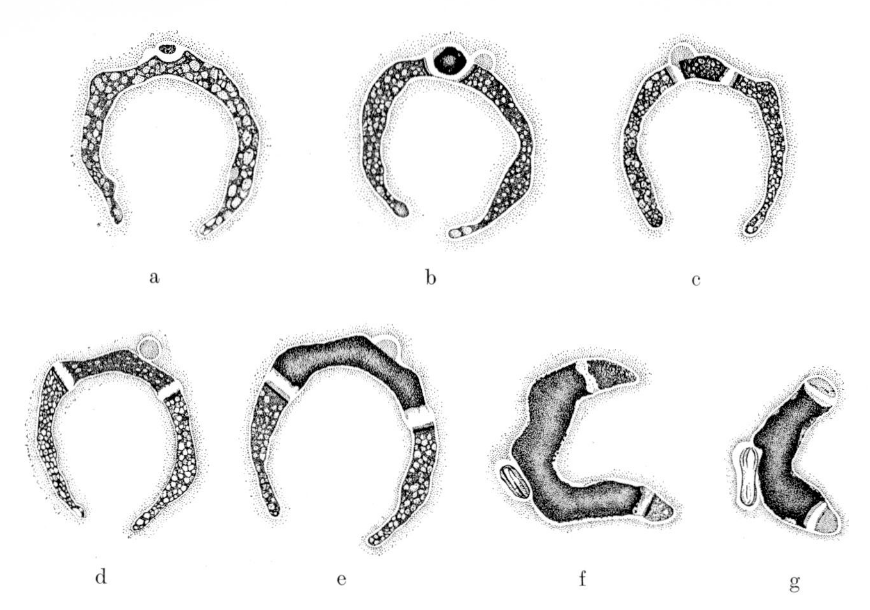

Fig. 102. *Aspidisca lynceus* (Hypotricha). Formation and course of the "replication bands" before division of the macronucleus. Sublimate alcohol, iron hematoxylin (d Feulgen). × 1180. After SUMMERS, 1935 [*1684*]

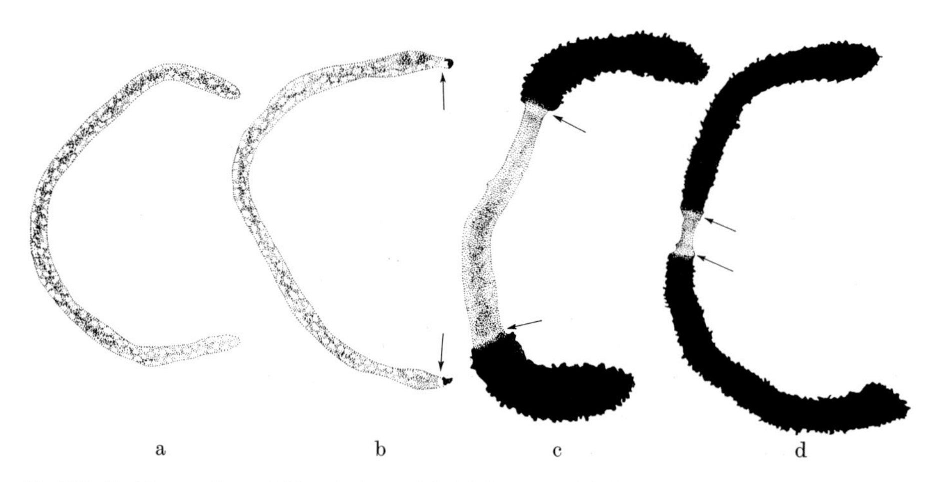

Fig. 103. *Euplotes eurystomus*. Different stages of isolated macronuclei after labeling with ^{3}H-thymidine. The arrows indicate the replication bands. a Before synthesis: no labeling. b Start of DNA synthesis: appearance of the replication bands on the ends. c Approximately 50% of the DNA of the macronucleus is reduplicated. d End of the synthetic phase: the replication bands meet in the middle. Exposure time of the autoradiographs was 12 hours. Length of the macronucleus approximately 140 µm. After micrographs by PRESCOTT, 1966 [*1348*]

macronucleus [*1388*, *1744*]. Even if these spindle fibers are not connected with the chromosomes themselves, there remains the possibility that their totality acts as a "stem-body" which functions during the phase when the macronucleus stretches.

In the hypotrichs and in some other ciliates, two light crossbands, the so-called *replication bands*, migrate along the whole macronucleus before division. They may either appear first in the middle of the macronucleus and then progress toward the ends (Fig. 102) or start at the ends and meet at the center. Sections passed by the bands are more dense and stain more intensely with the Feulgen reagent.

For the macronucleus of *Euplotes eurystomus* whose replication bands move towards the center (Fig. 103), it could be shown by autoradiography that they are involved in DNA and histone synthesis since the content of DNA and histone is doubled behind the bands. The identical course of both synthetic processes indicates a close relationship of histone to the genetic material. Possibly, other types of proteins are also doubled during the passage of the replication bands while RNA is continously renewed [*556*, *921*, *1348*, *1349*, *1433*].

In general, the volumes of macronuclei increase with the size of the cells. Larger species also have the most voluminous and often branched macronuclei. This relationship, whose variability is possible because of the polygenomic state, is an expression of the *metabolic activity* of the macronuclei.

If the macronucleus is removed by experimental means, the cell is no longer able to live. In the *am* mutant of *Paramecium aurelia* a high percentage of daughter cells without a macronucleus is continuously formed due to unequal nuclear distribution. Such cells cease to take in food, to grow and to divide. Even if they have enough micronuclei to correspond in mass to that of a macronuclear fragment capable of regeneration, they die within two days. A small fragment of macronucleus, on the other hand, is sufficient to keep them alive.

Amacronucleate cells of *Stentor* can survive 5–6 days. For about two days their behavior is entirely normal. In this time they can even regenerate their "foot" (holdfast) after excision. If an "anlagen field" (p. 133) is already present it can develop further. Even division and physiological regeneration can be completed, if development has reached a certain stage when the macronucleus is removed. However, no "anlagen field" can be initiated in an amacronucleate cell [*1713*].

Genetic studies have shown that the phenotype of ciliates is predominantly determined by the macronucleus (see p. 246ff.). Whether this phenotypic effect is due exclusively to the polygenomic state as a gene-dosis-effect or whether it is due to genetic inactivity of the micronucleus cannot be determined as yet. While earlier experiments indicated genetic inactivity of the micronucleus, signs of gene-dependent RNA synthesis have recently been found in the micronucleus [*1291a*].

Cytochemical studies show that cytoplasmic RNA is largely from the macronucleus. If RNA synthesis is inhibited, cytoplasmic protein synthesis will cease.

Removal of the micronucleus has, in general, no deleterious consequences. Strains without micronuclei have been cultured for many years and show no decrease of their ability to regenerate. In most species, with the exception of *Tetrahymena pyriformis*, cells without a micronucleus can even conjugate with a suitable partner, although they are naturally not able to form pronuclei which are the prerequisite of mutual fertilization [*1557*].

Although participation of the micronuclei in cellular metabolism cannot thus be wholly excluded, their actual importance must lie in the transmission of genetic information.

The problem of how it is possible for nuclei within the same cell to differentiate diversely (p. 100) can also be raised for the differentiation of the macronucleus and the micronucleus.

One hypothesis based on centrifugation experiments assumed that regional differences within the cytoplasm are responsible for determination [*1635*]. Recent experiments with the suctorian *Heliophrya (Cyclophrya) erhardi*, however, suggest that a feedback mechanism similar to that which exists in the heterokaryotic Foraminifera might also become effective in the determination of the macronucleus. By means of a micromanipulator, synkarya were dislocated to several regions of the cell. In all cases metagamic division of the synkaryon took place and one of the daughter nuclei differentiated into a macronuclear anlage. It can be concluded from these experiments that no special area of the cytoplasm is important for determination, but rather a "macronuclear differentiating factor" being effective throughout the cell, to which the nuclei react according to their varying "sensitivity". A nucleus once determined as macronucleus inhibits the determination of further nuclei, which then become micronuclei [*994b*].

c) Radiolaria

Aside from the polyploid macronuclei of ciliates, the *primary nuclei* of Radiolaria are the only protozoan nuclei with multiple division. However, their polygenomic nature is just as problematical as that of macronuclei, although in different respects. A satisfactory interpretation of their nuclear conditions has so far failed because knowledge of radiolarian development, which occurs in some cases at greater depths of the oceans, is only fragmentary. In addition, understanding of their nuclear division has long been hampered by the fact that Radiolaria are often infected with parasitic dinoflagellates which multiply within the central capsule and simulate by their own swarmer formation multiple division of the host (p. 356).

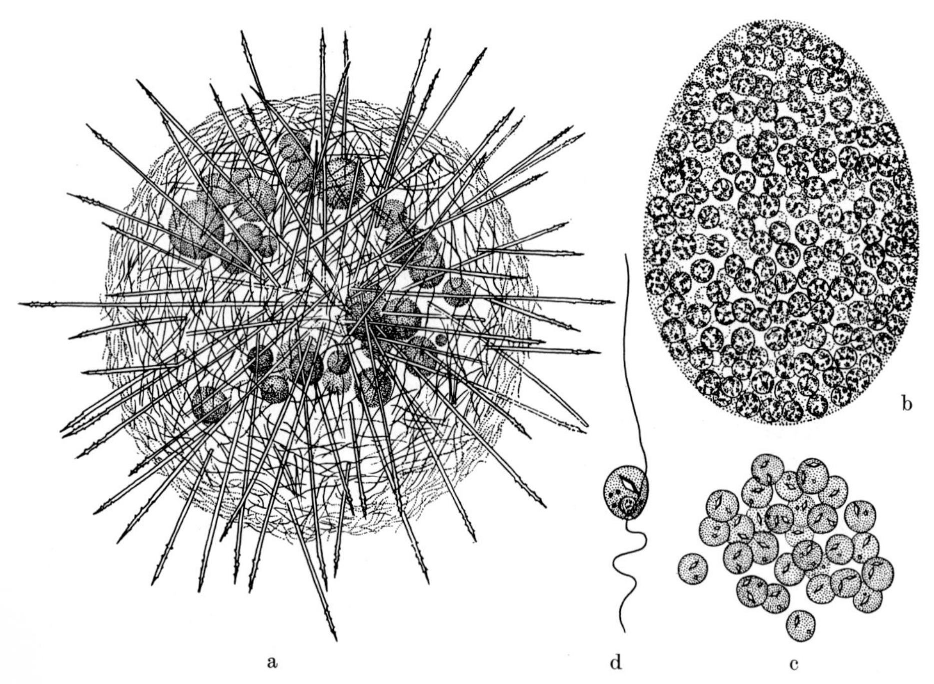

Fig. 104. *Aulacantha scolymantha.* Formation of crystal swarmers. a Survey view. Phaeodium dissolved, central capsule disintegrated into plasmatic spheres. × 70. b Section through a single plasmatic sphere with numerous cell nuclei. × 800. c Squash preparation of a plasmatic sphere (according to BORGERT: "Cell nuclei"). × 1200. d Individual crystal swarmer. a—c after BORGERT, 1909 [*151*], d after CACHON-ENJUMET, 1964 [*188*]

That Radiolaria also form swarmers preceded by the "disintegration" of the primary nucleus into numerous small secondary nuclei can nowadays be considered certain. These swarmers are characterized by highly refractile inclusions and are therefore called "*crystal swarmers*". How these become fully grown Radiolaria is unknown. Occasionally, growth stages with smaller nuclei have been found. It is therefore assumed that the large primary nuclei of fully grown Radiolaria are the result of endomitotic polyploidization [*655, 781, 789*].

In the *Collodaria*, as in *Thalassicolla nucleata* (Fig. 376) and *Thalassophysa sanguinolenta* (Fig. 126), the primary nucleus has a diameter of up to 180 μm. Its Feulgen reaction is weak and it contains large, sausage-like nucleoli. Shortly before swarmer formation, the nucleoli dissolve and from the fibrillar ground substance of the primary nucleus, groups of condensed chromosomes are differentiated, becoming the secondary nuclei.

The primary nuclei of *Tripylea*, in contrast to those of Collodaria, stain intensely with the Feulgen reagent. So far, swarmer formation is known from two species

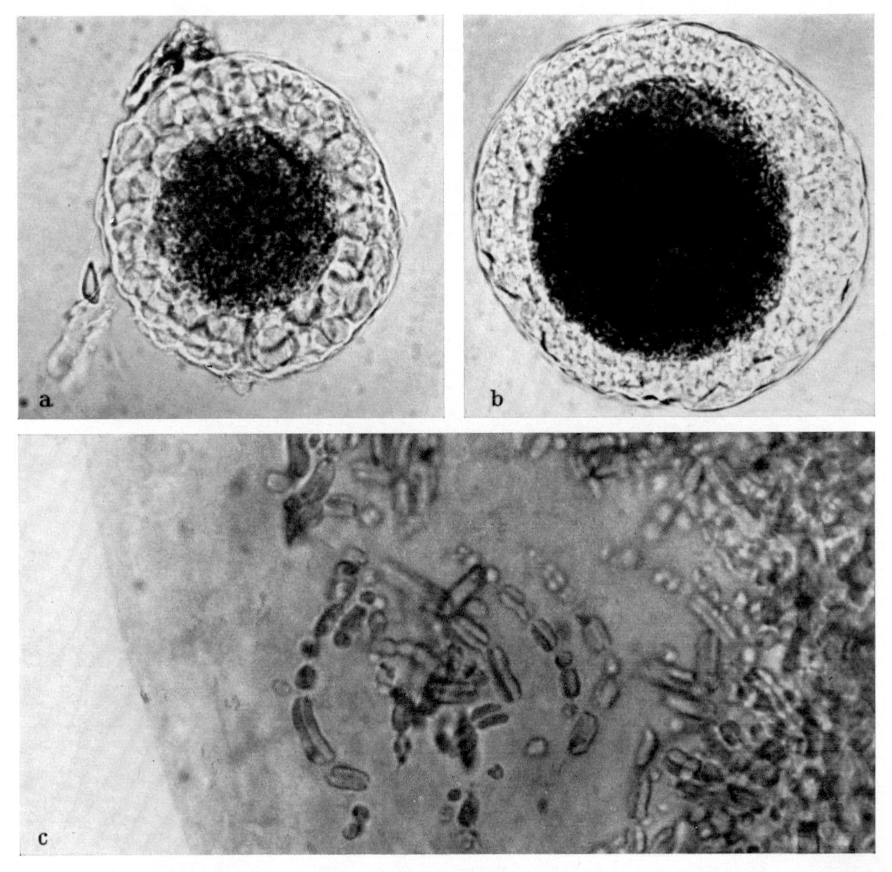

Fig. 105. *Aulacantha scolymantha*. a, b Central capsules with nuclei of different degrees of polyploidy. Sanfelice, Feulgen. × 220. c Section of a squashed central capsule: reduplication of the chromosomes. Aceto-carmine. × 1400. After GRELL, 1953 [*655*]

only, and only in the last stages, viz. *Coelodendrum ramosissimum* [*188, 1001*] and *Aulacantha scolymantha* [*151, 188*]. Some stages of the latter species are shown in Fig. 104. The phaeodium has disappeared and in the place of the central capsule there are several plasmatic spheres of different size (a). These already contain numerous "interphasic" secondary nuclei (b). It has been observed in life that the plasmatic spheres fall apart after some time into small cells with crystal inclusions (c). They are at first motionless but then form two long flagella and swim about as "crystal swarmers" (d).

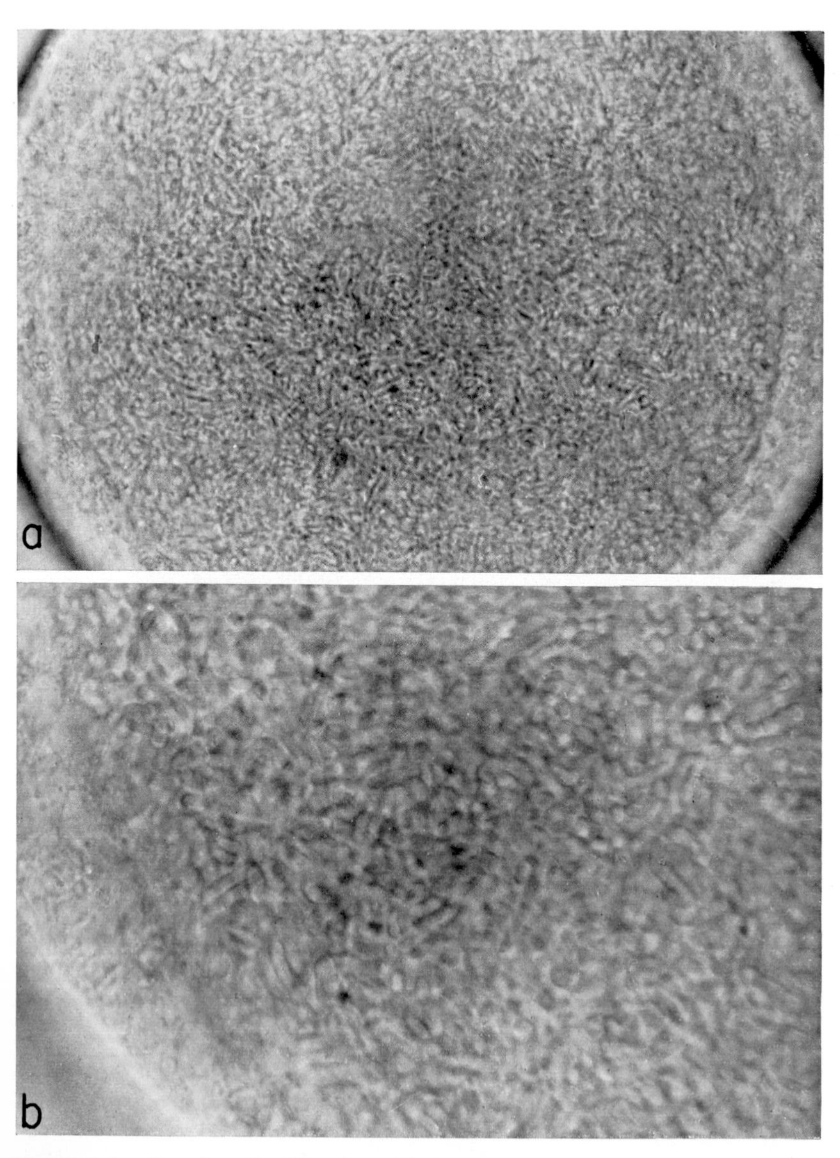

Fig. 106. *Aulacantha scolymantha*. Cell nucleus with chromosomes. a × 600. b Enlarged area of a, × 1600. Reduplication of the chromosomes. Photographed in life. After GRELL, 1953 [*655*]

In *Aulacantha scolymantha* younger growth stages have also been found (Fig. 105a) whose nuclei contain obviously fewer chromosomes than those of fully grown individuals (b). The primary nucleus is said to contain "more than a thousand", according to another estimate even two to three thousand chromosomes. In certain stages, interpreted as endomitoses (c), the chromosomes are double and consist of segments of different lengths. Perhaps this segmentation is an indication that we are dealing with "composite chromosomes". Each of these could correspond to a whole genome. One uncertain point about this interpretation is that constancy of segmentation has not been proved [*655*].

Photomicrographs of living cells in these stages give the impression of a very large number of chromosomes which far surpasses that of other protozoan nuclei with the exception of the polyploid macronuclei (Fig. 106). Within the nucleus the mass of chromosomes is in constant movement. This is especially striking in time-lapse microcinematography (film C 829).

The stages described precede the binary *division of the nucleus*. In dividing nuclei seen in life, a segmentation of the chromosomes is no longer recognizable. Although attempts have been made to interpret the division of the nucleus as a mitosis [*187*], it differs in essential aspects from a normal nuclear division. Instead of arranging themselves at the equator of a spindle, the chromosomes form a contorted "mother plate" within which they are oriented parallel to their future direction of movement and not at right angles to it (Fig. 107a). Even the daughter plates are at first still contorted as they move apart (b, c). It is only at the conclusion of this "anaphase movement" that they are arranged as plane parallel discs (d). Although microtubules have been identified between the daughter plates [*682*], their movement cannot depend upon a spindle mechanism as in mitosis because there are no recognizable spindle fiber attachments at the chromosomes and because spindle poles (centrioles) are undoubtedly absent. When the daughter plates have reached a certain distance from each other they are slowly tucked in at the margins so that the ends of the chromosomes which had been directed toward the interior of the central capsule are now directed toward the outside. The reconstructed daughter nucleus has therefore a radial structure (Fig. 108). Neither during division nor during the reconstruction of the daughter nuclei are there clear size differences among the chromosomes.

If it is assumed that the units of distribution are homologous composite chromosomes (genomes) rather than different individual chromosomes, some features of this nuclear division, such as the absence of an equatorial plate and the lack of regular chromatid distribution, become intelligible. However, nuclear division would then not be a mitosis but a case of "genome segregation" (p. 113).

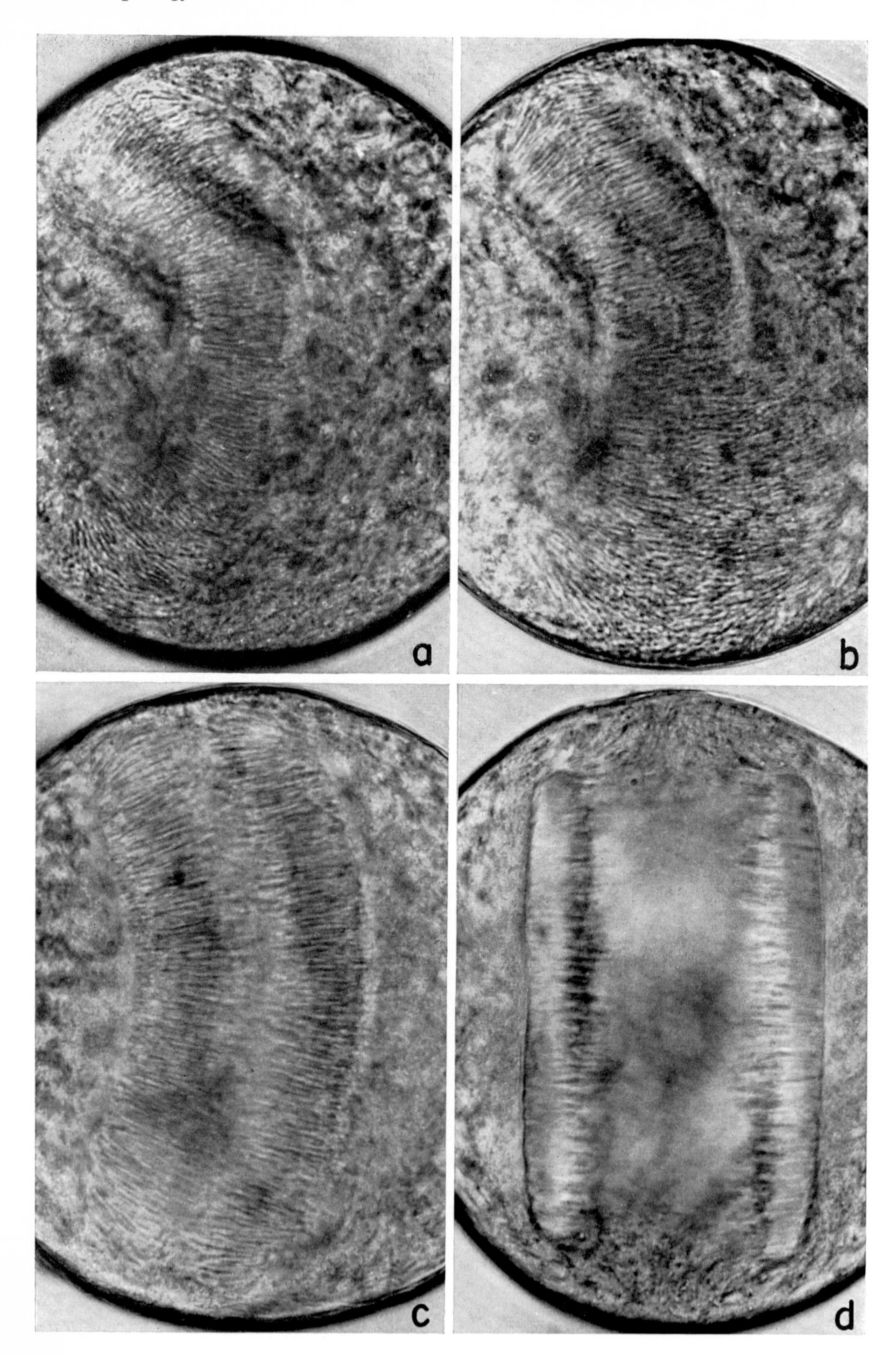

Fig. 107. *Aulacantha scolymantha*. Nuclear division. a, b The same central capsule. c, d Different central capsules. × 500. Photographed in life. After GRELL, 1953 [*655*]

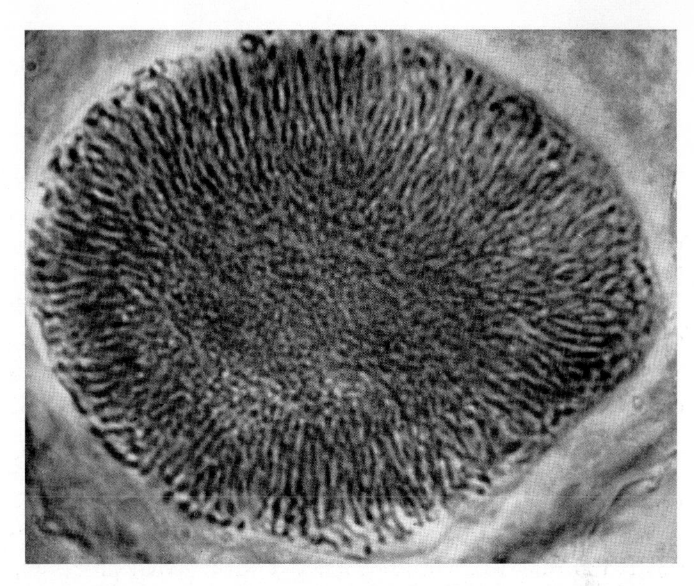

Fig. 108. *Aulacantha scolymantha*. Individual isolated daughter nucleus. Sanfelice, Feulgen. × 1200. After GRELL, 1953 [*655*]

D. Reproduction

The ability to reproduce is one of the basic characteristics of living systems. As soon as an organism has grown to a certain size, it starts reproduction. If, through continued amputation, an amoeba is prevented from reaching the size limit which is necessary for division, its reproduction can be completely suppressed [*719*]. Reproduction is therefore initiated by internal factors which are realized at a certain stage of growth.

In Protozoa, reproduction occurs by simple *cell division*. The daughter cells may either be of equal or of unequal size (*equal* and *unequal* cell division). The number of daughter cells may be two *(binary fission)* or many *(multiple fission)*. A variant of unequal cell division, which will be discussed under separate heading, is *budding* (p. 145).

In many cases, the daughter cells are capable of copulation (gametes). Since reproduction is thus intimately connected with sexual processes, we are speaking of *sexual reproduction* (p. 153).

An obvious result of each reproduction is *multiplication*. However, in sexual reproduction, multiplication may be restricted to a simple fission, which is then cancelled by the fusion of the gametes.

I. Binary Fission

The most elementary form of binary fission is found among *amoebae*. If they are uninuclear, the plane of division is usually at right angles to the axis of nuclear division. In *Amoeba proteus* (Fig. 109) the pseudopodia are withdrawn at the onset of division, the cytoplasm loses its transparency to a large degree and the pulsating vacuole disappears (spherical stage). After the beginning of telophase, the amoeba stretches and constricts in the middle. Large pseudopodia are formed at the poles of the divisional figure. They draw both daughter individuals in opposite directions. In amoebae with a firm ectoplasm, e.g. the moss amoeba *Amoeba sphaeronucleolosus*, the partition of the daughter cells is facilitated by counter-rotation of the two amoebae which seem to stem themselves apart through the pressure of broad pseudopods [*1814*].

Fission in *Heliozoa* is similar to that in amoebae (Fig. 110). Although they have a relatively constant body form, the plane of division is not predetermined by the organization of the cell because of their radial symmetry. In *Radiolaria*, on the other hand, whose basically radial symmetry is frequently modified, the plane of division may indeed be determined by the organization of the cell. *Tripylea*, for example, have a central capsule with one main pore (astropyle) and two minor ones (parapyles). A plane of symmetry through the main pore divides the cell into two mirror halves. In this case, the plane of division coincides with the plane of symmetry. It bisects the main pore, each daughter capsule receiving one minor pore. After constriction of the central capsule, the extracapsular formations (skeleton, phaeodium) are distributed equally between the two daughter cells.

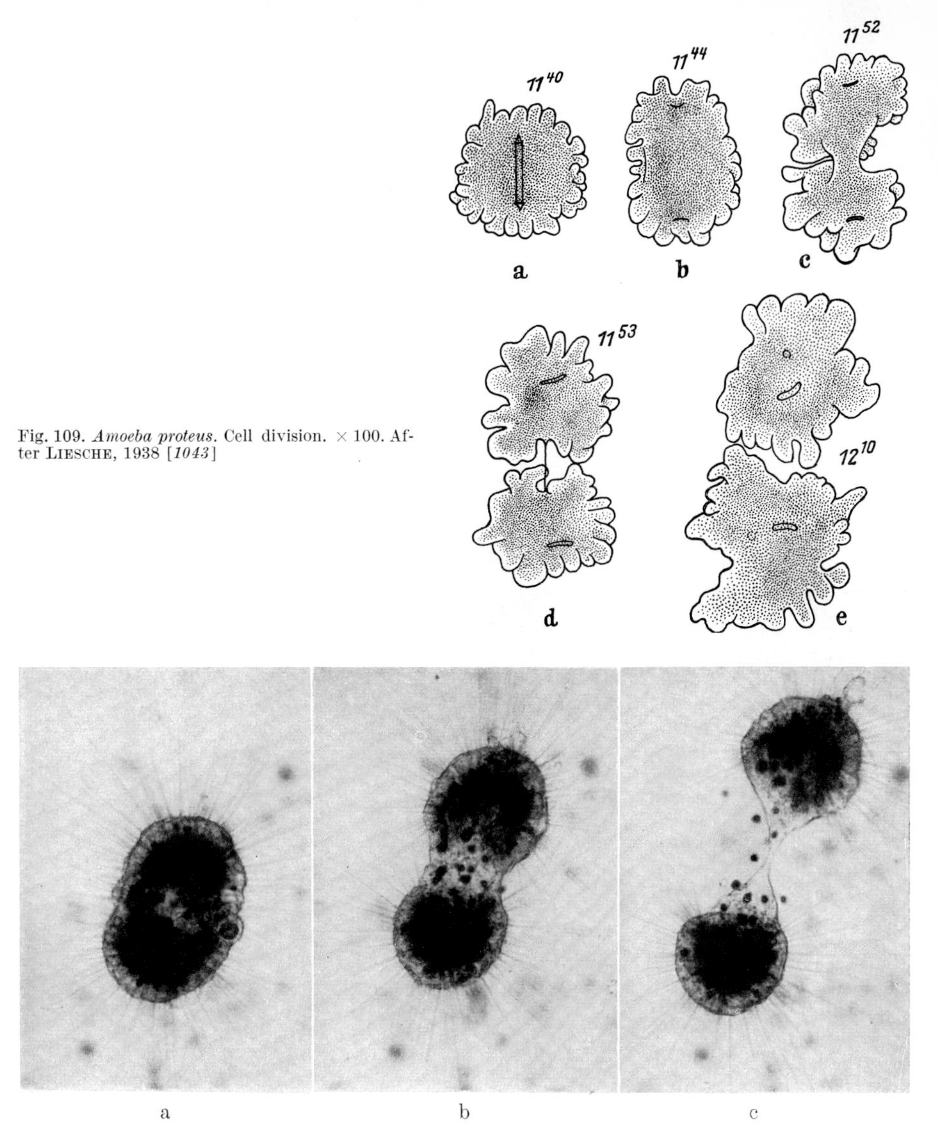

Fig. 109. *Amoeba proteus*. Cell division. × 100. After LIESCHE, 1938 [*1043*]

Fig. 110. *Actinosphaerium arachnoideum*. Binary fission. × 40. From the film E 648 (GRELL)

Most Protozoa have a *heteropolar structure*. Their cell bodies are stretched in the direction of the longitudinal axis and are variously differentiated at the anterior and posterior poles. The direction of division bears a definite relationship to the body axis.

In *Testacea*, polarity is determined by the opening of the shell from which the pseudopods protrude. The mode of division depends in this case upon the rigidity of the test. If it is of a soft texture, as in *Pamphagus hyalinus* (Fig. 111), division is longitudinal and the test is constricted into two halves. If it is rigid, however, a part of the cell body moves out of the opening and secretes a new shell at its

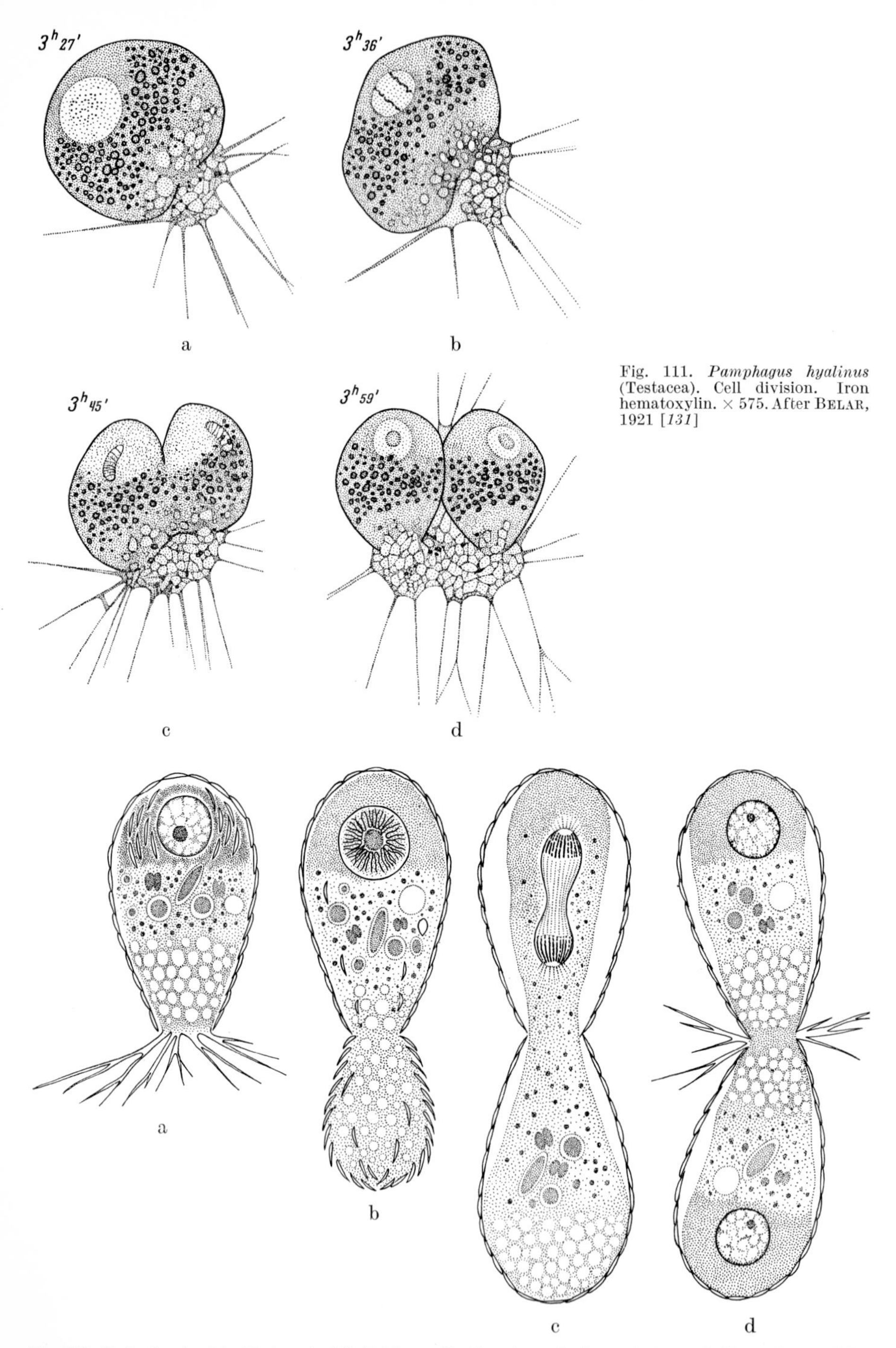

Fig. 111. *Pamphagus hyalinus* (Testacea). Cell division. Iron hematoxylin. × 575. After BELAR, 1921 [*131*]

Fig. 112. *Euglypha alveolata* (Testacea). Cell division. a Resting stage; in the posterior part: the nucleus and the reserve platelets for the shell. b Formation of a protoplasmic bud; the reserve platelets move to its surface. c Nuclear division; the reserve platelets have arranged themselves to form a shell. d Shortly before the separation of the daughter individuals. After SCHEWIAKOFF, 1887, from KÜHN, 1921

surface. In *Euglypha alveolata* (Fig. 112), whose test is made of small platelets, reserve platelets are laid down even before division within the cytoplasm. They are then moved to the surface of the protoplasmic mass which protrudes from the opening of the shell and are there arranged to form the new shell.

Also in *Arcella vulgaris* (Fig. 113) binary division starts with the protrusion of cytoplasm at the ventral opening of the shell. The protruded cytoplasm which lifts the shell from the substratum shows a hyaline peripheral zone (a). From this zone a cup-shaped protuberance arises, which encases the central mass completely (b, c). For a short time both cytoplasmic formations are clearly separated (d). Later on the central mass increases from cytoplasm of the "mother shell" and only a small interspace remains between both formations (e). After secretion of the "daughter shell" in this interspace the surrounding cytoplasm retracts and becomes reincorporated. Before both individuals separate (f) an equal distribution of cytoplasm occurs [*1229*].

In *flagellates* which are always heteropolar, division is usually longitudinal. Its course differs depending upon the number and arrangement of flagella and their associated organelles. Since the flagella themselves are incapable of division, they must be regenerated from basal bodies which arise in the vicinity of the old basal bodies. Multiplication of basal bodies usually precedes cell division.

Some flagellates divide only in an unflagellated stage. In species of *Chlamydomonas* the cytoplasmic body is cleaved lengthwise within the cellulose layer. Both daughter cells later become arranged one behind the other, giving the impression of dividing across. Before emerging from the cellulose layer they form flagella of their own.

In *euglenoids* the flagellum originates from an indentation at the front end. It was formerly thought to be a single flagellum with two roots. Close study showed, however, that one of the "roots" is actually a second short flagellum which grows before division to the length of the other flagellum (Fig. 320).

Division in *dinoflagellates* is usually more or less at right angles to the cell axis. This is due to the fact that the flagella which determine the plane of division are not located at the front end but at the side. In exceptional cases where the flagella originate at the front end, as in *Exuviaella marina* (Fig. 38b–d), fission is longitudinal. In species with an asymmetric armor (Fig. 114), the plane of division may be oblique because it follows the seams of certain armor plates. The two daughter cells are therefore unequal and have to reform the missing plates in different ways. In some species of *Ceratium* the daughter cells remain connected after division, leading to the formation of chains of individuals (Fig. 328).

In *trypanosomes* (Fig. 115) whose flagellum is transformed to an undulating membrane (p. 295), the newly formed basal body sprouts a fibrillar bundle during nuclear division; this becomes the margin of a new undulating membrane. One of the daughter cells will thus receive the old undulating membrane, the other a new one.

In *Polymastigina*, too, organelles of the mother cell are frequently handed on to one of the daughter cells while the other one forms new organelles. This holds true especially for flagella.

Fig. 113. *Arcella vulgaris*. Divisional stages. a—c Pseudopodium-like protrusion circumvallating a cytoplasmic bud. d, e Secretion of the new test. f Separation of daughter cells. From the film E 1643 (NETZEL) [*1229*]

It is also frequent that organelles of the mother cell such as axostyles and parabasal bodies are broken down and then formed anew after nuclear division. In some species the spindle connecting the centrioles stretches extremely after nuclear

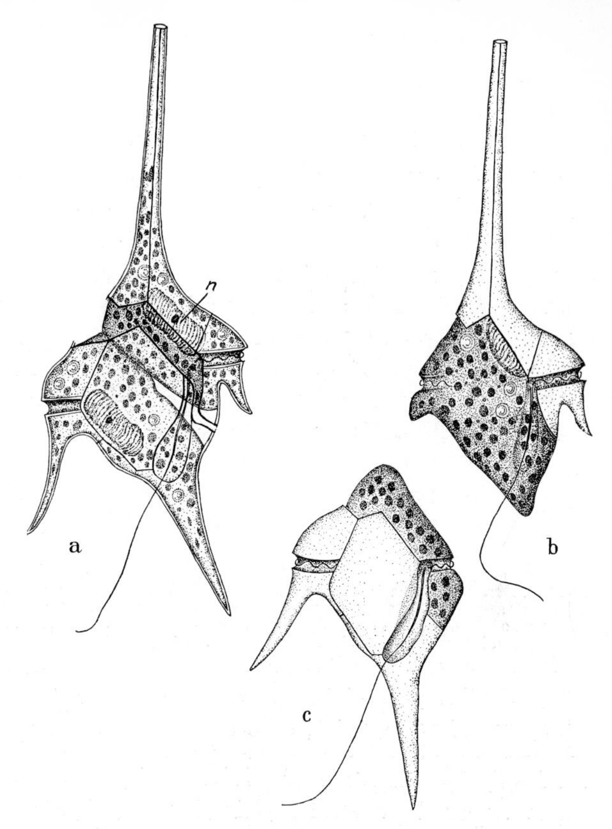

Fig. 114. *Ceratium hirundinella* (Dinoflagellata). Cell division. a Stage of division (the plane of division runs obliquely to the longitudinal axis between certain armor plates). b, c The two daughter cells after division. Combined after various authors, from KÜHN, 1921

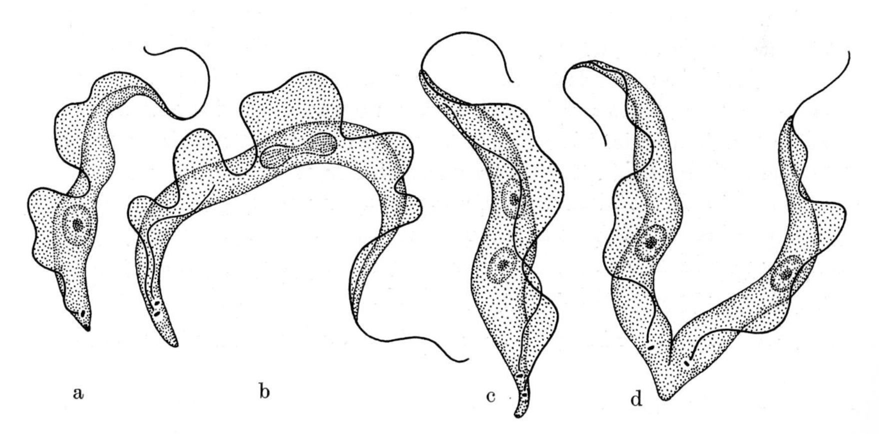

Fig. 115. *Trypanosoma brucei*. Binary fission. × 1500. After MACKINNON and HAWES, 1961

division and is then recognizable even in the daughter cells as a special structure next to the axostyle (Fig. 116).

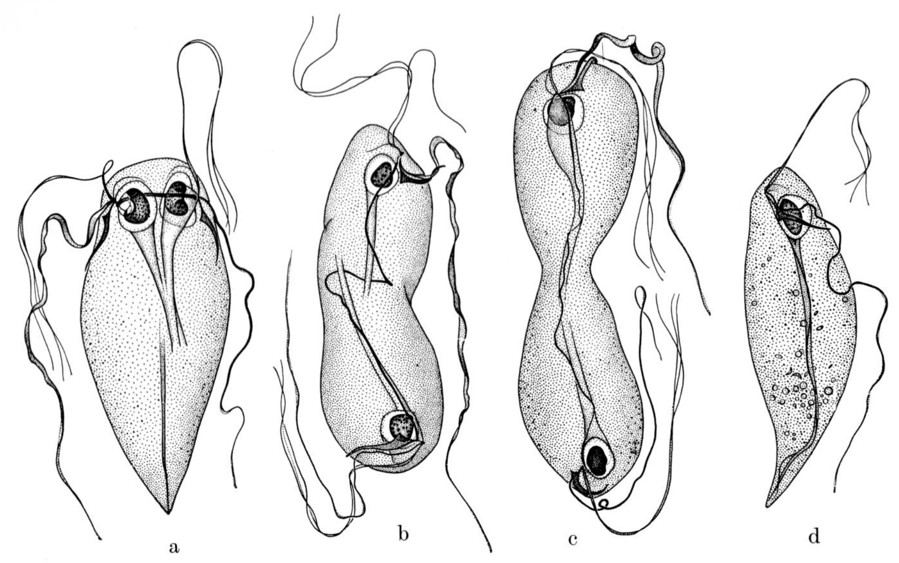

Fig. 116. *Devescovina lemniscata* (Polymastigina). Cell division. a Nuclear division is already complete, outgrowth of the new axostyles. b Daughter nuclei with flagella, axostyle etc. ("karyomastigont complex") moved to opposite cell poles. c Division of the cell body (extreme elongation of the spindle!). d Single daughter individual. Sublimate alcohol, iron hematoxylin. × 1200. After KIRBY, 1944 [*936*]

In the *Hypermastigida*, whose cells are the most highly differentiated among the flagellates, the mode of cell division can be very modified. This is especially so in the family *Spirotrichonymphidae* whose flagella arise from spiral bands (Fig. 341). There are usually several flagellar bands which begin close to each other at the front end, traversing the cell in approximately equal distance but at an ever increasing helical pitch toward the posterior end. The flagellar bands form so-called relational coils and can therefore not be freely moved apart in the transverse direction. The most simple way to circumvent this difficulty would of course consist in retaining all flagellar bands in one daughter cell, the other one forming all bands anew. This principle has, however, not been adhered to in the Spirotrichonymphidae. The most complicated way is the uncoiling of the flagellar bands by rotating the free ends. This has indeed been observed in life in some species. Others have developed different "methods".

Two species of the genus *Spirotrichonympha*, which are scarcely distinguishable morphologically, show completely different modes of division [*272*]. They are only identical with respect to the dissolution of the axostyle of the mother cell and its re-formation in both daughter cells.

Spirotrichonympha polygyra (Fig. 117) has four flagellar bands which originate in two groups at the anterior end. Division begins with the reduplication of the anterior pole whereby each daughter pole receives one group (a). As the distance between the two daughter poles becomes larger, the flagellar bands unwind but coil up again at once at the daughter poles. Between two bands of each group a flattened spindle is formed with fibers to the chromosomes which are attached to the nuclear envelope (b). After nuclear division the daughter poles move so far apart that they are eventually lying opposite each other (c, d). Only after cell

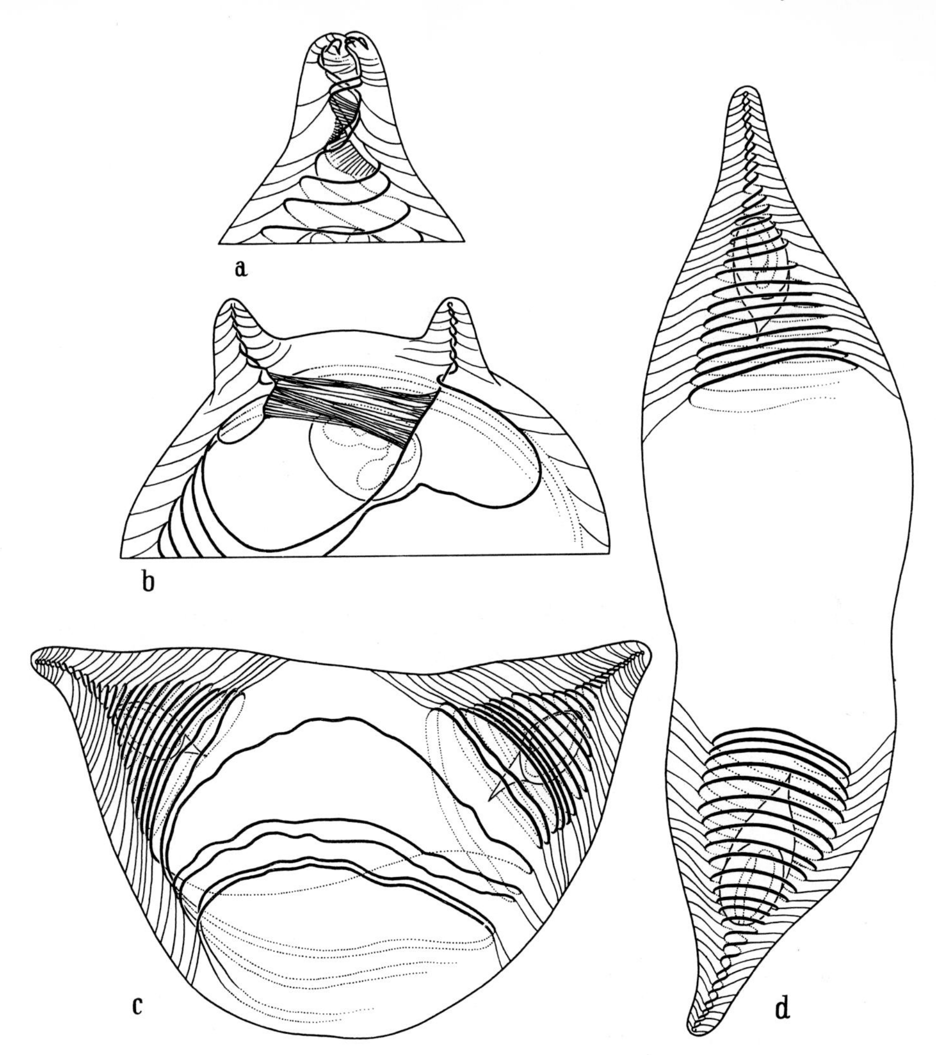

Fig. 117. *Spirotrichonympha polygyra*. Binary fission. Figures a and b show only the anterior end of the cell. Explanation in the text. a, b: × 1240; c,d: × 1000. After CLEVELAND, 1938 [*272*]

division, which in this case leads to daughter cells of equal size, does reduplication of the flagellar bands take place so that the original state is restored.

Spirotrichonympha bispira (Fig. 118) differs from the previous species essentially only in having two flagellar bands. However, they do not become separated from each other at the beginning of division. Instead, that portion of one of the flagellar bands which serves as spindle attachment reduplicates to form a new piece of flagellar band which remains connected with it by the spindle (a). During nuclear division this piece, which elongates beyond the section serving as the spindle attachment, moves further posteriad (b). As soon as the posterior end of the cell is reached, re-coiling of the new flagellar band commences, and a new "anterior pole" is thus formed (c). Cell division, which now begins, is unequal: while the large

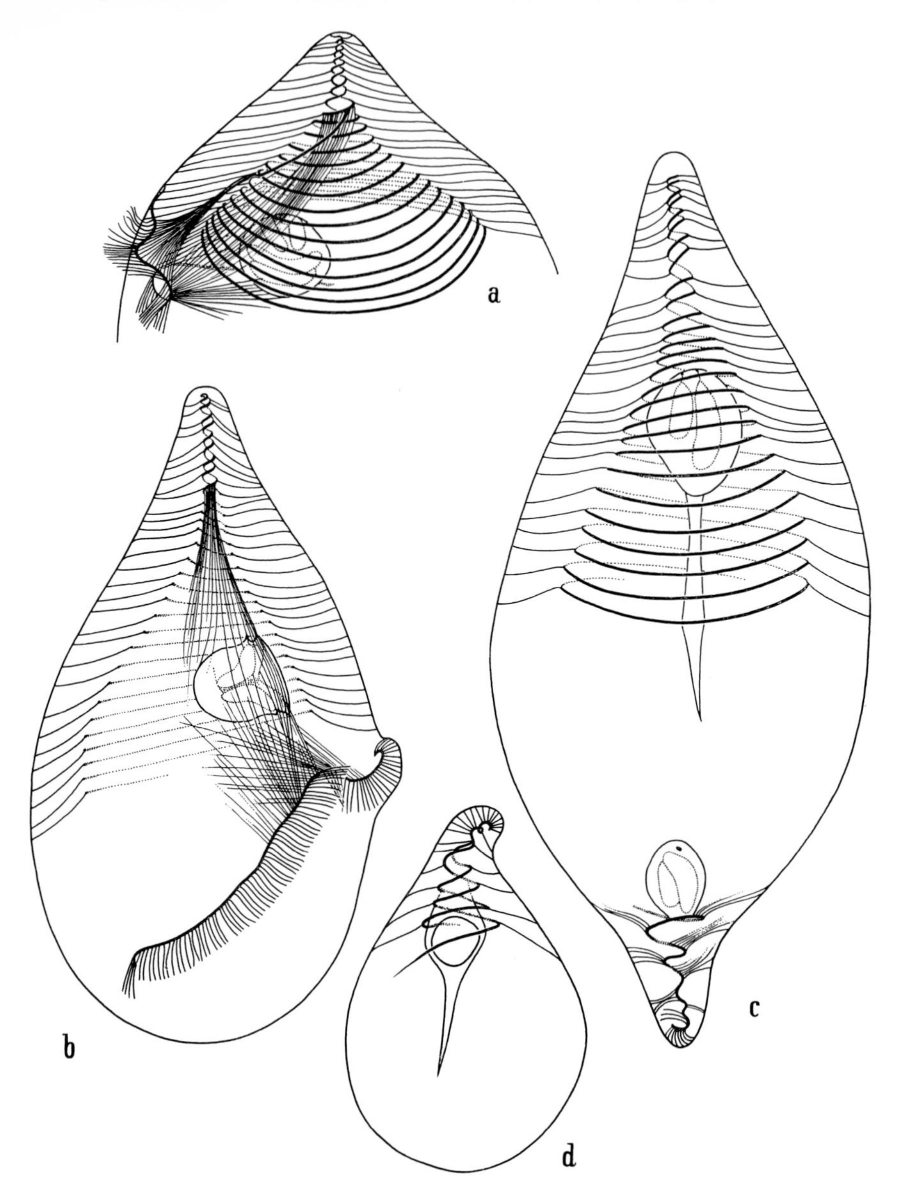

Fig. 118. *Spirotrichonympha bispira*. Binary fission. Figure a shows only the anterior end of the cell. Explanation in the text. × 1300. After CLEVELAND, 1938 [*272*]

anterior animal takes over all the "mother" organelles, a reorganization of the organelles, especially reduplication of the flagellar band (d), takes place in the small posterior animal.

The result of cell division is similar in *Holomastigotoides tusitala* (Fig. 119). No new flagellar band is formed, but that flagellar band to which the centriole of one of the daughter nuclei is attached breaks away from the others, unwinds completely and moves with the daughter nucleus to the posterior end of the cell (a), which has no

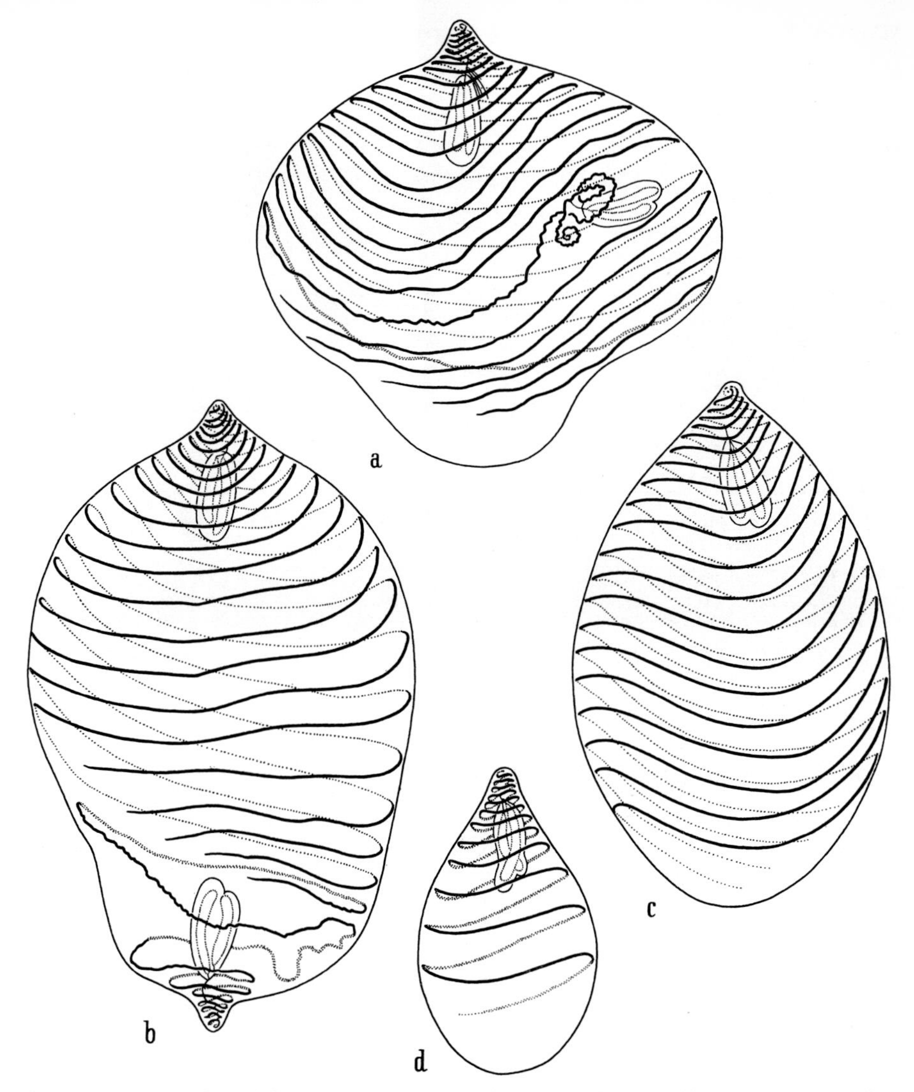

Fig. 119. *Holomastigotoides tusitala*. Binary fission. Explanation in the text. × 530. After CLEVELAND, 1949 [275]

flagellar bands. After re-coiling of the flagellar band, a new "anterior pole" (b) is formed as in the previous species. In this case the anterior animal (c) is again much larger than the posterior one (d).

Except for the fact that the two daughter cells are different in size, the mode of division in *Spirotrichonympha bispira* and *Holomastigotoides tusitala* also diverges from that of most flagellates in that it does not begin with reduplication of the anterior end but rather with fission at right angles to the body axis.

Transverse fission is also characteristic for most *ciliates*, the peritrichs being the only exception. Longitudinal division of the peritrichs is very likely an adaptation to their stationary way of life. However, it is equal only in colonial genera: both

Fig. 120. *Metaphrya sagittae* (Ciliata). Different stages of division in the body cavity of a species of *Sagitta* (from Messina). Photographed in life

daughter cells are alike and remain attached to the stalk of the mother cell (see p. 147).

In the simplest case (Fig. 120) transverse fission in ciliates begins with the appearance of an equatorial furrow which separates the surface cell layer or cortex into an anterior and a posterior half. A constriction, which is probably connected with contraction, leads then to the separation of the two daughter cells. Their form and structure frequently indicates from which half of the mother cell they have originated.

Depending upon the degree of differentiation of the species, division may involve far-reaching reorganizations. They consist partly in a transformation of preexisting structures whose position and size does not correspond to those of the daughter individuals, and partly in the formation of new structures. All of these events take place early in the divisional cycle.

Renewal of the ciliature begins with an increase in the number of basal bodies (kinetosomes). In some species this increase takes place in a "belt zone" corresponding to the position of the future division furrow (Fig. 121a–c). This "belt zone" gives rise to the ciliature of the new posterior half of the anterior daughter cell as well as that of the new anterior half of the posterior one. It can be assumed that the pellicular structures associated with each cilium (ciliary field, trichocysts, etc.) also originate in the course of those growth processes which follow the multiplication of the basal bodies (p. 45).

The cortex of ciliates is, however, not merely a mosaic of identical structural elements. In its entirety it presents a species-specific pattern of differentiation more or less determined by the form of the cell body, regional differences in ciliation and the appearance of special organelles.

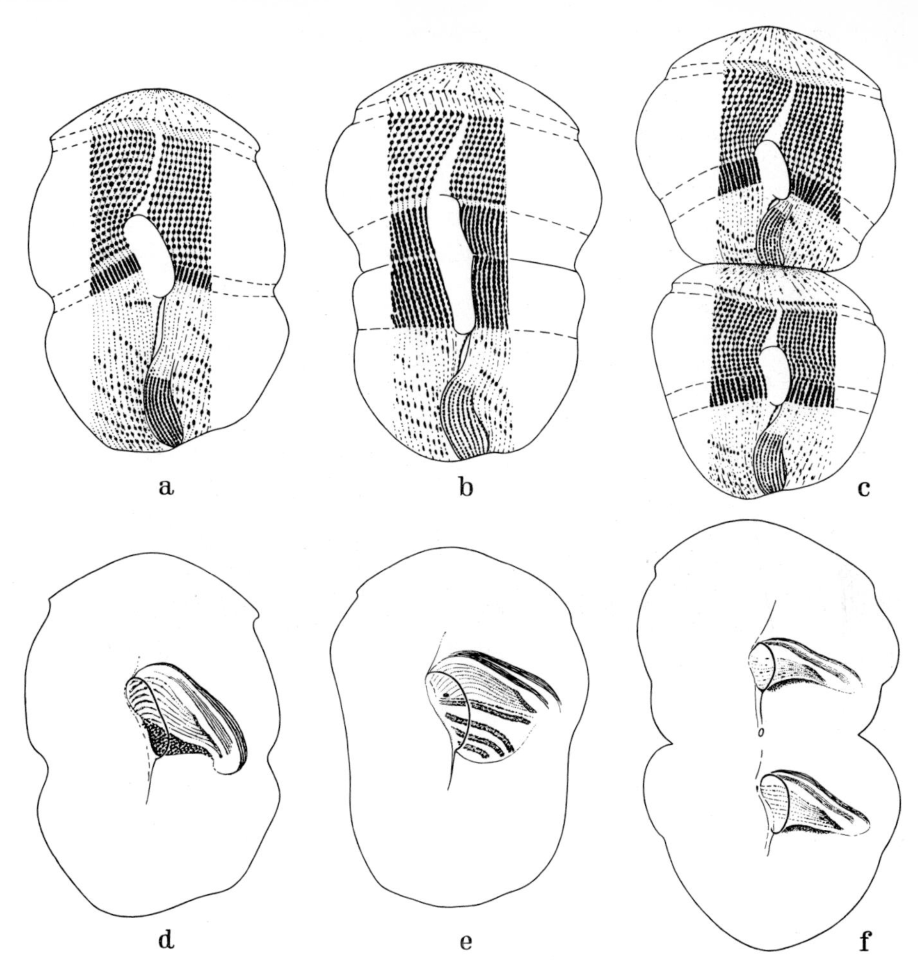

Fig. 121. *Urocentrum turbo* (Ciliata). Binary fission. In a—c the basal bodies of a middle zone of the body ciliature are shown, in d—f those of the rows of cilia which lead to the cytostome. Silver impregnation. After FAURÉ-FREMIET, 1954 [*501*]

The restoration of this pattern in the daughter cells has been analyzed by the methods of developmental physiology in a few species only [*1564*, *1642*, *1702*]. The results showed that differentiation frequently proceeds from a definite area in the cell cortex, the *primordial or anlagen field*.

As the example of *Urocentrum turbo* (Fig. 121 d—f) illustrates, the rows of cilia which lead to the cell mouth (cytostome) can arise in the anterior daughter animal from those of the mother animal, while those of the posterior animal develop from an anlagen field which originates at the beginning of division at the posterior margin of the mother's mouth region.

In *Paramecium* "reduplication" of the mouth region also takes place in a similar manner. It could be proved experimentally that a cell whose mouth region had been cut away was no longer capable of developing a new one [*714*].

In *Paramecium* the enlargement of the cortex which accompanies division does not take place uniformly over the whole cell. As in many ciliates (see above), a special growth zone is formed on both sides of what later becomes the division furrow (Fig. 122a, b). However, since pellicular growth is more intense at the "aboral" side, i.e. the side opposite the cytostome, than at the "oral" side, the original furrow separating the daughter animals is converted into the posterior suture (c) in the anterior daughter cell, and into the anterior suture in the posterior one (d). This explains why the image obtained by silver impregnation shows sutures at the oral surface (Fig. 35a).

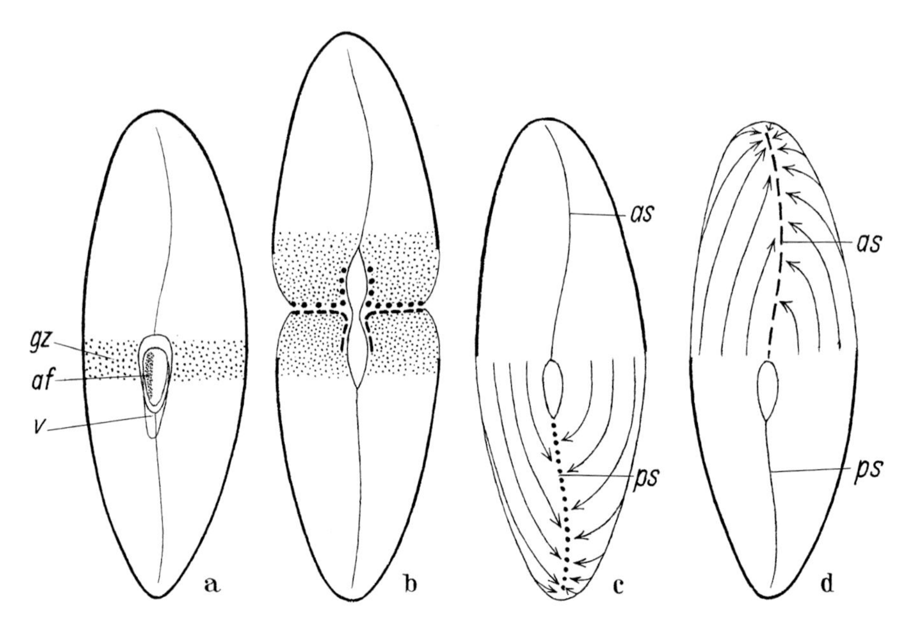

Fig. 122. *Paramecium.* Diagram of divisional growth; c and d the daughter cells after completion of divisional growth. The arrows show the direction of growth of the pellicle. *v* Vestibulum, *af* anlagen field, *gz* growth zone. *as*, *ps* Anterior and posterior sutures. After SCHWARTZ, 1963 [*1564*]

The meaning of a special anlagen field for the morphogenetic processes connected with fission is even more distinct in *Stentor coeruleus*. The cell is transformed at the apical end into an extensive mouth area (peristome) generating a whirlpool which is surrounded by a spiral band of membranelles. At the basal end an adhesive organelle designated as the "foot" is found. The outer cell layer contains a fine, bluish pigment which forms a regular pattern of stripes. The width of the stripes, which are separated by rows of cilia (Fig. 123a), shows a gradient following the circumference of the cell. On a definite longitudinal line, the so-called contrast meridian (*km*), the narrowest and the widest stripes are adjacent to each other. The anlagen field is found along the contrast meridian, at the side of the narrowest stripes.

The course of division shows that the anlagen field can be subdivided into areas of different prospective fates (b). The anlagen field (*aa*) and the prospective posterior end (*pa*) of the anterior daughter cell are located in front of the future divisional

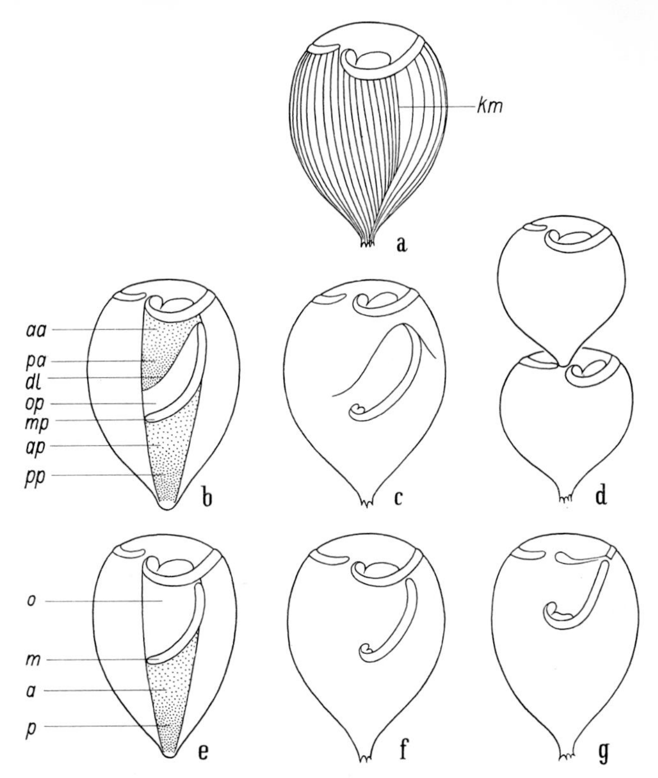

Fig. 123. *Stentor coeruleus*. a Diagram of the cortical pattern. The lines between the pigment stripes are shown. *km* Contrast meridian. b Differentiation of the anlagen field for binary fission (c, d). e Differentiation of the anlagen field for physiological regeneration (f, g). After SCHWARTZ, 1963 [*1564*]

line (*dl*). Then follow the primordia of the mouth- or oral-field (*op*), the adoral zone of membranelles (*mp*) and the anlagen field (*ap*) as well as the primordium of the posterior end (*pp*) of the posterior animal which is not sharply delimited from the former.

During division (c, d) the largest part of the anlagen field of the mother cell is thus allotted to the posterior animal. Growth and morphogenetic processes, which cannot be discussed here in detail, harmonize so perfectly that after division the two daughter animals cannot be distinguished from each other.

Occasionally a process called "physiological regeneration" takes place in *Stentor coeruleus* (also observed in *Blepharisma undulans* and *Spirostomum ambiguum* [*466, 1692*]). Its biological significance is unknown. The differentiation of the anlagen field becomes simpler (e). Its anterior region supplies part of the oral field (*o*) and a piece of membranellar band (*m*). The posterior region differentiates to form a new anlagen field (*a*) and a new posterior end (*p*). The piece of the old membranellar band adjacent to the anterior margin of the anlagen field is resorbed, while the newly formed piece advances to take its place (f, g).

It could be shown, by a variety of transection and transplantation experiments [*1556, 1702–1711, 1755, 1790*] that new formations within the anlagen field depend upon the meeting of wide and narrow stripes, the so-called "locus of stripe

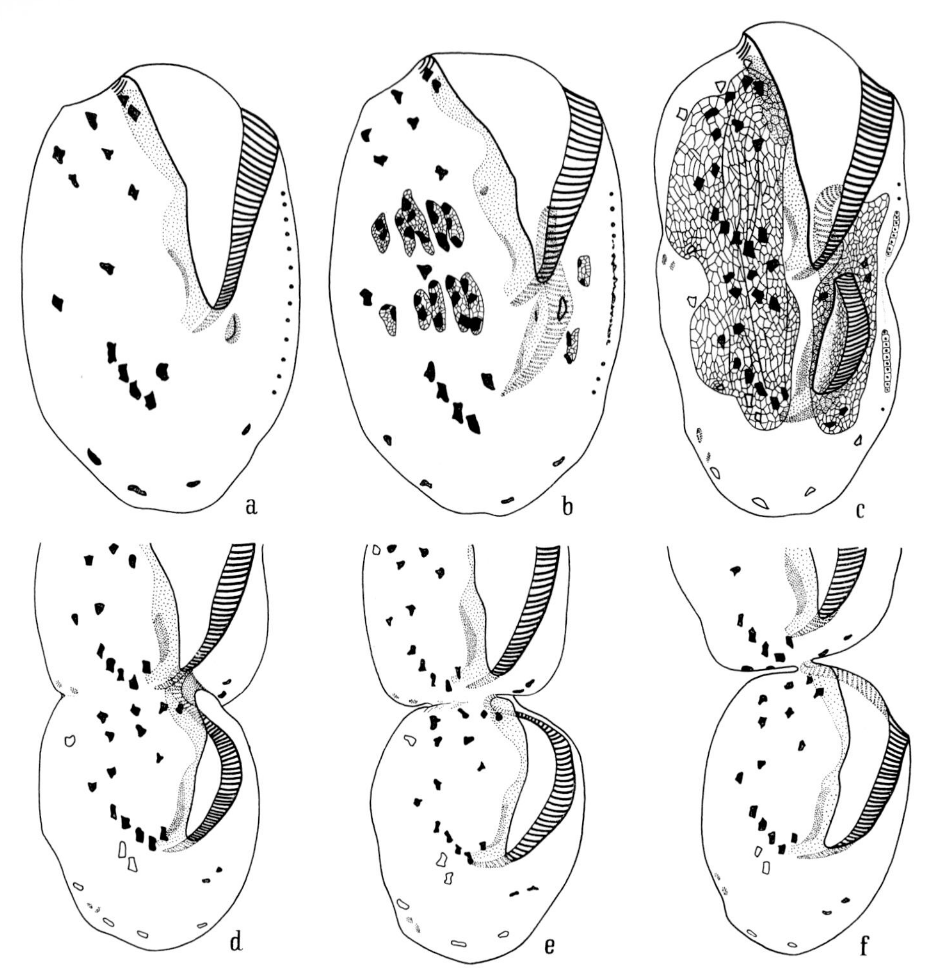

Fig. 124. *Euplotes eurystomus*. Stages of binary fission which show the reformation of the peristome for the posterior animal and the reorganization of the cortical pattern on its ventral surface. Silver impregnation. × 360. After WISE, 1965 [*1808*]

contrast". This is normally the case at the contrast meridian. Unlike *Paramecium*, excision of the anlagen field has thus no irreparable consequences. As soon as the margins of the wound have closed and a "stripe contrast" has been restored, a new anlagen field is built up again which can give rise to the new formations during division and physiological regeneration. Artificial stripe contrast meridians, which lead to predictable malformations, can be obtained by the transplantation of cell parts. Experiments of this type show that the capacity for new formations *(competence)* increases with declining stripe width, while the capacity to initiate the formation of an anlagen field *(induction)* increases with growing stripe width. Evidently there is a "circular" *gradient* around the cell which corresponds to the gradient of varying stripe width. Other experiments lead to the conclusion that there is also a gradient in the tendency to form basal structures. This gradient increases from the apical to the basal end. Differentiation during division and physiological regeneration is thus based upon the correlated interplay of a "circular" and a "basal-apical" gradient, whose cause must be sought in the gradient-like distribution of differences within the superficial cell layer (cortex). One may assume that these differences are of a cytochemical nature. Their analysis has just begun [*1569*].

The *hypotrichs* show, as is well known, a distinct dorsoventrality. At the dorsal surface there are rows (kineties) of short sparsely distributed cilia ("tactile bristles"). Before fission an increase of cilia takes place – similar to *Urocentrum turbo* (Fig. 121) – within a belt zone which borders the future division furrow. The ventral surface shows the peristome in front with the membranellar band and several groups of cirri, which are arranged in a definite pattern (Fig. 124). Of these differentiations, practically only the peristome remains before division. This is allotted to the anterior animal. The anlage of the peristome of the posterior animal arises from a granular condensation which may be traced to a group of basal bodies. It develops at first mainly below the pellicle and only during division of the two daughter animals does it open to its full extent toward the outside. The groups of cirri at the lower surface are resorbed. New groups of cirri are formed within definitely delimited fields which are distinguishable by silver impregnation from the remaining pellicle by the higher density of their net structure. Those of the anterior and the posterior animal are separated from the very beginning. Shortly before fission, these fields spread over an increasingly larger area of the ventral surface. In this manner the new groups of cirri ultimately reach the place prescribed by the proportions of the daughter animals.

Thus, while the dorsal surface is practically only being "built onto", the ventral surface undergoes a far-reaching reorganization of its cortical pattern.

II. Multiple Fission

In multiple fission, a single mother cell divides to form a multitude of daughter cells. It is usually preceded by nuclear multiplication within the mother cell, which then cleaves rapidly to form a corresponding number of daughter cells (*simultaneous* multiple fission).

In *flagellates*, this type of multiple fission is found in only a few species. Thus, in *Trypanosoma lewisi* it occurs in addition to binary fission. Not only the cell nuclei but also the kinetoplasts and the basal bodies from which the "marginal fibers" of the undulating membrane grow, multiply during the growth phase. In addition to binary fission, which is the most frequent mode of reproduction, multiple fission occasionally also takes place in *Noctiluca miliaris*. It leads to the formation of flagellated swarmers whose fate is unknown (Fig. 125).

Multiple fission is also found in many *Rhizopoda*. Some parasitic amoebae carry out multiple fission inside a cyst (*Entamoeba coli*, *E. histolytica*, Fig. 349). Some of the Heliozoa (e.g. *Actinosphaerium arachnoideum*) reproduce by multiple as well as binary fission (film E 648). Multiple fission leading to the formation of swarmers is also the normal mode of reproduction in Radiolaria (Fig. 126). As in *Noctiluca miliaris*, the further development of these swarmers is unknown. In Foraminifera, which are characterized by an alternation of generations, the agamonts, after

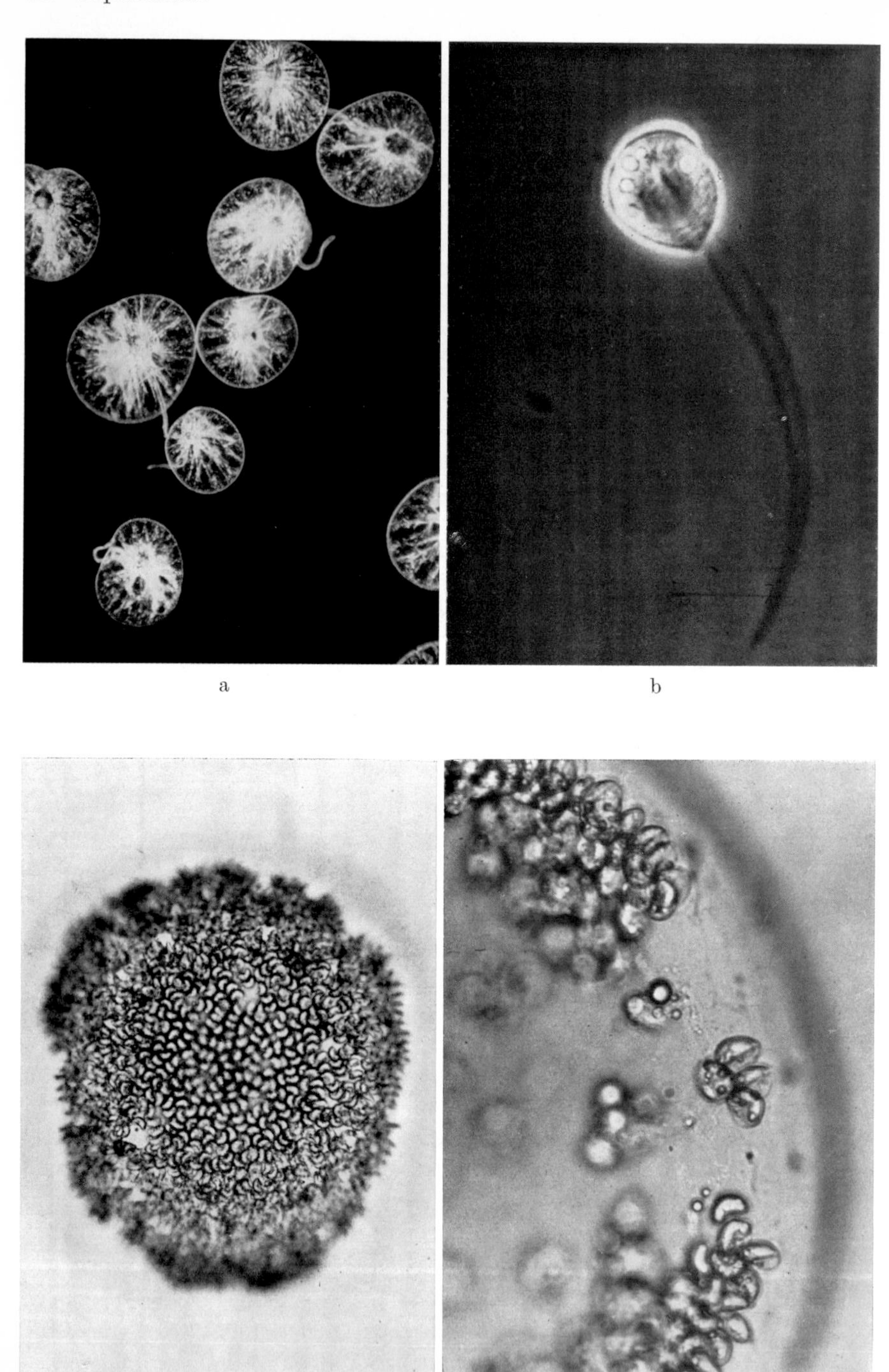

Fig. 125. *Noctiluca miliaris*. a Several cells in dark field. b Single swarmer (phase contrast). c Formation of swarmers. d The same, higher magnification. a × 40, b × 900, c × 190, d × 360. From the film C897 (UHLIG)

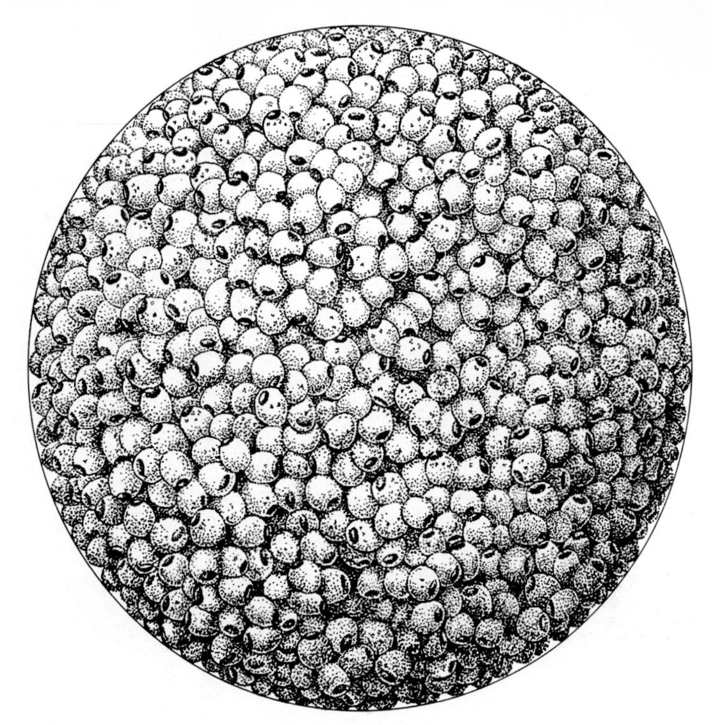

Fig. 126. *Thalassophysa sanguinolenta.* Central capsule with crystal swarmers. After HOLLANDE and ENJUMET, 1953 [*787*]

meiosis, as well as the gamonts reproduce by multiple fission. In many cases the protoplast emerges first from the shell and then subdivides outside to form young gamonts (agamont of *Rubratella intermedia*, Fig. 365) or gametes (gamonts of *Patellina corrugata*, Fig. 360).

In the life cycle of *Sporozoa*, multiple fission sets in after the cells have grown to a certain size (schizonts, gamonts and sporonts). In Coccidia (Fig. 127) the protoplast of the oocyst after a number of nuclear divisions is subdivided to form the spores. Within the latter, another multiple fission takes place leading to the formation of sporozoites. Frequently, not the whole protoplast is subdivided in the course of multiple fission in Sporozoa, a "residual body" being left over.

In some Protozoa, multiple fission resembles total cleavage of animal eggs. The mother cell is divided into smaller and smaller daughter cells without cell growth between the fissions. Each fission is accompanied by nuclear division (*successive* multiple fission). Among the *flagellates*, this type of multiple fission is especially widespread in *phytomonads.* Many species of Chlamydomonadidae divide within their cellulose wall into four or more daughter cells, which are thus smaller than the mother cell. In the colonial Volvocidae, all cells may be capable of forming small daughter colonies in this manner, or else they are differentiated into somatic and generative cells, only the latter being normally able to form daughter colonies (p. 8).

As shown in Fig. 128 (comp. Fig. 129) of the genus *Eudorina* (syn. *Pleodorina*), subdivision of the mother cell follows a fixed pattern. The first plane of division

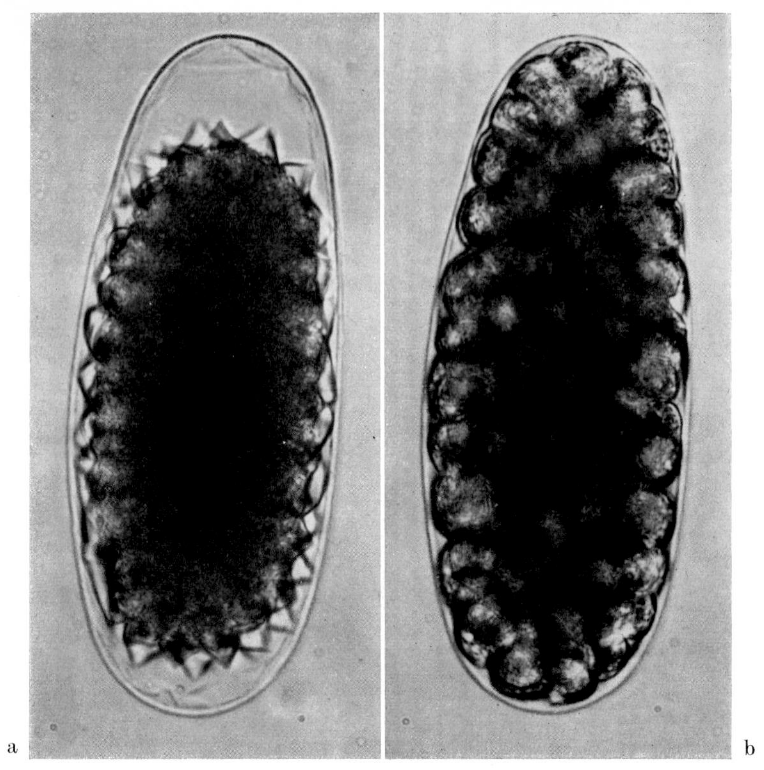

Fig. 127. *Eucoccidium dinophili.* Isolated oocyst. a Before, b after spore formation. Photographed in life. After GRELL, 1953 [657]

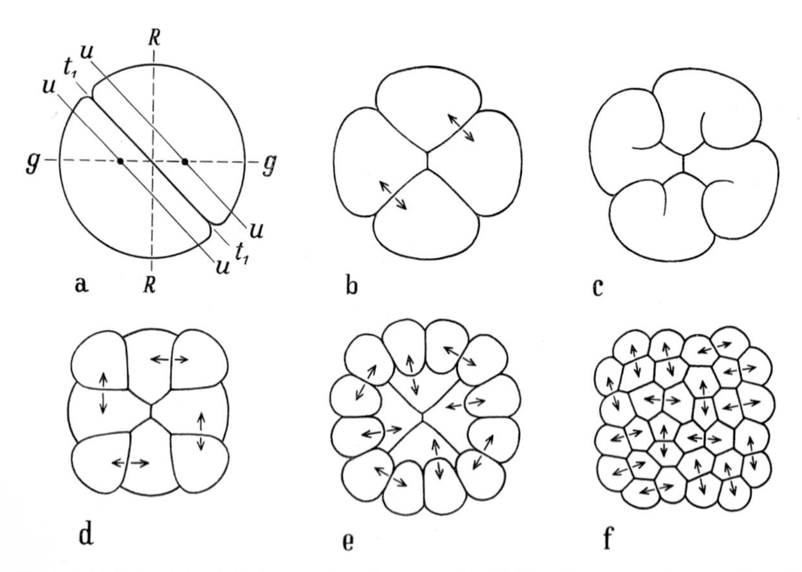

Fig. 128. *Eudorina (Pleodorina) californica.* Diagram of the divisional sequence of a generative cell. a First divisional step. t_1—t_2 plane of division, u—u planes of flagellar beating, g—g plane through the points of flagellar origin. R—R rotational axis of the colony. b 4-cell stage. c Third divisional step. d 8-cell stage ("*Volvox* cross"). e 16-cell stage. f 32-cell stage. The sister cells are joined by arrows. In accordance with GERISCH, 1959 [572]

(t_1—t_1) is parallel to the planes in which the two flagella beat (u—u). These planes are turned to the right by an acute angle relative to the plane connecting the two flagellar origins (g—g) (Fig. 128a). The second plane of division is at right angles to the first one, leading to the formation of four cells of equal size (b). The third division is unequal and dexiotropic (c). Thus, the typical "*Volvox* cross" of four cells touching in the middle and four others filling the interstices of the cross configuration is formed (d). Subsequent divisions also follow a constant pattern (e).

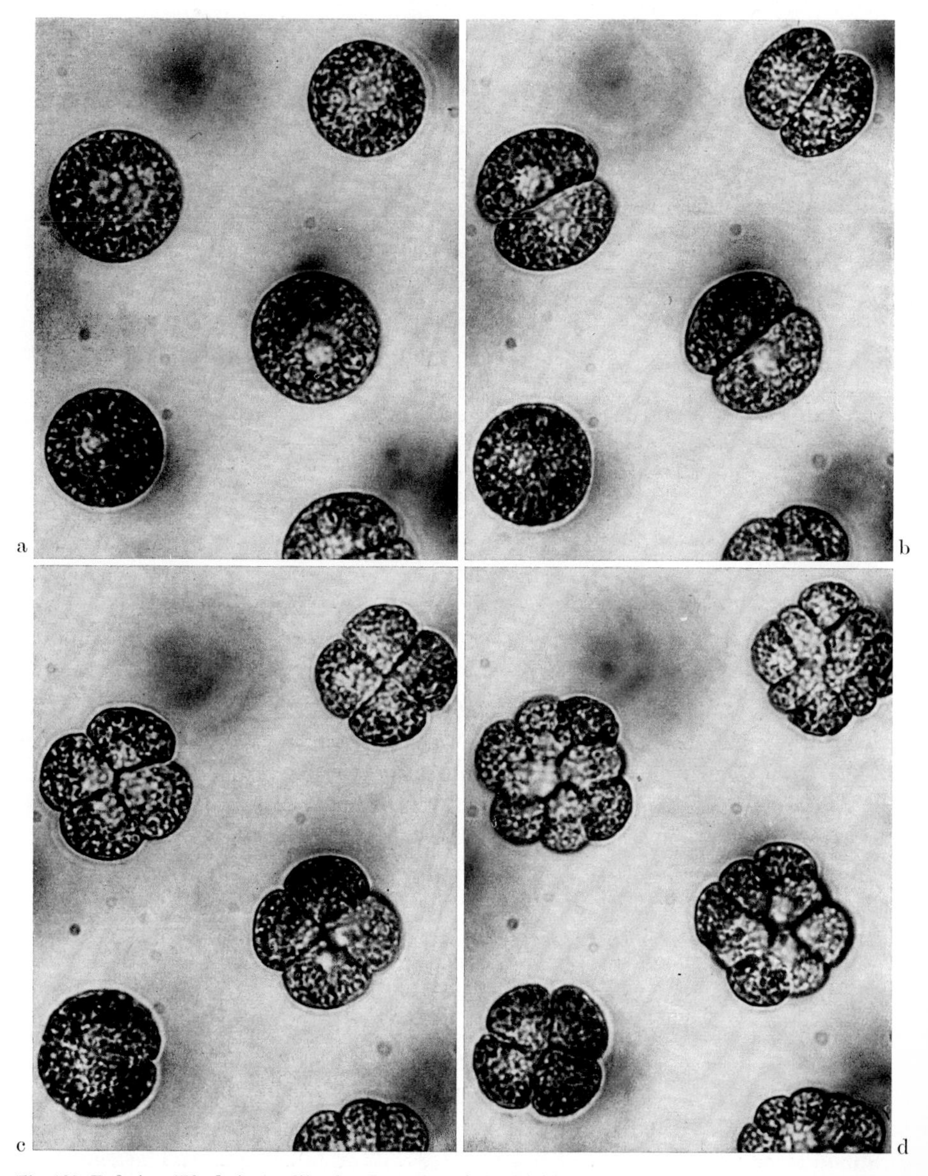

Fig. 129. ***Eudorina (Pleodorina) californica.*** Part of a colony. Division of the generative cells was photographed inside the colony (compare to Fig. 128). × 680. From the films C883 and E657 (GRELL)

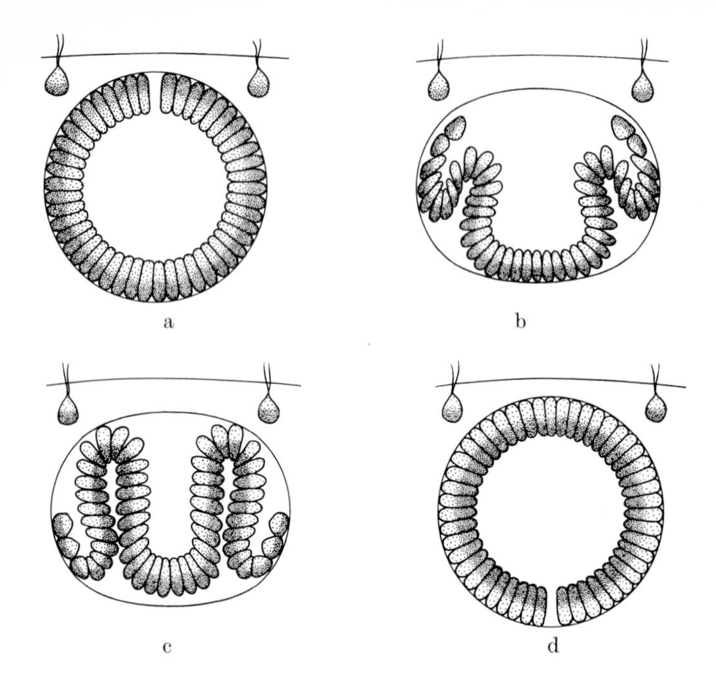

Fig. 130. *Volvox aureus*. Diagram of inversion. The polarity of the cells is indicated by varying density of stippling. Due to inversion the flagellar pole (light) is shifted to the outside. In accordance with ZIMMERMANN, 1921

Finally, a cup-shaped cell plate is formed (f). The flagella grow from the concave surface into the cavity.
At this stage, which is almost a closed sphere in the case of *Volvox* (Fig. 130), the flagella are directed toward the inside; therefore cleavage must be followed by an *inversion* to achieve direction of the flagella toward the outside. This process occurs rapidly. Its mechanism is not known as yet.
While the sequence of divisions is essentially the same in the genera *Pandorina* and *Volvox* as in *Eudorina (Pleodorina)*, it differs appreciably in *Gonium* [*572*].

Among other flagellates, this type of multiple fission is fairly common only in *dinoflagellates*. One pelagic species, *Dissodinium lunula* (Fig. 131), forms large, blister-like cells which give rise to 16 sickle-shaped ones by repeated fission. Within the latter, again several cell divisions take place, leading to a large number of small, *Gymnodinium*-like swarmers.
Successive multiple fission is peculiarly modified in those dinoflagellates which infest different marine animals either as ectoparasites *(Oodinium, Apodinium, Chytriodinium)* or as endoparasites *(Blastodinium, Syndinium, Haplozoon, Amoebophrya, Neresheimeria, Duboscquella)* [*186, 190, 193, 198, 199, 201, 228*].

The genus *Blastodinium*, whose species live in the intestine of pelagic copepods, is selected as an example (Fig. 132, 133). The host is infected by ingesting a small, *Gymnodinium*-like swarmer, the so-called dinospore, together with its food. The latter already has two nuclei and grows within the copepod gut to a large, immotile *germinative body*. It divides into two daughter cells of differing fates. The *trophocyte* (*T*) grows and again divides into two daughter cells (*T′*, *G′*), while the other cell,

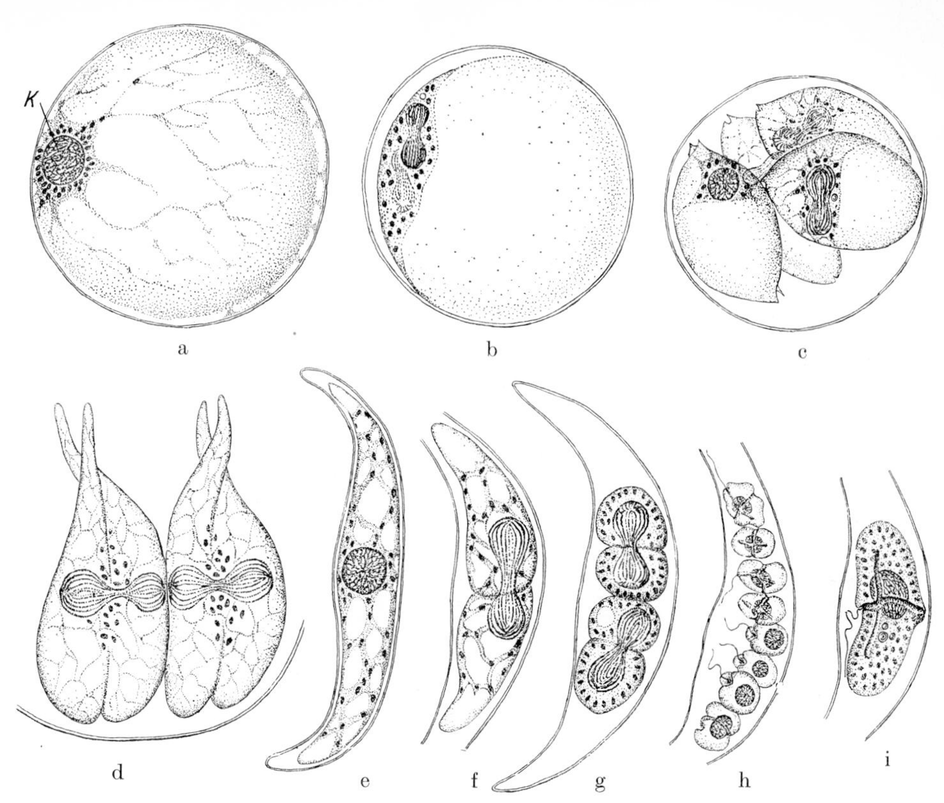

Fig. 131. *Dissodinium lunula* (Dinoflagellata). Multiple fission. a—c The first two cell divisions in the primary individual, which is surrounded by a cyst envelope. d Two of the eight cells after the third divisional step. Each cell divides once more (fourth step). e One of the 16 sickle-shaped cells. f—h Divisions within the same. i Occasionally the entire content of the sickle-shaped cell can change into a swarmer. × 500. After DOGIEL, 1906 from KÜHN, 1921

the *gonocyte* (G), forms a large number of swarmer mother cells (*sporocytes*, S, S') by a rapid succession of divisions. One of the two daughter cells of the second generation grows and divides into further daughter cells (T'', G''), while the other one gives rise to another set of swarmer-forming cells (S''). In some species this process can be repeated a number of times, leading to successive layers of swarmer mother cells which sheathe the central cells like peels of an onion. Sooner or later the outer casing bursts and the oldest swarmer-forming cells are released to the outside via the anus of the host. As long as the trophocyte continues to grow, this process of swarming is repeated at regular intervals — in some species every day.

Species of the genus *Amoebophrya* are intracellular parasites of other marine Protozoa such as dinoflagellates, radiolarians and ciliates. Their life cycle is very peculiar (Fig. 134). The uninucleated dinospore (a) infects a host cell, in which it grows into the so-called *Amoebophrya stage* (b—d). This growth phase is connected with an unusual morphological transformation: The posterior part of the cell (hypocone) forms a circumvallation which surrounds the anterior part of the cell (epicone) completely. In this way a cavity ("mastigocoel") arises, being connected with the outside by a small pore only. During this process a turning sets in, which leads to the formation of a helical ledge at the inner walls of the cavity. Mean-

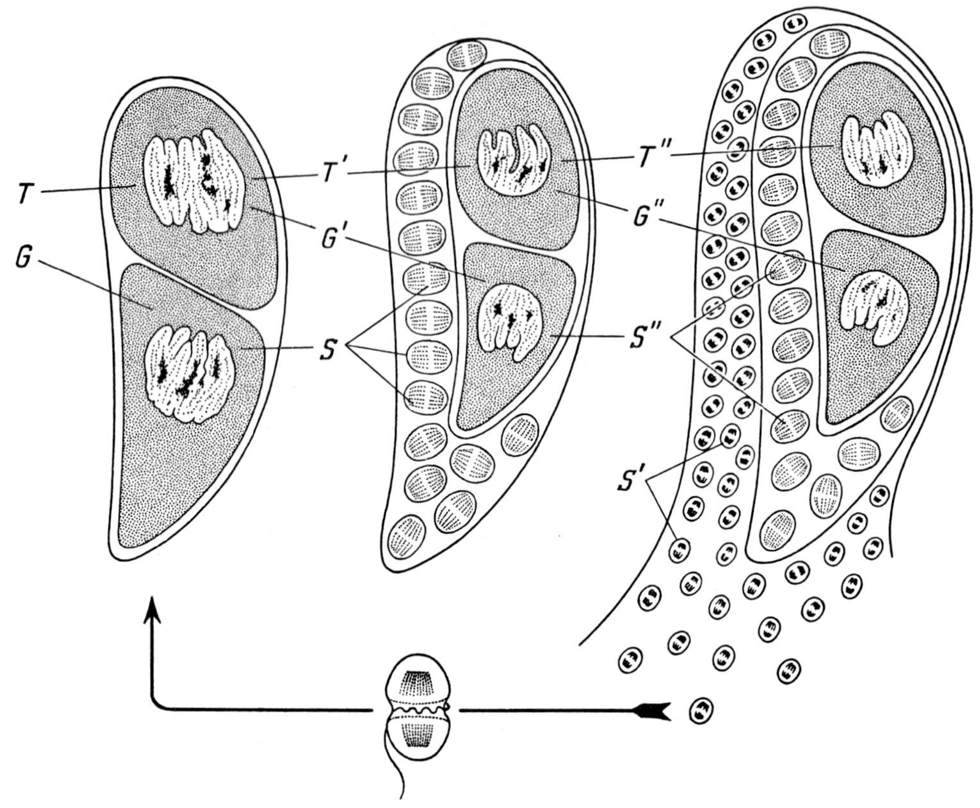

Fig. 132. *Blastodinium*. Diagram of the development of a germinative body. Explanation in the text

while the two basal bodies of the swarmer have multiplied along the helical ledge, outgrowing into hundreds of flagella which beat in the common cavity. In some species the nucleus of the swarmer multiplies at the beginning of the growth phase, in others at its end.

In any case hundreds of nuclei are present when the Amoebophrya stage transforms into the so-called *vermiform stage* (e). This process occurs in a few seconds. The originally anterior part of the cell starts to penetrate the pore, so that the inner walls of the cavity supplied with the helically arranged flagella become the outer walls (f). The vermiform stage which reaches a length of several mm swims about in the sea and disintegrates stepwise into single swarmers (g—i). It is interesting to note that the vermiform stage has a food vacuole within its posterior part in which nutritive material deriving from the host cell becomes digested during the pelagic life.

Successive multiple fission is also found in *ciliates*. It is usually preceded by encystment of the mother cell. Within this "reproductive cyst" the cell is then subdivided stepwise into even smaller descendants. This type of reproduction is especially characteristic of the genus *Colpoda* and the family Ophryoglenidae which also includes *Ichthyophthirius multifiliis*, a skin parasite of fresh-water fish (Fig. 414). Numerous small daughter cells are formed in the reproductive cysts of this species.

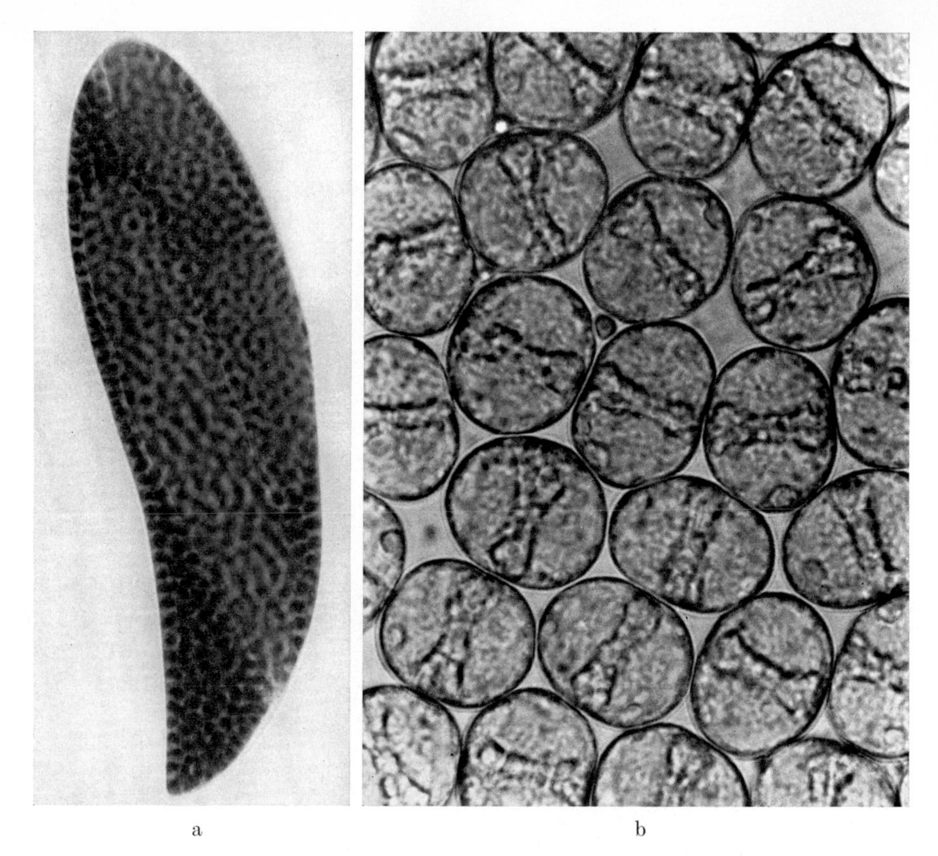

a b

Fig. 133. *Blastodinium*. a Germinative body from the intestine of a copepod stained with acetocarmine. b Swarmers in life

As mentioned above (p. 111), successive multiple fission is accompanied by progressive depolyploidization of the macronucleus.

III. Budding

While every unequal fission might be called "budding", it seems reasonable to reserve this term for those cases where the "mother cell" remains sessile and releases a "daughter cell" which serves as a motile "swarmer" in the dissemination of the species. Since the swarmer differs from the mother cell not only in its lower degree of differentiation but also in the presence of special organelles for locomotion and selection of a suitable substrate, its transformation into the sessile form is a true *metamorphosis:* the specific organelles of the swarmer are resorbed (regressive processes) and those of the sessile adult are formed anew (progressive processes). Budding and metamorphosis are found in most sessile *ciliates*. The *folliculinids*, which inhabit ampulla-like loricae, create currents for filter-feeding with the aid of two large peristomial wings (Fig. 421). The latter are resorbed before fission. The pigment, previously evenly distributed, accumulates in the anterior half of the cell

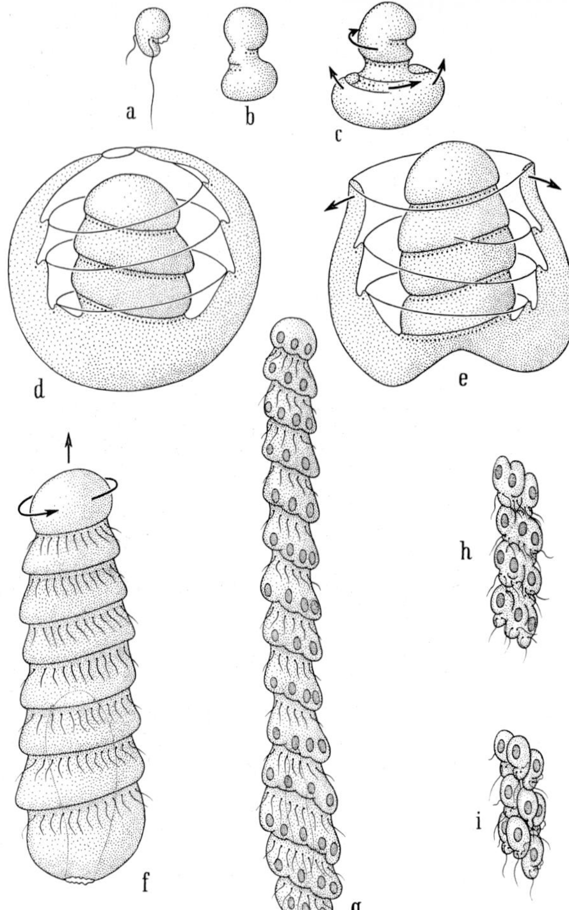

Fig. 134. *Amoebophrya*. Diagram of the life cycle. a—e Intracellular development of the parasite. Growth and transformation of the dinospore into the "*Amoebophrya*-stage". f, g Free-swimming "*vermiform*-stage". h, i Beginning of disintegration, formation of dinospores. Original drawings of J. CACHON (slightly modified)

Fig. 135. *Metafolliculina andrewsi*. The vermiform swarmer (anterior animal) crawls out of the lorica while the posterior animal (with peristomal wings) remains within it. × 150. From the films C 903 and E 649 (UHLIG)

which is budded off as a vermiform swarmer. The posterior half remains within the lorica and regenerates the peristomal wings (Fig. 135). Following attachment to a suitable substrate, the swarmer secretes a new lorica. This is accomplished by a series of coordinated steps [*1756*].

In the solitary *Peritricha*, one cell remains attached to the stalk after division while the other one forms a special circlet of cilia at the posterior end and swims away as the so-called "telotroch". Microconjugants are formed in a similar manner (Fig. 200). Ciliated swarmers are also characteristic for the small order *Chonotricha*.

In *Suctoria*, the difference between the "mother cell" and the bud which transforms into a swarmer is usually especially pronounced. *Exogenous budding*, which involves

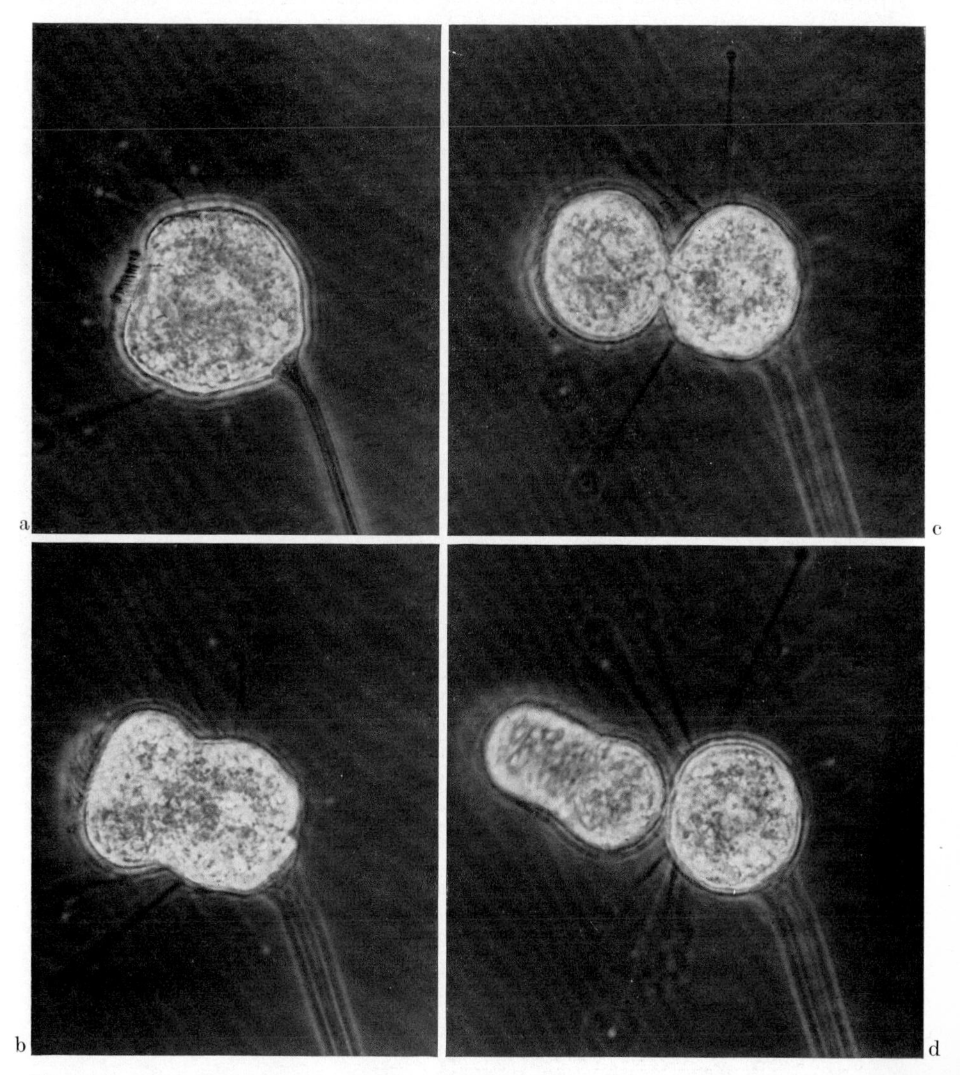

Fig. 136. *Paracineta limbata*. Exogenous budding. Only one swarmer of the same size as the "mother cell" is formed. In a the cilia of the swarmer are already recognizable. × 540. From the film C 913 (GRELL)

formation and separation of the bud toward the outside, may be looked upon as the primary type.

Paracineta limbata (Fig. 136) should be mentioned first because the bud is as large as the mother cell. Immediately before separation the bud stretches and thus becomes morphologically distinct from the mother cell.

In *Ephelota gemmipara*, an especially large suctorian, the mother cell forms several swarmers at once (Fig. 137, upper row). At its apical pole, small projections are noticed at first, which become ear-shaped in the course of time. At the concave side of each bud a ciliary field consisting of many rows of cilia is differentiated around the so-called *scopula*. The latter is a ring of dense cytoplasm and represents the primordium of the new stalk. In the course of its further transformations, the bud stretches and the ciliary field narrows down from two sides to form a groove. The completely formed swarmer thus assumes the shape of a coffee bean. After detaching from the mother cell, it crawls about for some time by means of its ciliary field. Once a suitable substrate is reached, it presses the scopula firmly to the ground and forms a stalk. Tentacles grow while the cilia disappear. Their initial stages are already recognizable in the swarmer within small grooves of the cell surface (Fig. 137, middle).

Ephelota gemmipara normally forms multiple buds only after the cell has grown to

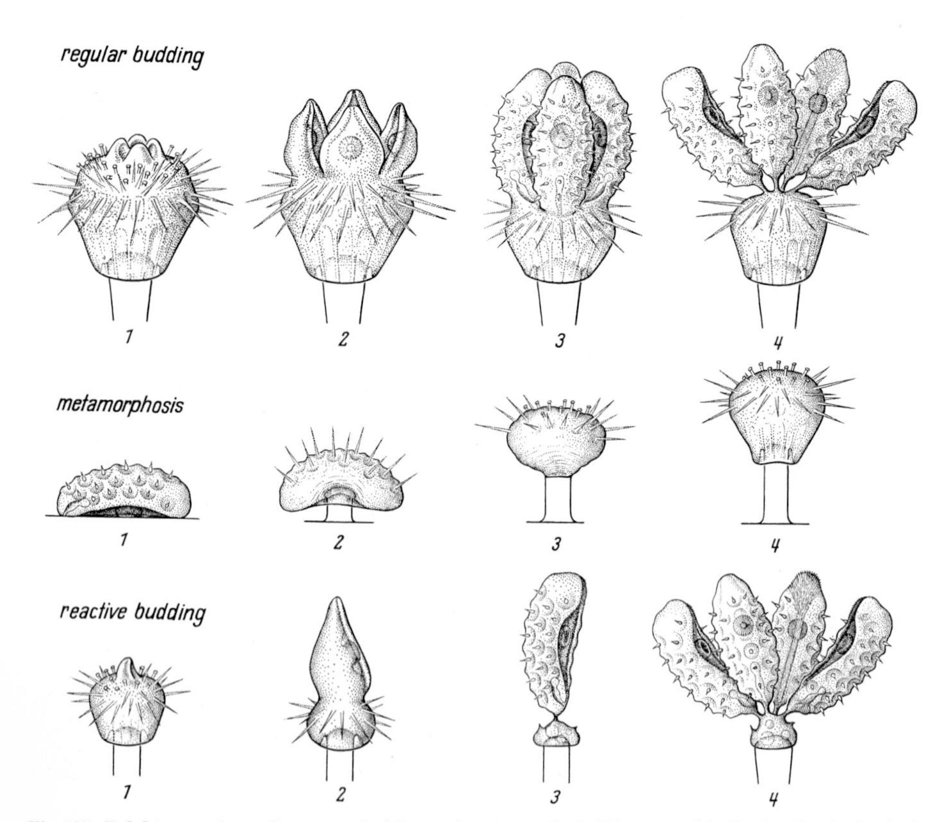

Fig. 137. *Ephelota gemmipara*. Exogenous budding and metamorphosis (diagrammatic). Explanation in the text

a certain size (*regular* budding). However, under unfavorable conditions, such as lack of oxygen, all cells, irrespective of size, form swarmers as a type of flight reaction (*reactive* budding). The individuals are then completely transformed into swarmers, unequal fission thus being replaced by equal fission. The number of swarmers depends upon the size of the cell. A small cell may be completely transformed into a single swarmer (Fig. 137, lower row, 1–3). "Reproduction" is then no longer associated with multiplication. Large cells form several swarmers (4) (films C 913 and E 1017).

A peculiar modification of budding is found in the free-living *Dactylophrya*-generation of *Tachyblaston ephelotensis* (see Fig. 431 for the developmental cycle). In this case the mother cell is subdivided by repeated budding and no intervening growth into the so-called *dactylozoites*. These have only a single tentacle but no cilia. After leaving the mother cell, which is enclosed in a cup-shaped casing, they attach themselves to the pellicle of *Ephelota gemmipara* and develop into the parasitic generation. Only the empty casing is left after subdivision of the mother cell (Figs. 138 and 432).

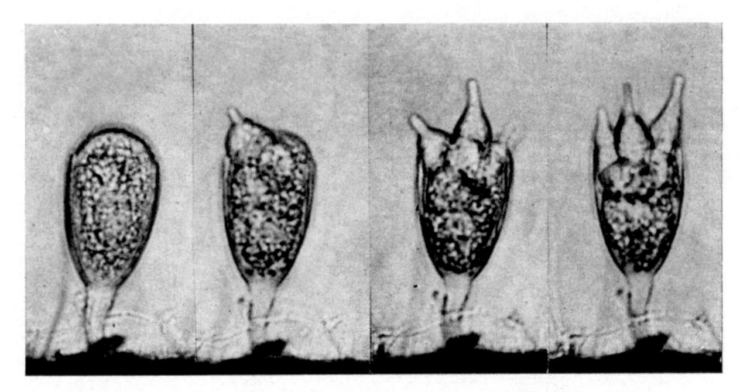

Fig. 138. *Tachyblaston ephelotensis*. Development of a *Dactylophrya*-stage (compare with Fig. 431). × 460. From the films C 907 and C 913 (Grell)

In most suctorian species the swarmer is formed inside the mother cell *(endogenous budding)*. As *Ephelota gemmipara* is an example for the simultaneous formation of several swarmers by exogenous budding, *Tokophrya quadripartita* exemplifies the same for endogenous budding. In *Acineta tuberosa* only one swarmer is formed. It escapes from the mother cell without leaving a wound (Fig. 139). This is the case because it differentiates within a preformed "brood pouch" which is connected to the outside by a pore (Fig. 140). The macronucleus is divided unequally in the course of bud formation.

A relatively transparent species, *Tokophrya lemnarum*, permits study of this process in life and by time-lapse microcinematography (Fig. 141). The chromosomal threads which fill the macronucleus are at first stiff and immotile (a). After some time, they lose their stiffness suddenly and are moved within the nuclear space as if thoroughly mixed (b). After the chromosomal threads have become stiff again, the macronucleus pinches off a bud (c) into the swarmer which has been formed in the meantime (d). The significance of these motions within the macronucleus is unknown. It seems likely that they also occur before division in other species

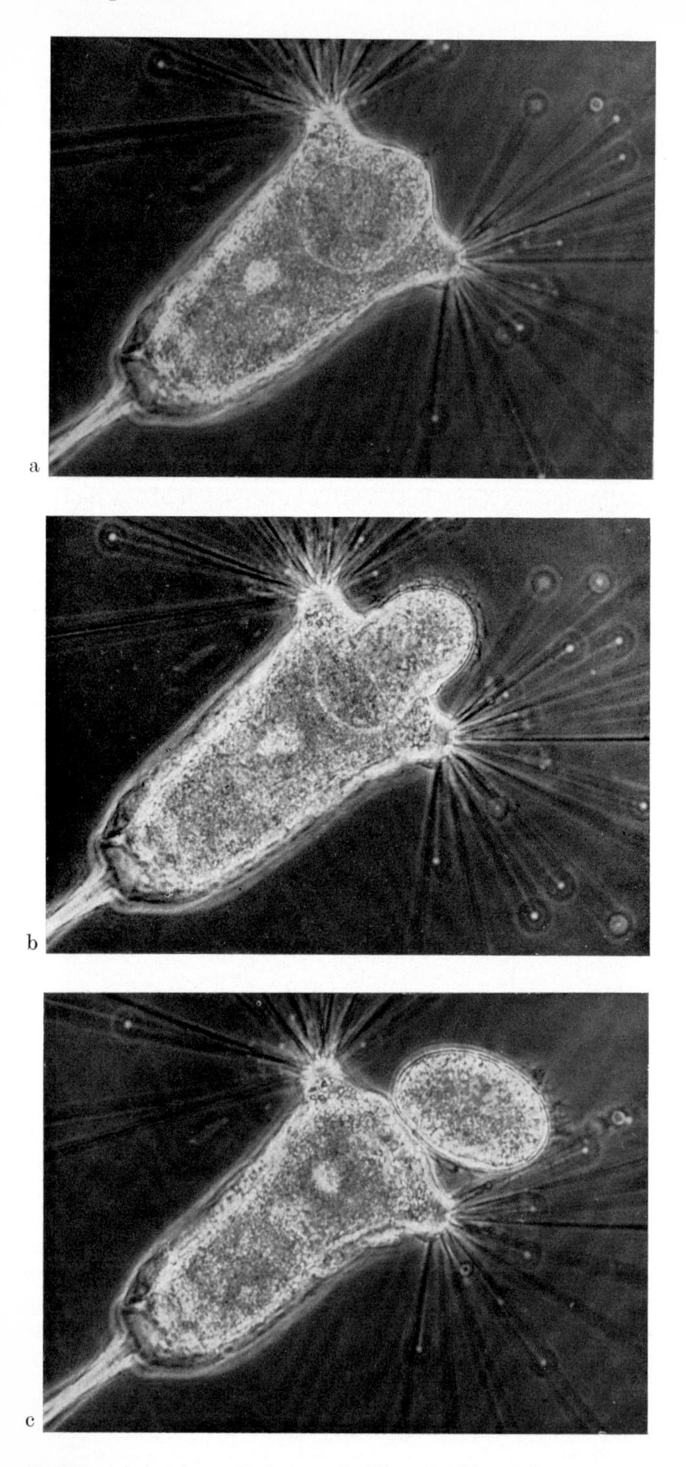

Fig. 139. *Acineta tuberosa*. Endogenous budding. Hatching of the swarmer. × 430. From the films C 913 and E 914 (GRELL)

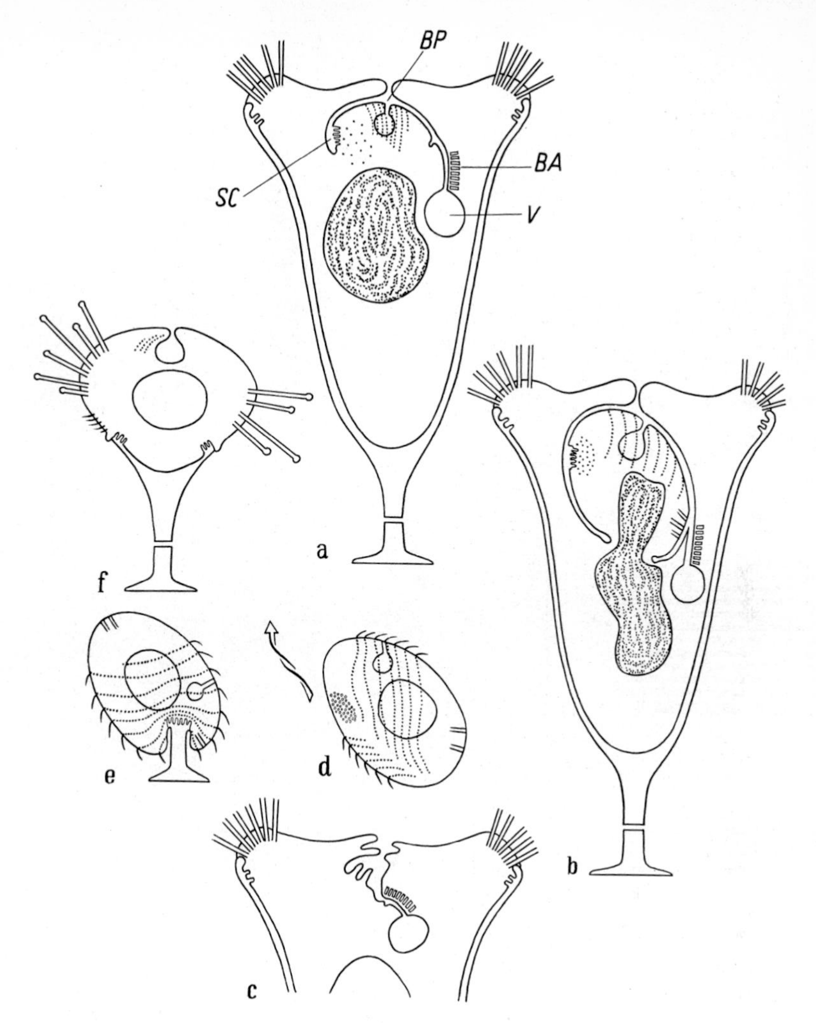

Fig. 140. *Acineta tuberosa*. Diagram of budding and metamorphosis. a Beginning, b end of swarmer formation (unequal division of macronucleus). c Upper part of mother cell after releasing of the swarmer. d, e, f Metamorphosis. *BP* Brood pouch. *BA* Barren basal bodies. *SC* Scopula. *V* Vesicle. After BARDELE, 1970 [*86*]

although they have not been previously noticed due to unfavorable optical conditions.

Since the brood pouch is formed by invagination of the cell surface, endogenous budding is probably derived from the exogenous mode of bud formation. It enables the mother cell to continue food capture and intake even during bud formation.

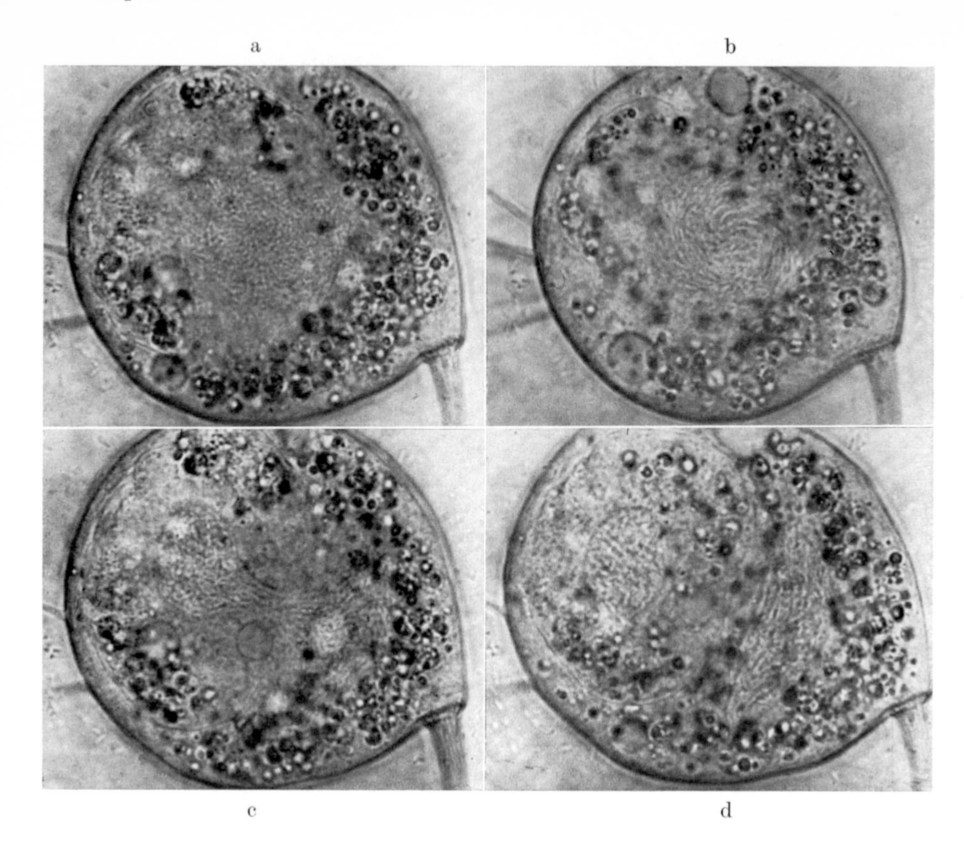

Fig. 141. *Tokophrya lemnarum*. Endogenous budding. The structural changes of the macronucleus are recognizable especially well in this case (compare with text). × 1100. From the films C 913 and E 913 (HECKMANN)

E. Fertilization and Sexuality

Fertilization, or copulation, is the fusion of two cells to form a *zygote*. Cells which have this ability are called *gametes*. One essential aspect of fertilization is *karyogamy*, i.e. the fusion of the two gametic nuclei (pronuclei) within the zygote to form a *synkaryon*. The doubling of the chromosome number which is connected with this is compensated by the nuclear events at meiosis (p. 78).

In all multicellular animals the gametes, which unite in pairs, are clearly differentiated from one another. They are called *eggs* and *spermatozoa* (or sperm cells) respectively. Eggs are formed by females and spermatozoa by males. Only in hermaphrodites are eggs and spermatozoa formed by the same individual.

The phenomenon of the diversity of the fusing gametes or the individuals producing the gametes is called *sexuality*.

Since the gametes of all multicellular animals are sexually differentiated, the connection between sexuality and fertilization seems so self-evident that the gametes are referred to as "sex cells". Among the Protozoa, however, there are cases where neither the gametes nor the individuals forming them are morphologically different from each other. It must therefore still be tested whether there are differences of any kind between them which might correspond to sexuality in multicellular animals.

If the gametes are not morphologically differentiated we speak of *isogamety*, and of *anisogamety* if they differ either in size, form or structure. Gametes which differ in size are called *macro-* and *microgametes*. In many Protozoa, the macrogametes resemble the egg cells of Metazoa while the microgametes resemble the spermatozoa. In this case anisogamety is referred to as *oogamety*.

In some Protozoa all cells can become gametes under certain conditions. However, as a rule a *gamont* is formed first which produces the gametes as a result of a special reproductive act *(gamogony)*. The latter may involve binary or multiple fission.

If gametes capable of fusing with one another are formed within the same clone, or if they arise from a common gamont, the organism is said to be *monoecious*. Species in which sexually compatible gametes arise from different clones or gamonts are called *dioecious*.

Sex determination can take place in various ways. If the realization of sex depends upon genes, we speak of *genetic* or *genotypic* sex determination. If the direction of differentiation is not due to genetic factors, sex determination is *modificatory*, or *phenotypic*. The potentialities of sexual differentiation are based upon the genetic "reaction norm" (p. 232) in either case.

One important variant of phenotypic sex determination is the production of two sexually different daughter cells or nuclei from a common progenitor by *differential cellular or nuclear division*.

Fertilization is of widespread occurrence among Protozoa. Even if we disregard doubtful cases where karyogamy and meiosis have not been proved to take place, there is no class of Protozoa without at least some orders where fertilization has been found. Among the flagellates, the Phytomonadina, the Polymastigina and the Opalinina are examples, among the rhizopods the Heliozoa and the Foraminifera. Fertilization is generally characteristic for the classes Sporozoa and Ciliata.

On the other hand, it should not be overlooked that fertilization has not been found in many flagellates and rhizopods in spite of thorough search. Some species which have been cultured for years in the laboratory have reproduced only asexually, i.e. without fertilization and meiosis. Despite widespread occurrence, there is thus no reason to look upon fertilization and sexuality as "basic phenomena of life" (HARTMANN).

Although all fertilization processes do lead to karyogamy sooner or later, the phenomena connected with them are of a surprising *diversity* among Protozoa. This includes the relation between gametes and gamonts, their mode of formation and their sexual differentiation as well as the occurrence of monoecy and dioecy.

On the basis of the relationship between gametes and gamonts, three types of fertilization processes can be distinguished in Protozoa, viz. *gametogamy*, *autogamy*, and *gamontogamy*.

I. Gametogamy

In the simplest case, gametes are formed which fuse as free-swimming cells and not within the gamonts. This mode, found in free-living Protozoa (Phytomonadina, Foraminifera) as well as in parasitic Protozoa (Polymastigina, Sporozoa) is called gametogamy.

Phytomonadina

In the phytomonads, fertilization has been proved for all well-known species and has been studied in detail in a number of forms. In all cases gametes fuse with each other of which at least one is free-swimming. Among the non-colonial species, especially those of the genera *Chlamydomonas*, *Dunaliella* and *Polytoma*, the gametes are often not formed by a special mode of reproduction but arise directly from vegetative cells *(hologamy)*.

The fact that cells become gametes, i.e. that they acquire the capacity to fuse with other gametes, is due to external factors which may be different in the various species and which may collaborate in different ways. In *Chlamydomonas*, the determination of gametes is induced primarily by lack of nitrogen [*1485*]. If distilled water is poured on an agar plate containing non-flagellated cells reproducing by fission, they form flagella and become gametes. Depending upon species and sex, illumination is also of importance in this connection [*527*, *1042*]. In *Dunaliella salina*, an inhabitant of salt pits, decrease of salinity induces the formation of gametes [*1026*].

External factors of this kind cause a *synchronization* so that many, if not all, cells of a population or a clone become sexually active at the same time.

In most phytomonads, however, external factors lead at first only to the formation of gamonts which, in turn, form gametes by successive multiple division *(merogamy)*. The nature of these external factors is not well known as yet.

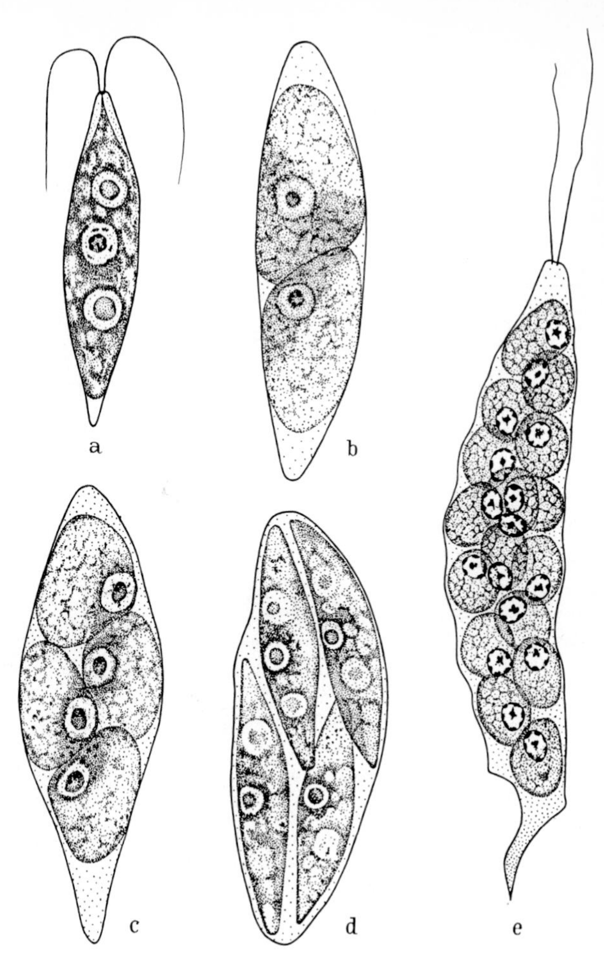

Fig. 142. *Chlorogonium elongatum* (Phytomonadina). a Normal cell. b—d Reproduction by two divisions within the cellulose envelope. e Gamete formation (16 cells). × 2200. After HARTMANN, 1956 [*721*]

The gamonts of phytomonads cannot be distinguished from ordinary cells by morphological criteria. In *Chlorogonium elongatum*, which normally reproduces by successive multiple division, the gamont carries out 2–3 additional fissions so that the gametes are smaller than the daughter cells of asexual reproduction (Fig. 142). Gametogenesis in *Chlamydomonas suboogama* (Fig. 143) is unusual enough to warrant special description. In this case, sexual reproduction begins with the determination of a gamont mother cell (progamont) which is enclosed in a gelatinous mass. It divides twice to form four daughter cells. Three of these always become macrogamonts while the fourth cell develops into a microgamont. The latter then forms four small microgametes. The macrogamonts do not divide but are directly transformed into macrogametes which are capable of fertilization. However, this change takes place only after the microgametes have left the gelatinous mass. The fusing macro- and microgametes are therefore normally descended from different gamont mother cells.

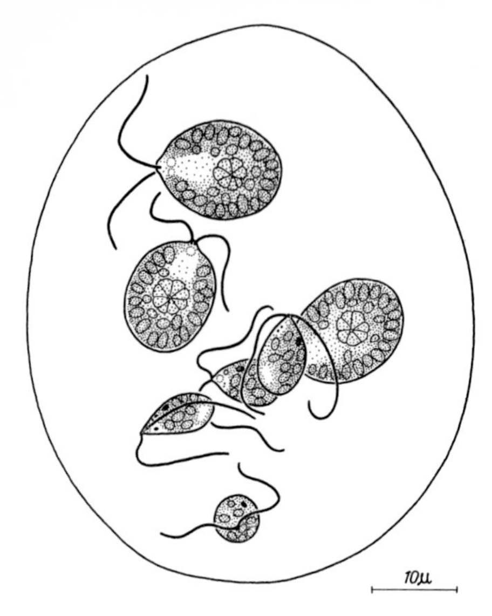

Fig. 143. *Chlamydomonas suboogama*. Gelatinous envelope of a gamont mother cell (progamont) with three macrogamonts and four microgametes. After TSCHERMAK-WOESS, 1959 [*1741*]

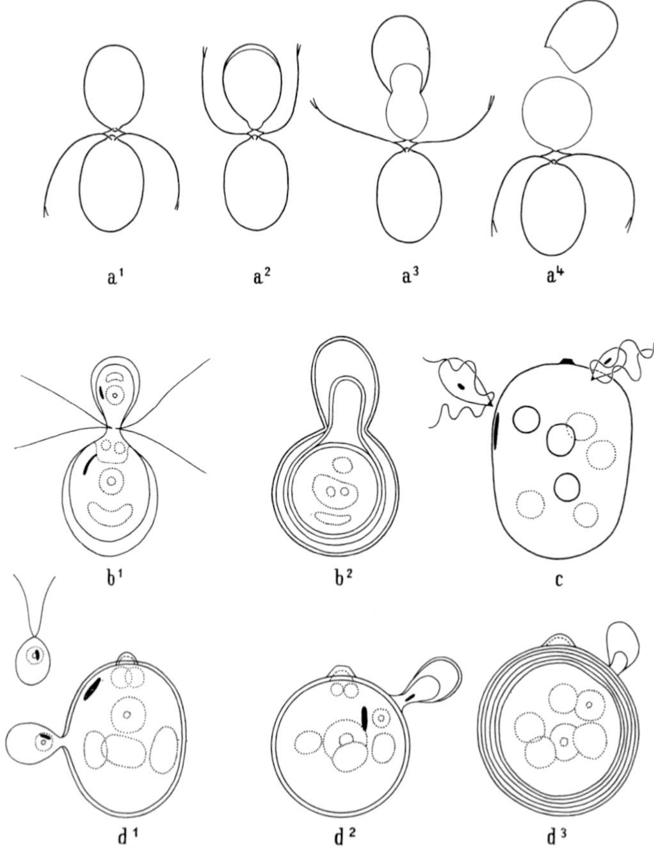

Fig. 144. Anisogamety and oogamety in various Chlamydomonadidae. a¹—a⁴ *Chloromonas saprophila*. After TSCHERMAK-WOESS, 1963. b¹, b² *Chlamydomonas braunii*. After GOROSCHANKIN, 1890. c *Chlamydomonas pseudogigantea*. After GEITLER, 1954. d¹—d³ *Chlamydomonas coccifera var. mesopyrenigera*. After SKUJA, 1949. From GRELL, 1967 [*675*]

In some colonial phytomonads all the cells of a colony can become gametes, in others special generative cells are formed which are the only ones capable of reproduction (p. 8). In some cases, e.g. *Pandorina morum*, it has been shown that gamete formation always sets in when colonies of different sex are brought together [*330*].

Sexual differentiation of gametes is of special interest in Phytomonadina because all transitions between isogamety and oogamety are found in this group. While some genera include only isogametic species *(Dunaliella, Pyramidomonas, Polytoma, Haematococcus, Brachiomonas)*, isogametic as well as anisogametic and oogametic species are known in the genera *Chlamydomonas* and *Chlorogonium*.

Most species of the large genus *Chlamydomonas* are isogametic. However, in some cases the fusing gametes exhibit differences which become manifest only during fertilization. The gametes of *Chlamydomonas eugametos syn. moewusii* swim about for hours after fusion. They lie opposite each other and are connected by a cytoplasmic bridge (Fig. 154b). Although both cells retain their flagella, they are active in one of the partners only. The flagella of the other gamete are turned back and only twitch occasionally. In *Chlamydomonas gymnogyne* and in *Chloromonas saprophila* one of the gametes sheds its cellulose casing after fusion (Fig. 144a^1–a^4).

Anisogamety in a morphological sense is of course only realized if both gametes are already different before pairing, as for instance in *Chlamydomonas braunii*

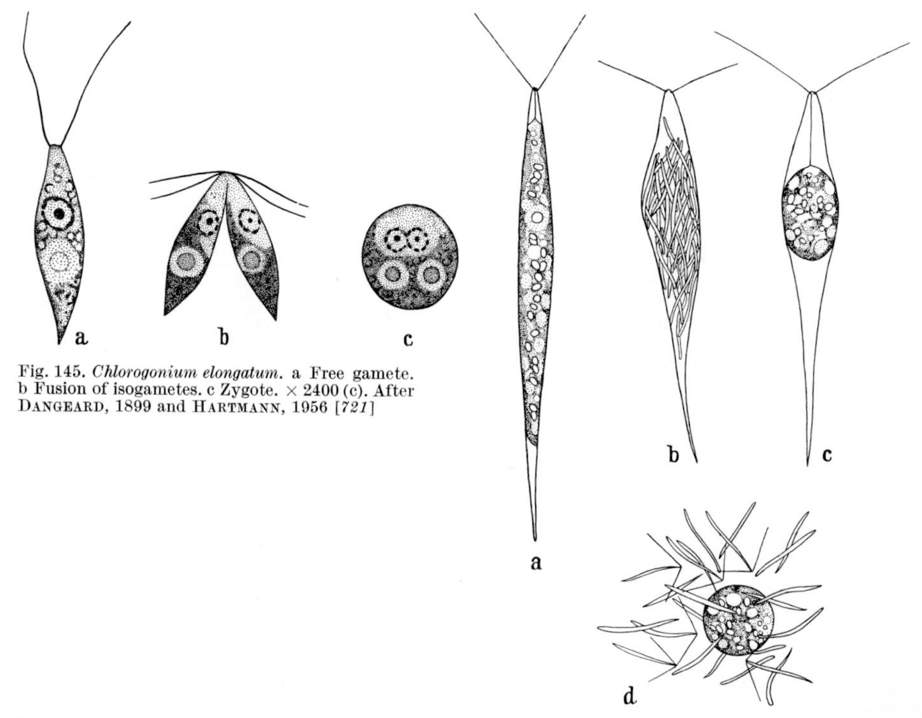

Fig. 145. *Chlorogonium elongatum.* a Free gamete. b Fusion of isogametes. c Zygote. × 2400 (c). After DANGEARD, 1899 and HARTMANN, 1956 [*721*]

Fig. 146. *Chlorogonium oogamum.* a Normal cell. b Gamont subdivided into numerous, slender microgametes. c Macrogamete within the cellulose envelope of the mother cell. d Hatched macrogamete surrounded by microgametes. After PASCHER, 1931

(Fig. 144 b^1, b^2). Some species such as *Chlamydomonas suboogama* (Fig. 143), *C. pseudogigantea* (Fig. 144c) and *C. coccifera* (d^1—d^3) can be considered oogametic since the macrogamete is a large, immotile cell, much bigger than the small motile microgametes. The microgametes of *Chlamydomonas pseudogigantea* contain a yellow pigment and lack chromatophores.

The small genus *Chlorogonium* also includes, besides isogametic species (e.g. *C. elongatum*, Fig. 145), a single oogametic form which was named *Chlorogonium oogamum*. The "male" gamont forms about 64 small flagellated microgametes whereas the cytoplasmic body of the "female" gamont contracts and is set free as a round "egg cell" after disintegration of the cell envelope (Fig. 146).

Among the colonial phytomonads, the genera *Stephanosphaera*, *Gonium* and *Pandorina* are isogametic, *Eudorina (Pleodorina)* and *Volvox* are oogametic.

Both *monoecy* and *dioecy* are found within small taxonomic groups. In the genus *Dunaliella*, for example, the species *Dunaliella parva* and *Dunaliella minuta* are monoecious while *Dunaliella salina* is dioecious. The genus *Chlamydomonas* also includes monoecious *(C. monadina, C. paupera, C. monoica)* as well as dioecious species *(C. eugametos syn. moewusii, C. reinhardi)*. *Haematococcus pluvialis* and *Stephanosphaera pulvialis* are monoecious. In the latter species, the eight cells of the mother colony carry out a variable number of fissions, giving rise to 4, 8, 16 or 32 gametes. These are spindle-shaped and are usually set free by rupture of the mother colony. Occasionally, especially if all cells of a colony have formed gametes, pairing occurs within the gelatine of the mother colony. Monoecy can thus lead to self-fertilization (autogamy, Fig. 147).

While *Gonium pectorale* is a dioecious species, *Gonium sociale* and *G. sacculiferum*, (both species with four cells in the colony), include dioecious as well as monoecious varieties or strains [*1655, 1659, 1660*]. Also in *Pandorina morum* a few monoecious clones could be isolated in addition to dioecious ones [*330*].

Fig. 147. *Stephanosphaera pluvialis*. Formation and fusion of the isogametes within the gelatinous envelope of a colony. After HIERONYMUS, 1884

The genus *Eudorina*, which now also includes the species formerly grouped together in the genus *Pleodorina*, shows high variability [*607*, *608*]. The species *Eudorina unicocca*, *Eudorina cylindrica*, *Eudorina (Pleodorina) illinoisensis* and the variety *elegans* of *Eudorina elegans* are dioecious. Some clones bring forth only female colonies and others only male colonies. While the cells of the female colonies are not distinguishable from normal cells, those of the male colonies carry out multiple fission. A packet of biflagellated microgametes or "spermatozoids" arises from each cell (Fig. 148). These sperm packets at first swim about as a unit. Only in the vicinity of the female colonies do they come to rest (b) and break up into the individual spermatozoids, which penetrate into the gelatinous mass and unite with the macrogametes or "eggs" (c).

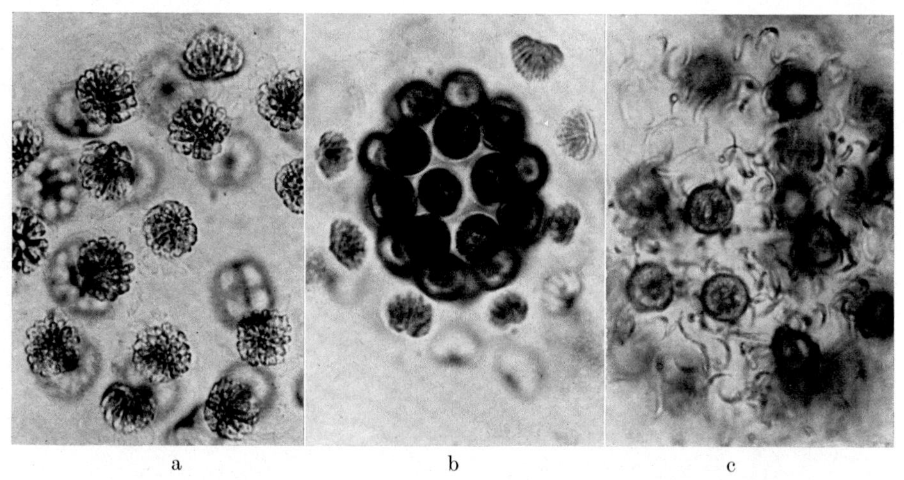

Fig. 148. *Eudorina elegans var. elegans*. Sexual reproduction. a Part of a male colony with packets of sperm. b Female colony, with sperm packets swarming around it. c The sperm have penetrated the gelatinous mass of a female colony. Photographed in life. × 300. After GOLDSTEIN, 1964 [*607*]

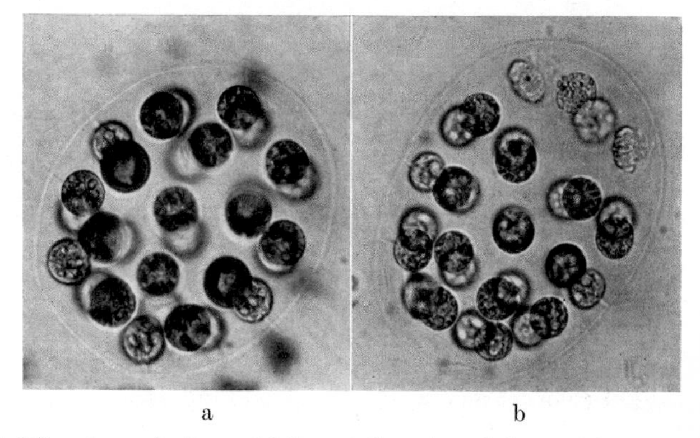

Fig. 149. *Eudorina elegans var. carteri*. Sexual reproduction. a Adult vegetative colony. b Sperm packets have formed from the four cells at the anterior pole. The remaining cells of the colony are "eggs". Photographed in life. × 320. After GOLDSTEIN, 1964 [*607*]

In monoecious species two possibilities can be discerned. In *Eudorina conradii* only the clones can be designated as monoecious while the individual colonies are either purely "female" or purely "male". Whether dioecy or monoecy is present can therefore be determined only by clone culture. In the other case, monoecy may also extend to the individual colonies, i.e. some of the cells become eggs while others form sperm packets. While the distribution of the sperm packets within a colony of *Eudorina elegans var. synoica* is random, in *Eudorina elegans var. carteri* the sperm packets are supplied only by the four cells at the anterior pole (Fig. 149). The polarity which manifests itself in the sexual differentiation of *Eudorina elegans var. carteri* is reminiscent of the polarity of *Eudorina (Pleodorina) illinoisensis*, where the four cells of the anterior pole are somatic, the remainder generative (Fig. 5). Also, of *Eudorina (Pleodorina) californica* at least two varieties exist in nature, one of which *(var. californica)* shows clonal monoecy (Fig. 150), the other *(var. tiffanyi)* colonial monoecy.

Very similar is the situation in the genus *Volvox*. As summarized in Table 2 (p. 161), from twelve species seven are monoecious, five dioecious [*1655a*]. The former species show either clonal or colonial monoecy.

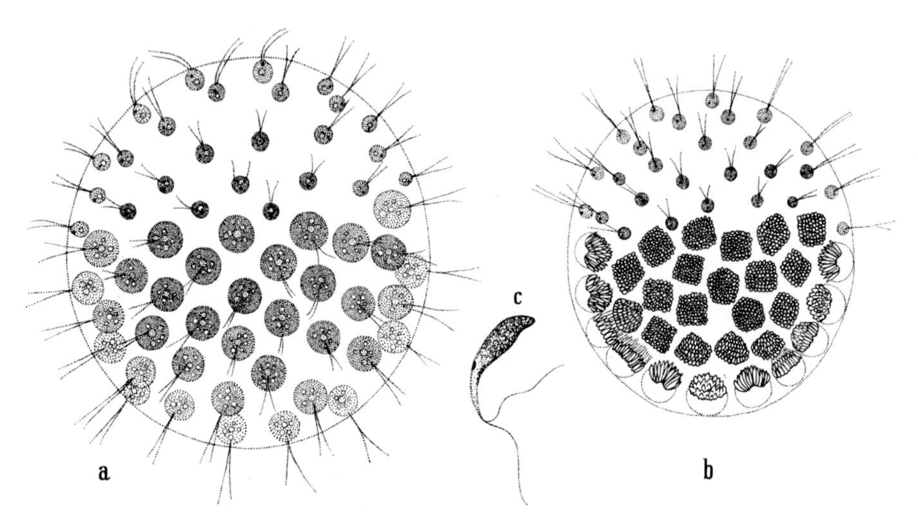

Fig. 150. *Eudorina (Pleodorina) californica.* Sexual reproduction. a Female colony. b Male colony with packets of sperm. c Single sperm cell. a, b: × 175. After CHATTON, 1911 [*227*]

As in *Eudorina (Pleodorina)*, the sperm packets swim about in the medium and become attached to the female colonies. A few minutes after attachment the sperm begin to penetrate and migrate within the gelatineous envelope (Fig. 151).

Recently, fusion of sperm and egg has been observed in *V. carteri* [*1657*]. It is unknown, however, whether the sperm can fertilize the eggs of the same colony in case of colonial monoecy (e.g. *V. globator*).

In several species of *Volvox* it could be shown that male colonies arise spontaneously under certain environmental conditions e.g. in old culture fluid. Initially only a few male colonies are formed. These produce an *inducing substance*, probably a protein, which causes the generative cells to form sexual colonies which contain sperm packets or eggs depending on the clone to which they belong. In other cases the effect is restricted to the differentiation of eggs in female clones only.

Table 2. *Volvox*. Distribution of monoecy (clonal, colonial) and dioecy in twelve species. After Starr (1968) [*1655a*]. Terminology changed as used in this context

Volvox	Monoecy		Dioecy
	clonal	colonial	
aureus	+		
barberi		+	
carteri			+
dissipatrix		+	
gigas			+
globator		+	
obversus			+
perglobator			+
powersii	+		
rousseletii			+
spermatosphaera	+		
tertius	+		

This induction system — though variable in the different species and strains — guarantees that numerous male and female colonies will be present simultaneously in a population and increases the chance for sexual reproduction.
It could be demonstrated experimentally that the inducing substances are species-specific, or even strain-specific *(V. carteri)*: cross-induction is unsuccessful [*365, 366, 952, 1655a, 1656, 1657*].

In all phytomonads with anisogamety or oogamety it is at once evident that sexual differentiation is confined to only two sexes. This *bipolarity*, familiar to us from Metazoa, holds for dioecious as well as monoecious species and varieties.

In the case of isogamety, however, only experimental methods can establish whether the copulating gametes are differentiated and if this differentiation is bipolar.

In *dioecious* species there is no doubt about sexual differentiation. That it is bipolar can be proved in many cases by combination experiments. If clones derived from the cells of a natural population or from the gones of a zygote are mixed together pairwise, fertilization will regularly take place in certain combinations and not in others. The pattern of combinations permits the assignment of all clones to one or another of two groups, designated as the "+" and "−" type.

Since it is highly improbable that this differentiation into two types is fundamentally different from sexual differentiation as expressed in cases of anisogamety and oogamety, the two types may also be looked upon as the "+" and "−" sexes.

The fact that a "+" gamete always fuses with a "−" gamete can be proved in some instances by *marking* one type of gamete. Marking may be phenotypic or genetic. If one of the two *Dunaliella salina* clones to be used in a combination test is reared in a medium poor in nitrogen and phosphorus, the cells become red after

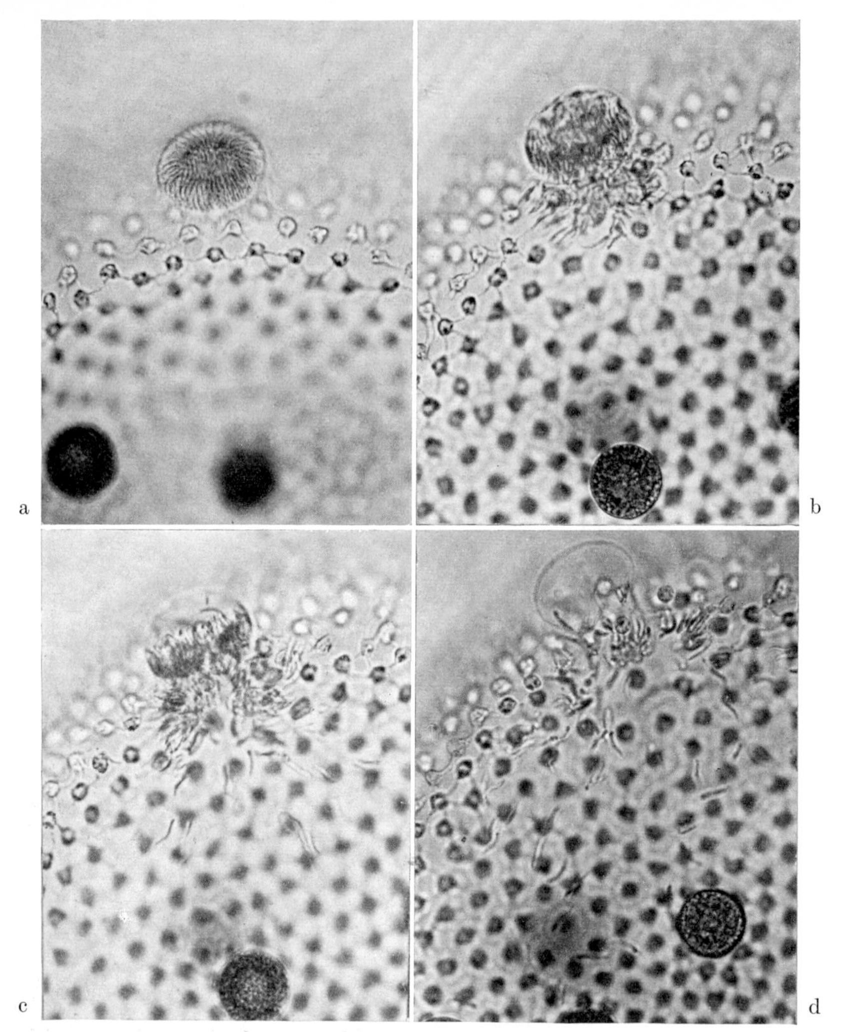

Fig. 151. *Volvox rousseletii.* a Packet of sperm attached to female colony. b Sperm begin to penetrate. c Sperm continue to enter female colony. d Most sperm inside female colony. After McCracken and Starr, 1970 [*1110*]

some time due to the deposition of carotene. If such "marked" isogametes are brought together with those of the other sex, pairing will always be between a red and a green cell (Fig. 152). If *Chlamydomonas reinhardi* is cultured on a nitrogen-deficient medium in the light, many starch granules accumulate in the cells. It is, therefore, easily possible to mark one of the gametic types (Fig. 153). If mutants are available, clones which are genetically different can be combined and the bipolarity of the gametes follows from the hybridization analysis (see p. 238).

In all dioecious species examined so far [*1040, 1477, 1536*] sex determination is genetic. It is based upon the separation of a pair of alleles (segregation), one being responsible for the +, the other for the − sex. Since meiosis is zygotic (p. 78), half of the gones which arise during division of the zygotes belong to one sex, half to the other. If only one gone emerges from the zygotic sheath, while the other

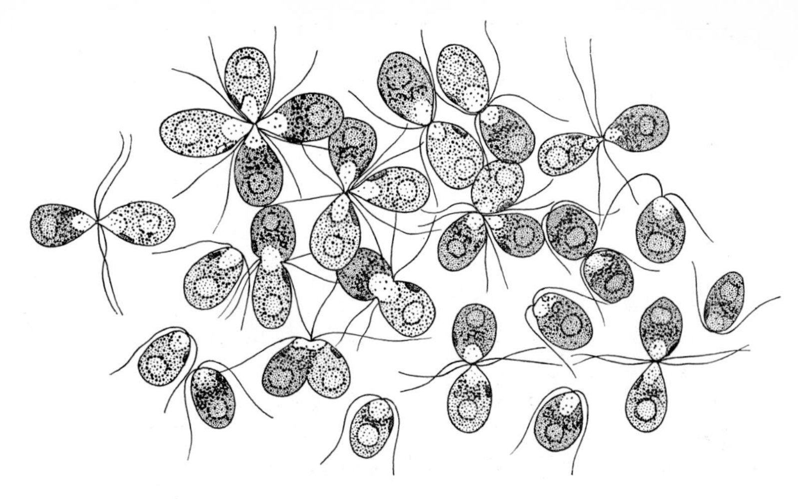

Fig. 152. *Dunaliella salina* (Phytomonadina). Fusion of isogametes which have been labeled due to different culture conditions. The darker ones shown are stained red, the light ones, green. After LERCHE, 1937 [*1026*]

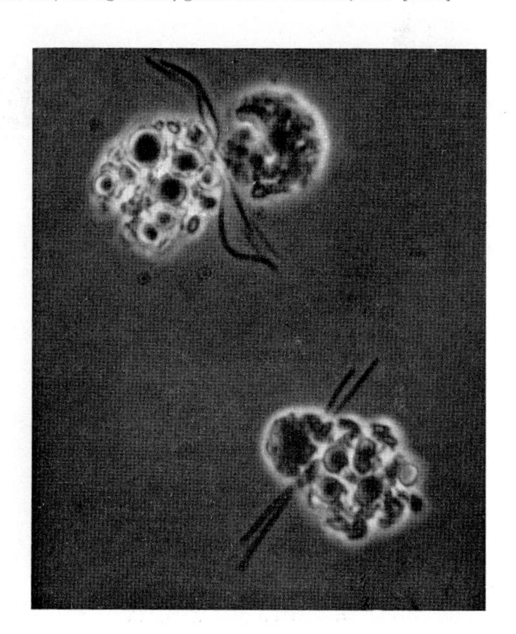

Fig. 153. *Chlamydomonas reinhardi.* Fusion of isogametes, one type containing starch granules (after culture in a nitrogen-deficient medium). Photographed in life (phase contrast). After SAGER and GRANICK, 1954 [*1485*]

gonal nuclei *(Pandorina)* or cells *(Eudorina, Volvox)* degenerate, then it is with equal likelihood either from the + or from the – sex [*330*, *607*, *952*].

Sexual differentiation and bipolarity are much more difficult to prove in the *monoecious* phytomonads with isogamety. One possible approach is the ,,Restgameten" (left-over gametes) method used successfully with *Acetabularia mediterranea* [*709*, *721*]. Corresponding experiments on phytomonads have not, however, led to conclusive results as yet. Should it turn out that even gametes which are not sexually differentiated can mate with each other, then one could, of course, not speak of "monoecy" [*1355*, *1356*].

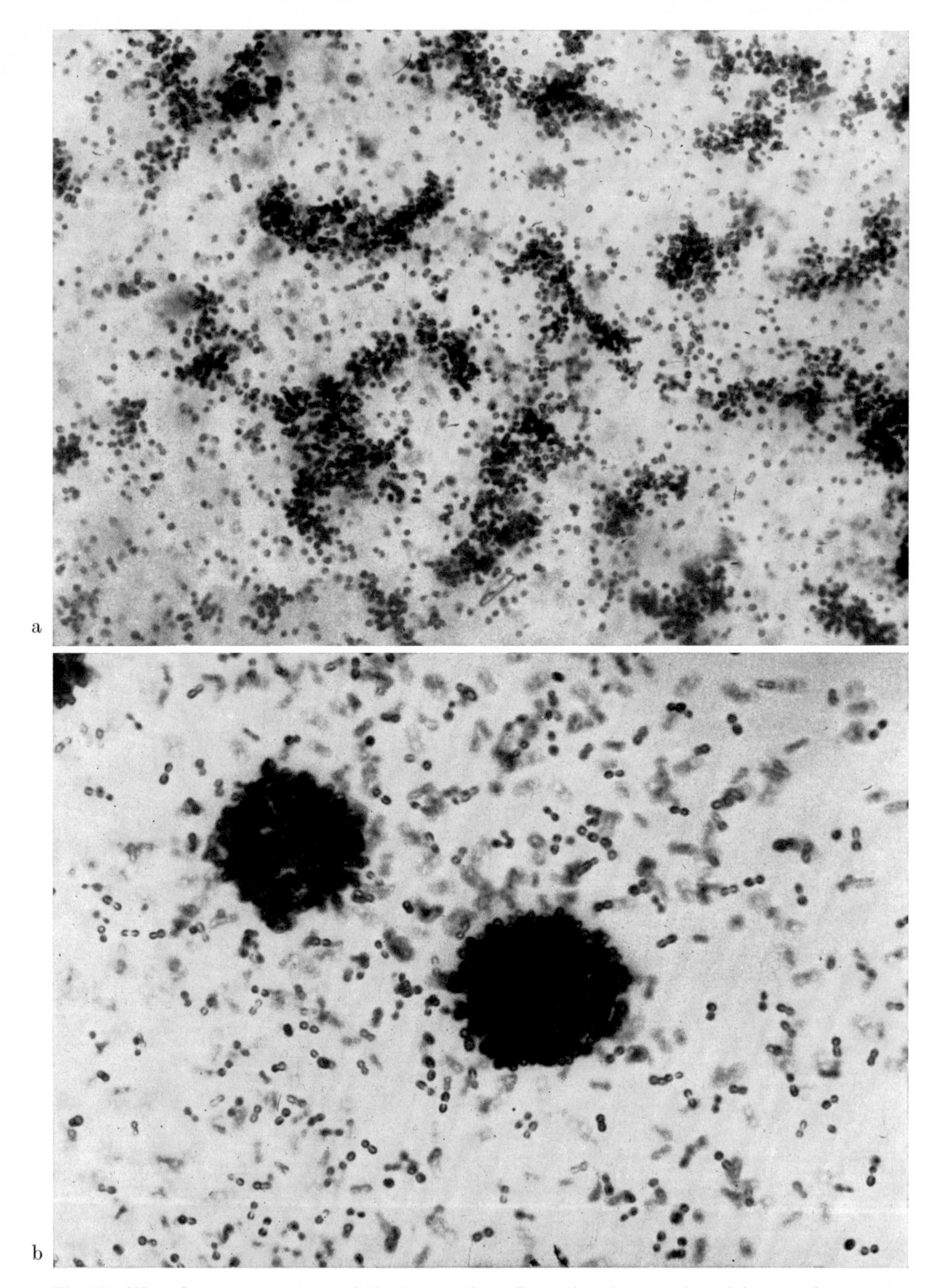

Fig. 154. *Chlamydomonas eugametos*. Agglutination reaction. a Formation of groups after mixing + and — gametes. × 160. b Two groups of agglutinating gametes and single pairs which are swimming about. × 400. From the film C 883 (GRELL)

An *agglutination reaction* takes place [*527–529, 1026*] if sexually different isogametes of dioecious species are brought together. Groups of many gametes – often several hundred — are thereby formed whose adhesion is due to stickiness of the flagellar tips. These groups remain intact for approximately half an hour in *Chlamydomonas eugametos syn. moewusii*, and for several hours in *Dunaliella salina*. Finally, they disperse into the individual pairs (Fig. 154).

In *Pandorina morum* [*330*] and *Astrephomene gubernaculifera* [*174*], agglutination begins when the cells are still within the colonies. Even after they have broken away from the colony, they still agglutinate for some time.

Since the agglutination reaction only takes place when + and — gametes are brought together, it must be based upon an interaction of sex-specific substances localized in the flagellar tips. Through this interaction, which can be thought of as a type of antigen-antibody reaction, the first contact between the gametes is established. In this respect, the sex-specific substances are reminiscent of the mating type substances of ciliates (p. 222).

However, while the mating type substances remain at the place of their action, the sex-specific

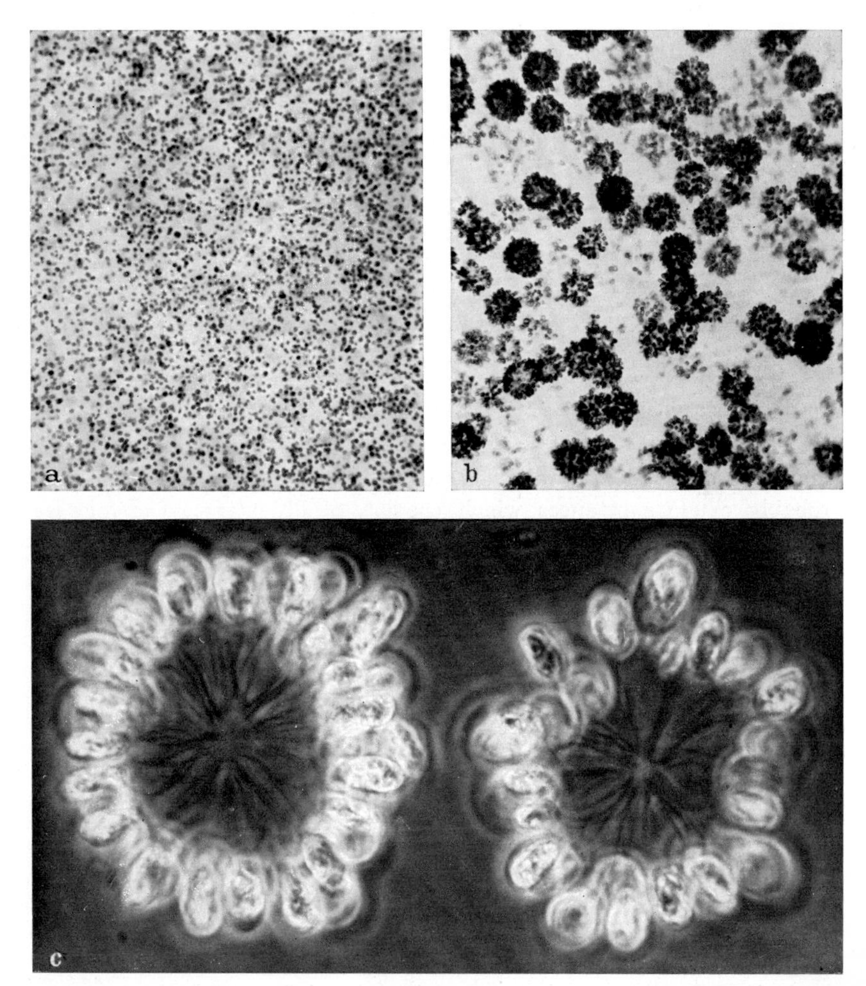

Fig. 155. *Chlamydomonas eugametos*. a "Male" gametes, untreated. b Isoagglutination of the "male" gametes after addition of a filtrate from "female" gametes. c Two groups of "male" gametes during isoagglutination (somewhat compressed by the cover slip). After Förster, Wiese, and Braunitzer, 1956 [*530*]

substances of *Chlamydomonas* can also be liberated into the culture solution. Although this may well be an "unphysiological" phenomenon which is meaningless for fertilization as such, it opens up the possibility to investigate the mode of action and nature of sex-specific substances more easily. Such investigations were carried out with *Chlamydomonas eugametos syn. moewusii* [*527—529*]. If the gametes of one sex are brought into a filtrate or centrifugate of gametes of the other sex, an *isoagglutination* takes place (Fig. 155). Thus, the substances liberated into the solution must be so-called polyvalent agglutinins, i.e. they must be capable of reacting with several "receptors", leading to the adherence of the flagellar tips of sexually identical gametes. It could be proved experimentally that they are identical with those substances which also permit the normal agglutination of the gametes [*1802*]. Chemical analysis showed them to be glycoproteins of high molecular weight [*530*]. The agglutinins of both sexes differ in their biochemical properties (temperature and pH-resistance) and in their nitrogen content. Moreover, the formation of the + agglutinine is light-dependent whereas the — agglutinine can also be formed in the dark. The substances are species-specific. Filtrates of *Chl. eugametos syn. moewusii* have no effect upon gametes of *Chl. reinhardi* and vice versa.

After contact of the flagella the gametes unite pairwise. In *Chlamydomonas eugametos syn. moewusii* the papillae which protrude between the flagella of both gametes are first to fuse so that the gametes lie opposite each other (vis-à-vis arrangement). In this manner the pairs swim about for hours, propelled by the flagellar beat of the + gamete while those of the – gamete remain inactive. The – gamete is probably immobilized by a substance formed in the + gamete and transferred to the – gamete by means of a cytoplasmic bridge [*1037*]. If the – clone is of a mutant which leads to "Siamese twins" as a result of a disturbance of fission, then the flagella of both cells are immobilized even though only one of them is combined with a + gamete.

In *Chlamydomonas reinhardi* both gametes become connected by a "fertilization tubule" which may be comparable to the bridge between both gametes in *Ch. eugametos syn. moewusii*. However, the zygote of *Ch. reinhardi* becomes motile only at a later stage when the protoplasts have nearly finished merging.
Electron microscopic investigations of pairing gametes showed that the + gamete contains a collar-shaped cytoplasmic organelle at the base of the fertilization tubule, the so-called "choanoid body". The – gamete has a similar structure, but much smaller and less conspicuous. There is some evidence that the gamete which produces the large "choanoid body" forms also the fertilization tubule, when + and – gametes come together [*549*].

In oogametic phytomonads, whose macrogametes are immotile "egg cells", agglutination cannot play a role in the contact of the sex cells. The macrogametes probably exude substances which chemotactically attract the microgametes. However, detailed investigations are not available as yet.

If the "egg cells" are not fertilized, they transform into "*parthenospores*" in some species (e.g. *Volvox aureus*). These are not distinguishable from zygotes: they have

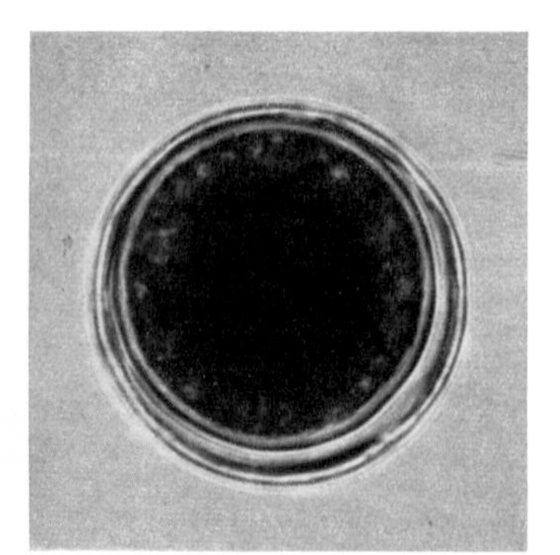

Fig. 156. *Volvox aureus*. Parthenospore. × 500

a double sheath and assume a reddish-brown color (Fig. 156). Under the same conditions which evoke the germination of zygotes, the "parthenospores" can also develop into colonies. Instead of "bisexual", a "monosexual" development or *parthenogenesis* thus takes place. In one strain of *Volvox aureus* parthenogenetic development is the rule [*1655a*].

Phytomonad species of which a fair number of clones have been isolated and cultured have been shown incapable of unlimited crossbreeding. Combination experiments indicated the existence of several "varieties" which are not distinct morphologically but are sexually isolated from each other: on combining gametes of different sexes belonging to two varieties, no mating takes place. In *Gonium pectorale* [*1659, 1661*], 12 dioecious "varieties" have been found so far, 15 in *Pandorina morum* [*330*], 6 in *Astrephomene gubernaculifera* [*174*], and 4 in *Eudorina elegans* [*607*]. It is probable that even more varieties exist in nature. Based on the same considerations discussed in the case of ciliates (p. 213), these "varieties" have been designated as *syngens*.

Polymastigina

Fertilization processes have also been described and thoroughly investigated in Polymastigina living in the intestine of roaches and termites [*276–284, 286–294, 299–301, 304, 307, 312–315, 318, 319, 324*]. At first it seemed that sexual reproduction would be restricted to species inhabiting the gut of the wood-feeding roach *Cryptocercus punctulatus*. More recently, fertilization was also described for some Polymastigina of termites. The possibility exists that it is widespread in this order of flagellates.

More than 30 species, of 14 genera, live in the intestinal tract of *Cryptocercus punctulatus*. However, the symbiont fauna of this roach, which is discontinuously distributed over the Southeastern and Northwestern United States, is not the same everywhere. *Macrospironympha xylopletha*, for example, occurs only in Californian populations of the roach.

Sexual reproduction of the flagellates is induced by the *molting hormone (ecdyson)* formed in the prothoracic glands of the host. As a rule, the roach molts only once a year. Fertilization of its flagellates is therefore limited to a comparatively short period of time. However, one must take into account that release of the molting hormone begins long before the actual molting act (ecdysis), i.e. before the old larval cuticle is sloughed off.

As several examples show (Fig. 157), sexual reproduction starts at different times in the individual genera. In *Barbulanympha*, it begins approximately 50 days before the molting act, in *Trichonympha*, on the other hand, only 5 days before. A certain concentration of the molting hormone, which differs in the individual genera and species, is apparently necessary to set the sexual processes into motion.

Table 3 (see p. 168) attempts to give a general survey of the diversity of the sexual processes which are even more complicated by the fact that some genera are haploid, others diploid.

In all cases, the effect of the molting hormone leads first to the determination of a gamont or progamont. In most genera, no special morphological changes are associated with this transformation. Only in *Leptospironympha* do the gamonts

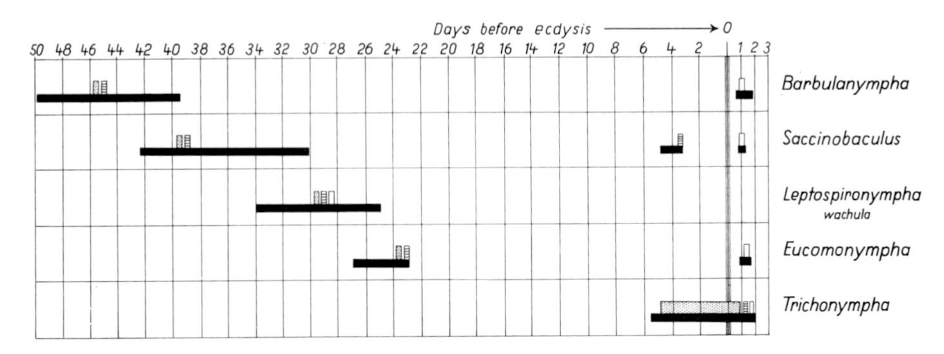

Fig. 157. Relationship between the sexual processes of some haploid Polymastigina of the roach *Cryptocercus punctulatus* and the molting period of the host. 50—0 days before, 0—3 days after molting (ecdysis). The vertical bars show the approximate duration of gamete formation (stippled), of fertilization (cross-hatched) and of meiosis (white) in a single individual. The horizontal bars (black) indicate the time-span for this phase when the results from several hosts are combined. Variants of sexual reproduction (autogamy, endomitosis) were not considered. In *Saccinobaculus* the symbol for fusion was entered twice because the pronuclei do not immediately unite to a synkaryon after gamete fusion. They remain side by side for several days and fuse finally just shortly before molting. Combined after CLEVELAND, 1957 [*294*]

Table 3. Sexual processes of Polymastigina living in the roach *Cryptocercus punctulatus*. Horizontal arrows indicate in which genera autogamy as well as gametogamy or gamontogamy has been observed

	Gametogamy	Autogamy	Gamontogamy
Haploids	*Saccinobaculus* ⟶	⟶	
	Oxymonas ⟶	⟶	
	Barbulanympha ⟶	⟶	
	Eucomonympha		
	Leptospironympha		
	Trichonympha		
Diploids	*Macrospironympha*	*Rhynchonympha*	
		Urinympha ⟵	⟵ *Notila*

differ conspicuously from ordinary cells. The flagellar bands which coil around the cell are broken down and the remaining flagellar tufts of the rostrum become transformed (Fig. 158). In species of the genus *Trichonympha* the gamonts round off and surround themselves with a cyst envelope. During encystment, a complete resorption of all extranuclear organelles (flagella, parabasal bodies) takes place; only the two centrioles, the nuclear cap and parts of the rostrum remain (Fig. 62). In *Saccinobaculus* and *Oxymonas* the axostyle is also resorbed.

In all haploid genera the gamont carries out a *differential division* which leads to the formation of a male and a female gamete. The peculiar chromosomal events which are connected with this division in *Barbulanympha*, *Leptospironympha* and *Trichonympha* have been pointed out above (p. 70). The various phases of gamete formation within the gamont cyst of *Trichonympha* take about 6 days and can be observed in life (Fig. 159). After the centrioles, between which an extranuclear spindle has been formed, have moved to opposite poles of the nucleus, new flagella are formed at the rostra (a). As soon as nuclear division is completed, the spindle

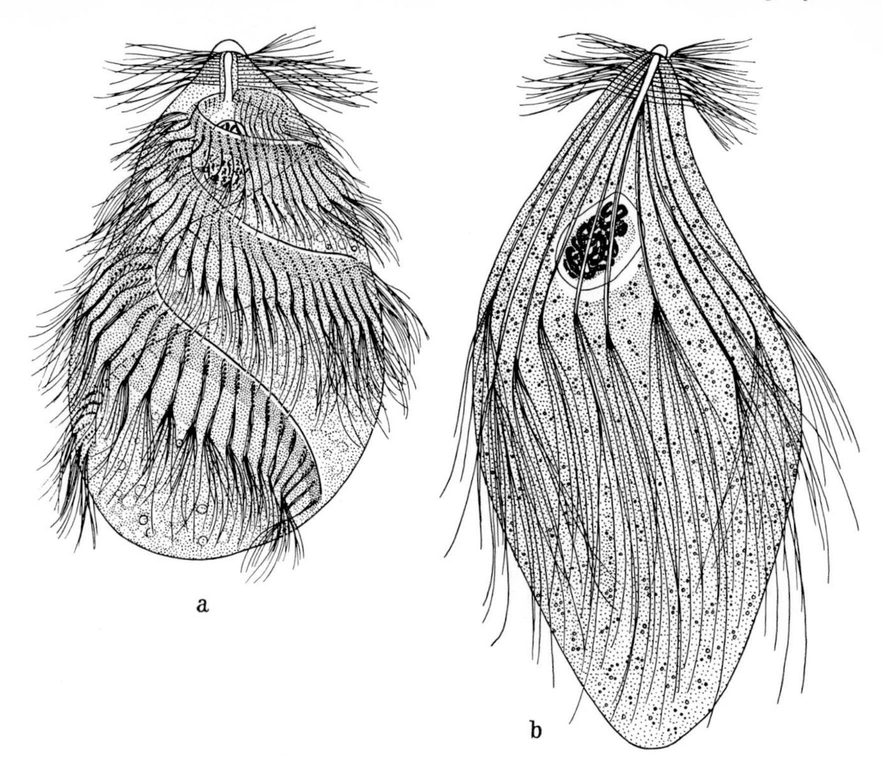

Fig. 158. *Leptospironympha wachula* (Polymastigina). a Normal cell. b Gamont. After CLEVELAND, 1948

is dissolved in the cytoplasm. Both daughter nuclei are now connected to their rostra by their own nuclear caps which are descended from that of the mother cell. The rostra can thus move independently within the cyst (b) together with their cell nuclei. With the elongation of the flagella their movement becomes more lively. At the same time the formation of the new parabasal bodies takes place (c). Eventually, the cytoplasmic bodies of both gametes are completely separated (d). Soon thereafter the gametes leave the gamont cyst whose wall is locally dissolved.

With respect to gamete differentiation, there are all intermediates ranging from isogamety to anisogamety. Oogamety, however, does not occur. While there is no difference between the gametes in *Saccinobaculus* (Fig. 160), *Oxymonas* and *Leptospironympha* (Fig. 163), the pairing gametes of *Eucomonympha* (Fig. 161), *Barbulanympha* (Fig. 162) and *Trichonympha* (Fig. 164) are usually of different size. There are also structural differences in *Trichonympha* which normally become apparent only after emergence from the cyst. The larger gamete forms numerous small granules which become arranged in a ring at its posterior end. Within this ring is the "fertilization cone", a hyaline region which can be withdrawn and extended. The smaller gamete has granules scattered in its cytoplasm, but no special differentiations at its posterior end.

The course of fusion is also quite different. In *Saccinobaculus* (Fig. 160), *Oxymonas*, *Eucomonympha* (Fig. 161) and *Barbulanympha* (Fig. 162) the cell bodies of the gametes can fuse at any point. It is characteristic of the first two genera that the

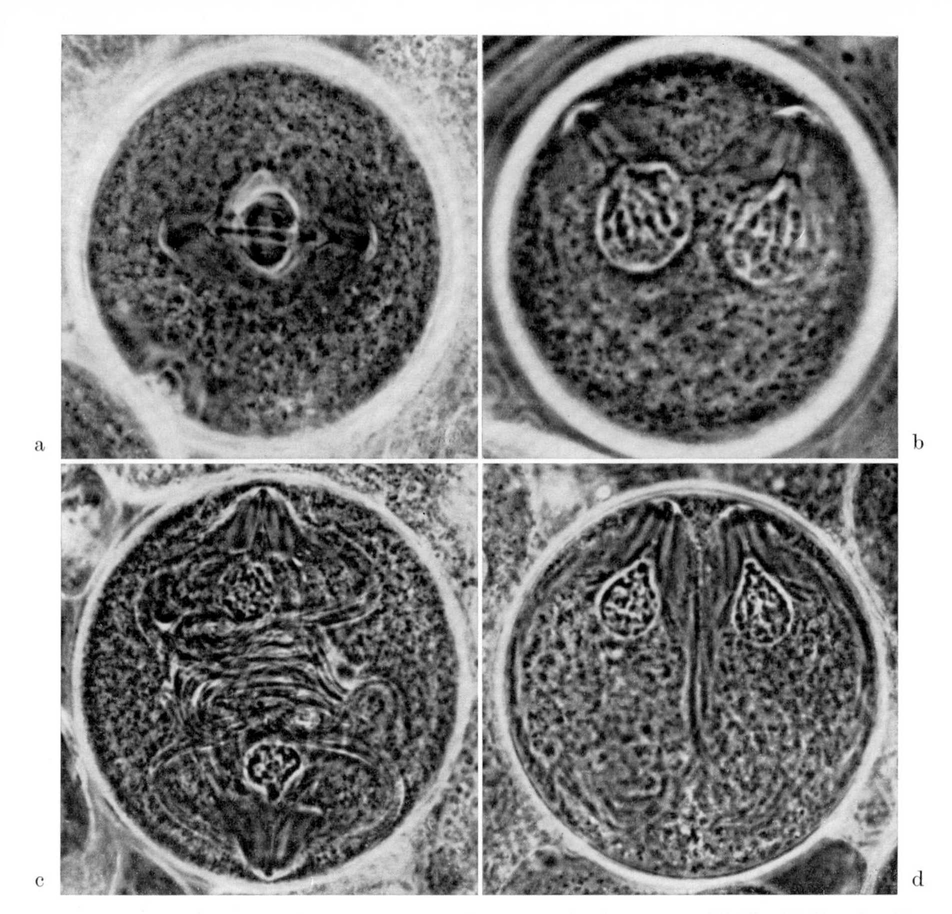

Fig. 159. *Trichonympha*. Stages of gamete formation. Gamogony mitosis (compare with Fig. 62). New daughter rostra have formed at the spindle poles. b The daughter rostra and the connected gamete nuclei move independently of each other. c The flagella originating at the daughter rostra have elongated. d The two gametes lie side by side and are still undivided in their lower halves. Photographed in life. × 600. After CLEVELAND, 1962 [*307*]

two axostyles also fuse after copulation. In *Eucomonympha*, the smaller gamete causes a local dissolution of the pellicle of the larger one. The gametes of *Leptospironympha* and *Trichonympha*, however, have a special zone of contact: one of the gametes attaches with its rostrum to the posterior end of the other one. In *Trichonympha* the fertilization cone is then withdrawn and the posterior gamete slips entirely into the anterior one.

Differences between the gametes are also expressed in the behavior of their nuclei. With the exception of *Saccinobaculus* and *Oxymonas* the nucleus of the smaller, respectively the posterior, gamete detaches itself from the organelles of the rostrum with which it is connected by the nuclear cap and migrates toward the other nucleus. The freed organelles are then resorbed in the cytoplasm of the zygote. This uniform behavior justifies a homology: the gamete whose organelles remain intact can be looked upon as "female" and the other one, whose organelles become resorbed, as "male". Once the pronuclei have touched, karyogamy follows. This takes a variable amount of time. In *Saccinobaculus*, the pronuclei lie next to each other

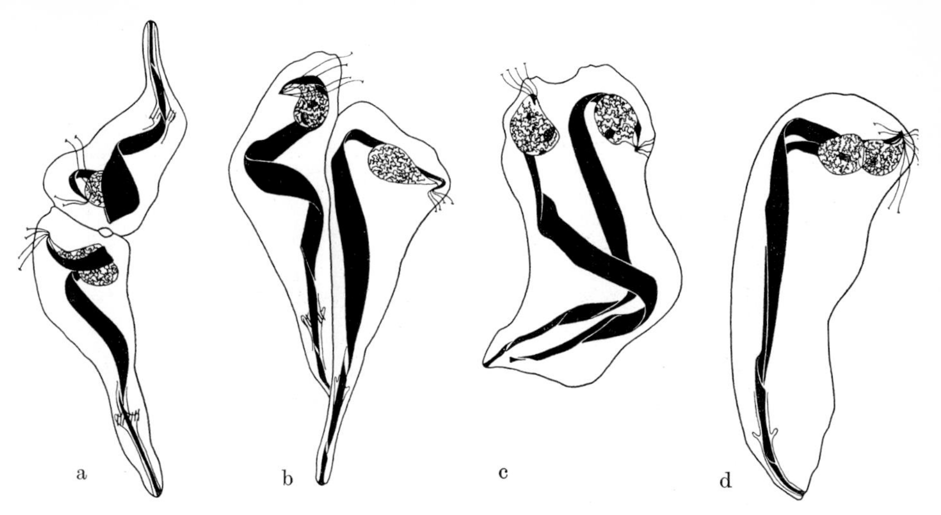

Fig. 160. *Saccinobaculus ambloaxostylus*. Different stages of fertilization. a The gametes fuse with the anterior ends. b Lateral union. c Fusion of the gametes completed. d Fusion of the axostyles, gamete nuclei united. × 530. After CLEVELAND, 1950 [*278*]

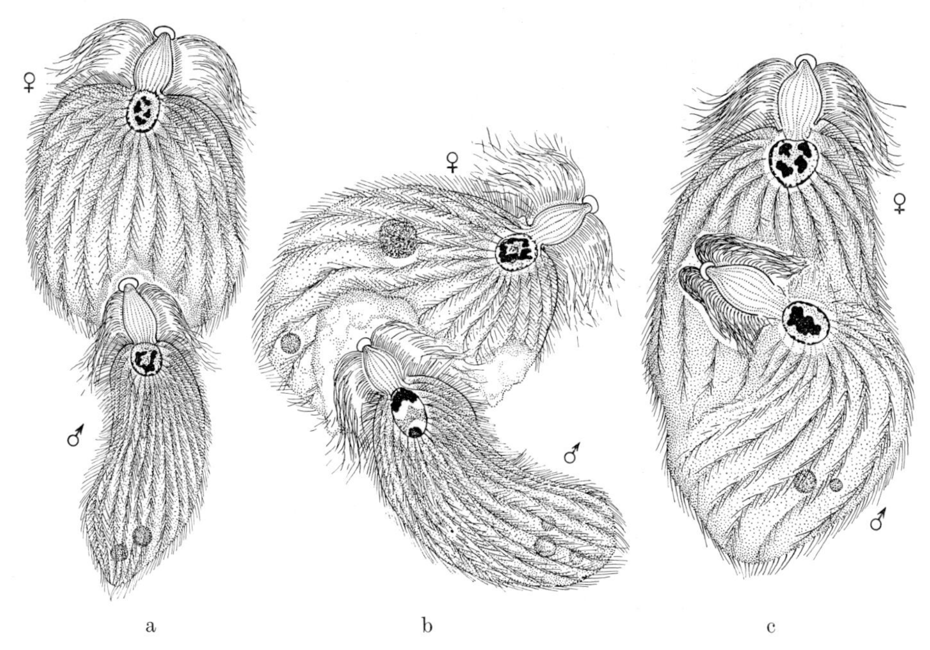

Fig. 161. *Eucomonympha imla*. Different stages of fertilization. × 460. After CLEVELAND, 1950 [*280*]

for several days before they finally fuse (Fig. 160d). In others, fusion takes place within a few hours.

While in haploids meiosis follows fertilization, it takes place in diploids prior to the formation of gametes. *Macrospironympha xylopletha* is, indeed, the only diploid polymastigote of *Cryptocercus punctulatus* which forms free-swimming gametes. In this case, sexual reproduction is initiated by the determination of a "progamont"

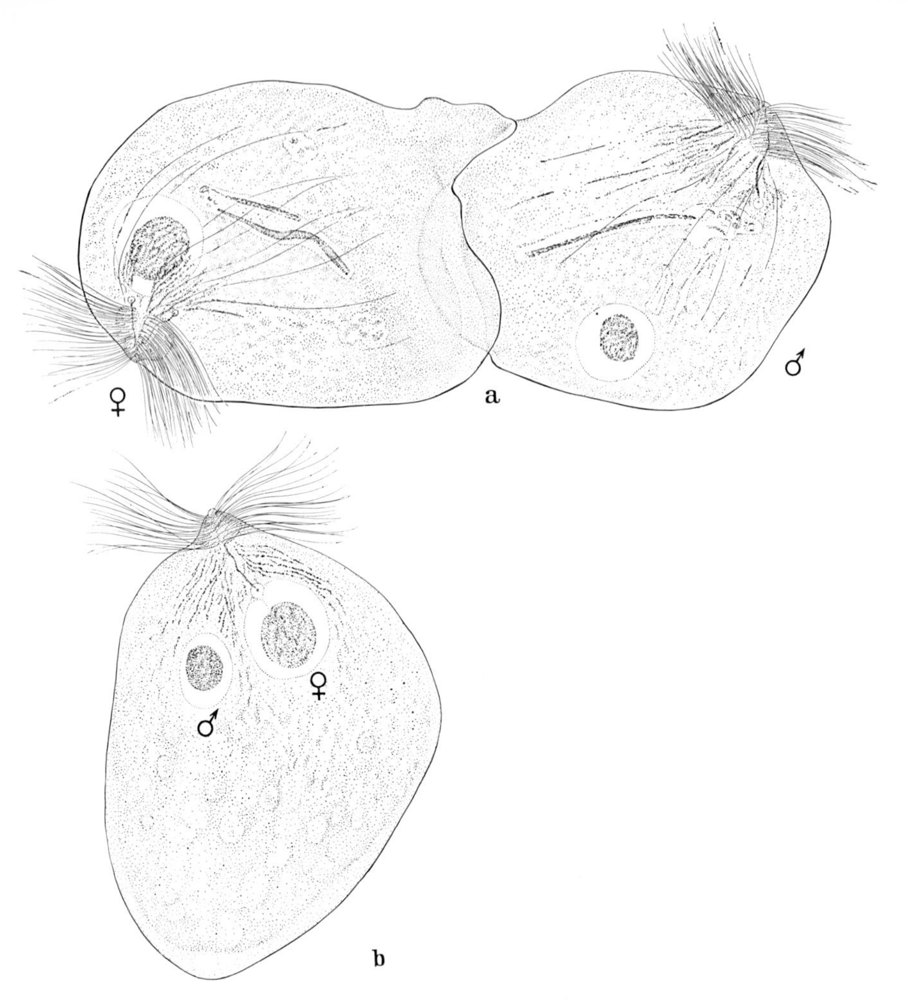

Fig. 162. *Barbulanympha*. a Fertilization. The nucleus of the male gamete has already given up its connection to the rostrum. b Zygote. The two pronuclei shortly before fusion. × 375. After CLEVELAND, 1953 [*284*]

which divides into two gamonts. This cell division is connected with the first division of meiosis. Then each gamont encysts and it is within the cyst that the gametes are formed by a differential division which corresponds to the second meiotic division. The two gametes of *Macrospironympha* differ only slightly in size, but even in this case the pronucleus of the smaller gamete separates from its organelles and migrates toward the pronucleus of the larger gamete.

As mentioned above, fertilization has recently also been observed among Polymastigina inhabiting the gut of termites. This holds for the genera *Trichonympha*, *Pseudotrichonympha*, *Deltotrichonympha*, *Koruga* and *Mixotricha*. It appears that

▶

Fig. 164. *Trichonympha*. Fertilization. a Male gamete with its rostrum attached to the protruded "fertilization cone" of the female gamete. b "Fertilization cone" of the female gamete withdrawn. Penetration of the male gamete into the female. c Complete fusion of the two gametes. The male gamete begins to lose its organelles. d Beginning fusion of the two pronuclei (note the difference in staining intensity!). × 375. After CLEVELAND, 1949 [*276*]

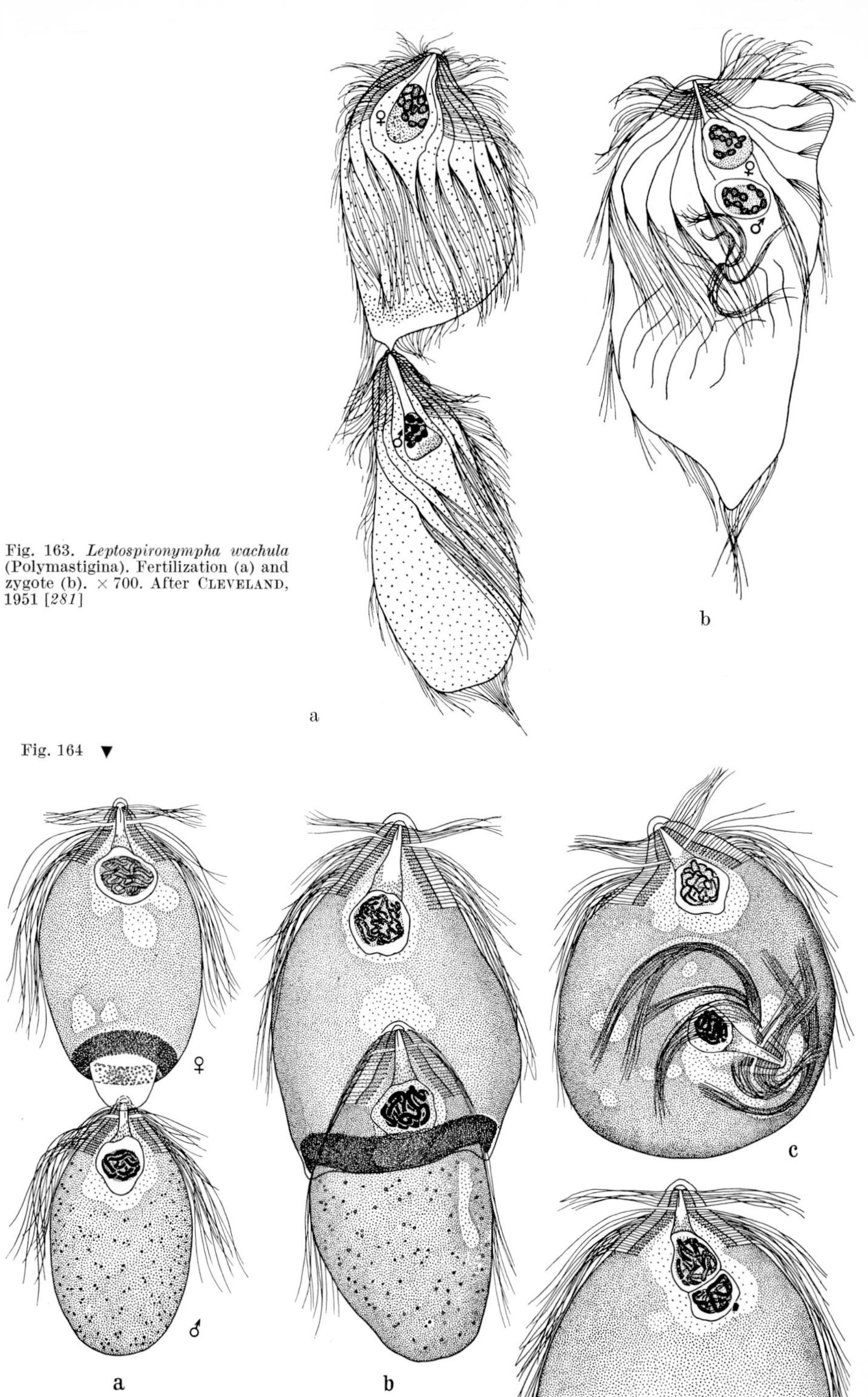

Fig. 163. *Leptospironympha wachula* (Polymastigina). Fertilization (a) and zygote (b). × 700. After CLEVELAND, 1951 [*281*]

Fig. 164 ▼

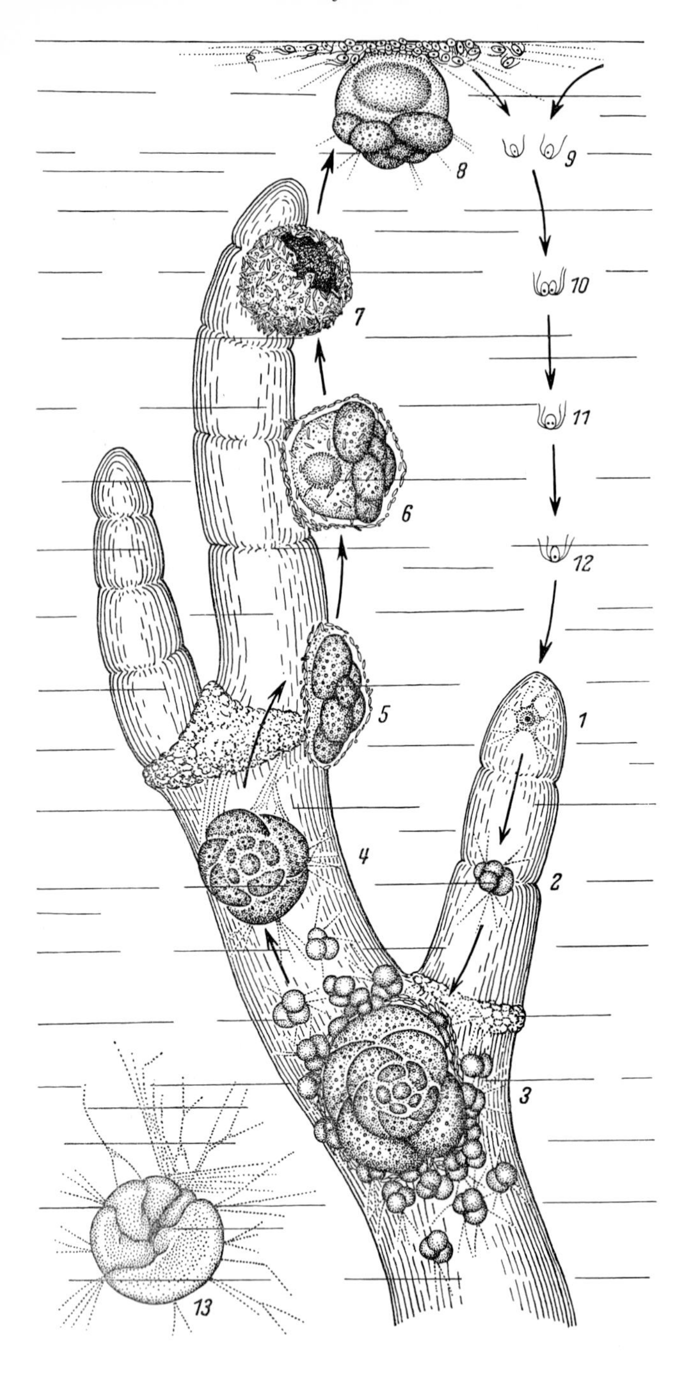

Fig. 165. *Tretomphalus bulloides*. Developmental cycle. *1* Amoeboid zygote. *2* Young (four-chambered) agamont. *3* Adult agamont forming agametes (gamonts). *4* Adult gamont. *5* The gamont surrounds itself with an envelope of detritus particles. *6* Formation of the "floating chamber" (compare with Fig. 353c). *7* Empty detritus envelope. *8* Gamont which has ascended to the surface of the sea, with swarming gametes. *9* Gametes. *10—12* Stages of fusion. *13* Gamont shown from below. After MYERS, 1943 [*1170*]

the sexual processes are also correlated with molting of the host in these cases [*312–314*].

Foraminifera

Among rhizopods the occurrence of free-swimming gametes was established conclusively in a few Foraminifera only. Since Foraminifera with sexual reproduction have an alternation of generations (p. 403), the gamonts representing the sexually reproducing generation, are not, as in Phytomonadina and Polymastigina, determined by external influences but arise by multiple fission of the agamonts. After the gamonts have reached a certain maximal size, they in turn undergo multiple fission. The entire cytoplasmic cell body is subdivided into gametes in the course of this fission and the empty shell is left behind.

The formation of free-swimming gametes was first observed in *Elphidium crispum (Polystomella crispa)*, the species where alternation of generations was first discovered. Later on, it became apparent that gametogamy is not restricted to polythalamic genera (*Peneroplis, Planorbulina, Discorbis, Orbulina, Tretomphalus* et al.) but that it also occurs in some monothalamic ones (*Myxotheca, Iridia* et al.).

In *Tretomphalus bulloides* (Fig. 165) the gamont, during its growth phase, creeps about at first on the substrate, e.g. an algal thallus. After it has reached a certain size, it surrounds itself with a sheath of detritus particles and forms a gas-filled floating chamber. Its specific weight thus decreased, it rises to the water surface where it releases the gametes. The empty shells of the gamonts (Fig. 353c) are occasionally seen drifting in large numbers in the water near the seashore and may form a regular seam where they are washed ashore.

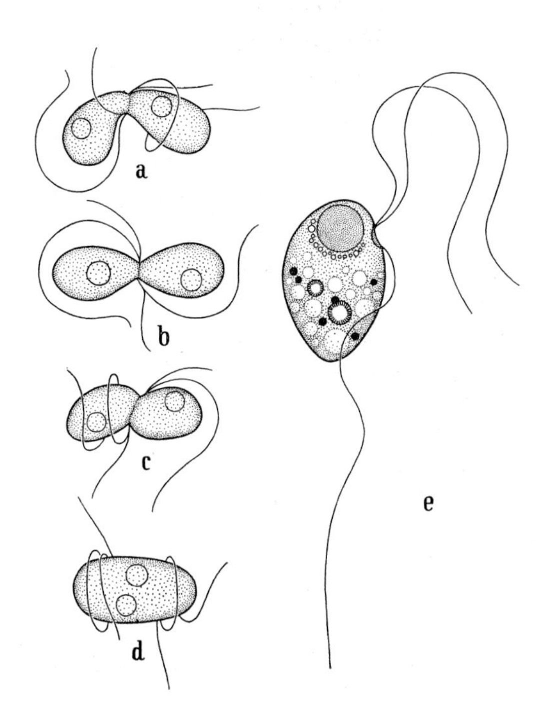

Fig. 166. Flagellated gametes of Foraminifera. a—d *Iridia lucida*. Stages of fusion. e *Discorbis mediterranensis*. Triflagellated gamete (this species is gamontogamous, see p. 185). e × 2100. After LE CALVEZ, 1938 and 1950 [*1003, 1004*]

The gametes formed by the gamonts of gametogamous species are very small (2–5 μm) and probably always biflagellated. Nothing is known concerning their sexual differentiation. It is likely that they are always isogametes. Their copulation has been described for *Iridia lucida* (Fig. 166a–d). The gametes formed by a single, isolated gamont of *Myxotheca arenilega* (Fig. 359) can copulate with each other [*658*]. The gamonts cannot therefore be sexually differentiated. In order to decide whether the gametes themselves are sexually differentiated, "left-over gametes" would have to be combined (see p. 163).

Sporozoa

With the exception of the Adeleidae, the formation of free-swimming gametes in Sporozoa is restricted to the Coccidia. The gamonts arise either directly from the sporozoites (Eucoccidia) or from the merozoites of a previous schizogony (Schizococcidia). While the "macrogamonts" are directly transformed into large, immotile macrogametes, the "microgamonts" undergo simultaneous multiple fission resulting in numerous small microgametes. *Oogamety* is thus characteristic for Coccidia. Sexual differentiation is often already recognizable in gamonts which may differ in size, form or structure *(anisogamonty)*. In *Eucoccidium dinophili* (Fig. 395), the macrogamonts become large and cigar-shaped in the course of growth but assume an oval shape when they transform into fertilizable macrogametes. The microgamonts transform soon into rounded cysts within which the microgametes originate.

While the microgametes of *Eucoccidium dinophili* and *Coelotropha durchoni* are cup-shaped and biflagellated, they have an elongated shape in most Schizococcidia. Species of *Eimeria* have microgametes with three flagella. As an example shows (Fig. 167), one flagellum may be more or less attached to the cell body, whose

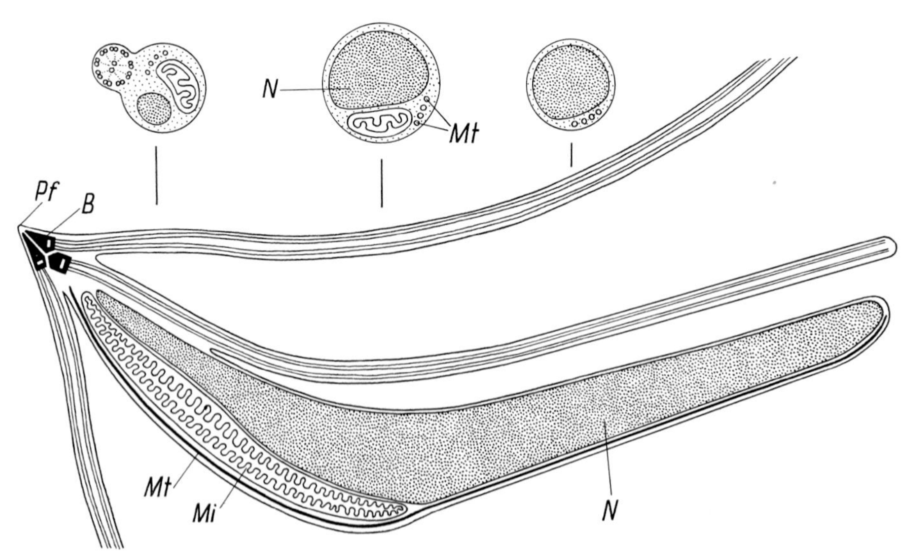

Fig. 167. *Eimeria maxima*. Diagram of microgamete in a longitudinal section and three cross sections. *N* Nucleus. *Mi* Mitochondrion. *Mt* Subpellicular microtubules. *Pf* Perforatorium. *B* Basal body. After SCHOLTYSECK, MEHLHORN, and HAMMOND, 1972 [*1532*]

space is occupied mostly by the nucleus (*N*) and a large mitochondrion (*Mi*). Microtubules (*Mt*) extend posteriorly beyond the mitochondrion. In *Aggregata eberthi* (Fig. 398) one of the three flagella is transformed into an undulating membrane. The thread-like microgametes of the Haemosporidae, on the other hand, seem to consist essentially only of a flagellum with the nucleus lying between the fibrillar bundle and the flagellar sheath [*564*].

In order to reach the macrogametes, the microgametes have to traverse a certain distance. It is very probable that the macrogametes release a substance which attracts the microgametes chemotactically. In *Eucoccidium dinophili*, where fertilization takes place within the body cavity of the host, some of the chemotactic substance extruded by the macrogamete is probably still clinging to the envelope of the oocyst, since it is usually surrounded by microgametes in lively motion (Fig. 168). Evidence for a chemotactic substance released by the macrogametes, has also been given for *Coelotropha durchoni* [*1326*].

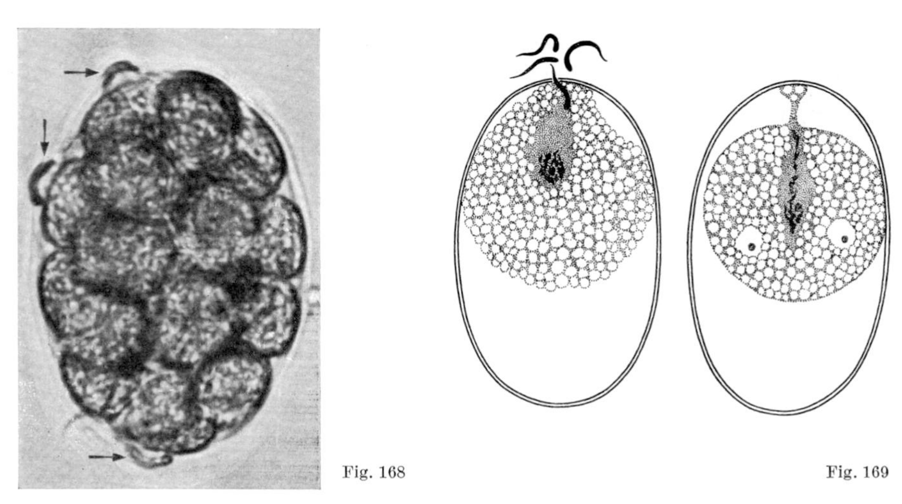

Fig. 168 Fig. 169

Fig. 168. *Eucoccidium dinophili.* Oocyst with spores, surrounded by microgametes (arrows!). The oocyst was pressed out of the body cavity of the host and lies free in the seawater. Photographed in life

Fig. 169. *Eimeria propria.* Fertilization. After SIEDLECKI from DOFLEIN-REICHENOW, 1949

Prior to fertilization the nucleus of the macrogamete moves to the periphery of the cell. At this point, where the cell envelope is temporarily dissolved, the whole microgamete penetrates into the macrogamete (Fig. 169). After karyogamy, the synkaryon becomes elongated. This stage, formerly referred to as "fertilization spindle", initiates the prophase of meiosis (p. 95). Meiosis in turn leads to sporogony.

As in phytomonads (p. 166), *parthenogenesis*, i.e. development of macrogametes without fertilization and meiosis, seems to take place occasionally in Coccidia. It has been proved to occur in *Eucoccidium dinophili* whose development can be followed in life (Fig. 170). When the macrogamont transforms into a macrogamete it rounds off and shortens (a, b). The nucleus then moves to the periphery of the cell where it looks like an indentation of the cytoplasm if viewed from one side (c).

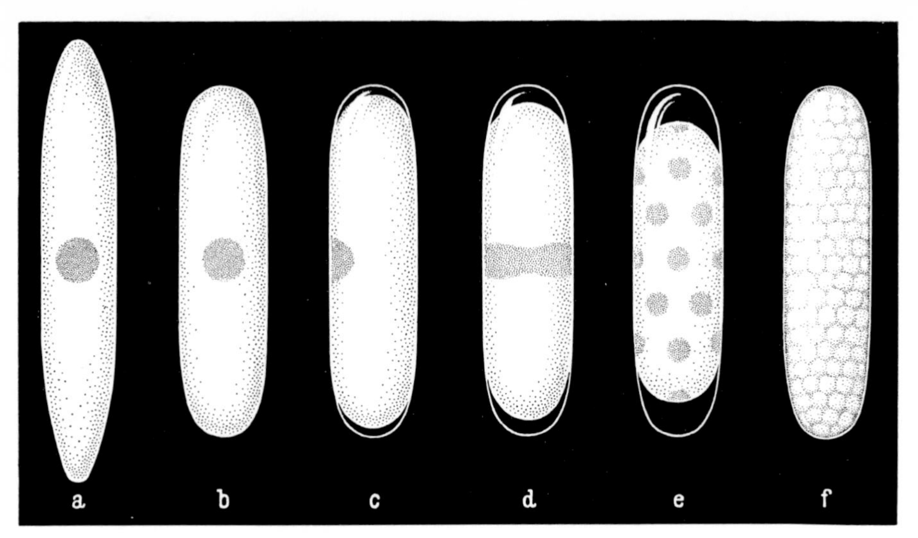

Fig. 170. *Eucoccidium dinophili*. Parthenogenetic development of the macrogamont. Life observations using incident light. After GRELL, 1953 [*657*]

Without assuming the characteristic shape of a "fertilization spindle", it divides first into two daughter nuclei (d), thus initiating sporogony which leads to the formation of spores (e, f). As after fertilization, the cytoplasm contracts in parthenogenetic development at the onset of sporogony, leading to a clearly recognizable space below the cellular envelope.

Parthenogenesis is a common mode of reproduction in *Eucoccidium dinophili*. It takes place whenever the host has been infected with only one or with few spores. In *weak* infections all the sporozoites within the host's body cavity develop into macrogamonts. It is only after a *heavy* infection that microgamonts are also formed, thus permitting bisexual reproduction. Even if the spores formed parthenogenetically from a single sporozoite are used to obtain a heavy infection, both sexes appear in the body cavity of the newly infected host. The determination of sporozoites into microgamonts must therefore be due to conditions which arise within the body cavity of the host only in the case of a heavy infection. Even though the nature of these factors is not known as yet, it is certain that *sex determination* is *phenotypic*. If it were genetic, the offspring of a single sporozoite could certainly not produce both sexes.

II. Autogamy

Autogamy is a process of fertilization in which only the gametes or gamete nuclei derived from the same gamont participate. Autogamy is thus a matter of obligatory monoecy. Whether there is any sexual difference between the gametes or gamete nuclei has to be tested in each individual case.

The possibility that a gamont gives rise to *gametes*, i.e. whole cells which fuse with each other, is realized in some Heliozoa and Foraminifera.

The only *Heliozoa* in which fertilization has been described in detail thus far are

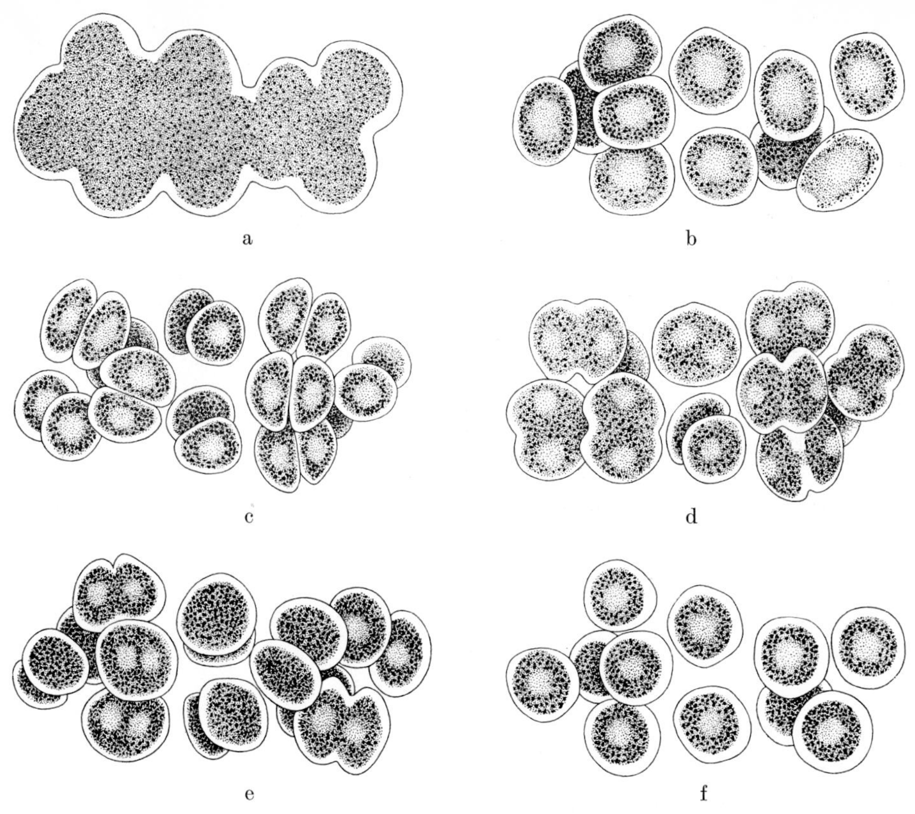

Fig. 171. *Actinosphaerium eichhorni*. Sexual reproduction (autogamy). a Flattened mother cell, lying on the bottom. Start of multiple fission. b Daughter cells (6 hours later). c Progamic division (15 hours later). d Gamonts during meiosis (8 hours later). e Fusion of the gametes (16 hours later). f Zygotes (9 hours later). Observations in life. After R. HERTWIG, 1898 [*754*]

the two fresh-water species *Actinosphaerium eichhorni* and *Actinophrys sol*. Sexual reproduction takes place when food suddenly becomes scarce after a period of multiplication.

Actinosphaerium eichhorni, a large, multinuclear species, then sinks to the bottom, withdraws its axopodia and flattens to an irregularly shaped mass which is surrounded by a gelatinous covering (Fig. 171a). Most of the numerous (up to 500) nuclei of the floating form degenerate. In an ensuing simultaneous multiple fission the cytoplasm is subdivided into as many single cells as there are surviving nuclei (b). Each cell becomes surrounded by a cyst envelope within which it divides into two daughter cells (c, d). The daughter cells, which may be called gamonts, now undergo meiosis and thus become gametes. These fuse within the cyst to form a zygote (e, f). The zygote is a resting stage. Under favorable conditions the young *Actinosphaerium*, already multinucleated, emerges from the zygote.

Sexual reproduction takes a simpler course in *Actinophrys sol* since this species is uninucleate. Each individual behaves as a cell of *Actinosphaerium eichhorni* after multiple division (Fig. 171b): it encysts and divides within the cyst into two daughter cells, the gamonts (Fig. 172a, b). The nucleus of each gamont then carries

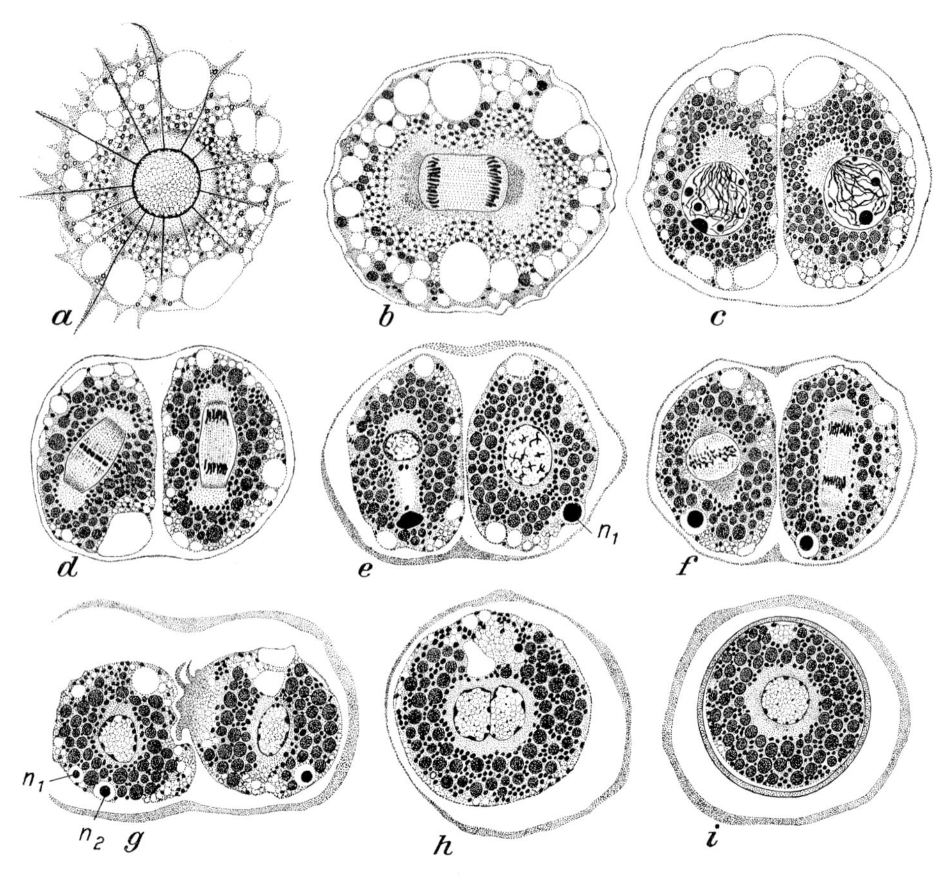

Fig. 172. *Actinophrys sol.* Sexual reproduction (autogamy). a Single cell. b Encystment and division of the two daughter cells. c—e First meiotic division (degeneration of one of the daughter nuclei, n_1). f, g Second meiotic division (degeneration of one of the daughter nuclei, n_2) and sexual differentiation of the gametes (pseudopodia formation in one of the gametes). h Karyogamy. i Zygote. × 830. After BELAR, 1922 [*132*]

out two meiotic divisions, whereby one of the two daughter nuclei becomes pycnotic after each division and degenerates (c—f). Although the gametes are morphologically identical (isogamety), they differ in behavior. One of the two gametes — probably the first to complete meiosis — forms a pseudopodium while the other one withdraws somewhat at this spot (g). The zygote is formed by fusion of the cells and karyogamy (h). It is also in this species a resting stage (i).

That the formation of the pseudopodium is a specific property depending on divergent differentiation of the gametes and not a matter of chance is indicated by observations in life. If the pseudopodium is formed by way of exception at the side away from the partner, it is resorbed and reformed at the "correct" place. Occasionally, two individuals will join at the beginning of sexual reproduction and proceed to form a common cyst. If the gametes formed by them happen to be arranged in series with active and passive gametes alternating, the resulting fertilization may be allogamic instead of autogamic (Fig. 173).

Gametes which fail to meet a partner or are incapable of fusion for other reasons

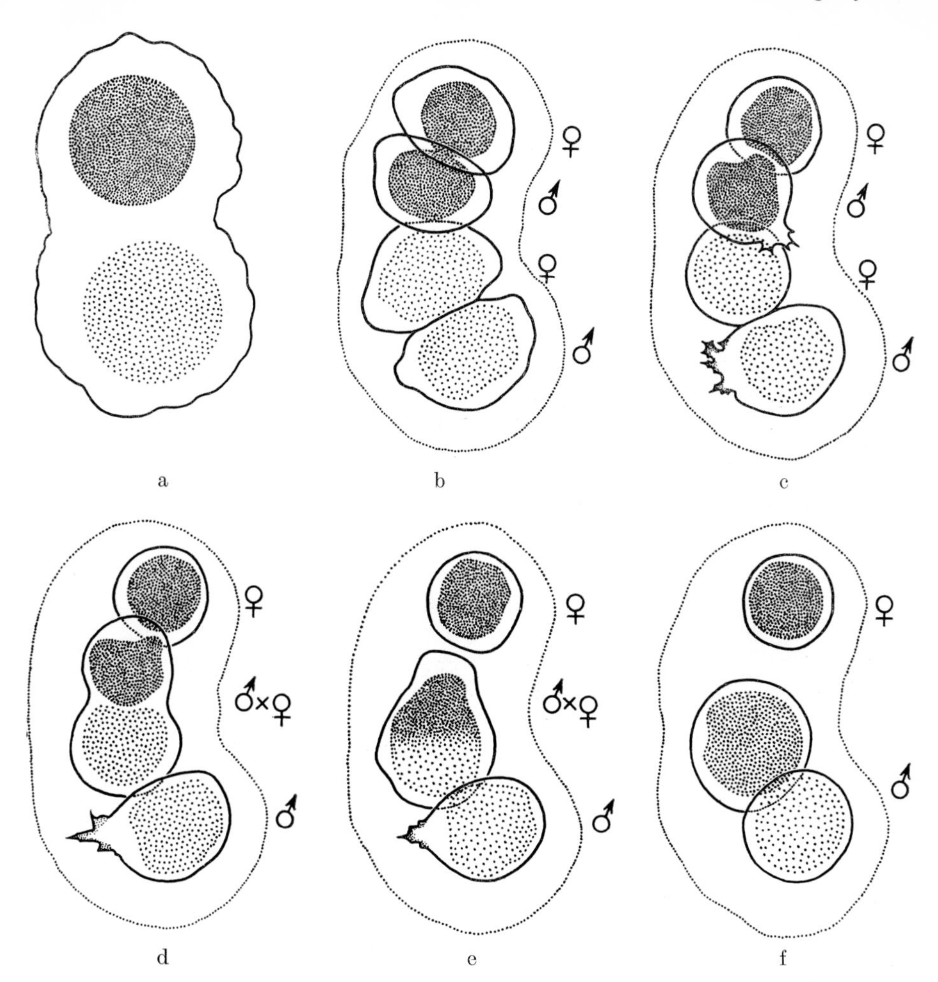

Fig. 173. *Actinophrys sol*. Fusion of two gametes, descended from different previously fused mother cells (diagrammatically indicated by dense and light stippling). Observations in life. × 600. After BELAR, 1922 [*132*]

can surround themselves with a cyst envelope of their own. In the case shown in Fig. 174 three individuals are joined; two of them carried out autogamy while the gametes of the third one have encysted independently.

The encysted gametes can develop further by *parthenogenesis*. Although only a fraction germinate and of these many die off, indefinitely multiplying clone cultures can be obtained from the survivors. It is likely that the chromosome number in the survivors is regulated to diploidy.

If sexual differentiation of the gametes is accepted as a fact, then sex determination must be connected with the progamic division which follows encystment. Thus, this could be looked upon as a differential division.

Among the *Foraminifera* one monothalamic species, *Allogromia laticollaris*, has been shown to be autogamic. Details of gamete formation and sexual fusion are not known [*67*].

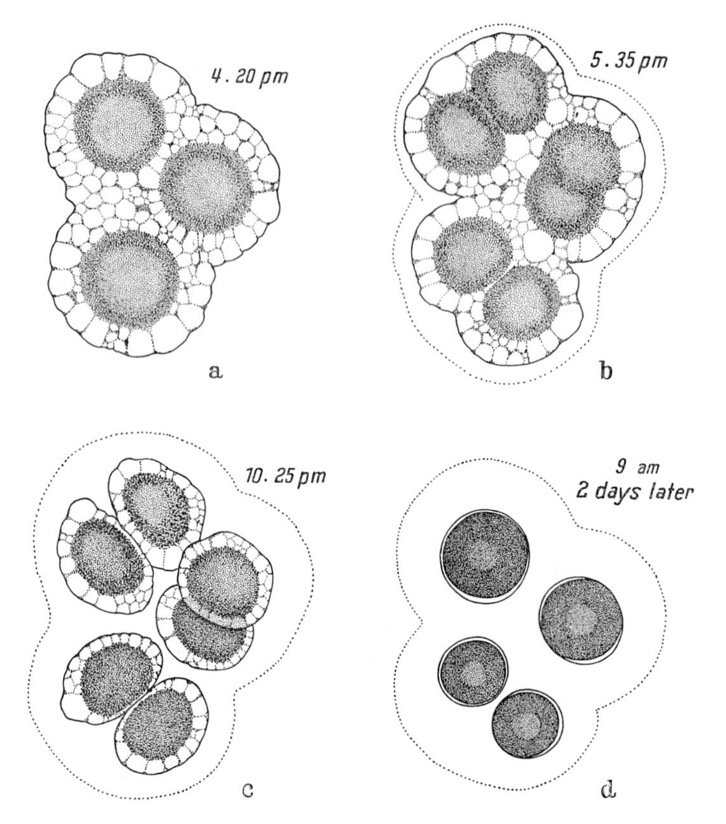

Fig. 174. *Actinophrys sol.* a Three cells are joined with their ectoplasmic layers ("plasmogamy"). b Each cell divides into two daughter cells. c After meiosis, a gamete originates from each daughter cell. d While the gametes of the two top pairs fuse and produce large zygotes surrounded by a cystic envelope, the gametes of the lower pair have not fused. However, they encyst also and develop further parthenogenetically. Observations in life. × 300. After Belar, 1922 [*132*]

More detailed studies are available concerning autogamy in the small polythalamic species of *Rotaliella*. As in all Foraminifera, the gamont is uninucleate. Gamogony consists of two phases and proceeds after the gamont has reached a certain critical size (Fig. 175). In the first phase the nuclear divisions are strikingly asynchronous (a—c). Eventually, all nuclei reach a resting stage (d). Their number may either be even or uneven. In the second phase there is one synchronous nuclear division (e) which leads to an even number of gamete nuclei (f). Subsequently, the gamont cleaves into a corresponding number of gametes (Fig. 176a) which are morphologically alike. They exhibit amoeboid motion and fuse in pairs to form zygotes (b, c).

The gamete nuclei in gamonts of *Rotaliella heterocaryotica* differ occasionally during karyogamy in their degree of condensation, one being always more condensed than the other (Fig. 177). It can only be assumed that this represents an expression of sexual differentiation. If this is indeed the case, it would seem likely that sex determination occurs in the last (synchronous) mitosis of gamogony. This would then take on the character of a differential division just as the progamic mitosis of Polymastigina and Heliozoa.

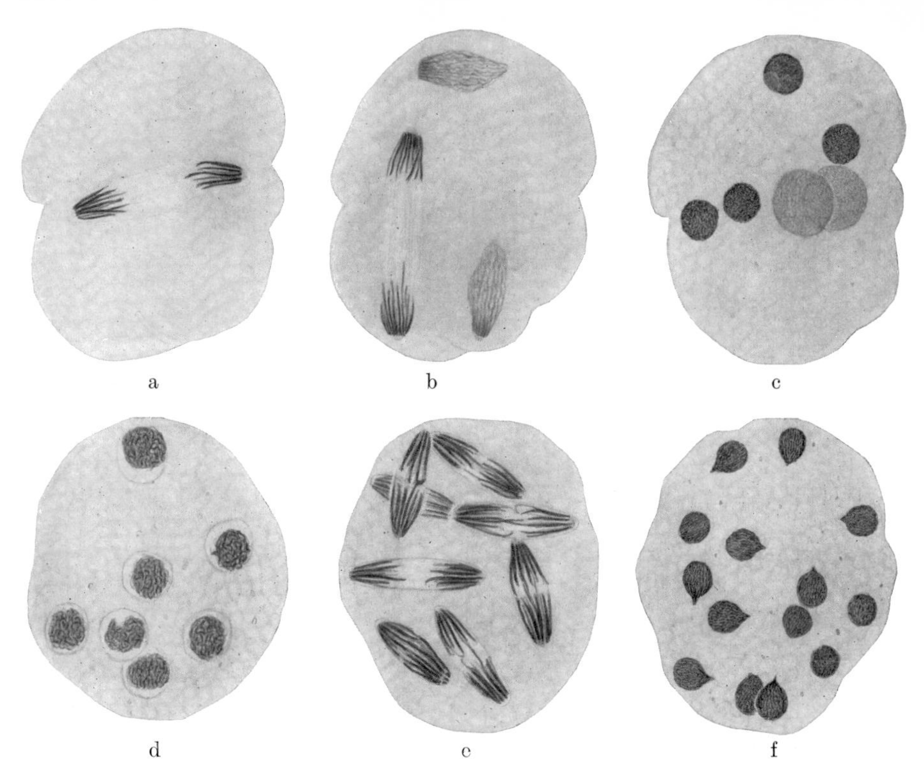

Fig. 175. *Rotaliella roscoffensis.* Gamogony. a—c Phase of asynchronous nuclear divisions. d Resting stage of the nuclei. e Synchronous division of all nuclei. f Gamete nuclei. Bouin-Duboscq, Feulgen staining. × 900. After GRELL, 1957 [*663*]

In a second mode of autogamy the gamont forms only *gamete nuclei* which unite pairwise to form synkarya. Besides gametogamy, this type of autogamy is frequently found in the haploid polymastigid genera *Saccinobaculus*, *Oxymonas* and *Barbula-*

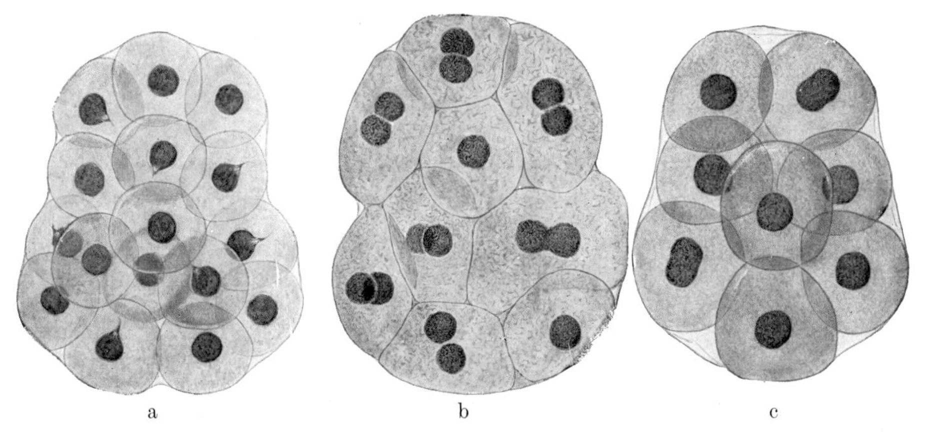

Fig. 176. *Rotaliella heterocaryotica.* Autogamy. a Gametes. b Karyogamy (two of the gametes still unfused). c Zygotes. Bouin-Duboscq, Feulgen staining. × 900. After GRELL, 1954 [*659*]

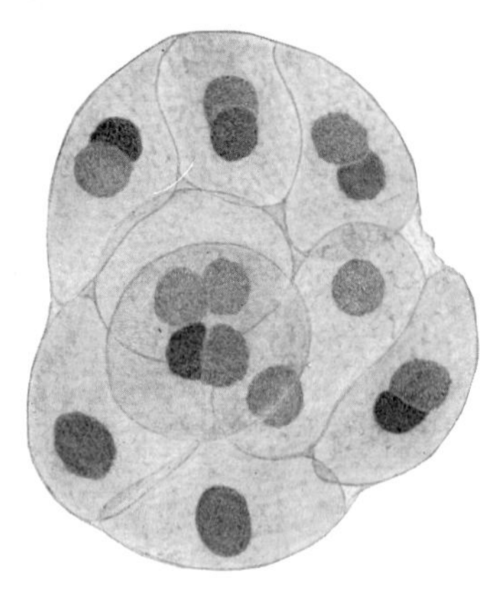

Fig. 177. *Rotaliella heterocaryotica*. Gamont with zygotes in karyogamy. The gamete nuclei show different degrees of condensation. Bouin-Duboscq, Feulgen staining. × 800. After GRELL, 1958 [*666*]

nympha from the roach *Cryptocercus punctulatus* (see Table 3, p. 168). Cell division seems suppressed in these cases and only a differential nuclear division takes place. Both gamete nuclei will then fuse within the same cell.

In *Barbulanympha* (Fig. 178) although the rostrum divides, one of the two daughter nuclei gives up its connection to the daughter rostrum with which it would otherwise remain in contact and migrates toward the other daughter nucleus. The two pronuclei thus exhibit during autogamy the very same difference in behavior as the gamete nuclei during copulation (p. 170).

A further simplification (which can scarcely still be considered "normal") consists in the total omission of nuclear division. What remains is the reduplication of the chromosomes without spindle formation, i.e. an *endomitosis*. This is evidently always the case if one of the centrioles degenerates prematurely. The "synkaryon" formed in this manner will later on carry out meiosis just as after gametogamy or autogamy [*286*, *288*].

In the diploid genera *Rhynchonympha* and *Urinympha* autogamy is the only mode of sexual reproduction. In *Rhynchonympha*, meiosis consists of two divisions, the first one being connected with cell division while the second (differential) division

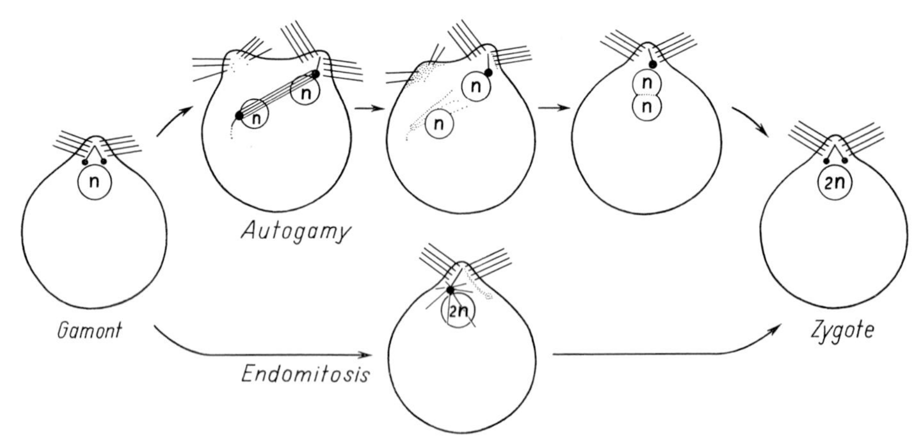

Fig. 178. *Barbulanympha*. Modifications of sexual reproduction due to omission of cell division (→ autogamy) and of nuclear division (→ endomitosis) of the gamonts. Based on a diagram by CLEVELAND, 1956 [*292*]

is only nuclear. The two nuclei then fuse in autogamy. In *Urinympha*, meiosis involves only one division. The two gamete nuclei fuse right away to form a synkaryon [*282, 283*].

In *ciliates*, where conjugation is the predominant mode of fertilization, autogamy is also occasionally encountered. The micronucleus undergoes the same divisions as in conjugation; however, the migratory nucleus is not transferred into another cell but fuses with its stationary sister nucleus. In most cases (e.g. *Paramecium aurelia, Euplotes minuta*) autogamy occurs along with conjugation [*395, 1595*]. Occasionally two cells attach as in conjugation but fail to exchange migratory nuclei, each partner carrying out autogamy. This modification of autogamy has been called "cytogamy" (Fig. 193). Both autogamy and cytogamy have been observed in *Paramecium polycaryum*, but conjugation has not [*402, 403*]. Some ciliates, e.g. *Tetrahymena rostrata*, carry out autogamy within a cyst [*338*].

Further studies will probably show that autogamy occurs in many ciliate species. Proof of autogamy requires close cytological control in clone cultures.

III. Gamontogamy

If sexual reproduction is initiated by the union of the gamonts, we can speak of gamontogamy. Either the gamonts form gametes by multiple fission or the gamonts fuse directly and form only gamete nuclei.

In cases of gamontogamy, the question of sexual differentiation arises both with respect to the gamonts and the gametes or gamete nuclei.

1. Gamontogamy with Gamete Formation

Foraminifera

In many polythalamic Foraminifera fertilization is initiated by the union of two or more gamonts to form an *aggregate*. After gamogony, the gametes fuse in the common space of the aggregated gamont shells.

It is highly probable that gamontogamy in Foraminifera has developed from gametogamy. In the genus *Discorbis* there are gametogamous *(D. vilardeboanus)* as well as gamontogamous species *(D. opercularis, D. patelliformis, D. mediterranensis)*. The latter form several hundred flagellated gametes which swim about within the shells of the gamonts. Indeed, the gametes of all gamontogamous species have three flagella (Fig. 166e). This also holds for *Glabratella sulcata* (Fig. 367) which is related to the genus *Discorbis*.

The remaining gamontogamous Foraminifera studied thus far form, on the other hand, but a few amoeboid gametes and are in this respect like the autogamous species (p. 182). They include *Patellina corrugata* (Fig. 360), *Spirillina vivipara, Metarotaliella parva* (Fig. 363), *Metarotaliella simplex* and *Rubratella intermedia* (Fig. 365).

In the simplest case *only two* gamonts mate. They creep toward each other, touch with their ventral surfaces and extrude a substance which firmly cements them together (Fig. 179 and 180). In the course of gamogony the inner walls separating the chambers of the gamonts dissolve, providing a common space within which the

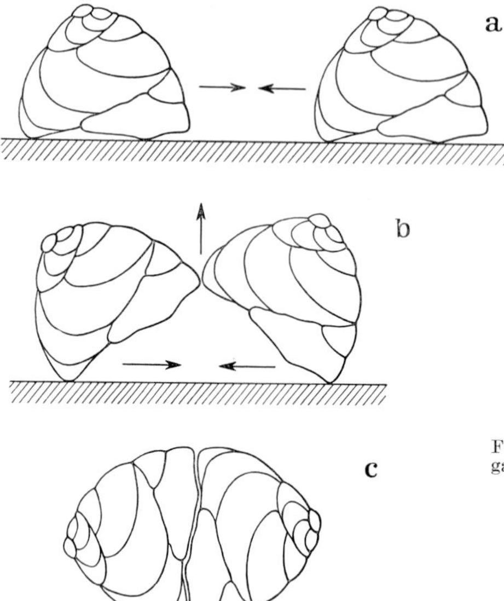

Fig. 179. *Discorbis mediterranensis*. Aggregation of two gamonts. After LE CALVEZ, 1950 [*1004*]

gametes can pair. In many species aggregates of *several* gamonts are regularly found. In *Patellina corrugata*, where the number of aggregating gamonts ranges from 2–14, the gamonts cover each other by the rims of their tests and form a thin organic membrane which fastens them to each other and to the substrate. All further events take place in this space roofed over by the gamonts (Fig. 181).

As soon as the gamonts have united (Fig. 360) some asynchronous nuclear divisions take place. In this way each gamont acquires between two and five nuclei depending upon its size. At this stage the protoplasts emerge from their shells and round off. Each protoplast is then subdivided into as many parts as the number of nuclei. Each of these parts divides to form two gametes.

The gametes are drop-shaped and thus clearly polarized. They move about for a time in the space provided by the empty gamont shells before fusing in pairs to form zygotes. As in all other Foraminifera studied, they are isogametes.

Many details of gamete formation and fertilization can be observed in life in *Patellina corrugata*. Fig. 182 shows the copulation of two gamete pairs taken from a time-lapse movie. In this case six zygotes were formed and two gametes did not fuse.

Such left-over gametes are frequent in *Patellina corrugata*. They are ingested by the zygotes. Only the nuclei are usually not ingested (see p. 91 for exceptions) and remain outside as pycnotic bodies.

Gamogony takes a wholly different course in *Metarotaliella parva* (Fig. 183). Usually, only two gamonts unite, aggregates of three being rare. The first phase of gamogony takes place in the initial chamber (a, b); even distribution of all nuclei throughout the cytoplasm follows (c). The last mitosis of gamogony, though synchronous in each gamont, begins a little earlier in one of the gamonts (d). This

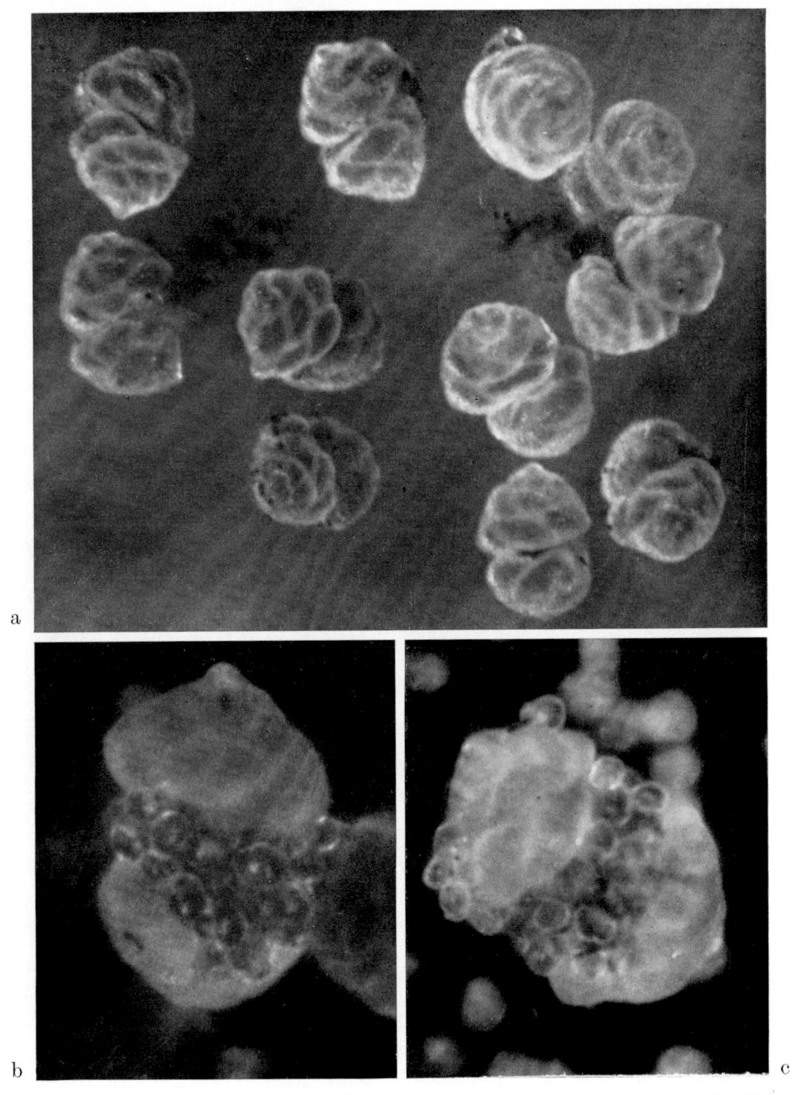

Fig. 180. *Glabratella sulcata*. Mating of the gamonts. Photographed in life. a Several gamont pairs. × 140. b, c Hatching of young agamonts. × 220. After GRELL, 1958 [*665*]

is also the gamont which begins with the amoeboid form changes which take place within the common space of the shells. Each gamont spreads into the shell of its partner, only a part of the plasmatic body remaining in its own shell. In this manner the gametes of both gamonts become thoroughly mixed (e–h). They then fuse in pairs to form zygotes (i).

The diagrammatic representation in Fig. 183 shows gamonts of equal size, each forming the same number of gametes. However, in *Metarotaliella parva* the gamonts are often of different size and form different numbers of gametes. Some gametes may thus be left without a mate, as described for *Patellina corrugata*

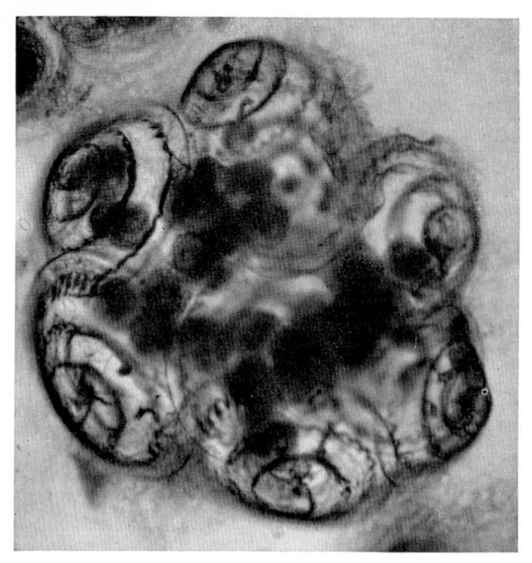

Fig. 181. *Patellina corrugata*. Aggregate of nine gamonts (three outside the focal plane). Photographed in life. × 120

Since neither the aggregating gamonts nor the copulating gametes are morphologically different in Foraminifera, the problem of their sexual differentiation can only be solved by indirect methods.

Combination tests with the gamonts of *Discorbis mediterranensis* [*1004*] and *Patellina corrugata* [*667*] led to the result that the gamonts are differentiated into two types designated as the "+" and "−" sex respectively. When two gamonts are put into a culture dish they either move toward each other and attempt to form an aggregate or they behave indifferently. After separation of the aggregating gamonts, their behavior toward other gamonts can be tested. By combination experiments of this type it can be decided whether all the gamonts derived from the same agamont are of one or the other sex.

In aggregates of *Patellina corrugata*, which frequently comprise several gamonts, both sexes are always present, though in varying proportions. Aggregates of three must of course include two gamonts of the same sex (Fig. 360).

The assumption that the gametes of *Patellina corrugata* are also unable to combine randomly is supported by the observation that in many of the aggregates left-over gametes are found which do not fuse with each other. Through analysis of time-lapse motion pictures the developmental history of zygotes and residual gametes could be traced back to the gamonts. In the triple aggregate shown in Fig. 184 six zygotes were formed and two gametes were left over. Analysis showed that gamont 1 had formed six gametes whereas gamonts 2 and 3 had given rise to four gametes each. Every gamete of gamont 1 took part in the formation of the six zygotes, besides all four gametes of gamont 2 and two of the gametes of gamont 3. The two left-overs were from gamont 3. Thus, if gamont 1 is of the "−" sex and gamonts 2 and 3 of the "+" sex, only gametes from gamonts of different sexes have united, the two left-overs belonging to the same sex.

In an aggregate of two, if one separates both gamonts by glass fibers, the gametes

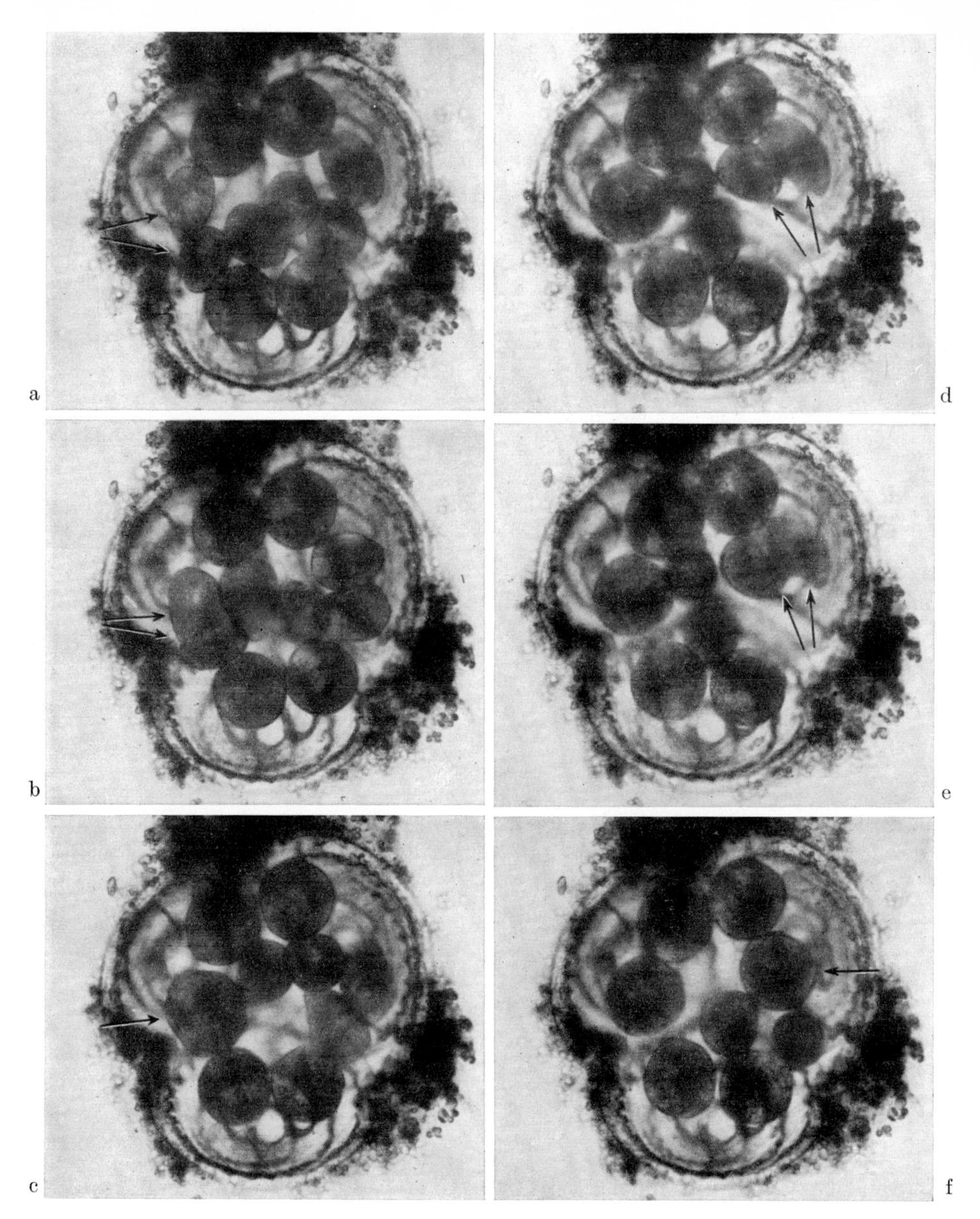

Fig. 182. *Patellina corrugata.* Aggregate of three gamonts. a Four zygotes, six gametes. b—e Fusion of two gamete pairs (arrows). f Six zygotes, two left-over gametes. × 120. From the films C802 and E258 (GRELL)

being produced by them cannot copulate – neither those of the different gamonts nor those of the same gamont. After some creeping around all gametes of each gamont finally unite to form a multinucleated plamodium which later dies.

Hence, *Patellina corrugate* is a dioecious species, probably with genetic sex determination [*140*].

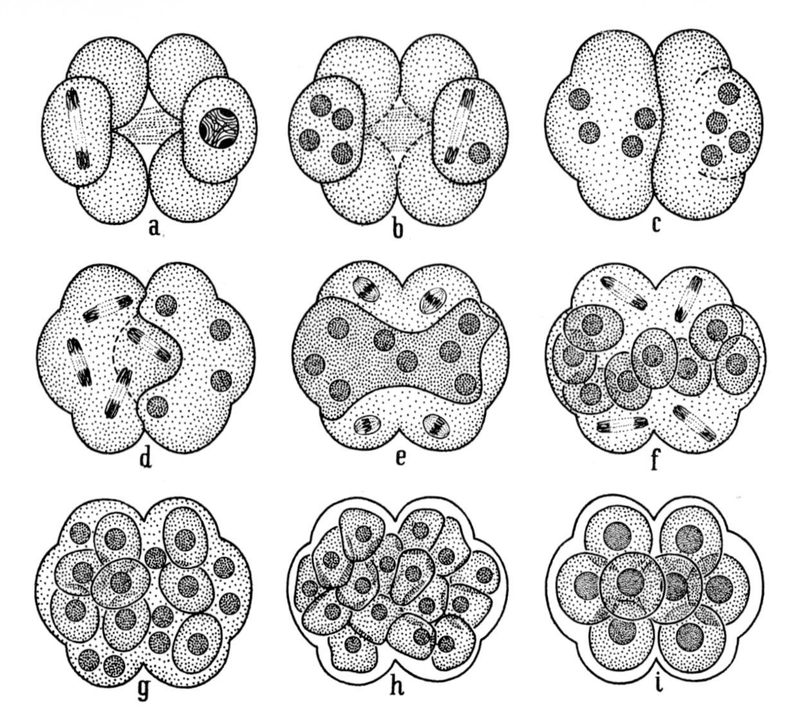

Fig. 183. *Metarotaliella parva*. Diagram of gamogony and fertilization. Explanation in the text. After WEBER, 1965 [*1789*]

There is also evidence in *Rubratella intermedia* that a gamete cannot mate with every other gamete. This foraminiferan forms only aggregates of two. The gamonts may be of equal or of unequal size (Fig. 185). In the first case, the gametes and their nuclei are also of equal size (a—d). In the second case, a mechanism of regulation becomes effective: Although both gamonts form the same number of gametes, those of the little gamont are smaller and their nuclei are more condensed. At karyogamy, a large and a small nucleus fuse (a′—d′). Thus, it is evident that mating takes place only between the gametes of different gamonts.

However, conclusions regarding the sexual differentiation of gametes cannot simply be extended to include the gamonts, as shown by experiments with *Metarotaliella parva* [*1789*].

It could be established conclusively that gamonts of this foraminiferan are *not sexually differentiated* before mating. Any two gamonts which are put into the same culture dish can form an aggregate.

The possibility of combinations is thus not restricted by the differentiation of types.

The question arises whether each gamete can also fuse with every other gamete. It cannot be solved by cytological study since a regulation, as described above for *Rubratella intermedia*, is absent in *Metarotaliella parva*. If gamonts of different size fuse, they form different numbers of gametes, some of which are left without a partner at pairing. Statistical analysis of the numbers of zygotes and residual

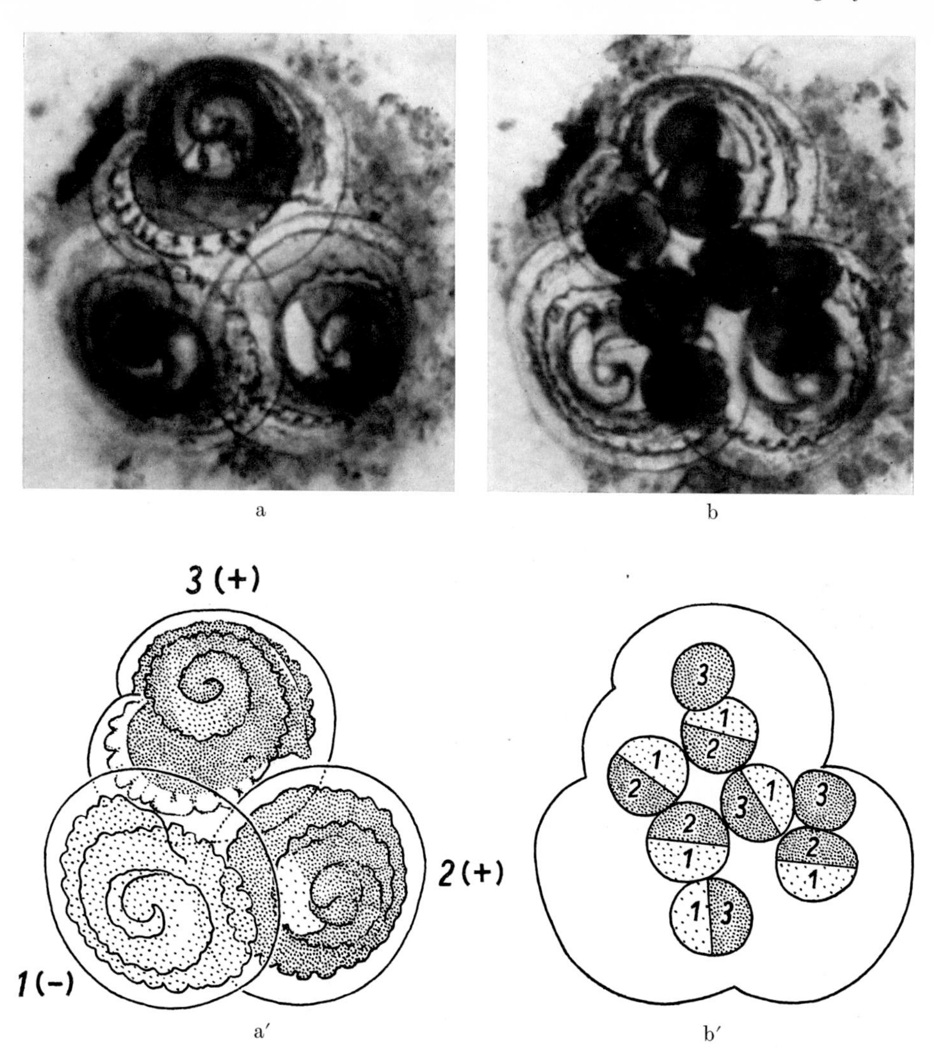

Fig. 184. *Patellina corrugata.* Aggregate of three gamonts. a Protoplasts still within the gamont shells. b The same aggregate after formation of 6 zygotes and 2 left-over gametes. a′, b′ Interpretation of sexual differentiation of gamonts and gametes, based upon frame by frame analysis of a motion picture (see text). × 120. After GRELL, 1960 [*668*]

gametes in aggregates showed, however, that only the gametes of different partners can pair with each other. Two residual gametes from the same gamont can never fuse. Gamont-specific differentiation of the gametes is also indicated by the results of combination experiments. If gamonts of different size are brought together to mate, the number of zygotes (or agamonts) formed depends solely on the smaller partner. If large gamonts of different sizes are combined with smaller partners of constant size, the number of emerging agamonts also remains constant.

Although the gamonts which form a pair are not sexually differentiated before mating, fertilization normally takes place between gametes from different gamonts *(allogamy)*.

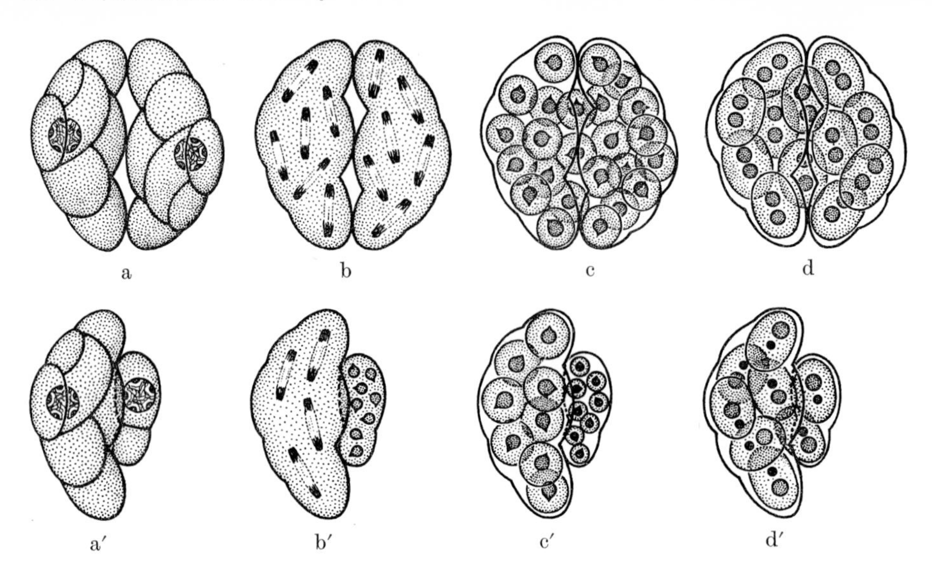

Fig. 185. *Rubratella intermedia.* Diagram of sexual reproduction when the two gamonts are of the same (a—d) and of different sizes (a′—d′). a, a′ Mating of the gamonts. b, b′ Gamogony. c, c′ Gametes. d, d′ Zygotes. After GRELL, 1958 [*664*]

Two possibilities may be considered in interpreting the sexual differentiation of the gametes of *Metarotaliella parva:*

1. The gametes of one gamont all belong to the "+" sex, those of the other to the "—" sex. This would imply that the gamonts are dioecious and that sex determination takes place after mating by mutual induction.
2. Each gamont forms "+" as well as "—" gametes. However, due to self-sterility, the "+" gametes of one gamont can only fuse with "—" gametes of the other one. If this is the case, the gamonts would be monoecious and sex determination might be connected with the last mitosis of gamogony, as discussed with respect to autogamy (p. 182).

Experiments with aggregates of three indicate that the second possibility is realized in *Metarotaliella parva.*

Combination experiments with gamonts of *Metarotaliella parva* show that they pass through three different stages in their individual development in which they differ in their mating behavior. These may be designated as the juvenile (I), the mature (II) and the old (III) stage.

Figure 186 illustrates these differences in graphs. When *gamonts of the same stage* are combined, every stage can be characterized by a definite type of mating curve (a).

It is characteristic of the *juvenile stage* (type 1) that in a I × I combination a latency period (t_0) elapses before mating sets in. The latter commences only after one of the partners has reached an age of at least 2 days.

In the *mature stage* (type 2), which lasts from about the second to the seventh day, pair formation commences as soon as the partners are brought together (II × II).

Old gamonts (type 3) differ from those in the mature stage in a progressive decrease of the speed of mating in III × III combinations. At the same time, the proportion of non-mating gamonts increases steadily.

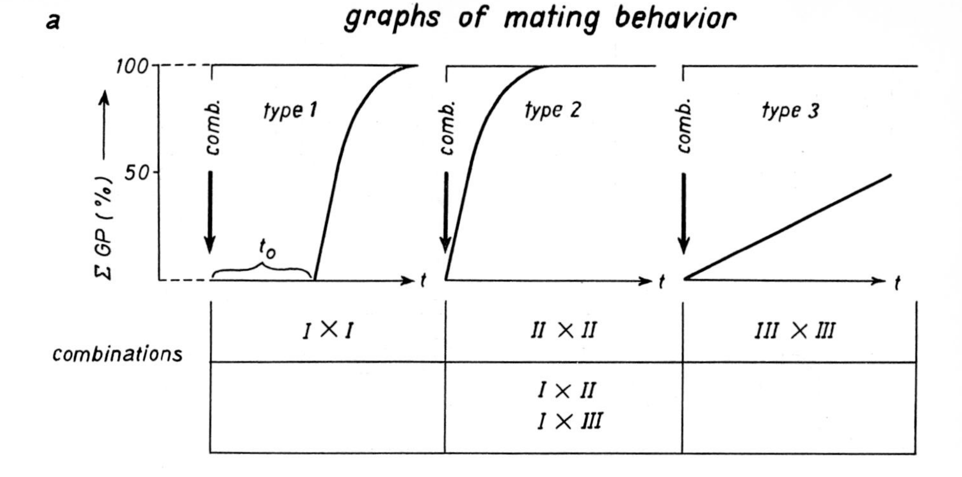

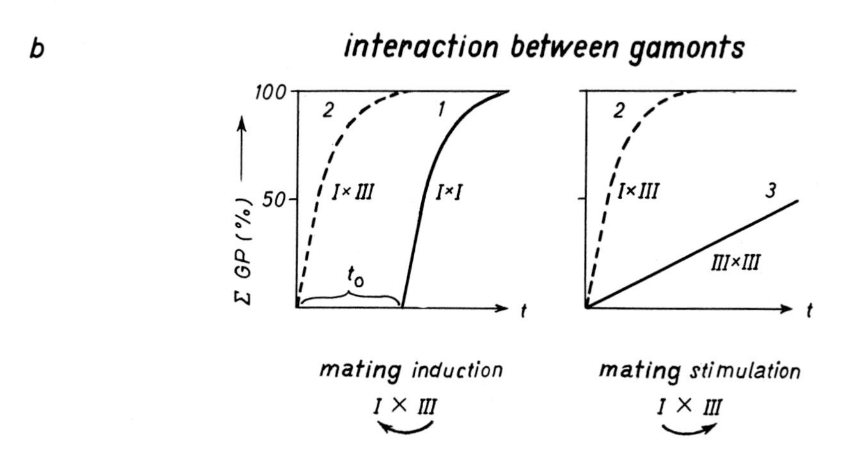

Fig. 186. *Metarotaliella parva.* Diagram of the mating behavior of gamonts of different stages of age (a) and the reciprocal interaction in a combination of young and older partners (b). Explanation in the text. After WEBER, 1965 [*1789*]

When gamonts of *different stages* are combined (I × II or I × III), the mating behavior corresponds to that of mature gamonts (II × II).

This result indicates a *reciprocal interaction* between gamonts of different stages. In a combination of young and old gamonts (I × III) both partners must influence each other (b). The old gamont effects the disappearance of the latency period (t_0) in its juvenile partner. Mating begins immediately upon combination at a time when the young gamont would not yet be capable of mating with a juvenile partner of type 1. This influence of the old gamont is called *mating induction.* On the other hand, the juvenile gamont also exerts an influence upon its old partner. Since the gamonts of stage III become more and more reluctant to mate (type 3), this effect of the juvenile gamont is all the more clearly expressed the older its partner is. This effect can be called *mating stimulation.*

Since the combination of young and old gamonts leads to the same course of mating

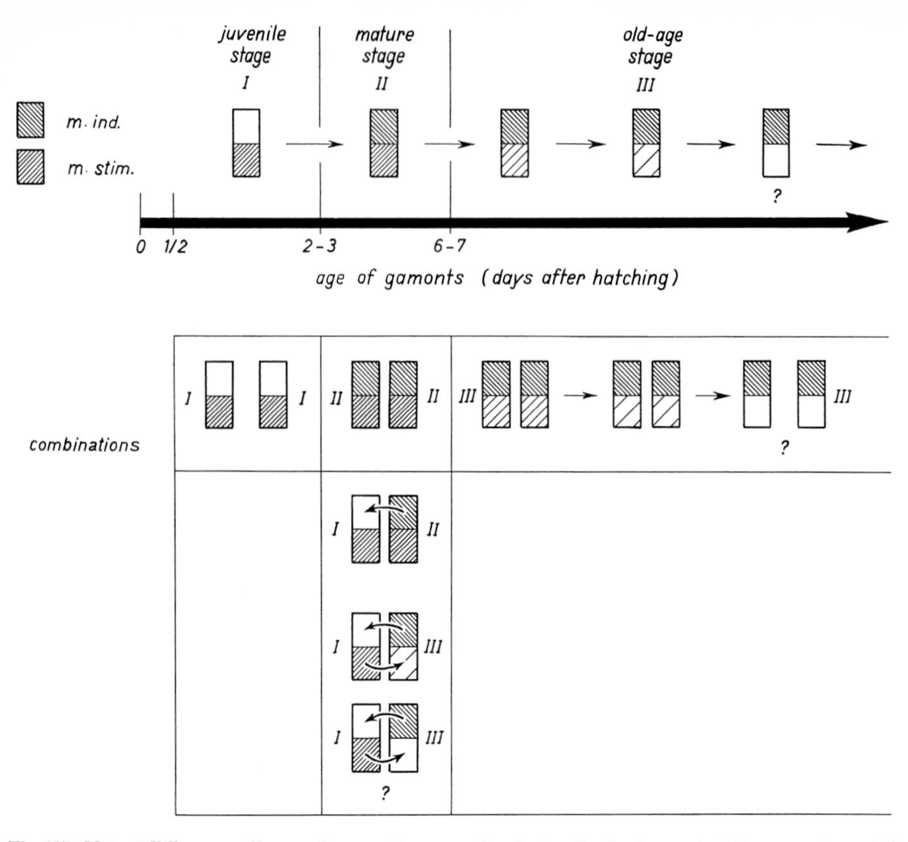

Fig. 187. *Metarotaliella parva*. Change of aggregation properties during the development of the gamonts. *m. ind.* = ability of mating induction. *m. stim.* = ability to stimulate mating. Curved arrows = proved partner effects. For the combinations in which mating is possible, the partial functions of the two partners are shown close together; for the combinations in which mating is not possible, they are shown far apart. After WEBER, 1965 [*1789*]

as the combination of two mature gamonts, it must be assumed that we are dealing with a *phenomenon of complementation* due to the effects of each partner upon the other.

Based on mating behavior we can thus recognize two properties of the gamonts which appear in succession in the course of development (Fig. 187).

In juvenile gamonts only the ability to stimulate mating is present. Therefore, mating between two gamonts is not yet possible in this phase. With transition into maturity the gamonts acquire the ability of mating induction, i.e., they can now mate with each other and with older as well as younger gamonts. In the old stage the ability to stimulate decreases steadily; the gamonts thus become more and more reluctant to mate. The ability to induce mating, however, remains unaffected.

In order to obtain mating in an aggregate, both properties must be present. However, it is not necessary that each partner has both properties as is the case in a mating of mature gamonts (II × II). Even if one partner is deficient in one characteristic (I × II), or if the properties of both supplement each other (I × III), mating can take place as between gamonts of the mature phase.

Sporozoa

Among the *Sporozoa* both the gregarines and the adeleids show gamontogamy. While the gamonts of eugregarines arise directly, by growth, from sporozoites, those of schizogregarines and of adeleids develop from the merozoites af a previous schizogony.

Mating of gamonts can take place in *eugregarines* either before or after termination of the growth phase. These differences seem to depend upon the species. It must still be clarified to what extent readiness for mating depends upon internal factors such as the growth stage reached and which external conditions might be involved. It has been shown in some cases that the hormonal situation of the host plays a role in the mating of gamonts [*465, 1541, 1777*].

The gamonts may attach side by side, as in many Monocystidae, or lengthwise. In some species, e.g. in *Stylocephalus longicollis* (Fig. 389), the gamonts attach by their protomerites. In most cases, however, one gamont fastens by its protomerite to the deutomerite of the partner. Thus, an anterior and a posterior gamont (designated as primite and satellite, respectively) can be distinguished. Chains of more than two gamonts or several satellites attached to one primite can also occur (Fig. 188). The further development of such aggregates is not known.

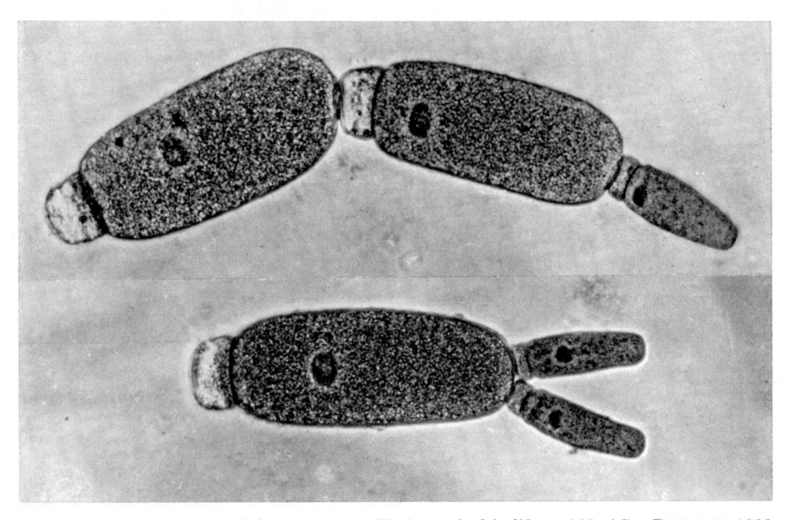

Fig. 188. *Gregarina sericostomae*. Aggregates of three gamonts. Photographed in life. × 280. After BAUDOIN, 1966 [*102*]

In *schizogregarines* the mating of gamonts is delayed until a certain degree of infection of the host is attained by way of schizogony. Sexual reproduction is therefore probably triggered by conditions which are brought about by infestation with the parasites. Since the duration of the gamontic growth phase is variable in schizogregarines, the number of gametes produced varies accordingly (compare Fig. 391 and Fig. 393).

After uniting, the gamonts flatten against each other, thus assuming a hemispherical shape and surround themselves by a common cyst envelope. The cysts of

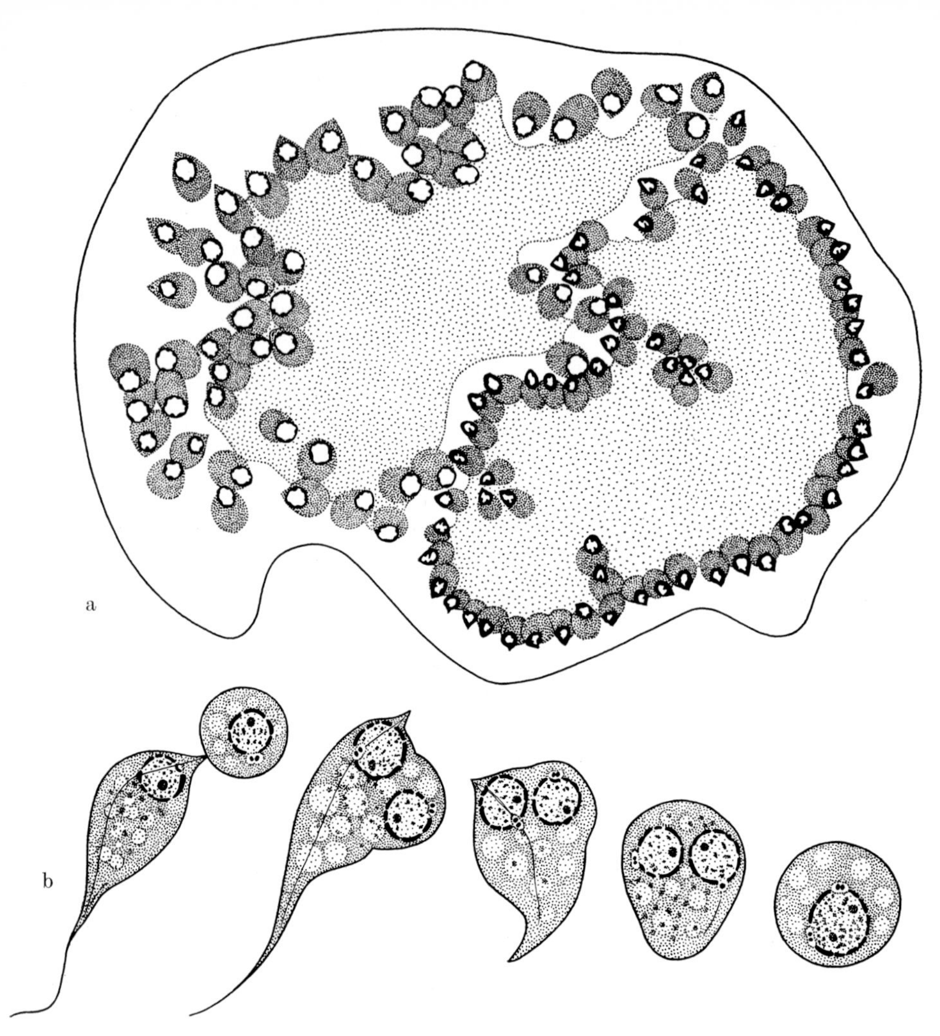

Fig. 189. Fertilization processes in eugregarines. a Section through the gamont cyst of a species of *Monocystis*. After BRASIL, 1905. b Fusion of the gametes of *Stylocephalus longicollis*. After LÉGER, 1904 [*1018*]

intestinal parasites are soon passed with the feces of the host. Species which inhabit the body cavity or tissues continue development within the host up to sporogony.

Encystment is followed by gamogony, which is initiated by a phase of nuclear multiplication. Most nuclei finally move to the surface of the cytoplasm, which has frequently been subdivided before into single portions (cytomeres). During gamete formation, which takes place synchronously in both gamonts, the whole cytoplasm is not generally used up. The remaining masses fuse in the center of the gamont cyst to form the so-called residual body. It may still contain a few nuclei.

In many gregarines, especially the schizogregarines, neither the aggregating gamonts nor the copulating gametes are morphologically different from each other. In other cases, the gamonts differ clearly in form, cytoplasmic structure and staining with vital dyes [*1153*]. Anisogamety, if present, is easily recognized in gregarines because the two gamonts form hundreds of gametes lying side by side within the same cyst (Fig. 189a). Occasionally, there are only minor differences in the form or size of the nuclei. However, there are also oogametic species where the gametes of one type are immotile while those of the other type can move about. A very peculiar differentiation is found in the gametes of *Stylocephalus longicollis*, the "egg cell" being smaller than the motile gamete. Their copulation is shown in Fig. 189b.

Although there is no doubt that sexual differentiation in gregarines is bipolar, the mode of sex determination is by no means clarified. Since meiosis takes place within the spore, the method of single-spore infection does not yield clear results as in Coccidia (p. 178).

Although the *Adeleidae* differ from other Coccidia with respect to gamontogamy, the direct transformation of the macrogamont into the macrogamete and the occurrence of multiple division only in the microgamont follows the typical coccidian pattern. The gamonts may either unite at the onset of the growth phase (*Karyolysus*, Fig. 404) or after its termination (*Klossia*, Fig. 403). In either case the paired gamonts are of different size. The lens-shaped microgamont adheres to the macrogamont and gives rise to only two or four microgametes, one of which

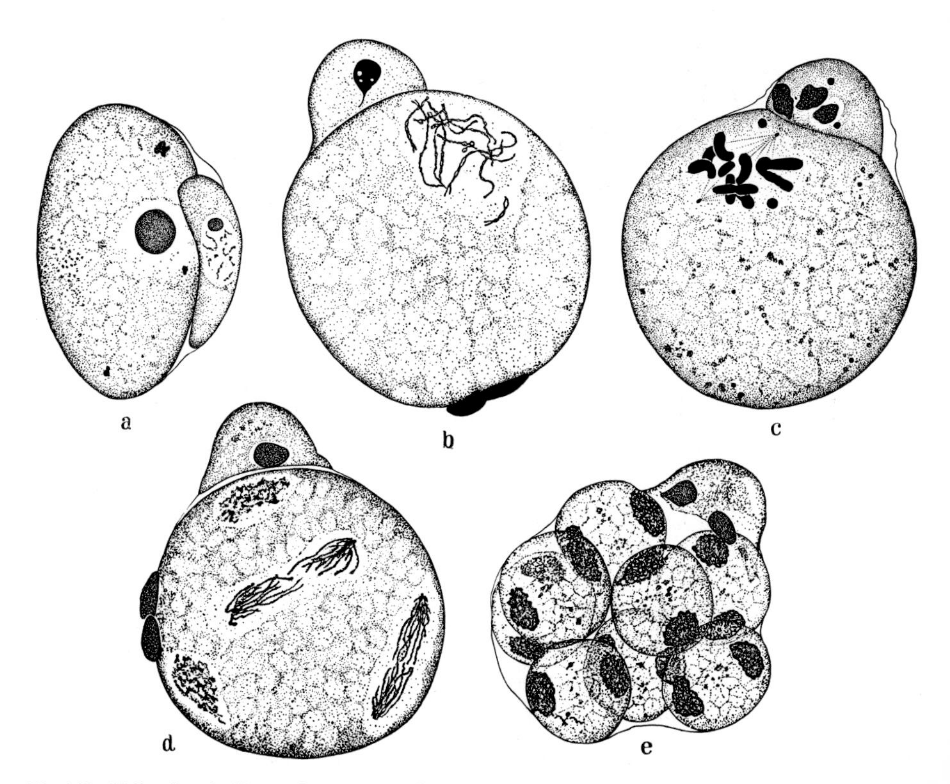

Fig. 190. *Adelina deronis*. Stages of gamogony and sporogony. a Uniting of macro- and microgamont. b Synkaryon in meiotic prophase (pairing of homologous chromosomes). c Meiotic metaphase. d Sporogony divisions. e Oocyst with 8 spores. In b and d two microgametes which have not succeeded in fertilization are attached to the oocyst. After HAUSCHKA, 1943 [*726*]

unites with the macrogamete. Meiosis of the synkaryon and the onset of sporogony follows immediately (Fig. 190).

The adeleids are therefore not only oogametic, which is characteristic for all Coccidia, but there is a difference in size of the gamonts as well (anisogamonty).

2. Gamontogamy without Gamete Formation

In a special type of gamontogamy the mating gamonts do not produce gamete cells but only gamete nuclei.

Among the *flagellates*, this has only been described for the diploid polymastigid, *Notila proteus* (Fig. 191). The gamonts which arise by differential cell division

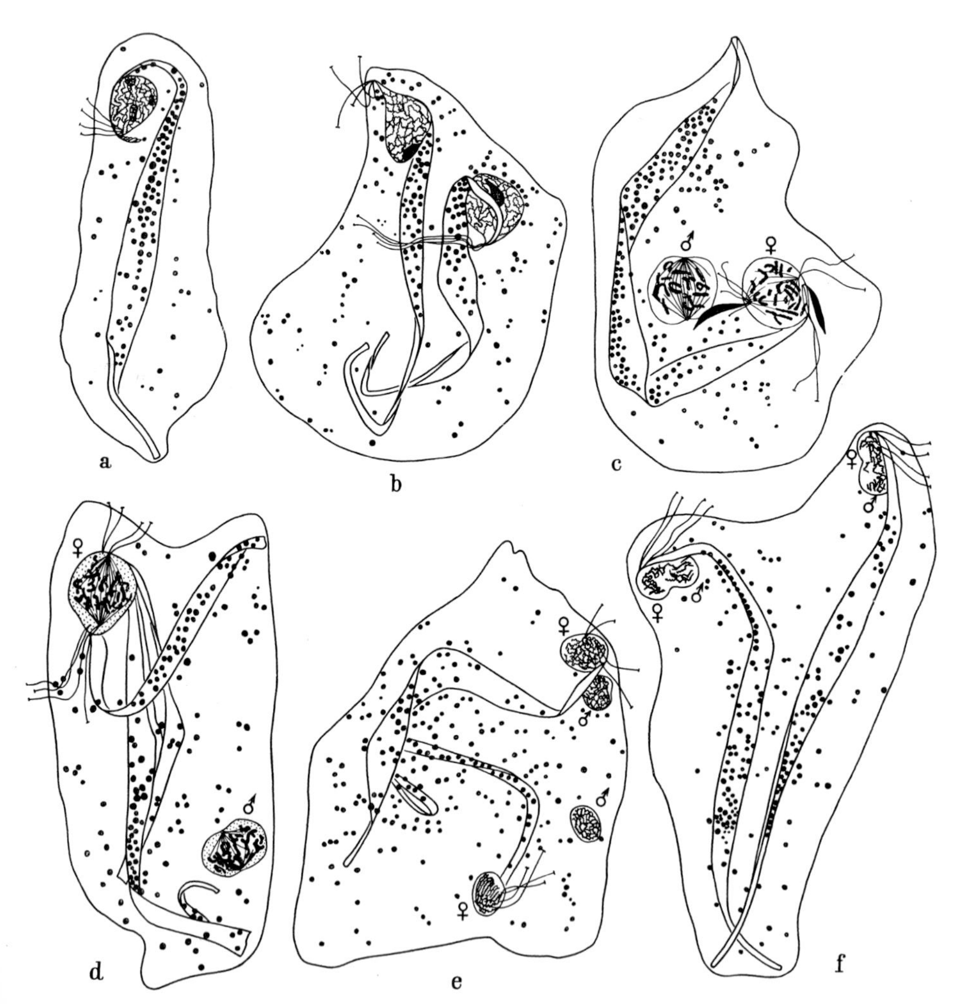

Fig. 191. *Notila proteus* (Polymastigina). Gamontogamy. a Ordinary diploid cells. b Fusion of two diploid cells (gamonts). d The nucleus of the male gamont has separated from its associated organelles (flagella, axostyle) anp moves about freely. New flagella and axostyles originate at the centrioles of the female gamont nucleus. d One-step meiosis of the male and female gamont nuclei. The old axostyle which has been formed as a fusion product disappears; instead, new axostyles are formed. e The male pronuclei migrate to the female pronuclei. f Karyogamy (followed by division of the "double zygote"). × 640. After CLEVELAND, 1950 [*279*]

from a "progamont" are not different from ordinary cells (a). Sexual differentiation, however, is expressed in the behavior of the nuclei. As soon as the "male" and "female" gamonts have fused with each other (b), the axostyles also fuse just as in gametes of *Oxymonas* and *Saccinobaculus* (Fig. 160). At the same time the nucleus of the male gamont separates from its associated organelles (axostyle, flagella). A new set of organelles is formed from the intranuclear centrioles of the nucleus of the female gamont (c). The old axostyle which has been formed as a fusion product is resorbed in the cytoplasm. Each gamont nucleus now divides and reduces the chromosome number (one-step meiosis, p. 83). Two female gamete nuclei, each connected with axostyle and flagella, and two motile male nuclei are formed in this manner (d, e). The latter migrate toward the female gamete nuclei and fuse with them, forming two synkarya (f). The "double zygote" then undergoes cytokinesis, leading to the formation of two diploid cells.

3. Conjugation

In ciliates, the process of fertilization is called "conjugation". Actually, this term denotes two rather different phenomena: in one case (isogamonty) both gamonts are morphologically identical; in the other case they are either different from the beginning or become different in the course of mating (anisogamonty). Both possibilities will be discussed separately.

a) Isogamonty

In the majority of ciliates the conjugants are morphologically identical (Fig. 192). The cells touch first by their anterior ends and then lie side by side along their whole length. The first change in the micronuclei is an enlargement (a). While the nuclear volume increases (compare Fig. 43 and Fig. 84), the chromosomes undergo the processes which are generally characteristic of meiotic prophase. Reduction of the chromosome number is achieved in the course of two micronuclear divisions which follow in rapid succession (b–d). In this manner four haploid daughter nuclei are formed, three of them becoming pycnotic to be resorbed in the cytoplasm (e). The remaining nucleus now divides once more (f). This nuclear division, which is designated as postmeiotic, gives rise to the two gamete nuclei (pronuclei). One of these, the *stationary nucleus*, remains within the same conjugant while the other one, the *migratory nucleus*, wanders into the mating partner. In each cell, the stationary nucleus fuses with the migratory nucleus of the other conjugant to form a synkaryon (g). Conjugation thus involves *mutual fertilization*, the mating partners separating afterwards. The cells are now referred to as *exconjugants*. In the latter, the synkaryon divides into two daughter nuclei, one becoming the micronucleus while the second develops into a new macronucleus (h–k). The old macronucleus dissolves in the course of conjugation. Usually it disintegrates into numerous fragments which are sooner or later resorbed in the cytoplasm.

In contrast to *Notila proteus* (Fig. 191) whose two "male" nuclei of one gamont migrate toward the two "female" nuclei of the other gamont, the ciliates form two gamete nuclei of different behavior in each gamont. Only one of these, which may be designated as "male", migrates into the partner and fuses with the "female"

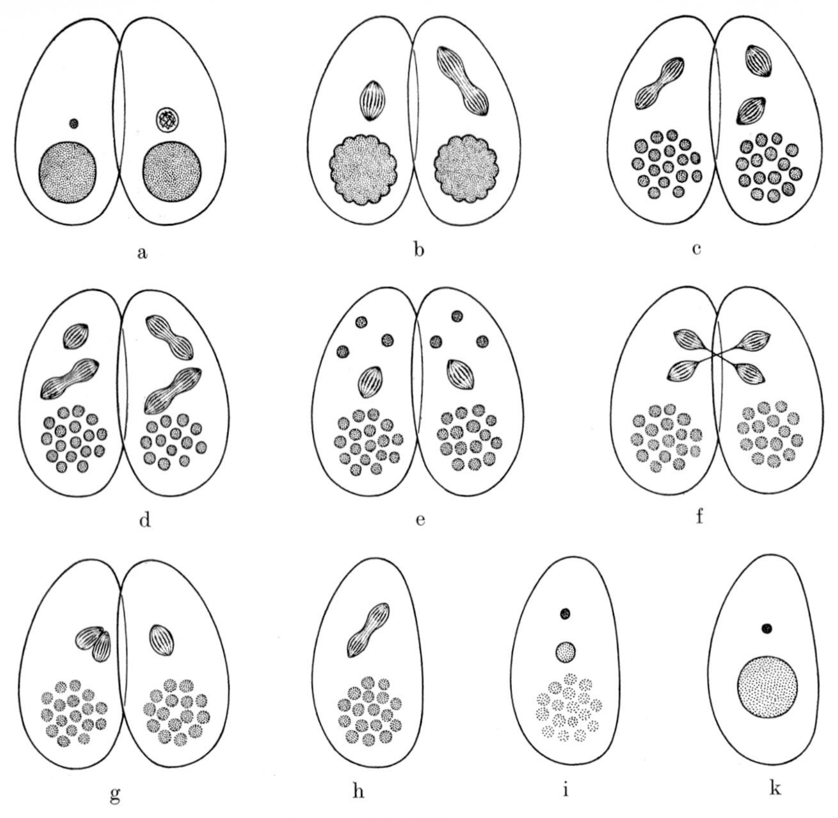

Fig. 192. Diagram of conjugation (isogamonty) in ciliates

gamete nucleus to form the synkaryon. According to the behavior of their gamete nuclei, the isogamontic ciliates can be looked upon as *monoecious*.

There are numerous variations of the normal type of ciliate conjugation as described above, both with regard to events before karyogamy (progamic) and following fertilization (metagamic).

The variability of *progamic* events is largely due to the fact that many ciliates have several micronuclei instead of just one. In all such cases the meiotic divisions are begun by all micronuclei and are frequently carried through even if their number, as in *Bursaria truncatella*, is very high [*1317*]. As a rule, only one of these micronuclei survives after the first two divisions.

In *Paramecium aurelia* (Fig. 193a–k), a species with two micronuclei, the two meiotic divisions give rise to eight daughter nuclei, seven of which become pycnotic. As mentioned above (p. 185), *autogamy* regularly occurs besides conjugation in this species. It entails fusion of the gamete nuclei within the same cell (l–n) [*395*]. After autogamy, the cells are called exautogamonts. Occasionally, two cells pair as in conjugation although no mutual fertilization takes place [*1795*], each individual carrying out an autogamy (o). The biological meaning of this "*cytogamy*" remains obscure. but it has to be taken into account when analyzing the results of genetic crosses (p. 247).

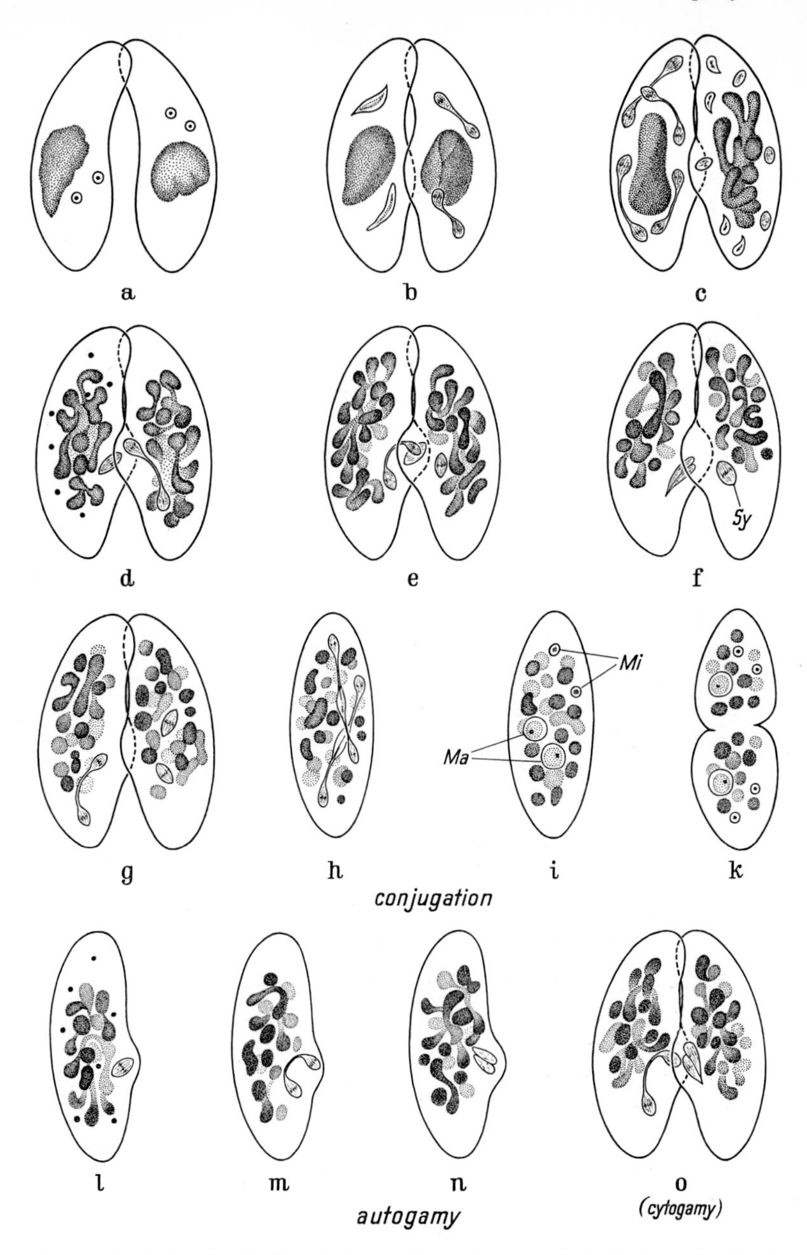

Fig. 193. *Paramecium aurelia*. Conjugation (a—k) and autogamy (l—o, o "cytogamy"). In the stages of conjugation the right partner is always depicted somewhat more advanced in its development than the left one. Partly diagrammatic. Combined from various authors

A most remarkable variation of progamic events is found among species of *Tracheloraphis* and *Trachelocerca*. Recent studies have shown that several of the gone nuclei formed by the two meiotic divisions can carry out a postmeiotic division [*450*, *967*, *1385*, *1392*].

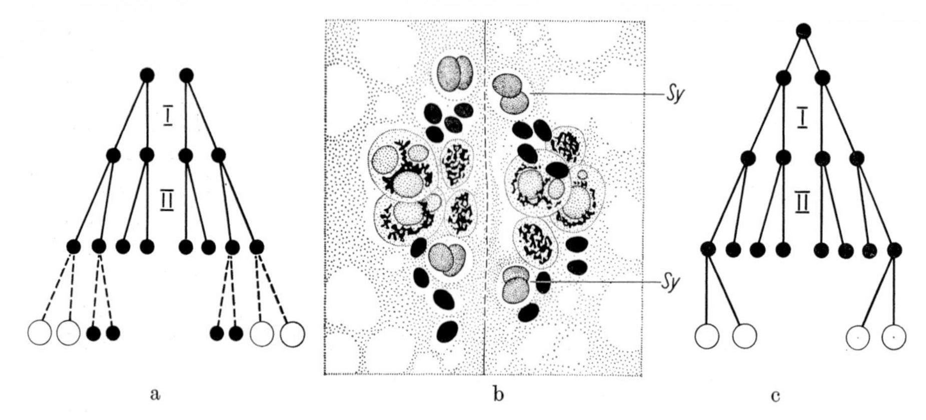

Fig. 194. Modifications of progamic nuclear processes. a, b *Trachelocerca coluber*. a Diagram of the nuclear division: formation of two stationary nuclei and two migratory nuclei (white circles), eight degenerating gone nuclei (black circles). b Section of the area of fusion of two conjugants after karyogamy; two synkarya (*Sy*). After RAIKOV, 1963 [*1392*]. c *Euplotes vannus.* Diagram of the nuclear divisions: of four gamete nuclei (white circles), two degenerate. In accordance with HECKMANN, 1963 [*735*]

In *Tracheloraphis phoenicopterus*, which has six micronuclei, a variable number of gamete nuclei originate in each conjugant. Some of these differentiate into migratory, others into stationary nuclei. After exchanging migratory nuclei, each conjugant forms several synkarya; all but one, however, are destined to become pycnotic. In *Trachelocerca coluber* the two micronuclei of each conjugant give rise to eight gone nuclei, four of which divide again. Four of the eight descendants of this division become pycnotic while the remaining four differentiate into two migratory and two stationary nuclei (Fig. 194a). Two synkarya are thus formed in each conjugant (b).

In *Euplotes vannus* (c) the micronucleus of each conjugant first undergoes a "premeiotic" division. After meiosis, there are then eight gone nuclei, two of which participate in a postmeiotic division leading to four gamete nuclei. Two of these degenerate, however, while the remaining two differentiate into a migratory and a stationary nucleus. Genetic studies [*735*] showed that the two functional nuclei can either be daughter nuclei of the same division or descendants of different gone nuclei. At least in *Euplotes vannus* it is therefore improbable that the occurence of a postmeiotic division is a prerequisite for the differentiation of a stationary and a migratory nucleus.

That the migratory nuclei are not simply pushed into the partner by the elongating spindle [*1560*] is indicated by cases where the postmeiotic division is parallel to the border between the conjugants. One example of this is provided by the suctorian *Heliophrya (Cyclophrya) erhardi.* As in all Suctoria, conjugation starts with the formation of lump-like swellings which elongate and connect both conjugants with each other (Fig. 195a). The gone nucleus which survives and gives rise to the gamete nuclei always moves into the left corner (seen from above) close to the border between both conjugants (b). Its division leads to the formation of a stationary nucleus at left and a migratory nucleus at right (c, d). One gets the impression

►

Fig. 195. *Heliophrya (Cyclophrya) erhardi.* Conjugation. a Clonal culture with different specimen forming lump-like swellings ("conjugational projections"). b Gone nuclei. c Postmeiotic divisions. d Stationary and migratory nucleus in each conjugant. e Karyogamy. f One synkaryon in each conjugant. Courtesy of N. LANNERS [*994a*]

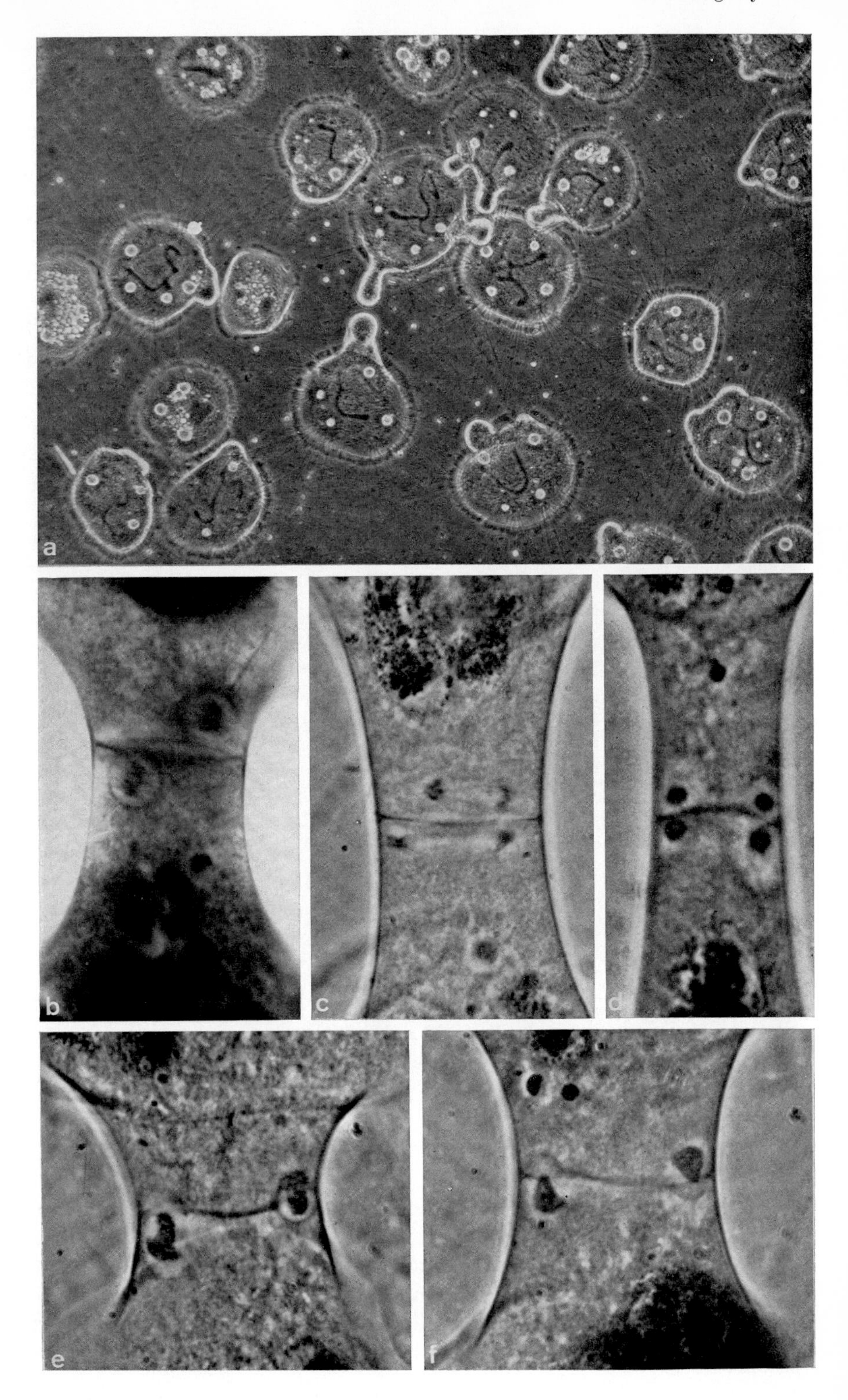
a
b
c
d
e
f

that influences of the membrane separating both cells are involved in the determination of both gamete nuclei. In addition to their different behavior, the gamete nuclei also show occasional structural differences as in some other ciliates (Fig. 197). After local dissolution of the membrane in front of each migratory nucleus, it fuses with the stationary nucleus of the same side while the membrane is reconstructed behind it (e, f).

In the free-swimming ciliates no uniform mode of attachment of the conjugants exists. In *Paramecium aurelia* the cells attach after the agglutination reaction described on p. 222, at first in a definite region of their anterior ends, the so-called holdfast region (Fig. 193). At this stage of mating they can still be separated with ease. Final fusion follows in a region close to the cytostome where the so-called *paroral cone* is formed during meiosis. The paroral cones of both conjugants lie on top of each other and are the sites at which the pellicles first dissolve and through which the migratory nuclei pass.

Detailed observations showed that the gone nucleus, from which the two gamete nuclei originated, migrates into the paroral cone. If this does not happen, as is regularly the case in a certain mutant of *Paramecium aurelia* [*1635*], it becomes pycnotic like the seven remaining gone nuclei. The region of the paroral cone seems to protect the nucleus from a physiological change in the remaining cytoplasm at the end of the progamic division which leads to degeneration of the gone nuclei.

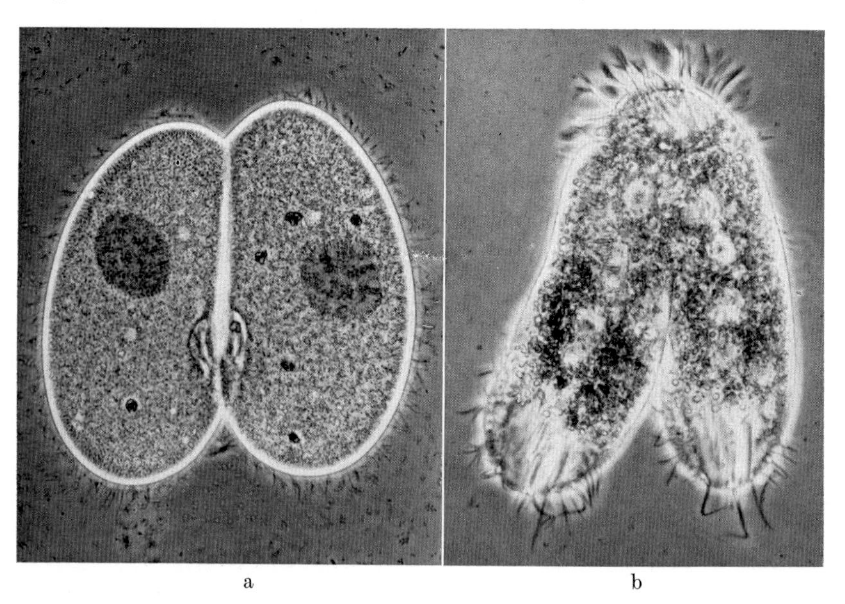

Fig. 196. Pairs of conjugants. a *Paramecium trichium*. × 500. b *Stylonychia mytilus*. × 200. Micrography by D. Ammermann

Whereas in *Paramecium* a wide cytoplasmic connection is only occasionally formed between the two conjugants (p. 262), this can be the usual manner of fusion in other ciliates (Fig. 196). Paroral cones are then not formed. Frequently, the entire region of the peristome partakes in fusion of the partners. In other cases the peristomes remain free during conjugation. In *Cycloposthium bipalmatum*, a ciliate living in ruminants, the two conjugants fuse only at the rims of their peristomes

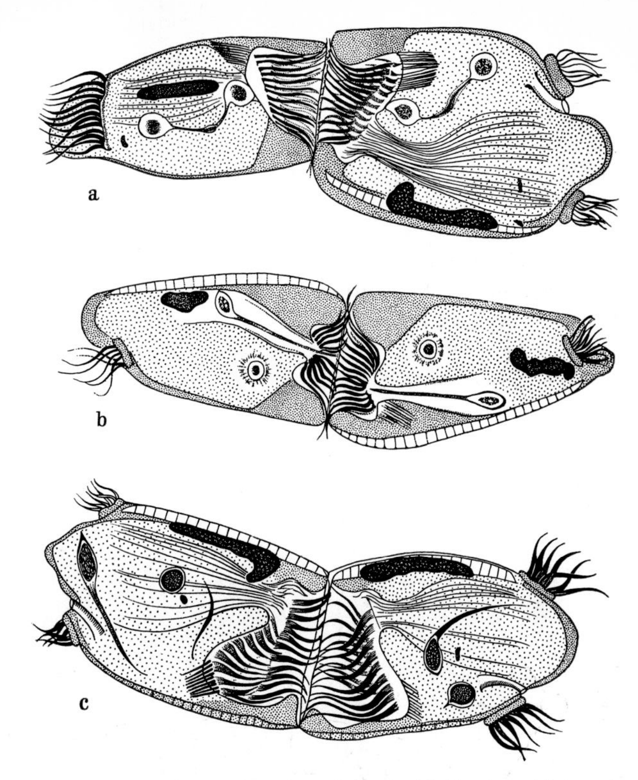

Fig. 197. *Cycloposthium bipalmatum.* Stages of conjugation. Both conjugants touch with the rims of their peristomes. a Third micronuclear division. b, c Transfer of the migratory nuclei with their cytoplasmic appendages through both peristomal spaces to the partner. After DOGIEL, 1925 [*429*]

(Fig. 197). The migratory nucleus has a long cytoplasmic appendage and must traverse the closed space between the peristomes in order to reach the other partner. It is similar in behavior and appearance to the spermatozoon of an animal.

The *metagamic* events are likewise variable (Fig. 198). It is by no means the rule that the first division of the synkaryon gives rise to a micronucleus and a macronuclear anlage as in *Chilodonella uncinata.* Frequently, this differentiation takes place only after the second metagamic mitosis. If a cell division occurs simultaneously, as in *Paramecium aurelia* (Fig. 193), each daughter cell receives a micronucleus and a macronuclear anlage. The micronucleus then divides again. In other cases, two of the four daughter nuclei degenerate *(Euplotes patella).* It can also happen that two macronuclear anlagen fuse with each other *(Didinium nasutum).* In *Paramecium bursaria,* three metagamic divisions take place; however, the first one is meaningless since one of the two nuclei degenerates. The three metagamic divisions of *Paramecium caudatum,* on the other hand, lead to the formation of eight daughter nuclei. Three of these die while one becomes a micronucleus and four develop into macronuclear anlagen. In two successive cell divisions the four anlagen are distributed to four cells while the micronucleus divides once with each cell division. Every offspring of the exconjugant thus receives finally one macronuclear anlage and one micronucleus.

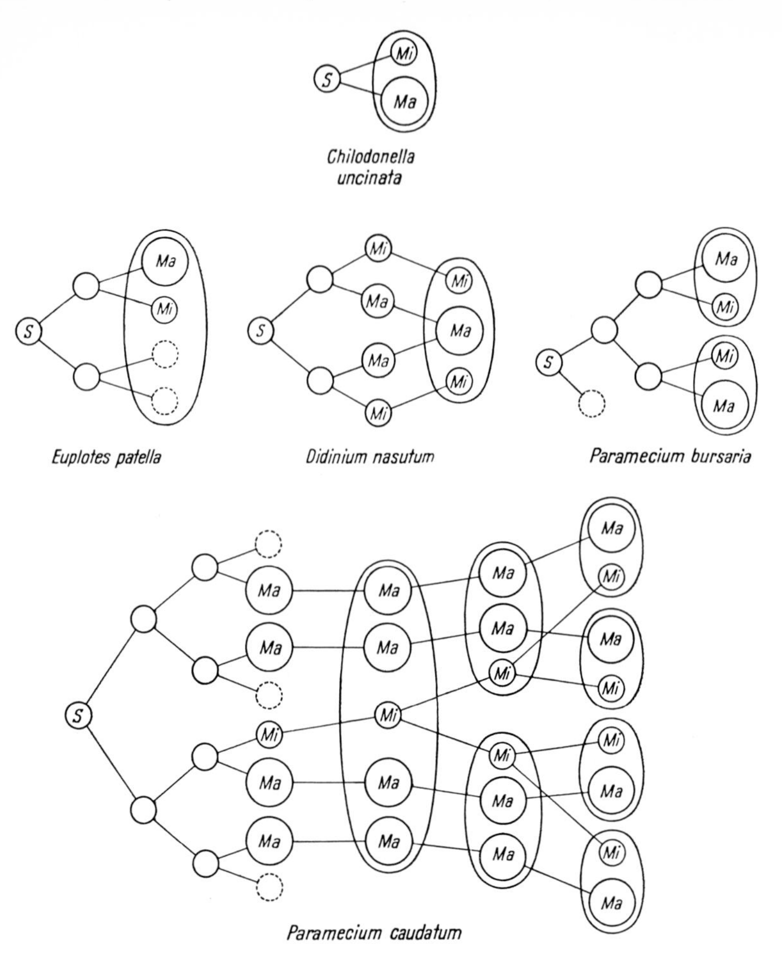

Fig. 198. Various possibilities of metagamic nuclear divisions with regard to the differentiation of micronuclei (*Mi*) and macronuclear anlagen (*Ma*). *S* Synkaryon

Even in one and the same species variations from the typical course of conjugation have frequently been observed. Thus, the third progamic micronuclear division may not be carried out in *Paramecium trichium*, two daughter nuclei of the second division becoming the gamete nuclei [*398*]. One race of this species is said to have only one progamic division [*399*]. In *Paramecium caudatum* more than three metagamic divisions of the synkaryon may take place before determination of micronuclei and macronuclear anlagen. The number of the latter may also vary from two to ten. It is also claimed that one of the two daughter nuclei of the synkaryon degenerates while the other one carries out three metagamic divisions [*396*, *400*].

In some ciliates fragments of the old, disintegrated macronucleus are exchanged at conjugation.

While the exconjugants are usually incapable of renewed conjugation, "re-conjugations" are occasionally found between individuals whose nuclear apparatus has not yet reformed and normal individuals (Fig. 199).

The duration of conjugation varies considerably among ciliates. In most cases it takes several days. The time needed for the various phases is tabulated below for *Paramecium bursaria* [*1796*]:

1st micronuclear division	14 hours
2nd micronuclear division	1 hour
3rd micronuclear division	3 hours
Exchange of pronuclei and karyogamy	about 20 minutes
1st, 2nd, and 3rd division of the synkaryon	2 hours each
Reorganization and division of the exconjugant	72 hours
Total duration of conjugation	4 days

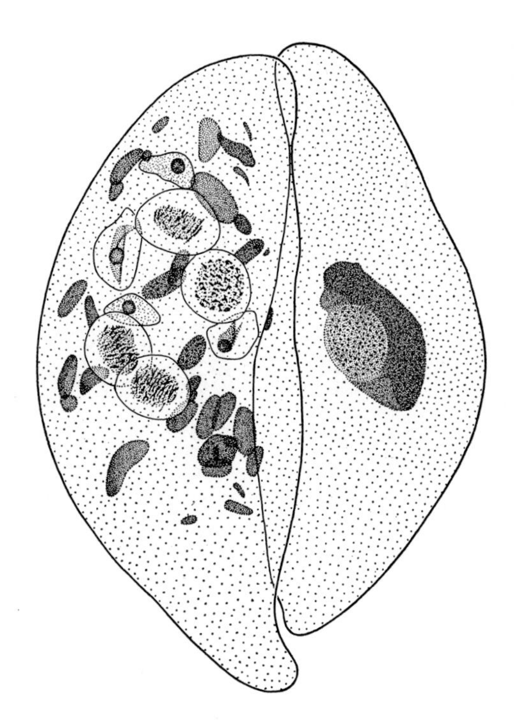

Fig. 199. *Paramecium caudatum.* Re-conjugation. Reconjugant (left) with four micronuclei which are in meta- or anaphase and four degenerating macronuclear anlagen. Normal partner (right) in prophase of the first progamic division. × 750. After DILLER, 1942 [*397*]

b) Anisogamonty

In some ciliates there is a constant difference in the size of the partners which is either recognizable before mating or becomes manifest in the course of mating. By way of analogy with other Protozoa such as the gamontogamous Adeleidae (p. 197) we can designate this as anisogamonty and speak of the two partners as the *macro-* and *microgamont* respectively.

In all closely studied cases, this anisogamonty is connected with the fact that a synkaryon is formed only in the macrogamont while the microgamont is resorbed by the macrogamont during mating. Instead of mutual fertilization inherent in the concept of conjugation, fertilization of only one partner takes place.

Anisogamonty is especially pronounced in the *peritrichs* where the microgamonts are the result of a special reproductive act. This can occur in two different ways.

In the *solitary* genera (e.g. *Vorticella, Opercularia, Urceolaria, Opisthonecta, Lagenophrys*) the gamonts are formed by a *differential* division which is unequal (Fig. 200a). The larger cell may become the macrogamont while the smaller one differentiates into the microgamont. The latter forms a special circlet of cilia in

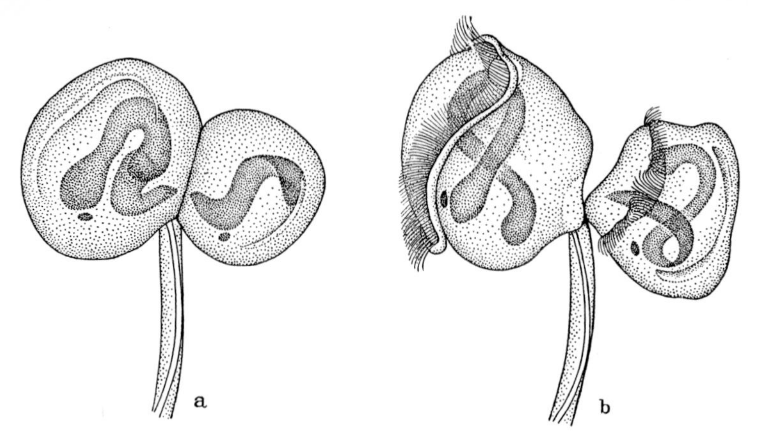

Fig. 200. *Vorticella campanula*. Unequal division, leading to the origin of a macroconjugant (left) and a microconjugant (right, with aboral circlet of cilia). After MÜGGE, 1957 [*1152*]

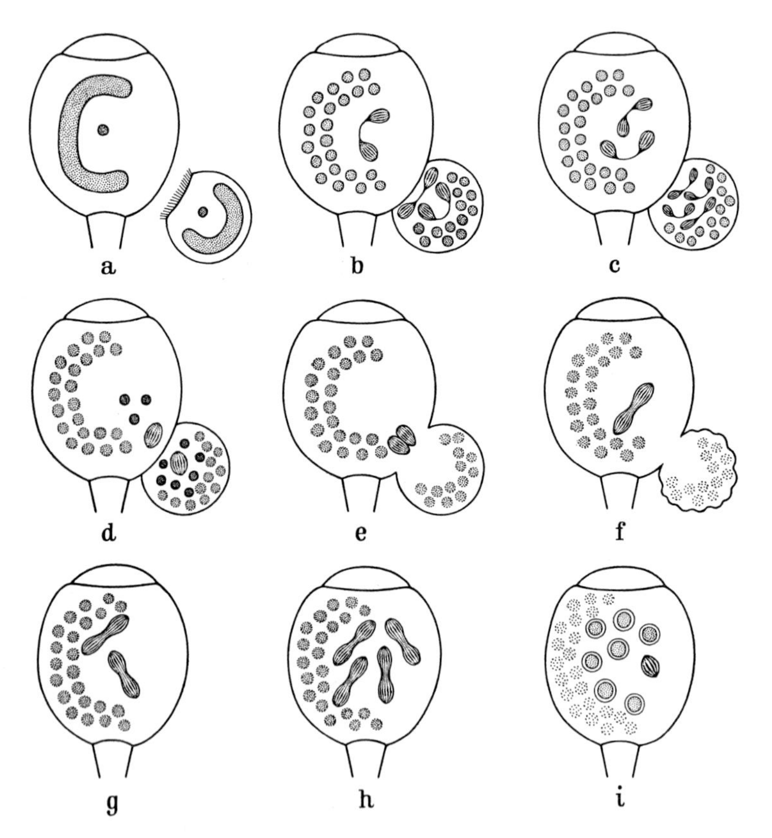

Fig. 201. Diagram of sexual processes (anisogamonty) in peritrichs

the posterior region of the cell(b) and corresponds in this respect to the "telotroch" of a normal cell division (p. 147).

If the gamonts are isolated to prevent conjugation, the microgamonts die, while the macrogamonts become normal vegetative cells capable of division. However, their daughter cells are no longer able to conjugate. A new differential division has to take place in order to obtain reactive macro- and microgamonts [*516*, *517*].

In the *colonial* genera (e.g. *Epistylis*, *Carchesium*, *Zoothamnium*) a cell divides once, twice or thrice to form the microgamonts.
While the macrogamonts remain attached to their stalks, the microgamonts are free-swimming. They seek out the macrogamonts and attach themselves near the origin of their stalk.
Progamic nuclear events do not take a uniform course. In most species they seem to take place as shown in Fig. 201. While the macronuclei disintegrate, the micronucleus divides twice in the macrogamont and three times in the microgamont (a—c). Three of the four daughter nuclei of the macrogamont and seven of the eight daughter nuclei of the microgamont are resorbed in the cytoplasm. Only a single nucleus is thus left in each gamont (d). In most cases this does not seem to divide again. After the pellicle separating the two cells has been dissolved, the nucleus of the microgamont migrates toward that of the macrogamont and fuses with it to form the synkaryon (e).
In some species (e.g. *Vorticella campanula*) it has been observed that a postmeiotic division takes place in each gamont. However, one of the two nuclei becomes pycnotic, leaving a "stationary nucleus" in the macrogamont and a "migratory nucleus" in the microgamont [*1152*]. Of one species (*Vorticella monilata*) it is said that both sister nuclei remain and that a mutual nuclear exchange takes place. A synkaryon is formed in the macrogamont only; the two pronuclei of the microgamont will not fuse and degenerate thereafter [*1109*].
These modifications of progamic events indicate that unilateral fertilization is derived from mutual fertilization.
As the microgamont is resorbed by the macrogamont, the synkaryon gives rise to eight daughter nuclei in the course of three metagamic divisions (Fig. 201f—i). One of these becomes the micronucleus and the remaining seven become macronuclear anlagen. These are then — similar to *Paramecium caudatum* — distributed to the descendants of the macrogamont in the course of subsequent fissions, each cell receiving finally one macronuclear anlage (Fig. 202).

Although it may seem reasonable to consider the type of fertilization found in peritrichs as an adaptation to the sessile mode of life, it should not be overlooked

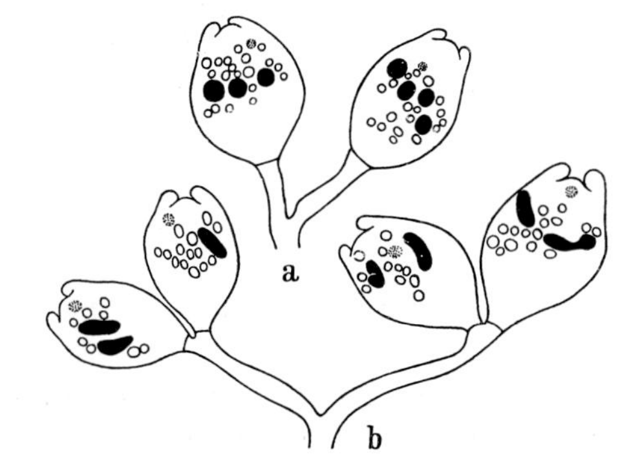

Fig. 202. *Epistylis articulata*. Metagamic divisions of the exconjugants. a After the first, b after the second division (macronuclear anlagen black). After DASS, 1953, somewhat modified [*367*]

that anisogamonty also occurs among free-swimming ciliates. It has been observed in species of the genera *Metopus* [*1240*] and *Urostyla* [*737*] that both partners are at first of equal size and indistinguishable by morphological criteria. However, after fusion one of the mating partners is gradually resorbed by the other one. In *Metopus sigmoides* a residue remains which detaches and dies, but in *Urostyla weissei syn. hologama* resorption of one partner by the other is complete. In both species resorption is followed by encystment.

In most *Suctoria* conjugation is not essentially different from that of other ciliates. After exchanging migratory nuclei, the two partners separate again. Only in *Tokophrya cyclopum*, in *Choanophrya infundibulifera* and in *Ephelota gemmipara* has conjugation been modified to unilateral fertilization, one partner being torn from its stalk and resorbed by the other one.

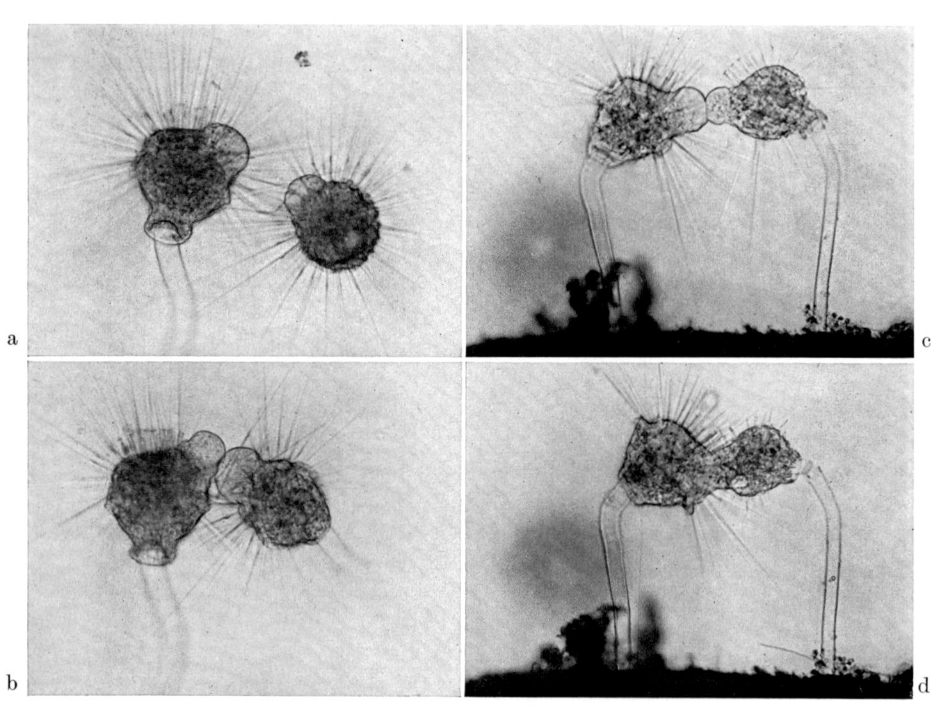

Fig. 203. *Ephelota gemmipara*. Start of mating. a, b Formation of the "mating lumps" and contact of the partners. The attachment point of the stalk of the right partner is outside the focal plane. c, d Withdrawal of the mating lumps and final fusion. In d the right partner detaches from its stalk. × 130. From the films C 913 and E 1017 (GRELL)

Both partners of *Ephelota gemmipara* may be of equal or of unequal size. The gamonts form at first lump-like swellings at their adjoining surfaces which carry out rhythmic pumping movements. As in other suctorians (Fig. 195), they are recognizable before the partners touch each other. Evidently, their formation must be due to chemical interaction at a distance. Initial contact is established by the food-capturing tentacles and it is by their aid that the gamonts approach in a hand-over-hand fashion (Fig. 203a, b). Connection is finally made at the "mating lumps" (c, d). After some hours one of the partners, in cases of unequal size probably always the smaller one, detaches from its stalk and is resorbed by the other (Fig. 204). In every case a size difference becomes manifest during mating. One partner can thus be designated as the macrogamont, the other

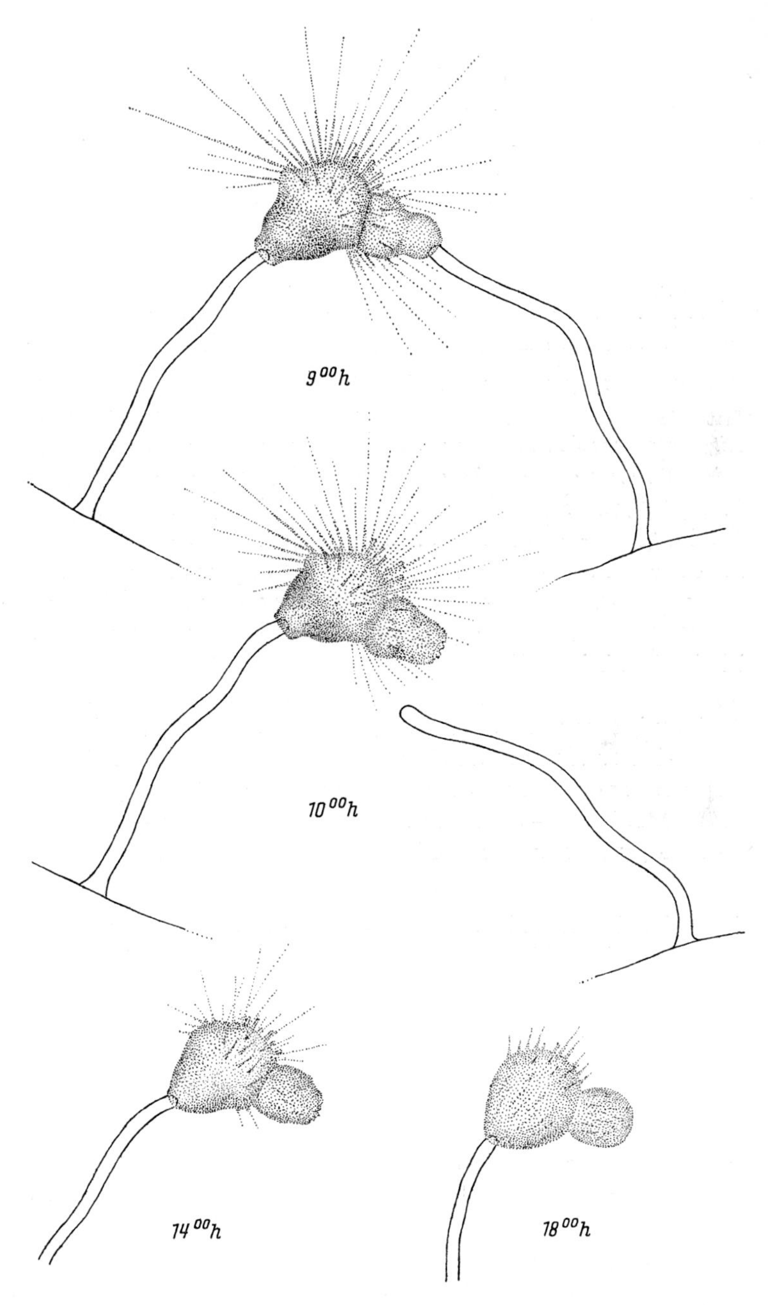

Fig. 204. *Ephelota gemmipara.* Anisogamonty. The right partner is torn away from its stalk. Drawn after observations in life. After GRELL, 1953 [*653*]

as the microgamont. Each cell normally forms only one gamete nucleus. The two gamete nuclei fuse to a synkaryon which divides within the macrogamont into two daughter nuclei, one of which becomes the micronucleus, the other a macronuclear anlage. Continued micronuclear divisions lead to many micronuclei while the macronuclear anlage develops into a complex, branched macronucleus.

c) Mating Types

Although in isogamontic ciliates the two partners cannot be distinguished morphologically and show the same behavior in the course of conjugation, it is found in many species that not every gamont can conjugate with every other gamont. The two partners must be of different *mating types.*

The existence of different mating types is proved in the same manner as sexual differentiation in the case of isogamety (p. 161). Cells of a natural population are reared in clone cultures. As a rule, conjugations do not occur within a clone. If certain clones are mixed, however, conjugation always takes place under suitable conditions (p. 221). Such clones belong to different mating types.

That pair formation involves partners of different types can best be shown by the so-called *split pair method:* shortly after mating, the pairs are separated from each other and allowed to reproduce in different culture dishes. There is no conjugation within the resulting clones. However, if the two clones obtained from a single pair are mixed, conjugation will always follow, given suitable conditions. The same proof can be obtained by marking the clones if the split pair method fails. One of the clones may for instance consist of "doublets", an abnormality which may be carried through many cell generations (p. 274). Occasionally, the individuals of two clones differ markedly in size (Fig. 205). If differences of this kind are not available, it may be possible to label the clones with different food organisms or with vital dyes such as neutral red or Nile blue sulphate.

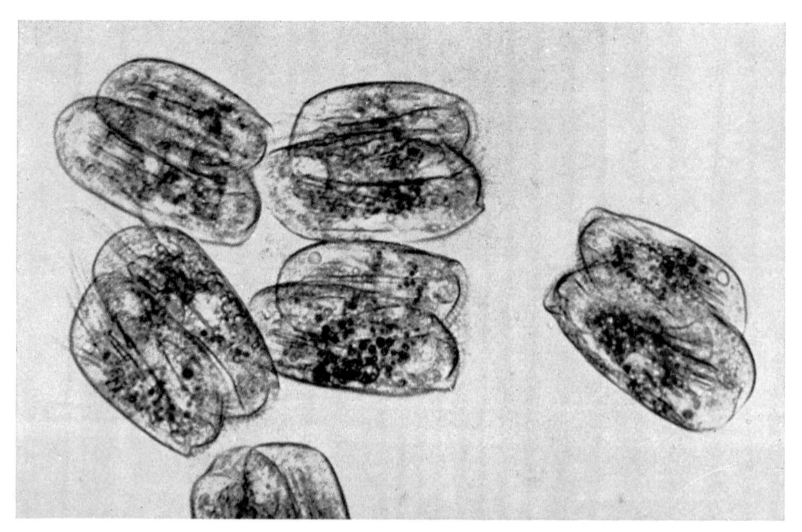

Fig. 205. *Euplotes vannus.* Conjugation between animals of two clones whose individuals differ in average size. Photographed in life. × 250. After HECKMANN, 1963 [*735*]

In all closely studied ciliates it has been found as in phytomonads (p. 167) that the "species" of the taxonomist is not a group of freely interbreeding populations in the sense of the usual species concept but includes a more or less large number of "varieties" which cannot be crossed with each other. Strictly speaking, these "varieties" should therefore be classified as different species. Since their morphological differences are insignificant compared to those which are usually employed to characterize species and since the concept of variety includes the ability to interbreed, the term "*syngen*" is preferred in this case. A syngen is defined as a group of individuals capable of an exchange of genes [*1637*].

The number of mating types within a syngen can be very different. Two kinds of mating systems can be distinguished on formal grounds.

Bipolar Systems

Bipolar systems are characterized by the existence of *only two* mating types in each syngen.

One example of this is furnished by *Paramecium aurelia* where the phenomenon of mating types was first discovered [*1628*]. About 330 strains of this species have thus far been collected all over the world and been taken into culture. Tests of their mating relations led to the recognition of 14 syngens (1–14) with 28 mating types (I–XXVIII). As shown in Table 4, most syngens can be allotted to two groups (A and B). Reasons for this distinction will be given below. The special position of syngen 13 which falls into a group of its own (C) will be discussed in the section on the genetics of mating types (p. 251).

Aside from differences which prevent interbreeding (incompatibility), the syngens differ in the form and size of the individuals, the rate of fission, temperature dependence, the spectrum of realizable antigen properties (serotypes, p. 255), the behavior at conjugation and other characteristics. Most differences can only be defined statistically or by means of special tests. Although the geographic distribution is only imperfectly known, it is already clear that certain syngens (e.g. 1 and 2) are cosmopolitan and others are restricted to certain areas (e.g. 9 to Europe).

The syngens are in general strictly isolated from each other. Normal conjugation which involves the major part (about 95%) of the individuals and which leads to viable clones of exconjugants occurs only if mating types of the same syngen are combined. Combinations of mating types from different syngens usually do not lead to any reaction. In some cases mating is incomplete and does not lead to an exchange of migratory nuclei (Inc.) or a certain number of pairs is formed (10% or even 95%). However, exconjugants or the clones derived from them die out sooner or later.

From the way in which the mating types react with each other in such cases it can be concluded that *homology* exists between the mating types of the different syngens involved. This is expressed by classifying all mating types *into two general types* designated as "—" and "+" respectively (Table 4). Mating types belonging to the "—" group (uneven numbers) react only with the (evenly numbered) "+" mating types of other syngens and vice versa (for example, I with X and XVI, II with V, IX, XIII and XV, and so forth). The bipolar expression of the mating types within a single syngen thus seems to be based upon a bipolarity of the whole system which has become varied in the course of evolution.

The mating type usually remains unchanged in the course of asexual reproduction. It is handed on over many cell generations and must therefore be looked upon as *inheritable* in this respect. In spite of this, the mating type can be *changed*.

Table 4. *Paramecium aurelia. Mating types.* Numbers within the table indicate in percent how many pairs were maximally formed on mixing the mating types in question. Inc. indicates that the reaction was incomplete and observed in a few individuals only. Based on data from SONNEBORN *[1644]*

Group			A												B														C		
	Syngen		1		3		5		9		11		14		2		4		6		7		8		10		12		13		General Type
		Mating Type	I	II	V	VI	IX	X	XVII	XVIII	XXI	XXII	XXVII	XXVIII	III	IV	VII	VIII	XI	XII	XIII	XIV	XV	XVI	XIX	XX	XXIII	XXIV	XXV	XXVI	
A	1	I	0	95	0	0	0	40	0	0	0	0	0	0	0	0	0	0	0	0	0	0	0	Jnc.	0	0	0	0	0	0	−
		II		0	1	0	40	0	0	0	0	0	0	0	0	0	0	0	0	0	10	0	Jnc.	0	0	0	0	0	0	0	+
	3	V			0	95	0	0	0	0	0	0	0	0	0	0	0	0	0	0	0	0	0	40	0	0	0	0	0	0	−
		VI				0	0	0	0	0	0	0	0	0	0	0	0	0	0	0	Jnc.	0	0	0	0	0	0	0	0	0	+
	5	IX					0	95	0	0	0	0	0	0	0	0	0	0	0	0	0	0	0	0	0	0	0	0	0	0	−
		X						0	0	0	0	0	0	0	0	0	0	0	0	0	Jnc.	0	0	0	0	0	0	0	0	0	+
	9	XVII							0	95	0	0	0	0	0	0	0	0	0	0	0	0	0	0	0	0	0	0	0	0	
		XVIII								0	0	0	0	0	0	0	0	0	0	0	0	0	0	0	0	0	0	0	0	0	
	11	XXI									0	95	0	0	0	0	0	0	0	0	0	0	0	0	0	0	0	0	0	0	
		XXII										0	0	0	0	0	0	0	0	0	0	0	0	0	0	0	0	0	0	0	
	14	XXVII											0	95	0	0	0	0	0	0	0	0	0	0	0	0	0	0	0	0	
		XXVIII												0	0	0	0	0	0	0	0	0	0	0	0	0	0	0	0	0	
B	2	III													0	95	0	0	0	0	0	0	0	0	0	0	0	0	0	0	
		IV														0	0	0	0	0	0	0	0	0	0	0	0	0	0	0	
	4	VII															0	95	0	0	0	0	0	95	0	0	0	0	0	0	−
		VIII																0	0	0	0	0	60	0	Jnc.	0	Jnc.	0	0	0	+
	6	XI																	0	95	0	0	0	0	0	0	0	0	0	0	
		XII																		0	0	0	0	0	0	0	0	0	0	0	
	7	XIII																			0	95	0	Jnc.	0	0	0	0	0	0	−
		XIV																				0	0	0	0	0	0	0	0	0	+
	8	XV																					0	95	0	0	0	0	0	0	−
		XVI																						0	Jnc.	0	Jnc.	0	0	0	+
	10	XIX																							0	95	0	0	0	0	−
		XX																								0	Jnc.	0	0	0	+
	12	XXIII																									0	95	0	0	−
		XXIV																										0	0	0	+
C	13	XXV																											0	60	
		XXVI																												0	

In syngens of *group A* such a change occurs frequently after conjugation or autogamy. It is remarkable in this connection that the change does not take place in the exconjugant or the exautogamont itself but in cells derived from the first division after conjugation or autogamy. Thus, the two daughter cells of an exconjugant or an exautogamont of a cell belonging to mating type I may both be either mating type I or II, or they may form different mating types, i.e. I and II. The distribution of mating types among the two daughter cells is random, as shown in the example of Fig. 206. Of 139 exconjugants, both daughter cells belonged to

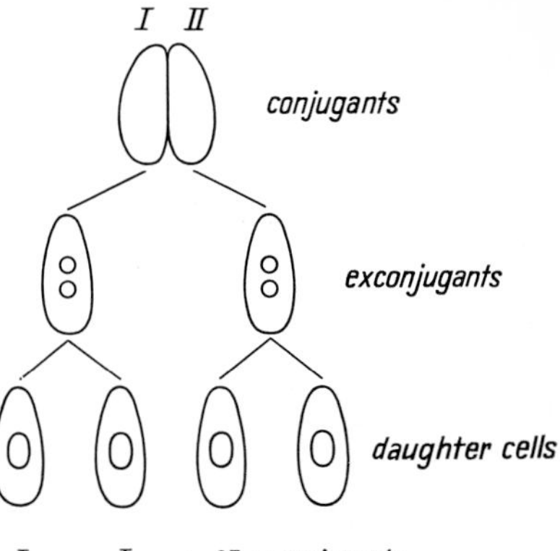

Fig. 206. *Paramecium aurelia* (group A, syngen 1). Change of mating type after conjugation. Below the two daughter cells of the left exconjugant the possible distribution of mating types and the frequency of this distribution in 139 examined exconjugants are presented. After data from SONNEBORN, 1937 [*1628*]

mating type I in 35 cases, to type II in 34 cases and in 70 exconjugants the two daughter cells belonged to different types. A clone which is pure with respect to mating type can therefore not be isolated by starting with exconjugants or exautogamonts. Instead, one of the two daughter cells must be isolated and allowed to reproduce asexually.

As the two macronuclear anlagen of *Paramecium aurelia* are distributed to the two daughter cells in the course of the first metagamic division (Fig. 193), it seems reasonable to assume that *determination of the mating types* takes place during the development of the new macronuclear anlagen. This assumption has been verified in a number of ways.

In one of the strains of syngen 1 more than two macronuclear anlagen are frequently formed in each exconjugant. Final fixation of the mating type is accordingly connected with a later fission when all macronuclear anlagen have been distributed and when they grow into new macronuclei in the descendants. The connection between mating type differentiation and macronucleus is also indicated by the fact that a cell with a macronucleus regenerated from a fragment of the old macronucleus which disintegrated during conjugation or autogamy (p. 200) always has the same mating type as the conjugant or autogamont from which this fragment originated.

The ratio of the two mating types formed after conjugation or autogamy can to a certain extent be influenced by external conditions. At the normal rearing temperature of 19° C the two types occur with equal frequency. Raising the temperature leads to a change in favor of the "+" types (II in syngen 1, VI in syngen 3, etc.). These studies showed that a relatively brief *sensitive phase* is passed through in the course of macronuclear development. It is only during this phase that the determination of mating types can be influenced by external conditions.

A change of mating type after conjugation or autogamy in the syngens of *group B* is a relatively rare event. As a rule, the clone resulting from an exconjugant or an exautogamont shows the same mating type as the conjugant or autogamont from which it originated. If, for example, mating types VII and VIII (syngen 4) are allowed to conjugate with each other, an exconjugant from the VII-conjugant will bring forth a clone belonging to mating type VII, and from the VIII-conjugant an exconjugant clone of mating type VIII will again arise.

However, in a relatively small number of cases a sudden change of mating type after conjugation or autogamy can also take place in syngens of group B. Further investigation proved that this change is regularly connected with the development of the new macronuclear anlage and can be influenced by temperature.

The difference between the two groups can therefore not be a basic one [*1187*]. The higher instability of mating type differentiation within group A is evidently due to the fact that cytoplasm plays only a minor role as a determination factor, whereas in group B a definite cytoplasmic condition is evoked by the macronucleus which then, for its part, determines a certain mating type for the developing macronuclear anlage. The fact that such cytoplasmic conditions can be maintained over many cell generations but can also on occasion suddenly change into another state will be treated further in the discussion of the so-called antigenic properties (p. 255ff.). At any rate, determination of mating types in both groups is by way of modification. As the exconjugants and their descendants have the same set of genes (p. 247), determination of the mating type cannot be due to genetic differences.

Four syngens could so far be isolated from *Paramecium multimicronucleatum*, which is closely related to *P. aurelia* [*92, 590, 1637*]. In addition to strains of syngen 2 in which the mating type of the daughter cell clones remains constant (III or IV), so-called cycler strains occur whose clones show a *daily periodic change* of the mating type. Some realize mating type III at night and mating type IV during the day or vice versa [*93, 94*].

The mating type of the "cyclers" at a given time can be tested with clones whose mating type remains constant ("non-cyclers"). The results of crossbreeding indicate that the difference is due to a single pair of alleles (*C* for cyclers, *c* for non-cyclers).

Diurnal periodicity is controlled by the alternation of light and darkness (timer); however, it persists in continuous light or dark culture (endogenous rhythm).

Of *Paramecium caudatum* 16 syngens are known so far, which differ partly in the size of the cell and the nucleus [*600—602*]. Autogamy is lacking but can be artificially induced [*1256*]. Instead, selfing (see below) takes place frequently in the clones, which impedes analysis of the mating system and genetic investigations. In some syngens (3, 12), determination of the mating types seems to be due to genetic factors [*761, 766, 768*].

Paramecium woodruffi and *P. calkinsi* also have bipolar mating systems [*42, 1637*]. Determination of the mating types is by way of modification.

Multipolar Systems

Most of the ciliates which have been examined thus far show a multipolarity of their mating system: in all syngens more than two mating types occur, every combination leading to conjugation.

In *Paramecium bursaria*, where this mode of differentiation was first discovered

[*149, 860*], four mating types have been found in some syngens (1, 3), and eight in others (2, 4, 5, 6) (Table 5).

Table 5. *Paramecium bursaria*. Reaction scheme of mating types in syngens 1 and 3 (a), resp. 2, 4, 5 and 6 (b). Based on data from BOMFORD, 1966 [*149*]

	I	II	III	IV
I	—	+	+	+
II		—	+	+
III			—	+
IV				—

a

	I	II	III	IV	V	VI	VII	VIII
I	—	+	+	+	+	+	+	+
II		—	+	+	+	+	+	+
III			—	+	+	+	+	+
IV				—	+	+	+	+
V					—	+	+	+
VI						—	+	+
VII							—	+
VIII								—

b

In *Paramecium bursaria* it can be easily proved, by *marking* one conjugant, that only individuals from different mating types conjugate. This species is stained green by zoochlorellae. In dark culture or after X-ray treatment the zoochlorellae can be made to disappear, leading to colorless cells. If a colorless clone of one mating type is brought together with an untreated clone of the other type, a pale individual will always conjugate with a green one.

As discussed earlier (p. 53), races of different degrees of polyploidy occur in *Paramecium bursaria* which can be distinguished from each other by the size of their micronuclei. If two mating types whose micronuclei are of different size combine, then one partner will always have a small micronucleus, the other a large one. Sometimes a third individual also attaches to a conjugating pair (Fig. 207). However, it never attaches to the conjugant with the same micronuclear size. The "odd" conjugant also carries out the progamic micronuclear division. The two pronuclei which arise in this manner fuse autogamically to a synkaryon [*245, 248*].

The mating system of *Tetrahymena pyriformis* has also been very thoroughly investigated. Until 1965, 12 syngens could be established, each comprising a definite number of mating types. Fifty mating types were found altogether. Syngen 1 with 7 mating types is cited as an example.
Multipolar systems are also found in hypotrich ciliates of the genera *Oxytricha* [*1585*], *Stylonychia* [*41, 436*], *Euplotes* [*735, 736, 907, 909, 916, 1234, 1799, 1800*] and *Uronychia* [*1425*]. The largest number of mating types within a syngen, a total of 48, was found in *Stylonychia mytilus* [*41*].
Detailed studies on the determination of mating types have only been carried out with some syngens of the species mentioned above. In syngen 1 of *Tetrahymena pyriformis*, it takes place in a manner similar to that of *Paramecium aurelia* (group A and B), i.e. through determination of the developing macronuclear anlage by way of modification. Since autogamy does not take place in *Tetrahymena pyriformis*, a change of mating type can, as a rule, only take place after conjugation. As in the syngens of group A of *Paramecium aurelia* (p. 215), the daughter cell clones in which different macronuclear anlagen develop can also realize different mating types. If a sufficiently large number of clones of such daughter cells are available, it can be shown that they are distributed among seven mating types. The individual types occur with fairly constant frequency under the same conditions. Thus, determination involves essentially the selection of one mating type from a spectrum of seven possible ones [*1186, 1211–1213*].

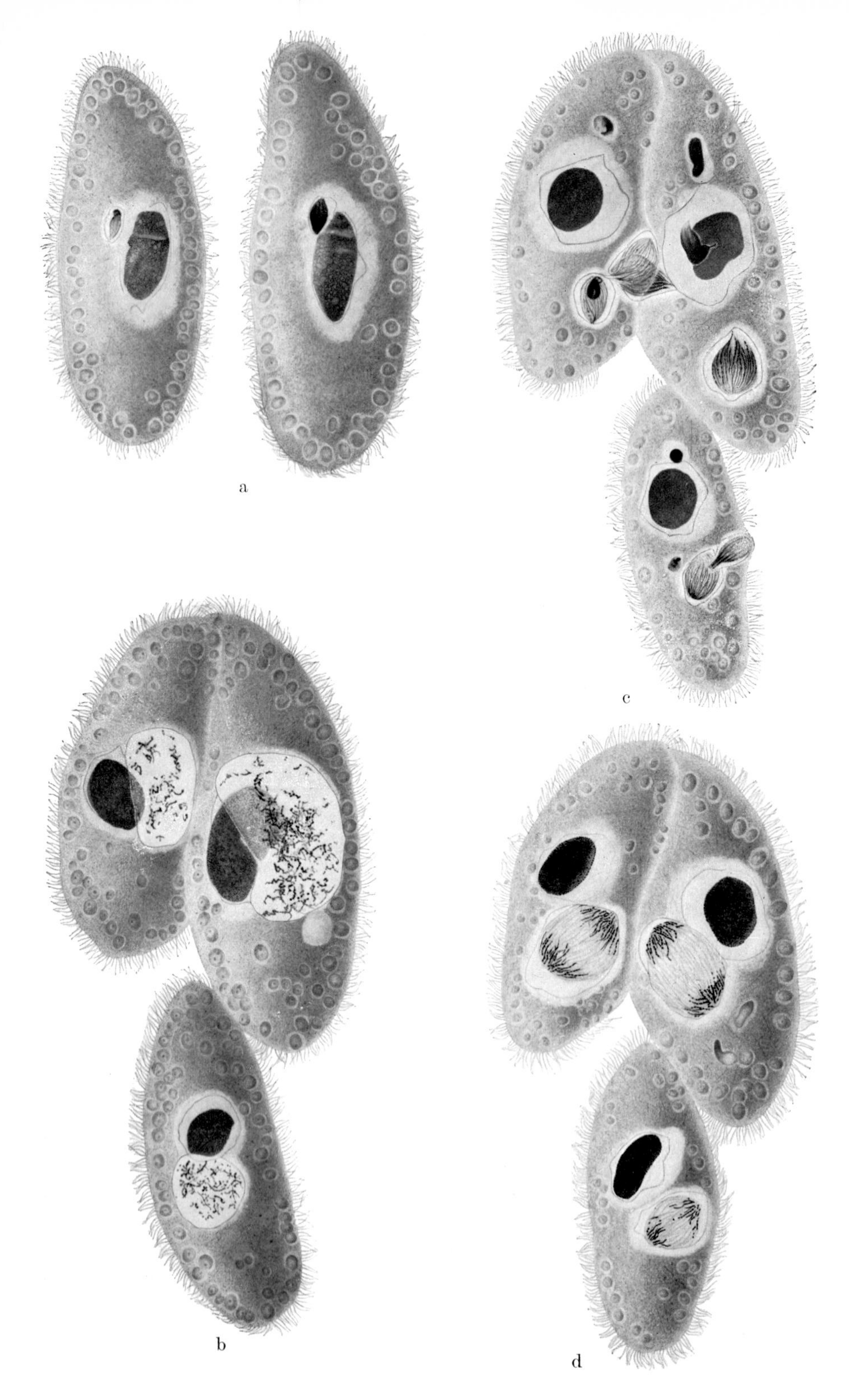
a
b
c
d

Fig. 207. *Paramecium bursaria*. Mating of three individuals. The odd "conjugant" attaches itself to the partner of the opposite mating type (distinguishable by the size of the micronuclei) and carries out an autogamy. a Individuals of the two races which mate with each other: left, race *Fd* with a small, probably not polyploid, micronucleus; right, race McD_3 with a large polyploid micronucleus. b Conjugation triplet. Late prophase of the first progamic division. c Exchange of pronuclei between the two actual conjugants. d Anaphase of the first division of the synkaryon. Sublimate alcohol, iron hematoxylin. × 610. After CHEN, 1946 [*248*]

◀

It has, however, been proved for certain syngens of *Paramecium bursaria* (1, 2) and *Tetrahymena pyriformis* (2, 8), a species of *Glaucoma*, all species of *Euplotes* which have been investigated so far, and *Uronychia transfuga*, that the mating types are determined directly by genetic factors and not by modification. The way this is achieved is different in the individual species and will be discussed in detail in the chapter on genetics.

Selfing

Although, as a rule, conjugation is only possible when clones belonging to different mating types are mixed together, in certain strains or under certain conditions intraclonal conjugation can also take place. This phenomenon, designated as *selfing*, still has to be examined thoroughly.

It may at first seem that the incompatibility barrier of mating type differentiation is abolished in the event of selfing, so that each cell can conjugate with any other one. However, a more detailed analysis of selfer clones showed that intraclonal conjugation is preceded by differentiation of a complementary mating type and that the partners belong to different mating types. The problem thus narrows down to how a complementary mating type can appear in a daughter cell clone (p. 215) which would normally differentiate uniformly.

In some strains of *Paramecium aurelia* intraclonal conjugations could be shown to be due to the fact that daughter cells of exconjugants or exautogamonts changed their mating type only after a few fissions and not right away. Since their descendants differed in this respect, the same clone contained individuals which belonged to the old mating type and others which had already expressed the new mating type [*915*, *1630*].

However, selfing cannot, as a rule, be explained as a phenomic lag of this sort. Instead, it is due to the fact that a differentiation which had been constant over many cell generations has become unstable. If single cells are isolated from "selfer clones", selfing occurs again in the subclones.

It could be shown that selfer clones of syngen 7 of *Paramecium aurelia*, which formerly were of mating type XIII, suddenly changed to mating type XIV. Since the change never took place in the opposite direction, it is thought that the mating type substance (p. 222) corresponding to mating type XIII represents a precursor of the substance corresponding to mating type XIV [*1719*, *1720*].

That irreversibility of this kind is not the general rule is shown by the example of the "cyclers" of *Paramecium multimicronucleatum* whose mating type changes with a diurnal periodicity (p. 216). Since during the time of change individuals of either mating type occur side by side, selfing is possible.

While with such rapid changes of mating type the assumption that the state of determination of the macronucleus changes wholly can scarcely be avoided, studies of selfer clones of syngen 1 of *Tetrahymena pyriformis* have led to a totally different interpretation. If single cells are isolated from selfer clones and allowed to multiply, selfing takes place again under good nutritional conditions. In starved cells selfing is suppressed and the subclones finally become "pure" with respect to a certain mating type. It was shown that only two mating types are formed in most selfer clones and that certain combinations occur preferentially. Since mating type differentiation is bound to the macronucleus, it seemed likely that the appearance of different mating types in selfer clones is due to *heterogeneity of the macronucleus*. The subunits of the macronucleus (p. 113) might become differentially determined during growth of the macronuclear anlagen. Stabilization of subclones would then be due to progressive *assortment of the subunits* and concomitantly increasing uniformity of the macronuclei. On the basis of this hypothesis and quantitative data obtained in the analysis of selfer clones it has been estimated that a macronucleus of *Tetrahymena pyriformis* should contain about 45 subunits after fission [*33, 1210, 1211, 1505*].

Although this hypothesis is attractive to explain the situation in *Tetrahymena pyriformis,* it certainly cannot be used to account for the phenomenon of selfing in other cases. In *Euplotes patella,* cells of certain mating types release specific substances which induce selfing among the cells of other mating types (p. 252). If clones of different mating types are mixed, cells of different as well as of the same types will conjugate with each other as shown by labeling experiments. The concept of "mating type" thus is different in this case from its original definition [*916*]. Curiously enough, the closely related *Euplotes eurystomus* shows precisely the opposite behavior. Intraclonal conjugations are the rule in this species. However, if clones of different "mating types" are mixed together, selfing between cells of the same type is suppressed and mating occurs predominantly between individuals of different types [*907*].

Since it is highly improbable that substances which either induce or suppress selfing play any role at all in the natural biotope of the species, the suspicion arises that we are dealing merely with "laboratory effects". Their analysis is, however, of no less importance in order to understand the physiology of conjugation (see below).

In *Euplotes crassus,* selfing can take place if clones heterozygous with respect to the mating type alleles have reached a certain age [*738*]. This phenomenon will be discussed later (p. 273).

While it is common to all cases of selfing discussed so far that differentiation as such remains constant despite an intraclonal change of mating type, the so-called *chemically induced conjugation* seems to be due to total abolishment of the incompatibility barrier. If paramecia are treated with certain substances such as K, Mg, heparin, EDTA, cells of the same type, even those of different syngens or species of the *aurelia*-group, can be brought to conjugation. The agglutination reaction (p. 222) does not take place in this case. Aside from the fact that all possible "irregularities" occur with respect to the relative positions of the conjugants, conjugation as such takes a perfectly normal course.

However, the exconjugants or the cells derived from them are not capable of living [*1142–1145*].

Abolishment of the incompatibility barrier is probably due to a change of the ciliary envelope. EDTA (ethylene diamine tetraacetic acid) might, for example, bind calcium incorporated in the ciliary envelope [*762*].

Even though there is no doubt that mating type differentiation is of widespread occurrence among ciliates, the possibility cannot be excluded that species without this differentiation exist. In such species each cell should be able to mate with any other one provided that all external and internal conditions otherwise required for conjugation are realized.

d) Physiology of Conjugation

Even if two clones which are combined belong to appropriate mating types, conjugation does not always follow. Besides "complementarity" as expressed in mating type differentiation, certain conditions must be fulfilled before mating is possible. Light and temperature are among the *external factors* which influence the readiness to conjugate. Even the time of day may play a role. A definite state of nutrition is indispensable since starved or overfed individuals generally will not conjugate.

In addition to these conditions whose importance may vary in the different syngens or strains, the readiness for conjugation depends on a more or less prolonged *immaturity period* in most ciliates. This means that the descendants of exconjugants are not immediately ready for another conjugation but must pass through a certain number of cell divisions before they enter the *period of maturity*.

The immaturity period of *Paramecium aurelia* is relatively short. There are strains whose clones are ready for conjugation again as soon as reorganization of the nuclear apparatus is completed. Even in the other strains the immaturity period lasts maximally 9 days (= 35 cell divisions). Once the state of maturity is reached, it may be interrupted briefly by autogamy but cannot be abolished by it.

In *Paramecium bursaria*, whose immaturity period may last for weeks or months, the state of complete maturity is preceded by a so-called *period of adolescence*. During this phase clones cannot yet conjugate with all other mating types of the cor-

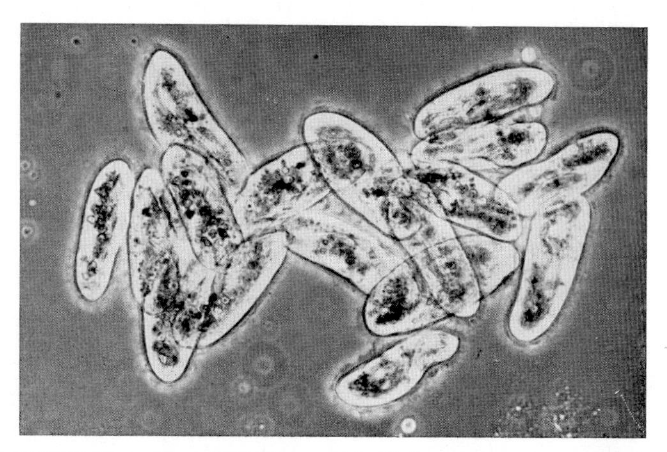

Fig. 208. *Paramecium woodruffi*. Agglutination reaction. × 150. Micrograph by D. Ammermann

responding syngen but only with some of them. In syngen 1, where this phenomenon has been studied in detail [*1588–1592*, *1594*], clones of the adolescence period conjugate with mature clones of only two (instead of three) mating types. An adolescent clone could thus belong to one or the other complementary mating type (e.g. $\overline{\text{I–II}}$). Only after transition to complete maturity can it be decided to which of the two types (e.g. I or II, Table 6) it belongs.

Table 6. *Paramecium bursaria*. Reaction scheme of two clones which cannot be distinguished in adolescence and which realize mating types I and II when mature

Immaturity				Adolescence				Maturity					
I	II	III	IV	I	II	III	IV	I	II	III	IV	←	Test clones
—	—	—	—	—	—	+	+	—	+	+	+	→ I	Final
—	—	—	—	—	—	+	+	+	—	+	+	→ II	mating type

If all conditions are optimal, species of *Paramecium* always react to a mixing of clones of different types with an *agglutination reaction* (Fig. 208) which reminds one of the reaction between gametes of different sexes (p. 164). Groups of many – often several hundred – individuals are formed which finally dissociate into single pairs.

Agglutination is due to the interaction of substances which are specific for the different mating types and are therefore termed *mating type substances* [*1120*]. Experiments with enzymes have shown that they contain protein. Work with inhibitors (puromycin, actinomycin D) indicated that these proteins are formed continuously and that their synthesis is RNA-dependent [*143*].

Their localization within the cilia is proved by mixing isolated, centrifuged cilia with living cells [*491*]. If the cells belong to a complementary mating type, the cilia remain attached to the peristomal region where contact normally takes place in the course of the agglutination reaction. In *Paramecium bursaria,* whose readiness to conjugate follows a diurnal periodicity, it could be shown that also the cilia would agglutinate only at the corresponding time of day (i.e. at noon) and that the region at which they stick enlarges and becomes smaller periodically [*326–328*, *1594*]. The mating type substances are thus not constant components of the cilia but are temporarily incorporated into them. These substances retain their reactiveness even if the cells themselves are no longer alive. Killed individuals can still agglutinate with cells of the complementary mating type [*1119*, *1120*].

In syngens with two mating types (e.g. *Paramecium aurelia*), the assumption of one substance per mating type suffices. In each syngen two substances are complementary ($\underline{A}_1$ and $\underline{a}_1$ in syngen 1, $\underline{A}_2$ and $\underline{a}_2$ in syngen 2, and so forth), while the homology of mating types (p. 213) might be due to structural similarity of the substances (e.g. $\underline{A}_1$, $\underline{A}_2$ . . . and $\underline{a}_1$, $\underline{a}_2$ . . .).

The peculiar phenomenon that the number of mating types in the syngens of *Paramecium bursaria* follows the geometric progression 2^n led to the hypothesis that there are two pairs of complementary substances ($\underline{A}$–$\underline{a}$, $\underline{B}$–$\underline{b}$) in syngens with four mating types (1, 3) and three such pairs ($\underline{A}$–$\underline{a}$, $\underline{B}$–$\underline{b}$, $\underline{C}$–$\underline{c}$) in syngens with eight mating types (2, 4, 5, 6) [*1120*].

In the syngens with four mating types, for instance, the following combination of substances would be possible: AB, aB, ab, Ab. If we assume that a reaction always takes place if either A and a (α-reaction) or B and b (β-reaction) are distributed to different individuals, the following scheme results (Table 7):

Table 7. *Paramecium bursaria*. Reaction scheme of a syngen with four mating types. According to the hypothesis of METZ, 1954 [*1120*], the reactions are due to the interaction of two pairs of complementary mating type substances: if the cells differ in A-a, an α-reaction takes place, and if they differ in B-b, a β-reaction occurs

Mating types		I	II	III	IV
	Substances	AB	aB	ab	Ab
I	AB	—	α	α, β	β
II	aB	α	—	β	α, β
III	ab	α, β	β	—	α
IV	Ab	β	α, β	α	—

This seemingly very formal hypothesis is supported by the *analysis of adolescence effects*. During adolescence the clones would be able to produce only one of the two substances and thus be capable only of an α- *or* β-reaction. They should, in any case, be able to react with two of the four clones used for the testing. In the maturity period, the other substance would be formed and both the α- *and* the β-reaction should be possible.

There were indeed no adolescent clones which reacted with only one of the four test clones. In addition, only the adolescent clones shown in Table 8 were observed while combinations I—III and II—IV, which would be impossible on the basis of the hypothesis, were not observed.

Table 8. *Paramecium bursaria*. Syngen 1. Reaction of adolescent clones with mature test clones. After SIEGEL and COHEN, 1963 [*1594*]

		Mature test clones			
		I	II	III	IV
Adolescent clones	Substances	AB	aB	ab	Ab
I—IV	A	—	α	α	—
II—III	a	α	—	—	α
I—II	B	—	—	β	β
III—IV	b	β	β	—	—

The assumption of two pairs of complementary substances is further supported by the fact that the reactions between adolescent clones were restricted to two possibilities, viz. $\overline{\text{I—IV}} \times \overline{\text{II—III}}$ (α-reaction) and $\overline{\text{I—II}} \times \overline{\text{III—IV}}$ (β-reaction), while other reactions did not take place (Table 9).

Table 9. *Paramecium bursaria*. Syngen 1. Reaction among four adolescent clones. After Siegel, 1961 [*1588*]

Adolescent clones		$\overline{\text{I—IV}}$	$\overline{\text{II—III}}$	$\overline{\text{I—II}}$	$\overline{\text{III—IV}}$
	Substances	$\underline{\text{A}}$	$\underline{\text{a}}$	$\underline{\text{B}}$	$\underline{\text{b}}$
$\overline{\text{I—IV}}$	$\underline{\text{A}}$	—	α	—	—
$\overline{\text{II—III}}$	$\underline{\text{a}}$	α	—	—	—
$\overline{\text{I—II}}$	$\underline{\text{B}}$	—	—	—	β
$\overline{\text{III—IV}}$	$\underline{\text{b}}$	—	—	β	—

If we assume that three pairs of complementary substances are effective in syngens with eight mating types, the following combinations result: $\underline{\text{A}}\,\underline{\text{B}}\,\underline{\text{C}}$, $\underline{\text{A}}\,\underline{\text{B}}\,\underline{\text{c}}$, $\underline{\text{A}}\,\underline{\text{b}}\,\underline{\text{C}}$, $\underline{\text{A}}\,\underline{\text{b}}\,\underline{\text{c}}$, $\underline{\text{a}}\,\underline{\text{B}}\,\underline{\text{C}}$, $\underline{\text{a}}\,\underline{\text{B}}\,\underline{\text{c}}$, $\underline{\text{a}}\,\underline{\text{b}}\,\underline{\text{C}}$, $\underline{\text{a}}\,\underline{\text{b}}\,\underline{\text{c}}$.

This is supported to a certain extent by the observation that reactions between different syngens are restricted to four of the eight mating types. One might think that of the substances of one syngen only one (e.g. $\underline{\text{A}}_2$ of syngen 2) has the degree of complementarity to a substance of the other syngen (e.g. $\underline{\text{a}}_4$ of syngen 4) which is required for a reaction.

Final proof of the correctness of the hypothesis could, however, only be obtained through *genetic analysis* (p. 254). It showed that the formation of the substances is controlled in syngen 1 by two, in syngen 4 by three loci with two alleles each (*A—a, B—b, C—c*).

The agglutination reaction *activates* the cells, i.e. they unite in pairs at the holdfast region and lose the ability to agglutinate with other cells. Further processes (formation of the paroral cones, start of micronuclear meiosis, disintegration of the macronucleus) follow necessarily. That this sequence of reactions is also gene-dependent is demonstrated by the mutant *CM* ("can't mate"). Although these cells show a typical agglutination reaction with those of the other mating type, all other phenomena of conjugation fail to occur. Even the ability to fuse with their partner at the holdfast region is blocked. Besides, they remain permanently capable of agglutination. Although they themselves cannot be activated, *CM*-individuals can activate other cells. That only the sequence of reactions leading to conjugation is blocked is shown by their ability to undergo normal autogamy [*1121*]. The very occurrence of autogamy shows that union at the holdfast region is not the prerequisite of the other phenomena of conjugation. The interrelationships of the latter are not yet clarified. The only thing which seems certain is that meiosis is not necessary for the formation of the paroral cones and for the disintegration of the macronucleus, as these processes can also be evoked in "conjugants" without micronuclei.

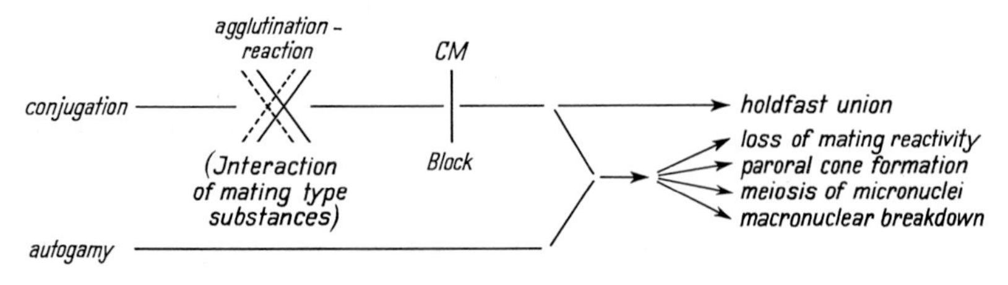

According to Metz, 1948 [*1119*]

Whatever is known so far about the interrelationships of sequential processes in conjugation and autogamy of *Paramecium aurelia* is summarized on the opposite page (below).

The agglutination reaction, which, in species of *Paramecium*, follows immediately upon mixing of clones of different types, is by no means observed in all ciliates. If clones of different types of *Tetrahymena pyriformis* or the above mentioned hypotrichs *(Oxytricha, Euplotes, Stylonychia)* are brought together, 1 to 2 hours will elapse before a reaction can be observed. There are experimental indications that chance contacts between different cells in the course of this "waiting period" induce an enhanced formation of mating type substances. This makes final fusion of the partners possible [*741*]. That the release of substances demonstrated for *Euplotes patella* and *E. eurystomus* probably represents a "laboratory effect" without importance in nature has been pointed out above. Perhaps the function of the "mating play", which precedes fusion of the partners in *Stylonychia mytilus* (Fig. 280) and other hypotrichs, is simply to stimulate the formation of further mating type substances.

In *Blepharisma intermedium*, the complementary mating types (I and II) were found to excrete two kinds of "gamone" into the medium. Gamone II ("blepharismin") is responsible for inducing mating type I to conjugate. Recently, it could be isolated in crystalline form and structurally elucidated [972a, 1145a].

IV. Retrospect

In a comparative survey of events connected with fertilization it must be kept in mind that a wide gulf exists between the situation in Bacteria and that in Protozoa (p. 5).

The so-called "*conjugation*" of *Bacteria* leads to the transfer of a more or less extended segment of the "chromosome" from the donor to the recipient cell. Within the recipient, which thus becomes a "*merozygote*", the transferred segment can now exchange pieces with the corresponding segment of the "chromosome" at hand.

In all *eukaryotes* fertilization leads to karyogamy. In its course both partners combine their whole chromosome sets, leading to the formation of a "*holozygote*". Segmental interchange between homologous chromosomes takes place at *meiosis*, which may be zygotic, gametic or intermediary (p. 78).

Karyogamy is the prerequisite for meiosis and thus also for any recombination of the parental genotypes. Hence, sexual processes are of importance for evolution because they favor the formation of new adaptations. On the other hand, it cannot be denied that this does not "explain" the biological meaning of sexual differentiation. A search for "sexual reproduction without sexuality" is therefore legitimate. However, no case has so far been described which would seem to be sufficiently verified by experimental methods [*721*]*.

Sexual differentiation of isogametes, on the other hand, could be demonstrated in numerous species. No case of more than two sexes has ever been found. To our present knowledge, the same *bipolarity* of sexual differentiation holds for isogamety

* Geneticists who view the phenomena of life from a more theoretical standpoint have often looked upon all processes as sexual which lead to recombination (including, for example, the so-called transduction). However, this extension of a biologically clearly defined concept (p. 153) is misleading since it gives the impression that sexuality must necessarily be connected with processes of recombination although it represents in reality a separate phenomenon with its own range of problems.

as for anisogamety and oogamety. This justifies the interpretation that all types of gametic differentiation can, in principle, be looked upon as *homologous*, although the designation "male" or "female" may be problematic in a given case.

Bipolar sexual differentiation of gametogamous Protozoa has not only been found among dioecious but also in monoecious species. The fact that sexual differences may even be recognizable in autogamically fusing gametes or gamete nuclei is especially impressive.

While the situation is completely uniform at the level of the gametes, the gamonts may represent different patterns depending upon their monoecy or dioecy.

If *gametogamy*, i.e. the formation of free-swimming gametes which copulate outside the gamonts, is taken to represent the basic mode of sexual reproduction, the relationships to autogamy and gamontogamy as the two remaining forms of sexual reproduction can be summarized as in Fig. 209.

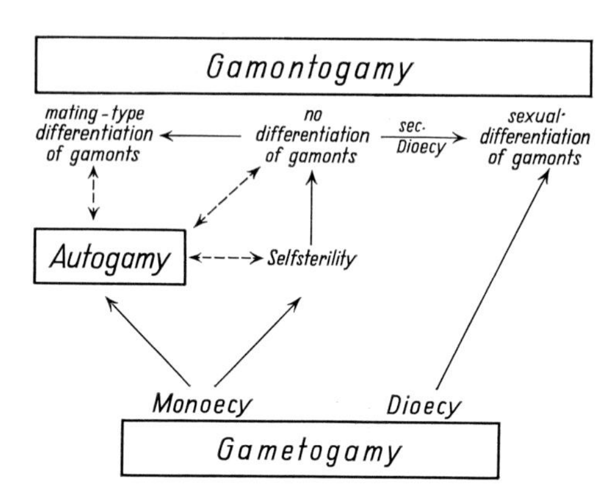

Fig. 209. The various types of sexual reproduction and differentiation in Protozoa and their interrelations. After GRELL, 1967 [*675*]

In *dioecy*, sexual differentiation of the gametes is, in a manner of speaking, extended to include the gamonts. The Coccidia may be taken as an example. Even in the gametogamous Eimeridae and Haemosporidae the gamonts may be clearly different. Pronounced anisogamonty prevails especially among the gamontogamous Adeleidae where the microgamont forms only a few microgametes and is thus much smaller than the macrogamont.

In the case of *monoecious* gamonts evolution can lead either to autogamy or to gamontogamy. In some species, e.g. the foraminiferan *Rotaliella heterocaryotica*, both forms of sexual reproduction are found side by side. Allogamy is only then ensured if the mating gamonts are *self-sterile*, i.e. if gametes or gamete nuclei from one gamont can only fuse with those from the other gamont, but not with each other.

The self-sterility barrier must temporarily be lifted if autogamy and gamontogamy exist side by side, while it is permanent in exclusively gamontogamous species. With bipolar differentiation of gametes or gamete nuclei, only cells or nuclei of different sex of both gamonts can of course fuse with each other.

The example of *Metarotaliella parva* provides evidence for the existence of gamontagamous Protozoa whose *gamonts are not differentiated into two types*. Every gamont

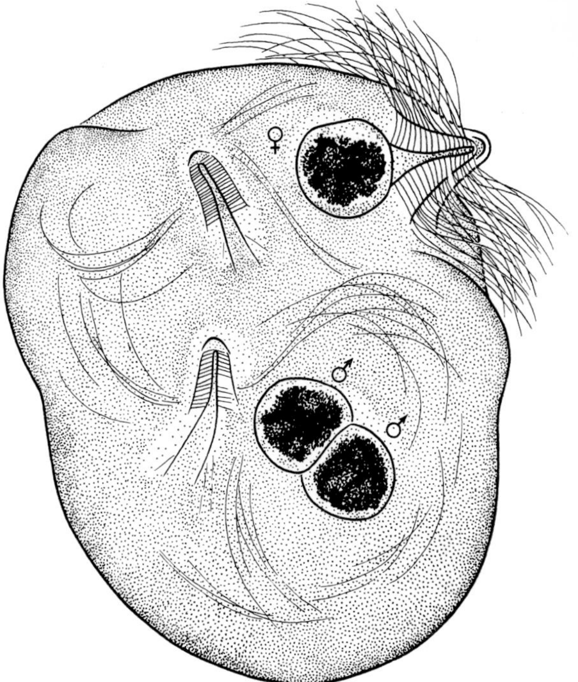

Fig. 211. *Trichonympha*. Abnormal fertilization stage. Two male gametes have penetrated into a female gamete (compare with Fig. 164). The nuclei of the two male gametes have fused. × 800. After CLEVELAND, 1957 [*293*]

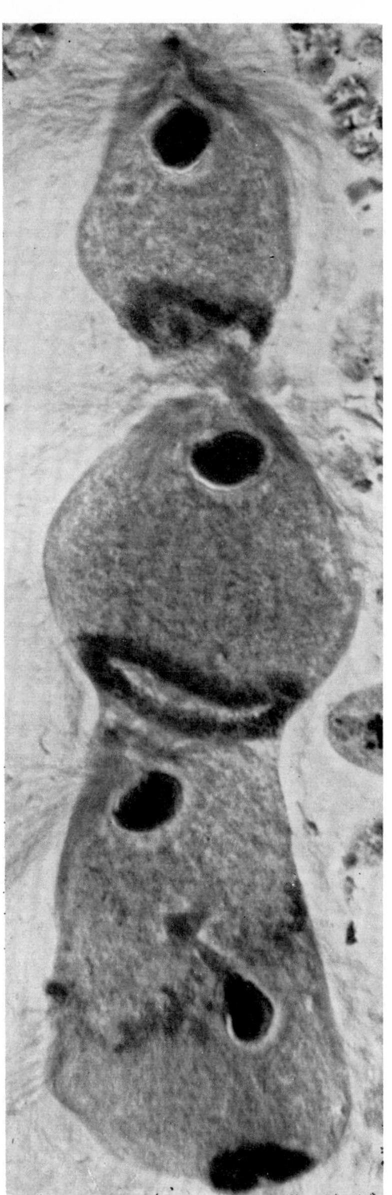

Fig. 210. *Trichonympha*. Chain of four gametes. Original micrograph by L. R. CLEVELAND

can mate with any other one in this foraminiferan species. Allogamy is nevertheless ensured.

This latter case may indeed provide the key for understanding the situation in *ciliates*. One could imagine that even in ciliates which must be regarded as monoecious on the basis of nuclear behavior a type-differentiation of gamonts was originally nonexistent. It cannot be excluded that even today ciliates may exist where any individual may conjugate with any other under suitable conditions (p. 221).

From this viewpoint, the differentiation into *mating types* represents a secondary phenomenon which has evolved on the basis of monoecy and is similar in this respect to the phenomenon of self-sterility in flowering plants. Although mating type differentiation shows some points of analogy to sexual differentiation (contact mechanisms, phenotypic and genetic determination), it is not advisable to designate it, too, as "sexual".

"*Anisogamonty*" is in ciliates undoubtedly a form of secondary dioecy which evolved from monoecy by suppression of the alternative sex. The micro- and macrogamonts of Peritricha are just as sexually differentiated as those of the Adeleidae.

Whether this secondary dioecy was preceded by a state without mating type differentiation (see Fig. 209), or whether suppression of the alternative sex occurred in the presence of two complementary mating types is a question of minor importance.

Sexual differentiation, which has reached such a varying degree among Protozoa, raises many questions whose solutions have only begun. It follows two divergent directions from the common basis of a bisexual reaction norm.

Although we know that its direction is determined both by genetic factors and by modification, the actual mechanisms of regulation are still completely unknown.

Some insights are gained whenever an abnormal course of fertilization indicates *incomplete or faulty differentiation* of gametes. In the polymastigid, *Trichonympha*, gametes of different sexes are derived from a differential division (p. 168). It often happens that the posterior gamete of a copulating pair also forms a fertilization cone which can then serve for attachment of another gamete. Fig. 210 shows four such gametes, of which the last two have already fused to a large extent.

Cases of this kind have been interpreted as "relative sexuality" [*721*]. However, the only certain fact is that the gametes in the middle behave as if they were "male" at the front and "female" at the posterior end.

A more simple interpretation would be that the cells are of opposite differentiation at their poles, and that they thus become cellular gynandromorphs [*276*].

Two male gametes of *Trichonympha* penetrate, on occasion, into a female gamete. Usually, the two male nuclei separate from their extranuclear organelles and both migrate to the female nucleus. Sometimes, this migration does not take place, and the nuclei of the male cells fuse with each other (Fig. 211).

One might, therefore, raise the question whether the gamete nuclei are indeed sexually differentiated. The possibility remains that following copulation the cytoplasmic conditions are such as to favor nuclear fusion. This view is contradicted, however, by the situation in *Notila* where of the four gamete nuclei of a pair of gamonts only the male ones fuse with the female ones, and never the male nuclei with one another (Fig. 191).

F. Alternation of Generations

This term is applied if in one species two or more generations* with different modes of reproduction alternate.

In contrast to higher organisms which *grow* by cell division, protozoan individuals *reproduce* by cell division. Their alternation of generations is therefore designated as *primary*.

Since the formation of gametes is usually connected with a special mode of reproduction, most sexually differentiated Protozoa alternate between sexual (gamogony) and asexual reproduction (agamogony). This type of alternation of generations is therefore the most frequent one. Alternation between different modes of asexual reproduction, on the other hand, is exceptional.

The alternation of reproductive modes is in most cases not strictly determined (*facultative* alternation of generations). Asexual reproduction can be repeated indefinitely, and it is only through a change of environmental conditions that sexual reproduction is triggered.

In many cases such a change is brought about automatically, if, for instance, one reproductive mode brings about those conditions which initiate the other mode. In many Sporozoa asexual reproduction (schizogony) has a natural limit given by the size of the host, the infected organ or tissue. After a definite number of asexual reproductive acts, sexual reproduction (gamogony) is therefore *regularly* initiated. It concludes the life cycle in that particular host.

It is only in a few cases that the rhythm of alternative modes of reproduction is strictly determined because *internal* conditions arising in the course of cell growth lead to the formation of daughter cells which can only reproduce in a different fashion (*obligatory* alternation of generations).

Besides the kind of alternating reproductive modes, the relative position of chromosome reduction in the life cycle can also be used to characterize the different types of alternation of generations (Fig. 212). If the change of reproductive modes is not connected with chromosomal reduction, i.e. if both modes take place in the same nuclear phase, we speak of a *homophasic* alternation of generations. If reduction is intermediary, i.e. if one generation is diploid and the other haploid, we have a *heterophasic* alternation of generations. We can thus distinguish between three types of the primary alternation of generations:

1. In the *haplo-homophasic* alternation of generations only the zygote is diploid; all reproductive processes take place in the haploid phase.

In the *Phytomonadina* the alternation of generations is facultative. Agamogony may consist of binary or multiple fission.

The alternation of generations found in *Sporozoa*, on the other hand, is more or less

* A generation is that developmental phase of a species which lasts from one reproductive act to the next. This is in contrast to usage in genetics where "generation" denotes the whole developmental cycle of a species. The complete life cycle of a sporozoon involving one "alternation of generations" would then represent *one* generation.

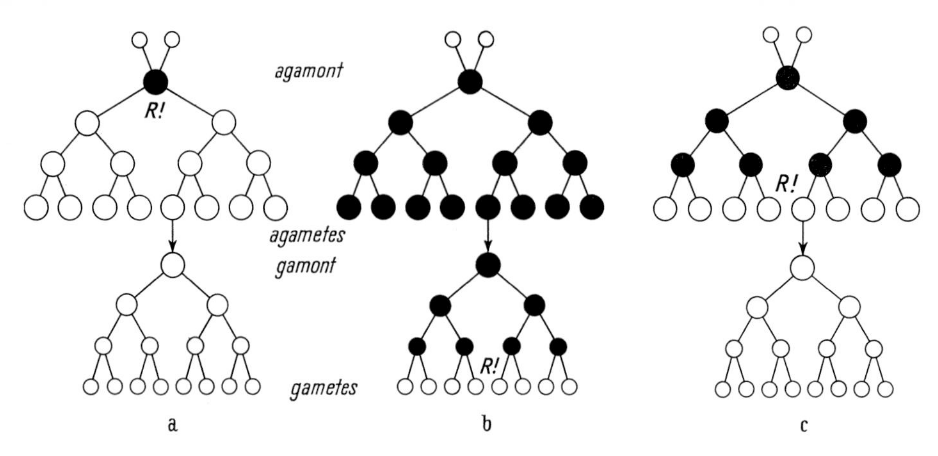

Fig. 212. Diagram of the different types of alternation of generations in Protozoa. Haploid cells: white. Diploid cells: black. R! Meiosis. a Haplo-homophasic, b diplo-homophasic, c heterophasic alternation of generations

determined by their parasitism. One mode of asexual reproduction found in all Sporozoa, *sporogony*, is a multiple fission. It leads to the formation of sporozoites which are transmitted via the spore to another host. Many Sporozoa (Schizogregarinida, Schizococcidia) have still another mode of multiple fission, *schizogony*. We have thus three different modes of reproduction in the life cycle of a species, i.e. sporogony, schizogony and gamogony. The change from schizogony to gamogony is facultative, but that from gamogony to sporogony is obligatory.

2. In the *diplo-homophasic* alternation of generations, the gametes represent the only haploid phase. All reproductive processes thus take place in the diploid phase.

The facultative type of this alternation of generations is found only among the sexually differentiated *Heliozoa (Actinophrys, Actinosphaerium)* and in *ciliates*. Binary fission is the usual mode of asexual reproduction.

3. In cases of *heterophasic* alternation of generations the agamont is diploid and the gamont haploid.

While heterophasic alternation of generations is widespread in the plant kingdom, the haploid gametophyte representing the sexual and the diploid sporophyte the asexual generation, it is found among Protozoa only in the *Foraminifera*. Agamogony and gamogony are by multiple fission in this case. Their alternation is obligatory.

The concept of alternation of generations is, in general, restricted to cases where one generation reproduces sexually. It can, however, happen that *two different asexual* modes of reproduction alternate. One example is provided by the suctorian, *Tachyblaston ephelotensis* (Fig. 431) with a regular alternation of a parasitic and a free-living *("Dactylophrya")* generation.

This case is of special interest because it illustrates how an alternation of generations might arise. If we take the free-living generation as the original one, the parasitic generation must have been due to swarmers becoming parasites with the ability to reproduce. As food intake became restricted to the parasitic generation, the free-living one proceeded to successive multiple division with no growth phases between the acts of budding.

Alternation of generations in *Metazoa* might frequently have arisen by way of metamorphosis. This pathway is directly recognizable in many cases (e.g. digenic trematodes, *Echinococcus granulosus*, and others).

G. Heredity

The totality of the *hereditary factors*, or *genes*, forms the *genotype*, the totality of the *characters* or *phenes*, the *phenotype*. There is no direct relationship between the hereditary factors and phenotypic traits. Instead, the genes represent the *reaction norm* which leads to the manifestation of a given phenotype only through the interplay with definite internal and external conditions. The genes are bound to structures which are handed on from cell to cell and have the ability for identical reduplication. The chromosomes of the cell nucleus play a key role among these structures. Their regular distribution in the basic processes of cellular reproduction (mitosis, karyogamy, meiosis) accounts for the inheritance of most traits. In haploid cells every gene is present only once, in diploid cells twice. Corresponding genes of homologous chromosomes are called *alleles*.

The material equivalents of the genes are the *nucleic acids*. The genetic information is determined by the sequence of nucleotides in them. Their ability for identical replication is the basis of the transmission of genetic information without changes.

Not all of the characters which comprise the phenotype of a species can be attributed to chromosomal genes. The cytoplasm can also contain specific constituents which transmit genetic information. Protozoa have been found especially suitable to demonstrate "extrachromosomal inheritance".

I. Mutability

All sudden changes of the hereditary material are called *mutations*. The geneticist, who wants to gain information about the inheritance of a trait, must endeavor to have as many strains or races with different mutations (mutants) available as possible. He is aided by the fact that the normally rather low mutation rate can be enhanced considerably by treatment with ultraviolet light (UV), ionizing radiations (e.g. X-rays) and the so-called mutagenic substances (e.g. nitrous acid, urethane, antibiotics, carcinogens).

If cross-experiments are impossible, the *nuclear transfer* method can give some information about the site of the mutational change. In *Amoeba proteus* mutations have been induced by the carcinogen N-methyl-N-nitrosourethan, which led to changes recognizable in the light microscope. After tranferring the nucleus from a mutant strain into an enucleated normal amoeba, the latter produced a clone showing the characteristics of the mutant strain. Hence, the nucleus, especially its chromosomes, must be the site of action of the mutagen. After transferring a mutant nucleus into a normal amoeba which retained its nucleus, it was the latter which determined the phenotype [*1253*, *1254*, *1834*].

Cross-experiments are necessary, however, if the geneticist wants to know whether a mutational change depends on a single hereditary factor. Such mutations are called *gene-mutations*.

In *haploids*, where the mutated gene is not accompanied by an allele which might suppress or weaken its effect, the change of a single hereditary factor must immediately become manifest in the phenotype. In most cases the disturbance caused by a gene mutation is extensive enough to cause the death of the affected cells *(lethal mutation)*. Even if they remain alive and divide, the mutation will frequently lead to a defect which decreases viability *(defect mutation)*.

Mutability has been studied extensively, particularly in species of *Chlamydomonas*. Under suitable conditions of illumination, they can be cultured on agar plates. As "wild types" they need only a minimal medium of some inorganic salts (NH_4NO_3, KH_2PO_4, $MgSO_4$, $CaCl_2$) and trace elements.

Spontaneous mutants as well as those induced by UV-irradiation, treatment with antibiotics or by other ways show various metabolic defects. The so-called *auxotrophic* mutants can only thrive if the minimal medium is supplemented by "growth factors" such as arginine, thiamine, p-aminobenzoic acid or niacin. The gene mutation must therefore have led to a block in the biosynthesis of these substances. Compared to Bacteria and *Neurospora*, relatively few amino acid-deficient mutants have been found in *Chlamydomonas*.

Other strains are no longer able to carry out normal photosynthesis and need sodium acetate as a carbon source (acetate-mutants). Some of these strains have been extensively investigated. The ac-20 strain of *Chlamydomonas reinhardi* for example is the result of a single gene mutation (linkage group XIII). It turned out that the gene exerts a control over chlorophyll-DNA transcription: the ability is reduced to synthesize chloroplast ribosomal RNA. The expression of the gene is different, however, under different growth conditions. Though "mixotrophically" grown cells (minimal medium with acetate) are capable of much more rapid growth than "phototrophically" grown cells (minimal medium without acetate), they have a highly disordered chloroplast membrane organization, usually no pyrenoids and only 5–10% as much ribosomes as the wild type, while the "phototrophic" cells are more normal in this respect. The biochemical pathways in both cases are under examination [*611, 612, 1034, 1035*].

While the "acetate-mutants" still have chlorophyll and retain their green color, other mutants show defects in chlorophyll synthesis. They are only of a faint green color or they become yellow in the dark while the wild type remains green even in darkness.

Many mutants show a more or less pronounced resistance towards substances which are lethal to the wild type at certain concentrations e.g. towards antibiotics. These mutants can be selected by cultivation in the substances concerned.

In many cases the strains can only be characterized by general physiological traits. One mutant can grow only if the osmotic pressure in the nutritive medium is increased [*619, 702*]. Mutants whose ability to swim is partially or totally affected have recently been shown to lack one or both central fibrils in the flagella [*1407, 1784*]. Some have abnormal flagellar length. Another mutant with greatly decreased sensitivity to light has no stigma [*722*].

In *diploids* the effect of the mutated gene can be weakened or suppressed by its nonmutated allele. In ciliates, the only diploid Protozoa whose mutability has been studied, there is the additional fact that the phenotype is predominantly determined by the polyploid macronucleus. Gene mutations, which appear in the course of asexual reproduction, are thus without any recognizable effect because of the large number of other, nonmutated alleles, which are present in the macronucleus. The fact that ciliates can withstand even very high dosages of mutagenic radiation may be directly connected with this. Nothing is known concerning mutability in ciliates with diploid macronuclei (p. 101).

Multiplication of a mutated gene in a cell can only take place while a new macronuclear anlage is developing. If irradiated cells of *Paramecium aurelia* are permitted to undergo autogamy (p. 200), all possible types of defects ranging from a slight decrease of the fission rate to strong lethality are found in part of the exautogamont clones. The assumption is therefore obvious that these defects are due to micronuclear mutations, which become homozygous at autogamy and then become manifest via the new macronucleus formed from the homozygous synkaryon. This assumption could also be supported by crosses of deficient exautogamont clones with normal ones [*918–920*].

Although thermosensitive and "behavioral" mutants have been induced by different agents recently (p. 264 and 307), most mutants of ciliates, in contrast to *Chlamydomonas*, were not induced by mutagenic radiation or substances. They have either been isolated from natural habitats or they appeared spontaneously in the cultures. In general, these were not "defect mutants" but variants with differences in their reaction norm, which did not affect viability. It cannot usually be decided, which of these variants is to be considered the "wild type".

Changes in the linear architecture of the chromosomes are termed *chromosome mutations*. The simplest type of such a change consists in breakage of a chromosome at a certain point. Breaks of this kind occur spontaneously from time to time, but they can of course only be recognized if the chromosomal set is sufficiently distinct. This is, for example, the case in the hypermastigid, *Holomastigotoides tusitala* (Fig. 213). Broken chromosome ends have the tendency to fuse with each other but they never seem to attach to the ends of unbroken chromosomes. If a broken

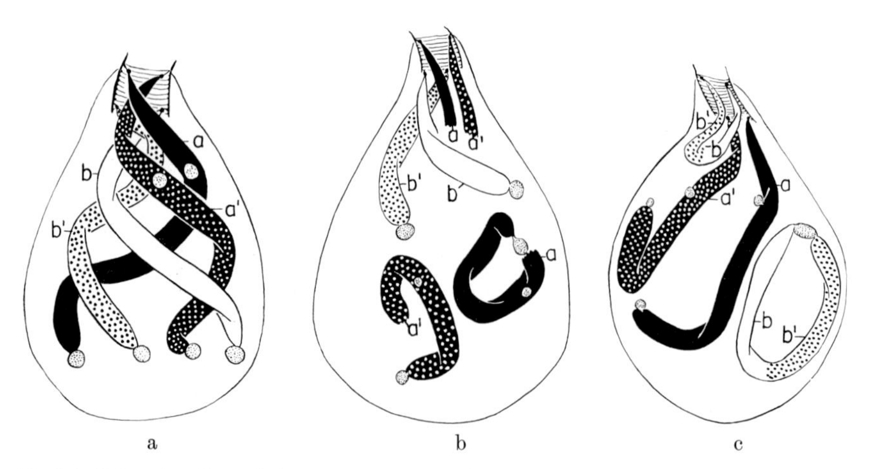

Fig. 213. *Hologastigotoides tusitala* (Polymastigina). Nucleus with normal chromatids (compare discussion on p. 55). The chromatids of the longer chromosome, which has a lateral nucleolus, are black (a) and black with white dots (a′); those of the shorter chromosome are drawn white (b) and white with black dots (b′). b Spontaneous breakage of the chromatids of the longer chromosome: two kinetic and two akinetic fragments. c Breakage of the chromatids of the shorter chromosome: the kinetic fragments have fused to form a bridge, the akinetic ones form a ring due to fusion of the broken surfaces at one end and fusion of the terminal nucleoli at the other. After CLEVELAND, 1953 [*285*]

chromosome divides into two chromatids, the breakage surfaces of the chromatids may fuse with each other. This leads to formation of a fragment with two points of spindle fiber attachment (dikinetic fragment) and an akinetic fragment without spindle attachment.

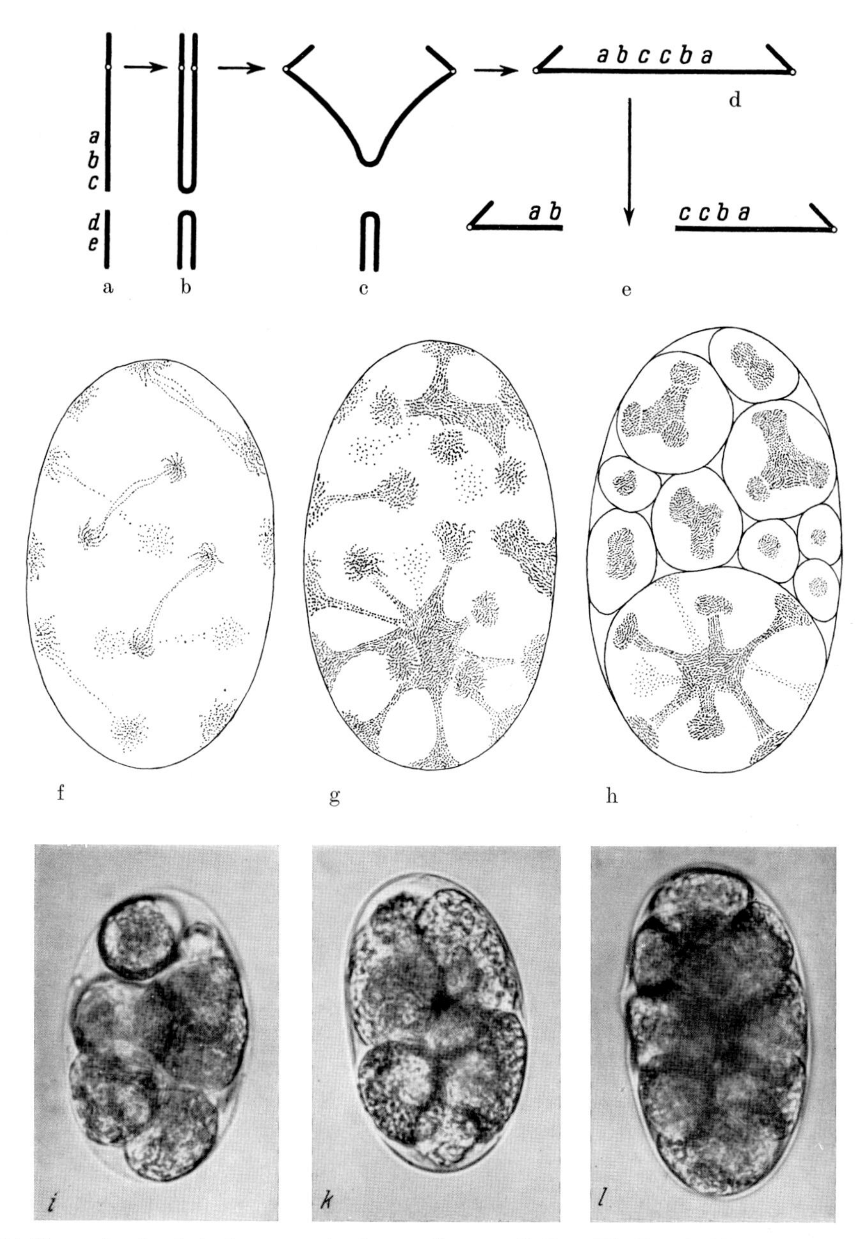

Fig. 214. The results of a single chromosome break, a—e diagrammatically. a Kinetic and akinetic fragments (kinetochore as white circle). b Reduplication of the chromosomes and fusion of the broken surfaces of the chromatids. c, d Bridge formation in anaphase. e Breakage of the bridge. f—h *Eucoccidium dinophili*. Bridge formation and nuclear fusion during sporogony after irradiation of the spores with X-rays. Whole oocysts are shown (in f and g the oocystic envelope has been omitted, in h not the entire content was depicted). Feulgen reaction. × 1090. i—l Live oocysts showing X-ray damage. × 730. After GRELL, 1955 [*660*]

The consequences of this type of chromosome breakage in successive mitoses are shown in Fig. 214a–e. While the akinetic fragment degenerates because it lacks spindle attachment, the dikinetic fragment stretches as a "bridge" between the two daughter plates. The bridge may either remain, or it will break at some point. In the latter case each daughter nucleus again receives a fragment with a newly broken surface. The process of bridge formation is therefore repeated in the succeeding and in all following mitoses (breakage-fusion-bridge-cycle). Since the bridges will not break at those points where the chromatid fragments had once fused, the genetic composition of nuclei with such single chromosome breaks is bound to become more and more disturbed as divisions progress. Descendants of cells with single breaks are therefore destined to perish sooner or later.

Chromosome breaks can easily be obtained by irradiation with X-rays, breakage frequency increasing with the dose of radiation. Sporozoites derived from irradiated spores of the coccidian, *Eucoccidium dinophili* (Fig. 395), develop perfectly normally at first. Chromosome breaks due to X-rays become apparent only during the nuclear divisions associated with microgametogenesis and sporogony (Fig. 214). Bridges are clearly recognizable between the daughter plates (*f*). They usually rupture in the first divisions of sporogony when the daughter plates move even further apart. Later on, when the pattern of nuclei at the surface of the oocyst condenses, they remain and thus lead to more or less extensive complexes of nuclei which cannot separate and will finally fuse (*g*). A certain mass of cytoplasm corresponding to the size of the fusion masses is then delimited around each nuclear complex (*h*). If the whole nuclear material has fused into one continuous complex, the cytoplasm is not subdivided at all. If chromosome breaks are caused by irradiating sporozoites, the oocysts derived from them are thus subdivided into cytoplasmic portions of unequal size (*i*—*l*). These portions are usually larger than normal spores. Because of genetic disturbances they cannot carry out the second phase of sporogony, i.e. they can form no sporozoites. They will finally perish.

If two or more chromosome breaks occur within the same nucleus, proper combination of the fragments may yield viable chromosome segments provided that each chromosome has a point of spindle attachment. Depending on the nature of the interchange, different viable chromosome mutations (translocations, inversions and duplications) can be distinguished.

Changes in the number of chromosomes are called *genome mutations*. The term *euploidy* is applied if whole chromosome sets are present in multiple, *heteroploidy* if only single chromosomes of a set are increased or decreased in number. Such cases have been discussed above (p. 53).

II. Crossing Experiments

In order to gain information about the nature and action of genes and the composition of the hereditary make-up from the inheritance of traits, it is necessary to cross individuals which differ in certain characteristics.

In Protozoa, crosses have been possible only to a limited extent thus far. The difficulties are due to the fact that among those Protozoa which are sexually differentated and thus suited for genetic analysis, only a few could be cultured under controlled conditions and be induced to reproduce sexually. Among haploids, only the *phytomonads* could be used for genetic analysis, among diploid Protozoa only the *ciliates*.

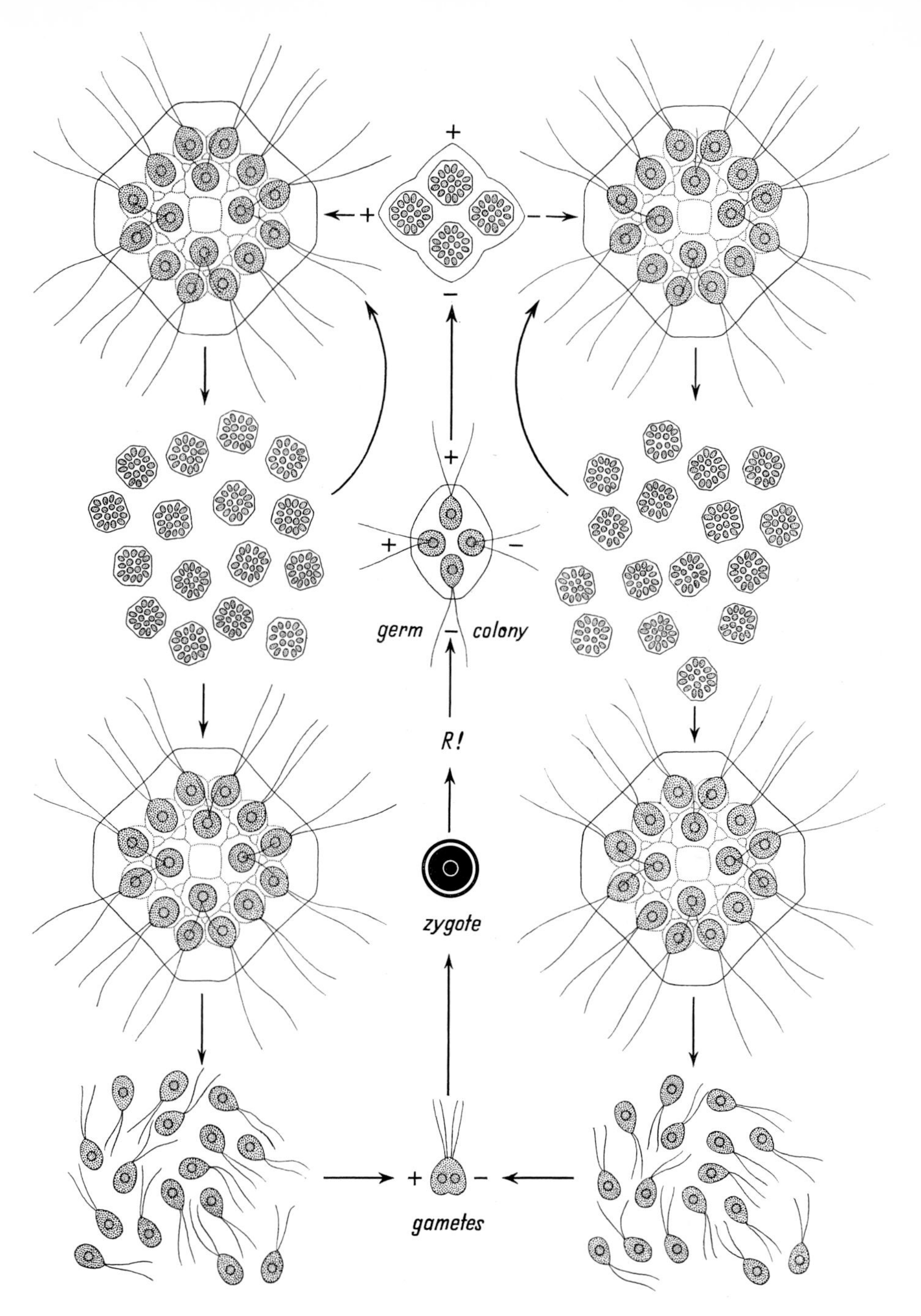

Fig. 215. *Gonium pectorale.* Sexual and asexual reproduction. A four-celled germ colony hatches from the zygote. Of the four gones, two belong to the + sex and two to the —sex. After the investigations of SCHREIBER, 1925 [*1536*] and STEIN, 1958 [*1659*]

1. Haploids

The zygote is the only diploid stage in the life cycle of a haploid organism. Meiosis takes place when the zygote germinates. The daughter cells formed by meiosis are called *gones*.

If two races of a haploid differing in one or several pairs of traits are crossed, only the zygote can therefore show *hybrid characteristics*, i.e. only it can be heterozygous with respect to the pairs of alleles, which are responsible for the different traits of the two races. Dominance of one trait over the other can thus only be expressed in the phenotype of the zygote, provided that the genes concerned manifest themselves phenotypically in the zygote.

When two races which differ in *one* pair of characters are crossed (*monohybrid* cross), the alleles responsible for this difference must be distributed by the chromosomes at meiosis in such a manner that one half of the gones receives one, the other half the other allele (Fig. 69). The alleles, which are combined in the synkaryon, are separated again in the course of meiosis. This phenomenon is called *segregation*. It can be demonstrated by *gone analysis*, i.e. by separate culture of the gones derived from a single zygote and by testing the clones obtained from multiplication of these gones.

In the dioecious phytomonads, the sex inheritance can be attributed to the segregation of two "sex alleles". This was first demonstrated in *Gonium pectorale* (Fig. 215) where a + and a – isogamete fuse during copulation. A germ colony of four gones hatches from the zygote. The gones give rise to four 16-celled colonies, which could be isolated and cultured as clones. Combination of clones showed that two gones belong to the + sex, the other two to the – sex [*1536, 1659*].

Crosses of mutants have so far been carried out almost exclusively in species of *Chlamydomonas (C. eugametos syn. moewusii, C. reinhardi)* [*467–472, 498, 499, 619, 1040, 1477, 1483*]. In most cases the traits segregate at meiosis in a 2:2 ratio.

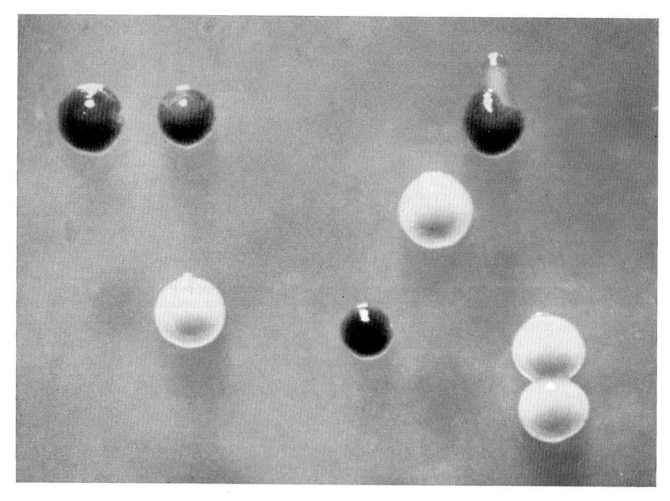

Fig. 216. *Chlamydomonas reinhardi*. Agar plate cultures derived from the eight gones of a zygote which arose from a cross between the normal strain and a chlorophyll-deficient strain. Four of the gones are again normal (dark-green) and four are chlorophyll-deficient (light-yellow). After SAGER, 1955 [*1473*]

Occasionally, eight instead of four cells emerge from a zygote if a mitosis took place after meiosis. One example is shown in Fig. 216.

If two races differing in *two* pairs of traits are crossed *(dihybrid cross)*, meiosis can lead to a *recombination of parental traits*. Three different types of zygotes can appear:

1. *Parental ditypes*, where two gones show the combination of traits from one parent, the other two those of the other parent.
2. *Nonparental ditypes*, where pairs of gones show the parental traits in new, reciprocal combination.
3. *Tetratypes*, where there are four different gones. Two of these correspond to the parents and two are new combinations.

The numerical ratios of these different types of zygotes permit conclusions as to whether the genes controlling the two traits are located on the same chromosome *(gene linkage)* or on different chromosomes.

If they are on the *same* chromosome (Fig. 217), parental ditypes will predominate. The small proportion of tetratypes observed is due to segmental interchange between two chromatids. In exceptional cases also nonparental ditypes are produced. They are due to an additional segmental interchange between the remaining two chromatids of the tetrad.

If two genes are located on *different* chromosomes (Fig. 218), nonparental ditypes appear just as frequently as parental ditypes since the probability that the chromosomes from different parents migrate to the same pole is just as large as the probability that the chromosomes of each parent move to the same pole.

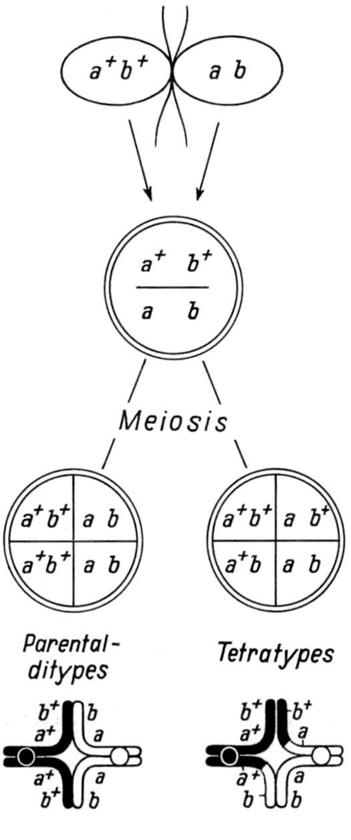

Fig. 217. Dihybrid cross in a haploid. Diagram of the possible gone combinations if the genes are on the same chromosome (linkage). In the tetrad diagrams, only that case is depicted in which a single exchange of chromatid segments took place in the tetrad

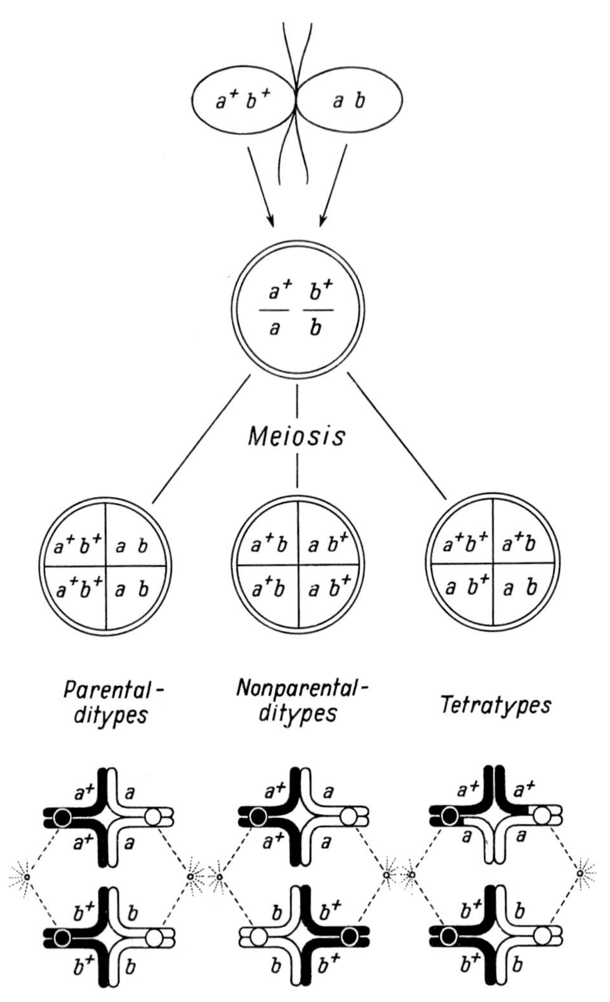

Fig. 218. Dihybrid cross in a haploid. Diagram of the possible gone combinations if the genes are located on different chromosomes. In the tetrad diagrams, only that case is depicted in which a single exchange of chromatid segments took place in one of the two tetrads

In any particular case this result may be modified because the frequency of parental or nonparental ditypes can be changed by even other segmental exchanges. However, with statistically significant numbers, a clear preponderance of parental ditypes always indicates gene linkage.

In race crossings which involve *more than two* pairs of traits the frequency of zygotic types must be determined separately for every two pairs of traits.

An example of a trihybrid cross involving gone analysis of 126 zygotes is summarized in Table 10.

The frequency of the zygotic types indicates that the genes *arg-2* and *pab-2* are on the same chromosome while the gene *ac-31* belongs to a different chromosome.

With the aid of statistical methods which shall not be described here in detail, it is also possible to calculate *genetic maps* which indicate the sequence of genes, their

Table 10. *Chlamydomonas reinhardi*. Example of a trihybrid cross: *arg-2, pab-2,* + x +, +, *ac-31*. PD parental ditypes, NPD nonparental ditypes, T tetratypes. After data of EBERSOLD and LEVINE, 1959 [*471*]

		PD	NPD	T
arg-2,	*pab-2*	88	0	38
arg-2,	*ac-31*	30	36	60
pab-2,	*ac-31*	19	23	84

relative positions (map distances) and their distances from the point of spindle fiber attachment (Fig. 219). In the case of *Chlamydomonas reinhardi* more than 80 genes have been mapped so far and assigned to 16 linkage groups [*472, 1034, 1483*].

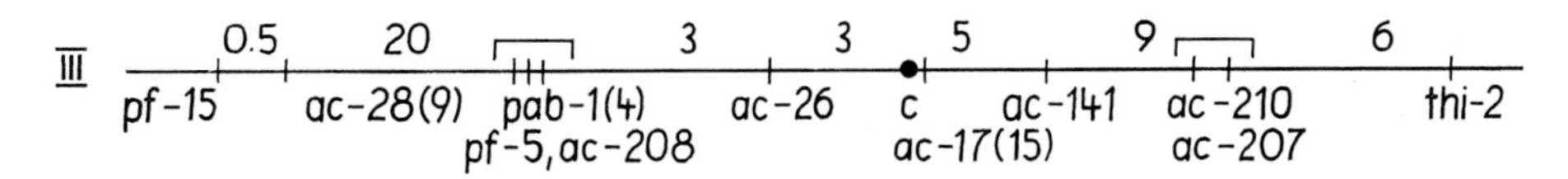

Fig. 219. *Chlamydomonas reinhardi*. Example of a genetic localization map (linkage group III). Figures in parenthesis after the name of the locus indicate the number of alleles known at that locus. Numbers above the line are map distances. The brackets above the group of markers indicate that their relative positions are uncertain or unknown. Abbreviations: *C* spindle attachment point; *pf* paralyzed flagella; *ac* acetate requiring; *pab* p-aminobenzoate requiring; *thi* thiamine requiring. From SAGER, 1972 [*1483*]

Chloroplast Genetics

Most mutants of *Chlamydomonas reinhardi* are due to genes whose inheritance can be explained by recombination of nuclear chromosomes and segmental exchange at meiosis. In the last ten years, however, more and more mutants with a different pattern of inheritance were found in crossing experiments. As will be shown later, they are due to "cytoplasmic genes" probably located in chloroplast DNA.

The majority of these mutants did not appear spontaneously but were induced by the antibiotic streptomycin [*1472, 1476, 1483, 1495, 1496*]. Some show properties which bear a direct relationship to the inducing agent: they are resistant to a certain concentration of streptomycin (e.g. to 50 or 500 mg/ml) or cannot grow without a definite concentration of streptomycin. That streptomycin is generally mutagenic is proven by the appearance of mutants, which show other biochemical defects. They may be unable to carry out photosynthesis and grow only if the medium is supplemented by a certain concentration of acetate depending upon the particular mutant strain (leaky or stringent acetate mutants). Other mutants induced by streptomycin were resistant to other antibiotics (e.g. erythromycin, spectinomycin, carbomycin, spiramycin, cleosine, oleandomycin) or they were temperature-sensitive.

While streptomycin is not detectably mutagenic toward nuclear genes, some cytoplasmic mutants could also be isolated after treatment with UV and nitro-

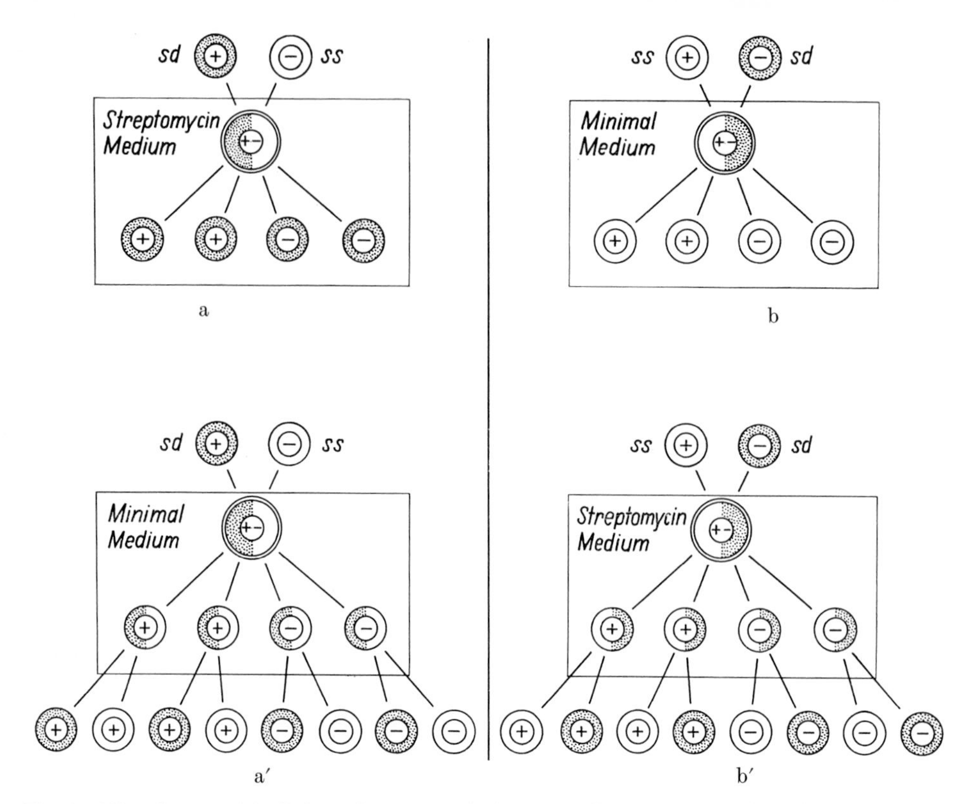

Fig. 220. *Chlamydomonas reinhardi*. Cross of a streptomycin-dependent (*sd*) with a streptomycin-sensitive (*ss*) clone. Generally (a, b), the gones derived from the zygotes manifest the trait of the + parent. Exceptional zygotes (a′, b′) give rise to gones whose descendants can manifest partly the trait of the + parent, partly that of the — parent. Based upon SAGER, 1965 [*1478*]

soguanidine [*593, 595, 1688*]. They show properties similar to or different (e.g. resistance to neamine) from those of the streptomycin-induced mutants.

In all cases the mutants were found to be stable. Some have already been cultured for many years.

Crosses of these mutants with the wild type or among themselves always lead to the same result: most zygotes give rise to gones, which show the trait of the + parent and also retain it in clone culture. If, for instance, the + parent is streptomycin-dependent (*sd*) and the — parent streptomycin-sensitive (*ss*), the zygotes mature only in a streptomycin medium and all the gones derived from them are dependent upon streptomycin (Fig. 220a). In the reciprocal cross, the zygotes mature only in the streptomycin-free minimal medium of the wild type and all of the gones derived from them are sensitive to streptomycin (b).

While the traits controlled by chromosomal alleles segregate at meiosis in a 2:2 ratio, the pairs of traits mentioned above show *uniparental* inheritance, i.e. the zygotes hand on only the trait received from the + parent.

In this respect the situation reminds one of the so-called "maternal inheritance" where certain traits are handed on to the F_1-generation only via the egg cell and not by the sperm. The usual interpretation of "maternal inheritance", i.e. that the zygote receives most of its cytoplasm from the egg and only a small amount from the sperm, can certainly not be true for *Chlamydomonas reinhardi* where both isogametes contribute equally to the formation of the zygote. Evidently the + gamete eliminates enzymatically the cytoplasmic genes of the − gamete. In accordance with this, it has been found that the chloroplast DNA of the −gamete disappears in the zygote [*1483, 1494*].

It is not less puzzling that a small proportion of the zygotes transmits traits of both parents to the gones so that we can speak of *biparental* inheritance [*1489, 1490*]. If zygotes of the crosses shown in Fig. 220 are transferred to media in which only cells with the trait of the − parent can live, about 0.1% will germinate and the clones derived from them are viable (a′, b′). By increasing the temperature of the cultures from 25 to 37° C, the numbers of these exceptional zygotes can even be increased tenfold.

Isolation of cells derived from the various gones shows that a portion of these again bears the trait of one parent, a portion that of the second parent. Segregation of the parental traits thus takes place in the course of postmeiotic divisions.

This result led to the hypothesis that normally uniparental traits which are inherited biparentally in exceptional zygotes depend upon separable determinants. Since they cannot be localized in the chromosomes of the nucleus, they have been termed *cytoplasmic genes*.

The *demonstration of recombinability* constitutes further support of this hypothesis. In the cross shown in Fig. 221 both parents differ — aside from sex (+, −) — in two pairs of traits (*ar-as*: resistance or sensitivity toward actidione; *mr-ms*: resistance or sensitivity toward methionine-sulphoximine) which are due to unlinked chromosomal genes. Zygotes of the tetratype (p. 239) can thus give rise to four genetically different gones by segregation and recombination, viz. *ar mr*, *as mr*, *ar ms*, *as ms*.

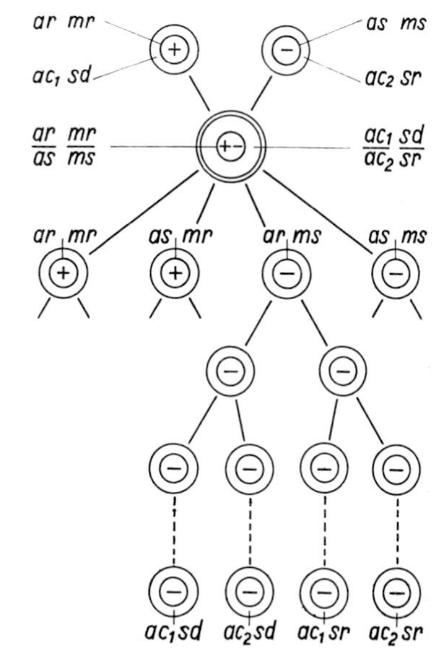

Fig. 221. *Chlamydomonas reinhardi*. Cross between two clones which differ with respect to two pairs of phenotypic traits due to chromosomal genes (*ar mr-as ms*) and two due to cytoplasmic genes (ac_1sd-ac_2sr). The gones derived from the zygote reflect the possible combinations of chromosomal genes. Exceptional zygotes give rise to gones whose descendants may represent recombinants of cytoplasmic genes. Based upon SAGER, 1965 [*1479*]

However, both parents also differ with respect to two traits with normally uniparental inheritance. In this case all gones show the traits of the + parent (ac_1, sd).

Analysis of exceptional zygotes which inherit the traits of both parents showed that cells derived from the gones have not only the combination of the parental traits (ac_1sd, ac_2sr) but also new combinations (ac_2sd, ac_1sr). A total of sixteen different genotypes can thus arise from a single zygote.

The gones of exceptional zygotes must therefore have obtained the cytoplasmic genes of both parents, and their segregation and recombination must have taken place in the course of the postmeiotic divisions.

If only recombinants of the types shown in Fig. 221 would occur, one might suggest that ac_1 and ac_2, sd and sr, respectively are unlinked. It turns out, however, that occasionally two additional types of recombinants appear which either correspond to the wild type or to the reciprocal double mutant. In crosses of $ac_1 \times ac_2$ for instance, the wild type "ac^+" ($ac_1^+ - ac_2^+$) and in crosses of $sd \times sr$, the wild type "ss" ($sd^+ - sr^+$) arose. By special methods also the double mutants ($ac_1 - ac_2$, $sd - sr$) could be identified. Hence, neither ac_1 and ac_2 nor sd and sr are alleles but *linked* genes which become separated or recombined by exchange events. The prerequisite for such exchanges is a linear structure, probably DNA. How these types of recombinations arise on the basis of the DNA concept is shown in Fig. 222 for the acetate system.

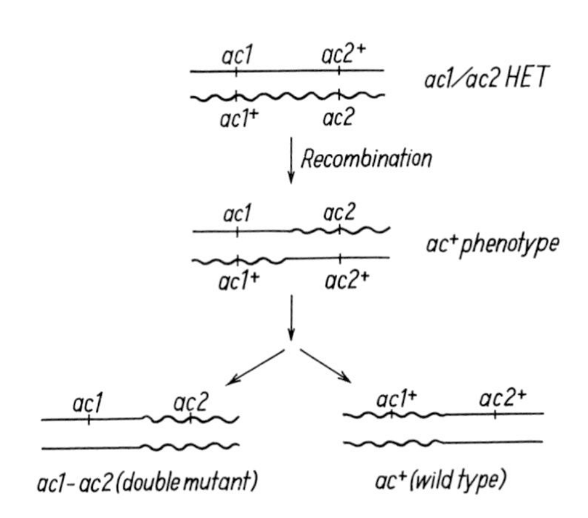

Fig. 222. *Chlamydomonas reinhardi*. Recombination of two closely localized acetate-genes (see Fig. 223) in a ac_1/ac_2 "heterozygote" or "HET"-cell (ac^+ phenotype) which leads to the ac_1-ac_2 double mutant and the ac^+ wild type. From SAGER, 1972 [*1483*]

Pedigree analysis, i.e. the isolation, culturing and phenotypic characterization of cells after different numbers of postmeiotic divisions, could be carried out on a much larger scale when a new discovery was made [*1491*]. If the + gametes are irradiated with a weak dose of UV (about 40 sec), biparental transmission of the cytoplasmic genes can be increased up to 50%. At a higher dose of UV irradiation only the traits of the − gamete are to be found under the gonal progeny, i.e. uniparental inheritance is completely reversed.

It seems likely that a weak dose of UV interferes with the transcription of a messenger RNA required for an enzyme which otherwise leads to the elimination of the cytoplasmic genes of the — gamete. The higher dose of UV evidently damages the genome of the + gamete so much that it is unable to replicate. Further evidence that its molecular equivalent is DNA can be taken from the fact that the damage is partially reparable by "photoreactivation", a repair mechanism only known from DNA [*1491*].

The *segregation* of the cytoplasmic genes begins directly after meiosis. While the gones are usually still "heterozygous" with respect to alleles in which the parents differ, in the course of the following divisions more and more cells arise which are "homozygous", i.e. pure for parental alleles.

The segregation may be *nonreciprocal*, i.e. while one sister cell is still "heterozygous", the other sister cell is already "homozygous", or it may be *reciprocal*, both sister cells being "homozygous" for either one or the other parental allele.

While the first type of segregation whose molecular basis is not yet understood (probably a miscopying process) occurs with approximately equal frequency for all genes, the second type is different for the different pairs of alleles: they segregate with a characteristic frequency.

This polarity provides evidence that segregation is not the result of a sorting-out process leading to a reduction in the number of DNA strands but the result of true exchange events, probably at a "four strand" stage after replication but before cell division. The polarity leads also to the postulation of a kinetochore-like *attachment point* which governs the distribution of the DNA strands at cell division. Based on the assumption that exchanges occur at random along the DNA strands, one can suggest that frequency of segregation increases with increasing distance from the postulated attachment point.

Without going into further details, it should be mentioned that the evaluation of pedigree data led to the conclusion that all cytoplasmic genes belong to *one linkage group* which – at least from the genetic point of view – is *circular* [*1483*].

On the basis of segregation and recombination frequencies between genes after multifactor crosses, a preliminary map could be constructed which shows the locali-

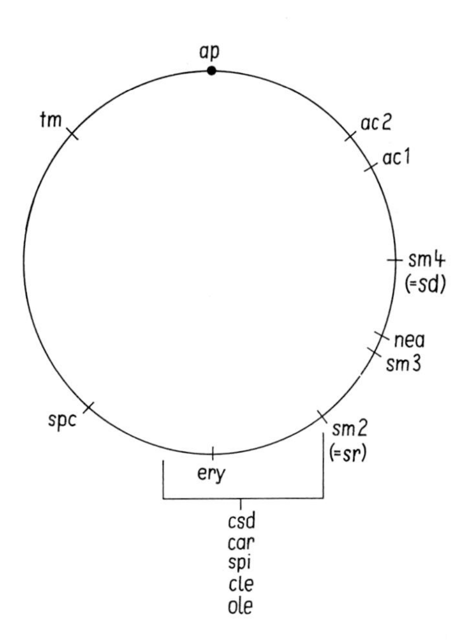

Fig. 223. *Chlamydomonas reinhardi*. Preliminary circular map of chloroplast genes. The bracketed genes all map close to *ery* and no linear order has been established. *ap* hypothetical attachment point; ac_2 and ac_1 acetate requirement; sm_4 streptomycin dependence; *nea* neamine resistance, sm_3 low level streptomycin resistance; sm_2 high level streptomycin resistance; *ery* erythromycin resistance; *csd* conditional streptomycin dependence; *car* carbomycin resistance; *spi*, spiramycin resistance; *cle* cleosine resistance; *ole*, oleandomycin resistance; *spc* spectinomycin resistance; *tm* temperature sensitivity. From SAGER, 1972 [*1483*]

zation of several cytoplasmic genes (Fig. 223). Since the gene *ery* shows the highest rate of segregation it has been assigned a position 180° from the attachment point.

The pedigree data suggest that the circular linkage group is present in *two copies*, correlating with cytological observations of two Feulgen positive regions in the chloroplast. Hence, while the cell as such is haploid, its chloroplast may be regarded as "diploid".

Not much is known as yet about the interrelationships of nuclear and chloroplast genes in cellular metabolism. Both control the same or similar traits. One nuclear gene designated as an "amplifier" [*1495*] increases the resistance of a nuclear gene mutant strain as well as that of a chloroplast gene mutant strain to such a degree that both strains can withstand more than 2 mg/ml of streptomycin. However, the amplifier gene influences the course of inheritance of neither the nuclear nor that of the chloroplast gene.

2. Diploids

The *ciliates* are the only diploid Protozoa, which have thus far been used in crossing experiments. Large-scale studies of this kind became possible only after discovery of the mating types (p. 212).

The course of inheritance to be expected on the basis of the special modes of reproduction in ciliates is shown diagrammatically in Fig. 224.

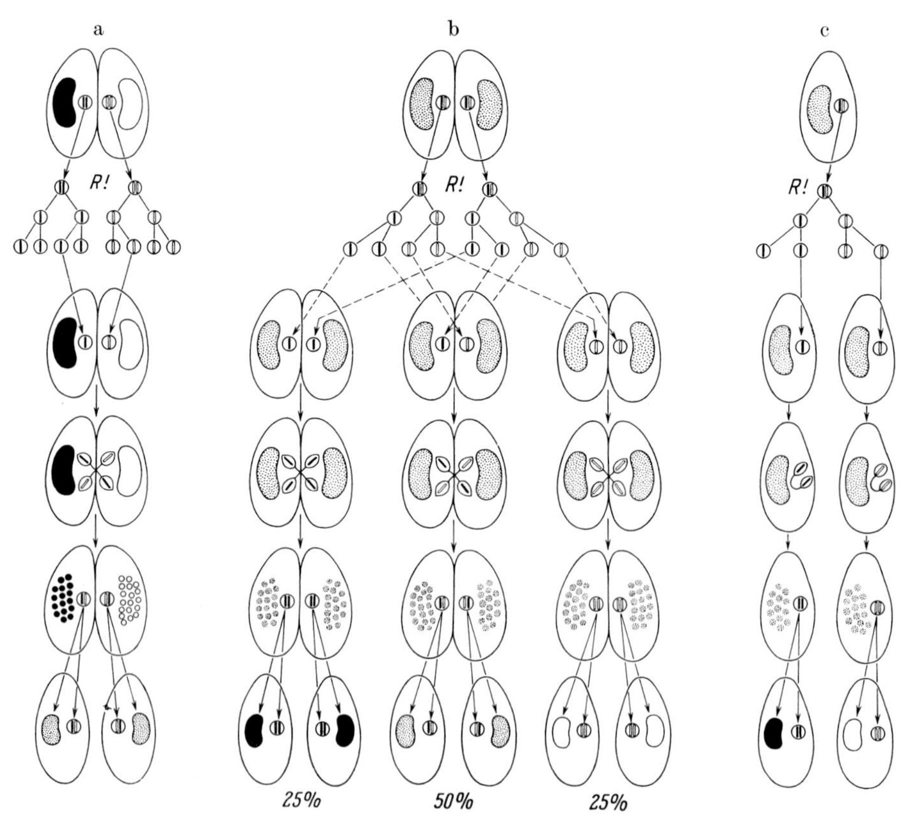

Fig. 224. Diagram of the course of inheritance of a pair of chromsomes in ciliates. a Cross between two homozygous individuals (black/black × white/white). b Cross between two heterozygous individuals (black/white × black/white). c Autogamy of a heterozygous individual (black/white) R! Meiosis

For the sake of simplicity it has been assumed that only one micronucleus and only one macronucleus is present. In addition, the inheritance of only one chromosome is shown, which has been drawn in black for one race and in white for the other. No chromosomes are shown in the polyploid macronucleus; the homozygous constitution is indicated by black or white and heterozygosity by stippling.

Conjugation of homozygous individuals (a). Since ciliates are diploid, the micronucleus of each race contains every chromosome in duplicate. In our diagram one conjugant has thus two black chromosomes, the other two white ones. After meiosis (*R*!) each chromosome is present only once. The third progamic micronuclear division gives rise to two gamete nuclei in each individual, one of which passes as a migratory nucleus into the mating partner. A synkaryon with one black and one white chromosome is thus formed in each conjugant. This constitution is handed on to its two daughter nuclei, viz. the new micronucleus and the new macronucleus.

If the two parental races differ in one or several pairs of alleles, the two exconjugants and the clones derived from them must be *genetically identical* or *isogenic*, in this case heterozygous.

If differences in phenotype persist in spite of the genetic identity of the two exconjugants or the clones derived from them, these cannot be genotypic, i.e. due to genes located in the chromosomes*.

It must then be assumed that such differences have been transmitted by way of the cytoplasm from the conjugant to the exconjugant. The importance which crossing experiments with ciliates have had for modern genetic research lies primarily in the possibility to distinguish easily between both modes of transmission.

Conjugation of heterozygous individuals (b). Since it is a matter of chance, which daughter nucleus is retained after meiosis, a gamete nucleus with either one or the other chromosome is formed in each conjugant when F_1-individuals conjugate. Each partner shown in our diagram will thus with equal probability contain either a black or a white chromosome after meiosis.

If we assume both F_1-individuals to be heterozygous for one pair of alleles (*Aa*), 25% of the exconjugant pairs of such a cross must be homozygous with respect to one allele (*AA*), 50% must be heterozygous (*Aa*) and another 25% must be homozygous for the other allele (*aa*). Among themselves, the exconjugants of any pair are of course isogenic again.

Isogenicity can, however, only be expected if conjugation follows the normal course, (e.g. *Paramecium aurelia*, Fig. 193), i.e. if the two gamete nuclei are derived from the same gone nucleus. Whenever the two gamete nuclei are derived from different gone nuclei as is frequently the case in *Euplotes vannus* (p. 202), one of the gamete nuclei may contain one allele (*A*), and the second the other one (*a*). If such a cell conjugates with another one, which forms gamete nuclei with identical alleles (e.g. *a-a*), one of the exconjugants will be heterozygous (*Aa*), the other one homozygous (*aa*).

Autogamy (c). If autogamy takes place in a heterozygous clone, the gone nucleus will, after meiosis, contain either the *A*-or the *a*-allele. If the two gamete nuclei which fuse in autogamy are derived from such a gone nucleus, the resulting ex-autogamont must be homozygous with respect to either one of the two alleles (*AA* or *aa*). Both cases are equally probable.

---

* In the exceptional case of "cytogamy" (p. 200) instead of true conjugation, the descendants of the mating partner remain, of course, genotypically different.

Autogamy is therefore genetically important: *It leads to segregation and homozygosity.*

Most crosses have been carried out with *Paramecium aurelia*. This species has the advantage that it can be cultured with ease and can be brought to autogamy at any time. It has only been in recent times that other ciliates such as *Paramecium bursaria*, *Tetrahymena pyriformis* and species of *Euplotes* have been used successfully for genetic analysis.

In a comparative survey it seems reasonable not to proceed from species to species but from traits whose inheritance has been studied. Special testing procedures are usually necessary in order to recognize these traits.

Mating Types

It was emphasized earlier (p. 219) that mating types may either be determined by modification or by genetic factors.

If determination occurs by way of *modification*, the reaction norm, in this case the possible numbers and kinds of mating types, is genetically fixed. In the syngens of *Paramecium aurelia* the reaction norm consists of a spectrum of two mating types, and in syngen 1 of *Tetrahymena pyriformis* it involves a total of seven mating types.

In some mutants all mating types of the "wild type" can no longer be formed.

The discovery of such mutants in *Paramecium aurelia* provided the first possibility of crossing two stocks which differed with respect to one pair of alleles [*1630*].

While most of the stocks of syngen 1 of *Paramecium aurelia* (p. 213) are able to form both mating types I and II, and are therefore referred to as "two type"-stocks, the "one type"-stocks can only form mating type I. Conjugation can therefore occur after autogamy in a "two type"-clone but not in a "one type"-clone since all exautogamonts retain mating type I.

A cross between a "two type"- and a "one type"-stock (Fig. 225) shows that this difference is due to a single pair of Mendelian alleles, II+ indicating the ability, II– the inability to form mating type II*. All exconjugant pairs (the F_1-generation) give rise to clones, which can form both mating types. Thus, the "two type"-trait is dominant over "one type".

When heterozygous (II+/II–) F_1-individuals are crossed, segregation takes place as expected. In one case involving clones raised from 120 exconjugant pairs, 88 showed the "two type"- and 32 the "one type"-characteristic. This corresponds approximately to a 3:1 ratio as expected with one dominant allele. The result of a backcross of heterozygous F_1-individuals (II+/II–) with the recessive parent (II–/II–) is also in accord with expectation. In one such backcross, 81 of 158 exconjugant pairs showed the "two type"- and 77 the "one type"-characteristic. This corresponds to the 1:1 ratio characteristic of backcrosses. Further backcrosses showed that $^2/_3$ of the "two type"-individuals of the F_2-generation are heterozygous and $^1/_3$ homozygous with respect to the "two type"-allele.

It should also be pointed out that in syngen 1 of *Paramecium aurelia* two mutants have been obtained through heat shock and UV-irradiation respectively, which only gave rise to mating

* More recently, the "two type"-allele has been referred to as $mt^{I,II}$ and the "one type"-allele as mt^{I}.

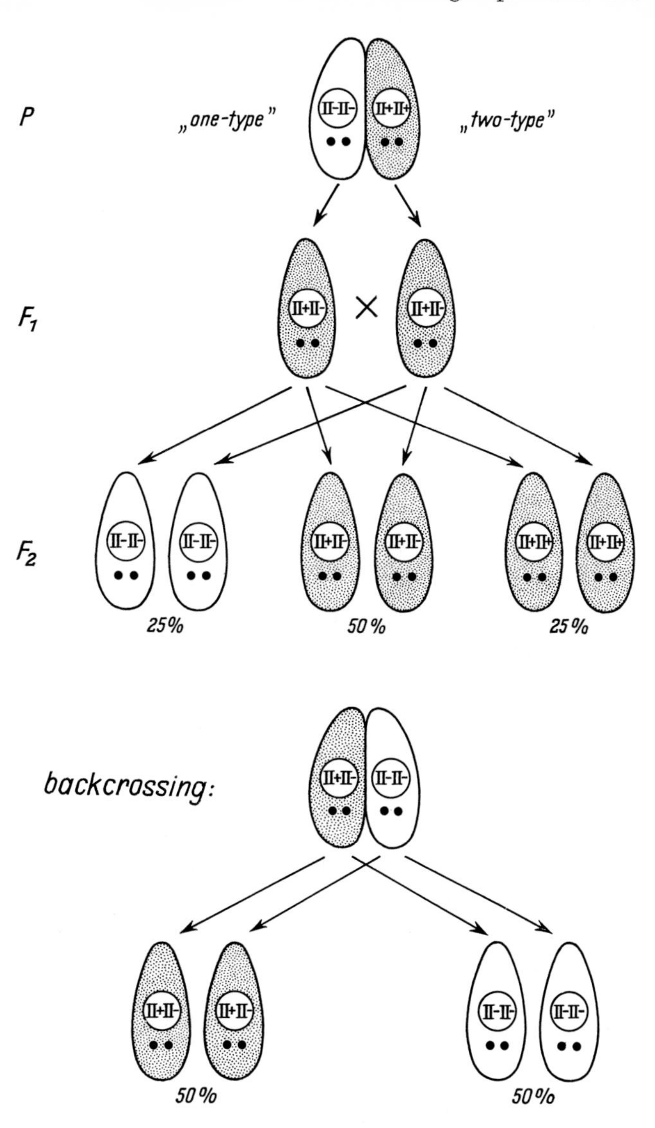

Fig. 225. Cross of a "one-type" (*II—II—*) with a "two-type" stock (*II + II +*) of *Paramecium aurelia*. After investigations of SONNEBORN, 1939 [*1630*]

type I following autogamy. Genetic analysis showed that inheritance is the same as in "one type"-stocks. In addition, evidence has been presented for the existence of a gene which increases the probability of mating type I determination in "two type"-stocks [*183*]. "One type"-stocks have also been found in syngen 7 (group B), which can no longer form mating types XIII and XIV but only XIII. In crosses of "one type"- and "two type"-stocks the exconjugant of the "one type"-stock will manifest mating type XIV although it is a characteristic of syngens of group B that the mating type of the exconjugant depends upon the partner from which it obtained its cytoplasm (p. 215). The action of the "one type"-allele is obviously not confined to narrowing the reaction norm down to one mating type, but it also influences the determinative properties of the cytoplasm in a specific manner [*1715—1718*].

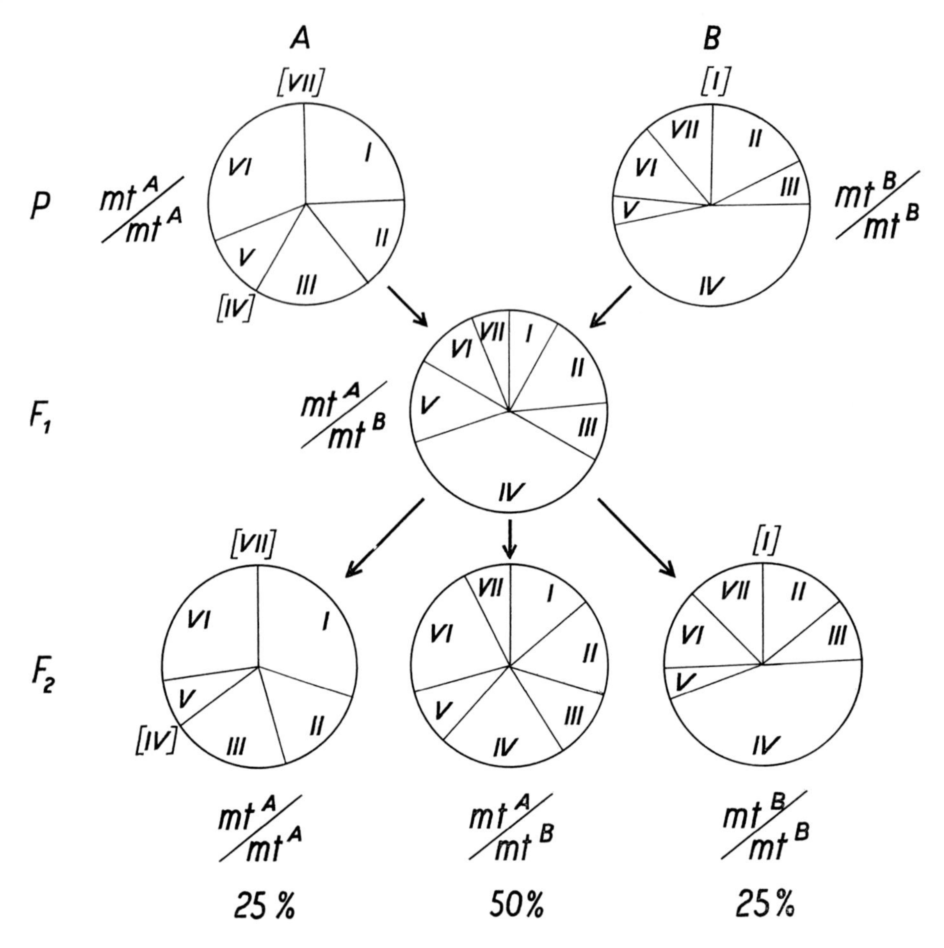

Fig. 226. *Tetrahymena pyriformis* (syngen 1). Cross between the mutants A and B. The sectors of the circles show the relative frequencies of the mating types. Mutant A (mt^A mt^A) is not able to realize mating type IV and VII, mutant B (mt^B mt^B), is not able to express mating type I (shown in brackets). All seven mating types of the F_1-generation (mt^A mt^B) can be realized. In the F_2-generation, the expected segregation due to a monofactorial course of inheritance takes place. Based on NANNEY, CAUGHEY, and TEFANKJIAN, 1955 [*1213*]

Two mutants have been found in syngen 1 of *Tetrahymena pyriformis*, which can no longer manifest the complete spectrum of seven mating types. One mutant (A) fails to form the mating types IV and VII, the other (B) is without type I. Since the first one can still form five and the other six of the mating types, conjugations can be obtained within the mutated stocks. All realizable mating types appear with fairly constant frequency among the exconjugant clones although this can be influenced by external factors such as temperature.

If A- and B-stocks are crossed (Fig. 226), the resulting F_1-generation is phenotypically identical with the wild type, i.e. it can form all seven mating types (phenomenon of complementation). It becomes evident that it is not genotypically identical with the wild type but heterozygous with respect to the alleles responsible for the phenotypic differences of the parents if their descendants are crossed: the

F_2-generation segregates into three classes, one corresponding to stock A, one to the F_1-generation and the third to stock B [*1213*].

Other mutants whose spectrum of realizable mating types is altered have not been discovered so far. Mutants have, however, been found which express the same spectrum as stock A but with a different frequency distribution of the mating types [*1191*].

The results of these crosses can formally be ascribed to two alleles (mt^A for mutant A and mt^B for mutant B, see Fig. 226).

It is, however, probable that the *mt*-locus does not correspond to a single functional unit (cistron) but represents in reality a "*complex locus*" of as many closely linked functional units as the number of realizable mating types (I, II in syngen 1 of *Paramecium aurelia;* I, II, III, IV, V, VI, VII in syngen 1 of *Tetrahymena pyriformis*).

Determination of a mating type by *modification* would then be due to a differential activation of one of these functional units, and the *mt*-mutants would only be deficient in those units, which correspond to the nonrealizable mating types.

In cases of *genotypic* determination, the mating type is directly fixed by the genetic make-up. If the mating system is *bipolar,* the possibility exists that one mating type is determined by the recessive allele, the other by the dominant one. One mating type would therefore always be homozygous (mt^1mt^1) while the other could be heterozygous (mt^2mt^1) or homozygous (mt^2mt^2) if autogamy takes place.

This possibly most basic mode of mating type determination has recently been found in syngen 13 of *Paramecium aurelia* – a species which otherwise has only type determination by modification. Syngen 13 can, for this reason, be assigned neither to the A- nor the B-group (Table 4). A similar dual system has also been discovered in *Paramecium caudatum*, syngen 3 [*766*].

In *multipolar* mating systems, determination of the mating type can be due either to combination of multiple alleles of a single locus or to combination of two alleles of several loci.

Multiple Alleles

This term denotes the appearance of *several* alleles (instead of just two) at the same gene locus. Depending upon their number, different combinations of these alleles in pairs are possible.

Extensive crossing experiments with three marine species of *Euplotes* (*E. vannus, crassus* and *minuta*) led to the conclusion that every allele is responsible for the realization of one mating type [*735, 736, 1234*]. The alleles can be grouped in a series on the basis of their mutual relationships. An allele with the higher index number is dominant over all those of lower index (e.g. $mt^5 > mt^4 > mt^3 > mt^2 > mt^1$). Individual mating types can thus be due to different numbers of genotypes as expressed in Roman numerals (Table 11).

Since mating type I is homozygous and the allele mt^1 is recessive to all others, it is especially suited for analysis of a clone of unknown genetic constitution: if the clone which is to be analyzed is heterozygous (e.g. mt^5mt^2), the F_1-generation will segregate into two different types (e.g. mt^5mt^1 and mt^2mt^1) from which the genetic constitution of the original clone can be directly deduced.

Table 11. *Euplotes crassus*. Mating types and genotypes. After HECKMANN, 1964 [*736*]

Mating type	Genotypes				
I	mt^1mt^1				
II	mt^2mt^2,	mt^2mt^1			
III	mt^3mt^3,	mt^3mt^2,	mt^3mt^1		
IV	mt^4mt^4,	mt^4mt^3,	mt^4mt^2,	mt^4mt^1	
V	mt^5mt^5,	mt^5mt^4,	mt^5mt^3,	mt^5mt^2,	mt^5mt^1

A system of multiple alleles which can be ordered serially for dominance has also been found in the syngens 2 and 8 of *Tetrahymena pyriformis* [*1255*] and in a syngen of a *Glaucoma* species [*259*].

The mating type system of a fresh-water species, *Euplotes patella*, is also due to a series of multiple alleles, in this case without dominance relationships. One of the syngens includes six mating types; three of these are determined by the homozygous and three by the heterozygous combinations of three alleles (Table 12).

Table 12. *Euplotes patella*. Induction of selfing. A heavy margin is drawn around the experimental results which indicate selfing within a mating type clone after treatment with a cell-free filtrate of a clone from another mating type. The remaining part of the table shows the interpretation of the results. After investigations of KIMBALL from SONNEBORN, 1947, slightly changed [*1632*]

			Cells					
		Genotypes	mt^1mt^1	mt^2mt^2	mt^3mt^3	mt^1mt^2	mt^1mt^3	mt^2mt^3
		Substances	1	2	3	1,2	1,3	2,3
	Substances	Mating type	IV	VI	III	I	II	V
Filtrates	1	IV	—	+	+	—	—	+
	2	VI	+	—	+	—	+	—
	3	III	+	+	—	+	—	—
	1,2	I	+	+	+	—	+	+
	1,3	II	+	+	+	+	—	+
	2,3	V	+	+	+	+	+	—

In contrast to the marine species of *Euplotes* mentioned above, *Euplotes patella* liberates *specific substances* into the culture fluid which induce "selfing", i.e. intraclonal conjugation, among cells of other mating types.

The effect of these substances can be tested by adding the cell-free filtrates to the cells. A definite system of reactions was found which can be understood in terms of the following assumptions:

1. Each of the three alleles is correlated with a specific substance. Genetically homozygous mating types can thus form one of the three substances, while heterozygotes form two such substances.
2. Selfing takes place whenever the filtrate contains a substance which cannot be formed by the cells on which the filtrate is acting.

Conjugations induced by mixing two clones of different mating types must therefore take place not only between cells of different but also of the same mating type. However, as mentioned above (p. 220), it is doubtful if the liberation of these substances plays any role under natural circumstances.

In *Uronychia transfuga*, a marine hypotrichous ciliate, mating types are also determined by a system of "codominant" alleles which cannot be seriated on the basis of dominance relationships. One of the syngens includes 10 mating types based on as many genotypes from pairwise combinations of four alleles. Six of the mating types are heterozygous and four are homozygous as shown in Table 13. Strange to say, homozygous mating types react with only those heterozygous mating types which differ from them in both alleles. Heterozygous mating types, on the other hand, can react with each other in any combination even if they have one allele in common.

Table 13. *Uronychia transfuga*. Reactions of the mating types and their genetic constitution. After REIFF, 1967 [*1425*]

Geno-types	mt^1mt^2	mt^3mt^4	mt^1mt^3	mt^2mt^4	mt^2mt^3	mt^1mt^4	mt^1mt^1	mt^2mt^2	mt^3mt^3	mt^4mt^4
Mating types	I	II	III	IV	V	VI	VII	VIII	IX	X
I	−	+	+	+	+	+	−	−	+	+
II		−	+	+	+	+	+	+	−	−
III			−	+	+	+	−	+	−	+
IV				−	+	+	+	−	+	−
V					−	+	+	−	−	+
VI						−	−	+	+	−
VII							−	+	+	+
VIII								−	+	+
IX									−	+
X										−

Analysis of this mating system has been possible by the aid of clones which had lost the micronucleus as a consequence of irradiation with X-rays. If normal and amicronucleate animals of complementary mating types are brought to conjugation, each of the "conjugants" receives a gamete nucleus of the normal partner. The haploid gamete nucleus becomes a "hemikaryon" in the "exconjugant", i.e. it divides into two daughter nuclei, one of which differentiates into a micronucleus, the other into a macronucleus. The micronucleus is haploid and the macronucleus is polyploid.

The genetic constitution of such "exconjugants" is called *hemizygous* since only one of the alleles of each locus is present, though in a large number in the macronucleus.

Such hemizygous descendants will also realize a definite mating type. Those from the conjugation of a normal animal of mating types VII to X with an amicronucleate partner have the same mating type as the normal animal. However, if the normal animal belonged to one of the mating types I to VI, the hemizygous descendants are with about equal frequency of any one of two different mating types of series VII to X. The same type of *segregation* as in autogamy (p. 247) is thus found. This can be explained by the assumption that mating types I to VI are heterozygous since hemizygous descendants contain either one or the other allele. The mating types are like those of normal animals which are homozygous for the same allele. Descendants of mating type I (mt^1mt^2), for instance, belong either to mating type VII or VIII.

Several Loci

The mating types of *Paramecium bursaria* are also directly determined by the genetic constitution. It is, however, not due to combination of pairs of multiple alleles of the same locus but to the combination of dominant or recessive alleles of two (syngen 1) or three different loci (syngen 4).

The genotypes of the four mating types (p. 217) of syngen 1 are listed in the table below. It is evident from the general formulae that mating type I is realized if dominant alleles of both loci are present. In mating type II a dominant allele is at the *B* locus, in mating type IV at the *A* locus. Mating type III is homozygous recessive for both loci.

Table 14. *Paramecium bursaria*, syngen 1. The genotypes corresponding to the four mating types and the mating type substances formed by the mature clones. After SIEGEL, 1961 [*1588*]

Mating types	Genotypes	General genetic formulae	Mating type substances	
I	*AABB* *AABb* *Aa BB* *Aa Bb*	*A-B-*	$\underline{\text{A}}$	$\underline{\text{B}}$
II	*aa BB* *aa Bb*	*aaB-*	$\underline{\text{a}}$	$\underline{\text{B}}$
III	*aa bb*	*aabb*	$\underline{\text{a}}$	$\underline{\text{b}}$
IV	*AAbb* *Aa bb*	*A-bb*	$\underline{\text{A}}$	$\underline{\text{b}}$

The validity of these genetic formulae has been ascertained by extensive crossing experiments [*1589, 1590, 1597*]. Mating type III is especially suited for the analysis of clones whose genotype is unknown.

The hypothesis discussed on p. 222 that the reactions between the four mating types of syngen 1 are due to the interaction of two pairs of complementary mating type substances has thus been fully verified. There is a clear correlation between the general genetic formulae and the hypothetical mating type substances

(Table 14): the dominant and recessive alleles at both loci determine which mating type substances appear in the cell surface of a *Paramecium*.

This viewpoint permits also a genetic interpretation of the various periods in the "life cycle" of exconjugant clones. During the period of immaturity both loci are still inactive. One of the loci becomes active in the period of adolescence, permitting the formation of one of the two substances. An adolescent clone can therefore react with two test clones which have the complementary substance (Table 8). The second locus becomes active in the period of maturity, permitting reaction with the third test clone which has the substance complementary to it.

The various states of differentiation through which the cells pass in the course of the "life cycle" of a clone can thus be traced back to *successive gene activation*.

It depends upon the genotype whether the *A* or the *B* locus is first to be activated. The temporal sequence of gene activation is determined by *regulator genes*. Moreover, a recessive allele of the *B*-locus (b^{-A}) has been found which suppresses manifestation of the *A*-locus. Cells homozygous for it can only produce the $\bar{b}$-substance. They behave therefore like permanently adolescent clones of the type $\overline{\underline{\text{III—IV}}}$ [*1591*].

Antigenic Properties

Thorough study has been made of those properties of ciliates which can only be demonstrated with *serological* methods.

If paramecia are injected into the ear vein of a rabbit, a certain substance within the paramecia acts as an *antigen*, i.e. it is responsible for the appearance of a specific *antibody* in the blood. The latter can readily be demonstrated by the effect of diluted blood serum of the rabbit on paramecia of the same clone as those injected to initiate antibody production. The reaction between antibody and antigen, which takes place over the entire cell surface [*115, 118*], causes the cilia to adhere to each other and the paramecia to sink motionless to the bottom *(immobilization test)*.

In actual practice the injection is repeated several times within a span of a few days. The minimum concentration of the serum which immobilizes the paramecia within two hours is called its *titer*. Such serum samples can be kept for months at low temperature without losing their effectiveness.

A close study of the various stocks of *Paramecium aurelia* [*106—108, 1649*] demonstrated that each stock can produce a large number of different antigens. Each antigen is strictly *specific*, i.e. the paramecia producing it can only be immobilized by the corresponding antiserum. They are not affected by serum gained with paramecia which produce another antigen. However, it is frequently found in the course of an immobilization test that not all paramecia of a clone are immobilized by the corresponding antiserum. While most individuals produce antigen A and are consequently immobilized by antiserum A, some form antigen B which can be demonstrated, after they have multiplied, by testing with antiserum B. Some may even remain motile in this test because they no longer form antigen B but antigen C instead (Fig. 227).

It has been found in this manner that each stock can form a whole series of different antigens, some appearing frequently, others rarely. A spectrum of 12 realizable antigens has been demonstrated in one particular stock [*1102*].

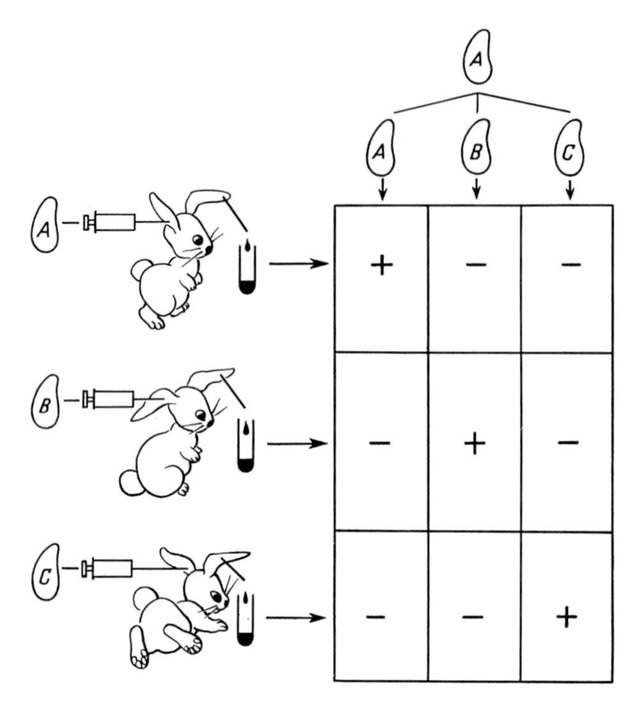

Fig. 227. Diagram of reciprocal interactions between paramecia of three clones descended from the same individual and the antisera obtained by injection of their sister clones into rabbits, + indicates immobilization in the homologous antiserum, — indicates no immobilization. Based on SONNEBORN, 1950, from EPHRUSSI, 1953 [*493*]

Although a given antigenic property may remain constant during many cycles of cell division, a sudden change to another property can always take place *(transformation)*. Which new antigenic property is acquired depends to a certain degree upon external conditions. A single cell can normally realize only *one* antigenic property.

A comparison of different stocks of the same syngen showed that some antigens can only be formed by certain stocks while others are common to all stocks. However, even the common or *homologous*, antigens of the various stocks can differ more or less clearly with respect to titer.

The genetic study of these relationships has led to far-reaching general conclusions. When two mating types of the *same* stock with different antigenic properties (e.g. A and B) are crossed, the properties of the parents reappear in the F_1-generation. The A-conjugant gives rise again to an A-clone, the B-conjugant to a B-clone. They retain the original property even after autogamy of F_1-individuals. This indicates that the antigenic properties are due to a specificity of the cytoplasm and that they are, therefore, directly transmitted to the next generation even at conjugation or autogamy (Fig. 228a).

Crosses of *different* stocks make it evident that nuclear genes are responsible for the realization of antigenic properties. This can be illustrated by a cross of stocks 51 and 29 (syngen 4) whose A-antigens can be distinguished by their titers: if the clones descended from the exconjugants carry out autogamy, segregation takes

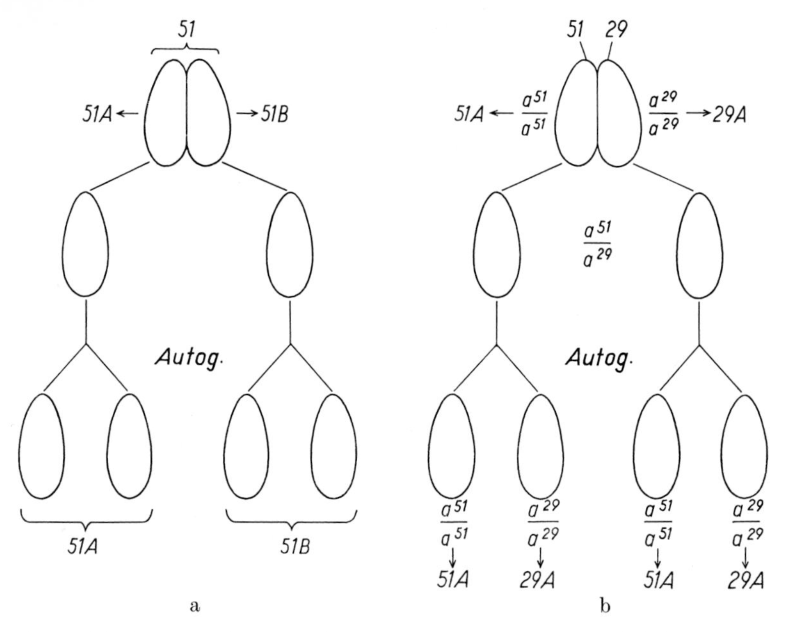

Fig. 228. Inheritance of antigen properties (syngen 4). a Cross between two individuals of stock 51, one forming antigen 51A, the other antigen 51B: no segregation after autogamy. b Cross of an individual of stock 51 which forms antigen 51A with an individual of stock 29 which forms antigen 29A: segregation after autogamy. Based upon studies by SONNEBORN and LESUER, 1948 [*1649*]

place among the exautogamonts (Fig. 228b). The difference between homologous antigens must therefore be due to different alleles [*1649*].

The most extensive studies of the inheritance of antigenic properties are those with syngen 1 where specificity is of the highest order and where the dependence of antigen formation upon external conditions is most clearly expressed [*106*].

Stocks of syngen 1 with the numbers 41, 60, 61 and 90 have the S-, G- and D-antigens in common. However, different homologous antigens can be clearly distinguished by their titer.

While of the S-antigens only that of stock 61 can be distinguished from others, the G-antigens of all stocks are characteristically different. The D-antigens of stocks 41 and 60 are identifiable while those of stocks 61 and 90 are not distinguishable from each other but from the two other antigens.

In all four stocks the S-antigens are formed predominantly at low (18°) temperatures, the G-antigens at intermediate (25°) and the D-antigens at higher temperatures (29–33°). These relationships are summarized in the following table.

Table 15. (Explanation in the text)

Stock	Antigens at			Genes
	18° C	25° C	29—33° C	
41	41 S	41 G	41 D	$s\ \ g^{41}\ d^{41}$
60	60 S	60 G	60 D	$s\ \ g^{60}\ d^{60}$
61	61 S	61 G	61 D	$s^{61}\ g^{61}\ d$
90	90 S	90 G	90 D	$s\ \ g^{90}\ d$

If paramecia of stock 41, for example, are cultured for prolonged periods at low temperature, they form antigen 41 S. If transferred to an intermediate temperature, they change after a number of cell divisions at an ever increasing rate to antigen 41 G and can thus no longer be immobilized by the specific antiserum to antigen 41 S. After prolonged culture at high temperature another change sets in, this time to antigen 41 D. The transformation of antigenic properties in response to temperature changes is similar in the other stocks.

Even though we find individuals in a clone with either one of the antigens at temperatures which are in between those mentioned above, the change from production of one antigen to another is, in any given individual, sudden and without intermediates.

Crosses of different stocks led to the result that each antigenic property depends upon a single gene and that homologous antigens are due to series of multiple alleles. Antigen 41 D, for example, is due to gene d^{41} and antigen 60 D to gene d^{60}. Since the D-antigens of stocks 60 and 90 cannot be distinguished by serological methods, it must remain open whether their d-alleles are different. Both are therefore designated as d. If the g- and s-alleles are also listed in like manner, the genetic constitution of each stock can be given, as in the right-hand column of Table 15.

The results of all crosses are in agreement with the genetic formulae. For example, if stocks 90 and 60 are crossed, the F_1-individuals produce *both* antigen 90 G and 60 G at the intermediate temperature (25°), when G-antigens become manifest, i.e. they are immobilized by both antisera. When transferred to the high temperature ($>29°$ C) suitable for the formation of D-antigens, they form both antigens 90 D and 60 D. This shows that the F_1 is heterozygous with respect to the two allele pairs $\left(\frac{g^{90}}{g^{60}}\frac{d^{90}}{d^{60}}\right)$. No dominance of one allele over the other can be observed. It is, however, remarkable that the reaction with the corresponding antiserum is clearly weaker in heterozygous (F_1-) individuals than in the homozygous parents ("gene dosage effect").

That different genes are indeed phenotypically manifested in the various temperature ranges is seen from the F_2-generation: if the F_1-individuals are permitted to undergo autogamy, the two new combinations $\left(\frac{g^{90}}{g^{90}}\frac{d^{60}}{d^{60}} \text{ and } \frac{g^{60}}{g^{60}}\frac{d^{90}}{d^{90}}\right)$ appear in addition to the parental types $\left(\frac{g^{90}}{g^{90}}\frac{d^{90}}{d^{90}} \text{ and } \frac{g^{60}}{g^{60}}\frac{d^{60}}{d^{60}}\right)$.

If the paramecia are transferred to another temperature range, the change to the other antigenic property does not take place immediately but after a large number of fissions (about 50). It is therefore possible to cross two stocks which form the antigens typical for different temperature ranges. One such cross is shown in Fig. 229. The stock 90-conjugant forms antigen 90 G because it derives from a culture to be raised at 25° C and therefore expresses gene g^{90}. The stock 60-conjugant produces antigen 60 D as it is from a culture at 29° C. Gene d^{60} is therefore still active. A test of the F_1-exconjugant clones (after about five fissions) led to the surprising result that both g-alleles come to expression in descendants of the 90-exconjugant and both d-alleles in descendants of the 60-exconjugant. If the 60-exconjugant clone is kept sufficiently long at a temperature of 25° C, the two g-alleles also become manifested in it, but if both clones are transferred to a temperature of 29° C, they both form the D-antigens of stocks 60 and 90.

An external condition such as temperature must therefore lead to a definite cytoplasmic state which is handed on over many fissions and even remains for some

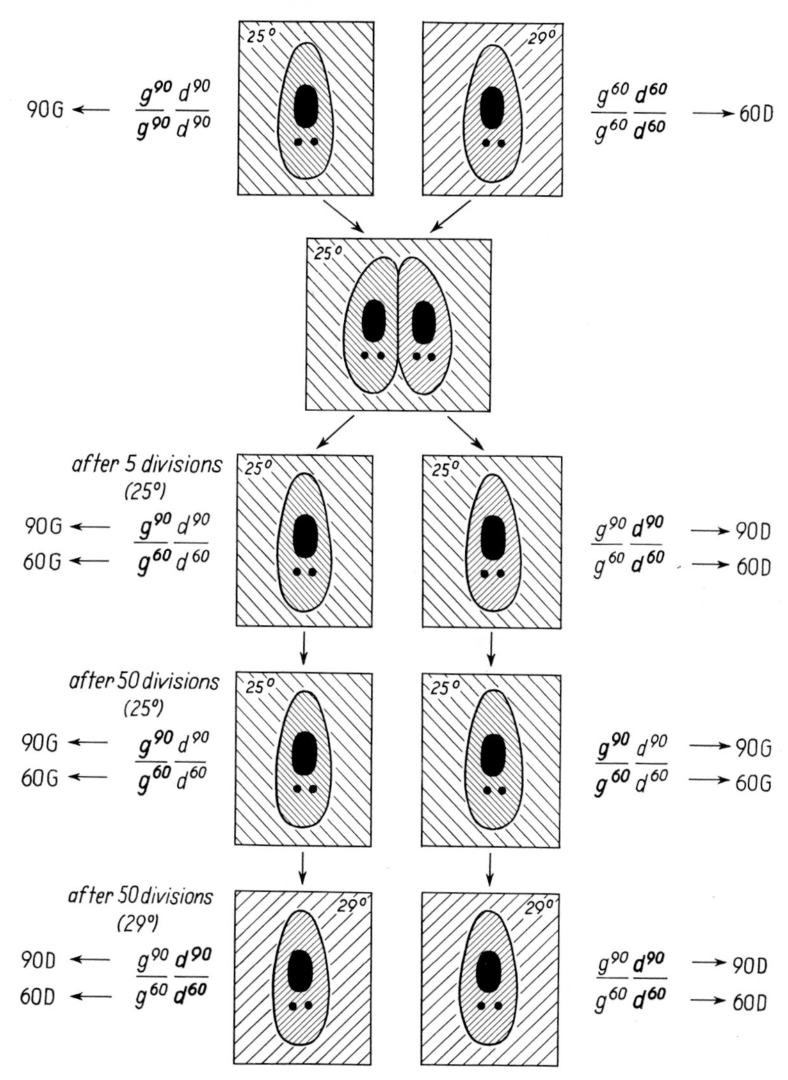

Fig. 229. Cross between stocks 90 (left) and 60 (right) of *Paramecium aurelia* (syngen 1). Stock 90 is from a culture kept at 25° (G-antigens), stock 60 from a culture kept at 29° (D-antigens). At 25°, the stock 60-exconjugant clone forms at first the two antigens 90D and 60D, later the two antigens 90G and 60G. After prolonged exposure to 29°, the exconjugant clones of both stocks form both D-antigens. In the genetic formulae the *s*-alleles were omitted. After investigations by BEALE, 1952 [*106*]

time after the cell has been transferred to another temperature. This state induces the activity of a definite gene and also activates alleles of this gene, which have been brought into the cell by crossing.

Tetrahymena pyriformis also forms different antigens depending upon the temperature [*1045, 1103*]. Stocks of syngen 1, which can be crossed with each other, form the L-antigens at low (10° C), H-antigens at intermediate (20–35° C) and T-antigens at high temperatures (40° C).

Some of these stocks form serologically different H-antigens, others different T-antigens. Research on *Paramecium aurelia* (see above) made it seem probable that these differences are controlled by multiple alleles of a H- as well as a T-locus. This was verified by appropriate crosses [*1192, 1195, 1198, 1214, 1300, 1301*]. Four such alleles could be demonstrated at the H-locus and three at the T-locus.

It was surprising, however, that the phenotypic manifestation of the antigens in heterozygous exconjugant clones is unlike that of *Paramecium aurelia*. If, for example, stocks which form antigens Hc and Hd at an intermediate temperature are crossed, the exconjugant clones with the genetic constitution $H^C H^D$ form *both* antigens only in the initial period of multiplication. Later on, more and more subclones can be isolated from the exconjugant clones, which form only one or the other antigen.
Crosses between such F_1-subclones revealed that the genetic constitution had not changed because the expected segregation (1:2:1) took place in the F_2-generation. Heterozygous clones could, on the other hand, be identified phenotypically only during the initial phase of multiplication. Later on, they "differentiated" into subclones which manifested either one of the antigens.
One of the two alleles of a heterozygous locus is thus alternatively "repressed", a phenomenon called "*allelic repression*". Exposure of "differentiated" heterozygotes to a variety of metabolic inhibitors failed to induce a single reversal of "differentiation". Hence, "allelic repression" must be regarded for the time being as irreversible nuclear differentiation [*880*].

Since the phenotype of ciliates is largely determined by the macronucleus, it seems possible that the increasing phenotypic "unification" of heterozygous exconjugant clones might be due to progressive assortment, in the course of successive fissions, of macronuclear subunits which became "differentiated" with respect to "allelic repression". Mathematical analysis showed that the kinetics of assortment correspond surprisingly well with that in selfer-clones (p. 220): it can be understood on the assumption of the presence of 45 subunits in a macronucleus just after fission.
In heterozygous exconjugant clones from stocks forming different T-antigens, assortment does not take place right after conjugation but about 30 fissions later. Since all crosses and clone cultures have been made at 25° C, and since the test for realizable antigens was carried out only with samples transferred to 40° C, "differentiation", i.e. the restriction of macronuclear subunits to manifest only one of the two antigens, must be independent of the actual formation of the antigen. In other words, although "allelic repression" forbids the expression of one of the two alleles, it does not necessarily lead to the manifestation of the other one. The actual "activation" of the allele in question requires the temperature factor.

Killers and Mate-Killers

Some stocks of *Paramecium aurelia* contain bacterial *symbionts* in the cytoplasm whose multiplication depends upon genes of the host cell. These symbionts, whose nature shall not be discussed here (comp. p. 360), confer upon their host cells the property to kill other stocks which lack such symbionts.
The effect of the so-called *kappa* symbionts has received most attention. Cells containing them are referred to as "*killers*". They continuously exude particles into the surrounding medium which lead to the death of cells lacking *kappa* symbionts. These latter cells are therefore called "*sensitives*" [*1334, 1631, 1638, 1640*].

It has been shown that the particles originate from the symbionts themselves and that one particle suffices to kill a cell of a sensitive stock. The particles are prob-

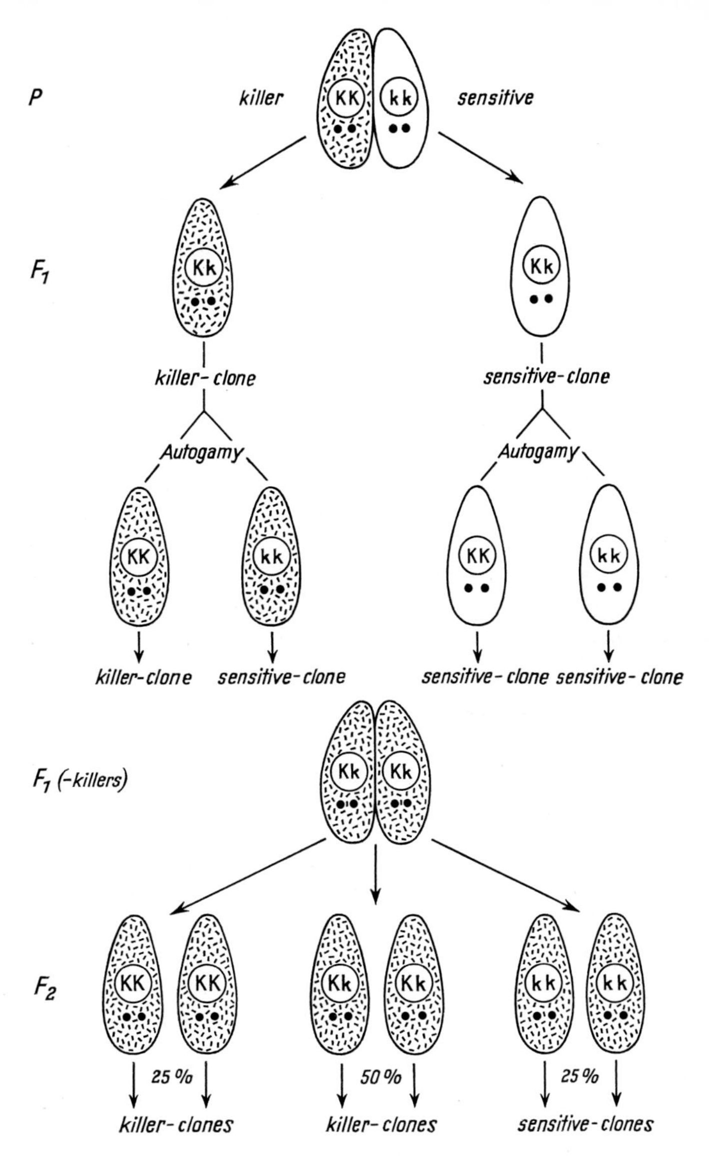

Fig. 230. Cross between a homozygous killer stock (*KK* with *kappa*) and a sensitive stock (*kk* without *kappa*) with subsequent autogamy in the exconjugant clones. Below: cross between heterozygous killers (*Kk* with *kappa*). After investigations by SONNEBORN, 1943 [*1631*]

ably taken up by the sensitives in the same way as food. During conjugation, the cells are protected from the particles because feeding ceases during this time.

This temporary "resistance" permits crossing a killer stock with a sensitive one (Fig. 230). If only migratory nuclei are exchanged between conjugants, the exconjugant clones are "phenotypically" like their parents. Dependence of *kappa*-multiplication upon the genotype of the host cell becomes evident only at autogamy. While exautogamonts derived from the sensitive exconjugant are again

sensitives, those from the killer exconjugant segregate. One half of them remain killers, the other half become sensitives, though only after a certain number of cell divisions.
The difference between the two stocks must therefore be due to a single pair of alleles, which is dominant (*KK*) in the killer stock and recessive in the sensitive stock (*kk*). After autogamy, the symbionts can only multiply in exautogamonts bearing the dominant allele.
Crosses between F_1-animals lead to the expected segregation: 75% of the exconjugant clones remain killers and 25% become sensitives. After autogamy, $^2/_3$ of the F_2-killers are seen to be heterozygous and $^1/_3$ to be homozygous.

Under certain conditions which can be brought about artificially, a broader than normal cytoplasmic bridge is formed between the conjugating partners. This permits exchange not only of migratory nuclei but also of considerable amounts of cytoplasm. If one of the partners is from a killer, the other from a sensitive stock (Fig. 231), *kappa*-symbionts can be transmitted to the sensitive conjugant. The resulting exconjugant also gives rise to a killer-clone.

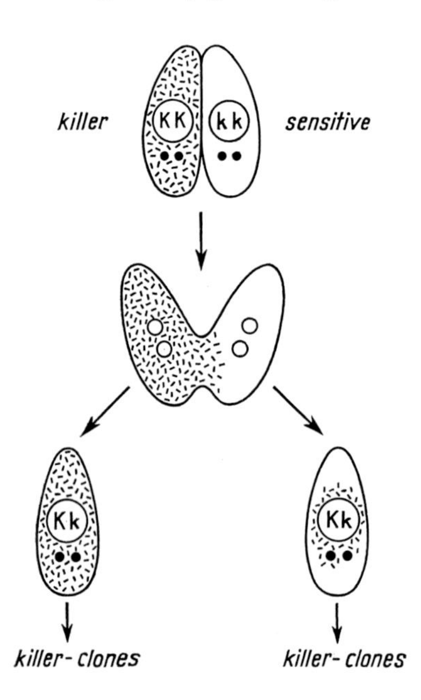

Fig. 231. Cross between a homozygous killer stock (*KK* with *kappa*) and a sensitive stock (*kk* without *kappa*) with a broad cytoplasmic bridge during conjugation (transfer of *kappa* to the sensitive partner). After investigations by SONNEBORN

The close relationship between the *K*-gene and the *kappa*-symbionts is also expressed in the fact that homozygous killers (*KK*) contain twice as many symbionts as heterozygous (*Kk*) ones [*218*].
However, killer cells can also lose their *kappa*-symbionts, e.g. when *S*-genes of certain sensitive stocks are acquired by crossing. Two nonlinked loci of such *S*-genes have been identified so far. If the paramecia are homozygous for the dominant alleles of both loci ($KKS_1S_1S_2S_2$), the symbionts disappear from the killer cells, but if they are present at one locus only ($KKS_1S_1s_2s_2$), they lead to loss of *kappa* only in a few individual cells of the clone [*76, 77*].

Disappearance of *kappa* can also be due to nongenetic factors, e.g. an increase of fission rate and temperature. Cells with the killer gene (*K*) can, however, be reinfected if exposed to a homogenate of *kappa*-containing cells (p. 360).

In some stocks of *Paramecium aurelia* the cells contain the so-called *mu-symbionts* in the cytoplasm. They are similar to *kappa*-symbionts in some respects but do not exude particles in the surrounding liquid. They kill the cells of sensitives only after they have conjugated with them and are therefore called *mate-killers*.

Since a normal exchange of nuclei takes place when a mate-killer conjugates with a sensitive, the two stocks can be crossed. The exconjugant derived from the sensitive partner dies.

The inheritance of mate-killing in stocks of syngen 8 is similar to that of the killer-phenomenon [*1027*, *1583*, *1584*]. Maintenance of the symbionts requires a gene M which corresponds to the K-gene.

It has been shown that mate-killer stocks of syngen 1 [*586*] contain two nonlinked loci, usually in homozygous combination (M_1M_1, M_2M_2). Crosses demonstrate that the cells remain mate-killers as long as they have one dominant allele at one locus. Sensitive clones arise only from cells which are homozygous recessive for both loci.

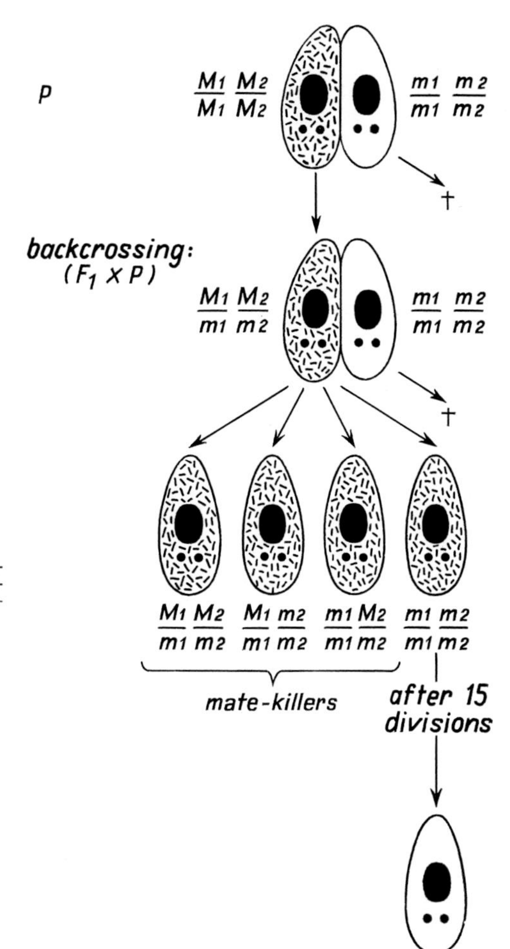

Fig. 232. Cross of a mate-killer stock with a sensitive stock. Backcross of the mate-killer F_1-exconjugant stock with the sensitive P-stock. After investigations of Gibson and Beale, 1961 [*586*]

In a backcross between a F_1-mate-killer and a sensitive, for example, the ratio of mate-killers to sensitives in the F_2 is 3:1 (Fig. 232).

Cells derived from killer or mate-killer exconjugants which became homozygous recessive (e.g. *kk*, $m_1m_1m_2m_2$) by crosses do not lose their symbionts at once. It is only after a varying number of fissions that more and more cells are found without symbionts. For any given cell the loss of symbionts is not gradual but sudden. The symbionts must therefore be capable of reproduction as long as they are present in the cell [*218*, *587*].

This phenomenon, known as *phenomic lag*, can be explained on the assumption that the genotype of the conjugant, which gave rise to the exconjugant exerts, over a number of cell divisions, an *after-effect* via the cytoplasm. It is not known, however, how this after-effect is brought about [*185*, *587*].

Thermosensitive Mutants

In *Paramecium aurelia* (syngen 4) thermosensitive mutations could be induced by various mutagenic agents (X-rays, UV, nitrosoguanidine). The mutants grow normally at temperatures up to 28° C but die at higher temperatures (35—36° C) at which wild type cells survive and multiply. The mutants, whose thermosensitivity is inherited as single recessive genes (*ts*), can be distinguished by the time of their survival at higher temperature and by their morphological appearance before dying.

Genetic analysis revealed that ts_{21m}, a complex thermosensitive and morphological mutant gene, which segregates normally and independently of ts_{111} and m_1 (a morphological mutant gene), is closely linked to ts_{401}. Frequency of crossing over is less than 10%. This was the first case of linkage so far reported in *Paramecium aurelia* [*129*].

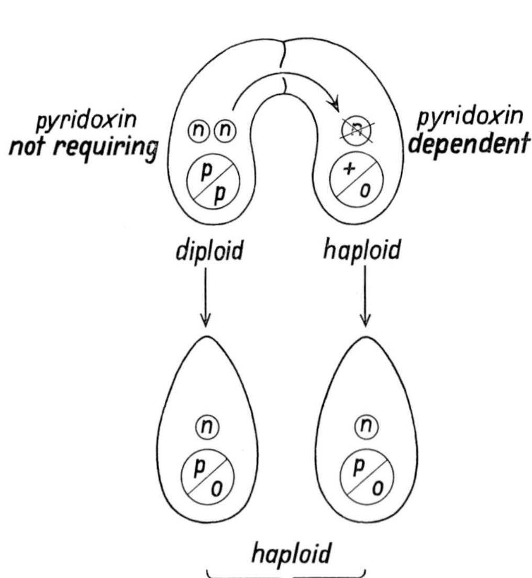

Fig. 233. *Tetrahymena pyriformis* (syngen 2). Cross of a diploid clone which can synthesize pyridoxin (*p*/*p*) with a haploid clone which is pyridoxin-dependent (+/*0*). Explanation in the text. Based on ELLIOTT and CLARK, 1958 [*488*]

Biochemical Mutants

While "deficiency mutants" known from *Neurospora* and bacterial genetics (compare also *Chlamydomonas*, p. 233) can no longer form certain compounds, two of the stocks of *Tetrahymena pyriformis* can synthesize substances which must be added to the culture medium of the "wild type". One stock can synthesize pyridoxin (vitamin B_6), the other the amino acid serine. Crosses showed both properties to be due to genes which are recessive to the wild type alleles [*488, 489*].

If a diploid mutant (e.g. "pyridoxin not requiring") is mated with a haploid wild type obtained by X-irradiation (e.g. "pyridoxin dependent"), the haploid micronucleus of the latter degenerates in the course of meiosis, and one gamete nucleus from the mutant becomes a "hemikaryon" (p. 253) in each "exconjugant". Both descendants are therefore hemizygous and show the phenotype of the mutant (Fig. 233).

Many studies in recent years have been concerned with the inheritance of enzyme systems (esterases) in *Tetrahymena pyriformis*, which can be demonstrated by starch gel electrophoresis. Details cannot be discussed here [*18—30, 34*]. It is, however, worth mentioning that the first case of *gene linkage* in ciliates has been described in the course of these studies. The *frequency of crossing over* between the two linked loci (the *mt*-locus and an esterase locus) was 25% [*21*].

Mitochondrial Genetics

Since it could be established that mitochondria contain DNA, questions have been raised, whether mitochondrial genes exist, what their functions are and how they cooperate with nuclear genes. The general idea was that mitochondrial genes code for mitochondrial transfer-RNA, ribosomal RNA and proteins, but the small amount of mitochondrial DNA suggested that only a small part of mitochondrial proteins could be coded in this way.

It turned out that *Paramecium aurelia* is as well a suitable tool in which to approach such questions [*1—4, 111, 116, 949, 950, 1693*]. Based on experiences with yeast, mutations which make paramecia resistant to certain antibiotics could be detected.

One mutation, resistance to erythromycin, arose spontaneously while another mutation, resistance to chloramphenicol, could be obtained after treatment with a mutagenic substance (nitrosoguanidine).

Resistance to both antibiotics is inherited through the cytoplasm at conjugation: a resistant conjugant produces a resistant exconjugant and a sensitive conjugant produces a sensitive exconjugant. If exchange of cytoplasm (p. 262) between a resistant and a sensitive conjugant occurs, both exconjugants produce a drug-resistant progeny.

Cytoplasmic mixing between erythromycin-resistant and chloramphenicol-resistant conjugants results in the production of "mixed" exconjugants. Clones deriving from them may retain their capacity to grow in the presence of either antibiotic when grown in a drug-free medium for a hundred or more fissions, but lose their resistance to the other antibiotic after growth in one antibiotic for a few fissions.

By microinjection of extracts which contained mainly mitochondria from resistant paramecia into sensitive cells of the same syngen a high percentage of the recipients became resistant.

Hence, the mitochondria are the sites of mutations in these cases. The "mixed" paramecia contain populations of both mutated types at the beginning, one population being eliminated during clonal growth by the respective antibiotic.

There is some evidence that the mutational change from sensitivity to resistance involves the alteration of proteins associated with the mitochondrial ribosomes.

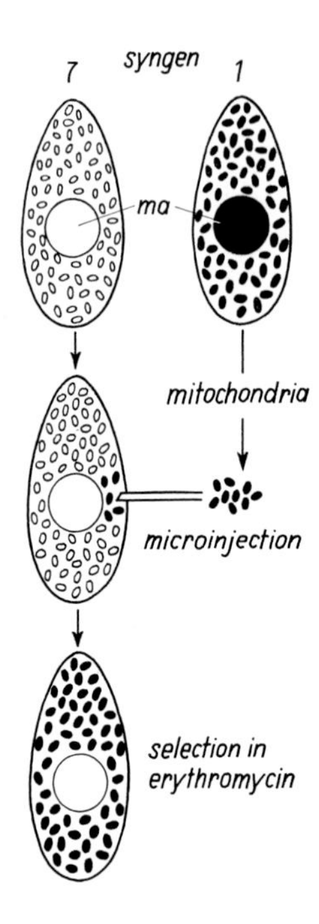

Fig. 234. *Paramecium aurelia.* Transfer of mitochondria by microinjection between two syngens whose mitochondria differ by sensitivity (syngen 7) or resistance (syngen 1) to erythromycin. After investigations of KNOWLES and TAIT, 1972 [*950*]

The microinjection method has also been used to demonstrate that a mitochondrial enzyme is not under mitochondrial but under nuclear control (Fig. 234). It was known that syngen 1 and syngen 7 differ with respect to the electrophoretic mobility of the soluble enzyme fumarase. Erythromycin-resistant mitochondria from syngen 1 were injected into paramecia from syngen 7 which contained erythromycin-sensitive mitochondria. By culturing the paramecia in an erythromycin solution the replication of the sensitive mitochondria could be inhibited. Finally, the syngen 7 paramecia contained only the resistant mitochondria from syngen 1. Nevertheless, the electrophoretic mobility of their fumarase corresponded to that of syngen 7. Hence, the mitochondrial enzyme fumarase is not coded by the mitochondrial but by the nuclear DNA [*949*].

In contrast to yeast, no case of recombination between the mitochondrial "genes" has as yet been found in *Paramecium*.

III. Modificability and Cell Heredity

We speak of *modificability* when individuals of the same genotype can manifest different phenotypes. Protozoa are especially suitable for studies of modificability because a progeny, a clone, can be obtained from any individual by asexual reproduction.

Any clone of *Paramecium caudatum*, for example, includes individuals of varying sizes which can be determined by measurements of body length (Fig. 235). The range of modification for this particular trait is given by the length of the smallest and that of the largest individual. Within these *extremes* (minimum and maximum), a continuous gradation of this trait is found. Most individuals are of intermediate length. The curve falls steadily off at both sides of this intermediate value.

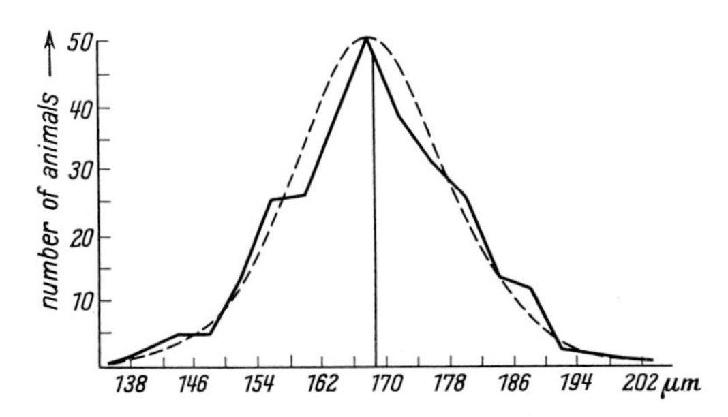

Fig. 235. Distribution of body length of 300 paramecia of a clone. The vertical line indicates the mean of the length in the distribution curve (= 168.5). The dotted line = chance distribution about the same mean. After data of JENNINGS from KÜHN, 1965 [*978*]

The curve coincides to a large extent with the *normal distribution curve* of statistics. This is due to the random action of external conditions upon individual cells, some acting to enhance, others to inhibit a trait, in this case body length. For any given individual, the net effect of inhibiting factors may exceed that of enhancing ones and vice versa. In most specimens, both are about equally effective. Every combination of factors leads thus to a definite expression of the trait in question.

Figure 236 shows two races of *Dunaliella salina* which differ in body shape. This can be expressed as the ratio of body length to width (δ). The two curves which illustrate the modificability of this trait overlap in a wide range. In a mixed population of both races it would therefore be impossible to decide to which of the races many individuals belong. After isolation and clone culture a decision can be made on the basis of modificability, which differs in the two races. Crosses have shown that the difference is due to a single pair of alleles.

The reaction norm as determined by the genotype thus permits *continuous* modificability with respect to certain traits, depending upon conditions of development.

Modificability is often clearly adaptational. Oocysts of the sporozoon *Eucoccidium dinophili* can be of very different sizes and contain a correspondingly different number of spores (Fig. 237). While the largest oocysts may form up to 250 spores, this number may be reduced in the smallest ones to four or even two. Modificability represents an adaptation to the strength of infection in this case. In a weak infec-

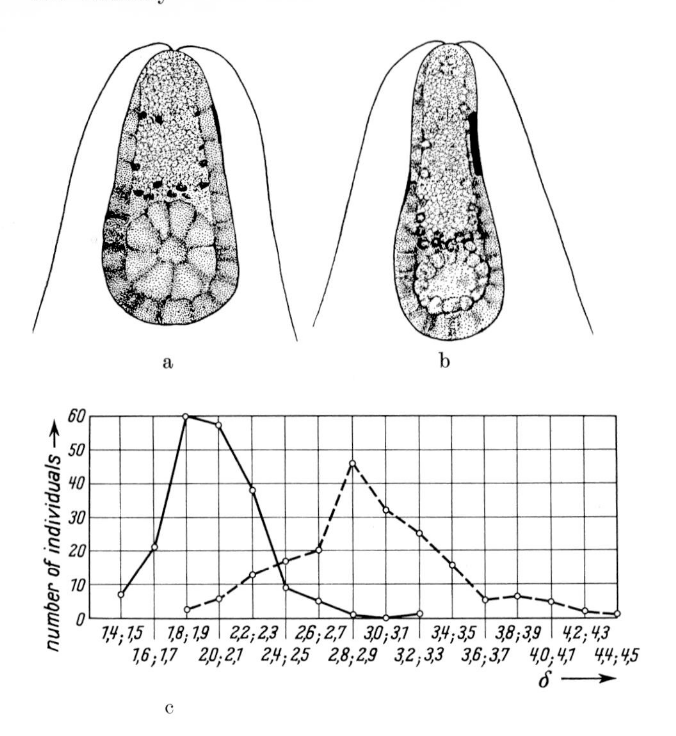

Fig. 236. *Dunaliella salina* (Phytomonadina). a Normal shape, b oblong shape (mutant). × 2080. c Modificability of body shape $\delta = \frac{\text{body length}}{\text{body width}}$ of the normal shaped race (solid curve) and the oblong shaped race (dotted curve). After LERCHE, 1937 [*1026*]

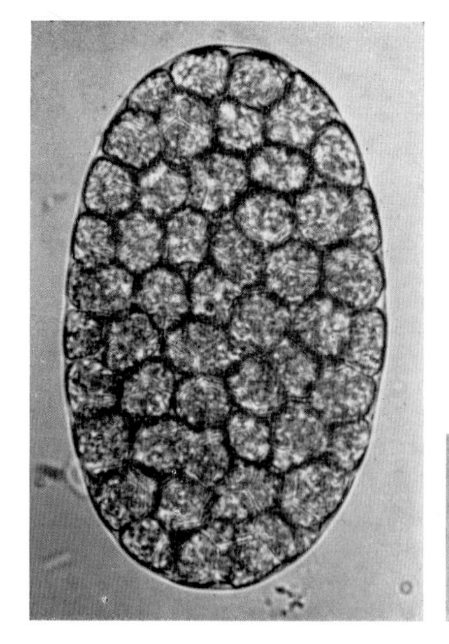

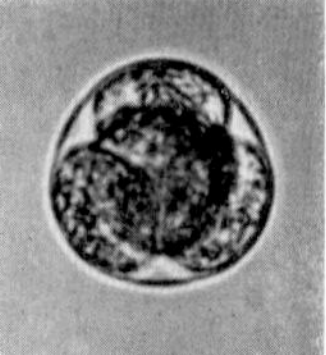

Fig. 237. *Eucoccidium dinophili* (Coccidia). Modificability of oocyst size at constant spore size. Photographed in life. × 820. After GRELL, 1953 [*657*]

tion, e.g. with a single spore, the growing macrogamonts have an ample supply of nutriments available in the body cavity of the host. They can therefore reach the maximal size (about 150 μm in length) given by the reaction norm. With strong infections, the nutriments are used up more rapidly and the macrogamonts transform earlier into macrogametes, i.e. sporogony sets in at an earlier stage of growth. All individuals can therefore finish development, in spite of lack of nutriments. Although the size of the oocyst varies, that of the spores remains constant.

Different external conditions can, in many cases, lead to *different cell forms*. The amoeba *Stereomyxa angulosa*, for example, is attached and more or less roundish when feeding. When the food, in this case diatoms, is exhausted, it detaches from the substrate and changes to the slim, angularly bent "floating form". It attaches and rounds off when it comes again into contact with a surface covered with diatoms. The related *Stereomyxa ramosa* does not detach as a "hunger form" but it branches extensively so that floating food organisms can easily be captured (Fig. 244a).

The term *cell differentiation* is generally only applied if the various cell forms of a species differ in structural characteristics. The so-called "*amoeboflagellates*" provide a well-known example. Since they appear either as a creeping amoeboid form or a flagellated swimming form, one might be uncertain whether they should be regarded as rhizopods or flagellates. Some species have the ability to encyst as well. *Naegleria gruberi*, which feeds on bacteria but can also be cultured axenically (p. 328), multiplies only in the amoeboid form (Fig. 238a). When the food supply is exhausted or the nutritive medium is diluted, it transforms within an hour to the flagellated form. When the swimming form contacts a suitable substrate, it forms pseudopodia, the flagellated pole becoming the posterior end (b, c). The flagella eventually disappear completely. Their basal bodies can no longer be detected by electron microscopy [*1544*, *1545*, *1807*]. Even during mitosis no centriole-like structures can be observed [*553*]. Under certain conditions, the amoeboid form, but not the flagellated form, encysts and from the cyst, it is the amoeboid form which hatches.

A pronounced *polymorphism* is also found in *trypanosomids*, which can appear as the *Leishmania*-, *Leptomonas*-, *Crithidia*- or *Trypanosoma*-form depending upon the host species. An especially far-reaching change of metabolism is naturally connected with the transformation from the intracellular, unflagellated *Leishmania*-form into the extracellular, flagellated *Leptomonas*-form [*343*, *1600*, *1737*, *1738*].

Cellular transformations also take place when sexual reproduction sets in, as described earlier (p. 153), not only when cells change directly to sex cells but also when they differentiate to gamonts which form gametes. If sex determination is by way of modification, gametes of different sexes can arise within a clone or from the same gamont. In many cases this alternative determination is connected with a differential nuclear or cell division. The causal relationship between both processes is unclarified as yet.

The differentiation connected with the change from one cell type to another can be *reversible* or *irreversible*. Sex cells provide examples for both possibilities. While isogametes of *Chlamydomonas* can change back into asexually reproducing cells, isogametes of Foraminifera must carry out fertilization or die. In cases of oogamety,

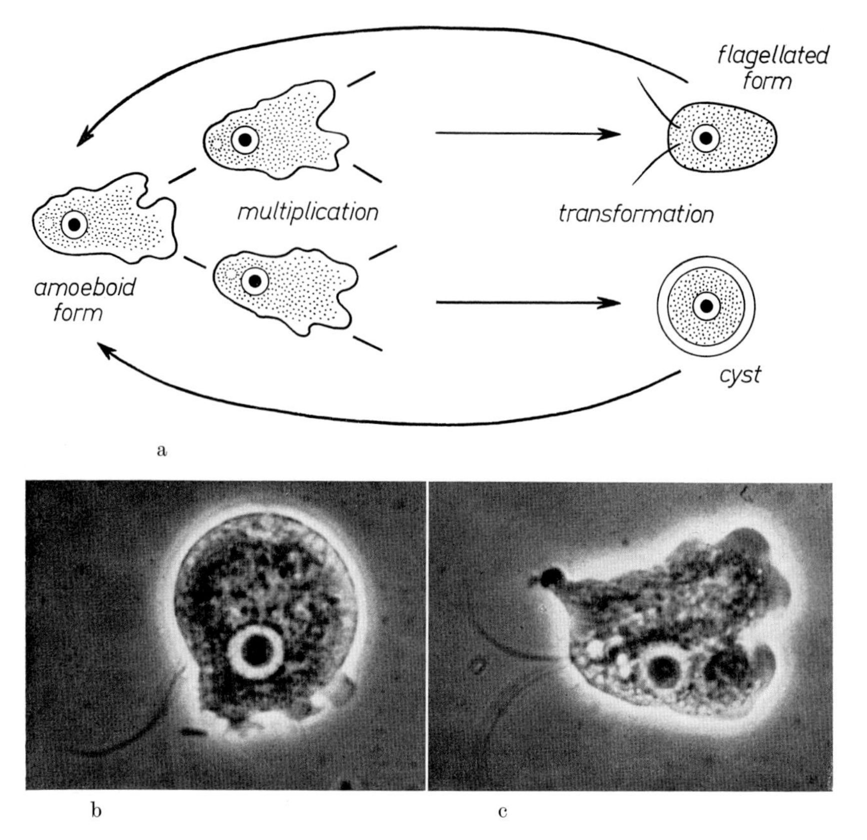

Fig. 238. *Naegleria gruberi*. a Diagram of multiplication and transformation. b, c Transformation of the flagellated swimming form into the amoeboid creeping form. From the films C942 and E1170 (GRELL)

unfertilized macrogametes can often develop parthenogenetically *(Volvox, Eucoccidium)*.
Cell differentiation on the basis of the same genotype can, according to present-day concepts, only be understood when *differential gene activation* is assumed. Models of gene activation at the molecular level are furnished by bacterial genetics [*734*, *834*]. Research on polytenic giant chromosomes led to the demonstration of microscopically visible changes of structure due to gene activity [*125*, *126*].

It has been shown in the preceding section that crossing experiments with ciliates can also furnish impressive examples of differential gene activation (Fig. 239). It can only be surmised that mating type determination by modification is due to activation of a functional subunit (cistron) of a complex locus (a). Successive activation of loci controlling mating type substances has been proved for *Paramecium bursaria* (b). Manifestation of different antigenic properties could also be attributed to gene activation (c). "Allelic repression" (d) remains as a peculiar special case.
Another important result of ciliate genetics is the proof of *cell inheritance:* phenotypic traits due to modification, for example, the property of forming certain mating type substances or antigens, are retained over many cell division cycles. If

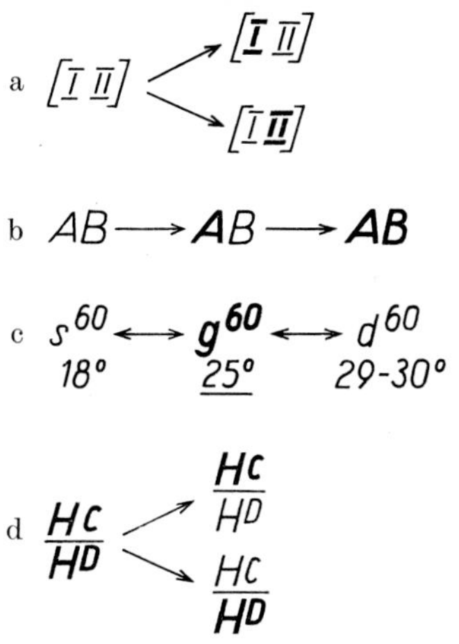

Fig. 239. Examples of differential gene activation in ciliates. Explanation in the text

their manifestation is due to gene activation, the possibility must exist that the pattern of active genes can be fixed to a certain degree.

Not much is known as yet about the regulatory mechanisms which enable such a fixation. We only know that it is restricted in many instances to the macronucleus or its subunits, and that it leads in other cases to the stabilization of a cytoplasmic state, which may perhaps represent a dynamic equilibrium (steady state system).

At any rate, the conception that phenotypic differences due to modification are not heritable requires qualification with respect to the phenomenon of cell inheritance: besides changes, which can only persist in the presence of the modifying influence, there are others, also due to modification, which can be handed on unaltered from cell to cell even after the inducing factor is no longer acting.

In an intermediate position, as it were, are those cases of prolonged changes in clones which are designated as "dauermodifications" and life cycle phenomena.

Dauermodifications arise gradually in response to external conditions. They remain over many cell generations after the inducing factor has ceased to act but then gradually disappear.

Paramecia, for instance, can become accustomed to poisons such as arsenious acid if the concentration is slowly increased [*869*]. In order to achieve this, a clone of *Paramecium caudatum* was left for some time in a concentration of arsenious acid which could just be tolerated by the cells. Most paramecia died after a slight increase of the concentration. The survivors were allowed to undergo multiplication and put again into a stronger solution. The final survivors after several repetitions of this treatment tolerated a 0.005-*N* solution, while a 0.001-*N* solution was originally lethal for most individuals. This increased resistance persisted during prolonged culture in a solution free of arsenic. Nevertheless, it was gradually lost. However, it took $8^1/_2$ months to reach the original level of low resistance.

In ciliates, the phenomenon of "dauermodifications" might be connected with the special organization of the macronucleus. Since there are many alleles of all genes in a polyploid macronucleus, it is not improbable that in the many macronuclei in a clone an allele might mutate to a form which is more suited to the experimental conditions to which the clone is exposed. In the given case, the mutated allele might confer an increased resistance towards arsenious acid. An individual with such an allele would be at a definite, though perhaps slight, selective advantage compared to its sisters in the clone. Through random distribution the allele could be multiplied in the macronuclei in certain descendants of this cell. Since a quantitative relation between the number of alleles and phenotypic effect has also been observed in other cases (p. 258), the formation of more and more resistant cell lines in this manner must be taken into account. Gradual fading of resistance later on could simply be due to a lack of selective advantage in the absence of arsenious acid. Cells with predominantly and later exclusively normal nonmutated alleles in their macronucleus can thus multiply again. The fact that resistance toward arsenic disappeared at once after conjugation, when a new macronucleus is built up, is in accordance with this interpretation.

In other cases, e.g. in the so-called calcium resistance of *Paramecium*, the dauermodification persists after conjugation or autogamy although the clones from daughter cells of exconjugants and exautogamonts differ clearly in resistance. This cannot be explained on the basis of a mutation. One can, however, imagine that selection acts upon the pattern of gene activation in the macronucleus rather than on mutated alleles. Maintenance of activation patterns in spite of sexual processes might be ensured as in the inheritance of mating types in group B of *Paramecium aurelia* (p. 215), i.e. due to the persistence of a cytoplasmic state which was induced by the macronucleus and which leads in turn to a fixation of the developing macronuclear anlage [*568*].

Life-cycle phenomena have already been mentioned (p. 221). Although clones from an exconjugant or one of its daughter cells are already determined with respect to mating type, they cannot as yet conjugate. They must at first pass through an immaturity period which may vary in time depending upon stock and syngen. They become capable of conjugation only after entering the maturity period which, in *Paramecium bursaria*, is preceded by a period of adolescence.

Analysis of the adolescence phenomenon (p. 254) led to the conception that the genes determining the mating type are not yet active during the period of immaturity and that mating type substances are consequently absent in the ciliary envelopes. The genes controlling these substances are within the macronucleus. Paramecia which do not receive a new macronuclear anlage after conjugation or autogamy but regenerate their macronucleus from a fragment of the old one will not become immature [*1589*]. The mating type genes, respectively the subunits of the *mt*-locus (p. 251), obviously remain irreversibly active after disintegration and regeneration of the macronucleus.

The duration of the immaturity period is gene-dependent as shown from crosses of two stocks of *Paramecium caudatum* which differed with respect to this trait [*764*].

In many ciliates, clones which had no opportunity to conjugate exhibited a steadily decreasing fission rate. They finally died even though conditions were optimal for life. This phenomenon has been termed *senescence*. Its cause is unknown.

In *Paramecium aurelia* it could be shown [*1636*] that senescence of clones does not set in if autogamy takes place from time to time (normally every 20—30 divisions). However, if lines are isolated which did not undergo autogamy, their division rate becomes lower and mortality higher, the older they are since the last autogamy. One such line died after the 324th division, 123 days after the last autogamy. However, even after the first month, mortality amounted to 1.7%, towards the end of the second month to 15.6% and to 25% during the last nine days. Although autogamy normally prevents senescence and thus "rejuvenates" the clones, decreased fission rate and higher mortality is also observed among a great many exautogamonts, depending upon the age of the lines when entering autogamy. If autogamy is induced in a subculture shortly before the clone dies, all of the exautogamonts will die. If a clone homozygous for the *am*-gene (p. 115) is aged, one can obtain macronuclear regeneration instead of autogamy at any time. This makes it plain that senescence can be prevented by timely regeneration of the macronucleus. As in autogamy, its rejuvenating effect becomes less and less probable as the clones advance in age. Lines whose macronuclei are derived from the various fragments of the same original macronucleus can exhibit great differences with respect to fission rate and mortality. The older the cell from which these lines were derived, the larger is the proportion of lines with a lowered fission rate and increased mortality. In very old clones all descendants die in spite of macronuclear regeneration.

In *Euplotes crassus*, where mating type determination is by way of a system of multiple alleles (p. 251), selfing occurs with increasing age in all clones heterozygous for the *mt*-locus [*738*]. The possibility that selfing was preceded by an autogamy could be excluded. It could be shown with the split-pair method (p. 212) that one conjugant had retained the original mating type while the other had the mating type corresponding to the recessive allele. The proportion of selfers can be considerably increased by transferring heterozygous clones which had reached a certain age to a lower temperature (20° C → 12° C). This observation indicates that the appearance of a new mating type is not due to segregation of the alleles. In addition, the newly appearing mating types proved to be unstable. They might revert to the original type after only a few fissions, and then back again to the type corresponding to the recessive allele. This suggests that dominance of a *mt*-allele can only be retained for a limited number of cell divisions (about 500). Later on, a condition, which abolishes the stability of this interallelic relationship, asserts itself in the cells. The originally recessive allele can thus also manifest itself. The nature of this condition is unclarified.

Selfing is of similar importance for heterozygous clones of *Euplotes crassus* as is autogamy for *Paramecium aurelia:* it enables the beginning of a new life cycle and saves the clones from the fate of dying out.

The phenotype of a cell is thus not only under the influence of conditions which prevail during its individual growth, but also of "historical factors", which led in one of its predecessors to the induction of a definite state in nucleus or cytoplasm. Although the regulatory mechanisms which enable the persistence of such a state may be very different, the genetic information stored in nuclear genes seems to play a role in all such cases.

After the discovery of "*organelle genes*" (pp. 241 and 265) which consist of DNA molecules in the organelles concerned (chloroplasts, mitochondria), the possible role of extranuclear genetic information for cell differentiation has to be clarified. The integration of the different genetic systems within the cell may be one of the main topics of cell biology in the future. At the present, however, our knowledge is not sufficient for any speculations.

In this connection the *basal bodies* or *kinetosomes* cannot be left unnoticed. According to an older hypothesis, based on light microscopic observations [*1066*], they were not only considered capable of binary division and growth to flagella or cilia but also of differentiation into other structures such as trichocysts under suitable conditions. Although electron microscopic investigations could not confirm this hypothesis, they do not exclude the possibility that basal bodies are able to "induce" the formation of new basal bodies or other structures, as will be discussed in the next chapter (p. 292).

To a certain degree, the outer cell layer or *cortex of ciliates* can be regarded as a "semi-autonomous" construction, at least with respect to the position and arrangement of its structural components. This is expressed in the perpetuation of disturbances of pattern over many cell generations.

Double animals containing all cortical differentiations in duplicate are occasionally found in many ciliate species. Usually they have only one, though correspondingly enlarged, macronucleus. Such malformations can be due to a variety of causes. In *Paramecium aurelia*, for instance, they are due to a failure of conjugants to separate. In the first metagamic division the anterior halves are pinched off as single animals but the posterior ones fuse to form a double animal (Fig. 240a).

Such double animals divide normally and can be taken into clone culture. Although they separate occasionally into singles, they remain fairly stable under constant conditions.

Experiments with *Paramecium aurelia* [*1641*] have shown that they retain this property even when crossed with single animals. Although their genetic make-up after conjugation is the same as that of singles, the doubles do not revert to the

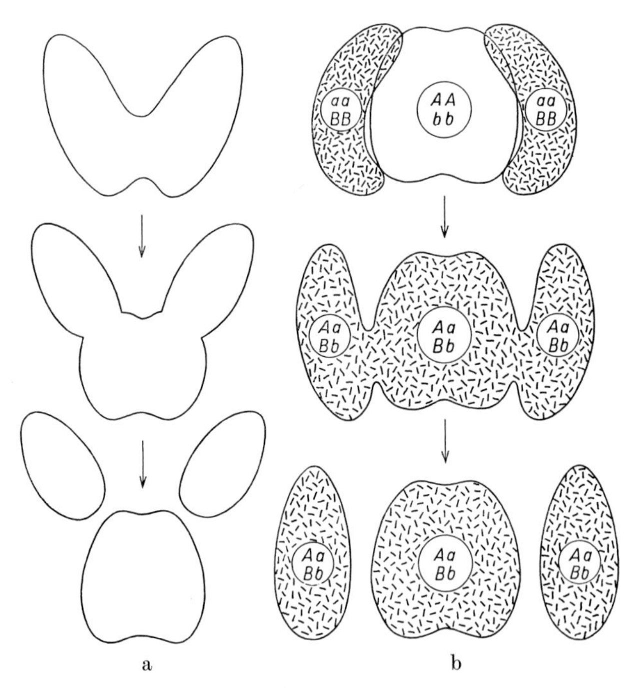

Fig. 240. *Paramecium aurelia*. Double animals. a Origin of a double animal. b Cross of a double animal with single animals: in spite of isogenization and exchange of cytoplasm (as indicated by the transfer of *kappa*), the trait "double animal" is inherited further. Diagram. After investigations by SONNEBORN, 1963 [*1641*]

single state. Even an exchange of cytoplasm at conjugation, as indicated by the transfer of *kappa*, does not lead to normalization (b).

In *Paramecium aurelia*, disturbances in the arrangement of single areals are also heritable. All of the fibrils which originate at the basal bodies run anteriorly at the right side of a ciliary row, as described earlier (p. 45). By artifice, cells can be obtained in which one or some of the 70 ciliary rows are inverted, the fibers running posteriorly at the left side (Fig. 241). One such abnormality persisted over 800 cell generations although many conjugations and autogamies took place in the meantime [*130, 1642*].

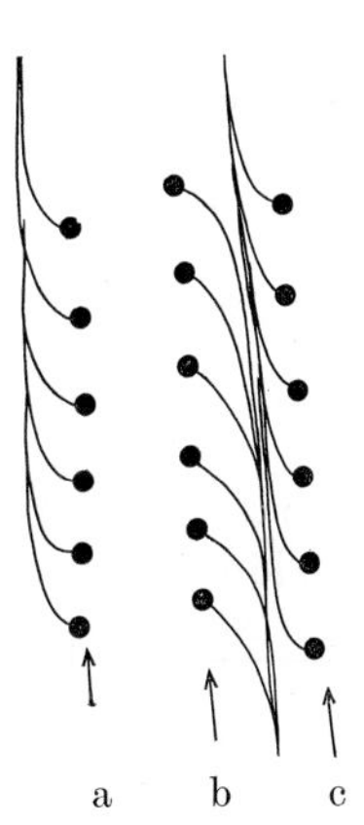

Fig. 241. Diagram of an experiment in which a part of a ciliary row (b) had been inverted. After SONNEBORN, 1964 [*1642*]

Cortical organization is thus obviously determined to a certain degree by the preexisting pattern of differentiation which had been artificially changed in the cited case.

Studies on *Tetrahymena pyriformis* revealed the existence of numerous *variations of cortical pattern* in natural populations. These differ in the number of ciliary rows (kineties), the relative positions of their parts and other characteristics as well. The various cortical patterns (*"corticotypes"*) are inherited in the course of the cell generations, irrespective of their own modificability. Inheritance is in this case independent of the genetic constitution of the cell nuclei or any determinants of the cytoplasm. "Heritability" must therefore be due to the "supra-molecular" organization of the cell cortex itself [*1201–1209*].

However, these examples should not be misinterpreted to indicate that cortical organization is entirely due to self-differentiation or that it is independent of the genes. Even in species with far-reaching "cortical autonomy", *neighborhood-relations* do play an important role, e.g. in the induction of the cytopyge (*Paramecium*) or the "pores" of pulsating vacuoles (*Tetrahymena*). Regulations which abolish malformations are also possible to a limited extent. The process of normalization takes several cell generations in many cases.

There are surprising differences among ciliates in this respect. *Stentor*, for example, is capable of nearly unlimited regulation. Disturbances of the stripe pattern are usually adjusted in the same cell generation. Missing parts can be replaced, supernumerary ones can be melted down. Aside from the ability to build up an anlagen field and to specifically organize it for division or physiological regeneration (p. 135),

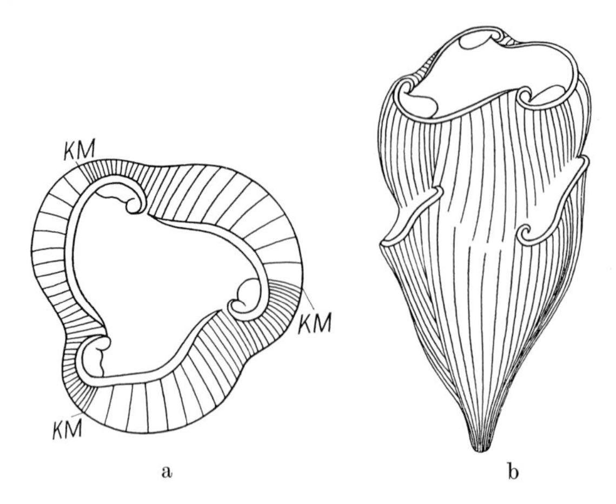

Fig. 242. *Stentor coeruleus.* Diagram of a triple animal. a Viewed from above, b from the side. *KM* contrast meridian (compare with Fig. 123). After experiments by UHLIG, 1960 [*1755*] from SCHWARTZ, 1965

the cell must have at its disposal a control system which registers malformations and initiates the corresponding steps for developmental regulation [*1564*]. Normalization seems to fail only if the disturbance does not lead to a disharmony of the induction processes connected with division. *Triple animals* could be made of *Stenor coeruleus* by surgical means. These had three contrast meridians, and they differentiated three new anlagen fields before division (Fig. 242). In one case a clone of such triple animals has been cultured for almost two years [*1755*].

The processes of resorption and regeneration associated with every division (p. 132ff.) show clearly that the pattern of cortical differentiation can only be autonomous to a more or less limited degree, depending upon the species. Even in physiological regeneration and during encystment and metamorphosis the cortical changes form only a part of the whole transformation of the cell. If the developmental cycle of a species involves different cell types, a specific pattern of differentiation corresponds to each of these cell types. This structural change cannot be autonomous but must depend upon changing conditions in the whole cell, which are in turn controlled by a sequence of gene activations.

We have no insight as yet into the nature of these regulatory processes.

H. Motility

Motility is a general property of protoplasm. Intracellular *cytoplasmic streaming* is found in all Protozoa. It is usually readily recognizable by microscopy or in microcinematographic films taken at normal speed; but occasionally it can only be detected by time-lapse cinematography. The streaming cytoplasm often follows definite pathways. Its motion can also be correlated in specific ways with certain phases of reproduction.

Motility is more noticeable if it leads to *locomotion* or to a *change of the shape* of cells.

I. Locomotion

Locomotion is a property of all Protozoa, even though in some parasitic groups as in Sporozoa it is restricted to definite stages of development. As a rule, movement is due to special *organelles of locomotion*. The latter may be temporary formations (pseudopodia) or permanent differentiations of the cell such as flagella and cilia. Some Protozoa carry out a *gliding movement* without the aid of special locomotor organelles.

1. Pseudopodia

Pseudopodia are extensions of the cell which can either be formed at any point or in a definite region and be melted down again at any time. They are especially characteristic for the Rhizopoda but can also appear in certain flagellates *(Cercomonas, Chromulina)*.

There is much diversity in the forms of pseudopodia. Several types have been distinguished which are occasionally difficult to delimit.

The so-called *lobopodia* are characteristic for most *Amoebina*. They are lobe-shaped, more or less broad extensions. Several lobopodia can usually be formed simultaneously. Only in some of the small species, such as in the "*Limax*"-amoebae living in the bacterial surface skin, is the cell body extended into a single, not sharply delimited pseudopodium. However, even typical "polypodial" species like *Amoeba proteus* can change to a "monopodial" form under certain conditions (e.g. negative phototaxis).

Many amoebae show *polarity* when moving. Species of the genus *Trichamoeba* have a tail appendage of thin protoplasmic threads (uroid). Small "monopodial" species (e.g. *Flabellula mira, Hyalodiscus simplex*) move ahead with a broad, hyaline pseudopodium.

Amoeba proteus and some other fresh water species have so-called *tubular pseudopodia*. Like the whole cell body, these consist of a viscous ectoplasm and a more liquid endoplasm. The formation of a pseudopodium is always connected with streaming of the endoplasm, referred to as the *axial stream*. A part of this endoplasm is formed from ectoplasm and is transferred again into ectoplasm at the tip

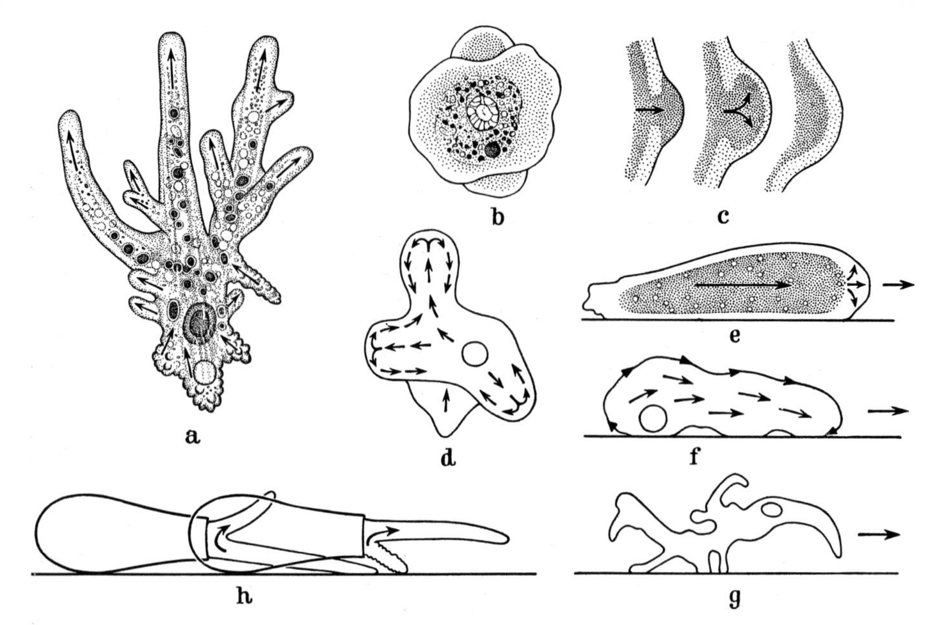

Fig. 243. Different types of amoeboid locomotion. a *Amoeba proteus* with tubular pseudopodia. b *Entamoeba histolytica* with broad, lobose pseudopodia. c Diagram of the origin of such pseudopodia by sudden rupture of the ectoplasm (light) and an outpour of endoplasm (dark). d An amoeba with fountain streaming. e Creeping motion of an amoeba of the *Limax* type. f Rolling motion. g Walking motion. h Locomotion in an "inchworm" fashion in a *Difflugia*. Combined after various authors

of the pseudopodium (*"ecto-endoplasmic process"*). The posterior end of a moving cell has a wrinkled surface (Fig. 243a).
In some amoebae, e.g. *Entamoeba histolytica*, the ectoplasmic layer is locally "liquified" as a pseudopodium forms. It seems to rupture finally and the endoplasm spreads like a droplet over the adjoining regions of the ectoplasm (b, c). The cellular envelope is, of course, not ruptured in the process.

In many cases there are no such pronounced differences of viscosity recognizable within a pseudopodium. If the protoplasm is highly liquified, the axial stream may divide near the tip of the pseudopodia and flow backward at the sides as a *fountain stream* (d). In some marine species the pseudopodia consist of a fairly viscous and hyaline protoplasm. Their movement is slow, but they are highly branched so that drifting food organisms may easily stick to them (Fig. 244). While the cells of *Stereomyxa ramosa* (a) form in a culture dish a dense mesh suggestive of a mesenchymatic tissue, a single cell of *Corallomyxa mutabilis* (b) grows to a net-like plasmodium from whose strands innumerable pseudopods originate.
Species of the genus *Thecamoeba* (e.g. *T. orbis*, Fig. 346b), which have a very thick longitudinally folded outer layer, carry out a *rolling* motion (Fig. 243). The amoeba moves in a caterpillar-tractor fashion, its surface attaching successively at the front end while contact with the substrate is given up at the rear. This type of locomotion is, of course, not connected with any formation of pseudopodia. In amoebae which do not lie with a broad surface area upon the substrate but form pseudopodia downwards, locomotion can give the appearance of "walking" (g). A type of walking motion is also carried out by *Pontifex maximus*. Contact with the

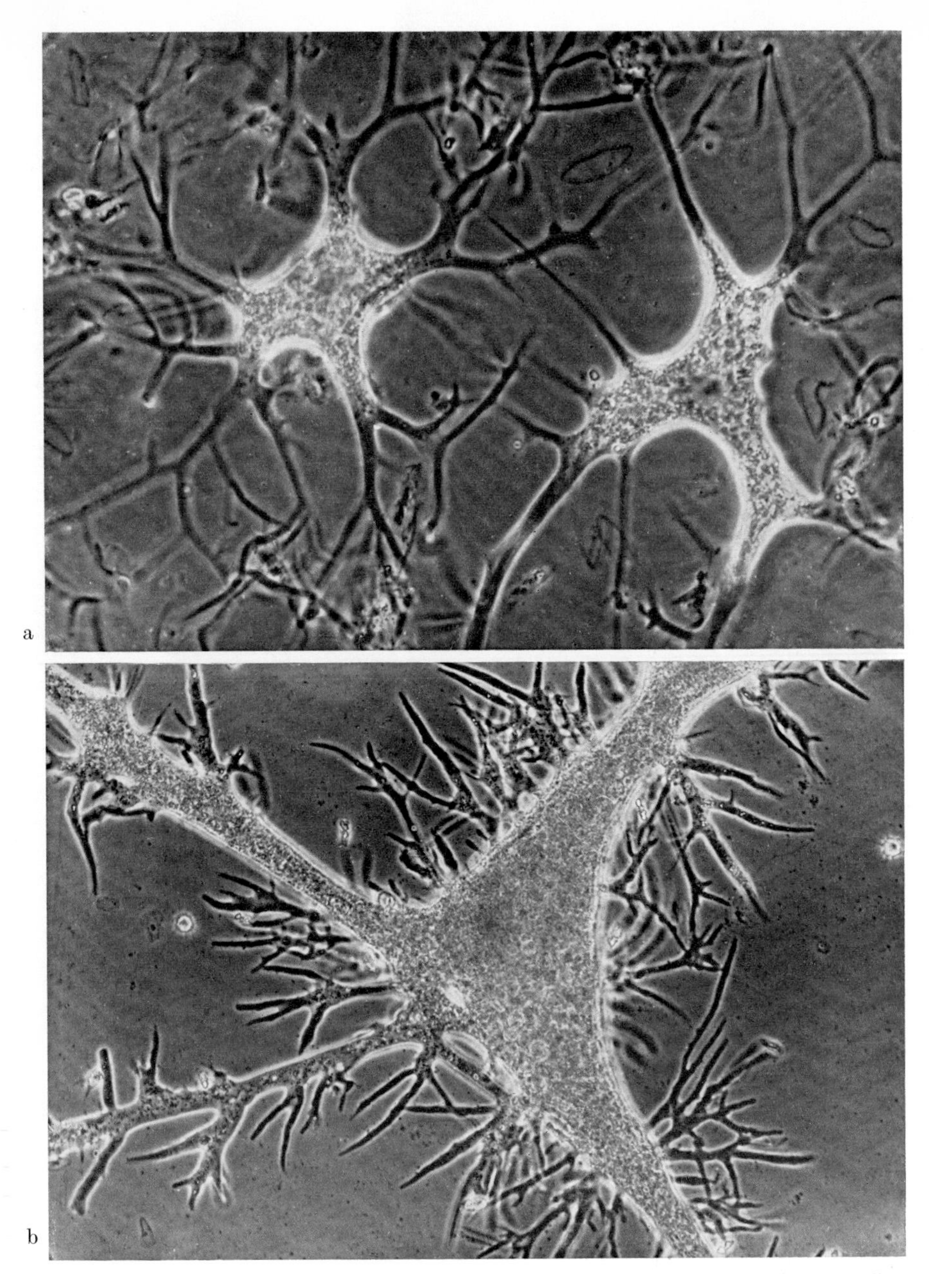

Fig. 244. Amoebae with branched pseudopodia. a Two cells of *Stereomyxa ramosa*. b Part of a net-like plasmodium of *Corallomyxa mutabilis*. Photographed in life. × 360. After GRELL, 1966 [*674*]

substrate is possible by way of special formations of the cell surface, the "adhesive projections" (Fig. 253, 346c).

The pseudopodia of *Testacea* are also predominantly hyaline. They may either be broad and lobular (Lobosa) or thin and thread-like (Filosa), in the latter case occasionally branched and connected by anastomoses. The pseudopodia originate frequently from a "pseudopodial cone" which protrudes from the shell. The movement of lobose testaceans resembles locomotion of the caterpillars of certain moths: a pseudopodium is extended and attached to the substrate. The cell body with the shell is then pulled forward and a new pseudopodium is formed (Fig. 243h).

In the freely floating *Heliozoa* and *Radiolaria* the pseudopodia radiate in all directions. They serve mainly as extensions for floating and food capture. Only when they contact a substrate can they also be used for locomotion.

The pseudopodia of *Heliozoa* are called *axopodia*. They contain a firm axoneme, which imparts a certain rigidity so that they can be bent to the side. Cytoplasmic flow along the axoneme is evident from the movement of inclusions (mitochondria, protein granules), which are carried along in both directions. The axonemes of the pseudopodia extend far into the interior of the cell, where they either end freely in the cytoplasm *(Actinosphaerium)*, at the cell nucleus (*Actinophrys*, Fig. 172a) or at the so-called central granule (*Acanthocystis*, Fig. 369). In *Echinosphaerium nucleofilum* they are implanted into depressions of the nuclear surface.

Axonemes are seen to be birefringent when viewed with polarized light [*1511*]. Electron microscopy [*718, 819, 942, 944, 1082*] showed that they consist of a

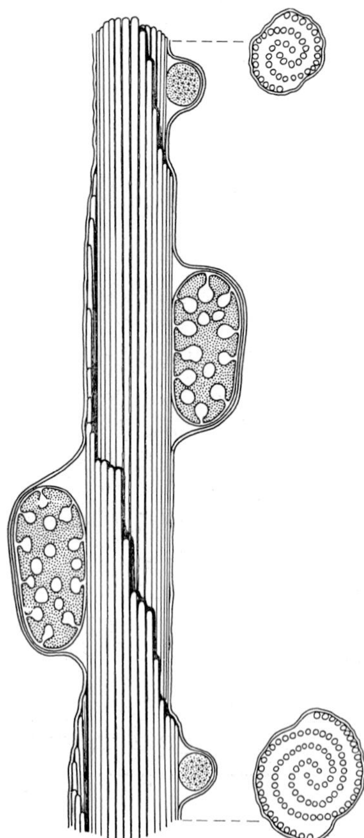

Fig. 245. *Echinosphaerium nucleofilum*. Reconstruction of an axopodium, based on electron micrographs. Beneath the cell envelope are two mitochondria and granules. The cross sections (right) illustrate the arrangement of the microtubules. After Tilney and Porter, 1965 [*1728*]

bundle of microtubules. In cross sections the microtubules are seen arranged as two interposed spiral lamellae wound about a central axis (Fig. 22c, d). While about 500 microtubules have been counted at the base, this number decreases more and more toward the tip of the pseudopodium, leaving finally only a few central tubules. Fig. 245 shows an attempt at reconstruction.

Agents which affect the integrity of microtubules (low temperature, hydrostatic pressure and colchicine), known as "antimitotic agents", lead to a breakdown of the axoneme and consequently to a retraction of the axopodium. After removal of the colchicine the axopodia begin

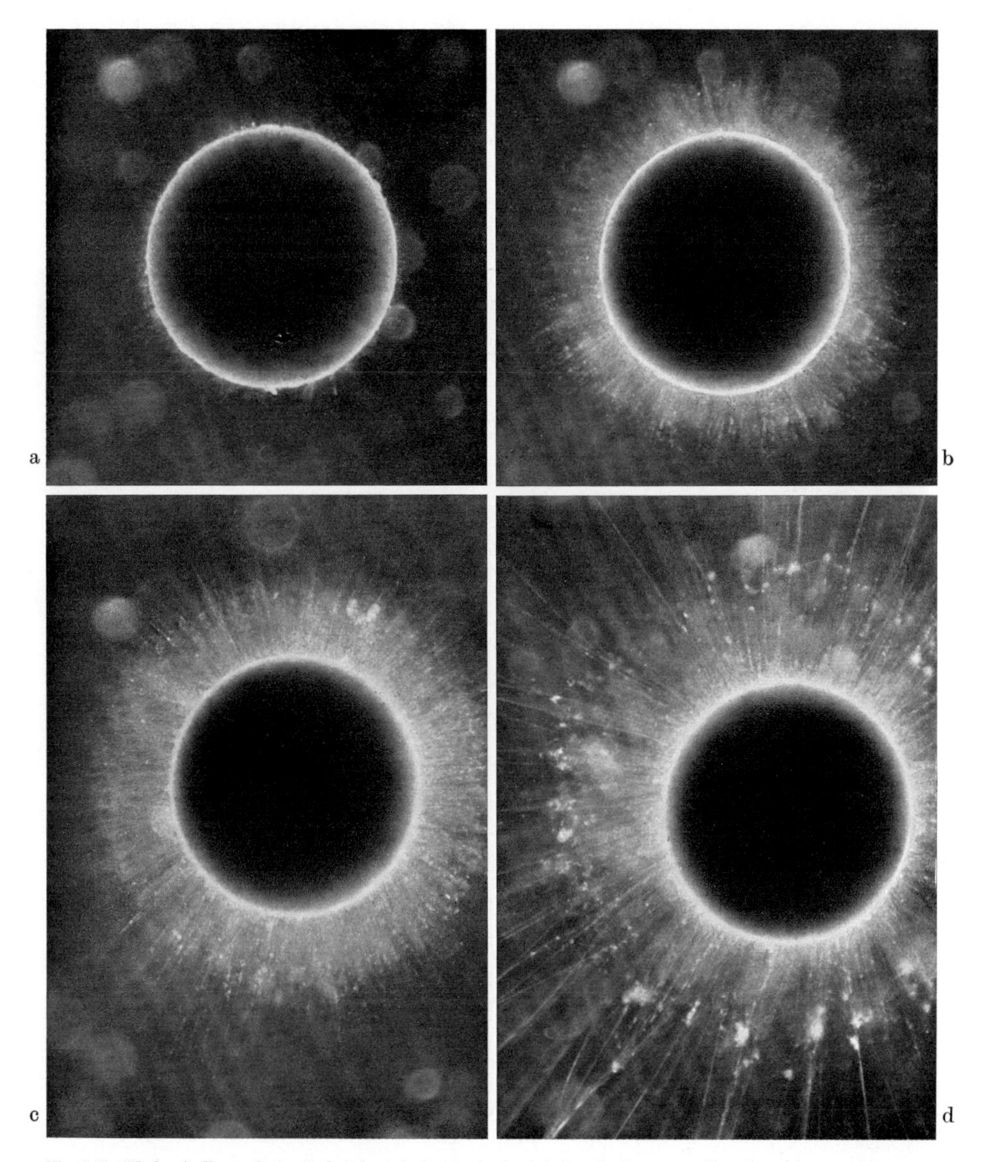

Fig. 246. *Thalassicolla nucleata*. Isolated central capsule (moist chamber) regenerating the extracapsulary cytoplasm. × 60. From the film C829 (GRELL)

to regrow and birefringent axonemes can easily be identified. Evidently, the microtubules are not only related to the maintenance of the axopodia but also to their growth [*1426*, *1725* to *1729*].

Few investigations of the pseudopodia of *Radiolaria* have as yet been carried out. In some cases it could, however, be shown with certainty that they contain bundles of regularly arranged microtubules [*189*, *200*, *203*, *206*, *779*]. They may thus also be termed axopodia. In Radiolaria, too, the rapidity of pseudopod formation is surprising. If the central capsule of *Thalassicolla nucleata*, an especially large species without skeleton, is transferred into a moist chamber, pseudopodia emerge through fine pores shortly thereafter (Fig. 246). Later on, the vacuolated layer of the "extracapsularium" is also regenerated (Fig. 376).

Colonial Radiolaria (Fig. 377) have many central capsules in one common gelatinous casing. While the colony extends filiform pseudopodia for food capture towards the outside, the central capsules are connected within the gelatinous mass by a cytoplasmic network which serves to distribute the food (Fig. 247).

Most *Foraminifera* are benthonic and form branched, rootlike pseudopodia connected by anastomoses. Pseudopodia of this kind are called *rhizopodia* or *reticulopodia*. Sessile species and others of restricted motility cover their surroundings with a dense network in order to *capture food organisms* (Fig. 248). If destroyed, the net can be repaired rapidly. Cut-off pieces can unite with the net again if they are not too far away.

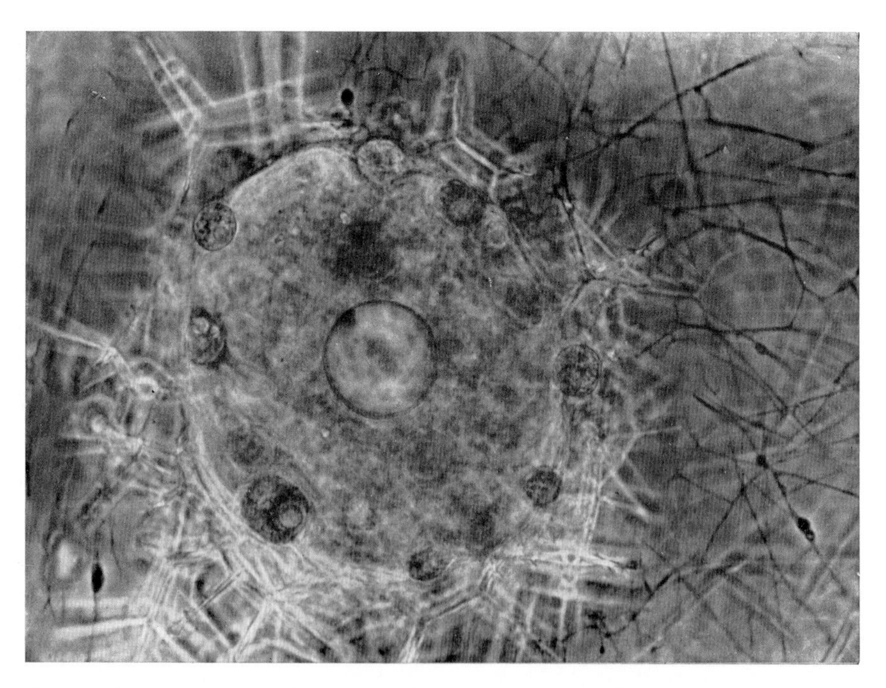

Fig. 247. *Sphaerozoum punctatum*. Single individual of a colony. In the middle of the central capsule is a large oil sphere, outside and adjacent to it are zooxanthellae and skeletal needles. At the right, rhizopods are recognizable, connecting the individuals with each other. Photographed in life (phase contrast)

Fig. 248. *Allogromia laticollaris.* Several individuals on a cover glass in a culture dish with their rhizopodia extended. Photographed in life

Especially striking is the lively cytoplasmic streaming in rhizopodia. In Foraminifera with transparent shells this is seen to be connected with the streaming in the cells interior. Granules (mitochondria, protein granules) are transported in both directions even in very delicate protoplasmic strands. It is, however, by no means true that all granules migrate to the tip of each protoplasmic strand and then back again [*849*]. Usually they only oscillate back and forth over a more or less extended stretch. At the most delicate sections of the rhizopodia, only extension and retraction can sometimes be observed. The impression of countercurrents might thus be due to a parallel displacement of functionally independent threadlike units. Electron micrographs showed that rhizopodia are complexes of fine cytoplasmic threads [*1817*]. The unidirectional transport of food particles can most easily be understood on the basis of a retraction of the cytoplasmic threads to which the particles became attached. Meanwhile, others extend, so that, on the whole, the rhizopodia change very little. On the other hand, there can be no doubt that the cytoplasmic threads can also expand and retract jointly. If a foraminiferan is detached mechanically from its substrate, the rhizopodia are immediately retracted.

Rhizopodia are not only active in food capture but also enable the cell to carry out *locomotion.* They extend into a definite direction and then drag the cell behind them. Fig. 249 shows that a "guide rhizopodium" may first be extended while the other rhizopodia follow like the tail of a comet. Even pelagic species, whose rhizopodia normally serve only in floating and food capture, can creep about if they come into contact with a substrate (Fig. 250).

In polythalamic species the rhizopodia also seem to take part in the *formation of the shell.* While it was formerly thought that the new chamber of a foraminiferan

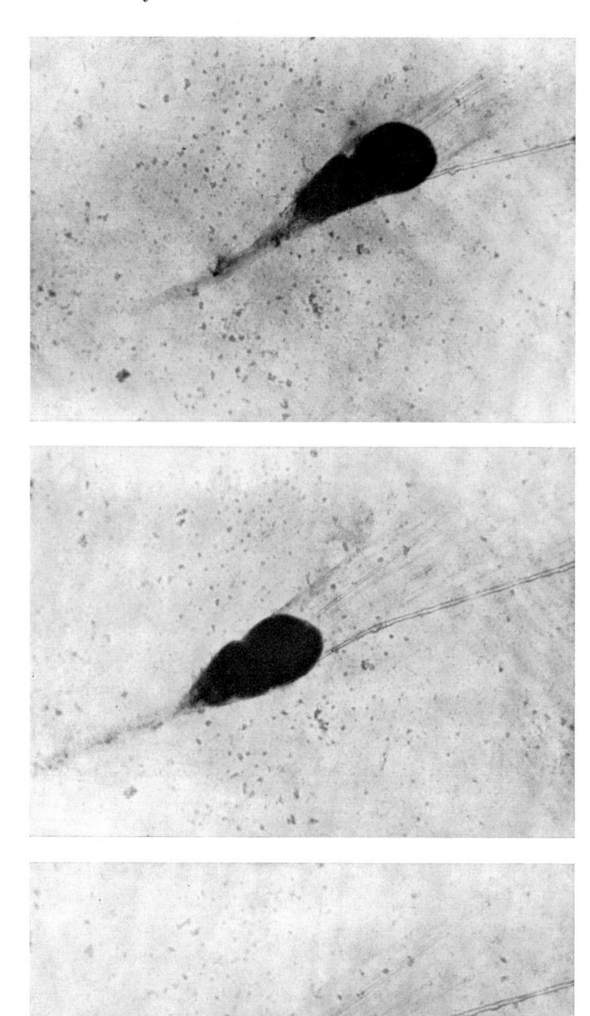

Fig. 249. *Iridia sp.* Locomotion of a newly hatched agamete (compare to Fig. 358)

is formed around a cytoplasmic droplet which protrudes from the shell, it is now clear from live observations that a new chamber does not contain protoplasm at first but is gradually filled with it. It seems that the rhizopodia prepare the new chamber (Fig. 251). In species which are attached to a substrate, the rhizopodia arrange themselves first in a fan-shape (a). Then they retract somewhat, leaving the accumulated particles of detritus behind (b) and secrete an organic membrane

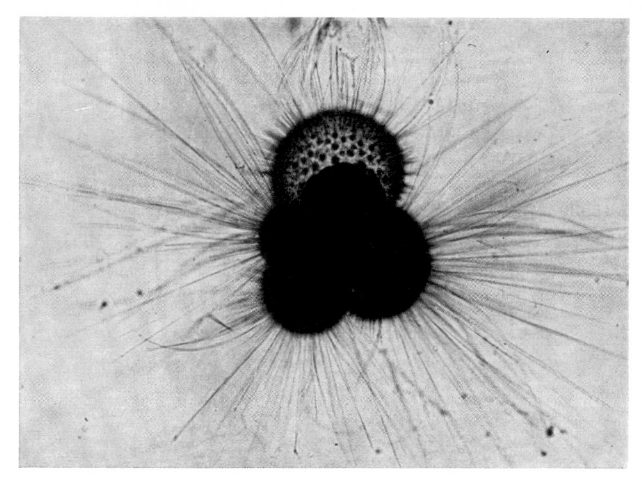

Fig. 250. *Globigerina bulloides*. Photographed in life. × 160. From the film C836 (GRELL)

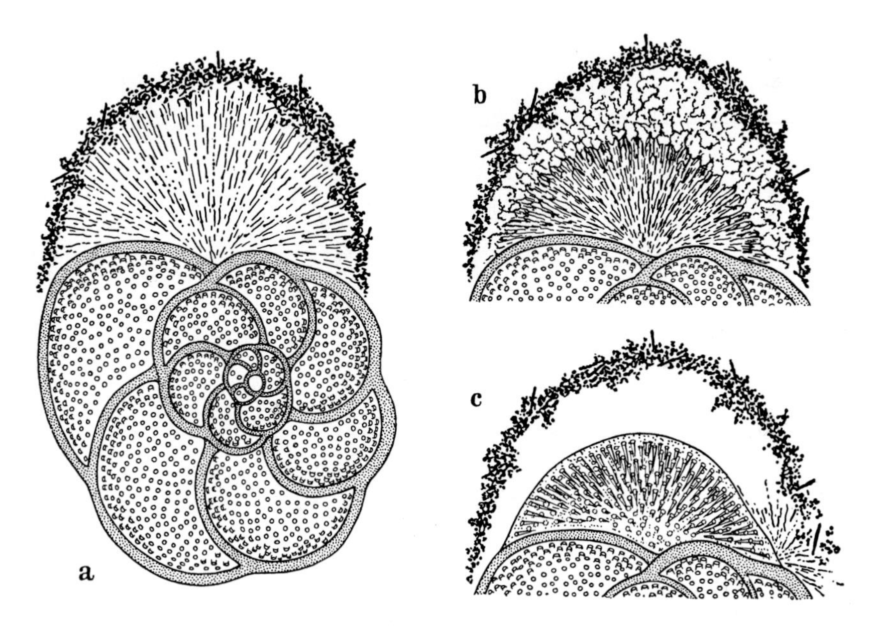

Fig. 251. *Discorbis bertheloti*. Formation of a new chamber. a Fan-shaped arrangement of the rhizopodia. b Retraction of the rhizopodia. c Secretion of a new chamber wall perforated by pores. After observations in life. × 55. After LE CALVEZ, 1938 [*1003*]

which is reinforced by incrustation or addition of calcium carbonate to become the wall of a new chamber (c).

In the monothalamous foraminifera *Iridia diaphana* and *Shepheardella taeniformis* bundles of microtubules have been found, parallel to the long axis of the rhizopodia but occupying limited areas in cross sections. They may be responsible for the rigidity or for the transportation of particles along the cytoplasmic strands [*744, 1104, 1113*]. With suitable techniques they will probably be found in other species too.

A peculiar mode of locomotion is found in the *Labyrinthulales*, a group of unicellulars whose taxonomic relationships are still controversial [*1316, 1513, 1674*]. Their cells are spindle-shaped and form extensive aggregates after exhausting the food supply. If such an aggregate is transferred to a dish with food organisms, the cells form fine hyaline extensions which are first mobile and resemble pseudopodia (Fig. 252a). However, they soon grow to long and branching thread-like paths which seem to be completely stiff. The paths of different cells are capable of fusion, leading finally to an extensive network. The cells which multiply by binary fission perform a gliding movement back and forth along these pathways (b). The mechanism of this gliding is still unclarified.

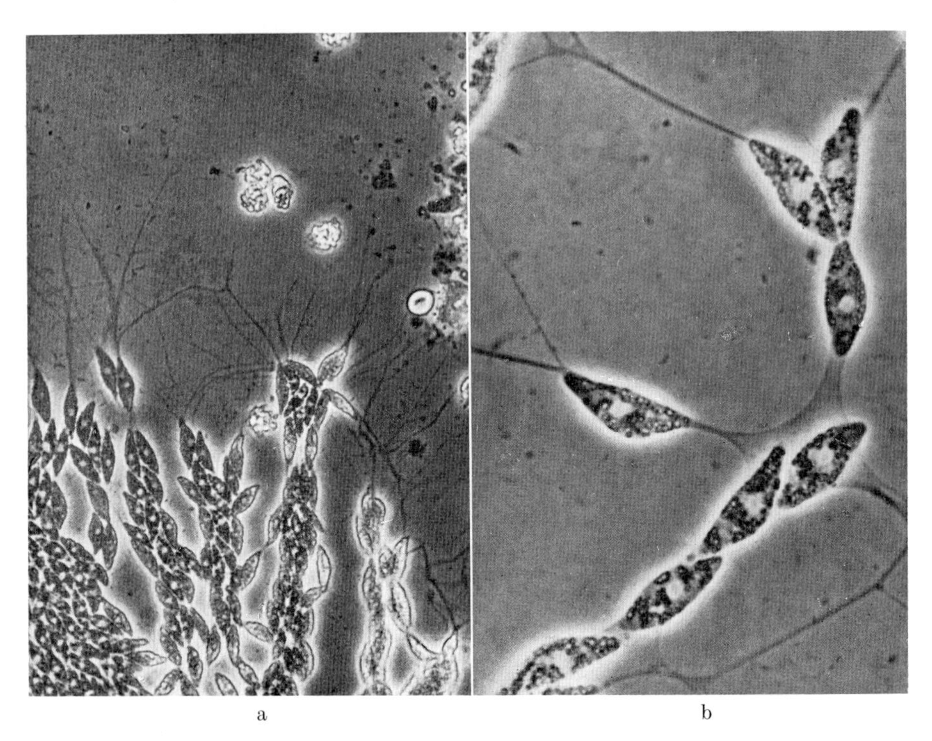

Fig. 252. *Labyrinthula coenocystis*. a The spindle-shaped cells of an aggregate form extensions. × 520. b The spindle cells glide within the thread-like pathways. × 1170. From the films C942 and E1172 (Grell)

Many investigations have been carried out in order to gain insight into the *physiology of pseudopodial movement* [*9–11, 14, 17, 372, 604, 638, 842, 844, 847, 960, 1241, 1817*]. Although the results will be valuable elements of a general theory, they are as yet insufficient to put forth such a theory.

Most investigations are restricted to one form of pseudopodial movement, i.e. locomotion of fresh-water amoebae.

The theory that pseudopod formation is due to a local decrease of *surface tension* rests upon too simplified physicochemical concepts and is no longer discussed seriously. Whenever a differentiation of ecto- and endoplasm exists as in the tubular pseudopodia of *Amoeba proteus* (Fig. 243a), one is naturally tempted to look upon the *ecto-endoplasmic process* as basic for pseudopodial movement. However, as the pseudopodia of many other amoebae have no such differentiation, it can only be a phenomenon of secondary importance not causally related to pseudopodial movement as such.

It has frequently been assumed that pseudopodial movement would be connected with a constant renewal of the cell envelope. Detailed investigations have, however, shown that no such connection exists. Although the cell surface is continuously renewed (p. 40), this renewal is much slower than expected if it were correlated with locomotion [*638*].

Recent hypotheses are all based upon a mechanism involving *contractility*. Disagreement prevails regarding the seat of the contractile forces. According to the so-called "pressure-flow hypothesis" the endoplasm is propelled during pseudopod formation by the contractility of other cell regions; streaming as such would then be a passive process. In normal locomotion the amoeba would contract at the posterior end and the forward flow of the endoplasm under pressure would lead to pseudopod formation. Simple observation shows, however, that a pseudopod can also be retracted, e.g. if it has been extended upwards and found no contact with the substrate. The endoplasm will then flow in the opposite direction (see film C942). According to the "pressure-flow hypothesis", streaming could only be caused by suction due to the expansion of other cell regions or by contraction at the tip of the pseudopodium.

Recent experiments [*16a*] indicate, however, that artificial changes of the internal hydrostatic pressure do not influence the extension or retraction of a pseudopodium. In other words: the cytoplasmic streaming is an active process not dependent on a pressure gradient as the "pressure-flow hypothesis" assumes. The possibility exists that the contractile forces are not localized, i.e. the amoeba can contract wherever it is "necessary".

Even though we know little concerning the details of these processes, it seems to be certain that the cellular envelope does not actively take part in the contractions and expansions of the cytoplasm but that it is folded and unfolded passively [*638*].

The assumption that motility depends in the last analysis on contractility is supported by the isolation of thin (50—70 Å) and thick (140—160 Å) filaments from cytoplasm. Thin filaments bind the muscle protein heavy meromyosin (HMM) to form polarized "arrowheads" typical for the actin-HMM-complexes in muscle [*1321, 1322*]. Hence, there is good evidence that the filaments consist of "contractile proteins" whose mode of action might be similar to that of the "sliding filament" system in muscle fibers.

In ultrathin sections of amoebae separate bundles or aggregates of filaments could be demonstrated, predominantly near the cell surface [*1320, 1817, 1818*]. They have been found also in "glycerinated", i.e. glycerin-extracted cells of *Amoeba proteus* which still show a definite though relatively small contraction after addition of ATP [*775, 1504, 1598*].

It is of course difficult to prove that the electron microscopically visible filaments correspond to those obtained by isolation. If pseudopodia of *Difflugia* get attached to the substratum (Fig. 243h), and are observed through a polarizing microscope, a positively birefringent fibrillar array can be seen to develop rapidly at the attachment point and to extend back to the cell body within the shell. The birefringent fibrils correspond to refractile fibrils visible with the Nomarski differential inter-

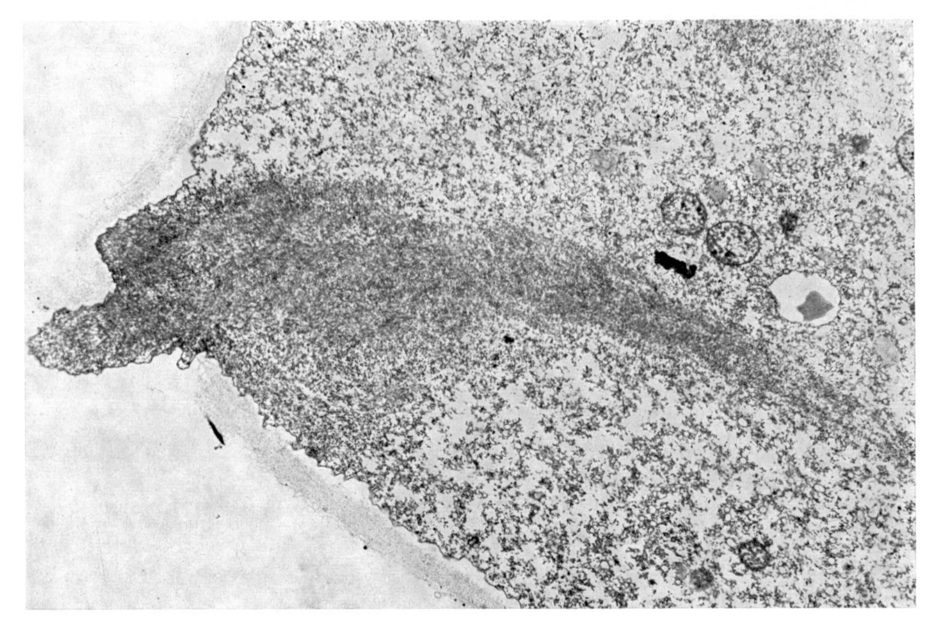

Fig. 253. *Pontifex maximus* (comp. with Fig. 346c). Part of the cell with "adhesive projection". × 10000

ference microscope and probably to bundles of intimately associated, aligned microfilaments to be seen on sections in the electron microscope. The fibrils are formed *before* pseudopod retraction occurs. After retraction, which leads to the result that the heavy shell is dragged forwards, they disappear [*1820*].

In *Pontifex maximus* where no differentiation in endo- and ectoplasm exists attachment is brought about by the already mentioned "adhesive projections". An electron microscopic picture (Fig. 253) shows a dense material of fibrillar nature directed into the cell's interior, just before retraction occurs.

2. Flagella and Cilia

Flagella and cilia are permanently formed, thread-like extensions which carry out irregular or periodic movements. This creates a water current which can move the cell about or carry food particles to it.

Electron microscopic studies revealed the same basic plan of fine structure in the flagella and cilia of all eukaryotes (Protozoa, Metazoa). This is an indication not only of phylogenetic relationship but also of a similar mode of action.

The essential constituent of every flagellum or cilium (Fig. 254a) is a *bundle of fibrils*. This includes a *basal body* or *kinetosome* (B) within the cell and a *shaft*,

▶

Fig. 254. Diagram showing the ultrastructure of flagella or cilia. a Longitudinal section and three corresponding transverse sections. b A basal body inducing spindle fibers (microtubuli) within a nucleus. c Diagram illustrating the assignment of numbers of the 23 rows of globular protein subunits of the peripheral subfibrils. *B* Basal body. *PF* Peripheral fibrils (*A B*). *CF* Central fibrils. *Ax* Axosome. *S* Septum. *Sf* Spindle fiber (microtubule). *Chr* Chromosome. Combined after several authors

Fig. 255. *Pseudotrichonympha.* Cross section through flagellar shafts (left) and basal bodies (right). The flagella originate in grooves of the pellicle, which are recognizable in the left half of the picture. × 76000. After GIBBONS and GRIMSTONE, 1960 [577]

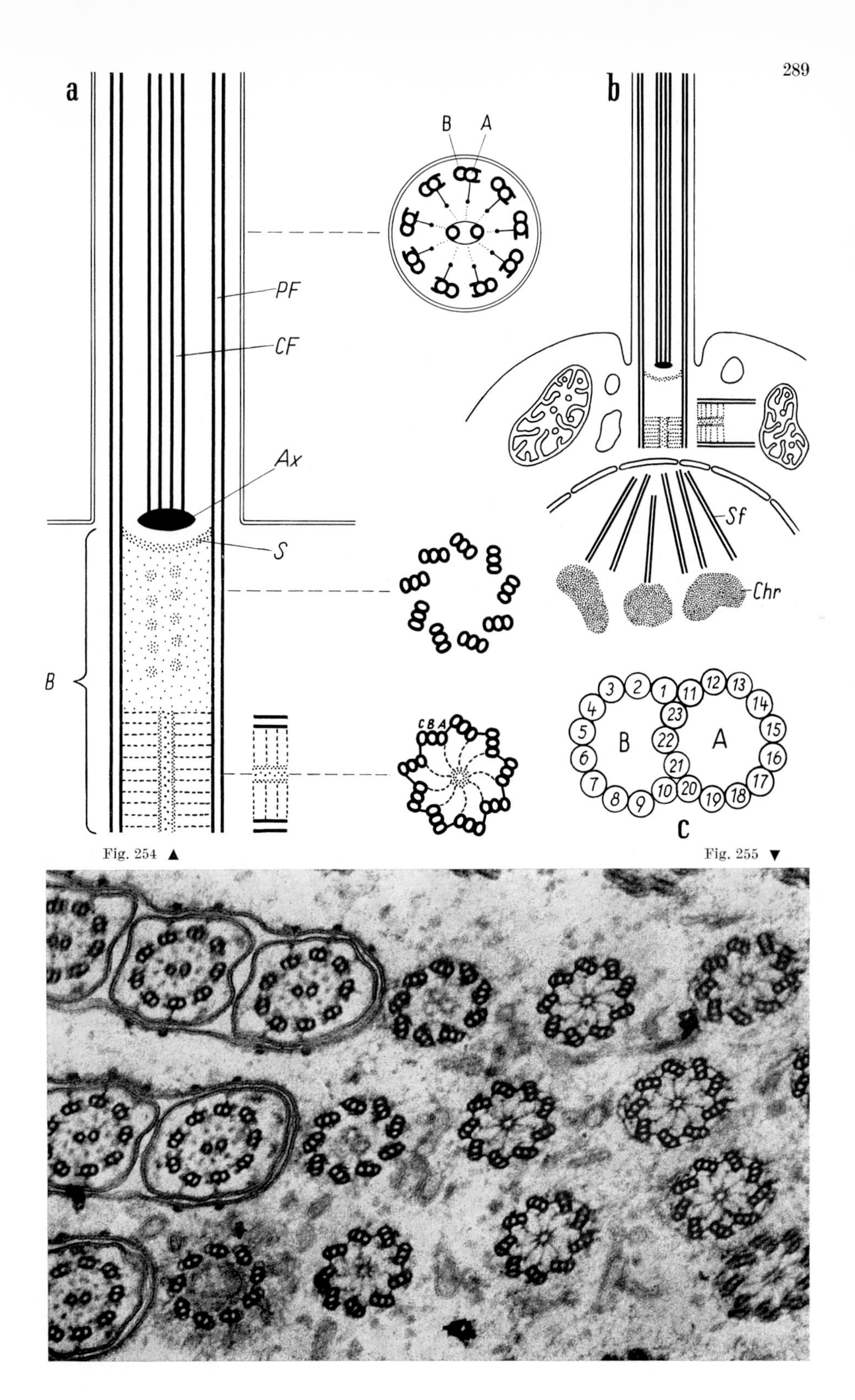

Fig. 254 ▲

Fig. 255 ▼

which extends from the cell. The latter is surrounded by a *membrane*, which is a continuation of the cell envelope. The fibrils are embedded in a *ground substance* and consist of *microtubules*. Their number and arrangement is surprisingly constant.

In all well-analyzed cases, the basal body and shaft are seen to differ in their cross-sectional views (Fig. 255).

The region of the *basal body* contains only peripheral fibrils arranged in nine groups around the center. Each group consists of three adjacent subfibrils, which share the walls where they are connected. These "triplets" are not arranged tangentially, but are set at an angle. The interior of the fibrillar cylinder is occupied by a variety of structures. Near the proximal end, a structure resembling the spokes of a wheel is frequently recognizable *("cartwheel-structure"):* each triplet is linked by a slightly bent "spoke" to a tubular axis. This is clearly of lower contrast than a subfibril and certainly not a microtubule. The triplets may also be connected to each other.

Within the *shaft*, the fibrils are arranged in the so-called *9 + 2-pattern*. The nine *peripheral fibrils* (PF) consist of two subfibrils only which represent a continuation of the two inner subfibrils of a triplet group (AB). Although these "doublets" are not set at such a conspicuous angle as the triplets, the fibrillar bundle as a whole also has a marked *asymmetry* as seen in cross sections of the shaft. Two *central fibrils* (CF), which do not touch, are in the interior of the shaft. Their dimensions are identical with those of the subfibrils. They are often surrounded by a common sheath.

The peripheral subfibrils are different from other microtubules in sharing a common wall. While single microtubules consist of about 13 rows of globular protein subunits (p. 29), doublets are made up of 23 rows (c). The protein is called *tubulin* and has an amino acid composition similar to that of the muscle protein actin.

In *Chlamydomonas*, two proteins called tubulin 1 and 2 have been identified in both the A and B tubules of the peripheral doublets. There is evidence that the rows of subunits ("protofilaments") consist either of tubulin 1 or tubulin 2 and are arranged in a regular manner according to their protein composition [*1813*].

From subfibril A two types of projections arise. The side *arms* occur in pairs and consist of a protein called *dynein* which — like the muscle protein myosin — has ATPase activity [*575, 576*]. The radial *spokes* which are attached at right angles to subfibril A extend into the lumen of the shaft in the direction of the central pair. Negative staining preparations reveal that the spokes also occur in pairs but along the length of the shaft. They terminate with a hammerhead-like attachment [*815*].

Longitudinal sections show that the interior of the basal body is frequently separated by a fine *septum* (S) from the shaft and that the central fibrils of the latter frequently end at a special structure, the so-called *axosome* (Ax). Further details of ultrastructure will not be discussed here.

The *formation* of a flagellum or a cilium is still largely unclarified. However, there is no doubt that it always originates from a basal body. This is heteropolar also in a morphogenetic sense and grows only at its distal end to form a shaft.

When *Naegleria gruberi* (Fig. 238) is exposed to conditions which induce transformation into the free-swimming stage, the first basal bodies appear about 55 minutes

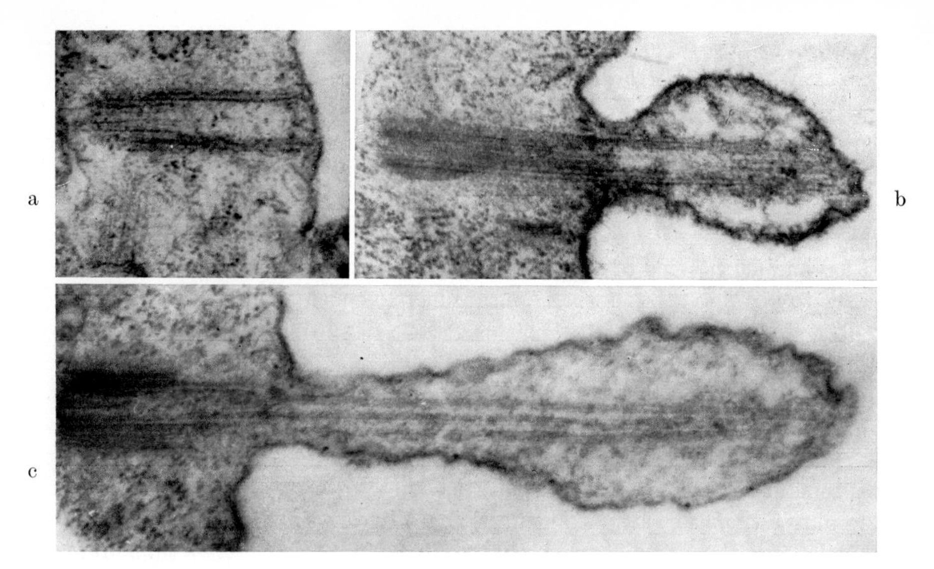

Fig. 256. *Naegleria gruberi.* Longitudinal sections through flagella during growth stages. × 36500. After DINGLE and FULTON, 1966 [*407*]

later. From the beginning they are oriented at right angles to the cell envelope (Fig. 256a). A few minutes later they sprout fibrillar shafts which "draw" the cell envelope along (b, c). The growth rate of all fibrils of a bundle is about the same (0.5 μm/min) [*407*]. If the cells are exposed to sublethal temperature shocks (about 38° C) during amoeba-to-flagellate transformation they develop about twice as many flagella as controls. Reverting to amoebae they lose both normal and supernumerary flagella [*406*].

Basal bodies may also occur without a shaft. In many cases a special physiological situation is apparently required before they give up their "waiting positions".

An electron microscopic study of the hypermastigid flagellates *Deltotrichonympha* and *Koruga* from the primitive Australian termite, *Mastotermes darwiniensis,* has shown that the cells contain more than a half million free basal bodies in the anterior cytoplasm. There are longer basal bodies scattered singly throughout the cytoplasm and shorter ones, which are arranged end-to-end in polarized chains of varying lengths. The number of free basal bodies is much greater than that of active flagella. The significance of this remarkable accumulation of free basal bodies is not known as yet [*1698*].

If cilia or flagella are detached by thermal or chemical treatment, they can be regenerated. For instance, the mitotic inhibitor chloralhydrate induces deciliation in *Paramecium* [*983*]. Electron microscopic studies show that the agent causes a breakdown of the ciliary shaft at the junction with the basal body. The latter retains its septum and its axosome. After removal of the agent the basal body regenerates the shaft in a short time [*912*].

There can thus be no doubt that basal bodies, at least from the morphogenetic point of view, possess a certain autonomy. This statement cannot be applied, however, with respect to their multiplication. As the case of *Naegleria* shows (p. 269) basal bodies — and the same seems to be true for the structurally identical centrioles —

may develop "de novo" in the ground cytoplasm or from precursors not yet traceable with the electron microscope.

On the other hand, in the majority of cases, a new basal body develops in the vicinity of an old basal body. This has been seen in an overwhelming number of flagellates and ciliates. Careful investigations [*409*] show that the new basal body develops close to the old one but perpendicular to it (Fig. 254a). Its synthesis consists of a precise succession of stages not to be treated in detail. The exact position and orientation of the new basal body in relation to the old one makes it difficult to imagine that this relationship has no substantial basis.

In this connection it has to be mentioned that the problem whether basal bodies contain DNA cannot yet be regarded as solved [*521, 771, 824, 1372, 1570, 1617, 1618, 1833*].

As exemplified above, basal bodies are often associated with various other structures (basal fibrils, parabasal bodies, costae, axostyles, etc.). Usually these structures remain connected with their basal bodies; in certain cases, however, they detach after their formation and participate in the construction of specific cell organelles, e.g. the cytopharyngeal basket of cyrtophorine ciliates (p. 335) [*1746*].

In some cases basal bodies serve as centrioles. Even if their position is extranuclear, spindle fibers (microtubules) radiate from them within the nucleus (Fig. 254b).

Hence, it is certainly a reasonable assumption that basal bodies are "nucleating sites" or "foci" for the assembly of components to be used for the formation of other basal bodies or of different – mostly microtubular – structures. The formation of these structures by artificially dislodged basal bodies would constitute an experimental evidence for this assumption.

An explanation for the asymmetry in fine structure and the constant number and arrangement of the fibrils in terms of their function cannot be given as yet.

Probably, the central fibrils are not directly involved in the mechanism of motility. The fact that flagella without the central fibrils are incapable of movement [*1407, 1784*], does not contradict this opinion. Even if they play a role as static or elastic components, they might well be essential for motility. Studies on cilia from *Opalina* have demonstrated that the plane of the central pair is always perpendicular to the direction of bending at any place of the cilium, though the form of beat is three-dimensional and the direction of the stroke is variable (p. 300). This supports the view that the orientation of the central pair is related to the direction of bending [*1700*].

Very likely, the mechanism underlying movement is differential activity of the peripheral fibrils. There is some evidence that the fibrils do not "contract", but slide past one another. This has led to the assumption that bending is based on a sliding-filament mechanism comparable to that of muscle contraction [*772, 773, 1502, 1503, 1611, 1686*]. The interaction between adjacent doublets seems the most likely since ATPase activity is associated with the arms which project from one doublet to the next (Fig. 254).

That motility of flagella depends on ATP is shown by glycerin-extracted or Triton X-treated "models", which perform rhythmic motion after addition of ATP [*172, 173*]. Even isolated flagella of *Polytoma uvella* can still move though with lower speed (10 µm/sec) and less amplitude than in the living cell (40–50 µm/sec).

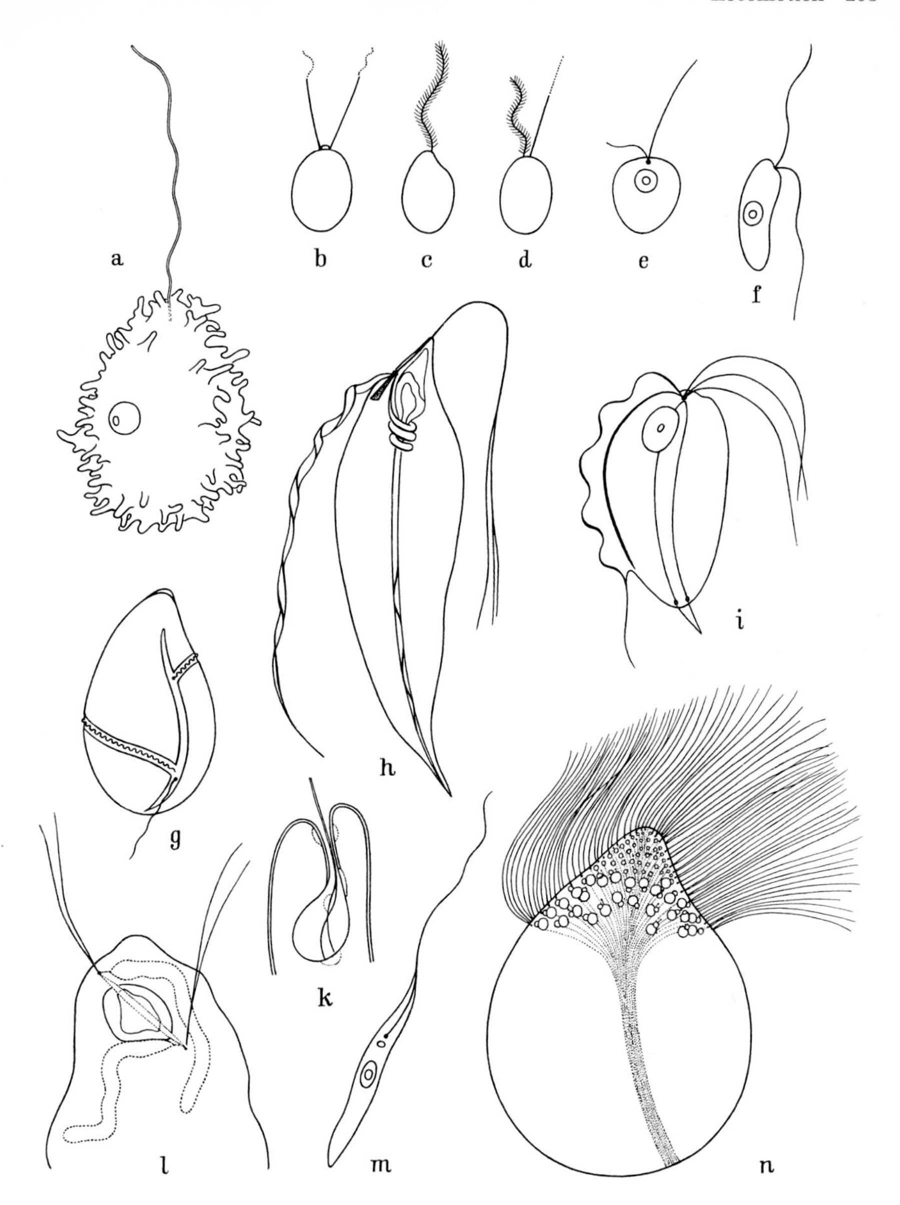

Fig. 257. Different types of flagella in flagellates. a *Mastigella vitrea*. b *Polytoma*. c *Mallomonas*. d *Synura*. e *Monas vulgaris*. f *Bodo ovatus*. g *Gyrodinium pepo*. h *Devescovina lepida*. i *Trichomonas*. k *Euglena ehrenbergi*, anterior end with flagellar sac. l *Devescovina striata*, stage of division. Two flagella and one parabasal body (dotted) originate from each of the two poles of the spindle. *m Leptomonas*. *n Calonympha grassii*. Combined after various authors

Flagella

Flagella are generally as long or longer than the cell and are usually present in low numbers only. They are characteristic of flagellates but they also appear in certain developmental stages of Rhizopoda, e.g. in swarmers of Radiolaria and the gametes of many Foraminifera, and, among Sporozoa, in the microgametes of Coccidia.

The morphological diversity of flagella is considerable (Fig. 257). Usually they are long threads either of uniform thickness (a) or with a thick basal part and a thinner terminal thread of different length (whip-like flagella, b). It was shown, in the case of *Chromulina pusilla*, that only the central fibrils extend into the terminal thread [*1084*].

Dark field observations and electron microscopy have shown a fine, hair-like covering (c, d) on the surface of many flagella, sometimes arranged in one or two rows, and in others all around the flagellum. These "hairlets", called *mastigonemes*, may also have a species-specific fine structure. Sometimes they are arranged in groups. In *Ochromonas danica* (Fig. 258), the mastigonemes of the anterior flagellum form two lateral unbalanced rows, each row on opposite sides. Each mastigoneme consists of lateral filaments of two distinct sizes attached to a tubular shaft.

In several cases it could be shown that the mastigonemes are formed in cytoplasmic vesicles and then transported to the cell surface for deposition on the flagella [*1015*].

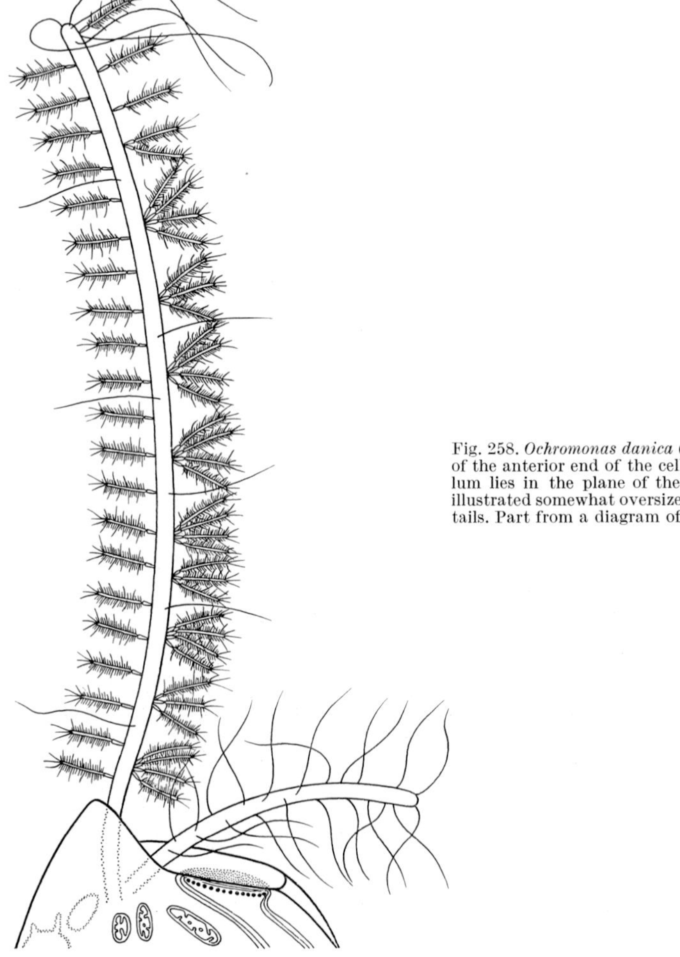

Fig. 258. *Ochromonas danica* (Chrysomonadina). Diagram of the anterior end of the cell with flagella. Short flagellum lies in the plane of the eyespot. Mastigonemes are illustrated somewhat oversized in order to include fine details. Part from a diagram of BOUCK, 1971 [*152*]

In *Ochromonas danica* the assembly of the shaft begins within the perinuclear cisterna enclosing the nucleus and the chloroplast (p. 25) while addition of the lateral filaments to the shaft and extrusion of the mastigonemes on the cell surface is mediated by the Golgi complex [*152*].

In flagellates with *several* flagella, the latter may differ with respect to position, size and form. Besides a long *major flagellum*, there may be one or more *minor flagella* (Fig. 257e). Some flagella are directed backward during movement and are therefore referred to as *recurrent flagella* (f). Dinoflagellates always have two flagella, one extending across the cell body, the other extending backward. The first one performs a helical motion within a groove which encircles the cell body like a belt, while the second flagellum undulates posteriorly (g). In many species of *Devescovina*, the posterior flagellum is more or less flattened and hence called a *ribbon flagellum* (h).

Even the so-called *undulating membranes* are basically ribbon flagella, loosely connected to the cell body with the possible exception of the tip which may be free-swinging (i). An electron micrograph (Fig. 259) shows the "undulating membrane" to be completely surrounded by a sheath of its own which corresponds to that of the shaft. Inside is the fibrillar bundle in the typical 9 + 2-pattern. Next to the fibrillar bundle is the so-called paraxial strand, an electron dense longitudinal structure of unknown significance.

A flagellum may arise either directly from the cell surface or from a special depression at the front end, the so-called *flagellar sac* (Fig. 257k). It originates directly from the centriole (l) or from a basal body. If the basal body is located deep within the cell, it is first continued as an intracellular fibrillar bundle, which then extends into the shaft of the free-swinging flagellum (m). In Polymastigina, flagella are especially numerous (n). In many cases the basal bodies are connected with special differentiations, the *flagellar bands*, which course around the cell in helices (Fig. 341).

Some flagellates, e.g. among chrysomonads *Chrysochromulina* [*1091, 1096*] and *Prymnesium* [*1085*], have, besides ordinary flagella, a so-called *haptonema*. Externally similar to a flagellum,

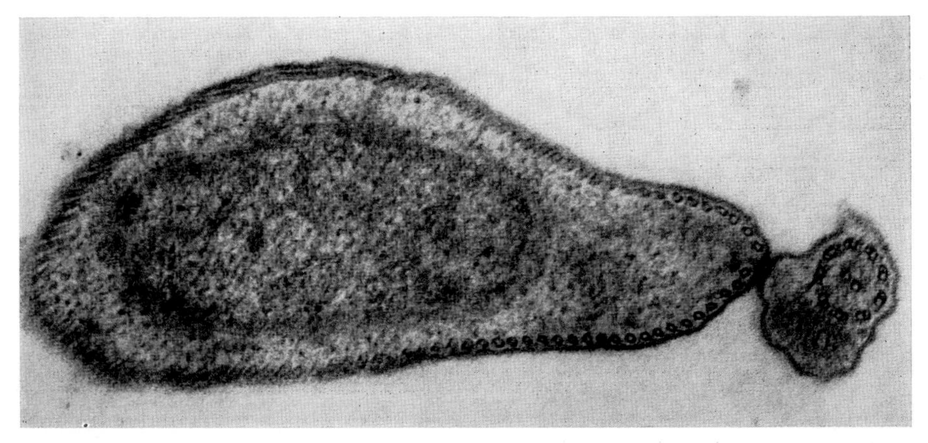

Fig. 259. *Trypanosoma lewisi*. Cross section through the anterior region of the cell (with cell nucleus). Beneath the pellicle: microtubules (some sectioned across, some obliquely). At the right the "undulating membrane", which is only loosely attached to the cell, with fibrillar bundle ((9 + 2-pattern) and "paraxial strand". × 17200. After ANDERSON and ELLIS, 1965 [*56*]

its primary function is to fasten the cell to the substrate. Although it does not carry out undulating motions, the haptonema can be wound up to a helix and be stretched out again. Electron microscopy showed a fine structure which differs from that of flagella. The interior contains single microtubules, sometimes arranged as a ring, sometimes less regularly. In *Prymnesium*, the base always seems to contain 9 microtubules. But some of these do not reach up to the tip of the haptonema, so that more distal cross sections show only 8 or 7 microtubules. The microtubular bundle is surrounded by three unit membranes, the two outer ones being due to a surrounding fold of the cell envelope. The haptonema is obviously derived from a flagellum whose structure has been simplified and changed in connection with its function as an adhesive organelle.

The *mode of flagellar motion* can be very different. Even within the same species there is no uniform pattern, and motility can be modified in many ways. Motility is frequently not even connected with locomotion.

Flagellar undulations, which lead to locomotion, i.e. to swimming in the water, are usually too rapid for visual analysis. High speed cinematography and stroboscopic methods must therefore be used. Model experiments can also substantially contribute towards an understanding of flagellar motion.

Most flagella carry out undulatory motions, either wholly within one plane *(uniplanar)* or in the form of a helix *(helicoidal)*.

The water pressure which propels the cell is created by sinusoidal or helicoidal waves (Fig. 260). They move distally if the flagellum inserts at the posterior end (a, e.g. animal sperms, *Ceratium*), and proximally in the case of an anterior flagellum (b, e.g. *Mastigamoeba*, trypanosomids). The widespread notion that the bending wave must always originate at the basal body is therefore erroneous [*842*].

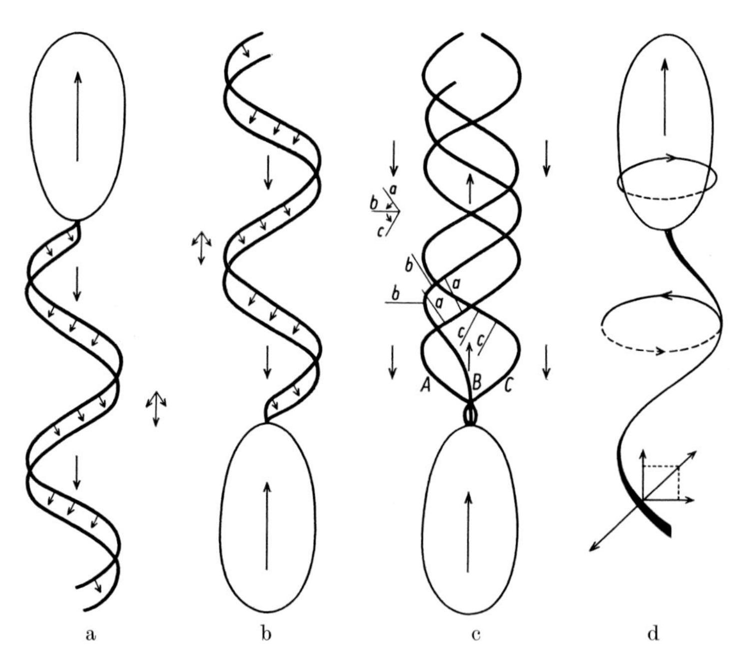

Fig. 260. Modes of flagellar motion. a, b, c planar-sinusoidal waves. a Flagellum posteriorly (sine waves: proximal → distal). b Flagellum anteriorly (sine waves: distal → proximal). c Flagellum anteriorly, with mastigonemes (only two are indicated); *A*, *B*, *C* three different phases of a sine wave progressing proximal → distal. After Jahn, Landman, and Fonseca, 1964 [*848*]. d Flagellum posteriorly, helicoidal wave. After Buder from Nultsch, 1964

High-speed motion pictures show that occasionally waves may move distally in spite of flagellar insertion at the anterior end. In the case of the chrysomonads *Ochromonas* and *Chromulina*, it could be shown that such flagella are covered by stiff, bristle-like mastigonemes (p. 294) which are moved backward in the course of wave motion like oars and thus reverse the hydrodynamic effect (c) [*848*].

In some sessile flagellates *(Actinomonas, Monas, Poteriodendron)* the waves also move distally although the water current which carries food particles is directed towards the cell. It is probable that mastigonemes are also present in this case [*1607*].

In others, e.g. in *Euglena* (Fig. 271), the waves do move towards the tip of the flagellum but the latter is directed backward during locomotion.

Frequently, no sharp distinction can be made between sinusoidal and helicoidal motion. In the latter case, a gyrational component is added to the longitudinal component. The flagellum thus progresses through the water in a screw-like motion (Fig. 260d).

It was formerly considered certain that the more or less stretched flagellum would gyrate around a cone-like space and thereby create suction. However, this mode is improbable for hydrodynamic reasons and has never been observed so far by high-speed cinematography.

In *dinoflagellates*, locomotion is primarily due to the longitudinal flagellum which is usually directed backwards. Stroboscopic studies of species of *Ceratium* showed uniplanar sinusoidal waves moving distally [*846*]. Helicoidal movements of the flagellum can probably also be carried out. Undulation of the flagellum encircling the body leads not only to a rotation of the cell but also contributes to its locomotion. Dinoflagellates follow a helical course while swimming, their body axis being inclined towards the axis of progression [*1299*]. The direction of movement can be changed by turning the longitudinal flagellum. The direction of rotation can also be changed while swimming.

Periodic beating is also found among flagellates. The colonial *Volvocidae* are especially interesting in this respect. It has been pointed out above (p. 141) that the plane of flagellar beating is determined by the asymmetric organization of the cells. The two flagella beat in parallel planes, i.e. from the upper left to the lower right, which leads to a clockwise rotation of the whole colony as seen from the anterior pole (Fig. 128). This rigid fixation of the plane of beating is essential for coordinated locomotion of the colonies. As a consequence, isolated cells cannot carry out locomotion but merely rotate on the spot [*572*].

A great diversity of motions can be expected among *Polymastigina* because the number and arrangement of flagella is extremely variable in this group. *Pyrsonympha* has four undulating membranes which carry out asynchronous wave motions. *Trichomonads* have recurrent flagella or undulating membranes directed backwards. Wave motions of the axostyle or the costa may also aid in locomotion (p. 32).

The trichomonad *Mixotricha paradoxa*, an inhabitant of the primitive termite, *Mastotermes darwiniensis*, exhibits a most unusual mode of locomotion. The flagella serve only in steering in this case. The whole cell surface has rows of small, console-like protuberances. A bacterium is fastened in front of each protuberance and a spirochaete behind it. While the bacterium has evidently no "function", that of the spirochaete is undoubtedly the propulsion of the cell. This is evident since all spirochaetes are directed backwards and carry out undulating motions, which are clearly coordinated. Because of this *Mixotricha* resembles a hypermastigid and was indeed for a time considered to be one. However, the presence of spirochaetes instead of flagella has been confirmed by electron microscopy. This "motility symbiosis" is one of the most curious examples of functional complementation of two entirely different organisms [*311, 322*].

In *Hypermastigida* the flagella may be restricted to bundles fastened at the front end of the cell. They may originate from helical bands or from longitudinal furrows (Fig. 255). For the most part, their motions seem to be undulating. Flagella inserted at the same level undulate synchronously, while those of a given longitudinal groove undulate in metachrony. Metachronal waves thus move continually from the front end backwards, their frequency and amplitude indicating the speed of propulsion. If the front end of the cell body is set apart as a rostrum, it enables the organism to change direction while moving and is frequently provided with longer flagella (Fig. 342).

Cilia

A sharp distinction between flagella and cilia can nowadays no longer be made. For the sake of convention, it would of course be expedient if flagellates could be said to have flagella and ciliates, cilia. But even this principle cannot be unequivocally defended since the opalinids are now usually classified as flagellates although their type of "ciliation" is not unlike that of ciliates.

From a phylogenetic point of view, cilia can be considered specialized flagella, which, in connection with their higher number and higher degree of functional differentiation, became shorter than the flagella of flagellates.

A dense ciliation which covers the whole cell body evenly is found only in the *holotrichous ciliates*. The Holotricha, usually considered the most primitive group, also have the largest number of cilia. In one especially large species, *Prorodon teres*, their number is estimated at 12000.

The manifold modifications of the ciliature in the other orders of ciliates, as well as the integration of fields of cilia to complex organelles of locomotion, will be discussed below (p. 304). Like flagella, cilia can also be used in various ways to capture food (p. 337).

The cilia of opalinids are set in rows between the anterio-posterior longitudinal folds of the pellicle (Fig. 32). The ciliary rows of holotrichs are not separated by folds; but their "meridional" course can be more or less "disturbed" by the form of the cell body, especially the position of the peristome, and the differential growth of the pellicle (p. 134).

The basal bodies of cilia, like those of flagella, also serve as points of origin of *fibrils*, whose course and relative position is subject to variation.

Those originating laterally at the basal bodies are called *kinetodesmal fibrils*. They are cross-striated in electron micrographs and always seem to originate at the right side of the basal body when the cell is observed from the front end (rule of desmodexy, [*233*]). In *Paramecium* the kinetodesmal fibrils of the individual basal bodies are all directed anteriorly (Fig. 34). They taper to a point and end somewhere below the fourth or fifth ciliary field of the same row. In this arrangement, they cover each other like shingles, thus creating the impression of a continuous "longitudinal fibril" (kinetodesma). Similar arrangements have been found in other holotrichs [*1310—1315*] and in the body ciliation of heterotrichs [*507, 519, 911, 1401, 1406, 1829*].

The ends of basal bodies may also serve as points of attachment for fibrils. They are called *nemadesmal fibrils* and differ from kinetodesmal fibrils in that cross striations are absent. Like the axostyles of Polymastigina, they consist of rows of microtubules (p. 31). Sometimes the fibrils arise from special plates which are attached to the ends of the basal bodies. In some of the ciliates of ruminants the fibrils are directed towards the wall of the cytopharynx or towards the so-called "concrement vacuole", thought to represent a mechanoreceptor [*54, 620*].

Bundles of microtubules are generally seen in the vicinity of basal bodies. Sometimes they connect the latter with the surrounding pellicle.

Since there is such variability both with respect to the occurrence and the course of these fibrillar systems, they shall not be discussed further here. Besides, only assumptions can be made about their *function*. They might serve in "anchoring" the cilia and in maintaining a definite cell shape. In some cases they might represent contractile elements which enable the cell to change its shape. As yet, there is no definite proof that any of the fibrillar systems described thus far serves in the

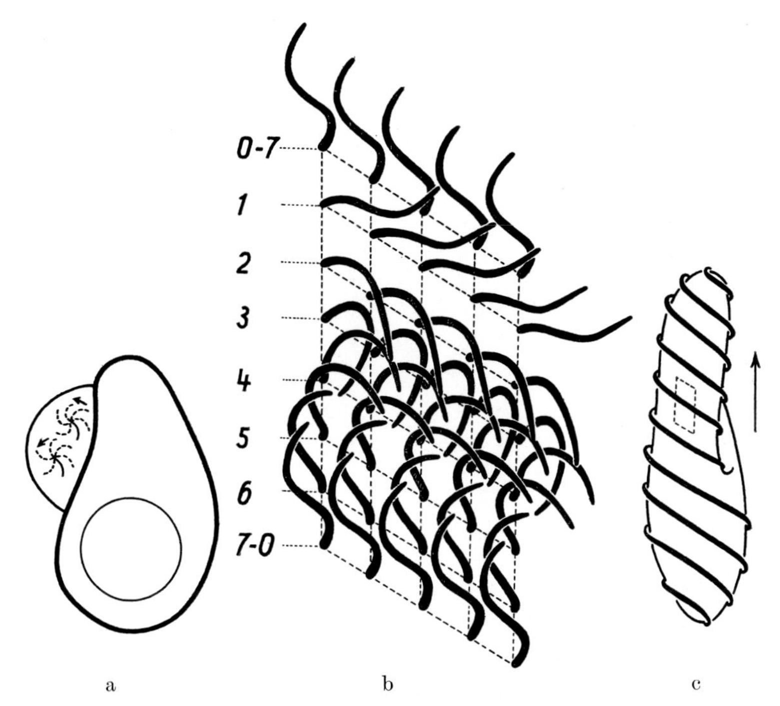

Fig. 261. Coordination of ciliary beating. a *Colpidium*. Hyaline bleb due to treatment with ammonia. On its surface two cilia, which are torn out of their point of insertion, carry out uncoordinated circular movements. b, c *Paramecium*. Form of the ciliary stroke and metachrony. b Diagram of an instantaneously fixed surface area with five ciliary rows. *0—2* effective, *2—7* recovery stroke. c Diagram of the course of the metachronal waves in forward movement (the rhombic section corresponds to the area shown in b). After PARDUCZ, 1954 [*1273*]

intracellular conduction of impulses, as frequently assumed in the older literature. The existence of such conductile fibrillar systems is also improbable on general physiological grounds [*179*].

The most impressive phenomenon exhibited by the cilia of opalinids and ciliates is the *coordination of their beating*. This is of greater plasticity than that of the flagella of a hypermastigid or a phytomonad colony and thus permits a much larger spectrum of reactions [*40, 1271–1287*].

In order to understand coordination, it must be recognized that even an isolated cilium which lacks connections to its neighbors can carry out motions. Treatment with ammonia vapors leads in some ciliates to the elevation of hyaline blebs. Cilia at the surface of such blebs, which are torn out of the pellicle, always carry out a continuous circular movement in a counterclockwise direction. The whole cilium thus outlines by its movement the surface of a cone (Fig. 261a). The same circular motion is observed when ciliates exposed to vapors of chloroform or to magnesium ions are close to death. The whole body ciliation participates in what is obviously an "autonomic" form of motion, independent of those influences which enforce coordination.

In contrast to this, the beat of a cilium in the course of locomotion is clearly *polarized*. It has a definite direction and includes the rhythmic repetition of a fast *effective stroke* and a slow *recovery stroke* which brings the cilium again into position for the next effective stroke. The time ratio of effective and recovery strokes varies from 1:6–1:2.5 in *Paramecium*. Because of the more straight form and higher angular velocity of the effective stroke, the cilium acts as an oar. This action is amplified by the large number of simultaneously beating cilia which propel the cell in the opposite direction.

Both the effective and the recovery stroke were formerly thought to take place in in the same plane as is the case in some ciliated epithelia of Metazoa. It has, however, been shown that this assumption is neither correct for the opalinids, which are usually cited as an example, nor for the ciliates. In *Paramecium*, where successive phases of movement could be analyzed with the aid of an instantaneous fixation method [*1271–1273*], the cilium carries out a gyrating motion in the recovery phase. It is held close to the cell surface and moves counterclockwise (Fig. 261b) as in the "autonomic" motion discussed above.

Coordination also expresses itself in the fact that the strokes of different cilia do not take place independently but in time sequence along definite directions. The rhythmic succession of strokes is thus connected with a phase difference between adjacent cilia, a phenomenon known as *metachrony*. After rapid fixation of a ciliate during locomotion, the cilia are seen arrested in various phases of movement which are passed by a single cilium in the course of its effective and recovery strokes (Fig. 261b).

The direction of the maximal phase shift can be recognized in the form of the so-called *metachronal waves*, which pass as zones of higher and lesser transient ciliary density over the cell. The increased density is due to the fact that the cilia converge during certain periods of their individual cycles.

▶

Fig. 262. Patterns of metachronal waves. a *Opalina ranarum*, ventral surface. b Enlargement of region indicated in a. c *Paramecium multimicronucleatum*, fixed during forward swimming (comp. Fig. 261c). Scanning electron micrographs. a and b after TAMM and HORRIDGE, 1970 [*1700*]; c after TAMM, 1972 [*1699*]

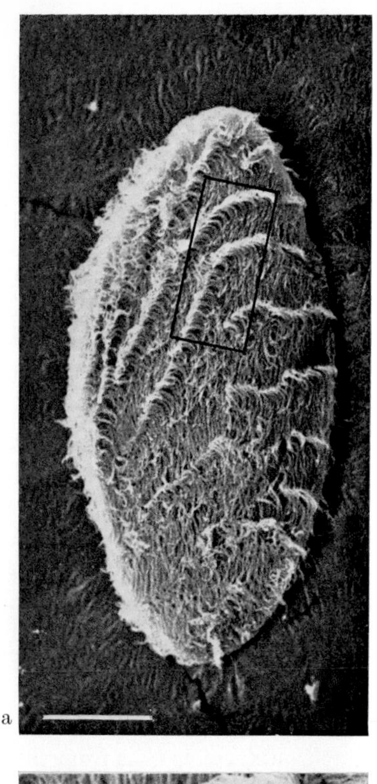
a

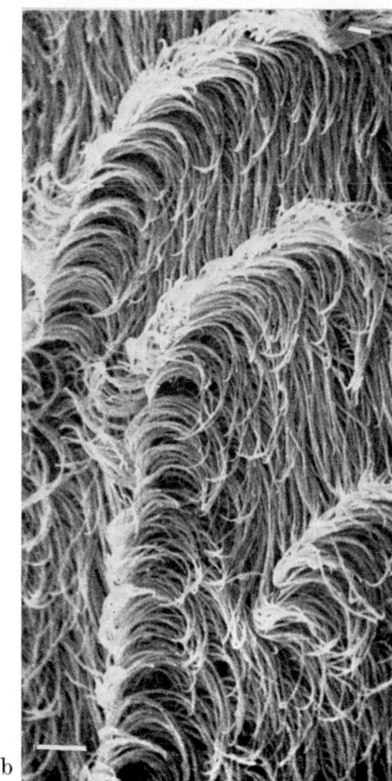
b

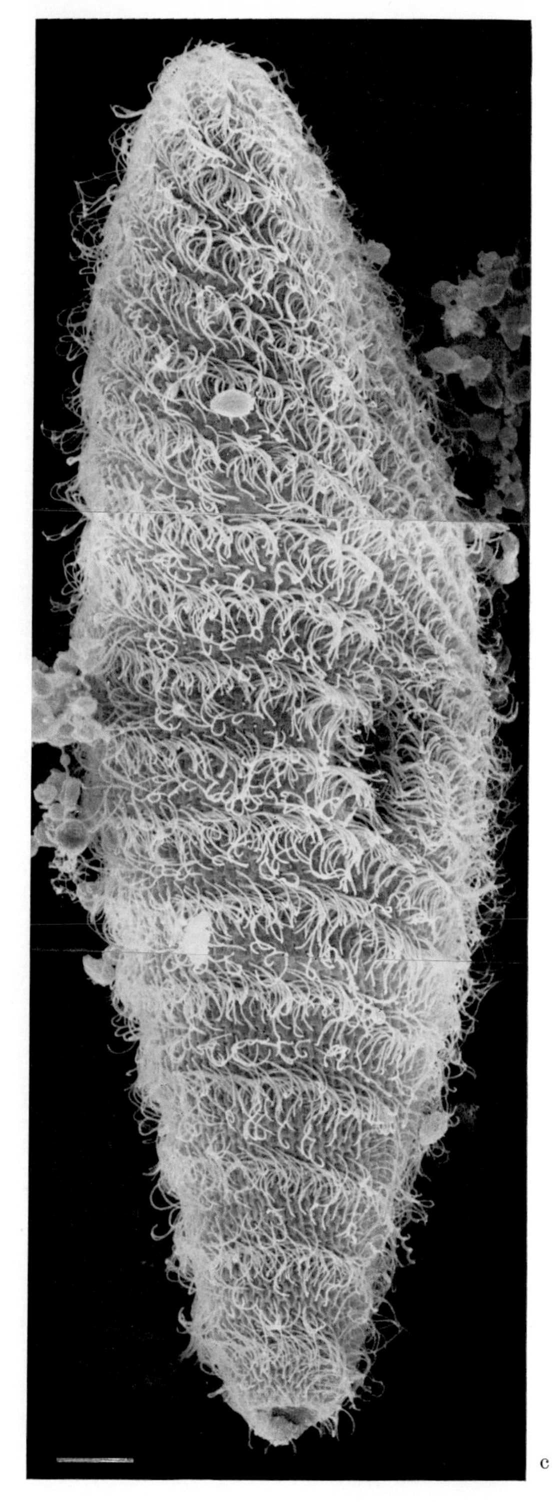
c

The instantaneous fixation method appears to preserve the metachronal waves in a remarkable true-to-life form as pictures illustrate taken with the scanning electron microscope (Fig. 262). There is no reason to believe that the appearance of metachronal waves is due to fixation as some authors suggest [*988, 989*]. In *Opalina ranarum*, metachronal waves move slowly enough to be observed in the microscope. In *Paramecium multimicronucleatum* where motion is much faster they can be demonstrated by electronic flash photography if viscosity of medium is low enough [*1080, 1081*].

Comparative studies [*948*] have shown that the direction of the metachronal wave can differ relative to the direction of the effective stroke (Fig. 263). We speak of *symplectic* metachrony if both directions coincide (a). Convergence of the cilia takes place therefore during the effective stroke, and the metachronal waves pass from the front end backwards as the ciliate moves ahead. In the other case, termed *antiplectic* metachrony (b), the effective stroke takes place in the opposite direction. The cilia converge in the course of the recovery stroke, and the metachronal waves pass from the posterior end forward as the ciliate moves ahead.

In *diaplectic* metachrony, the effective stroke is directed at right angles to the direction of the phase difference. It can be directed either to the right or to the left (*dexioplectic* or *laeoplectic*).

Symplectic metachrony, formerly thought to represent the prevailing mode, is characteristic for *Opalina*. Most ciliates show dexioplectic metachrony, and perhaps sometimes intermediates of the different types. An example of this possibility is *Paramecium*.

During *forward movement* of a *Paramecium* the metachronal waves pass from the posterior end forward. Rather than at right angles to the longitudinal body axis,

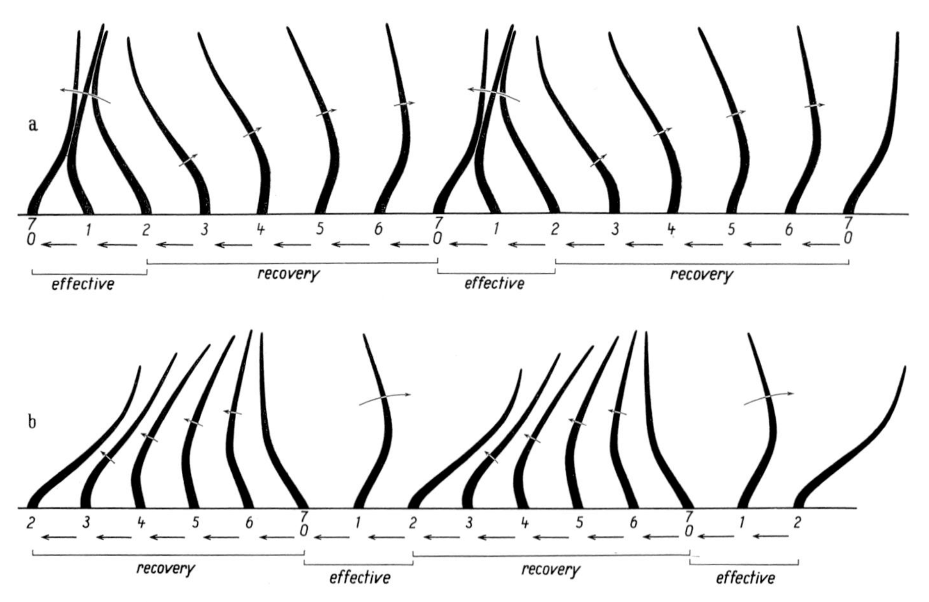

Fig. 263. Diagram of symplectic and antiplectic metachrony. *0—2* effective, *2—7* recovery stroke. In symplectic metachrony (a), the direction of the effective stroke corresponds to that of the metachronal waves, in antiplectic metachrony (b) the two directions are opposite. Courtesy of H. Machemer

the wave fronts are displaced towards the right at an angle of about 20–40 degrees (Fig. 261 c). The effective stroke is carried out in the direction of the wave front towards the right (dexioplectic), i.e. towards the right rear with respect to the body axis.

This mode of ciliary beating propels the cell along a left-handed helix rather than straight ahead. In each turn of the helix it rotates once about its longitudinal axis, the peristome being directed towards the axis of the helix.

In *backward movement*, as during an "avoiding reaction" (Fig. 276), a *Paramecium* follows a straighter course, in this case a right-handed helix with respect to the direction of movement. This is due to the fact that the effective stroke is carried out anteriorly and to the right. What takes place is therefore not simply a "reversal of stroke" but a change of the plane of beating by an obtuse angle. Accordingly, the metachronal waves move from the left anterior end towards the right posterior, but they are directed more meridionally than during forward movement (Fig. 261 c).

The wave patterns are subject to much *variability* depending upon the ever-changing stimuli impinging upon the cell. On occasion, *Paramecium* can also move ahead along a right-handed helix. The effective stroke is then directed posteriorly and towards the left and the wave-fronts are inclined accordingly. Rotation is decreased when it has to swim against a water current. The effective stroke is then directed meridionally from front to rear in a mode which was formerly considered "normal".

Not only the direction of the effective stroke as expressed in the wave pattern but also its *intensity* can be changed. A *Paramecium multimicronucleatum* of about 200 μm length generally moves with a speed of approximately 1300 μm/sec. It can, however, increase its speed to about 3500 μm/sec by increasing the ciliary frequency. Apparently, the frequency may also differ regionally. "Swimming along an arched course" may be understood in termes of locally modified direction and frequency of beating.

Characteristic changes of the wave pattern could be induced in *Opalina* [*1174, 1245–1247*] and in *Paramecium* [*1275*] by local stimulation. When a *Paramecium* is attacked by *Didinium* (p. 336), concentric wave fronts originate from the point of contact.

Although certain directions and intensities of the ciliary stroke clearly prevail, the direction of the effective stroke can be changed and its frequency can be varied within a wide range. However, the fact that waves are always formed, if sufficient frequency is given, indicates that ciliary activity is coordinated at all times.

Even among Holotricha, ciliation can be reduced to a considerable extent. The cell body of *Didinium nasutum*, for example, is encircled by only two ciliary belts (Fig. 287). They consist of short rows of cilia whose orientation is not meridional but oblique from the anterior left to the posterior right. Metachronal waves are seen to pass clockwise over both ciliary belts as the animal moves ahead, the cilia beating from the anterior right to the posterior left. It thus moves ahead along a right-handed helix. When swimming backwards, the waves are directed counter-clockwise and the cilia beat from the posterior left to the anterior right. In contrast to *Paramecium* (p. 303), the direction of the effective stroke thus changes by 180 degrees when the direction of movement is reversed [*1280*].

The transformation of groups of cilia to functional units like undulating membranes and membranelles, which is already indicated in Holotricha (Hymenostomata) and in Peritricha, has progressed further in Spirotricha. The *adoral band of membranelles*, which is typical for this order, serves primarily for the creation of a water current which carries food to the cytostome, but it can also play a role in locomotion.

Each membranelle is a platelet of triangular or trapezoidal shape containing 2 or 3 rows of cilia. While the basal bodies are firmly cemented to a plate of fibrils or amorphous material, the ciliary shafts seem to be only loosely connected. The membranelles, whose broad surfaces are arranged close to each other, beat in metachrony. In all cases studied thus far, the waves passing over the band of membranelles are seen to originate at the cytostome.

In addition to the band of membranelles, the *Heterotricha* also have longitudinal rows of body cilia like Holotricha.

In *Stentor* the band of membranelles encircles the funnel-shaped peristome and extends in a helical course towards the cytostome. The membranelles beat approximately at right angles to its direction towards the outside (dexioplectic).

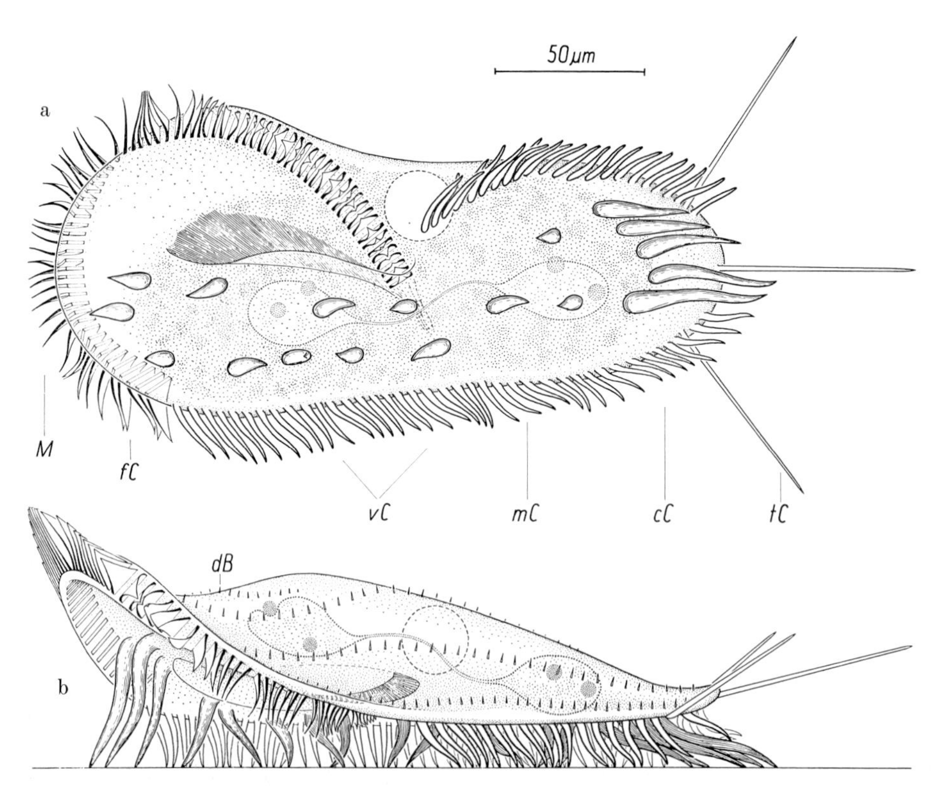

Fig. 264. *Stylonychia mytilus*. a Ventral view. b Side view. *M* Membranelles. *fC*, *vC*, *mC*, *cC*, *tC*: Frontal, ventral, marginal, caudal, terminal cirri. *dB* Dorsal "tactile bristles". Courtesy of H. MACHEMER

Usually, they cannot reverse their direction of beating [*1608*]. Bundles of nemadesmal fibrils originate at the basal bodies. They extend vertically into the cytoplasm and unite there in a longitudinal strand which follows the membranelle band to the full extent of its length [*1406*].

In Hypotricha, the body cilia are reduced to short bristles of unknown function at the dorsal surface. Ventrally, cirri are formed as complex organelles of motion. The membranelles beat primarily in the direction of the band.

It has been shown that the membranelles at both sides of the front end of a *Stylonychia mytilus* (Fig. 264) beat posteriorly. This leads to two mirror-image vortices which unite above the front end of the animal to a semi-circular ring (Fig. 265a). The membranelles do not beat as stiff plates: one edge seems to precede the other in time (turbine effect). The individual cilia composing a membranelle thus do not seem to become active all at once but in a fixed metachronal pattern (b).

Cirri are complexes of numerous rather long cilia closely grouped together and are round or polygonal in cross section. Their ends are pointed. They are always found in groups of relatively constant number and position.

In contrast to cilia and membranelles, the cirri do not exhibit a uniform activity of beating. Used even for "walking" on a substrate, their motions are often jerky,

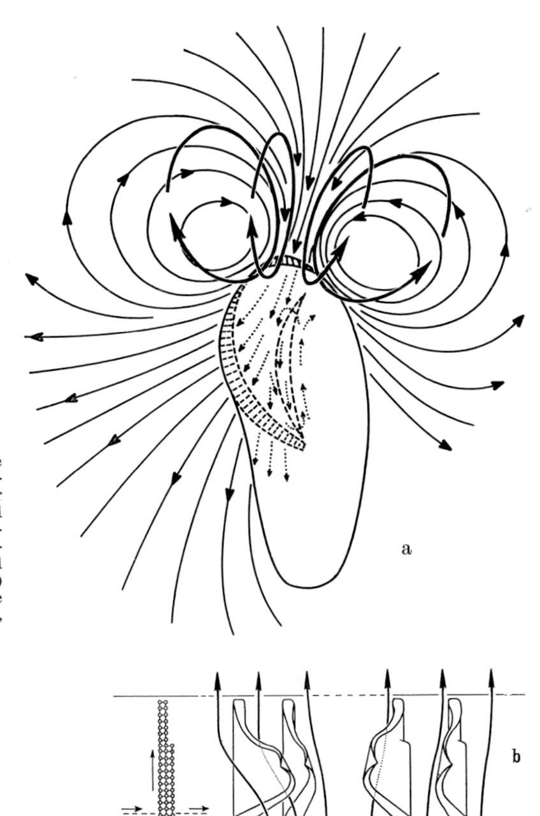

Fig. 265. *Stylonychia mytilus*. a Course of the water currents in the area of the band of membranelles (dorsal view). Some of the water streaming into the peristome region is deflected laterally. b Mode of membranelle activity (hypothetical): a continuous water current directed crosswise (long arrows) can be generated during both phases of the motion (bent arrows) if the membranelles are set obliquely like turbine blades while beating. After MACHEMER, 1966 [*1072*]

and sometimes in synchrony or metachrony. Groups of cirri and even single ones can also be moved independently. Analysis of these patterns of motion, which must involve a high degree of coordination, is still in its beginnings [*1070—1072*].

Electrophysiological studies have shown that electrical phenomena are associated with metachronal coordination. Since it has been possible to measure membrane potentials of single cells with microelectrodes, it could be shown that modifications of the direction and speed of the normal effective stroke are correlated with a change in membrane potential in *Opalina* [*929, 1174*] as well as in *Paramecium* [*1831*].

In *Opalina*, every change of the direction or speed of beating can be read directly from the metachronal waves. *Paramecium* exhibits one special case of directional change which is particularly suited for experimental analysis. This is the "reversal of beating" which enables the animal to swim backwards (p. 303).

Reversal can be induced by certain concentrations of cations in the surrounding culture fluid but is cancelled after some time. All cations inducing membrane depolarization in the presence of at least small amounts of Ca^{++} ions, are effective in reversing the cilia [*639, 641*]. In a galvanic current, reversal takes place at the side of the cathode (p. 324).

A suitable ratio of barium to calcium leads to *periodic reversal* in *Paramecium* with directional changes of effective strokes at intervals of 0.5—1 sec [*457*].

Since the same ionic effects are also observed in mechanically immobilized animals, the potential changes associated with reversal can be recorded. This correlation is especially impressive in cases of periodic reversal, each cilium beating in the opposite direction 22—36 msec after the beginning of depolarization [*930—933*].

Low concentrations of nickel (Ni^{++}) ions induce in *Paramecium* a forward movement along a right-handed helix [*984, 1282*], but at higher concentrations the cilia are immobilized. This state persists for about 30 min after the Ni^{++} ions are removed. If paramecia are exposed to K^{+} ions or to galvanic current during this phase, all cilia become directed to the anterior right as in swimming backwards. In a galvanic current, this is only observed near the cathode. It is only later that they resume rhythmic activity. This experiment supports the contention that the activity which leads to reversal of beating is not directly associated with ciliary beating as such.

Measurements of potential showed that a change of direction involves depolarization of the cell membrane to the same extent as with normally beating cilia [*1178*]. Directional change and depolarization are thus closely connected, while ciliary beating appears to represent an independent process as indicated by observations of isolated cilia (p. 300).

Recent investigations led to surprising results [*473, 474, 1179—1184*]. By extraction with the detergent Triton X-100 it was possible to obtain "models" of *Paramecium*. The cell interior of such models is freely exchangeable with the exterior, since the cellular envelope is disrupted. Nevertheless, the models could be reactivated to swim if they were transferred into a medium with ATP and Mg^{++} ions. While swimming the models showed typical metachronal waves whose frequency could be measured. It was essentially similar to that of unextracted live specimen.

It turned out that the direction of ciliary beating depended on the internal concentration of Ca^{++} ions. Whereas the cilia beat in a normal direction (towards the posterior right) when the Ca^{++} ion concentration was less than 10^{-6} mole/liter, they beat in the reversed direction (towards the anterior right) when Ca^{++} ion concentration was raised above 10^{-6} mole/liter. At the critical concentration of 10^{-6} mole/liter, the extracted paramecia spun about in one place without progress forward or backward, since the direction of the effective stroke was intermediate.

With respect to the electric phenomena one can suggest that depolarization increases the calcium conductance of the cell membrane, permitting an influx of extracellular Ca^{++} ions so that the subcortical space reaches a concentration sufficient to activate a shift in the direction of the effective stroke.

On the basis of these experiments, it is assumed that ciliary reversal of *live* paramecia is mediated by an increased cytoplasmic calcium concentration around the ciliary system through calcium dependent membrane responses to external stimuli. Evidently Mg^{++} ions are necessary for the activation of the cilia as such, while Ca^{++} ions – according to their concentration – determine the orientation of the ciliary movement.

A "behavioral" mutant of *Paramecium aurelia,* which differs from the wild type by a single gene, is of special interest in this connection. The cells of the mutant (called "*Pawn*") are unable to swim backwards even in the face of stimuli such as collision with an obstacle and high potassium concentration to which wild type cells respond by swimming backwards. "Models" of the mutant cells, however, swim backwards when the Ca^{++}-concentration was raised above 10^{-6} mole/liter. This result shows that the deficiency is most likely due to an alteration of the membrane (calcium conductance) rather than an impairment of the ciliary apparatus [*980—981*].

It is an open question, whether this concept, which assumes that ciliary activity is controlled primarily through electrical events of the cell membrane, may explain *all* motor responses of the ciliates in the future. It cannot be overlooked that the variation of directional changes is not yet understood. In any case cilia have to be regarded as "effectors" being sensitive to stimuli impinging on the cell membrane. At this time only a few events within a complicated receptor-effector system have been investigated.

It is not surprising, therefore, that other authors prefer to use provisional biological terms. They speak of "excitatory impulses", which cause the cilium to undergo an effective stroke, i.e. to perform a polarized form of movement instead of the "apolar" circular beating which occurs without "excitation" [*1287*].

In some cases such "excitatory impulses" seem to have a constant direction, which is fixed by the structural organization of the ciliary apparatus.

In *Stentor*, for instance, the "excitatory impulses" seem to originate at the cytostome (p. 304). If the membranelles had no part in their conduction, we should expect the length and speed of the metachronal waves to be uniform throughout, even if the spacing of the membranelles varies. Measurements made in *Stentor* showed this expectation not to be verified [*1603, 1604*]. In the vicinity of the cytostome, where the membranelles are smaller and more densely packed, the waves are shorter and have a lower velocity. The number of membranelles in a given wave is, however, the same throughout the length of the band of membranelles.

If a cut is made into the membranellar band, metachronal waves can start behind it. Their frequency is sometimes less and sometimes greater than that of the waves in front of the cut.

These observations have led to the *hypothesis* illustrated in Fig. 266. Every cilium – strictly speaking, every membranelle – is thought to have a certain degree of spontaneous excitation, which can build up to the level required for

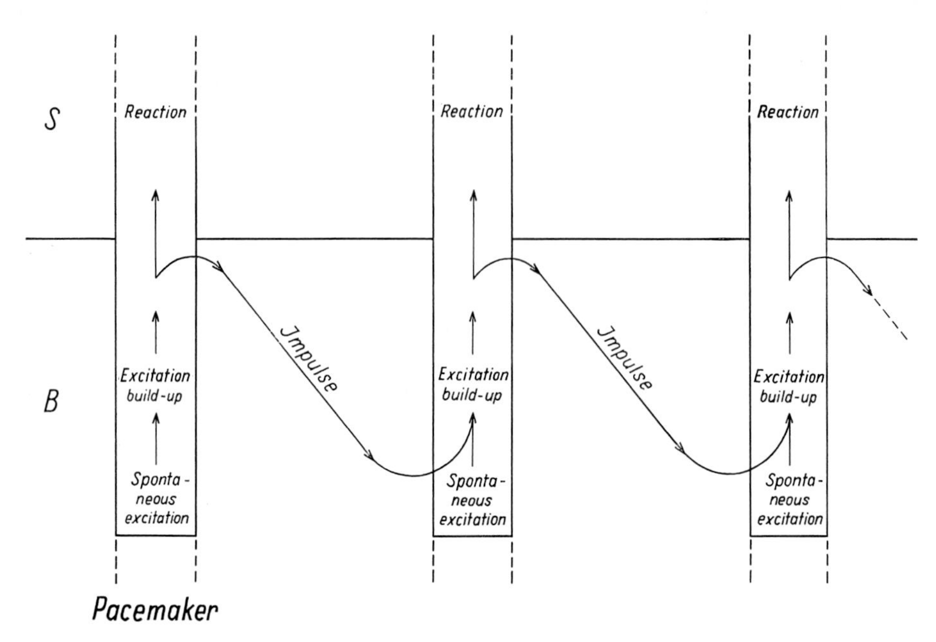

Fig. 266. Diagram of excitatory processes on three consecutive membranelles of *Stentor* (hypothetical). Explanation in the text. After SLEIGH, 1957 [*1604*]

autonomous beating. Before that, it is usually reached by an impulse which induces a coordinated stroke. As its excitation exceeds a given threshold, it transmits an impulse to a neighboring cilium. Impulse transmission is therefore delayed since the level of "excitation" has to be built up in every cilium. This means that the speed of the metachronal waves along a given path depends upon the number of cilia.

Every cilium can, however, become the *pacemaker* for the following cilia, e.g. behind a cut. When it has reached the necessary level of "excitation", it transmits an impulse to the next cilium and thus determines the frequency of strokes for the whole row.

This hypothesis makes it understandable why the velocity of metachronal waves is less than that of impulses along a nerve fiber.

In *Opalina* and *Paramecium*, where metachronal waves can pass in every direction over the cell, it is highly improbable that fibrillar systems, e.g. the kinetodesmal fibrils of *Paramecium* (p. 298), represent pathways for the conduction of "excitatory impulses". In *Euplotes eurystomus* the anal cirri are connected by nemadesmal fibrils with the distal end of the band of membranelles. Transection experiments seemed at first to indicate that the coordinated activity of cirri and membranelles, in jerking backwards, for example, is disturbed when this connection is severed [*603, 1721*]. However, repetition of these experiments has shown that such disturbances exist only shortly after the operation. Although the fibrils were transected, coordination was normalized after some time [*1248*]. Even in this frequently cited case, the intracellular fibers seem to be involved in impulse conduction as little as the "neurofibrils" of nerve cells.

3. Absence of Locomotor Organelles

While the gametes of *Sporozoa* can move with the aid of flagella, the *sporozoites* have no externally recognizable organelles of motion. They are nevertheless able to leave the spore and to find the final site for their parasitic mode of life. In the course of their migration they may have to pass through other tissues such as the intestinal wall. The zygotes of Hemosporidae, called "*ookinetes*", are also motile and pass through the peritrophic membrane into the intestinal epithelium while carrying out helicoidal movements.

Growing *gamonts* of eugregarines also move in a lively manner in the gut fluid of their host, even after they have linked together in "syzygy". They can actively change direction and twist through narrow passages. Pronounced snake-like motions are to be observed particularly in species of *Selenidium* in the gut of polychaetes.

If the intestinal contents are diluted with physiological saline, the motions of most gregarines are, however, restricted to a uniform *gliding*, interrupted at times by "periods of rest", which are often connected with a change of direction. Occasionally, the gamonts bend slowly to one side and quickly flip back to the original position. Motion pictures show at times wave-like motions of the cell surface, especially at the lower surface of the protomerite, and also reveal that locomotion is not correlated with cytoplasmic streaming. Particles adhering to the surface are always carried from the front backwards, even if locomotion of the cells is prevented by the micromanipulator. The speed at which the particles are transported corresponds approximately to that of the gliding motion [*979*, *1430*].

In many eugregarines the pellicle is differentiated into a system of longitudinal folds, which are uniformly distributed around the circumference. They are continuous from one end to the other, including the region between protomerite and

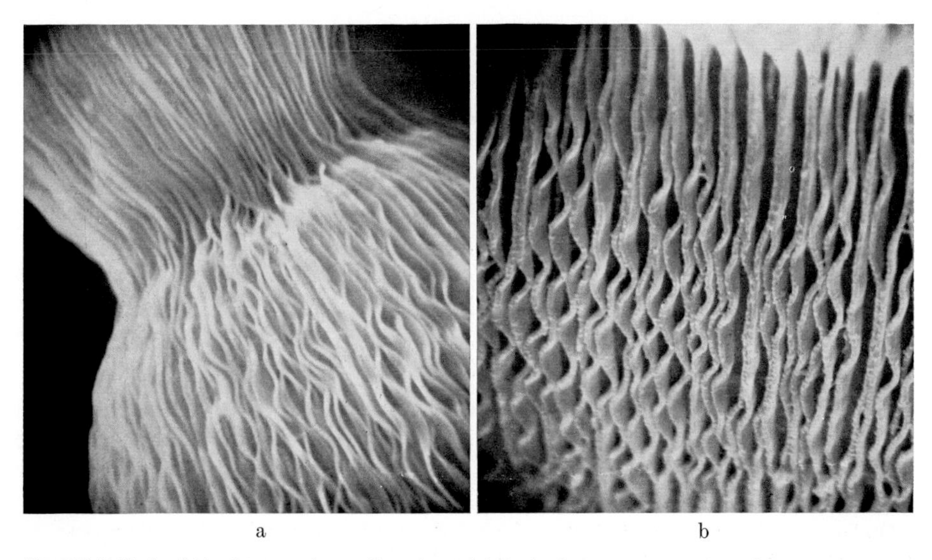

Fig. 267. Pellicular folds of eugregarines. a *Gregarina steini*. Region between protomerite and deutomerite. × ca. 6400. b *Gregarina cuneata*. Region of deutomerite. × ca. 7800. Stereoscan micrographs. After VAVRA and SMALL, 1969 [*1765*]

deutomerite. Cinematographic and ultrastructural observations [*1539, 1771–1773*] especially with the scanning electron microscope (Fig. 267) indicate that it is very probable that the type of movement known as *gliding* is brought about by undulations of these folds. Gregarines which do not glide (e.g. *Nematocystis*) either lack pellicular folds or the folds are permanently joined at their tips and are thus unable to undulate [*1765*].

There is no clarity as yet regarding the contractile elements of this cellular "surface muscle system". Microtubular fibrils have been demonstrated in sporozoites and ookinetes directly below the unit membrane [*558*], and the gamonts of some gregarines have a layer of "myonemes" below the pellicle. Future studies will have to clarify whether these elements are indeed contractile and how their activity is coordinated. It seems likely that all movements in Sporozoa which are not due to flagellar action are based on the same mechanism.

The pelagic Heliozoa and Radiolaria can sink or rise in the water by changing the gas (carbon dioxyde) content in their richly vacuolated outer cytoplasmic layer. Even Testacea like *Arcella* and *Difflugia* can rise from the bottom and float for prolonged periods by forming a gas bubble. The gamont of the foraminiferan *Tretomphalus bulloides* forms, shortly before gametogenesis, a special gas-filled chamber by the aid of which it rises to the water surface (Fig. 165 and 353c).

II. Changes of Shape

While an amoeba changes its shape constantly due to pseudopod formation, most other Protozoa have a definite cell shape which can, however, be actively changed in many cases. Such changes of shape are possible because of the presence of contractile fibrils which are sometimes arranged in bundles called *myonemes*.

In some euglenoids and monocystids the cell body widens and narrows rhythmically, suggestive of intestinal peristaltic waves. It is uncertain whether this "metaboly" is of any importance to the cell.

Changes of cell shape in *ciliates* are undoubtedly reactions which are meaningful to their way of life. Best known are contractions by which the animal extricates itself from harmful influences. The free-swimming *Spirostomum* suddenly shortens to one half of its length; the sessile *Stentor* contracts in seconds to a third of its length and assumes a spherical shape. Many ciliates (folliculinids, tintinnids, loricate peritrichs) have casings to conceal themselves when contracted. Besides simple contractions and expansions of the cell body, we also find complicated changes of shape in ciliates. Nothing is known about the coordination of these movements. The swinging movements of the proboscis of *Dileptus anser* (Fig. 289) may be cited as an example.

While it may often be very difficult to decide in these cases which of the fibrillar systems seen in the light or electron microscope are to be regarded as myonemes, there can be no doubt that the "*stalk muscle*" (spasmoneme) is the actual contractile element of the stalk of *Vorticellidae*. The stalk shortens by its contraction, as it is either bent into a zig-zag shape (e.g. *Zoothamnium*) or coiled to a helix (e.g. *Vorticella*). This contraction is especially impressive in the marine floating species,

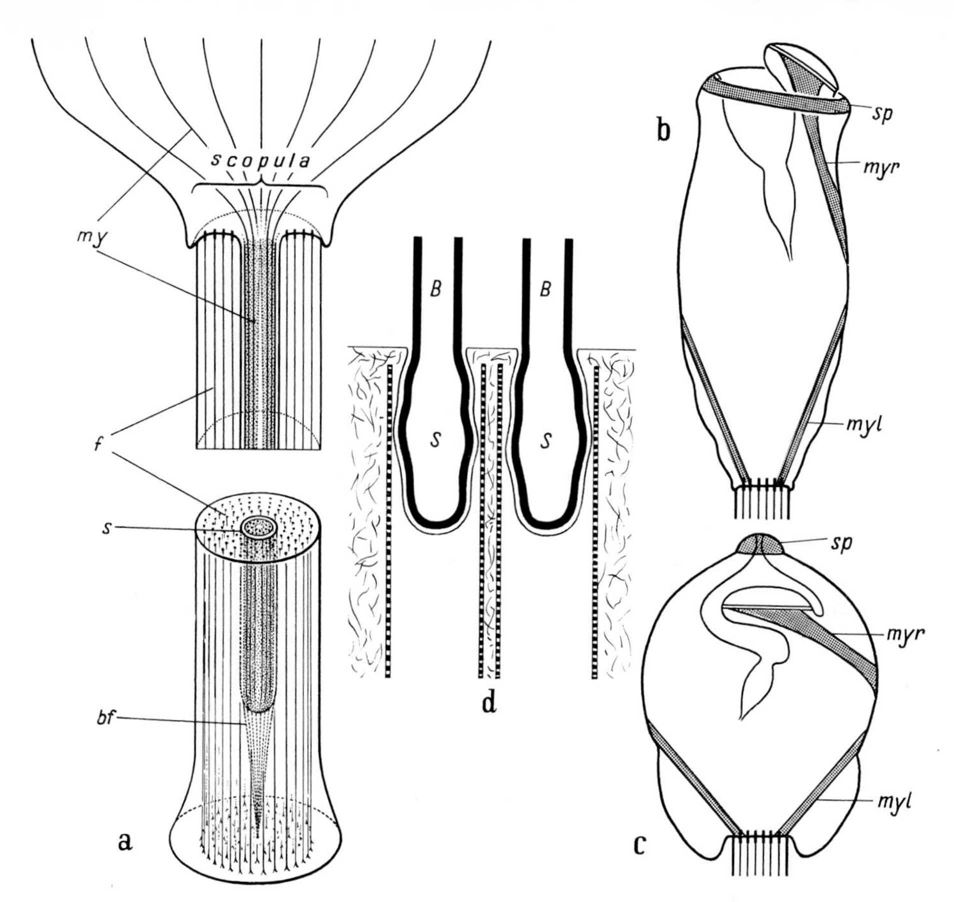

Fig. 268. Contractile fibrillar systems of peritrichs, schematized. a Stalk of a vorticellid. *my* Myonemes of the stalk muscle. *s* sheath of the stalk muscle, *f* stalk fibers, *bf* fibrillar bundle. b, c *Epistylis anastatica* (*b* elongated, *c* contracted), *sp* "sphincter", *myr* "retractor", *myl* lateral myoneme. Simplified after FAURÉ-FREMIET, FAVARD and CARASSO, 1962 [*506*]. *d Epistylis plicatilis*. Section of the scopula. *B* Basal body, *S* shaft. Schematized after an electron micrograph of RANDALL and HOPKINS, 1962 [*1405*]

Zoothamnium pelagicum, because all individuals (zooides) of the branched colony are connected by the common stalk muscle (film C836).

The stalk muscle is morphogenetically an evagination of the cell body into the stalk (Fig. 268a). It is bordered at the outside by a tubular sheath, which represents an extension of the cell envelope (*s*). Fine fibrils within a cytoplasmic ground substance extend as myonemes (*my*) far into the cell body and permit contractility. If the stalk muscle shortens, the cell body will also contract.

In *Epistylidae* (b, c), whose stalk is rigid and incapable of motion, a stalk muscle is absent. However, they do have myonemes (*myl*) which originate at the socket of the stalk and permit contraction of the cell body. In addition, many peritrichs seem to have a "retractor" (*myr*) to withdraw the peristomial disc and a "sphincter" (*sp*) to close the peristome cavity.

The stalk itself is an extracellular secretion product formed by a special zone, the so-called *scopula*. It consists of the wall of the stalk and a system of tubular fibers with a purely static or elastic function (*f*). In vorticellids, these stalk fibers are arranged as a cylinder around the stalk muscle which is connected by a fiber bundle (*bf*) to the base or wall of the stalk (a). In Epistylidae, the stalk fibers may be arranged cylindrically around a hollow space, perhaps reminiscent of a formerly present stalk muscle *(Opercularia, Campanella)*. In others, the stalk fibers fill the whole cross section of the stalk *(Epistylis)*.

Electron microscope studies [*506, 1405, 1456*] have shown that modified cilia are important structural parts of the scopula. The basal bodies are of normal structure but the shafts are more or less reduced (d). Since the tubular stalk fibers, which are seen to be cross-striated in the electron microscope, originate in the region of the shafts, it seems possible that the latter are centers of organization for the stalk fibers. This interpretation is also supported by the fact that the wall of the stalk fibers consists of 9 single elements in *Epistylis anastatica*.

J. Behavior

The way in which an organism establishes an active relationship to its environment can be called its behavior. It is largely determined by environmental influences to which the organism is subjected. Specific influences or stimuli call forth definite behavioral patterns or reactions *(reactive behavior)*. If the stimuli are within the framework of those conditions in which a species lives in its natural environment, the reactions are in a meaningful relation to the stimuli: they enable the organism to adapt to the situation it meets.

Stimuli are changes in energy and can be of a mechanical, chemical or electromagnetic nature ("optical stimuli", "thermal stimuli") within the natural environment of an organism. They lead to a reaction only if the amount of energy exceeds a definite *threshold of stimulation*. The repeated action of amounts of energy below this value *("subthreshold stimuli")* may, however, also lead to a reaction *(summation of stimuli)*. Whether an organism carries out a reaction depends also to a certain degree upon its physiological state *(susceptibility to stimulation)*. The susceptible state is often reached only after a definite course of development. The way an organism reacts can also be influenced by changes of its state due to earlier stimuli *(individual reaction base)*.

A stimulus induces a change in the physiological state of an organism. This is termed *excitation*. It spreads from the local point of stimulation, or it is conducted by special pathways to the point of reaction *(conduction of excitation)*. The energetic processes between stimulus and reaction are largely unknown.

In Protozoa, the reception of a stimulus and the response to it takes place within the same cell. Either the whole cell or definite regions or organelles may function as a receptor for stimuli. The response is in most cases a *reaction involving locomotion* or *taxis*, designated as phototaxis, mechanotaxis (thigmotaxis, rheotaxis), thermotaxis or chemotaxis depending upon the physical nature of the stimulus. It must, however, be taken into account that the same reaction can be elicited by very different stimuli. With respect to the relationship between the reaction and the direction of the stimulus we can distinguish between *phobotaxis* and *topotaxis*. In phobic reactions ("shock reactions"), no relationship to the direction of stimulation is recognizable. Such a relationship exists in topic reactions ("orientation reactions"): the cell moves either towards the source of the stimulus (positive topotaxis) or away from it (negative topotaxis). The following will show that some seemingly topic reactions are in reality of a phobic nature.

Phototaxis has received especially close study in this respect [*705, 724, 725, 1754*]. Locomotor reactions to light stimuli are not only found in all phototrophic flagellates, where their biological meaning is immediately evident, but also in some heterotrophic flagellates as well as in some rhizopods and ciliates.

In contrast to phototrophic purple Bacteria and Cyanophycea, which react only phobically to changes of light intensity, flagellates have the ability to carry out a *directed* movement towards the light source. In a unilaterally illuminated culture dish, they accumulate at the light side, i.e. their reaction is a positive phototaxis. This reaction can change to the reverse if the light intensity is increased: they show negative phototaxis by either swimming away from the light source, or they withdraw by repeated "shock reactions" from the region of greater light intensity.

Not only the intensity but also the quality of the light plays a role in the reaction. At first it might seem reasonable that the spectral ranges optimal for photosynthesis should evoke a positive reaction. Detailed study of the "action spectrum" showed, however, that no such coincidence exists. The most effective wave range for phototaxis is in the short-wave blue while the photosynthetic maximum is in the long wavelength red range of the spectrum.

In order to be effective, light must be absorbed by *pigments*. The absorbed energy starts a chemical process which in turn leads to the light-dependent reaction. Since blue light is most effective in phototaxis, the absorbing pigment must be yellow.

Mere light absorption is not sufficient to carry out a reaction related to the direction of the stimulus. The cell must be able to register *differences in intensity* of the incoming radiation.

Many phototrophic flagellates have *stigmata* or "*eyespots*". Although they cannot be regarded as the actual "photoreceptor organelles", they certainly do play an important auxiliary role in light perception. In many cases the stigmata are differentiations of the chloroplasts, located below the cell envelope at the front end (phytomonads). However, they can also appear separated from the chloroplasts as in euglenoids whose stigmata are closely applied to the wall of the flagellar sac. A swelling of the flagellar base, termed *paraflagellar body*, is in its vicinity (Fig. 320).

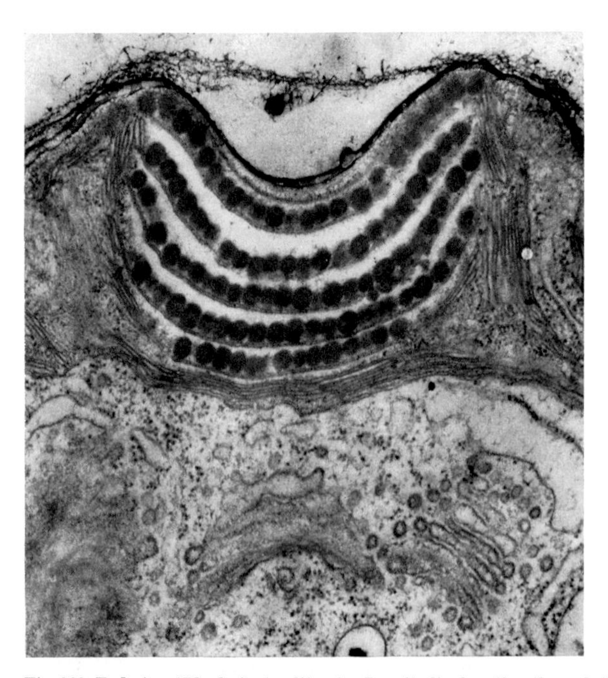

Fig. 269. *Eudorina (Pleodorina) californica.* Longitudinal section through the stigma of a somatic cell. The stigma lies within the chloroplast. Below: sectioned Golgi complexes. × 30000. Courtesy of G. SCHWALBACH

Stigmata have a yellow or red color in life. This is due to an accumulation of pigment granules of equal size, arranged as a single layer or in several layers on top of each other. The layers may be bent like a dish and are separated by membranes evidently connected with thylakoids of the chloroplast (Fig. 269). The chrysomonad, *Chromulina psammobia*, has a short flagellum above the stigma which is enclosed by an invagination of the cell surface. It is certainly superfluous for locomotion but may play a role in light perception [*1455*]. Absorption measurements after extraction have proved that the pigment of stigmata is a carotenoid [*618*, *705*].

The importance of the stigma for light perception seems to differ among flagellates. This follows from the fact that flagellates without a stigma (e.g. *Chilomonas*, *Bodo*) also exhibit phototaxis. Mutants of *Chlamydomonas reinhardi* without a stigma react less precisely than the wild type but can still orient towards the light [*722*]. Even in the case of *Euglena*, whose light reactions will be discussed in detail, it has been shown that the stigma plays only an auxiliary role in phototactic orientation.

On the other hand, the stigma seems to be essential for phototaxis in *colonial phytomonads*. In *Eudorina californica* only somatic cells which have a stigma (Fig. 269) can react to light. No reaction is shown by generative cells which lack a stigma [*572*].

The only reaction of somatic cells to alternating illumination and shadowing is a decrease in the frequency of flagellar beating. This inhibition sets in immediately after a change of light intensity but lasts only a few seconds when the flagella resume normal beating, although the light intensity remains the same. With

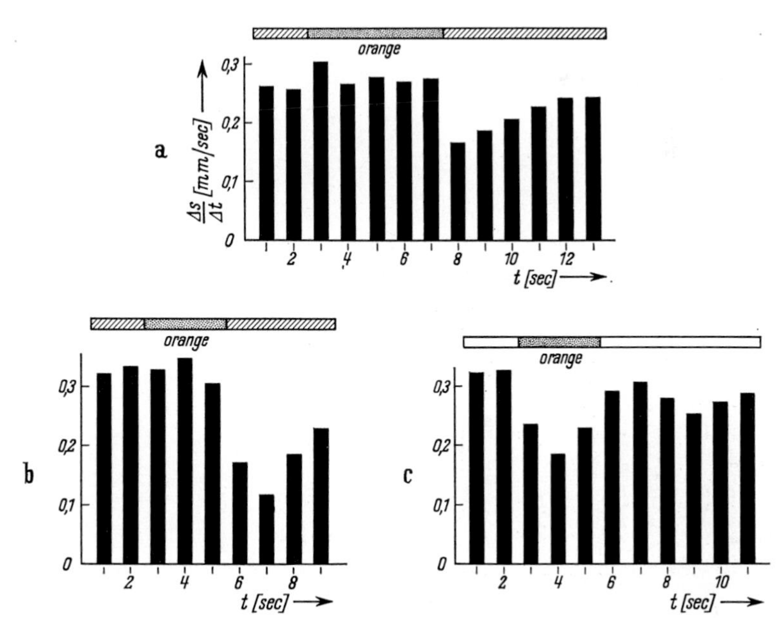

Fig. 270. *Eudorina (Pleodorina) californica* (a) and *Volvox aureus* (b, c). Change of swimming speed when darkened temporarily with an orange filter. a, b Positively phototactic colonies. c Negatively phototactic colonies. Explanation in the text. After GERISCH, 1959 [572]

sufficiently strong stimulation, the flagella go into a "blocked position" in which only brief twitches are carried out.

In positively phototactic colonies (low light intensity) inhibition follows upon an increase and in negatively phototactic colonies (high light intensity), a decrease of light intensity.

The reactions of individual cells as observed in artificially fixed colonies is in accordance with the speed of swimming of a colony in light/dark experiments. In positively phototactic colonies of *Eudorina californica* or *Volvox aureus* darkened by addition of an orange filter, a temporary decrease of swimming speed is observed when the filter is removed (Fig. 270a and b). If negatively phototactic colonies are darkened, a speed decrease sets in when the filter is put in place (c).

This concurrence provides the key for understanding the phototactic reaction. Since a swimming colony rotates because all flagella beat in a definite plane (p. 141), the individual cells are alternately exposed to light and darkness as long as a course deviating from the direction of light incidence is followed. The cells at the "light side" decrease the frequency of flagellar beating or are even blocked. As a consequence, the positively phototactic colony will turn towards the light. The reverse holds true in negative phototaxis: cells at the "shaded" side lessen the frequency of flagellar beating, and the colony moves away from the light in a straight line.

Whereas non-colonial phytomonads, like *Chlamydomonas*, for example, react topically, i.e. they swim in positive and negative phototaxis either towards the light source or away from it, the individual cells of colonial species carry out a phobic reaction. It has no relation to the direction of light incidence and is induced by a periodic change in the intensity of the stimulus. The nature of this reaction and the arrangement of the cells within the complex lead, however, to a topic reaction of the colony as a whole.

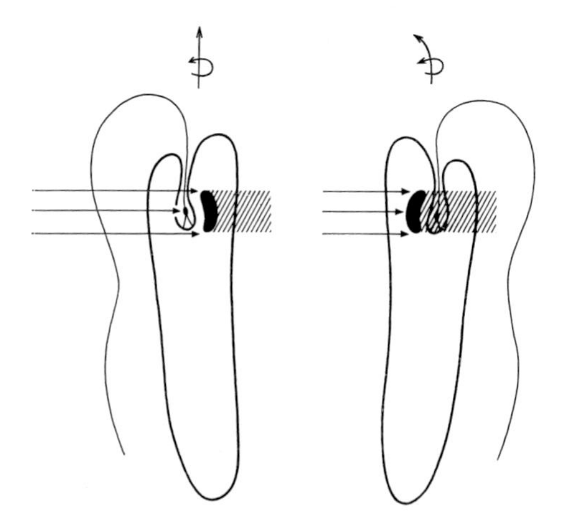

Fig. 271. *Euglena* illuminated from the left (arrows), in two different positions toward the light. The stigma (kidney-shaped in the picture) shadows the photoreceptor (thickening at the flagellar base) at the right position: the result is course change during forward movement (arrow over the cell). Rotation about the longitudinal axis is indicated by another arrow. After HAUPT, 1965 [*724*]

Studies made with *Euglena* [*180*, *181*, *618*] indicate that the stigma has only a "shadowing" function in this case. Light absorption, which is essential for phototaxis, takes place outside the stigma, probably at the local thickening of the flagellum which is known as the paraflagellar body.

In locomotion connected with *positive* phototaxis, a *Euglena* rotates about its longitudinal axis. With lateral light incidence, the shadow of the stigma is therefore periodically cast upon the paraflagellar body. The cell reacts by changing its course until it moves without the stimulation of periodic changes of intensity towards the light source (Fig. 271).
Since the periodic changes of flagellar beating necessary for course correction are not related to the direction of light incidence, they are to be classed as phobic reactions. The summation of these reactions leads to a seemingly topic (pseudotopic) response to the stimulus.
By contrast, *negative* phototaxis of *Euglena* is purely phobic, even with respect to the behavior of the whole cell. Light microscopy shows that when cells hit upon a part of the visual field with sufficiently high light intensity, they merely recoil. In an intensity gradient, however, they do not carry out any directed motion away from the light source.
Thus, we can assume that in *Euglena* negative phototaxis takes place without the aid of the stigma. The reaction is indeed also carried out by stocks lacking a stigma, while a stock which also lacks the paraflagellar body is wholly incapable of reacting to light.
If it is correct that the paraflagellar body represents the actual "photoreceptor" of *Euglena*, it should also contain the pigment whose light absorption is essential in phototaxis. However, there is no evidence regarding the nature of this pigment. It might be a carotene which differs chemically from that in the stigma [*705*].

The phototactic responses in *Euglena* are controlled by a temperature-independent biological clock with an endogenous rhythm which persists in continuous darkness.

Sense organelles for light reception, often called "ocelli", are found in one marine dinoflagellate family *(Warnowiidae)*. As they are not multicellular organs but organ-like differentiations of a single cell, i.e. organelles, the term *ocelloids* seems more appropriate.
Erythropsis pavillardi (Fig. 272), whose cell body is also equipped with a tentacle capable of extraordinary contraction and expansion, has an ocelloid of a striking size (about 40 μm). By light microscopy, a protruding lens-like, occasionally clearly mushroom-shaped, "dioptric apparatus" and a brownish-black pigment mass of variable extent are recognizable.
Electron microscope studies [*685*, *687*, *1148*] reveal a surprisingly intricate fine structure (Fig. 273). The part of the "dioptric apparatus" which protrudes above the cell surface forms a hemispherical "cornea" (*C*) containing a layer of large, flattened mitochondria (*M*). Below the "cornea", a large "crystalline body" is found. It represents the major mass of the "dioptric apparatus" and is composed of five vesicle-like strata (*K*). It consists of a highly refractile substance, completely transparent in life, which attains the hardness of an eye lens. The "crystalline body" is bounded at its middle by a triple layer of tapelike structures called "constrictors" (*k*).

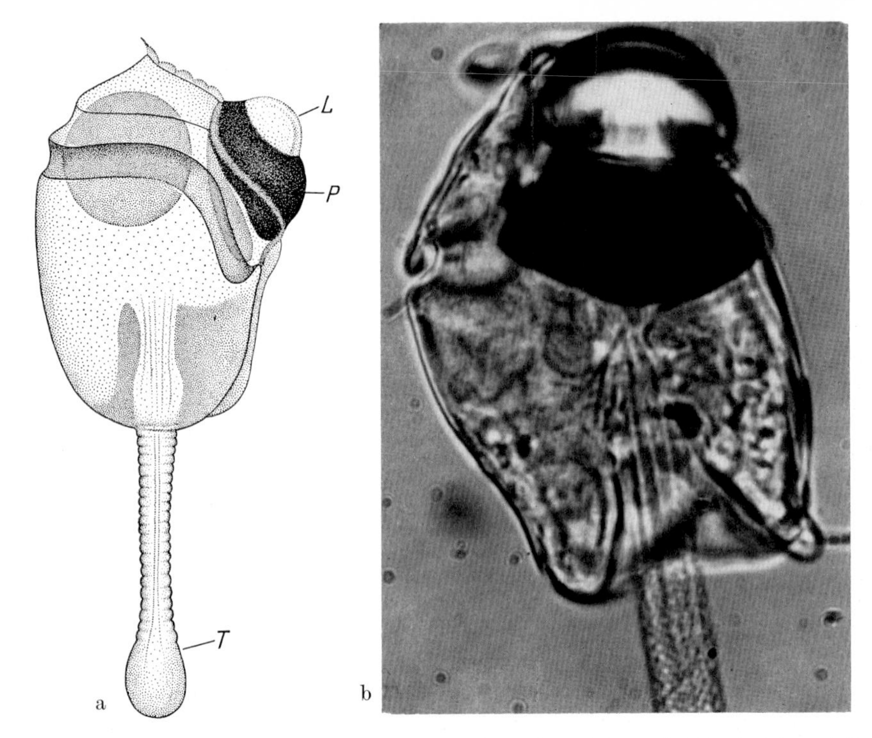

Fig. 272. *Erythropsis pavillardi*. a Survey picture. *T* Tentacle, *P* pigment, *L* so-called "lens" of the ocelloid. After KOFOID and SWEZY, 1921 [*956*]. b Photographed in life by C. GREUET (Villefranche-sur-mer)

The region below the "crystalline body" consists of two parts of which the proximal one is wholly enclosed by pigment (*P*). An outer layer of large carotene granules can be distinguished from an inner layer of small ones. The layer at the bottom of this pigment cup has only become known in detail through electron microscopy. It represents the actual photoreceptor and is therefore called "retina" (*R*).

A section at right angles to the longitudinal axis of the ocelloid shows that the "retina" has a "paracrystalline" fine structure (Fig. 274). It is composed of parallel double membranes and a thicker membrane in between each doublet in a regular wavy arrangement resembling corrugated board.

Since dinoflagellates with such ocelloids are relatively rare and prone to perish quickly, not much is known about their habits. Perhaps they generally live in the deeper regions of the sea where they depend upon optimal utilization of the light available. Since they are heterotrophic, light can only be of importance for orientation, e.g. in finding phototrophic food organisms.

Rhizopods and *ciliates* show generally only negative phototaxis unless they have phototrophic symbionts. *Amoeba proteus* creeps away from the light source. It reacts most strongly to light in the green range of the spectrum [*760*]. If illuminated from two directions with equal intensity, it orientates only to one of the two light sources. Apparently the same part of the cell remains the front end when the amoeba changes direction [*1155*]. Only *Stentor coeruleus* will be mentioned as an example

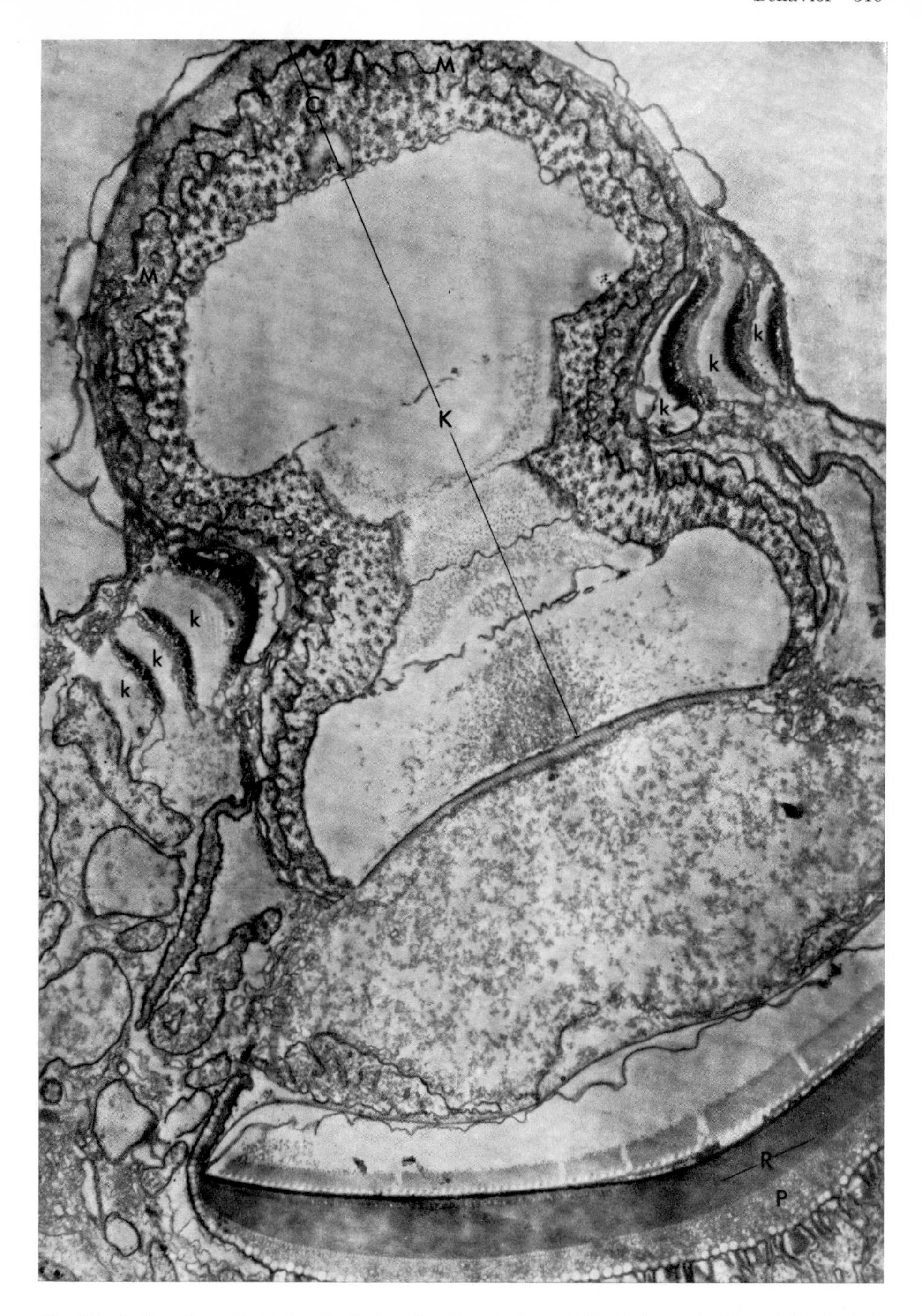

Fig. 273. *Erythropsis pavillardi.* Longitudinal section through the ocelloid. *C* "Cornea", *K* "crystalline body", *R* "retina", *P* pigment layer, *k* "constrictors", *M* mitochondria. × 4800. Courtesy of C. GREUET (Villefranche-sur-mer)

Fig. 274. *Erythropsis pavillardi.* Transverse section through the "retina". × 55000. Courtesy of C. GREUET (Villefranche-sur-mer)

of a phototactic ciliate. If cells are kept in a dish with varying light intensity, all of them will finally accumulate in the darker region. Ciliates, which otherwise show no phototactic reactions (e.g. paramecia), can be sensitized with fluorescent dyes (eosin, erythrosin): they react with negative phototaxis when illuminated from one side after they have taken up small amounts of dye ("induced" phototaxis).

However, light stimuli are generally of minor importance to heterotrophic Protozoa. Reactions to other modes of stimulation may be all the more pronounced.

Mechanical stimuli are pressure differences due either to touch or to water current. If the reactions in question are locomotive, those elicited by touch are called *thigmotactic* and those due to current are called *rheotactic.* However, the response to such stimuli can also be a change of shape, e.g. a contraction, especially in "haptic" species which attach temporarily or in those which are permanently sessile.

The "mechanoreceptors" of flagellates and ciliates are first and foremost the flagella and cilia. If touched, the cilia of *Paramecium* cease to beat. Finally all cilia except those of the peristome may come to a standstill.

While *Paramecium* usually swims about, *Dileptus cygnus* has a more sedentary mode of life. It lies on the bottom and performs oscillating movements with its proboscis at the base of which the cytostome is located (comp. Fig. 289). Experiments show that the body has two areas of reactivity (Fig. 275). Mechanical stimulation (touching, puncture) of the anterior part of the body leads to withdrawal ("backward response area"), stimulation of the posterior part to a start forward

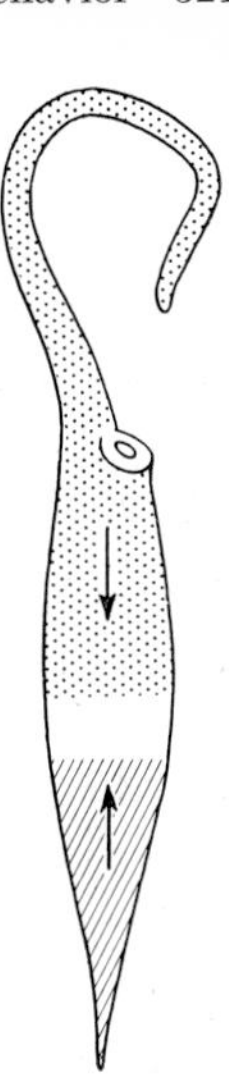

Fig. 275. *Dileptus cygnus.* Distribution of "backward response area "(anterior part) and "forward response area" (posterior part). After DOROSZEWSKI, 1970 [*434*]

("forward response area"). After bisection in the intermediate region both areas still behave in the same manner. The missing areas of reactivity become restored in the course of regeneration [*431–435*].
Sometimes cilia are transformed to "tactile bristles", i.e. they are no longer motile and are specialized for the perception of pressure differences. Whether the bristles to be found on the dorsal surface of hypotrichs are really "tactile" as is usually assumed needs to be proved experimentally. Regions of increased sensitivity to pressure might also be of aid to the motile stages of sessile species (telotrochs of peritrichs, swarmers of suctorians) in finding a suitable substrate for attachment. However, it is not known to what extent chemical stimuli might also play a role.

Chemical stimuli are undoubtedly of great importance for Protozoa although they lead only rarely to directed locomotor reactions. It is often difficult to decide at a glance whether a given reaction is thigmotactic or chemotactic, e.g. if an amoeba puts forth a pseudopodium at a spot touched by a *Paramecium* (film E1171). Only the fact that the amoeba reacts negatively to equally intense touch stimulation of other sources makes it probable that the positive reaction was due to the chemical make-up of the surface of the *Paramecium* and not to a pressure difference.
Strictly speaking, we are only dealing with chemotaxis if the reaction involves locomotion. In the diffusion gradient of a substance the reaction could either be phobic, e.g. involving a sudden recoil when a certain concentration threshold is exceeded, or it is topic and leads to a directed motion with respect to the center of the diffusion gradient. Even then the possibility often cannot be excluded that a seemingly topic reaction is actually composed of a sequence of phobic parts.

The so-called *chemotaxis experiments* have shown that many Protozoa react to certain substances either with positive or negative chemotaxis. The substance in question is filled into a capillary pipette which is then introduced into a suspension of the protozoan species to be tested, e.g. under a coverslip. The results of such experiments will not be discussed here. However, the importance of chemotaxis for microgametes seeking out macrogametes or for the affinity of parasites to a specific tissue should also be pointed out.

It could be shown in *Paramecium caudatum* that chemotaxis is often not elicited by a specific substance as such but by changes of pH due to its presence. The speed with which paramecia swim is highest in their optimal pH range of 5.4 to 6.4 [*453*, *454*]. A test of negative chemotaxis to various alcohols showed that the intensity of reaction increases with increasing molecular weight of the alcohol [*452*, *457*].

It is only natural that *temperature* influences all protozoan life processes to varying degrees. However, little is known concerning locomotor reactions evoked by *thermal stimulation*. It is nevertheless evident from many experiments that Protozoa do react negatively or positively to thermal stimuli and that they can seek out a species-specific preferred range when put into a temperature gradient. This preferred range is from 24—28° C in the case of *Paramecium*.

For a long time it was considered proved that in ciliates all mechanical, chemical and thermal stimuli call forth the same reaction, the so-called *avoiding reaction*. In *Paramecium*, where forward movement represents the "natural state" [*1273*], this reaction was thought to set in whenever the animal chances upon a mechanical

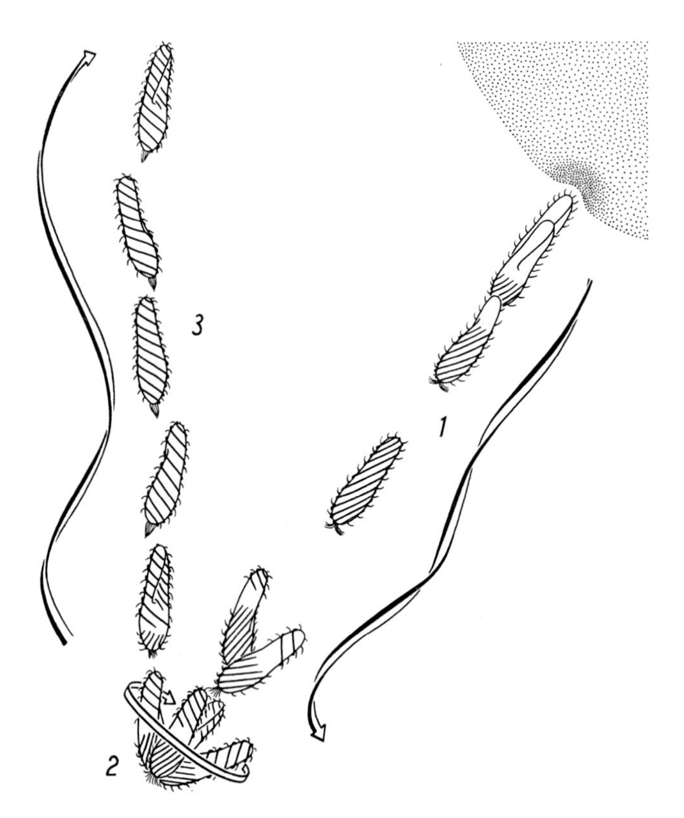

Fig. 276. *Paramecium caudatum*. Changes in metachronal wave pattern during the so-called avoiding reaction (scheme). *1* Backward motion after stimulation. *2* Cone swinging phase. *3* Forward motion. After PARDUCZ, 1959 [*1279*], adapted

obstacle or either a chemical or thermal intensity threshold. It entails three consecutive phases*, viz. a fast backward movement, then a circular movement where the cell body outlines a conical surface ("cone swinging phase"), and renewed forward movement (Fig. 276). According to the so-called "trial and error theory" [*859*], the physiological role of the cone swinging phase would be to obtain water samples from different directions, enabling the *Paramecium* to select that direction for forward movement from which a stimulus no longer impinges upon it.

The "avoiding reaction" is especially pronounced in a *gradient of stimuli*. An upper and a lower intensity limit at which an "avoiding reaction" is evoked exists with respect to many environmental conditions. The range within both limits (*preferred range* or *optimal zone*) leads to no reactions. Thus, paramecia may accumulate in the diffusion gradient of a substance within a definite ring-shaped zone around the center of diffusion (Fig. 277a). This may also be the case in a definite pH-range (p. 322). They swim at first into this zone without any reaction but are then retained in it as in a trap. The path of a single *Paramecium* has a zig-zag shape and consists of successive "avoiding reactions" induced by contact with the upper and lower intensity threshold (b).

Although the "avoiding reaction" is often observed as a mere succession of phases, closer analysis has shown that it is by no means a uniform reaction [*1276, 1277, 1279*]. The phases do not necessarily have to take place in the sequence described above.

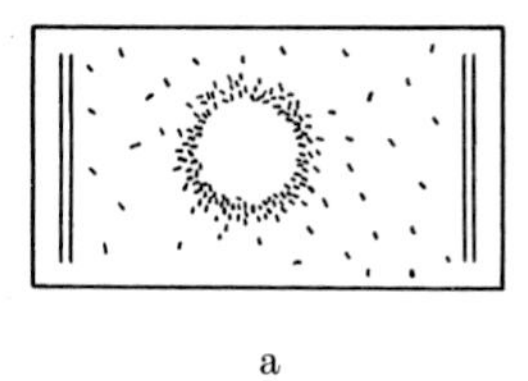

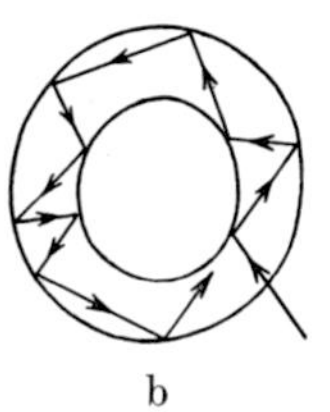

a b

Fig. 277. Behavior of *Paramecium* in a concentration gradient. Accumulation around a drop with a mixture of salt and acid. b Path of a single individual within an optimal zone (successive avoiding reactions). After JENNINGS, 1906 [*859*]

The "cone swinging phase" is simply explained as the physiological change to normal forward movement as discussed on p. 303 (Fig. 276). The role ascribed to it by the "trial and error theory" seems improbable if only for the reason that cone rotation is also carried out in the complete absence of stimuli. Backward movement due to a stimulus is often not even followed by cone rotation but changes directly to another mode of locomotion, such as movement of the front end to any side or mere rotation [*454*].

Although the tendency to reduce locomotor behavior in ciliates to a single principle is understandable, the cognition of the extraordinary diversity of orientation movements should not be ignored [*40, 1272–1287*]. *Paramecium* can, for instance, change direction while swimming and thus avoid noxious influences. It can swim along an arc with or without rotation, turn on the spot and also carry out other motions, which, though caused or influenced by external stimuli, cannot be wholly interpreted as abbreviated or modified "avoiding reactions".

* Backward movement is said to be preceded by a single synchronous stroke of all cilia towards the front end [*1287*].

Analysis of wave patterns (p. 303) shows this variability to be due to the ability of a *Paramecium* to respond to differences in stimulation over its cell surface with local modifications of ciliary activity (direction and intensity of the effective stroke).

With the exception of rheotaxis, where a *Paramecium* swims upstream against the water current, topic orientation movements as reactions to symmetrically impinging stimuli are relatively rare in *Paramecium*.
When put into a vertical tube filled with water rich in carbonic acid, the paramecia swim upwards, i.e. they react negatively to gravity. This reaction, referred to as negative *geotaxis*, was explained as due to pressure of inclusions within vacuoles upon the cytoplasma below. Accordingly, the cells move in a magnetic field away from a magnetic pole if previously fed with powdered iron [*953, 955*].

It is doubtful that geotaxis plays any role under natural circumstances. This is certainly true in the case of *galvanotaxis*. Many Protozoa carry out directed movements either towards the anode or the cathode under the influence of a galvanic current.
This phenomenon has been thoroughly studied in *Paramecium*, which swims towards the cathode when the current is switched on [*632, 633, 636, 838, 839, 859, 954*].
Physiologically, this reaction is due to augmented ciliary beating at the body region closest to the anode while a smaller population of cilia close to the cathode beats in the reversed direction. In this manner all paramecia carry out a forced orientation which directs the front end to the cathode, irrespective of the position they had when the current was switched on (Fig. 278).

The cilia of the peristome may also play a significant role in turning the animal to the cathode since they retain their direction, continuing to beat towards the cytostome [*640*].

In a medium of increased viscosity the cilia are largely immobilized but can be re-activated by a galvanic current (Fig. 279).
If the *Paramecium* happens to be oriented with its front end towards the cathode *(homodromic* orientation), at first, with low currents, only the cilia of the front end will start beating. The effective strokes are towards the cathode, i.e. to the right anterior. As the current is increased, the posterior cilia also start beating, in this case towards the anode, i.e. to the right posterior. Finally, both regions extend until they meet at a "dividing line" at right angles to the direction of the current (a).
In *antidromic* orientation, i.e. with the front end to the anode, the cilia of both body regions start to beat simultaneously. Those of the posterior end now beat away from the cathode, i.e. towards the right anterior, and those of the front end away from the anode, i.e. towards the right posterior (b).

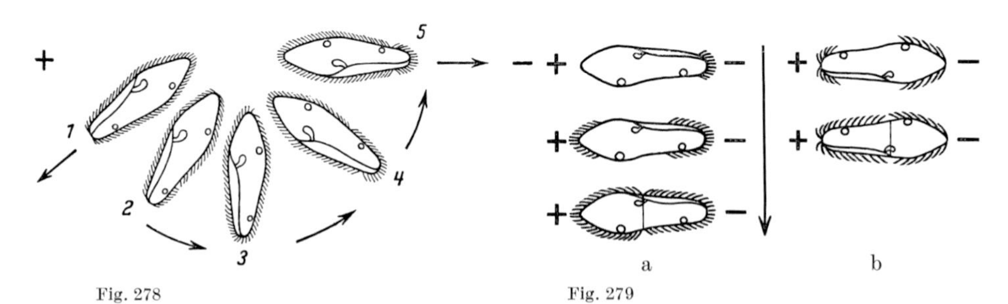

Fig. 278 Fig. 279

Fig. 278. Galvanotaxis in *Paramecium*. The *Paramecium*, which was swimming about in every direction, turns towards the cathode when the current is switched on. After JENNINGS, 1906 [*859*]

Fig. 279. Effect of an electric current on paramecia which were immobilized by increasing the viscosity of the medium. Only the re-activated cilia are depicted. a Homodromic orientation: anterior end oriented towards the cathode. b Antidromic orientation: anterior end oriented towards the anode. After investigations of KAMADA, 1931 [*893, 894*]

The activation of cilia is such as to induce "reversed" beating in the body region nearer to the cathode and augmented "normal" beating in the region closer to the anode.

This experiment shows that the influence of the galvanic current includes a cathelectrotonic as well as an anelectrotonic effect. The fact that the cathelectrotonic effect sets in earlier in the case of homodromic orientation might be due to higher sensitivity of the front end to stimulation [*893, 894*]. This interpretation is also supported by other experiments involving stimuli of different kinds.

That the dividing line between the cathodic and anodic influence is — at any position of the cells — always at right angles to the direction of current is also recognizable in galvanotactically reacting, free-swimming paramecia. Analysis of wave patterns by the instantaneous fixation method shows metachronal waves in the cathode range as in normal backward swimming and in the anodic range as in normal forward motion (p. 303). Ciliary coordination is thus not disturbed by a galvanic current.

According to recent investigations [*474*] the basis of galvanotaxis is the local potential difference across the cellular membrane by the applied current. The membrane on the side facing the anode is somewhat hyperpolarized by inward electronic current while the membrane facing the cathode is depolarized by outward electronic current. As mentioned before (p. 306) hyperpolarization elicits an increased frequency of beating in the normal direction (forward) while depolarization elicits ciliary reversal (backward).

Also in Protozoa, behavior is not limited to reactions to external influences. Ciliates especially show behavioral patterns which are due to internal conditions and must therefore be called *spontaneous*. They may appear of necessity at a definite developmental stage of the cell or maturity stage of the clone and form a regular sequence of phases.

As soon as the swarmer of *Metafolliculina andrewsi* (Fig. 135) has left the lorica of the mother cell, it selects a suitable substrate with the aid of a thigmotactic field of cilia. It rotates a few times in a circle at its future attachment point and comes finally to rest, assuming a very flattened shape. It attaches firmly to the substrate by a secretion from its base. Then it stretches a little and forms the ampulla-shaped part of the lorica which remains open only at the front end. After lifting the front end at a 45° angle from the substrate, the swarmer stretches further, rotating continuously towards the left at the same time. The neck of the lorica, which is supplied with a helical ridge, is secreted from a special pigmented zone in the course of this rotation. The collar-like end of the neck is formed when the front end widens to a mushroom shape. Metamorphosis of the swarmer, especially growth of the two peristomial wings, sets in only after the lorica is finished [*1756*].

In *Stylonychia mytilus* and other hypotrichs, conjugation is preceded by a curious "mating play" (Fig. 280). At first, both partners rotate in circles next to each other, with the peristomes facing down. Rotation is clockwise as seen from above and consists of single jerks covering angles of 45—60 degrees (a). Finally they rise up and touch with their peristomes (b). This play may be repeated several times before the peristomes finally become glued together and the cells merge (c).

Changes in behavior which can be interpreted as adaptations to a special stimulus situation are of particular interest. If *Stentor roeseli* is stimulated with a water current from a pipette, it withdraws like a flash into its tube of mucus. However, it soon extends again and starts to whirl with its cilia. If this is repeated several times, it no longer reacts to the stimulus. This becoming accustomed to a repeated innocuous stimulus *(habituation)* has also been observed in other ciliates [*926*].

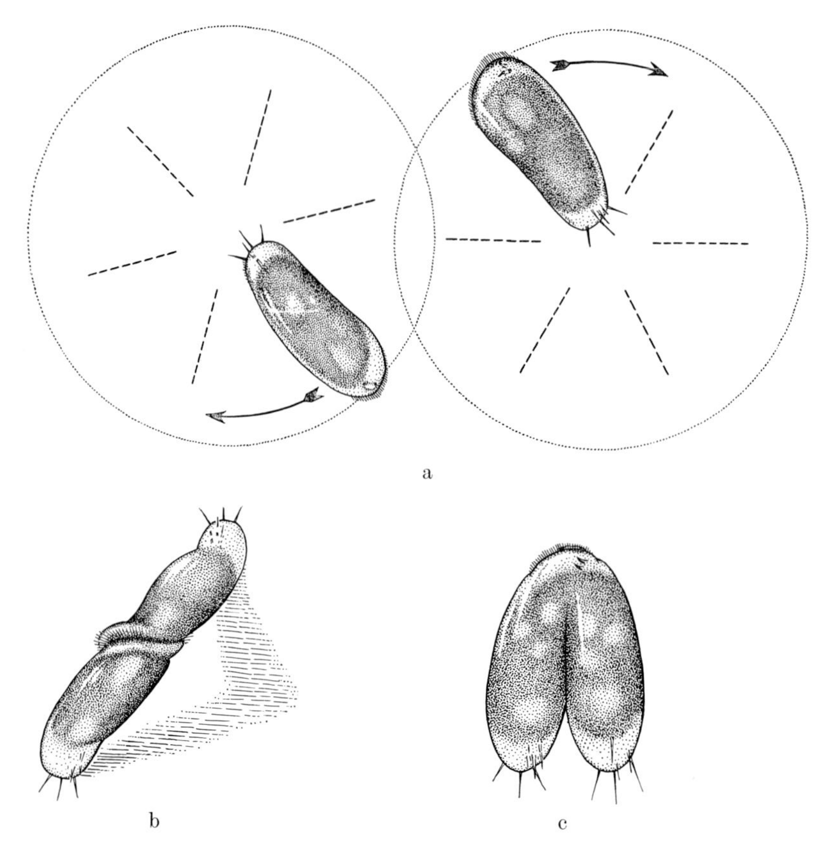

Fig. 280. *Stylonychia mytilus*. Mating play preceding conjugation. a Jerky rotation in circles by the two partners. b Touching of the peristomes. c Final fusion. After GRELL, 1951 [*652*]

In some cases the change in behavior is not habituation but a switch to a totally different behavioral pattern. If a suspension of India ink or carmine is directed at a *Stentor*, it soon turns its front end away from the source of stimulation. However, if stimulation continues, the body cilia reverse the direction of their beating, driving the cloud of particles away from the peristome. More and more extended contractions follow if stimulation persists. Finally, the *Stentor* detaches from its substrate and swims to another place. In this case, the cell responds to the same constant stimulus with a sequence of four different reactions [*859*].

Higher animals are known to be able to store excitatory states due to stimuli as "engrams". In the course of their individual lives, they can form firm connections *(associations)* between engrams of different stimuli and are therefore able to "remember" combinations of stimuli. For instance, if a circumstance which itself evokes no reaction appears long enough in combination with a stimulus, the organism can also respond to that circumstance even if it is no longer accompanied by a stimulus.

Many attempts have been made to demonstrate the ability to form such associations *("learning")* in ciliates too. In several cases it could be shown later that the experiments were not carried out carefully enough. What appeared as "learning" turned out to be "sensitization" or "habituation": constant repetition of the same stimulus led to a temporary lowering or elevation of the threshold value for the stimulus in question [*63*, *64*, *136*]. There are other results, however, which are more convincing. In *Tetrahymena*, for instance, a firm connection between the tendency to perform an "avoiding reaction" at stimulation with an electric current impulse and the excitation by a flash of light could be established. While the cells do not react to light under normal conditions, they did it after repeated combination of the "negative" and the "neutral" stimulus, i.e. they performed an "avoiding reaction" without the electric current impulse when exposed to a light flash only. This association persisted for some time even after cell division [*139*].

K. Nutrition

Every organism is dependent upon the uptake of certain *nutriments* from which it can build up new body substances or gain energy necessary for its life processes.

Autotrophic organisms need only inorganic nutriments from which they synthesize organic substances using chemical (chemoautotrophs) or radiant energy (photoautotrophs). Only some Bacteria can live autotrophically by the use of chemical energy. All autotrophic Protozoa, like green plants, utilize the sun's light as a source of energy. They can synthesize carbohydrates from carbonic acid and water with the aid of certain pigments [*614*] for the absorption of light energy. The pigments are bound to special carrier structures, the plastids (p. 23). This process is called *photosynthesis*. From the carbohydrates thus formed and certain water-soluble inorganic compounds, the most important being those containing nitrogen and phosphorus, they can form the remaining organic substances, especially proteins and lipids. Such Protozoa can therefore be cultured in a purely inorganic solution in the presence of light.

Culture experiments with pure nutriment solutions [*704, 828, 829, 1114, 1353, 1357*] have, however, shown that numerous flagellates, though capable of photosynthesis, are dependent upon an external supply of certain organic compounds which they cannot synthesize themselves. Although these substances are needed only in trace amounts, growth is impossible without them in many chrysomonads, euglenoids and dinoflagellates. Examples of such "*growth factors*" are thiamine (vitamin B_1), cobalamin (vitamin B_{12}) and biotin.

Many "phytoflagellates" can, on the other hand, grow without photosynthesis if forced to live in the dark for prolonged periods. They can then satisfy their demand for nutriments by the uptake of organic carbon or nitrogen compounds if these are available in their natural environment or in the culture liquid.

Flagellates which depend upon organic compounds in spite of their ability to carry out photosynthesis are called *mixotrophic*. If, when lacking light, they can switch from auto- or mixotrophic to purely heterotrophic nutrition, they may be called *amphitrophic*. Since it is in many cases unknown to which type a plastid-containing flagellate belongs, the general term *phototrophic* might be the best to comprise all nutritional types of photosynthesizing flagellates.

Most Protozoa are *heterotrophic*. They cannot carry out photosynthesis and are fully dependent upon organic nutriments.

In many cases it has been possible to culture photo- or even heterotrophic Protozoa *axenically*, i.e. in chemically defined nutritive solutions without any other organisms. This revealed surprising differences in nutritive requirements even of closely related species. *Chlamydomonas reinhardi* and *Euglena gracilis*, for example, can be cultured in the dark if sodium acetate is available as a carbon source. *Chlamydomonas eugametos syn. moewusii* and *Euglena pisciformis*, on the other hand, are unable to utilize acetate and perish in the dark.

Some ciliates which feed on bacteria under natural circumstances could also be adapted to synthetic media. *Tetrahymena pyriformis*, for example, can be cultured in a sterile medium containing 10 amino acids, 7 of the B-vitamins, guanine, uracil and some inorganic salts. While this species requires vitamin B_6 (with the exception of the pyridoxin mutant, p. 265), *Tetrahymena vorax* can also be cultured without this substance [*807*].

Different modes of food intake can be distinguished in Protozoa, although they can neither be sharply delimited nor does one mode preclude the other.

1. Permeation

The penetration of dissolved substances into the cell can be referred to as permeation.

Since the cell envelope consists of at least a unit membrane (p. 14), its ability to permit passage of only certain nutriments from a variety of external substances *(selective permeability)* must be due to structural, chemical and physical properties of the unit membrane. Smaller molecules are generally more easily taken up than large ones which might be due to the structure of the unit membrane itself, acting as an "ultrafilter". Preferential passage of lipid molecules can be understood on the basis of its lipid content. For transport of electrically charged ions it must be of importance whether the membrane itself maintains a certain resting potential. It seems that special carrier molecules take part in ion transport. Experience with Bacteria suggests that specific enzymes (permeases) might be available, which break down some substances into smaller molecules capable of permeation. Although knowledge of these processes is still fragmentary, it is certain that many nutriments are not taken up by passive transport, which can be explained by the laws of diffusion and osmosis, but by *active transport* involving energy expenditure. Even when the cell envelope consists of only a unit membrane, it is unlikely that its structure and permeability characteristics are the same everywhere. This applies especially when the cell envelope has a more complex structure because of additional unit membranes or other layers. In such cases permeability is probably restricted to special points which appear in the light- or electron microscope as "pores" (although these are naturally not perforations of the cell envelope).

2. Pinocytosis

Another process, called pinocytosis (literally: cell drinking), also seems to play a role in the uptake of liquids. This term denotes the abscission of vesicles from the unit membrane into the interior of the cell.

Pinocytosis has been most thoroughly studied in diverse amoebae (*Amoeba proteus*, *Amoeba dubia*, *Chaos chaos syn. Pelomyxa carolinensis*, and others), but there are indications that it also occurs in other Protozoa [*166, 168, 169, 219–224, 802, 803, 1675–1677*].

In *Amoeba proteus* (Fig. 281), pinocytosis involves first the formation of thin, tubular invaginations from the cell surface into the cytoplasm. These disintegrate then into single vesicles which can in turn pinch off more vesicles. The diameter of the vesicles thus varies within wide limits (2 μm – 0.01 μm), the largest being still recognizable in the light microscope and the smallest only in the electron microscope.

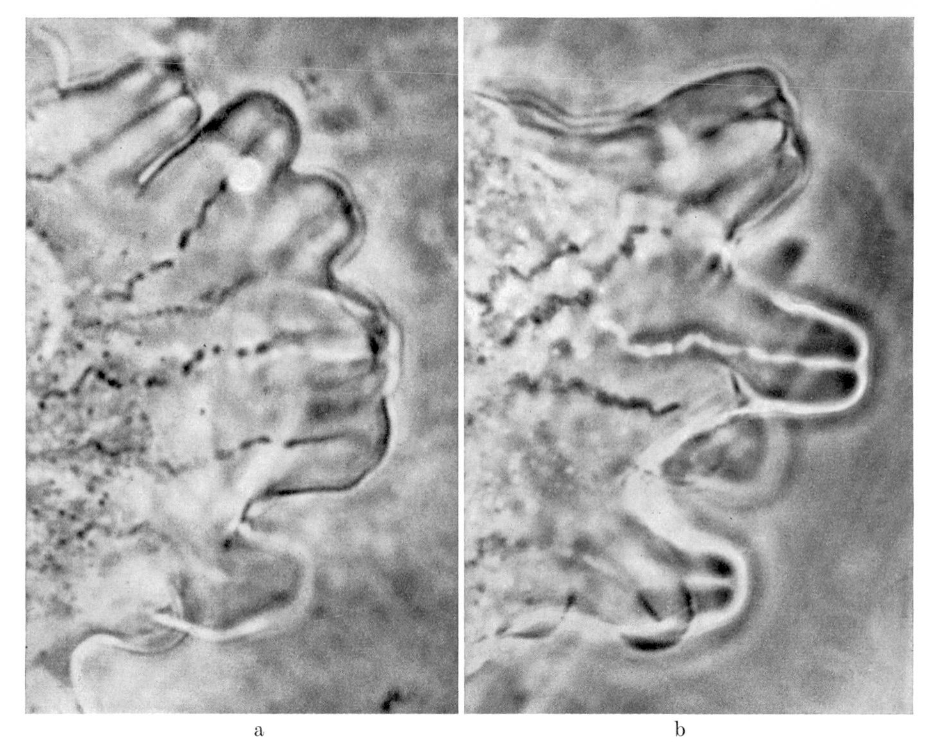

Fig. 281. *Amoeba proteus*. Formation of pinocytotic tubules which disintegrate into vesicles. Micrograph by D. M. PRESCOTT from HOLTER, 1959 [*802*]

As a rule, pinocytosis takes place at the tips of pseudopodia-like extensions, but in amoebae small vesicles can evidently also be pinched off at other points of the cell envelope.

It has been shown experimentally that macromolecules which cannot pass through the cell surface as such, can reach the cytoplasm by way of pinocytosis. Some substances, especially proteins and amino acids, call forth a lively pinocytotic activity. Such substances are first adsorbed to the cell envelope. The mucopolysaccharide layer (p. 40) probably plays a role in this. Other substances (e.g. carbohydrates) which do not by themselves induce pinocytosis, can in the presence of inducing compounds also pass into the cell. Experiments with labelled amino acids and sugars showed that these substances are distributed in the ground cytoplasm after pinocytosis. It seems that the wall of the vesicles becomes permeable to the adsorbed materials instead of breaking down right away once inside the cell.

Pinocytotic activity cannot last indefinitely. Once an *Amoeba proteus* has formed about 100 tubules in half an hour, pinocytosis stops suddenly and begins again only 3–4 hours later. If *Amoeba proteus* is fed with ciliates, it forms only about one tenth as many pinocytotic tubules after transfer to an inducing medium. Amoebae which have just completed a period of intense pinocytotic activity will, however, not accept food ciliates.

These observations show that pinocytotic activity, rather than representing merely a membrane phenomenon, is regulated by the cell itself.

Pinocytosis is probably widespread among species whose nutrition has so far been looked upon as "osmotic" ("osmotrophy"). It has been mentioned earlier (p. 43) that the opalinid, *Cepedea dimidiata*, constantly forms small pinocytotic vesicles at the base of the pellicular folds [*1238*].

3. Phagocytosis

The intake of solid food particles is termed phagocytosis. A sharp distinction from pinocytosis is naturally not possible because small particles may also be taken up in "cell drinking" and also because phagocytosis probably always entails uptake of some liquid.

Many Protozoa lack special organelles for food intake. Amoebae without definite polarity can take in food at any point of the cell surface. If they feed solely on bacteria, this process seems very simple indeed (Fig. 282): the amoeba flows, in a manner of speaking, around the bacterium and encloses it in a "food vacuole" formed in the same way as a pinocytotic vesicle, i.e. by infolding of the cell envelope. Enzymes which cause the eventual "lysis" of the bacterium and breakdown of its

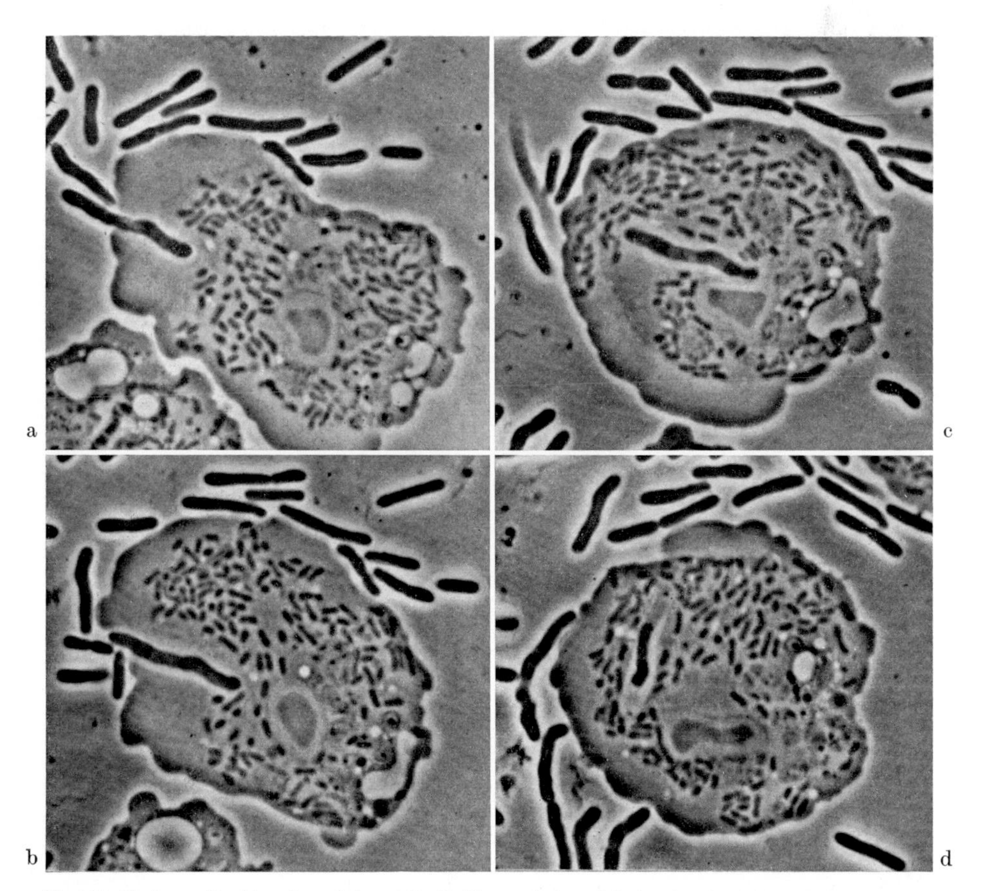

Fig. 282. *Hartmannella (Acanthamoeba) castellanii*. Phagocytosis and lysis of a bacterium. Cell nucleus and mitochondria are recognizable in the amoeba. × 1500. From the films C 943 and E 1169 (GRELL)

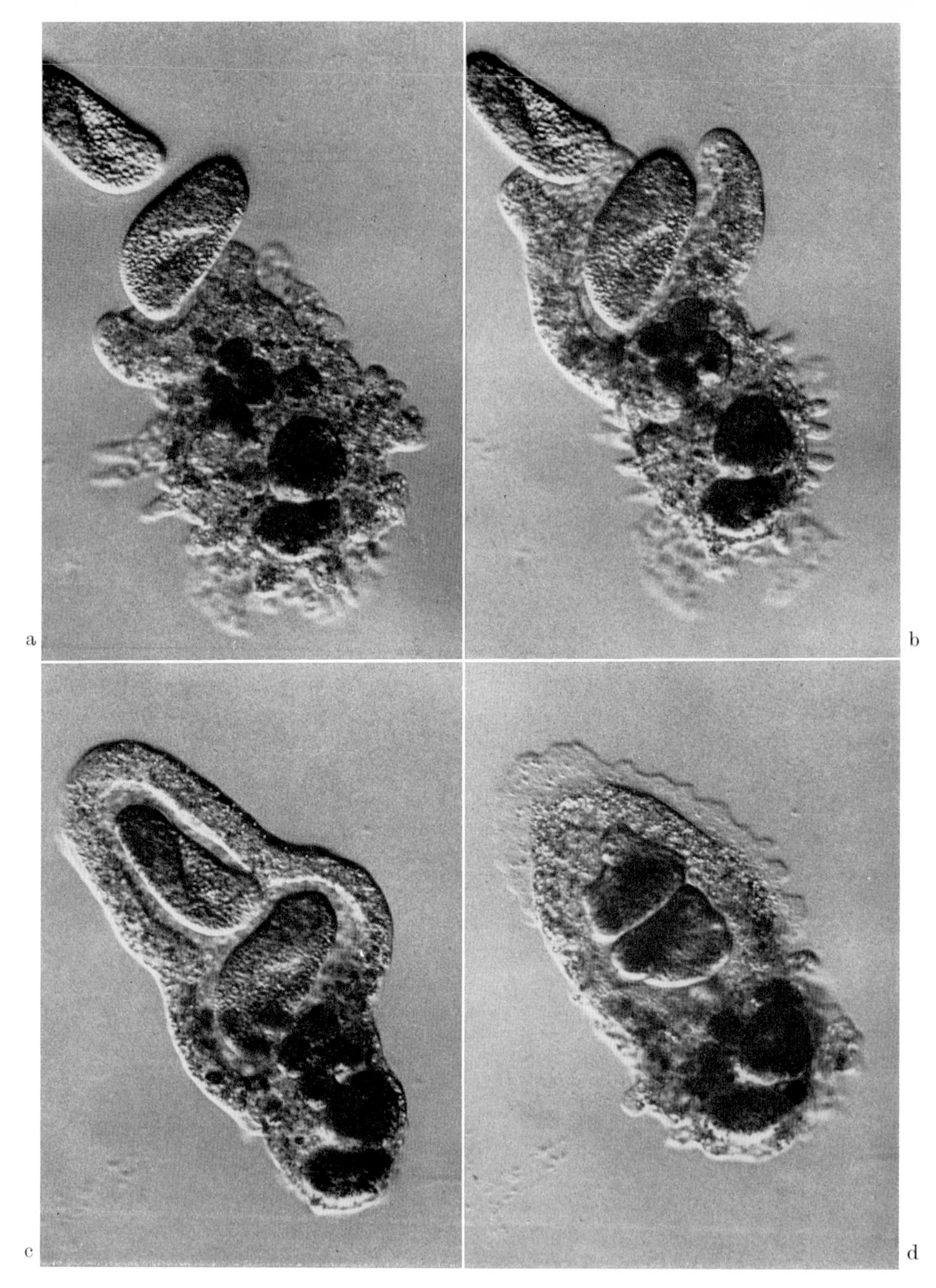

Fig. 283. *Amoeba proteus*. Phagocytosis of two cells of *Paramecium bursaria*. × 200. From the films C943 and E1171 (GRELL)

chemical substances into split products of lower molecular weight are discharged into the food vacuole by lysosomes. During digestion and resorption of the split products the food vacuole becomes smaller and smaller until it finally contains only indigestible material.

In an amoeba feeding on larger prey, the process of food intake is more complicated. Fig. 283 shows *Amoeba proteus* capturing *Paramecium bursaria*. While the paramecia remain completely immotile, they are gradually "encircled" by the amoeba which actively changes shape (circumvallation). An immobilizing influence obviously emanates from the amoeba which only touches the cilia of the paramecia. Only after the paramecia have been enclosed in the "food vacuole" will they attempt to escape. However, after a few violent motions they suddenly become rigid, assume a spherical shape and are subjected to the process of digestion.

Thecamoeba verrucosa is able to wind up a thread of the cyanophycean *Oscillatoria* of several hundred micra length in its cytoplasm. It is evident that such events require the coordinated activity of the whole cell.

Testacea enclose their food with pseudopods which emerge from the opening of the shell. In Heliozoa, Radiolaria, and Foraminifera the food organisms are not taken up directly into the cell body but stick to the axopodia or rhizopodia. They are then carried to the cell by surface cytoplasmic streaming or by contraction of the cytoplasmic strands composing the pseudopodia (p. 283). According to some descriptions, digestion can occasionally begin at the still extended pseudopodium.

In the heterotrophic flagellates food intake is usually restricted to a definite surface region due to the heteropolar organization of the cells. Many chrysomonads (e.g. *Chromulina*) pick up their food by means of a pseudopodium at the anterior end. *Peranema trichophorum* (Fig. 322a) attacks other euglenoids provided that they are stationary. The so-called rod organelle protrudes when the prey touches the predator: While the anterior end of the *Peranema* dilates, the rod organelle, through continual attachment and re-attachment, pushes the prey into the cell's interior. In some dinoflagellates (e.g. *Oxyrrhis marina*) the process of food uptake is not exactly known. Several pelagic species (*Noctiluca*, Fig. 325a; *Erythropsis*, Fig. 272) have motile tentacles acting like "lime-twigs" in food capture.

While choanoflagellates (Fig. 330) use their "collar" for capturing food particles, other "Protomonadina" may have a special "cytostome", sometimes continuing into a "cytopharyngeal tube" (e.g. *Ichthyobodo necator*, Fig. 331). Electron microscopic studies have shown that even members of the family Trypanosomidae, for a long time regarded as "osmotrophic", have a cytostome where pinocytotic vesicles are produced (e.g. *Trypanosoma mega*, *T. conorhini*). In *Trypanosoma raiae* the cell envelope forms a deep invagination near the flagellar pocket. Experiments with ferritin — an electron dense tracer — show that pinocytosis occurs from the blind ending of this invagination [*1352*]. In contrast to most other heterotrophic flagellates, hypermastigids living in the termite gut take up their food (small pieces of wood) at their posterior end.

"Micropores" found in the pellicle of various developmental stages of Sporozoa evidently serve also in the exchange of substances [*6–8, 82, 242, 1460, 1463, 1466, 1467*]. Erythrocytic stages (schizonts, gamonts) of *Plasmodium* species have regularly one such micropore whose function is undoubtedly to permit passage of host cell cytoplasm into the parasite (Fig. 284a). Erythrocytic cytoplasm is taken

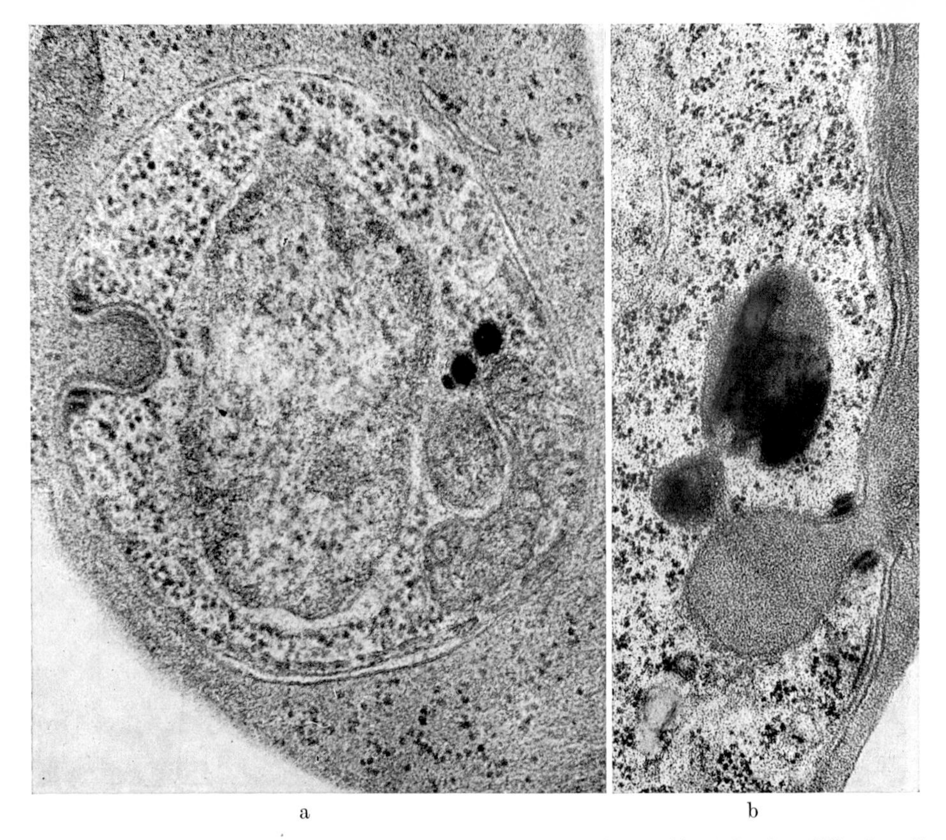

Fig. 284. *Plasmodium cathemerium*. Micropores (microcytostomes). a Young schizont. In the middle, the cell nucleus. × 56000. b Part of a gamont. Above the incorporated erythrocytic cytoplasm is a "food vacuole". × 43000. After AIKAWA, HEPLER, HUFF, and SPRINZ, 1966 [7]

up in a droplet-shaped invagination of the cell envelope (unit membrane) which is then pinched off as a "food vacuole". The whole process looks much like the formation of a pinocytotic vesicle. However, it cannot be simply called pinocytosis since the "micropore" is evidently a permanent differentiation of the stages in question. Its rim is formed by a structure of two electron dense rings. Hence, we are actually dealing with a kind of "cytostome".

In bird malaria plasmodia digestion takes place inside the food vacuole as evidenced by a dense deposition of hemozoin, the remnant of hemoglobin digestion (b). Newly formed food vacuoles are of the same density as host cytoplasm. Gradually with digestion progressing, they become less dense, but more of the hemozoin pigment aggregates in them. In mammalian plasmodia, however, small vesicles become detached from the food vacuole, later floating freely in the cytoplasm. The presence of the pigment inside them indicates that they are the sites of digestion [*1460*].

Possession of a *cell mouth* or *cytostome* is a characteristic especially of Euciliata, although some groups, such as the Astomata, have secondarily lost the mouth again. The euciliate cell mouth – strictly speaking, not an opening but a cortical region specialized for food intake – leads to a pharynx-like formation which is, however, filled with cytoplasm. It may be reinforced by rod- or thread-like differentiations (trichites) and is designated as *cell gullet* or *cytopharynx*.

The position of the cytostome and the differentiation of the functionally associated cell region is extraordinarily diverse in ciliates. A starting point is provided by the "Prostomata" (Holotricha, Gymnostomata) whose cytostome is located at the front end and at the surface (Fig. 405). In most ciliates the cytostome is either displaced from the front end or set deep into a "buccal cavity".

With respect to the mode of food intake two types can be distinguished, gulpers and swirlers.

Gulpers [*440*] feed on large prey taken in with the cytostome. It can therefore be greatly expanded and is located either at the front end or the side of the cell body. Some gulpers *(Nassula, Chlamydodon)* feed on filiform Cyanophycea which they can pinch off with the aid of a complex organelle, the so-called *cytopharyngeal basket*. Electron microscopic studies [*1745*–*1751*] show that it comprises dif-

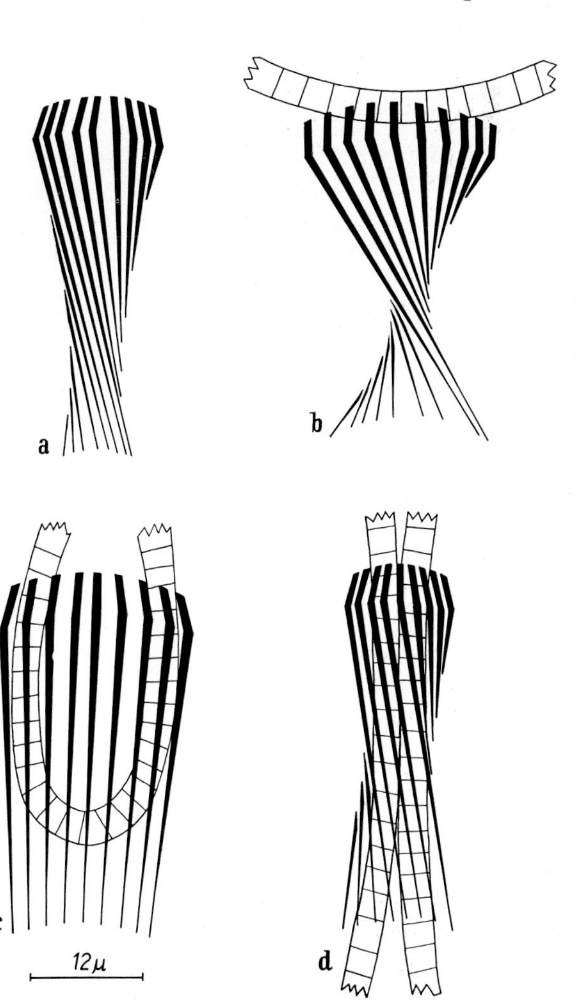

Fig. 285. *Nassula*. A series of scale drawings showing the arrangements of the rods (black) in the near side of a basket as they appear in a living specimen at different stages in the ingestion of an algal filament. After TUCKER, 1968 [*1745*]

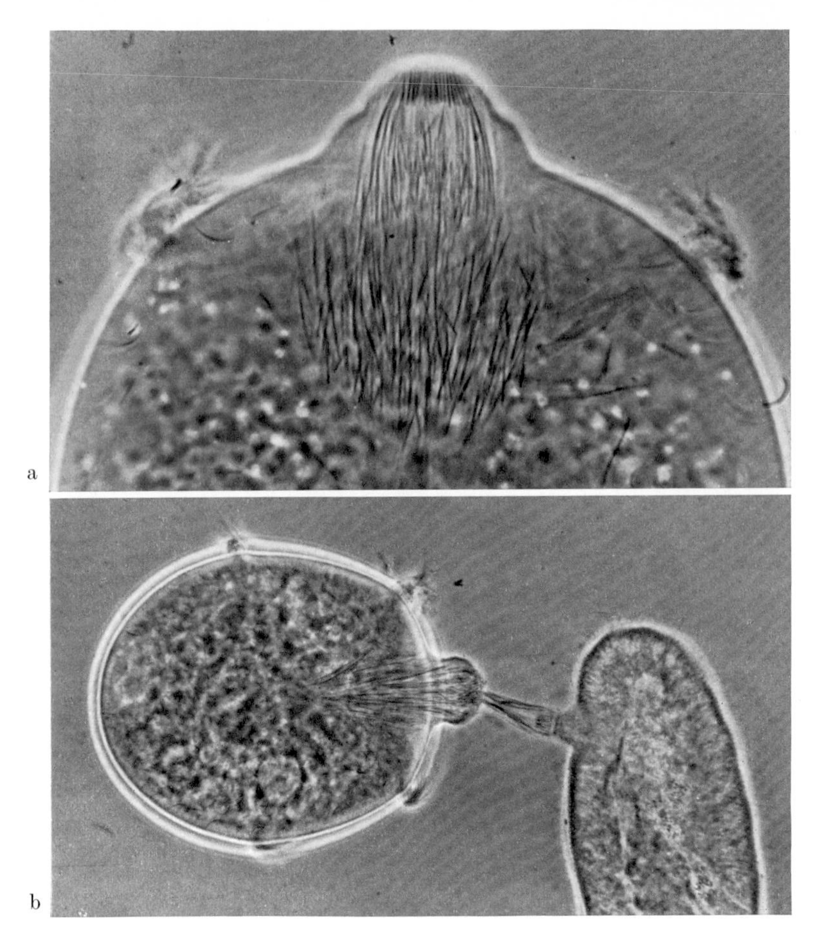

Fig. 286. *Didinium nasutum*. a Anterior end. Under the top of the oral cone: pexicysts. The longer rods are toxicysts. × 500. b *Didinium* and *Paramecium*, fixed shortly after contact has been established. During preparation the trichocysts of the *Paramecium* were lost. × 350. After SCHWARTZ, 1965 [*1566*]

ferent structural components which consist predominantly of microtubules. The most prominent component to be seen also with the light microscope is a circular palisade of rods. As Fig. 285 illustrates, ingestion of an algal filament is brought about by movement of the rods which perform different feeding positions. Displacement of the rigid rods is enabled by antagonistically acting contractile structures.

Another ciliate, *Coleps hirtus*, engulfs dead cells to which it is attracted by chemotaxis. Most gulpers, however, take live prey and are often specialized for certain species.

Didinium nasutum feeds exclusively on paramecia. They are not actively sought but are caught in accidental collisions [*1566*]. It can happen that several didinia chance upon the same *Paramecium*.

The front end of *Didinium* tapers to the so-called oral cone. If the latter contacts prey, it attaches immediately to it. Contact is established by short rods below the top of the oral cone, called pexicysts. The considerably longer and more peripherally located rods are toxicysts which are discharged only after contact has been established (Fig. 286a). The attacked *Paramecium* discharges its trichocysts. Even if it can push *Didinium* somewhat back in doing so, the pexicysts maintain contact but are occasionally only connected to the oral cone by a cytoplasmic thread (b). Experiments have shown that the following reaction, engulfment of the prey, is induced by a substance diffusing from the cytoplasm of the *Paramecium* which is at first only killed locally. The prey is finally pulled into the *Didinium* through streaming of the cytoplasm attached to it (film C881).

Pictures taken with the scanning electron microscope give an impression of the dramatic event when *Didinium* engulfs a *Paramecium*. Even if the predator is smaller than the prey, the former is able to ingest the latter completely (Fig. 287). After ingestion the temporary cytostome of the predator closes and the oral cone is reformed [*1793, 1794*].

Dileptus anser has a lateral cytostome, the tapering part of the cell body in front of it forming a movable "proboscis" (Fig. 288a). The latter contains numerous toxicysts at the oral side (b) which are immediately discharged when the proboscis touches a prey, e.g. *Colpidium*. The discharged threads of the toxicysts transfer a highly effective poison to the prey which usually dies and disintegrates right away. It seems that the poison itself causes only local lysis of the pellicle. The supply of toxicysts in the proboscis is sufficient to kill 70 colpidia. *Dileptus* needs about two hours to replenish its supply after all have been discharged. If prey is plentiful, it kills many more than it eats. When hungry, it glides along the cadaver and widens the cytostome, which is surrounded by a thick torus, even before the prey touches the mouth. In the process of "swallowing", the prey is taken into a large food vacuole (Fig. 289). The cytostome is finally closed, probably by a sphincter-like myoneme (film C881).

Swirlers create a water current which carries food to the cytostome. The latter is therefore always surrounded by a special region, the *buccal field* or *peristome*, whose position, size and form determines the shape of the cell body. The peristome frequently narrows in front of the cytostome, forming the so-called *oral funnel* or *vestibulum*.

The cilia of the peristome are employed mainly to create a current carrying food particles to the cell. Their beat can be autonomous with respect to the body cilia (p. 303). Even among *Holotricha* (Hymenostomata) an undulating membrane appears in the vicinity of the cytostome. It consists of fused ciliary rows and directs the food stream more effectively than singly beating cilia.

In some species (e.g. *Pleuronema*) the undulating membrane has the shape of a sail which extends far out of the peristome (Fig. 415).

Further morphological differentiations are found in the *Peritricha*. The whole front end of the cell body is flattened to a disc, covered by a spiral of cilia turning counterclockwise toward the cytostome. It consists of two or three rows of cilia of which the outer one has become transformed partly into an undulating membrane.

The adoral band of membranelles of *Spirotricha*, which typically surrounds the whole peristome, has already been mentioned (p. 304). In creating currents for

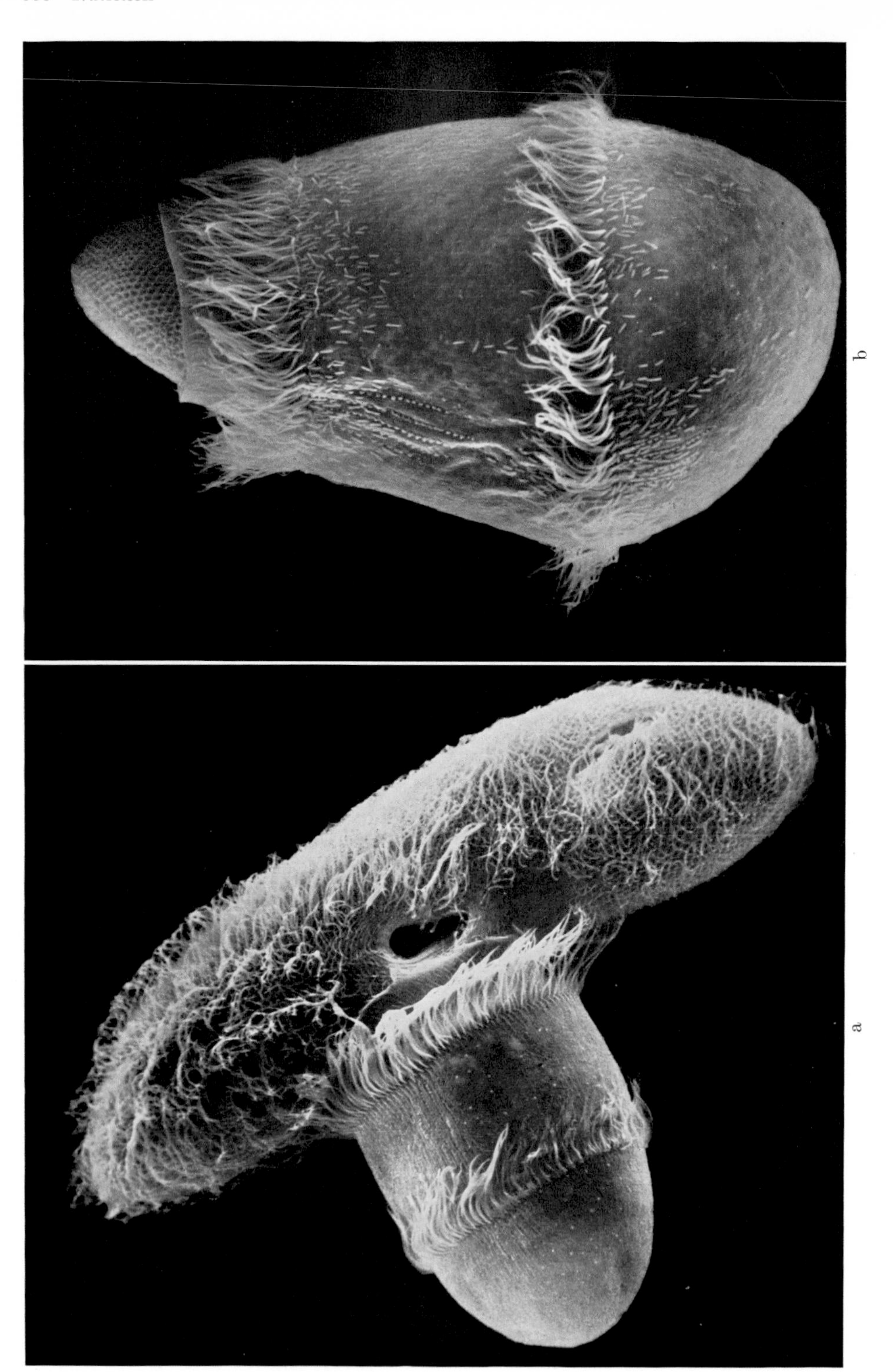
b
a

Fig. 287. *Didinium nasutum* ingesting *Paramecium multimicronucleatum*. a An early phase of ingestion showing the buccal opening of *Paramecium* just above the turned out oral rim of *Didinium*. × ca. 1200. b The *Paramecium* is almost swallowed and the *Didinium* is expanded. The oral rim has a smooth contour but does not touch the prey. × ca. 1260. Stereoscan-micrographs. a After WESSENBERG and ANTIPA, 1970 [*1794*], b Courtesy of the same authors

◀

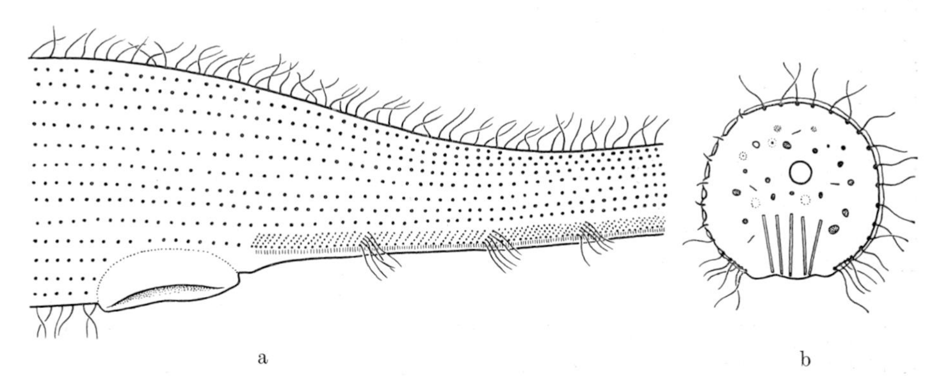

Fig. 288. *Dileptus anser*. a Region of the cytostome and proboscis. b Cross section through the proboscis. The toxicysts are located ventrally. After DRAGESCO, 1963 [*441*]

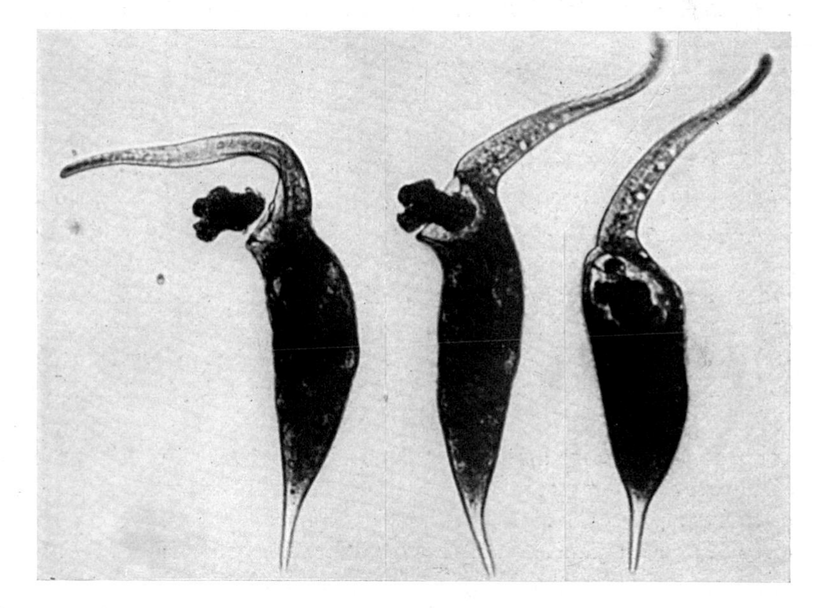

Fig. 289. *Dileptus anser*, devouring a previously killed prey. × 170. From the film C 881 (GRELL)

feeding, it is especially effective in semi-sessile *(Stentor)* and sessile species *(Folliculina)*.

Suctoria have *tentacles* rather than a single cytostome to take in food. Most species have many such tentacles, often arranged in two or more bundles. They arise as pin-shaped extensions from the cell surface and can be withdrawn and expanded (Fig. 429a, b). The tentacles of *Dendrocometes paradoxus* are of a special kind. They

Fig. 290. *Acinetopsis rara* tearing an *Ephelota* from its stalk with its prehensile tentacle. During contraction of its prehensile tentacle the *Acinetopsis* is still covered with remains of a previously captured *Ephelota*. × 130. From the films C 907 and C 912 (GRELL)

represent thick and stiff extensions of the cell which are branched into numerous points (Fig. 429c, d).

While the tentacles of most suctorians serve in food capture as well as food intake, the two functions are carried out by different types of tentacles in some cases.

Ephelota gemmipara, whose tentacles are not bunched but evenly distributed over the apical cell surface, has long, tapering *prehensile tentacles* and short, blunt *feeding tentacles* (Fig. 430a). Food organisms stick to the prehensile tentacles. These bend inward and hand the prey over to the feeding tentacles in the middle of the apical surface. The latter are also contractile, but they can only shorten or elongate, not bend to the side.

Another suctorian, *Acinetopsis rara* (Fig. 290), which feeds exclusively on *Ephelota gemmipara*, has usually only one (occasionally two) capturing tentacle. It can be extended to 1 mm length and carries out real "seeking movements". As soon as its button-like thickened end touches the pellicle of an *Ephelota*, it contracts with astonishing tensile force. Although *Ephelota* is so firmly attached that its stalk bends towards the predator, the latter will eventually succeed in tearing it away. Only after the fully contracted capturing tentacle has pulled the cell body of *Ephelota* onto its apical surface does *Acinetopsis* extend numerous short feeding tentacles which take in the cytoplasm of the prey. With every act of feeding, a swarmer develops within *Acinetopsis*. It often leaves the mother cell even before the prey has been devoured completely.

In Suctoria whose tentacles serve both in food capture and intake, the prey (chiefly ciliates) touches by chance the swollen capitate end of the tentacle (tentacle knob) and remains attached to it. Sometimes the ciliates succeed in escaping, especially if they are larger than the suctorian. Usually, however, they come into contact with more tentacles while struggling and escape becomes impossible. It has been observed in many cases that the prey becomes immobilized after attachment. If the prey is a ciliate, the cilia are successively immobilized, beginning at the point of contact. This effect need not be irreversible. If the prey is artificially detached shortly after paralysis, it can become motile again after some time [*1458*].

Electron microscope studies [*91*] have shown that contact is established by the *haptocysts* described earlier (p. 36), which are present only in the membrane of the tentacle knob (Fig. 291). It seems possible that substances which immobilize or lower cytoplasmic viscosity are transferred via the haptocysts into the prey. However, their main function must be to establish contact since they also occur in parasitic Suctoria (e.g. *Podophrya parameciorum*) which do not immobilize their hosts [*882*].

The suctorian *Phalacrocleptes verruciformis* which lives on the pinnulae of the prostomial cirri of the polychaete, *Schizobranchia insignis*, has numerous very short tentacles distributed all over the cell body. Each tentacle contains only a single haptocyst which establishes contact with a cilium of the host [*1054*].

As soon as the prey is attached, the tentacle contracts and becomes thicker (Fig. 292a). Its external surface is formed by the tentacle sheath which is continuous with the pellicle of the cell body and is drawn into folds in the course of contraction (b). It must be stressed, however, that the tentacle sheath corresponds

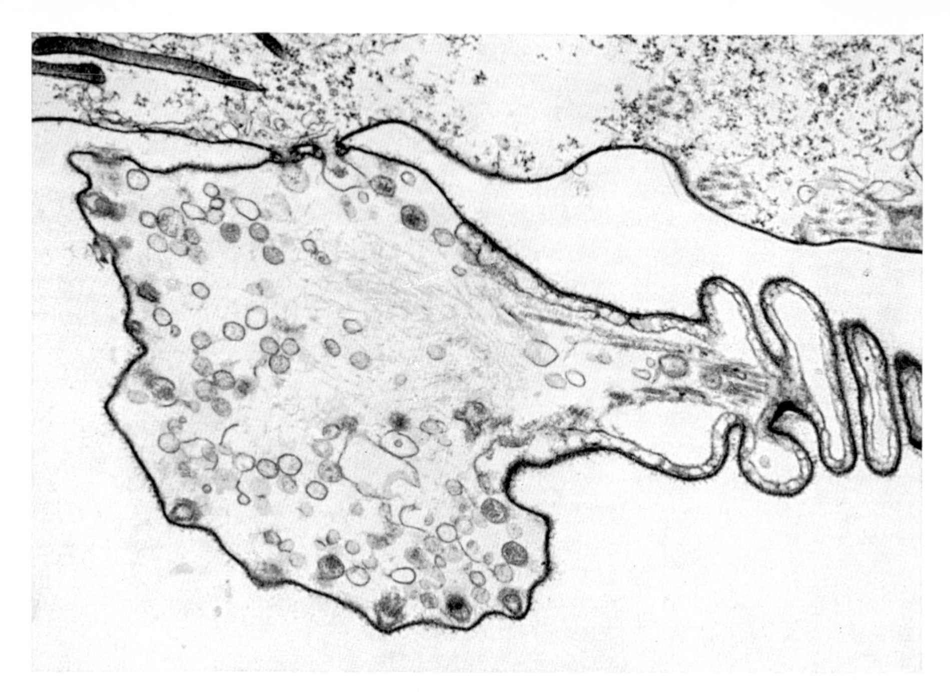

Fig. 291. *Acineta tuberosa*. Contact of a tentacle knob with prey *(Strombidium)*. Two haptocysts, which establish the connection with the prey, are recognizable in this section. Sections of inactive haptocysts below the membrane of the tentacle knob. × 40000. After BARDELE and GRELL, 1967 [*91*]

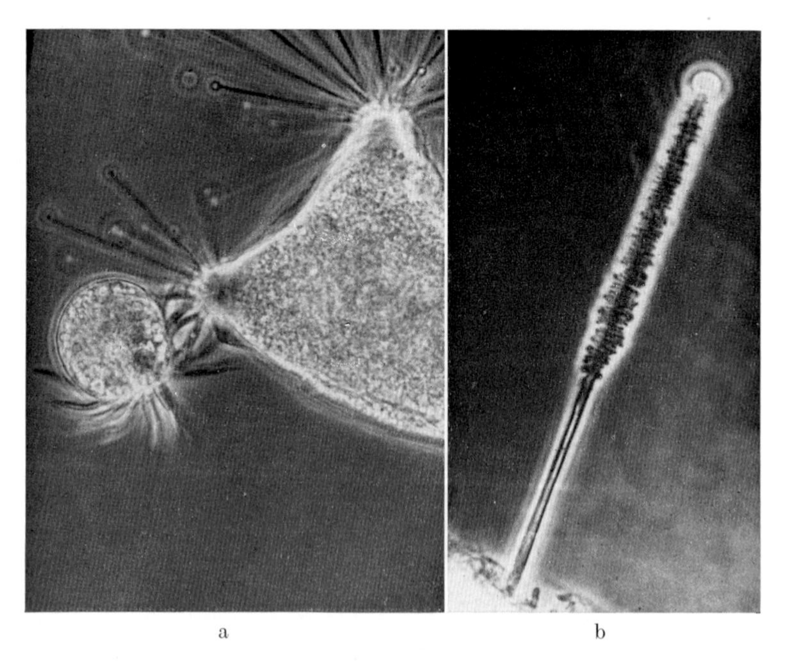

a b

Fig. 292. Suctorian tentacles. a *Acineta tuberosa*, capturing *Strombidium*. b Prehensile tentacle of *Acinetopsis rara* in a highly contracted state. × 560. From the film C912 (GRELL)

in structure to the pellicle only below the tentacle knob. The latter is bordered by a simple unit membrane.

It was formerly assumed that the tentacle would be traversed by a thin "tube" which merges distally with the wall of the tentacle knob. However, electron microscopy revealed that the "tube" is not formed by a continuous membrane. It consists of many separate microtubules whose number and arrangement vary from species to species. In tentacles serving both in food capture and feeding, the microtubules are arranged in a roughly circular pattern (Fig. 293a). The same holds true for the feeding tentacles of *Ephelota*. The prehensile tentacles of *Ephelota*, however, contain two or three continuous groups of microtubules separated by a septum (b). Since these tentacles cannot feed, but only carry out bending movements, it seems likely that the microtubules are responsible for these movements.

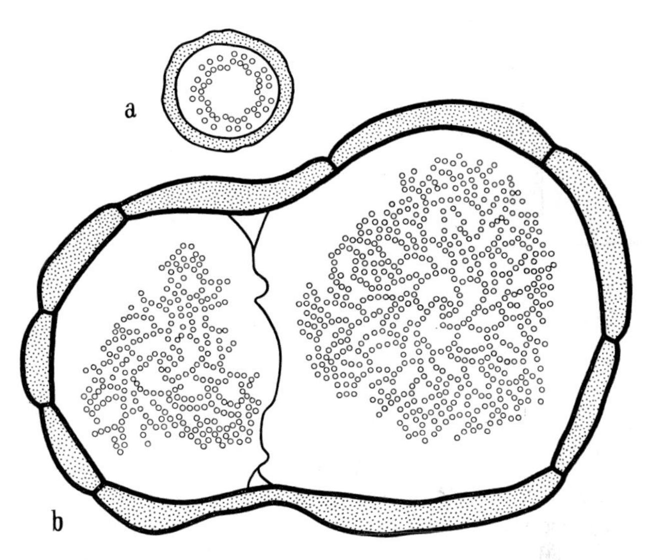

Fig. 293. Cross sections through suctorian tentacles. Drawn from electron micrographs. a Prehensile and feeding tentacle of *Tokophrya infusionum*. × 24500. After RUDZINSKA, 1965 [*1458*]. b Prehensile tentacle of *Ephelota gemmipara* with two groups of microtubules. × 22500. After BATISSE, 1965 [*95*]

Electron micrographs suggest that a feeding tentacle is closed when inactive (Fig. 294). In the act of feeding, it becomes a "tube", as the membrane of the tentacle knob is drawn along into the interior of the cell. The cylindrically arranged microtubules (Mt) may serve as abutment or are responsible for transportation of the membrane. Food vacuoles are pinched off where the microtubules terminate in the cytoplasm. According to this view, food intake by Suctoria is basically a specialized form of phagocytosis.

How Suctoria produce the force which enables them to take in large amounts of food in a short time through their tentacles is still a problem [*826, 827, 938*]. An important approach to solve this problem is the finding that the transportation of the membrane enclosed food seems to be mediated by microtubules [*90, 1751*].

Food materials are never taken directly into the cytoplasm but are always enclosed in a vacuole. In ciliates which feed on small particles, the vacuole forms at the

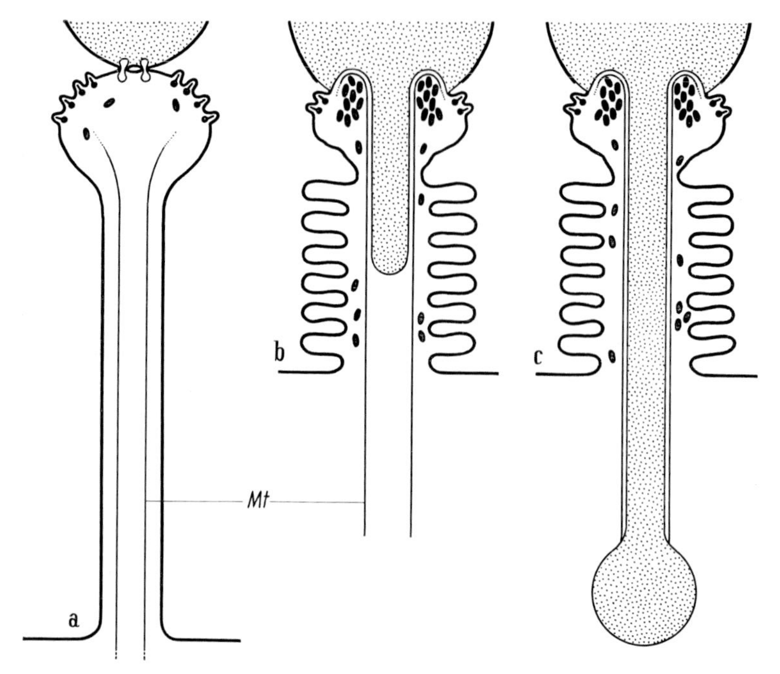

Fig. 294. Diagram of food intake through a feeding tentacle. a Contact with the prey by way of haptocysts. b. c Feeding process: the apical membrane of the tentacle knob surrounds the inflowing cytoplasm of the prey. It forms the wall of the food vacuoles which detach where the ring of microtubules (*Mt*) ends. In the tentacle knob are osmiophilic granules which probably play a role in the re-formation of membrane material. After BARDELE and GRELL, 1967 [*91*]

cytostome and is called "*receiving vacuole*". Since its formation can also be induced by indigestible particles (e.g. carmine or latex granules), it is obviously due to mechanical touch at the cytostome and not to a chemical stimulus. This is probably the reason why *Tetrahymena pyriformis* grows better if the axenic medium is improved by addition of insoluble particulate material.

When the "receiving vacuole" has become filled to a certain degree, it detaches from the cytostome and travels as a *food vacuole* within the cytoplasm. Often, e.g. in *Paramecium* and *Vorticella*, the food vacuole follows a definite path determined by cytoplasmic streaming. In the course of this travel *(cyclosis)*, *digestion* takes place within the food vacuole.

In *Paramecium* it can be shown with suitable indicator dyes, for example Kongo red, that the content of a food vacuole is at first alkaline, then acidic and later alkaline again. Acidity in the second phase corresponds to about 0.3% hydrochloric acid, sufficient to kill the bacteria.

The digestive enzymes (proteases, carbohydrases, esterases) secreted into the food vacuole seem to be available in the cell in a definite quantity rather than being synthesized when food is taken in. The content of cytochemically demonstrable acid phosphatases, at any rate, is the same whether food vacuoles are present or not [*1162—1167*].

Recent studies [*490*] have shown that digestive enzymes synthesized at endoplasmic cisternae accumulate in small vesicles, the so-called *lysosomes*. These surround the outer surface of food vacuoles in order to deliver their contents to them.

In *Tetrahymena pyriformis*, which forms fewer food vacuoles in an axenic medium (p. 328) than in one containing bacteria, enzyme activity increases with the number of food vacuoles. However, it is unimportant in this respect whether the food vacuoles contain any digestible material. The amount of acid phosphatases is the same whether the food vacuoles contain indigestible polystyrene latex granules or bacteria [*1164*].

The food vacuole becomes smaller and its contents become more concentrated as digestion proceeds. The vacuole content is finally liquified to a large extent, and the vacuole swells a little again. It has frequently been observed that small vesicles are pinched off from the wall of the food vacuole which become distributed in the cytoplasm [*508, 509, 881, 1163, 1446, 1516*].

In *Metafolliculina andrewsi* the wall of the food vacuole forms numerous folds leading to considerable surface increase which favors permeation [*1757*].

After resorption of breakdown products, indigestible residues are voided to the outside as feces. In ciliates, defecation takes place via a special differentiation, the *cell anus (cytopyge)*, frequently located within or in the vicinity of the peristome. It is not morphologically distinguishable, and therefore it usually becomes recognizable only in the course of defecation.

The *pulsating vacuoles* [*939*] found in many Protozoa differ from food vacuoles by the absence of solid particles and by their rhythmic activity. They increase periodically, taking up liquid from the surrounding cytoplasm *(diastole)*, and empty their contents to the outside *(systole)*.

In *amoebae* which constantly change shape, it is often difficult to decide whether the pulsating vacuole forms either at any spot or in a given region of the cytoplasm. *Thecamoeba verrucosa* (Fig. 346a), which has a definite polarity, always forms it in the posterior cell region (uroid).

The formation of the pulsating vacuole (Fig. 295) seems to proceed in some amoebae from small vesicles which fuse successively like droplets of liquid. New vesicles appear at the same place as soon as the pulsating vacuole has voided its contents to the outside. It can, however, be demonstrated in some of the larger species (e.g. *Thecamoeba verrucosa*), whose pulsating vacuole can be removed undamaged from the cell, that it is not simply an accumulation of liquid but is bounded by an elastic, semipermeable membrane. Electron micrographs show the latter as a unit membrane [*1268, 1269*]. Rhythmic appearance and disappearance must therefore be connected with continuous reformation and breakdown of membranes. Since the isolated pulsating vacuole of *Amoeba proteus* contracts if treated with ATP and Mg^{++} ions, the force for systole must be generated at least in part by the wall of the vacuole. However, the structural basis for contraction is still unknown in this case. In other species of amoebae fibrils bordering the vacuole have been reported [*1358*].

The boundary of pulsating vacuoles is less clearly defined in Heliozoa, whose outer cell layer is already highly vacuolated. In *Actinophrys* and *Actinosphaerium* some of these vacuoles are seen to open periodically to the outside. This is the only distinction from the many "sap vacuoles" of the cortical layer.

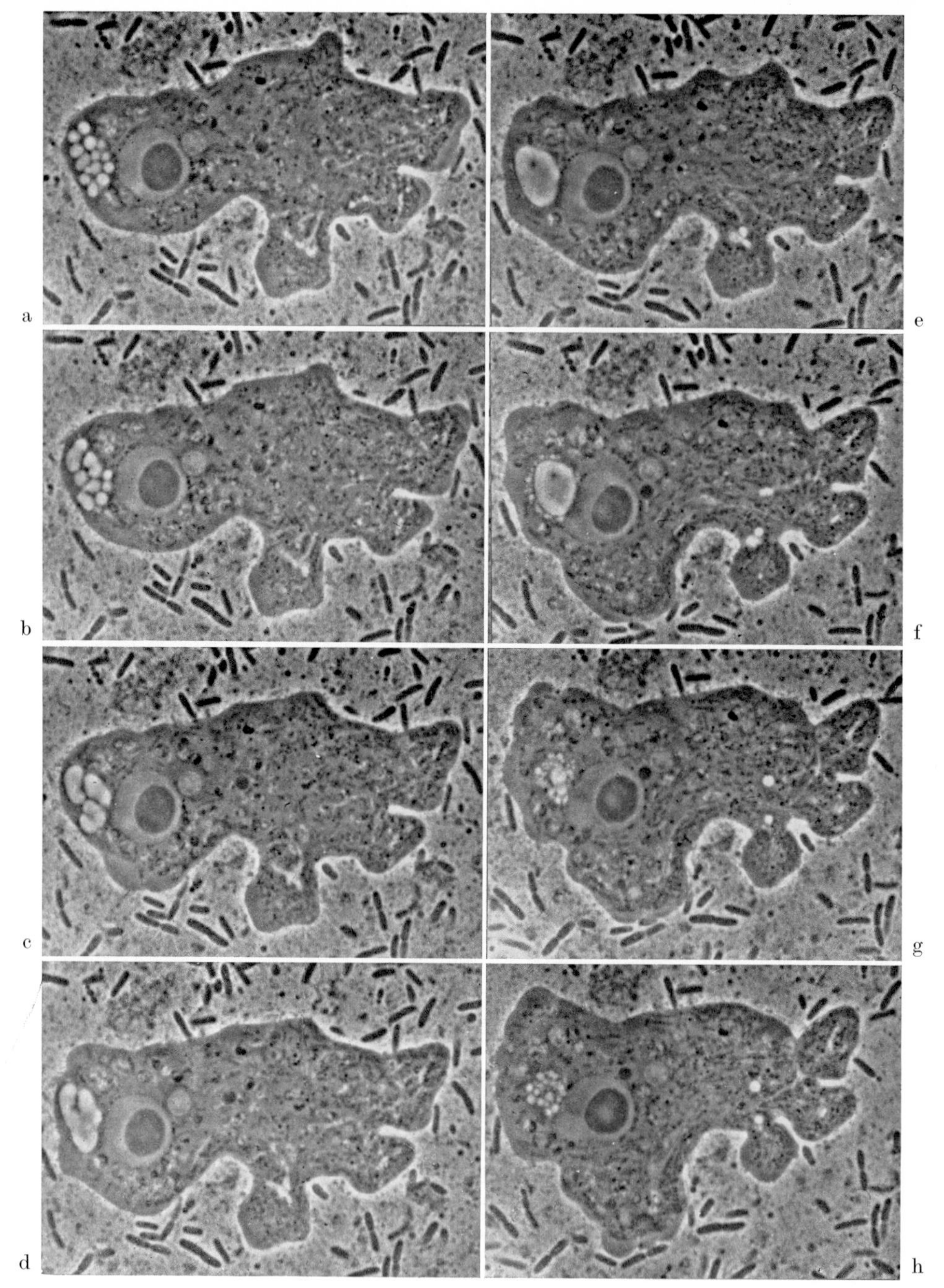

Fig. 295. *Naegleria gruberi*. Cycle of the pulsating vacuole (left of the cell nucleus). × 670. From the films C 942 and E 1170 (GRELL)

A definite position of pulsating vacuoles is characteristic for *flagellates*. Phytomonads often have two vacuoles at the front end (Fig. 313). In euglenoids the vacuole forms from many small vesicles (Fig. 320) and empties its contents into the flagellar sac.

The pulsating vacuoles of *ciliates* are of a special type. They appear usually in constant number and void their contents through a preformed pore in the pellicle.

Paramecium caudatum (Fig. 413) has two pulsating vacuoles which work alternately: while one is in diastole, the other is in systole (Fig. 296). The so-called *reservoir*, a large central bladder, is filled by radially arranged *feeder canals*. A feeder canal consists of a pointed end portion where liquid is taken in, an ampulla-like middle portion which swells up before diastole and a short, undilatable connective neck through which liquid is pressed into the reservoir. Light-microscopic observations indicated that the end portions of the feeder canals are surrounded by a specialized cytoplasm [*566*]. Electron microscopy showed this "nephridioplasm" to consist of a wickerwork of branched tubules in connection with endoplasmatic cisternae (Fig. 297, *NT*). During diastole (a) the connections between tubules and

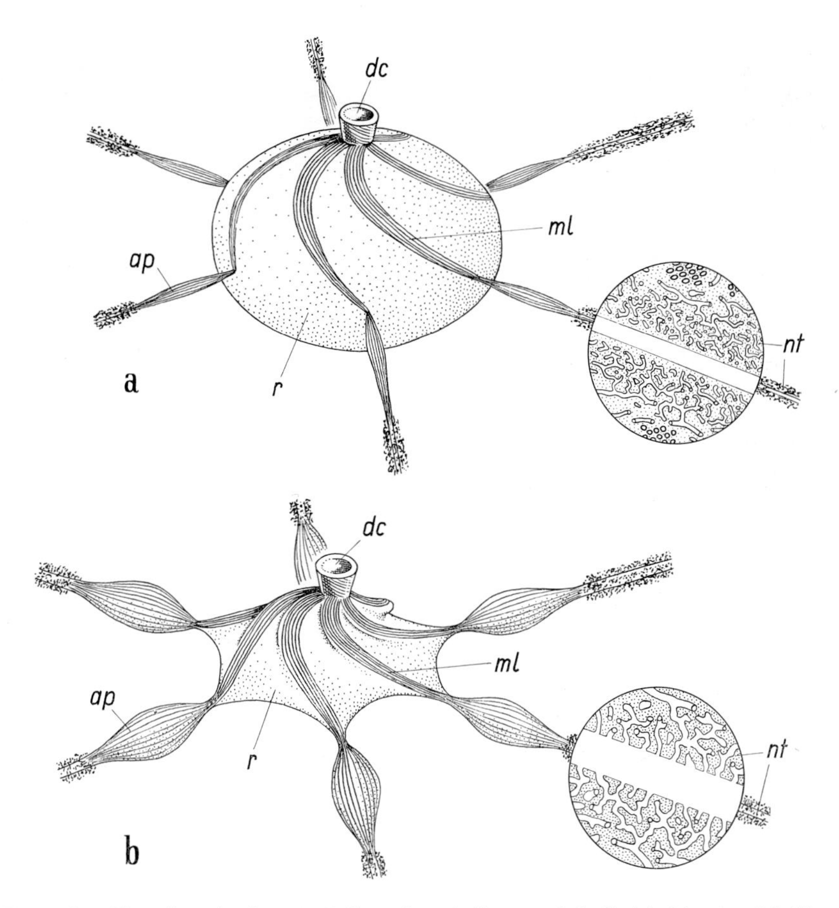

Fig. 296. *Paramecium*. Three-dimensional representations of a pulsating vacuole in diastole (a) and systole (b). *ap* Ampulla of a feeder canal. *r* Reservoir. *dc* Discharge canal. *ml* Bundles of fibrils running from the wall of the discharge canal down the wall of the reservoir and then along the outside of the feeder canal. Enlarged circular inset into a feeder canal shows the corresponding ultrastructure of the "nephridioplasm" (comp. Fig. 297). After JURAND and SELMAN, 1969 [*885*]

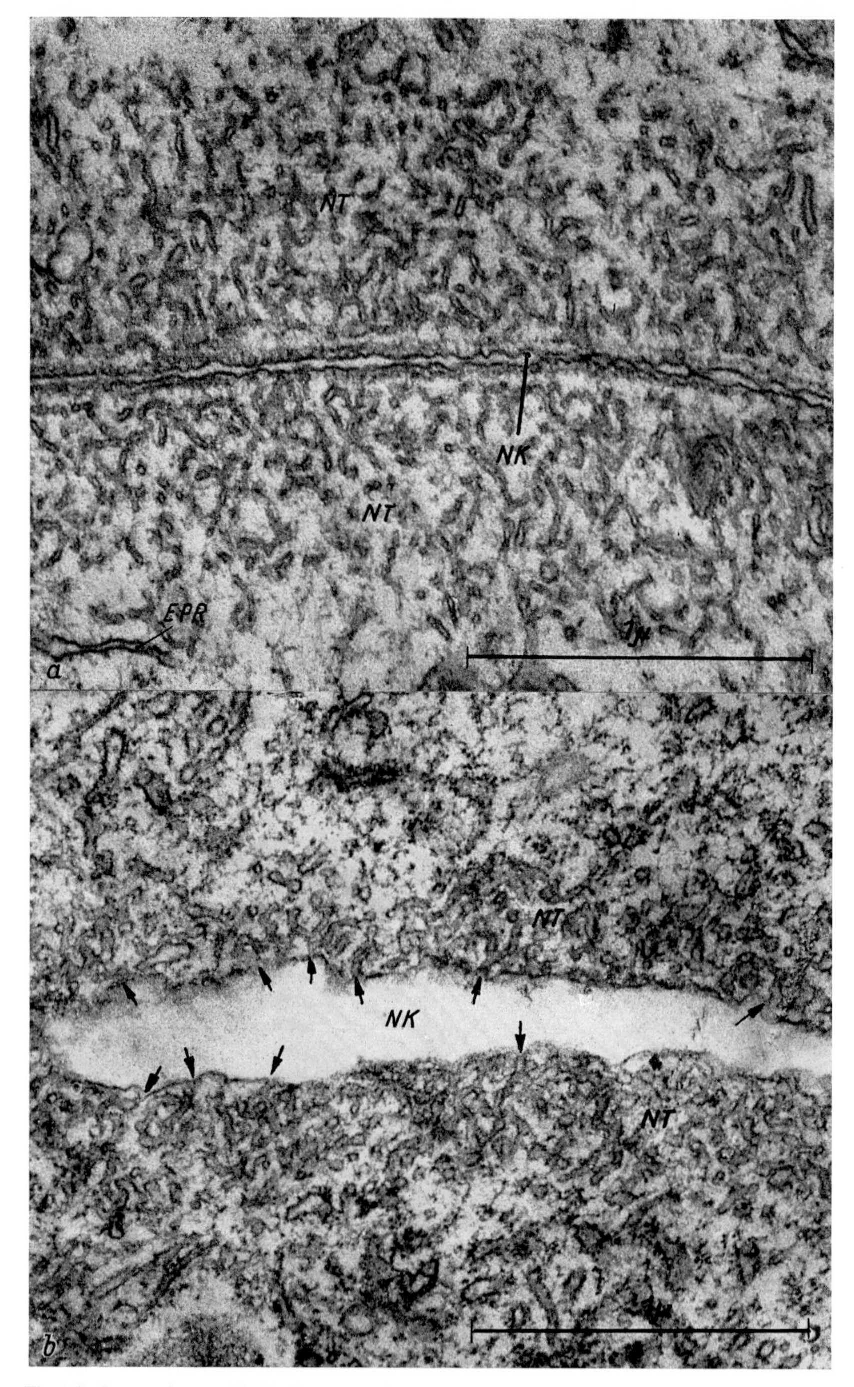

Fig. 297. *Paramecium aurelia.* Region of a feeder canal of the pulsating vacuole. a Diastole, b systole. In the latter case, the tubules of the "nephridioplasm" are open to the feeder canal (arrows). Magnification approximately 50000. After SCHNEIDER, 1960 [*1514*]

the feeder canal (*NK*) are closed. During systole (b) the communication is open (arrows!). Tapelike fibrillar bundles thought to be endowed with contractility are arranged at the connecting necks, the reservoir and the canal leading to the pore [*1514*].

The vacuole of *Spirostomum ambiguum* fills nearly the whole posterior end of the cell while the feeder canal which opens into it reaches to the front end (Fig. 418). *Stentor coeruleus* has a pulsating vacuole with two feeder canals, one directed longitudinally, the other encircling the margin of the peristome.

In most ciliates the pulsating vacuoles are of a simpler structure, i.e. they lack feeder canals. The tubules which have been demonstrated in several species [*485*] thus lead directly into the vacuole. The pore through which the vacuolar contents are discharged to the outside is located at the base of a pellicular invagination whose wall is connected to the membrane of the vacuole by fine fibers. In the suctorian, *Tokophrya infusionum,* a thin tube is inserted between vacuole and pellicular invagination.

While the vacuole is discharged (systole) in a fraction of a second, its filling (diastole) takes much longer. Considerable differences, partly due to vacuole size, exist between species. An *Amoeba proteus,* whose vacuole reaches a diameter of about 50 μm, diastole takes about 5 min. In *Paramecium caudatum,* whose reservoir swells up to 5–10 μm, it takes 5–10 sec. On the other hand, the large vacuole of *Spirostomum ambiguum* empties only every 30–40 min.

In a given species, pulsation frequency depends upon a variety of conditions, especially temperature and salinity of the environment. Size and pulsation frequency determine the efficiency of the vacuole. The volume of liquid discharged by some ciliates in the span of 10 min equals that contained in the whole cell.

Functionally, pulsating vacuoles seem to be primarily organelles for *osmoregulation*. Since the salt concentration within the cells is higher than that of the environment, water passes continuously inward through the cell envelope. This would lead to excess pressure and swelling of the cell if water were not pumped out continuously by the pulsating vacuoles. It is in accordance with this interpretation that pulsating vacuoles are found especially in species inhabiting fresh water and that they are usually absent in Protozoa living in media of higher salt concentrations (marine species, parasites and symbionts). Sometimes they are present but pulsate so slowly that their activity can only be discerned in time-lapse photographs. If transferred to a medium of higher salinity, Protozoa decrease the pulsation frequency of their vacuoles. Some ciliates (e.g. *Frontonia*), on the other hand, seem to be able to adapt their osmotic pressure within limits to variations in salinity without pronounced changes of pulsation frequency. A mutant of *Chlamydomonas moewusii* which is unable to form pulsating vacuoles can only be kept alive in a medium with artificially increased osmotic pressure [*702*]. The mutation has changed it to an obligatory inhabitant of brackish water [*1736*].

Whether pulsating vacuoles may also have an *excretory function* in voiding water-soluble end products of metabolism cannot be decided as yet.

L. Parasitism and Symbiosis

We speak of *parasitism* if one organism (parasite) withdraws body substances from another one (the host) and thereby injures it.

Unlike a *predator*, the parasite depends either temporarily or permanently upon the life of its host. Therefore, parasitic infection generally leads to the death of the host only if the parasite has already finished its development or if other circumstances add to the injury caused by the parasite. In many instances it is difficult to distinguish parasitism from mere *commensalism*. A commensal is an organism which lives on prey wastes or body juices of another organism but does not cause any significant harm in doing so.

Parasitic infection can be a passing event or it may lead to a permanent association of parasite and host (*temporary* and *stationary* parasitism). The parasite may be attached externally *(ectoparasitism)* or live within the host *(endoparasitism)*. In the latter case, the parasite may either inhabit various cavities, e.g. the alimentary canal or the coelom *(body cavity parasitism)*, or it may actually penetrate into tissue cells *(tissue parasitism)*.

One measure of adaptation of parasites is the degree of *host specificity*. While some parasites are specialized to a single species, others can infect several or even many host species. In most cases there is also an *organ* or *tissue specificity*, if the parasites or its developmental stages can infect only specific organs or tissues of a host. It is frequently the case that a parasite can infect different hosts, but within each host only definite tissues or organs.

We speak of *symbiosis* if two organisms living together supplement each other functionally instead of injuring one species. No sharp line can be drawn between parasitism or commensalism and symbiosis. If the parasite or commensal lives not only at the expense of its host but also contributes to keep it alive, it becomes a symbiont.

Parasitism and symbiosis are widespread among Protozoa, either as parasites or symbionts of other organisms or as hosts of parasites or symbionts. Each possibility will be treated separately.

I. Protozoa as Parasites and Symbionts

About one fifth of all Protozoa are either parasites or symbionts. Among *flagellates*, the otherwise predominantly phototrophic *dinoflagellates* exhibit the most diverse forms of parasitism [*186, 228*]. The parasitic mode of life has usually led to loss of the flagella and such extensive changes of cell organization that nuclear structure and the occurrence of *Gymnodinium*-like swarmers are the only indications of their dinoflagellate nature. Some forms *(Oodinium, Apodinium)* live as ectoparasites on pelagic marine animals (siphonophores, polychaetes,appendicularians,thaliaceans). Their cell bodies are supported by a stalk which sends rhizoid-like extensions into the host tissue.

Among endoparasites, the species of the genus *Blastodinium* (Fig. 298) are especially noteworthy. They live within the gut of marine copepods and grow there to large germinative bodies. Small swarmers formed by a periodically repeated reproductive act (successive multiple division, p. 142) are voided in groups through the host's anus to the outside. Although it is highly probable — but not proved — that *Blastodinium* withdraws substances from its host, it still has plastids and pyrenoids and undoubtedly carries out photosynthesis. Infected copepods are easily recognized in life since the yellow or brown germinative bodies are visible through the transparent host tissues.

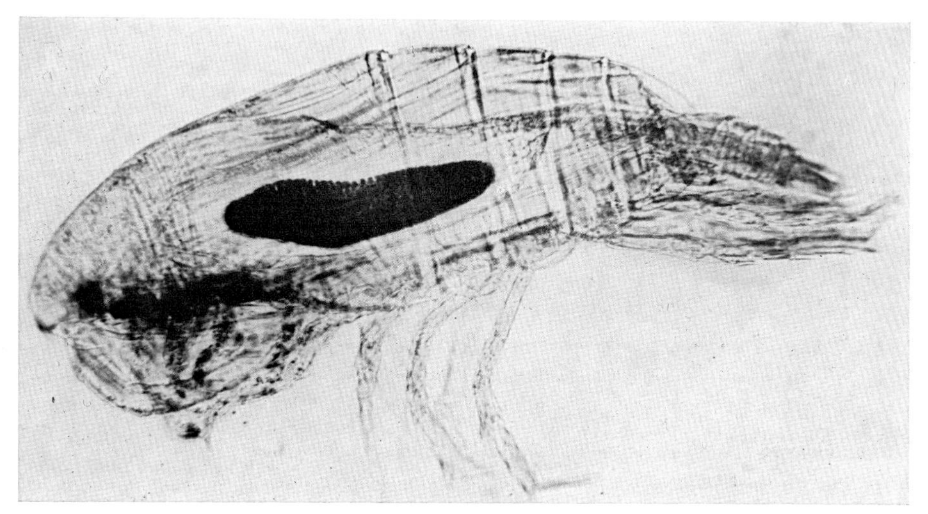

Fig. 298. Marine copepod, in its stomach a germinative body of *Blastodinium* (dinoflagellate). Sanfelice, Feulgen

By contrast, the *Syndinium*- species living in the body cavity of copepods have become true parasites. They grow to multinucleated plasmodia and finally disintegrate into numerous swarmers set free upon the death of the host. Other dinoflagellates have become intracellular parasites, e.g. species of *Ichthyodinium* living in fish eggs. Many infect other Protozoa (p. 356).

The *trypanosomids* of the order Protomonadina are also true parasites. They inhabit the blood stream of vertebrates and invertebrates (or the latex juice of plants) and have frequently become intracellular parasites (*Leishmania*-species, *Trypanosoma cruzi*, Fig. 334). Diplomonadina, Polymastigina *(Trichomonas)* and Opalinina, on the other hand, inhabit the intestinal tract almost exclusively and do not seriously injure their hosts.

Among *Rhizopoda*, the amoebae seem to be the only group with true parasites or commensals. They usually live in definite sections of the alimentary canal or its appendages (Fig. 349). Species of *Malpighiella* and *Malpighamoeba* have become specialized for the Malpighian tubules of insects, where they also encyst.

The *Sporozoa* are solely parasitic. Eugregarinida and Eucoccidia appear predominantly in body cavities. Intestinal Eugregarinida can temporarily penetrate into epithelial cells which they leave again when they associate in "syzygy"

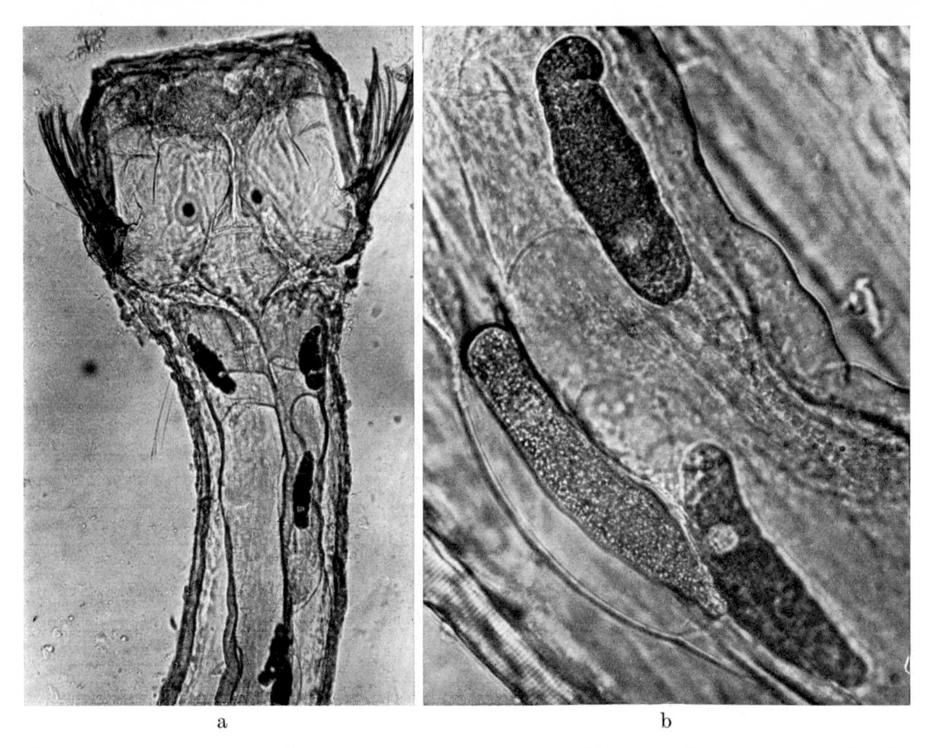

Fig. 299. Eugregarines in the body cavity of a *Sagitta* species (from Messina). a Survey, anterior end of the arrowhead worm. b Three single gregarines. Photographed in life

(p. 195). Many Eugregarinida penetrate the gut wall as sporozoites and continue development in the host's body cavity (Fig. 299). The same holds true for Eucoccidia.

Parasites of body cavities are expected to have a lower degree of host specificity than tissue parasites. *Eucoccidium ophryotrochae* which generally lives in the body cavity of the polychaete *Ophryotrocha puerilis* (Fig. 300a, b) can, for example, be transferred to the archiannelid *Dinophilus gyrociliatus*. However, only a fraction of the oocysts finishes sporogony completely (c).

The Schizogregarinida include many species which carry out the major part of their development within host cells. This tissue parasitism is the rule among Schizococcidia. Since intracellular parasites depend upon the size of the host cell, they are generally smaller than parasites of body cavities. Thus, Sporozoa which penetrate into a host cell generally form fewer spores than species growing in body cavities. The evolution of *schizogony* as an additional form of agamogony in Schizogregarinida and Schizococcidia might well be connected with this. It permits further multiplication of the parasite within the host and infection of many host cells. Decreased spore production due to the limited size of the host cell is compensated for in this manner.

Some Schizococcidia can induce hypertrophy of the host cell, i.e. enlargement beyond normal limits (Fig. 301). This also includes the nuclei of infected cells. Since the content of DNA is considerably increased compared with nuclei of

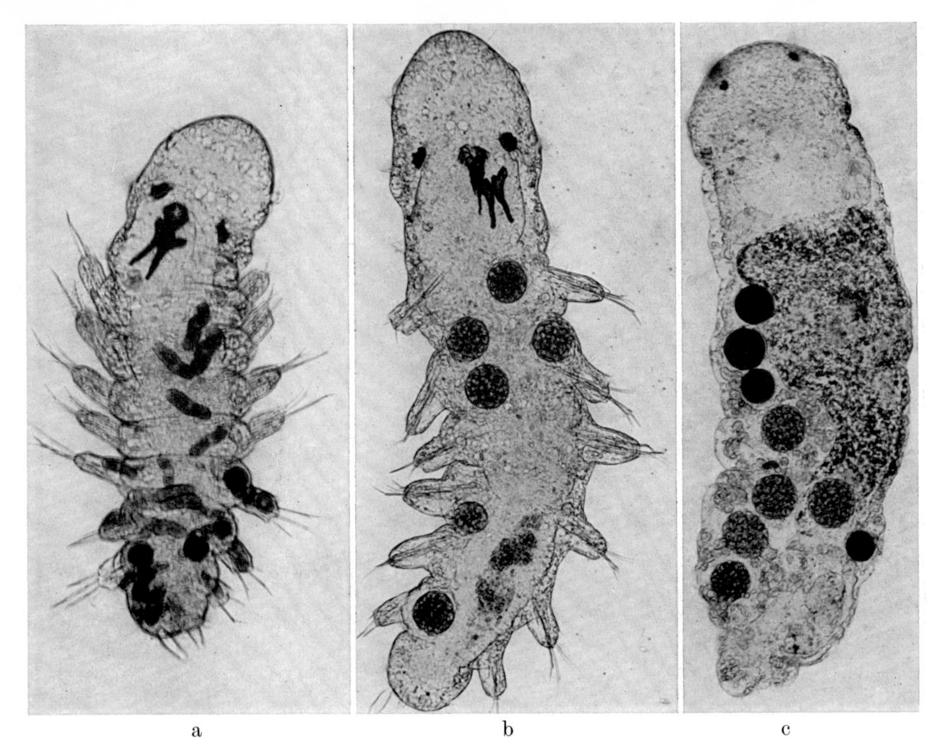

Fig. 300. *Eucoccidium ophryotrochae*. a, b In the body cavity of young individuals of *Ophryotrocha puerilis* (a growth stages of the macrogamonts, b oocysts). c In the body cavity of *Dinophilus gyrociliatus*: five oocysts have continued development and are forming spores, four have degenerated. b, c After GRELL, 1960 [*669*]

uninfected cells [*1022*], it seems likely that nuclear enlargement is due to endopolyploidization (p. 106). Evidently this is connected with increased metabolic activity of the host cell which withdraws more substance from the surrounding medium. The cell of the parasite thus forces the host cell into a definite direction of differentiation in this case. With respect to the physiology of metabolism, both cells have become a unit and represent together a separate system *(xenon)* within the host organism, just as a gall is a foreign body formed from the plant's own tissues around the egg of the gall insect.

Parasitism is also widespread among *ciliates*. Besides species attached as specific commensals to a variety of aquatic organisms, often at definite body regions, there are also true ectoparasites, e.g. *Ichthyophtirius multifiliis* on the skin of freshwater fish, or the *Apostomea* and *Thigmotricha* which live on marine animals. Endoparasitic species and endocommensals are found in the alimentary canal of many Metazoa. Especially noteworthy are the Entodiniomorpha living in the rumen of ruminants and the Astomata in oligochaetes.

Parasites differ extraordinarily in their *effects upon the host*. The parasite may merely withdraw body substances which are either directly taken up or first broken down enzymatically into resorbable split products. Usually, this causes little harm, or it becomes harmful only after the parasite has concluded its development in that particular host. However, parasitic infection can in many cases block the

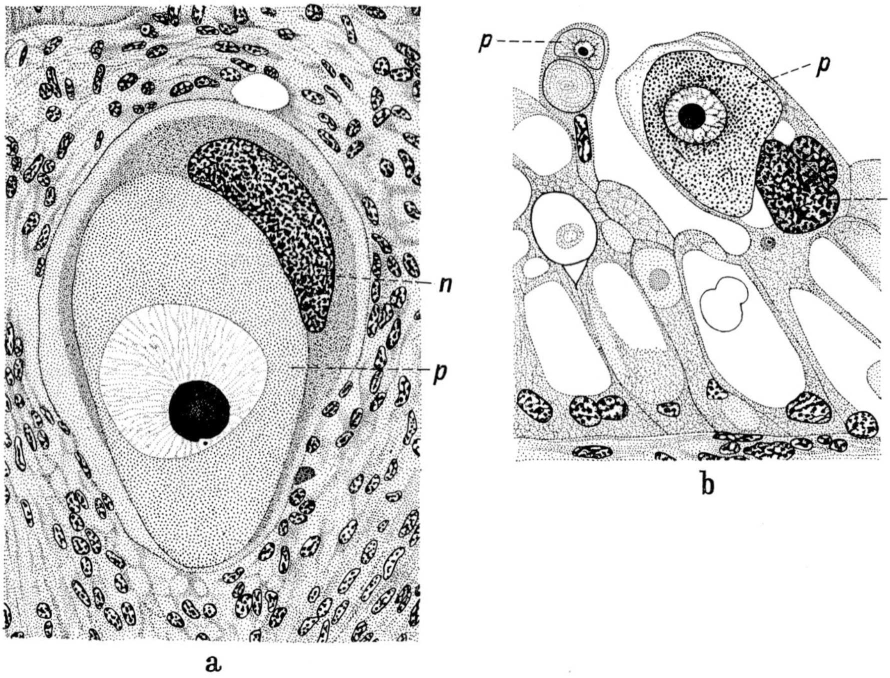

Fig. 301. Hypertrophy of the host cell and its nucleus. a Young gamont of *Aggregata octopiana* in the intestinal mucosa of *Octopus*. b Two gamonts of different age of *Klossia helicina* in cells of the kidney epithelium of *Cepaea nemoralis*. *p* Parasite (gamont), *n* cell nucleus of the host. Iron hematoxylin. a × 390, b × 700. After WURMBACH, 1935 [*1828*]

exchange of materials mechanically, or disintegration of infected tissues may lead to functional disturbances in the host organism. If metabolic end products excreted by the parasite act as poisons *(toxins)*, the host will usually become ill soon after infection. The host-parasite relationship depends largely upon the extent to which the host is able to mobilize its defenses against the parasite. Specific antibodies can be formed against blood parasites or their metabolic end products which either destroy the parasite or render its toxins ineffective. The parasites may be devoured by phagocytes. Hosts are often able to encapsulate the focus of infection (e.g. with connective tissue cells in the case of vertebrates, and hemocytes in insects) and thus prevent further spreading of the parasite.

Many Protozoa live as *symbionts* with Metazoa, both partners supplementing each other in their metabolism. Symbionts referred to as *zoochlorellae* and *zooxanthellae* are phototrophic. Their general role seems to be to supply oxygen released during photosynthesis and to bind the carbonic acid set free by their hosts. The zoochlorellae are Chlorococcales (relatives of phytomonads) and seem for the most part to belong to the free-living genus *Chlorella*. They live as intracellular symbionts in many fresh-water animals, e.g. in endodermal cells of *Chlorohydra viridissima*. The green symbionts of the marine turbellarian *Convoluta roscoffensis*, however, belong to another genus which also includes free-living species and are designated as *Platymonas convolutae* [*1290*].

The zooxanthellae are for the most part dinoflagellates. Within their host cells, they reproduce by binary or, occasionally, multiple fission followed by the formation of small, *Gymnodinium*-like swarmers which seek out other host individuals. Zooxanthellae are found in very different Metazoa (polychaetes, the giant clam *Tridacna*, and others). They are most widely distributed among Cnidaria. In the mangrove jelly fish, *Cassiopea andromeda*, (Scyphozoa), the zooxanthellae are found in the polyp as well as in the free-swimming larvae and the jelly fish itself. They live within certain "carrier cells" of the endoderm and the mesoglea. Recent studies have shown that the rate of strobilation increases with the number of zooxanthellae [*1063*]. Zooxanthellae are especially widespread among sea-anemones and corals (Anthozoa) of warm seas. In this case they are predominantly in the endoderm. Autoradiographic experiments have shown that they supply their host cells with assimilation products and reserve substances. In *Anemonia sulcata* over 60% of the carbon fixed in photosynthesis by the symbionts is transferred to the host [*1722*]. The zooxanthellae themselves take up nitrate and phosphate and perhaps also trace substances (vitamin B_{12}, thiamine, biotin) which are absent or scarce in open sea water. Since most reef corals feed only at night (plankton), symbiosis with zooxanthellae enables them to continue metabolic activity also during the daytime. They withdraw the calcium necessary for skeletal formation from the seawater just when the zooxanthellae are at their maximal rate of photosynthesis. It is obvious that zooxanthellae also influence the mode of reaction of their hosts. Planula larvae containing zooxanthellae react with positive phototaxis while those lacking such symbionts are indifferent towards light. Since each species of coral has its specific optimum of light intensity, the zooxanthellae must be of great importance for the ecology of a coral reef [*451*].

Zooxanthellae and zoochlorellae can be cultured without their hosts in suitable media. It is, however, uncertain whether the latter can exist permanently without their symbionts. It is probable that the excreted products of the symbionts are of value only as a supplement to the host's metabolism and that the primary source of nourishment comes from the exogenous food supply.

Protozoa can also live in symbiosis with other organisms without such cellular connection. The *Polymastigina* in the gut of termites and of the roach *Cryptocercus punctulatus* may serve as examples [*270, 323*]. The termites as well as the roach live exclusively on wood. If freed from their flagellate fauna – e.g. by transfer to high temperature or an oxygen atmosphere – they die after some time but remain alive if re-infected with flagellates in time. Termite and roach are incapable of synthesizing the cellulases required for the breakdown of cellulose. The flagellates, however, can take up small pieces of wood into their cell bodies and digest them. Since the flagellates excrete metabolic end products and occasionally disintegrate, their hosts receive substances in this manner which they can digest with their own enzymes. The Polymastigina, on the other hand, are unable to live outside their hosts. Termites lose their symbionts with every molt and become re-infected by eating feces of their hive-mates. The roach, on the other hand, retains its flagellates when molting. The latter enter a sexual cycle (p. 167) which, in *Trichonympha*, is connected with encystment.

Ciliates living in the rumen of ruminants can also take up and digest small pieces of wood. However, they do not seem to play an essential role in the metabolism of their hosts, in spite of their great number. The hosts show no ill effects if they grow up without ciliates. Ruminant ciliates must therefore be looked upon as commensals. Transmission from one host to another is not by cysts but directly through saliva (mutual licking and feeding at the same spot).

II. Parasites and Symbionts of Protozoa

Protozoa can also become infected with parasites or live in permanent association with symbionts. The rotifer *Hertwigella volvocicola* is specialized for species of the genus *Volvox* and does not attack other Volvocidae. Amictic females hatched from winter eggs penetrate into *Volvox* spheres (Fig. 302) and live on the peripherally arranged cells. They deposit small, smooth-shelled summer eggs into the gelatinous mass which fills the major part of the *Volvox* sphere. The amictic females which hatch from these eggs attack other *Volvox* spheres. When a certain density of infection is reached, mictic females hatch from the summer eggs. Their eggs give rise to small dwarf males without an intestinal tract which fertilize the mictic females. Fertilized winter eggs are larger than summer eggs and provided with a tough, spiny shell. They are set free when the *Volvox* spheres die and can survive for long periods, e.g. throughout the winter. The close adaptation of the parasite to its host is also expressed in the fact that the action spectrum of phototaxis of *Hertwigella volvocicola* is largely identical with that of *Volvox* but differs from that of other phototactic rotifers.

Most parasites of Protozoa are also unicellular organisms. Parasitic *dinoflagellates* can infect Radiolaria. They penetrate into the central capsule as swarmers and often develop at first exclusively in the cell nucleus. This can lead to the formation

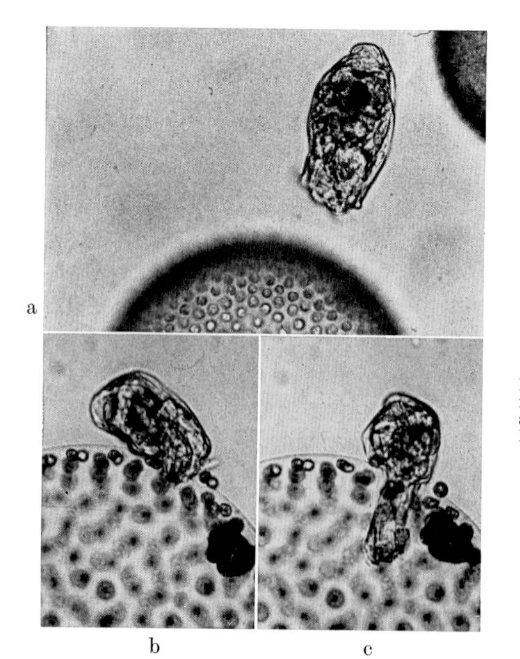

Fig. 302. *Hertwigella volvocicola* (rotifer). a Swimming amictic female. b, c Female penetrating into a colony of *Volvox aureus*. × 210. From the film E566 (GRELL)

of multinuclear plasmodia which disintegrate again into swarmers. In many Radiolaria (e.g. *Thalassicolla nucleata, Aulacantha scolymantha*) these stages were for a long time thought to represent a developmental phase of the host itself (p. 116). Some dinoflagellates *(Duboscquella, Amoebophrya)* infect pelagic ciliates (*Strombidium*, tintinnids) or other dinoflagellates (*Prorocentrum, Ceratium* and others) and have often become modified in the most peculiar manner in connection with intracellular parasitism (Fig. 134). Their developmental stages have, therefore, often been mistaken for organelles of the host cell or for "Mesozoa". "Hyperparasitism" is also found occasionally, for example, when a parasitic dinoflagellate (e.g. *Oodinium*) is in turn infected by a parasite (e.g. *Amoebophrya grassei*).

Infection of opalinids *(Zelleriella)* by amoebae, gregarines by microsporidians also represents hyperparasitism. Nothing is known about the degree of host specificity in these cases.

Euciliates which infect other Protozoa are relatively rare. Species of *Hypocoma* are provided with a sucking tube and live as ectoparasites on Suctoria and marine species of *Zoothamnium*. *Hypocoma acinetarum* can also penetrate into the cell body of *Ephelota gemmipara* and reproduce there by binary fission (film C 907).

On the other hand, parasitism is fairly frequent in *Suctoria* which seem to be predestinated for it because of their stationary and predatory mode of life. *Sphaerophrya* and *Podophrya* species infect free-swimming ciliates in whose cell bodies they reproduce by budding. Species of *Pseudogemma* live as ectoparasites on marine Suctoria. *Tachyblaston ephelotensis* is specifically a parasite of *Ephelota gemmipara* (Fig. 303). Although it lives in a deep invagination of the host cell, it is also considered an ectoparasite. It takes up host cell cytoplasm with the aid of a feeding tentacle and, in doing so, it forms one swarmer after another. However, the swarmers do not serve for re-infection but change to the free-living *Dactylophrya*-stages (Fig. 431).

Aside from certain *fungi (Sphaerita, Nucleophaga)*, Protozoa are infected especially by *bacteria*. Some bacterial species have become specific nuclear parasites. The marine amoeba *Paramoeba eilhardi*, for example, can fall prey to a bacterium which multiplies exclusively in the cell nucleus (Fig. 304) and induces its enormous enlargement (hypertrophy). After some time the amoeba ceases to feed and finally dies, liberating the bacterial mass. The bacteria are immotile as long as they are in the nucleus but form flagella once they get into sea water. There they remain alive for months without dividing. Tests showed that this bacterium cannot be transferred to other marine amoebae. The macronucleus of ciliates can also sometimes become infected with bacteria and occasionally with leptomonads [*598*].

In the case of *symbionts* of Protozoa it is often difficult to tell to what extent they are really necessary for the life of their hosts. The zoochlorellae found in the cell bodies of some amoebae *(Amoeba viridis)*, testaceans (*Difflugia lobostoma, Hyalosphenia papilio* etc.) and ciliates (*Paramecium bursaria, Stentor polymorphus* etc.) are closely related to free-living species of *Chlorella*.

The symbiosis of *Paramecium bursaria* has received most attention because it can easily be freed of its zoochlorellae (p. 217). The various stocks of *Chlorella* differ in infectivity and fission rate. Most of them can be cultured independently of their

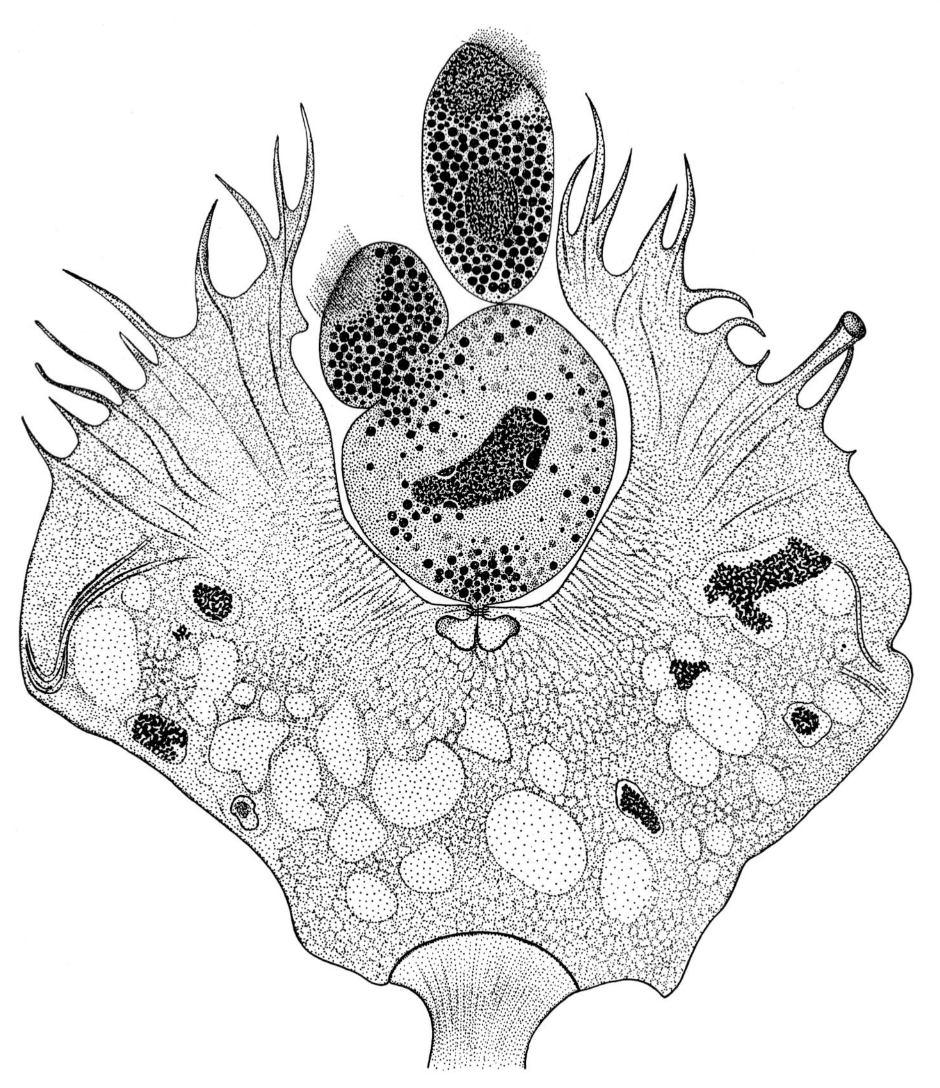

Fig. 303. *Ephelota gemmipara*, infected with *Tachyblaston ephelotensis*. The parasite (with bud and fully formed swarmer) takes up the cytoplasm of the host cell with its feeding tentacle. FLEMMING, iron hematoxylin. × 750. After GRELL, 1950 [*650*]

paramecia or even axenically. Paramecia without chlorellae can easily be re-infected not only with stocks isolated from them but also with zoochlorellae of *Chlorohydra viridissima* and with *Chlorella vulgaris*. While the fission rate of paramecia lacking symbionts does not differ significantly from that of green paramecia as long as the medium is rich in bacteria, the difference becomes pronounced when paramecia are reared in an axenic solution or one poor in bacteria. The zoochlorellae can thus compensate for a deficiency in bacterial food to a certain extent, probably because they provide their hosts with products of photosynthesis. It has, on the other hand, been shown that the zoochlorellae continue to divide in the dark if sufficient food bacteria are available to the paramecia. They must therefore also

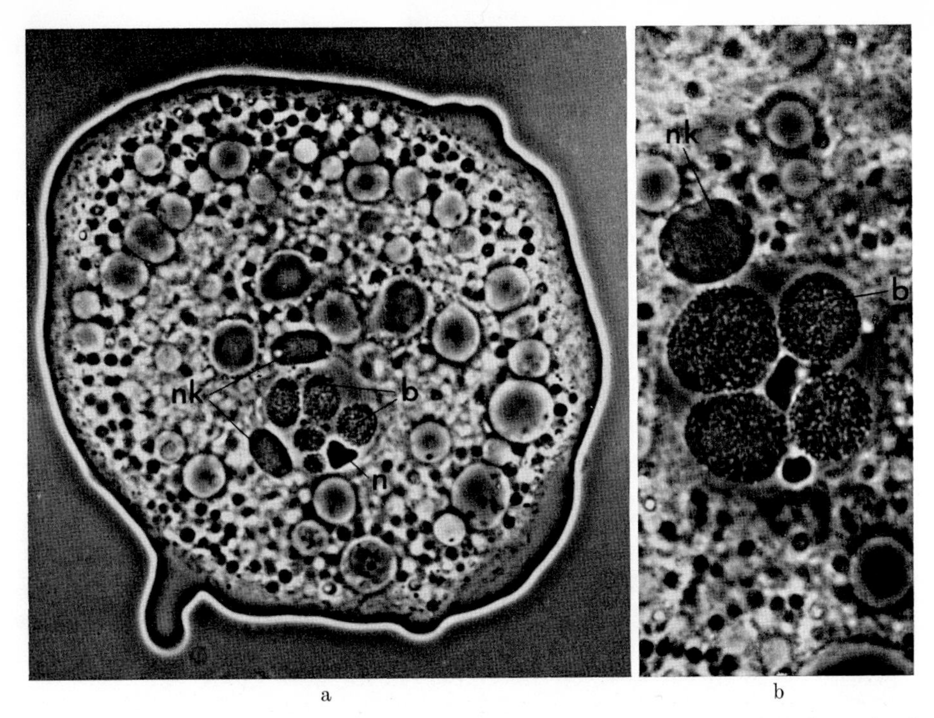

Fig. 304. *Paramoeba eilhardi*. Bacterial infection of the cell nucleus. a A whole cell. × 1300. b Nuclear region of another amoeba. × 2600. *nk* „Nebenkörper", *b* bacterial mass, *n* nucleolus. Photographed in life

obtain nutriments from their hosts and be capable of shifting to heterotrophic nutrition. The close integration of both partners is especially clearly expressed in the fact that their division rates are coordinated, maintaining a definite population density of zoochlorellae by regulation [*148, 903–906, 1587, 1596*].

Zooxanthellae are found in many Radiolaria and Foraminifera (*Peneroplis, Orbitolites, Globigerina, Heterostegina*). Some are dinoflagellates, some chrysomonads. Though it has been asserted that *Heterostegina depressa* is able to grow without uptake of food particles, not much is known about the metabolic relationships.

Cyanophycea are relatively rarely found in symbiosis with Protozoa. The flagellate *Cyanophora paradoxa* normally contains two or four cells of a cyanophycean which can, by electron microscopy, be identified and distinguished from chloroplasts [*1736*].

Spirochaetes and *bacteria* frequently live together with Protozoa. In many Polymastigina they are externally attached to the cell, often limited to certain areas or associated with special structures of the pellicle. The peculiar "motility symbiosis" of *Mixotricha paradoxa* with spirochaetes has been mentioned earlier (p. 297).

Electron microscope studies have shown that some free-living amoebae always harbor bacteria in their cytoplasm which are enclosed in vesicles. The multinucleate saprobiotic amoeba *Pelomyxa palustris* has two species, one usually in the vicinity of the cell nuclei [*362, 363*]. A danish strain contains even three different species of bacteria [*60*].

Bacteria are also found in the cytoplasm of many ciliates (e.g. *Euplotes patella, Oxytricha bifaria, Halteria grandinella*, several Astomata and Suctoria) without any recognizable effect upon the host cells. Only axenic culture of both partners could show whether this represents a metabolic symbiosis [*1108*].

The organisms designated as *kappa, lambda, mu-* and *gamma*-particles are gram-negative bacteria and can impart definite properties upon their ciliate host, *Paramecium aurelia*, which stimulated research on these systems [*114, 883, 1340, 1342, 1638*].

The multiplication of these organisms depends upon nuclear genes of the host (see p. 260); they are metabolically dependent upon them. On the other hand, they do not seem to be only useless commensals, as it has been shown by axenic culture that paramecia containing *lambda* can live without added folic acid which otherwise must be present in the nutritive medium. Upon treatment with penicillin *lambda*-containing paramecia lose their symbionts and die if folic acid is not added to the medium [*1620*].

The *kappa* particles present in most killer stocks of *Paramecium aurelia* are of two kinds (Fig. 305a). The so-called N-particles (nonbrights) contain no refractile inclusion and reproduce by division. N-particles isolated from the cells can infect other paramecia if these are of the genetic constitution required for the multiplication of *kappa* (b). A portion of the N-particles transforms continuously into B-particles (brights). The latter contain a refractile inclusion (R-body). B-particles are extruded into the surrounding culture fluids. Their ability to kill sensitive paramecia is due to the R-bodies (c).

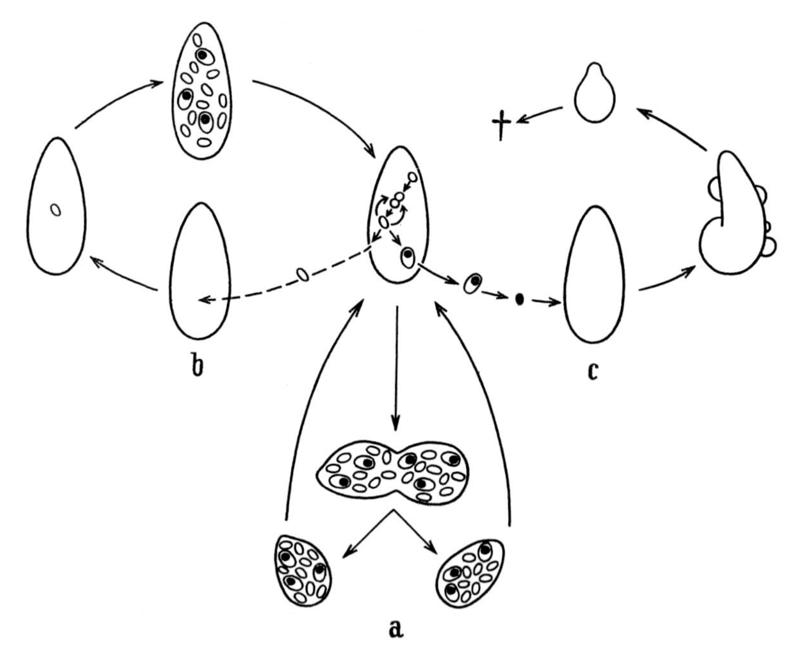

Fig. 305. *Paramecium aurelia*. Multiplication, mode of infection and killer effect of the *kappa* particles. N-particle without inclusion, B-particle with a refractile inclusion (drawn black in the diagram). After SONNEBORN, 1961 [*1640*]

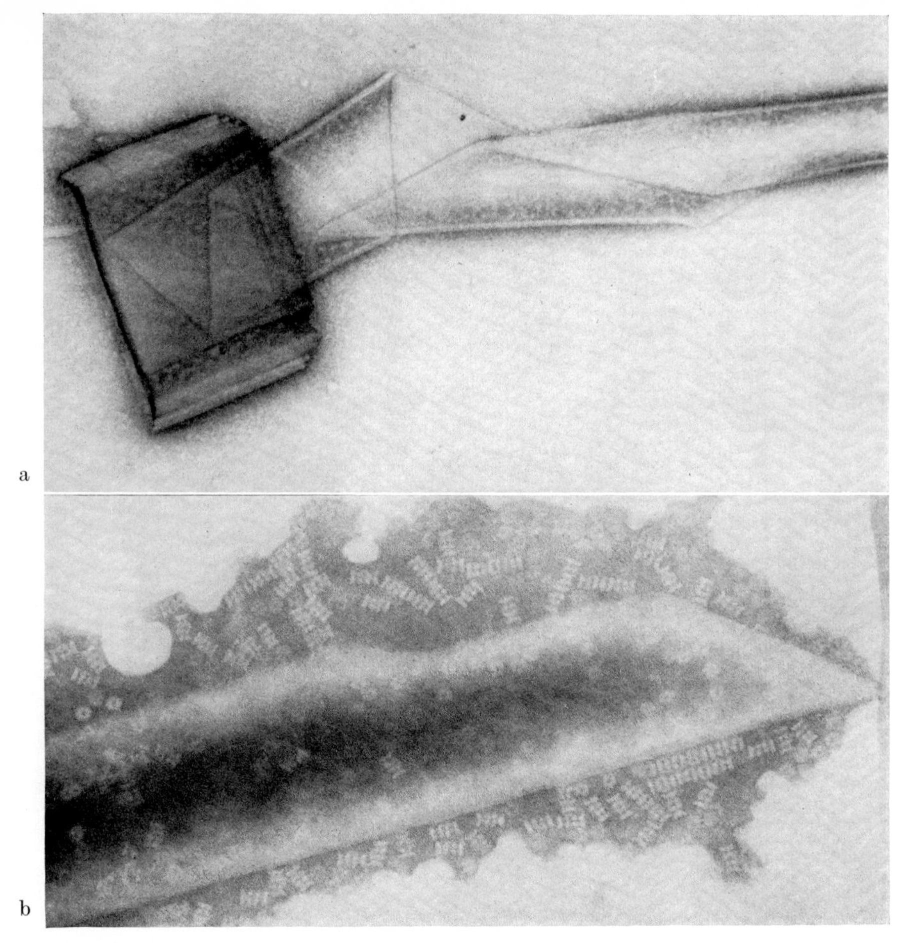

Fig. 306. *Paramecium aurelia.* a R-body, partly unfolded (stock 51). × 52000. After J. R. PREER, A. HUFNAGEL, and L. B. PREER, 1966 [*1337*]. b Inner end of an unfolded R-body with numerous virus particles (capsomere-like structures) present, many aggregated in rows (stock 7). × 150000. After L. B. PREER, A, JURAND, J. R. PREER, and B. M. RUDMAN, 1972 [*1342*]

In the electron microscope [*55*, *1337*, *1343*] the R-bodies are seen to be folded tapes which are, however, frequently more or less unfolded. In one stock of *kappa* (7) this unfolding always proceeds from the outside, in another (51) from the inside (Fig. 306a).

Particles, which have recently [*1338*, *1344*] been demonstrated at the ends of the tapes, seem to be *viruses* (b). They have even been isolated and shown to contain DNA [*1339*]. There are indications that these viruses are present within the N-particles as inactive pro-viruses. After transformation into the active form they change the N-particles to B-particles, i.e. they stop multiplication and induce the synthesis of R-bodies [*694*]. It is probable that the R-bodies, which consist of a specific protein, unroll in the food vacuoles of sensitive paramecia and break through the vacuolar wall in doing so [*884*]. In this manner the viruses could get into the cytoplasm and manifest their toxic action which leads to

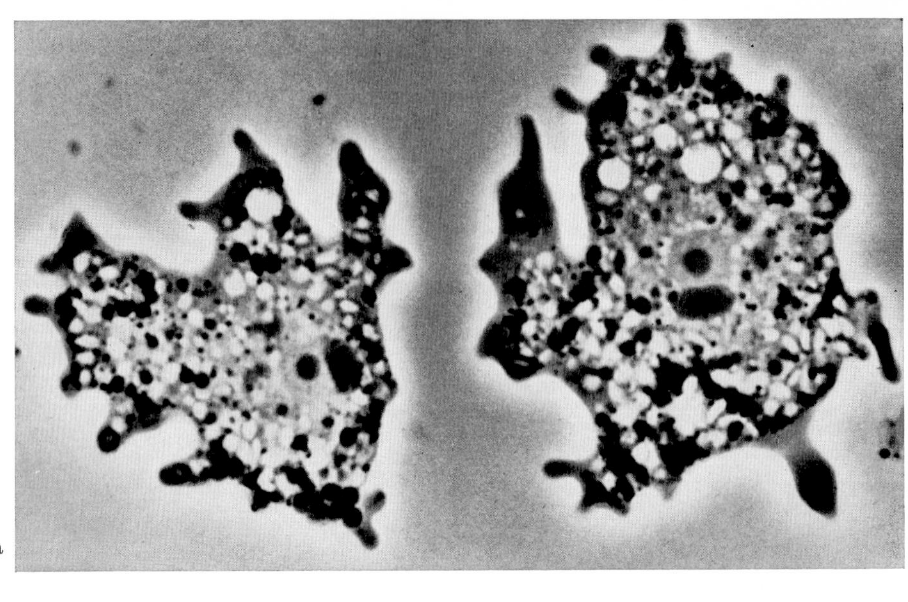

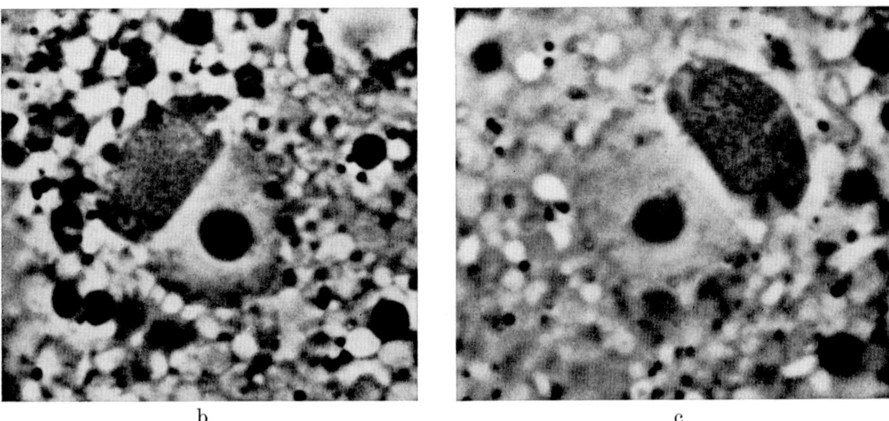

Fig. 307. *Paramoeba eilhardi.* a Two amoebae. × 1300. b, c Cell nucleus and ,,Nebenkörper". × 3000. Photographed in life. After GRELL, 1961 [*670*]

death. That this toxicity is actually due to the virus or a protein formed by it is indicated by the proof that R-bodies of stock 51 are not toxic if no viruses are attached to them.

The frequently discussed "killer phenomenon" would thus have a surprising "explanation": bacterial endosymbionts whose multiplication depends upon genes of the host have become vectors for viruses. An intracellular regulatory mechanism ensures that not all symbionts are destroyed in their function as transmitters. A part always remains to hand on the inactive pro-viruses.

In some cases the integration of symbiont and host has progressed to a stage where the symbiont appears as an "organelle" of the cell. Attached to the nucleus of *Paramoeba eilhardi* are structures called ,,Nebenkörper" (associated bodies;

Fig. 307). Some stocks have only one or two, others four or more such bodies. Each of these bodies shows a polar structure. It consists of a Feulgen-staining, filamentous middle portion and the two, more or less empty-appearing poles. At first sight, the „Nebenkörper" might seem to be a nucleus. Electron micrographs showed, however, a thin granular layer below the envelope of the „Nebenkörper". The latter might therefore be a cell whose amount of cytoplasm is considerably reduced, living as an intracellular symbiont in close and permanent association with the nucleus of the amoeba [*670, 678*].

„Nebenkörper" divide sometimes independently of the host cell nucleus, sometimes together with it (Fig. 308). Differences between stocks indicate that the numbers of such „Nebenkörper" are, within limits, subject to regulation by the host cell.

Amoebae without „Nebenkörper" have never been found in clone cultures, nor was it possible to obtain viable amoebae after the „Nebenkörper" had been irradiated by UV microbeam. It is therefore likely that the „Nebenkörper" has indeed become an indispensable "organelle" of the cell. Its DNA contains perhaps part of the genetic information necessary for the metabolism of the amoeba.

This case is interesting with respect to the so-called "*endosymbiosis theory*" which assumes that "true" cell organelles capable of autoreduplication and endowed with their own, though restricted, protein-synthesizing systems may derive from prokaryotic endosymbionts: mitochondria from Bacteria and plastids from Cyano-

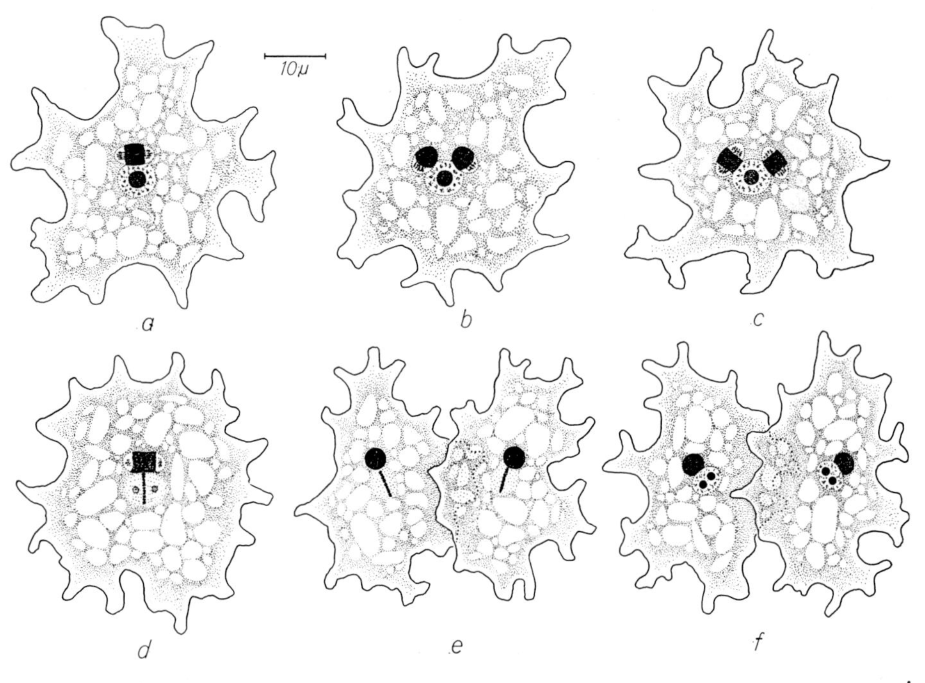

Fig. 308. *Paramoeba eilhardi*. Diagram of the processes of division (after preparation with iron hematoxylin) a—c Autonomous division of the „Nebenkörper". d—f Cell division (synchronous division of the nucleus and of the „Nebenkörper"). After GRELL, 1961 [*670*]

phycea [*1426*]. Though many arguments are in favor of this assumption, it will always remain a theory. The possibility cannot be excluded that these cell organelles derive from structures of prokaryotic ancestors e.g. mitochondria from "mesosomes" [*1383*].

M. Taxonomic Survey

As our knowledge of structure and reproduction of numerous groups is still fragmentary and their taxonomic relationships are unknown, it is difficult to draw up a satisfactory system of the Protozoa. In addition, it is left to a certain degree to the taxonomist's individual judgment which criteria he uses in his classification. Reform proposals published in recent times [*241, 810, 1376*] agree only in restricting the term Sporozoa to the "Telosporidia" (Gregarinida, Coccidia) and in classifying the Opalinina among the flagellates.

The following survey is not intended as a complete system, but rather as a taxonomic grouping of the better known forms which were also treated in the preceding chapters. Some of the smaller groups including those of uncertain taxonomic standing are not mentioned at all.

The customary subdivision of Protozoa into classes, which is retained in the following survey should not obscure the fact that these classes are very heterogeneous and that various relationships exist between them. There are flagellates which can form pseudopodia and rhizopods which can form flagellated stages. The so-called amoeboflagellates could be grouped into either one of the two classes equally well. There are indications that the Sporozoa may have originated from flagellates with a haplo-homophasic alternation of generations. The deep cleft which seemed to separate the ciliates from all other Protozoa is largely bridged today since it became known that their nuclear differentiation is not unique and that there is no fundamental difference between flagella and cilia. Quite problematic is the maintenance of the class "Cnidosporidia".

First Class: Flagellata

Fritsch, F. E.: The structure and reproduction of the Algae. 2. Ed. London: Cambridge University Press 1952.

Fott, B.: Algenkunde. Jena: Fischer 1959.

Grassé, P. P.: Traité de zoologie. **1**, Fasc. 1: Protozoaires (Généralités. Flagellés). Paris: Masson et Cie. 1952.

Hollande, A.: Etude cytologique et biologique de quelques flagellés libres. Arch. Zool. exp. gén. *83* (1942).

Kühn, A.: Morphologie der Tiere in Bildern. Heft 1, Teil 1: Flagellaten. Berlin: Bornträger 1921.

Oltmanns, F.: Morphologie und Biologie der Algen. 2. Aufl. Jena: Fischer 1922/23.

Pascher, A.: Flagellatae. In: Die Süßwasserflora Deutschlands. Heft 1. Jena: Fischer 1914.

— Flagellaten und Rhizopoden in ihren gegenseitigen Beziehungen. Arch. Protistenk. **38**, (1917).

Pringsheim, E. G.: Farblose Algen. Ein Beitrag zur Evolutionsforschung. Stuttgart: Fischer 1963.

Smith, G. M.: The fresh-water Algae of the United States. 2. Ed. New York: McGraw-Hill Book Comp. 1950.

Flagellates have one or more *flagella*. Many species can, however, temporarily change to an unflagellated state *(palmella stage)* in which they surround themselves with a gelatinous capsule. As such cells remain together after fission, more or less extensive complexes of individual cells are formed. Some species can also form colonies with a definite number and arrangement of the cells composing the colony. In this manner a variety of *intermediates to the Algae* are found.

Longitudinal fission is the usual mode of reproduction. The occurrence of sexual processes has only been proved in Phytomonadina, some Polymastigina and Opalinina.

Phototrophic flagellates (p. 328) can be colored green, brown or yellow due to the presence of plastids. Since flagellates capable of photosynthesis are restricted to certain orders (the 1st to the 5th), the latter are frequently referred to collectively as "*phytoflagellates*" in contrast to the following, exclusively heterotrophic orders (6th to 9th) called "*zooflagellates*". This difference cannot, however, be utilized in their taxonomic characterization since the "phytoflagellates" also include many species which became colorless due to loss of plastids (p. 26) [*1354*].

First Order: Chrysomonadina

HOLLANDE, A.: Classe des Chrysomonadines. In: GRASSÉ, P. P.: Traité de Zoologie. *1*, Fasc. 1. Paris: Masson et Cie. 1952.

PASCHER, A.: Chrysomonadinae. In: Die Süßwasserflora Deutschlands. Heft 2. Jena: Fischer 1913.

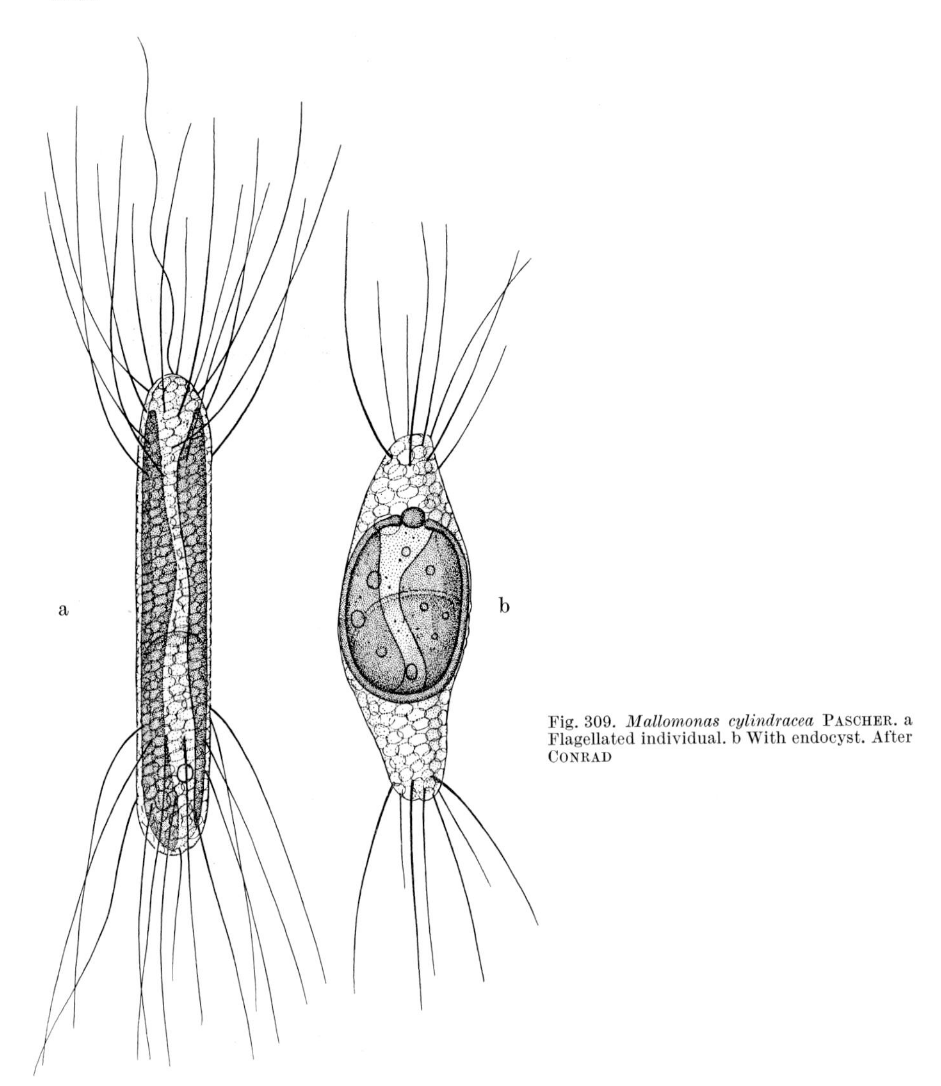

Fig. 309. *Mallomonas cylindracea* PASCHER. a Flagellated individual. b With endocyst. After CONRAD

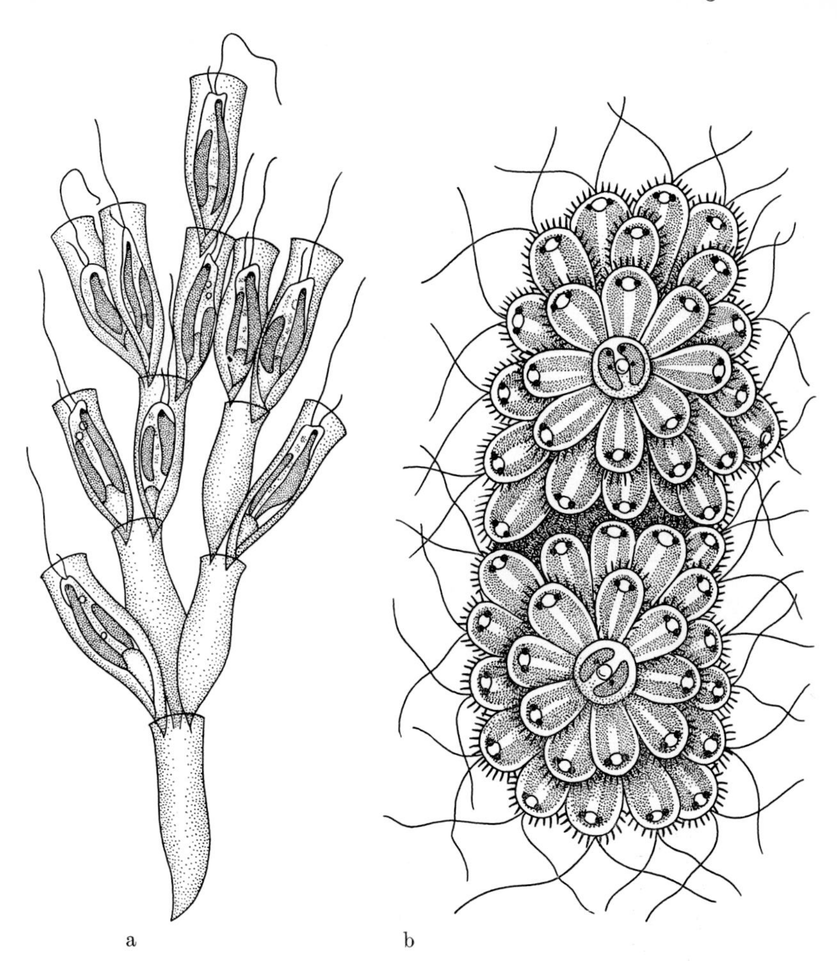

Fig. 310. Colonial chrysomonads. a *Dinobryon sertularia* EHRENBERG. After SENN. b *Synura uvella* EHRENBERG, in division. After KORSCHIKOFF

PASCHER, A.: Die braune Algenreihe der Chrysophyceen. Arch. Protistenk. **52** (1925).

The chrysomonads are small flagellates with yellow-brownish plastids. They usually have two flagella of unequal length, the longer one bearing mastigonemes (p. 294). In addition, a flagella-like haptonema (p. 295) is frequently present which serves as an organelle for attachment. The cell envelope of many chrysomonads is covered with small platelets of species-specific fine structure, often supplied with a thorn (Fig. 311). Leucosin (chrysolaminarin) and oil are found as reserve substances. Some species take in particulate food with the aid of pseudopodia.

So-called endocysts are often formed as persisting stages. They arise within the cytoplasm, contain silica and are supplied with an opening covered by a type of plug (Fig. 309 b). Reproduction by binary fission.
Chrysomonads are found both in fresh water and in the sea.

Chromulina. 1—2 brown, plate-shaped plastids. Amoeboid motility. Brown surface cover on ponds.

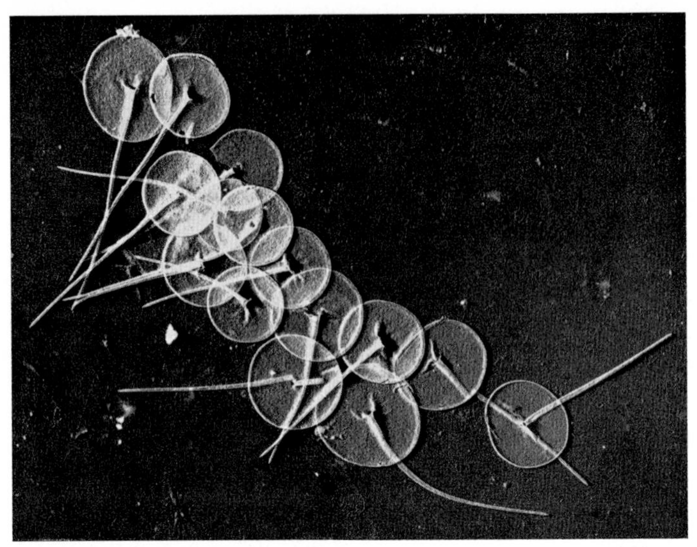

Fig. 311. *Paraphysomonas vestita* (STOKES) DE SAEDELER. Silicon platelets detached from the cell surface. × 7500. After MANTON and LEEDALE, 1961 [*1095*]

C. pascheri HOFENEDER, fresh water.
Mallomonas. Usually elongated. Silicous thorns originate at the cell envelope.
M. cylindracea PASCHER (Fig. 309), fresh water.
Ochromonas. Two unequal flagella, without test.
O. danica PRINGSHEIM (Fig. 16 and 258), fresh water.
Dinobryon. Single or branched, tree-like colonies whose component cells are attached by their tests.
D. sertularia EHRENBERG (Fig. 310a), fresh water (plankton).
Synura. Spherical colony of numerous (up to 50) cells.
S. uvella EHRENBERG (Fig. 310b), fresh water.

Palmelloid chrysomonads can form thread-like or branched complexes reminiscent of algae (e.g. *Hydrurus foetidus*, in brooks).
An important part of the so-called dwarf plankton or nannoplankton of the sea are the *coccolithophorids* which have a test composed of lime platelets or rods (coccoliths). The *silicoflagellates* with their skeletons of silica are also part of the marine plankton. These groups are often classified as separate flagellate orders.

Second Order: Cryptomonadina

HOLLANDE, A.: Classe des Cryptomonadines. In: GRASSÉ, P.P.: Traité de Zoologie. **1**, Fasc. 1. Paris: Masson et Cie. 1952.
PASCHER, A.: Cryptomonadinae. In: Die Süßwasserflora Deutschlands. H. 2. Jena: Fischer 1913.

Cryptomonads can be of very different colorations. They are frequently brown or red. The cell has bilateral symmetry. A pharynx-like invagination (vestibulum) opens at one side of the front end. The two flagella arise from the vestibulum, whose wall bears the highly refractile ejectisomes (p. 37). Starch is found as reserve substance. Cryptomonads reproduce by binary fission and abound both in fresh water and in the sea.

Chilomonas. Deep vestibulum, accumulations of starch, colorless, in infusions (bacterial feeder).
C. paramecium EHRENBERG (Fig. 312a and 28), fresh water.

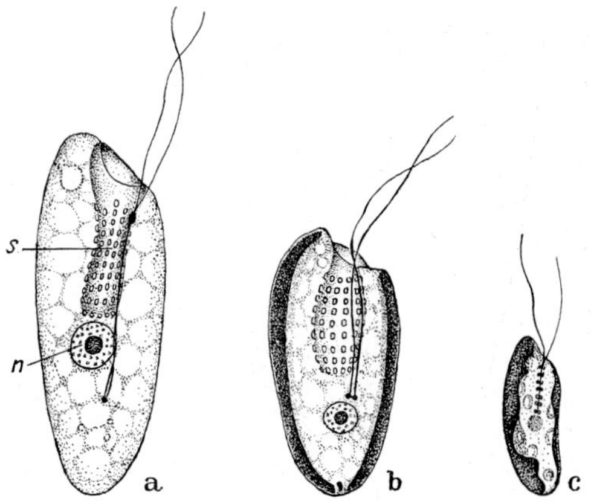

Fig. 312. Cryptomonads. a *Chilomonas paramecium* EHRENBERG. b *Cryptomonas ovata* EHRENBERG. c *Cryptochrysis commutata* PASCHER. *s* Vestibulum, *n* nucleus. × 800. a, b after DOFLEIN; c after PASCHER from KÜHN

Cryptomonas. Two dish-shaped plastids.
C. ovata EHRENBERG (Fig. 312b), fresh water.

Cryptochrysis. Vestibulum reduced.
C. commutata PASCHER (Fig. 312c), fresh water.

Third Order: Phytomonadina

ETTL, H.: Beitrag zur Kenntnis der Morphologie der Gattung *Chlamydomonas* EHRENBERG. Arch. Protistenk. **108**, 271—430 (1965).

JANET, CH.: Le Volvox Mémoire I—III. Paris: Masson et Cie. 1912, 1922, 1923.

PASCHER, A.: Volvocales — Phytomonadinae. In: Die Süßwasserflora Deutschlands. Heft 4. Fischer 1927.

PAVILLARD, J.: Classe des Phytomonadines ou Volvocales. In: GRASSÉ, P. P.: Traité de Zoologie. **1**, Fasc. 1. Paris: Masson et Cie. 1952.

Phytomonadina have two, four or eight flagella of equal length which originate at the front end. The cell envelope can be reinforced by a layer of cellulose with pectins. The major part of the cell is occupied by a large, usually cup-shaped chloroplast which imparts a green color to it. Starch is found as reserve substance. Reproduction by binary or multiple fission (p. 139). Sexual processes are widespread and some have been studied in detail (p. 154). The Phytomonadina are haploids with zygotic meiosis. As a rule, four gones arise from the zygote, which is often stained red by hematochrome (Fig. 215). In some genera *(Pandorina, Eudorina, Volvox)* only one gone hatches because the other gone cells or nuclei degenerate beforehand.

Phytomonads are found predominantly in fresh water. Most are single-celled, others form colonies either with all cells capable of reproduction or with differentiation into generative and somatic cells (Fig. 5).

a) Single cells

Dunaliella. Without a firm cellulose case.
D. salina TEODORESCO (Fig. 152), in salines.

Chlamydomonas (Fig. 313). Spherical or ovoid; two flagella arise near a papilla-like thickening of the front end. Numerous (more than 300) species. Fresh water and marine.
C. reinhardi DANGEARD, fresh water.

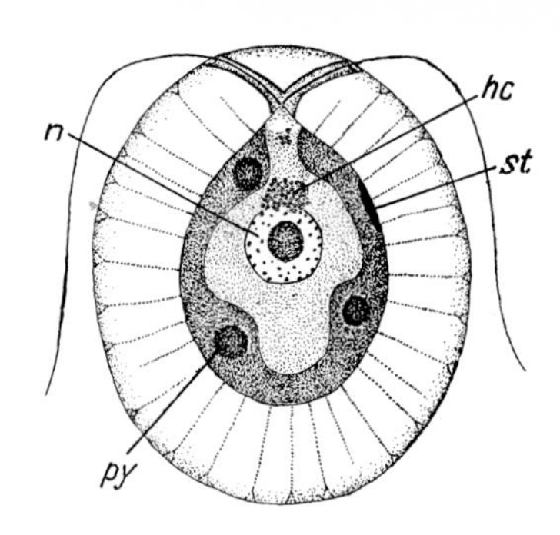

Fig. 314. *Haematococcus pluvialis* FLOTOW. *st* Stigma, *n* nucleus, *py* pyrenoid, *hc* hematochrome. After REICHENOW from KÜHN

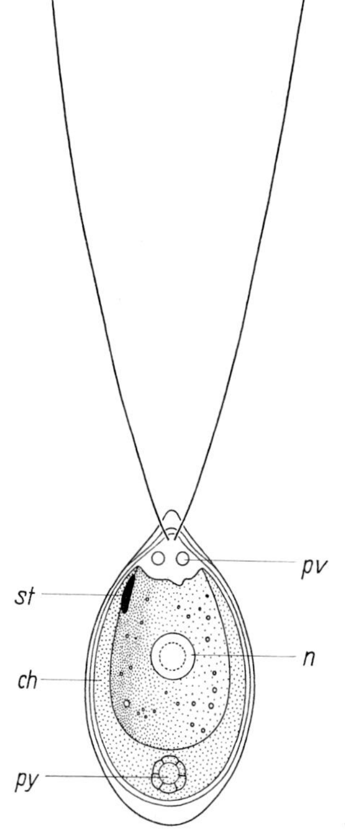

Fig. 313. *Chlamydomonas*. Diagram. *st* Stigma, *ch* chloroplast, *py* pyrenoid, *pv* vacuole, *n* nucleus. Adapted after PASCHER

Chlorogonium. Elongated, spindle-shaped.
  C. elongatum DANGEARD (Fig. 142 and 145).
  C. oogamum PASCHER (Fig. 146), fresh water.
Haematococcus. Cell body separated from cell membrane by a gelatinous layer traversed by cytoplasmic strands and two flagellar canals. Often stained red by hematochrome. Cause of the "pink snow".
  H. pluvialis FLOTOW (Fig. 314), fresh water.
Polytoma. Colorless, heterotrophic.
  P. uvella EHRENBERG, in stagnating fresh water.

b) Cell colonies

Chlamydobotrys. Colony of eight cells, with two flagella each.
  C. korschikoffi PASCHER (Fig. 315), fresh water.
Spondylomorum. Colonies of sixteen cells, with four flagella each.
  S. quaternarium EHRENBERG (Fig. 316), fresh water.
Stephanosphaera. Four, eight or sixteen cells arranged in a circle. Cells with pseudopodia-like but stiff extensions.
  S. pluvialis COHN, fresh water.
Gonium. Four or sixteen cells form a disc in a gelatinous envelope.
  G. pectorale MÜLLER, sixteen cells (Fig. 215), fresh water.
Pandorina. Spherical colony of eight or sixteen heart-shaped, closely applied cells, in a gelatinous covering.
  P. morum BORY, fresh water.
Eudorina (Pleodorina). Cells arranged loosely, usually in several successive tiers in a highly developed gelatinous mass. Colonies with polarity. Cells of the foremost tier *(E. illinoisensis)*

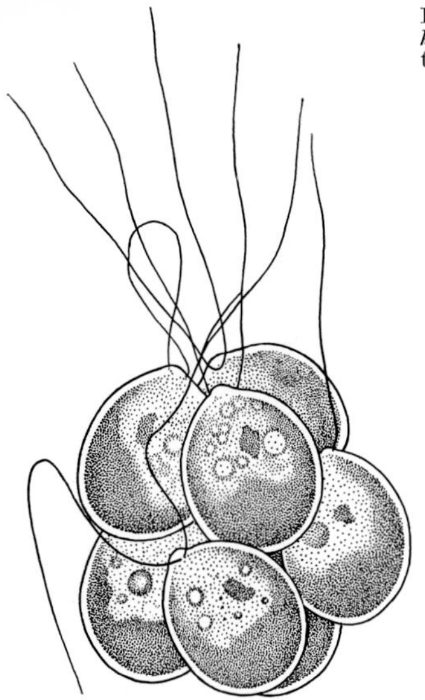

Fig. 315. *Chlamydobotrys korschikoffi* PASCHER. After KORSCHIKOFF

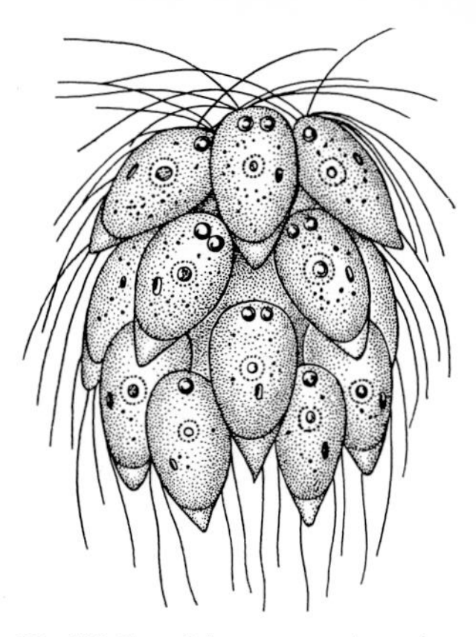

Fig. 316. *Spondylomorum quaternarium* EHRENBERG. After STEIN

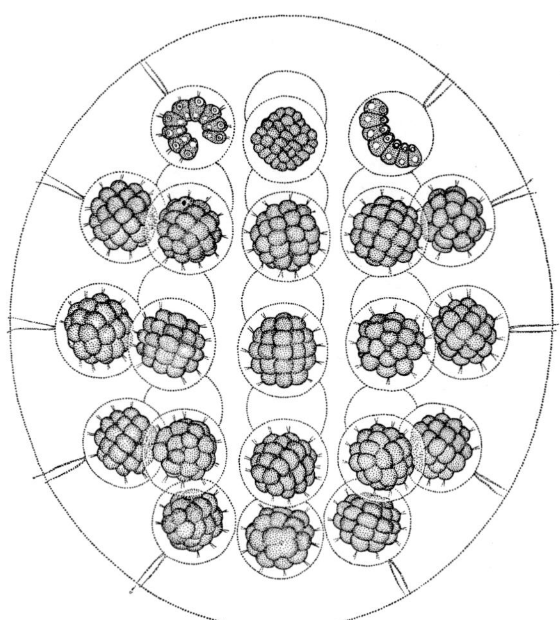

Fig. 317. *Eudorina elegans* EHRENBERG. Divisional stage. After HARTMANN

or the anterior half *(E. californica)* can be smaller than the others and differentiated as somatic cells (Fig. 5).

E. elegans EHRENBERG (Fig. 317), *E. (P.) illinoisensis* KOFOID.

E. (P.) californica SHAW (Fig. 318), all fresh water.

Volvox. The colony is a large, gelatinous sphere of hundreds of individual cells mostly connected with one another by cytoplasmic strands. They are peripheral and for the most part somatic. Only in the posterior half of the colony are single, scattered generative cells (Fig. 5).

V. globator EHRENBERG (Fig. 319a).

V. aureus EHRENBERG (Fig. 319b), both in fresh water.

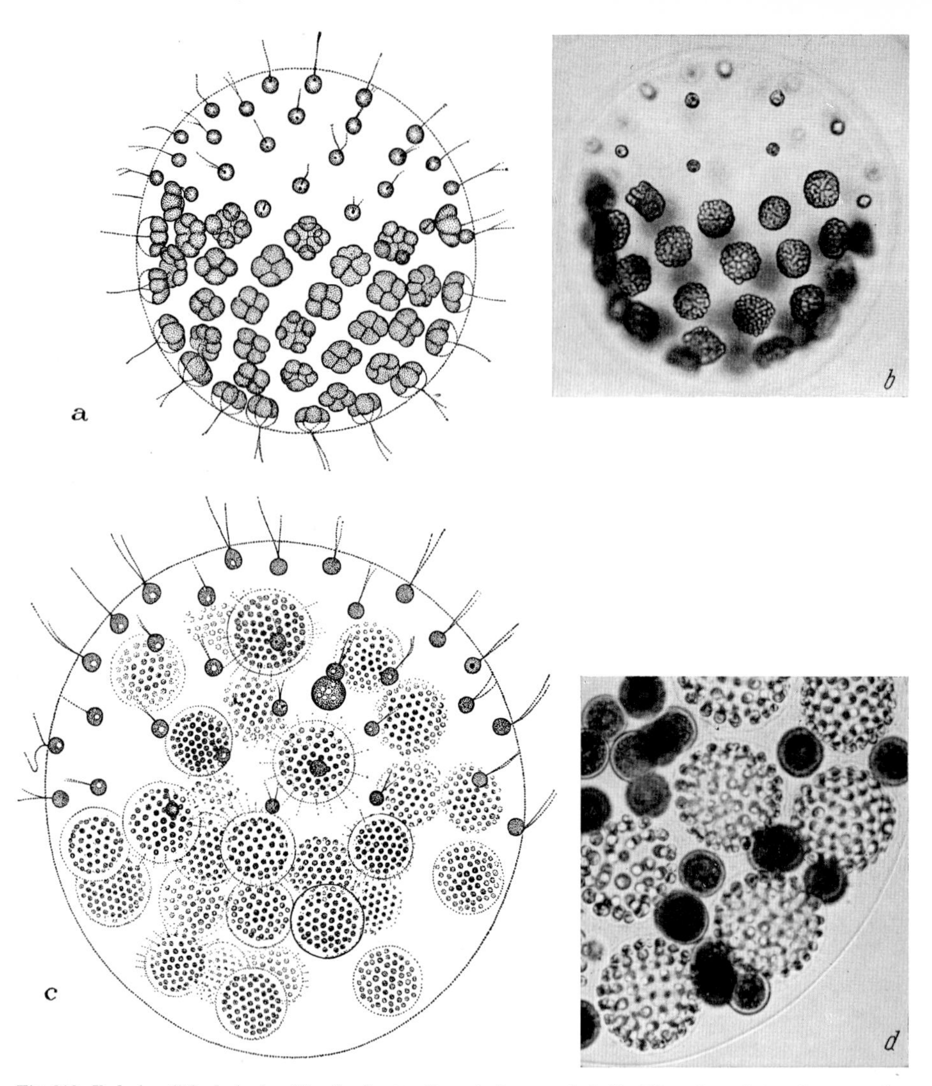

Fig. 318. *Eudorina (Pleodorina) californica* SHAW. Above is the somatic half of the colony, below the generative half (compare Fig. 5). a Early, b middle, c late stage of asexual reproduction. d Section of the posterior end of a colony with daughter spheres and zygotes (dark). a, c After CHATTON, 1911; b, d photographed in life

Fourth Order: Euglenoidina

BUETOW, D. E. (Ed.): The Biology of *Euglena*, Vol. I: General Biology and Ultrastructure; Vol. II: Biochemistry. New York and London: Academic Press 1968.

DANGEARD, P.: Recherches sur les Eugléniens. Botaniste 8 (1901).

GOJDICS, M.: The genus *Euglena*. Madison: The Univ. of Wisconsin Press 1953.

HOLLANDE, A.: Etudes cytologiques et biologiques de quelques Flagellés libres. Arch. Zool. exp. gén. **83** (1942).

— Classe des Eugléniens. In: GRASSÉ, P. P.: Traité de Zoologie. **1**, Fasc. 1. Paris: Masson et Cie. 1952.

JAHN, T.: The Euglenoid Flagellates. Quart. Rev. Biol. **21** (1946).

LEEDALE, G. F.: Euglenoid Flagellates. New Jersey: Prentice Hall 1967.

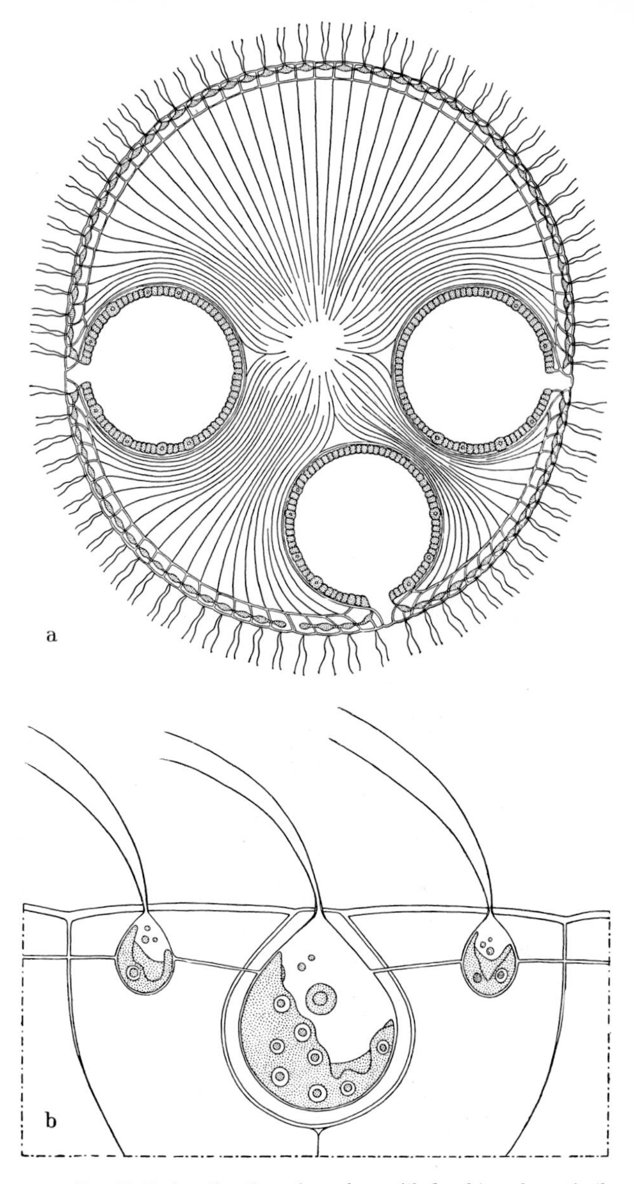

Fig. 319. *Volvox*. a *V. globator* EHRENBERG. Longitudinal section through a colony with daughter spheres (in the posterior half). v *V. aureus* EHRENBERG. A generative cell between two somatic cells (schematic). After JANET

LEMMERMANN, E.: Eugleninae. In: Die Süßwasserflora Deutschlands. Heft 2. Jena: Fischer 1913.
WOLKEN, J. J.: Euglena. 2. Ed. 204 pages. New York: Appleton-Century Crofts 1967.

Euglenoidina either have green plastids or they are colorless. They usually have two flagella originating from an invagination at the front end, the so-called flagellar sac. One of the two flagella is often too short to protrude from the flagellar sac, but it can also be long and dragged behind as a recurrent flagellum. The

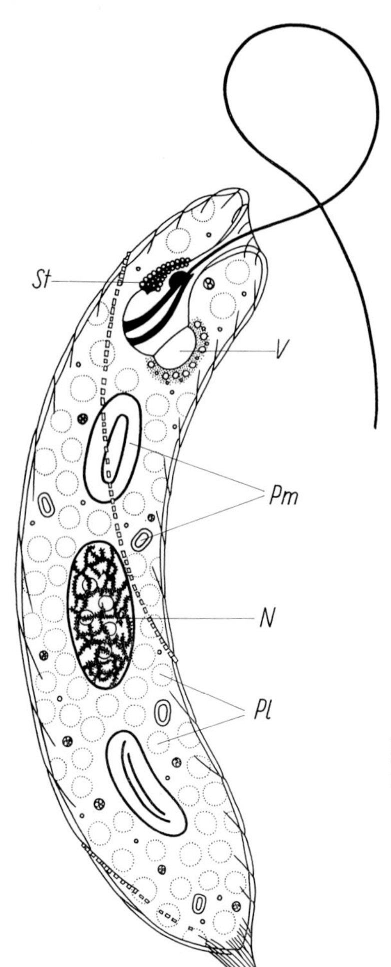

Fig. 320. *Euglena spirogyra* EHRENBERG. Diagram. *St* stigma, *V* vacuole, *Pm* paramylon, *N* cell nucleus, *Pl* plastids. × 1250. After LEEDALE, MEEUSE, and PRINGSHEIM, 1965 [*1016*]

pellicle of euglenoids is highly differentiated and frequently bears spiralling stripes. The stigma, which is not connected to the plastids, is in the vicinity of the flagellar sac. Reserve substance is paramylon, a starch-like carbohydrate deposited in granules or platelets. Reproduction by binary fission. Most euglenoids live in fresh water, a few are marine or parasitic.

a) Colored

Eutreptia. Both flagella directed anteriorly, cell body elongated.
E. marina DA CUNHA, marine.

Euglena. One flagellum short (within the flagellar sac), cell body elongated.
E. spirogyra EHRENBERG (Fig. 320), *E. viridis* EHRENBERG (Fig. 321 a), *E. oxyuris* SCHMARDA (Fig. 321 b), all in fresh water.

Phacus. Cell body flattened, stiff.
P. longicaudus DUJARDIN (Fig. 321 c), fresh water.

Trachelomonas. With a brown, sculptured shell.
T. hispida STEIN (Fig. 321 d), fresh water.

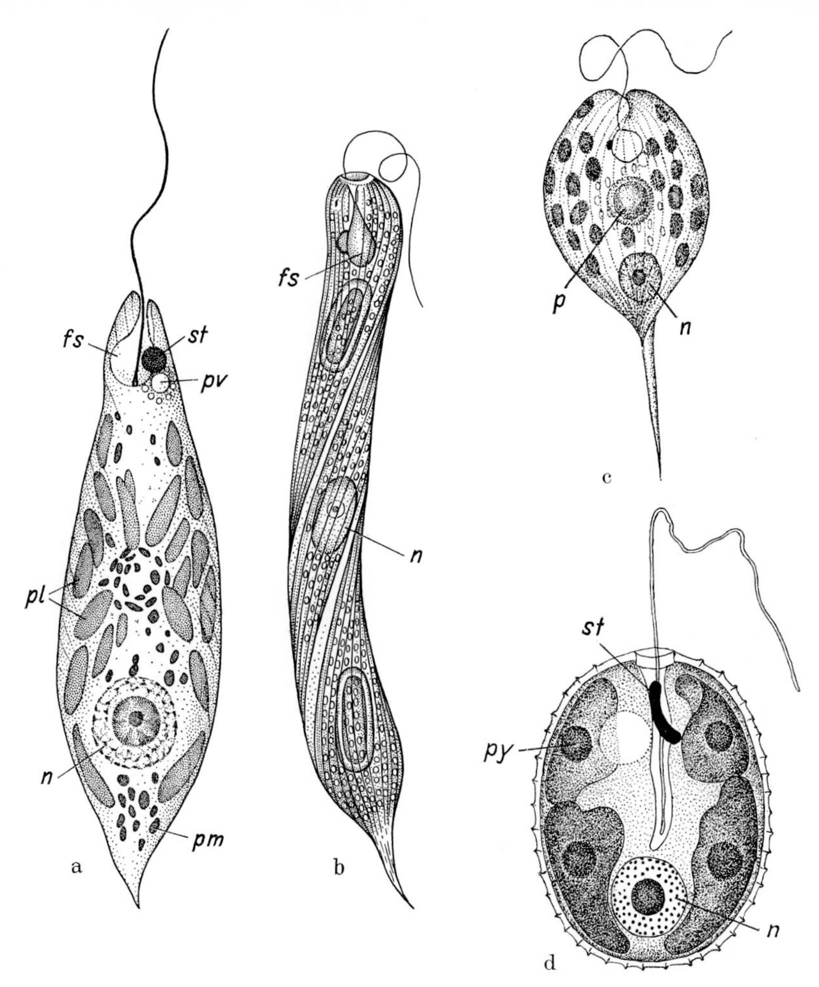

Fig. 321. Colored euglenoids. a *Euglena viridis* EHRENBERG, after DOFLEIN. b *Euglena oxyuris* SCHMARDA, after LEMMERMANN. c *Phacus longicaudus* DUJARDIN. × 800. After STEIN from KÜHN. d *Trachelomonas hispida* STEIN. × 730. After DOFLEIN from KÜHN. *fs* flagellar sac, *st* stigma, *p* paramylon, *py* pyrenoid, *n* nucleus, *pv* pulsating vacuole

b) Colorless

Peranema. Highly metabolic cell body. Closely applied to the flagellar sac is the so-called rod organelle which plays a role in food intake.
P. trichophorum EHRENBERG (Fig. 322a), fresh water.

Anisonema. Cell body stiff, with a short anterior and a long recurrent flagellum. The ventral side bears a groove.
A. costatum CHRISTEN (Fig. 322b), fresh water.

Petalomonas. One flagellum short (within the flagellar sac). Cell body flattened, with longitudinal folds.
P. hovassei MIGNOT (Fig. 322c), fresh water.

Fifth Order: Dinoflagellata

BIECHELER, B.: Recherches sur les Péridiniens. Bull. Biol. Fr. Belg. Suppl. **36** (1952).

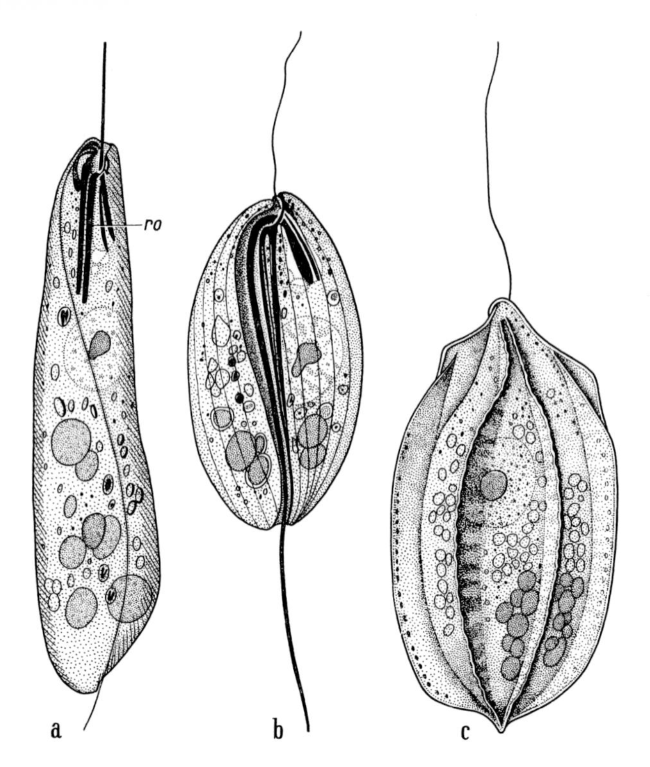

Fig. 322. Colorless euglenoids. a *Peranema trichophorum* EHRENBERG. *ro* so-called rod organelle. b *Anisonema costatum* CHRISTEN. c *Petalomonas hovassei* MIGNOT. After MIGNOT, 1966 [*1128*]

CACHON, J.: Contribution à l'étude des Péridiniens parasites. Cytologie, cycles évolutifs. Ann. Sci. Nat. Zool. (Paris) 12. Sci. **6** (1964).

CHATTON, E.: Les Péridiniens parasites. Morphologie, reproduction, éthologie. Arch. Zool. exp. gén. **59** (1920).

— Classe des Dinoflagellés ou Péridiniens. In: GRASSÉ, P. P.: Traité de Zoologie. **1**, Fasc. 1. Paris: Masson et Cie. 1952.

KOFOID, C. A., SWEZY, C.: The free-living unarmored Dinoflagellates: Berkeley: Mem. University California Vol. 5, 1921.

PETERS, U.: Peridinea. In: GRIMPE-WAGLER: Die Tierwelt der Nord- und Ostsee II. Leipzig: Akadem. Verlagsges. 1930.

REICHENOW, E.: Parasitische Peridinea. In: GRIMPE-WAGLER: Die Tierwelt der Nord- und Ostsee II. Leipzig: Akadem. Verlagsges. 1930.

SCHILLER, J.: Dinoflagellatae. In: RABENHORST: Kryptogamenflora, Bd. 10, Teil 3. 2. Aufl. Leipzig: Akadem. Verlagsges. 1933.

SCHILLING, A.: Dinoflagellatae (Peridineae). In: Die Süßwasserflora Deutschlands. Heft 3. Jena: Fischer 1913.

Dinoflagellates either have yellow-brownish plastids or they are colorless. One of their two flagella undulates longitudinally, the other one transversely. In most species the longitudinal flagellum originates in a longitudinal groove (sulcus) while the transverse flagellum undulates in a transverse groove (annulus) which encircles the cell body much like a belt. The part of the cell body in front of the transverse groove is called epiconus, the posterior part, hypoconus.

The name "armored flagellates" indicates that they frequently possess a firm cellulose casing differentiated either as an armor of numerous plates or as a shell of two valves.

Chromosomes are always recognizable in the resting nucleus. Extensive sap-filled spaces (pusules) are often found in the cell interior. Some species have stigmata or ocelloids (p. 317). The occurrence of trichocysts (p. 36) has been proven in many cases (nematocysts in *Nematodinium* and *Polykrikos*).

Reproduction by binary or multiple fission (p. 125 and 142). Pelagic species often form chains (Fig. 328).

As a group, dinoflagellates present manifold differentiations. They live in fresh water and in the sea. A large number of marine species are heterotrophic, many are ecto- or endoparasites (p. 350).

The subdivision of dinoflagellates into Adinida and Dinifera is based on a morphological difference which should not be overestimated: the Adinida are probably derived from Dinifera by reduction of the epiconus (compare *Amphidinium*, Fig. 324).

First Suborder: **Adinida**

Without grooves. Flagella originate at the front end. Longitudinal fission.

Prorocentrum. With a thorn-like protrusion at the front end (probably a remainder of the epiconus).
  P. micans Ehrenberg, marine.

Exuviaella. Without protrusion, shell of two valves.
  E. marina Cienkowski (Fig. 38 b—d), marine.

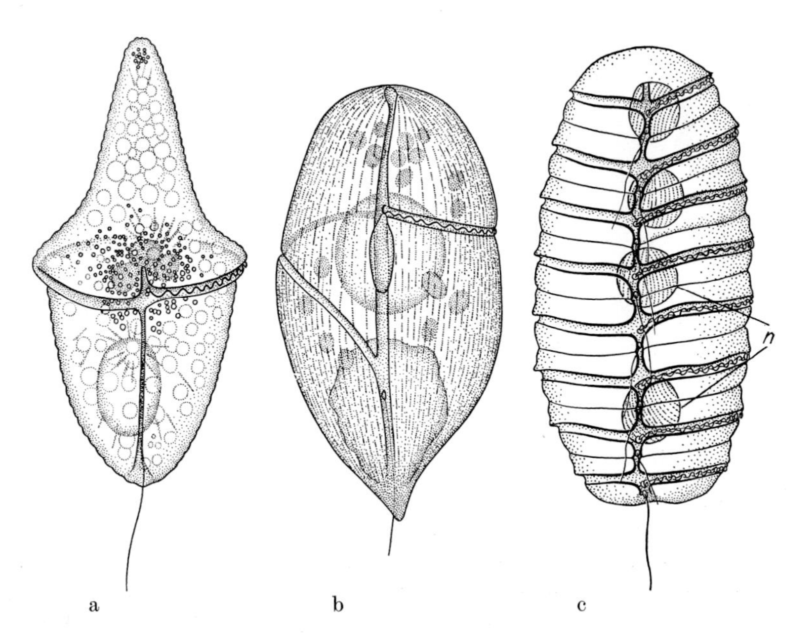

Fig. 323. Gymnodinida. a *Gymnodinium dogieli* Kofoid et Swezy. × 360. b *Gyrodinium postmaculatum* Kofoid et Swezy. × 780. c *Polykrikos schwartzi* Bütschli, *n* nuclei. × 450. After Kofoid and Swezy, 1921 [*956*]

Second Suborder: **Dinifera**

With grooves (exception: Dinophysida). Flagella originate at the side. Transverse fission.

1. Family Group: Gymnodinida

Without armor plates, but with longitudinal and transverse grooves.

Gymnodinium. Transverse groove in the middle of the cell body.
G. dogieli KOFOID et SWEZY (Fig. 323a), marine.
Gyrodinium. Helical transverse groove.
G. postmaculatum KOFOID et SWEZY (Fig. 323b), marine.
Polykrikos. Numerous nuclei, grooves and flagella. The number of nuclei is half as large as the number of groove systems.
P. schwartzi BÜTSCHLI (Fig. 323c), marine.
Amphidinium. Transverse groove close to the anterior pole of the cell. Epiconus therefore small, often tooth-like.
A. elegans GRELL et WOHLFARTH-BOTTERMANN (Fig. 324), marine.

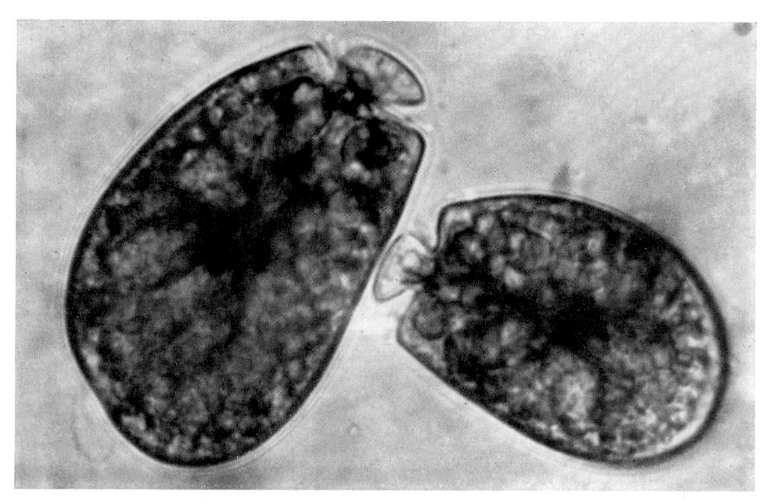

Fig. 324. *Amphidinium elegans* GRELL et WOHLFARTH-BOTTERMANN. × 1400. After GRELL and WOHLFARTH-BOTTERMANN, 1957 [*684*]

Dissodinium. Blister-like. Reproduction by multiple fission, swarmer *Gymnodinium*-like.
D. lunula SCHÜTT (Fig. 131), marine.
Noctiluca. Large gelatinous sphere, traversed by cytoplasmic strands which originate at the so-called central plasma. Small flagellum interpreted as rudimentary longitudinal flagellum. Tentacle for capturing prey. Binary and multiple fission (Fig. 125). Causes "marine luminescence".
N. miliaris SURIRAY (Fig. 325a), marine.
Pomatodinium. Shell shaped like a watch glass, loosely attached.
P. impatiens CACHON et CACHON-ENJUMET (Fig. 325b), marine.
Phytodinium. The alga-like resting cell has neither grooves nor flagella; it reproduces by binary fission. Occasionally, it transforms into a *Gymnodinium*-like swarmer.
P. marinum GRELL (Fig. 326), marine.

The *parasitic dinoflagellates* are also to be included in the Gymnodinida. Their morphology and reproduction are highly modified (comp. p. 350 and 356).

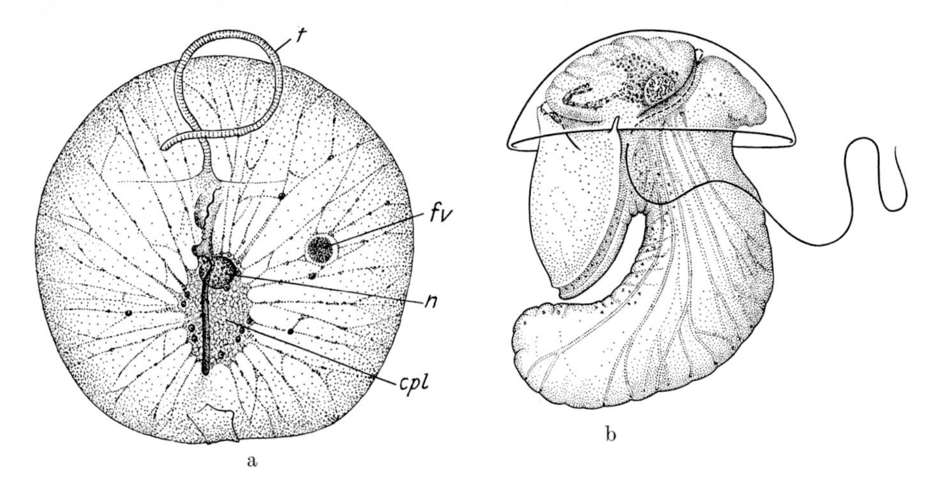

Fig. 325. Gymnodinida. a *Noctiluca miliaris* SURIRAY. *cpl* Central plasma, *n* nucleus, *fv* food, *t* tentacle. × 110. After PRATJE from KÜHN. b *Pomatodinium impatiens* CACHON et CACHON-ENJUMET with a shell shaped like a watch glass. × 200. After CACHON and CACHON-ENJUMET, 1966 [*191*]

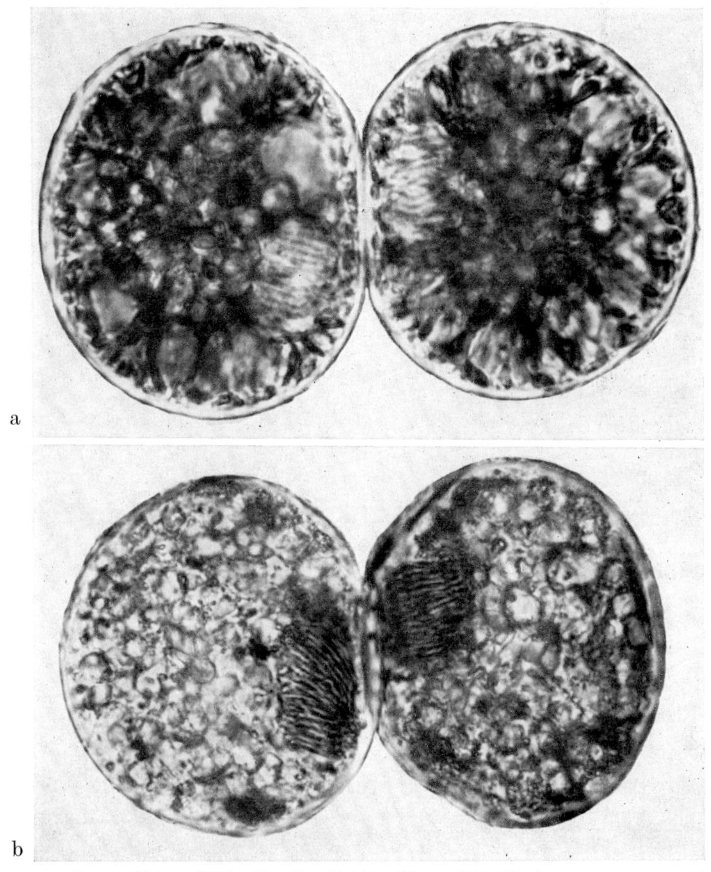

Fig. 326. *Phytodinium marinum* GRELL. Sister cells shortly after division. The nuclei with chromosomes arranged in parallel rows can be recognized at the point where the two cells touch. a In life. b The same cells after staining with aceto-carmine. × 1100. Photomicrographs

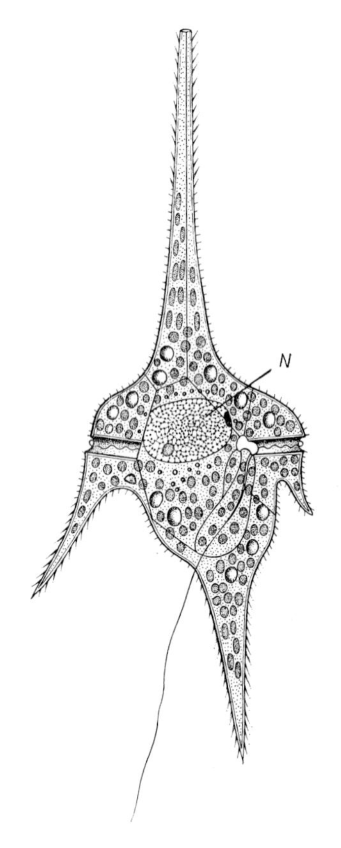

Fig. 327. *Ceratium hirundinella* Müller. *N* Nucleus. × 450. After Lauterborn

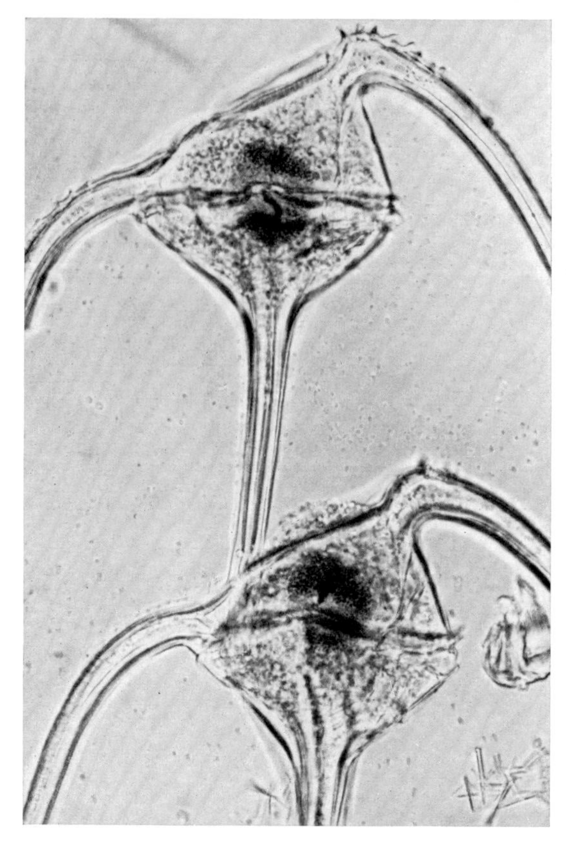

Fig. 328. *Ceratium tripos* Nitzsch. The posterior individual is connected to the anterior one by a posterior horn-like protrusion of the latter. Aceto-carmine

2. Family Group: Peridinida

Cell envelope reinforced by armor like plates of definite number and arrangement.

Peridinium. Rounded shape.
P. tabulatum Claparede et Lachmann, fresh water.

Ceratium. Flattened shape with horn-like protrusions.
C. hirundinella Müller (Fig. 327) fresh water, *C. tripos* Nitzsch (Fig. 328), marine.

3. Family Group: Dinophysida

Armor divided by a sagittal seam into two usually not quite symmetrical halves. Transverse and longitudinal "grooves" not deepened but bordered by "alar ledges" which may be drawn out to protrusions as an aid in floating.

Pelagic oceanic species (Fig. 329).

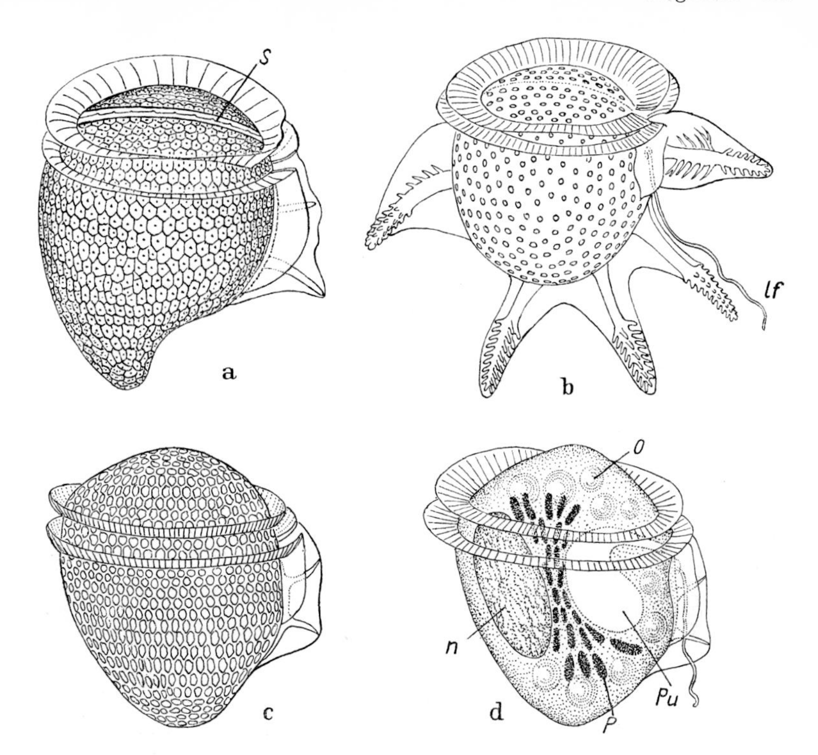

Fig. 329. Dinophysida. Species of the genus *Phalacroma*. a *P. mitra* SCHÜTT. b *P. jourdani* SCHÜTT. c *P. vastum* SCHÜTT, armor. d The same, cell contents. *S* Sagittal seam, *Lf* longitudinal flagellum, *O* oil droplet, *Pu* pusule, *P* plastids, *n* nucleus. Magnification in a, c, d × 575, in b × 470. After SCHÜTT from KÜHN

Sixth Order: Protomonadina

GRASSÉ, P.P.: Ordre des Trypanosomides. In: GRASSÉ, P.P.: Traité de Zoologie. **1**, Fasc. 1. Paris: Masson et Cie. 1952.

HOLLANDE, A.: Ordre des Choanoflagellés ou Craspédomonadines. In: GRASSÉ, P.P.: Traité de Zoologie. **1**, Fasc. 1. Paris: Masson et Cie. 1952.

LEMMERMANN, E.: Protomastiginae. In: Die Süßwasserflora Deutschlands. Heft 1. Jena: Fischer 1914.

All colorless flagellates with only one or two flagella which cannot be directly derived from any of the preceding genera simply by loss of plastids are classified as Protomonadina.

This order, certainly not a natural group, includes many free-living and parasitic flagellates. They live on bacteria or organic substances set free through bacterial decay or available directly from their hosts. Some of the free-living protomonads are temporarily or permanently sessile. They may either attach directly to the substrate, often by a filiform extension of the posterior end (e.g. *Amphimonas*), or they may be enclosed in a delicate shell (e.g. *Bicoeca*).

The so-called *choanoflagellates* bear a structure termed *collare* at the front end. It consists of long projections difficult to distinguish with the light microscope (Fig. 330). The similarity of the choanoflagellates to the choanocytes of sponges led to the assumption that the latter derive phylogenetically from the former. It

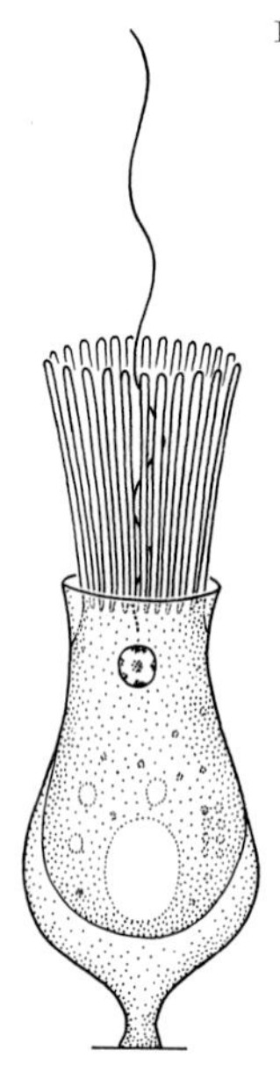

Fig. 330. *Salpingoeca amphoroideum* CLARK

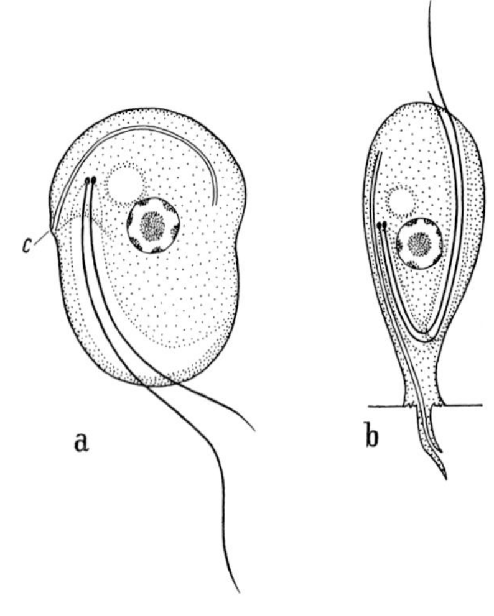

Fig. 331. *Ichthyobodo necator* HENNEGUY (syn. *Costia necatrix*). a Free-swimming form, ventral surface (*c* Cytostome), b fixed form, attached to host cell. After JOYON and LOM, 1969 [*877*], slightly modified

has to be pointed out, however, that choanocytes have been found also in the epidermis of enteropneusts. On the other hand, the choanoflagellates might have evolved from chrysomonads by loss of plastids since one chrysomonad species (*Stylochromonas minuta* LACKEY) has a similar collar [*1354*].

The predominantly free-living Bodonidae are now classed together with the Trypanosomidae as "Kinetoplastida" because they too, frequently have a *kinetoplast* (p. 22) [*810*].

To the *Bodonidae* belongs *Ichthyobodo necator (Costia necatrix)* which exists in a free-swimming and a fixed form (Fig. 331). In the latter case it is an ectoparasite of fishes forming a sucking organelle which penetrates the host cell.

As the *Trypanosomidae* include, besides harmless commensals, many pathogenic species, they are of great practical importance and have therefore received much study. They are characterized by *polymorphism*. Depending upon the host lived in, the same species can appear in various modification types. Essentially four different types can be distinguished:

1. The *Leishmania*-form: no flagellum, rounded cell body.
2. The *Leptomonas*-form: flagellum inserted at the front end.
3. The *Crithidia*-form: flagellum inserted at the middle of the cell (in front of the nucleus), directed anteriorly as marginal thread of a short undulating membrane.
4. The *Trypanosoma*-form: flagellum inserted at the posterior end, running up to the front end as marginal thread of a long undulating membrane.

The *Leptomonas*- and *Crithidia*-form are found predominantly in invertebrate hosts, the *Trypanosoma*-form in the vertebrate host. The *Leishmania*-form can

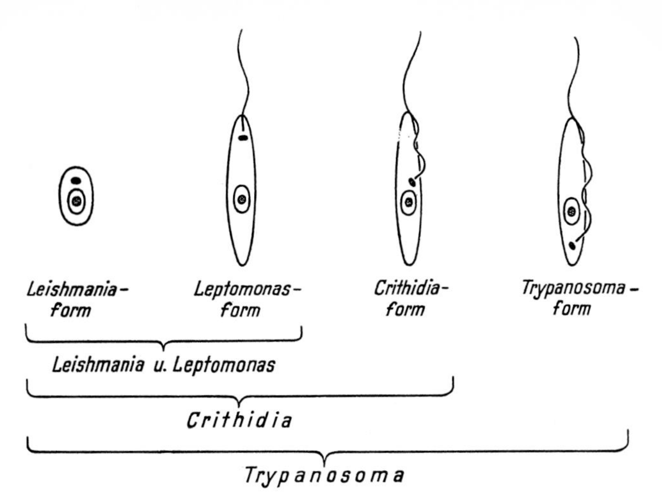

Fig. 332. Modification types in trypanosomids. The brackets indicate which modification types can be formed by the individual genera

appear in both, intracellularly in the vertebrate host and extracellularly (i.e. attached to cells) in the invertebrate host.

Which of the modification types can become manifest in the various genera is indicated in the preceding survey (Fig. 332):

Leishmania. Alternates between vertebrate host (*Leishmania*-form in cells of the reticuloendothelial system) and insect host (*Leptomonas*-form in the gut lumen). *L.*-species important as *pathogenic parasites of man* are all transmitted by butterfly-midges or sandflies (phlebotomids), in the Mediterranean area mainly by *Phlebotomus papatasii.*

L. donovani LAVERAN et MESNIL. Causative agent of visceral leishmaniasis (kala-azar).

Visceral leishmaniasis (kala-azar) is widespread in South Asia (especially India), South China, and in the Mediterranean countries. A major characteristic of the disease is a considerable enlargement of the spleen due to blockage of the reticuloendothelial system by the parasites. It is usually fatal if untreated.

L. tropica WRIGHT. Causative agent of skin leishmaniasis.

Skin leishmaniasis (oriental sore) is a disease endemic in warmer countries and its geographical distribution is much like that of visceral leishmaniasis. Parasitic infection is restricted to the endothelium of skin capillaries and leads to lump-like boils.

L. mexicana BIACI and *L. brasiliensis* VIANNA. Causative agents of the South American leishmaniasis of skin and mucoid membranes.

Leptomonas. Without alternation of hosts. In the gut of many insects and other invertebrates (*Leptomonas*-form free-swimming in the intestinal lumen, *Leishmania*-form attached to the intestinal epithelium).

L. jaculum WOODCOOK. In the water scorpion *Nepa cinerea.*

The genus *Phytomonas,* whose species live in the milk-sap of tropical plants and are transmitted by bugs, is distinguished from the morphologically identical genus *Leptomonas.*

Crithidia. Without alternation of hosts. In the gut of many insects.

C. gerridis PATTEN. In the water strider *Gerris.*

Trypanosoma. Alternates between vertebrate (*Trypanosoma*-form in blood) and invertebrate host (*Leishmania-*, *Leptomonas-* or *Crithidia*-form in the gut).

T. granulosum LAVERAN et MESNIL. In the eel. Vector: the leech *Hemiclepsis marginata.*

T. rotatorium MAYER. In the European green frog. Vector: the leech *Hemiclepsis marginata.*

T. lewisi KENT. In rats. Vector: fleas. This species can reproduce by binary or multiple fission.

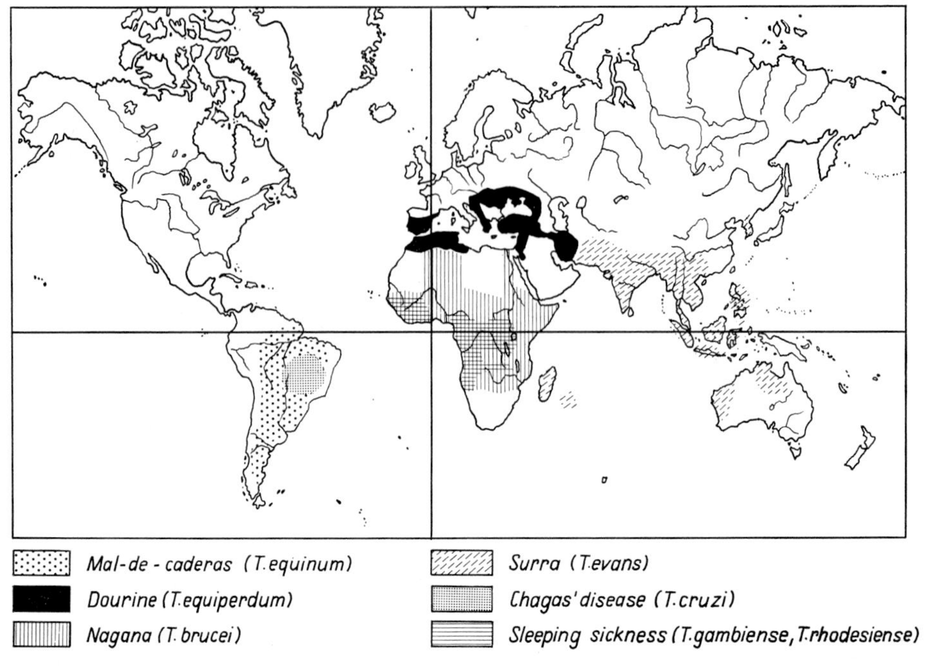

Fig. 333. Distribution of some diseases caused by trypanosomids. After DOFLEIN and REICHENOW

T. brucei PLIMMER et BRADFORD (Fig. 115). Causative agent of the nagana disease (Fig. 333). *Nagana* is a disease of African domesticated animals (horses, cattle) transmitted by the tsetse flies *Glossina morsitans* and *Glossina palpalis*. The African species of large game (zebras, antelopes etc.) play the role of reservoir hosts. *T. brucei* cannot be transmitted to man.

T. evansi STEEL, causative agent of surra (Fig. 333).

Surra is a disease of domesticated animals (horses, camels) in the North African and South Asian range (the name is Indian), transmitted by horseflies. (tabanids). The parasite does not undergo development in the tabanid; it is only mechanically transmitted.

T. equinum VOGES. Causative agent of mal-de-caderas (Fig. 333). The species lacks a kinetoplast.

Mal-de-caderas (croup paralysis) is a South American disease of horses, also transmitted by horseflies (tabanids).

T. equiperdum DOFLEIN. Causative agent of dourine (Fig. 333).

Dourine is a venereal disease of horses and donkeys, especially widespread in Mediterranean countries. In contrast to other *Trypanosoma*-species, transmission is not due to an insect vector but takes place as the mare is covered.

T. gambiense DUTTON and *T. rhodesiense* STEPHENS et FANTHAM. Causative agents of sleeping sickness (Fig. 333).

Sleeping sickness is a dangerous disease of man in tropical Africa. It is transmitted by tsetse flies, of which *Glossina palpalis* plays the major role. Within the fly the trypanosomes are predominantly in the *Crithidia*-form. It is only in the salivary gland that they change to the *Trypanosoma*-form. After they have passed with the fly's saliva into man, the trypanosomes multiply at first in the vicinity of the injection point, leading often to a furuncle-like primary symptom. After about a week they appear in the blood stream. Infestation of the lymph system leads to glandular swelling which is symptomatic for sleeping sickness (neck glands). After months, trypanosomes penetrate into the cerebro-

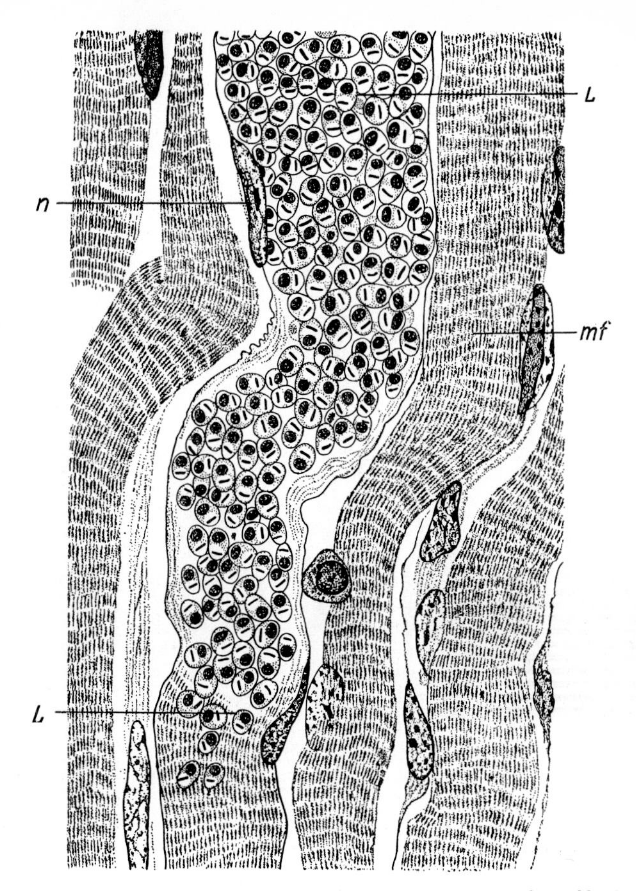

Fig. 334. *Trypanosoma cruzi* CHAGAS. Muscle fibers (*mf*) of the rat with *Leishmania*-forms (*L*). *n* nucleus of host cell. After BRUMPT

spinal fluid. They cause brain damage and bring about the lethargy characteristic for sleeping sickness. If untreated, the disease will usually lead to death.

T. cruzi CHAGAS. Causative agent of CHAGAS' disease (Fig. 333). In contrast to other species of *Trypanosoma*, *T. cruzi* multiplies as *Leishmania*-form in the vertebrate host.

Chagas' disease is widespread in South and Middle America. Especially children (but also dogs and cats) are affected. It is transmitted by bugs of the genus *Triatoma*. Within the bug, it multiplies as the *Crithidia*-form. Transmission to man is not due to the bug's bite, but through its feces. The trypanosomes penetrate through mucous membranes into different tissue cells (e.g. muscle fibers, Fig.. 334) where they change to the *Leishmania*-form and even multiply as such. They are released into the blood when the host cell bursts but cannot multiply in the blood. The bugs infect themselves when sucking blood and spread the infection by eating their feces. An acute infection in childhood is rarely fatal, but after a latency period of many years, severe subsequent injuries may result in the form of denervation and dilation of the heart and other hollow organs.

Seventh Order: Diplomonadina

Diplomonadina are bilaterally symmetric double individuals. Nuclei and flagellar groups are present in duplicate. Usually, eight flagella are present.

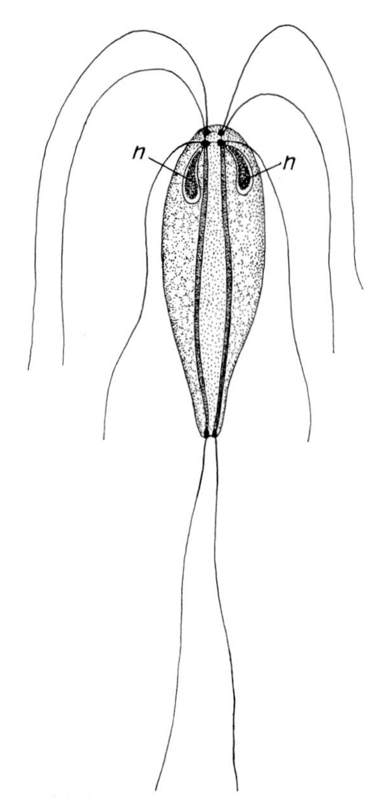

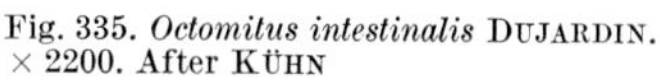
Fig. 335. *Octomitus intestinalis* DUJARDIN. × 2200. After KÜHN

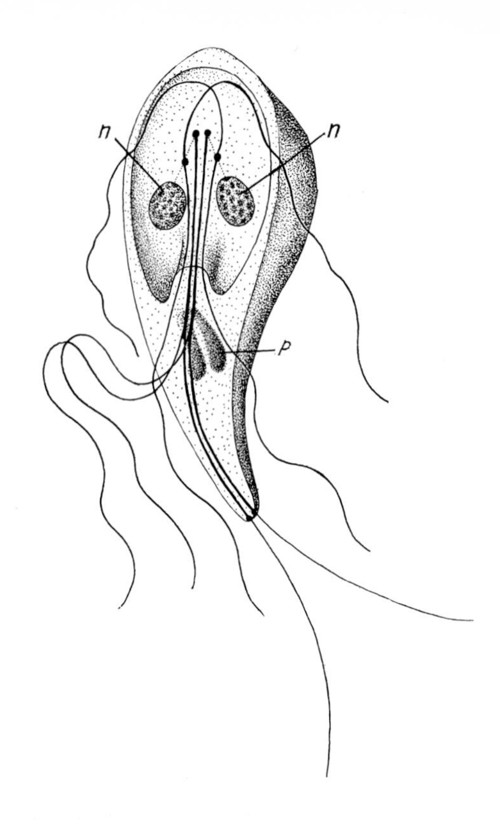

Fig. 336. *Lamblia intestinalis* LAMBL. *n* Nuclei, *p* so-called parabasal body. × 3200. After RODENWALDT from KÜHN

Octomitus. With two anterior flagellar groups of three flagella each and two tail flagella which arise from two thickened longitudinal fibrils.

O. intestinalis DUJARDIN (Fig. 335) in the frog intestine.

Lamblia. Dorsal side convex, ventral side flattened and deepened anteriorly to form a suction groove (for attachment to cells of the intestinal epithelium). The flagella form a system of intracellular fibrils before they leave the cell body. Two flagellar pairs leave the cell laterally, one at the ventral side and one at the posterior end (tail flagella).

L. intestinalis LAMBL (Fig. 336). Most frequent intestinal flagellate of man. Cysts are voided with the feces.

Eighth Order: Polymastigina

CLEVELAND, L. R., HALL, S. R., SANDERS, E. P., COLLIER, J.: The wood-feeding roach *Cryptocercus*, its Protozoa and the symbiosis between Protozoa and roach. Mem. Amer. Acad. Arts Sci. **17** (1934).

GRASSÉ, P. P.: Ordre des Trichomonadines. In: GRASSÉ, P. P.: Traité de Zoologie. **1**, Fasc. 1. Paris: Masson et Cie. 1952.

HONIGBERG, B. M.: Evolutionary and systematic relationship in the flagellate order Trichomonadida KIRBY. J. Protozool. **10** (1963).

KIRBY, H.: The devescovinid flagellates of termites. I—V. Univ. Calif. Publ. Zool. **45** (1941/49)

Polymastigina have usually four or more flagella and typical organelles (axostyles, parabasal bodies etc.) connected with basal bodies. Reproduction by binary fission. Sexual processes have been observed in species inhabiting the roach, *Cryptocercus punctulatus*, and also in some inhabiting termites (p. 167).

Most Polymastigina live as parasites, commensals or symbionts in the gut of arthropods or vertebrates.

There are some flagellates which have less than four flagella, but organelles typical for Polymastigina (esp. Trichomonadida). An example is *Histomonas meleagridis* SMITH, a parasite of the turkey and other birds. It can exist in an amoeboid and in a flagellated form. In the latter case it has usually one flagellum, an axostyle, a parabasal body and a structure called pelta.

First Suborder: **Pyrsonymphida**

Only one nucleus, intranuclear mitosis, with axostyles but without parabasal bodies.

Pyrsonympha. Four to eight flagella are connected with the cell envelope and turn posteriorly in a left-handed helix.

P. vertens LEIDY. In the termite *Reticulotermes flaviceps*.

The following species are found in the roach *Cryptocercus punctulatus:*

Oxymonas doroaxostylus CLEVELAND.

Saccinobaculus ambloaxostylus CLEVELAND (Fig. 160).

Notila proteus CLEVELAND (Fig. 191).

Second Suborder: **Trichomonadida**

Only one nucleus, extranuclear mitosis, with axostyle and parabasal body, four to six flagella, one directed backward as a recurrent flagellum.

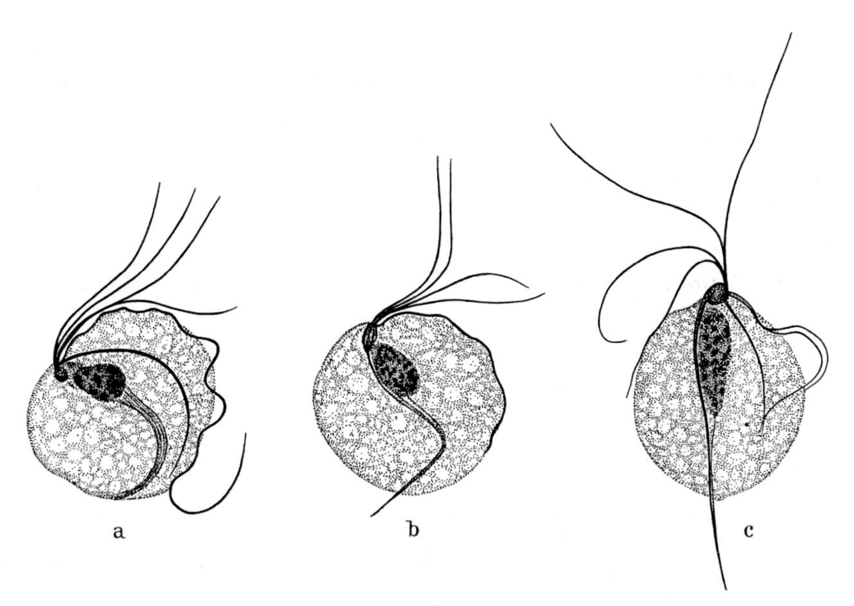

Fig. 337. *Trichomonas* species of man. a *T. hominis* DAVAINE. b *T. tenax* MÜLLER. c *T. vaginalis* DONNÉ. × 2160. After WESTPHAL

Devescovina. Free, usually tape-like recurrent flagellum. A parabasal body is frequently wound about the anterior part of the axostyle. Numerous wood-eating species in the gut of termites (Fig. 21 and 116).

Trichomonas. Recurrent flagellum modified into an undulating membrane.
T. termopsidis KOFOID et SWEZY (Fig. 11). In the termite *Termopsis angusticollis*.

Many species live in the alimentary tract of vertebrates and feed on bacteria:
T. lacertae BUETSCHLI. In lizards.
T. muris HARTMANN. In mice.

Trichomonas species in man (Fig. 337):
In the intestine:
T. fecalis CLEVELAND. Three free flagella.
T. hominis DEVAINE. Four free flagella.
T. ardin delteili DERRIEU et RAYMOND. Five free flagella.

In the mouth:
T. tenax MÜLLER. Four free flagella.

In the vagina:
T. vaginalis DONNÉ. Four free flagella.

Third Suborder: **Calonymphida**

Numerous nuclei, each of these may be equipped with a flagellar group, parabasal bodies and axostyles. Such complexes are often multiplied without an attached nucleus. In the gut of termites.

Calonympha. Numerous flagella originate from the anterior half of the cell.
C. grassii FOA (Fig. 257n). In *Cryptotermes grassii*.

Snyderella. Only the posterior end is free of flagellar bundles.
S. tabogae KIRBY (Fig. 338). In *Calotermes longicollis*.

Fourth Suborder: **Hypermastigida**

Numerous flagella, but only one nucleus. The latter is usually fastened at the front end in a special nuclear sac. The mitotic spindle is always formed outside the nucleus.

Due to their rich supply of parabasal bodies, axostyles, flagellar bands etc., the hypermastigids appear as the most highly differentiated colorless flagellates. As intestinal inhabitants of termites and roaches, they feed to a large extent on pieces of wood taken up at the rear end.

Joenia. With a flagellar bundle and an axostyle. Most primitive trichomonad-like hypermastigids.
J. duboscqui HOLLANDE (Fig. 339). In *Calotermes praecox*.

Lophomonas. With a flagellar bundle and an axostyle which widens anteriorly to a goblet-shaped calyx surrounding the nucleus.
L. blattarum STEIN (Fig. 340). In the rectum of the roach *Blatta orientalis*.

Spirotrichonympha. Flagella originate from flagellar bands which follow a helical course within the ectoplasm of the cell. With axostyle.
S. bispira CLEVELAND (Fig. 341 and 118). In *Calotermes simplicicornis*.

Holomastigotoides. Similar to the preceding genus, but without uniform axostyle.
H. tusitala CLEVELAND (Fig. 119). In *Prorhinotermes*-termites.

Eucomonympha. Front end (rostrum) broad and filled with a hyaline mass. Flagella originate partly at the rostrum, partly in longitudinal rows along the cell body.
E. imla CLEVELAND (Fig. 161). In *Cryptocercus punctulatus*.

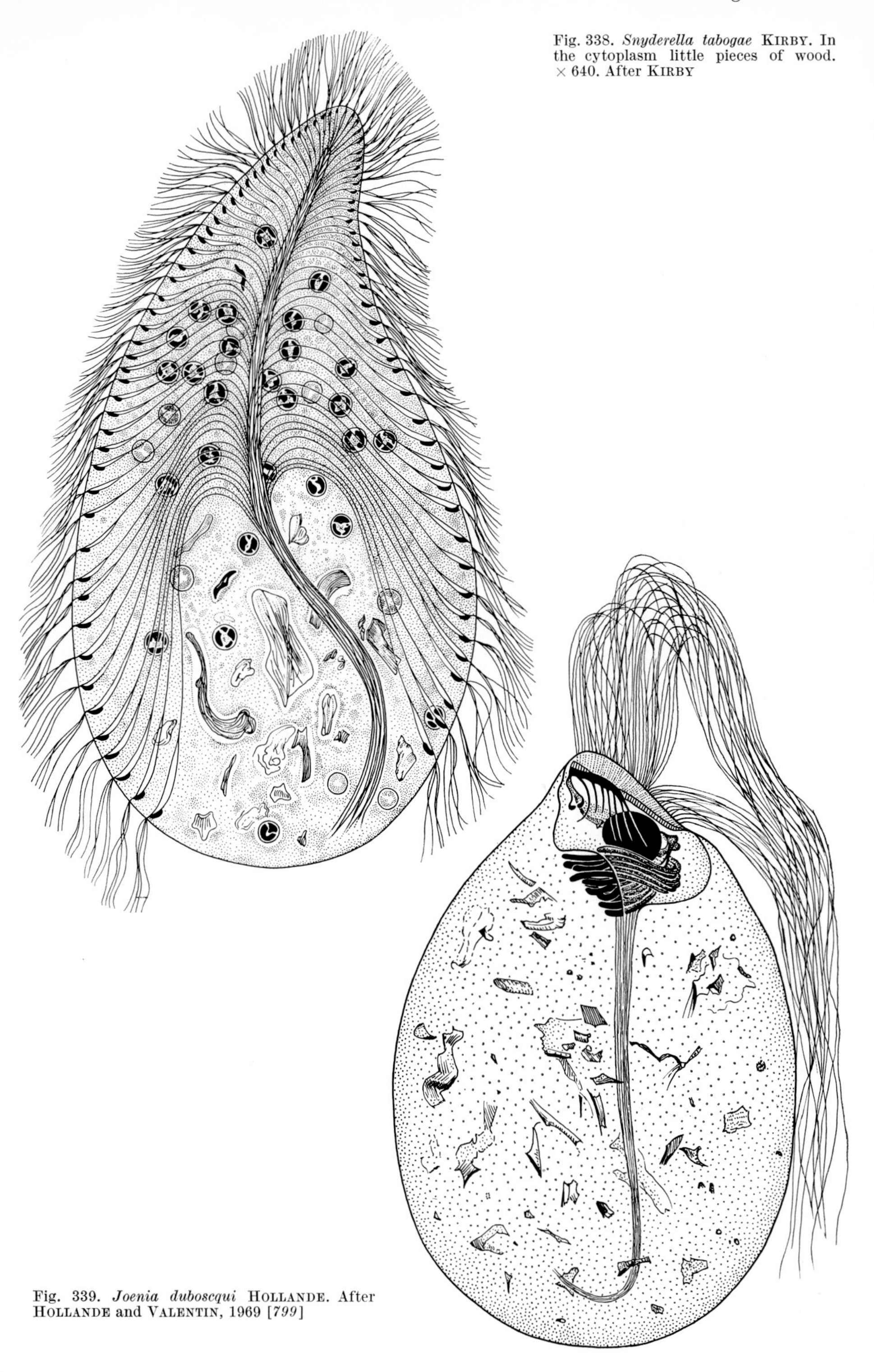

Fig. 338. *Snyderella tabogae* KIRBY. In the cytoplasm little pieces of wood. × 640. After KIRBY

Fig. 339. *Joenia duboscqui* HOLLANDE. After HOLLANDE and VALENTIN, 1969 [*799*]

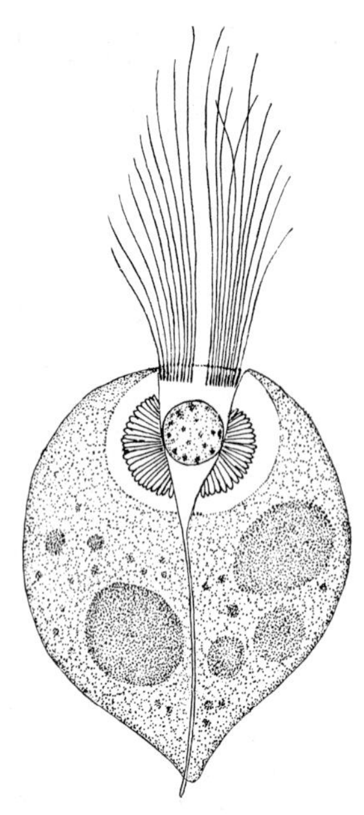

Fig. 340. *Lophomonas blattarum* STEIN. × 660. After JANICKI from KÜHN

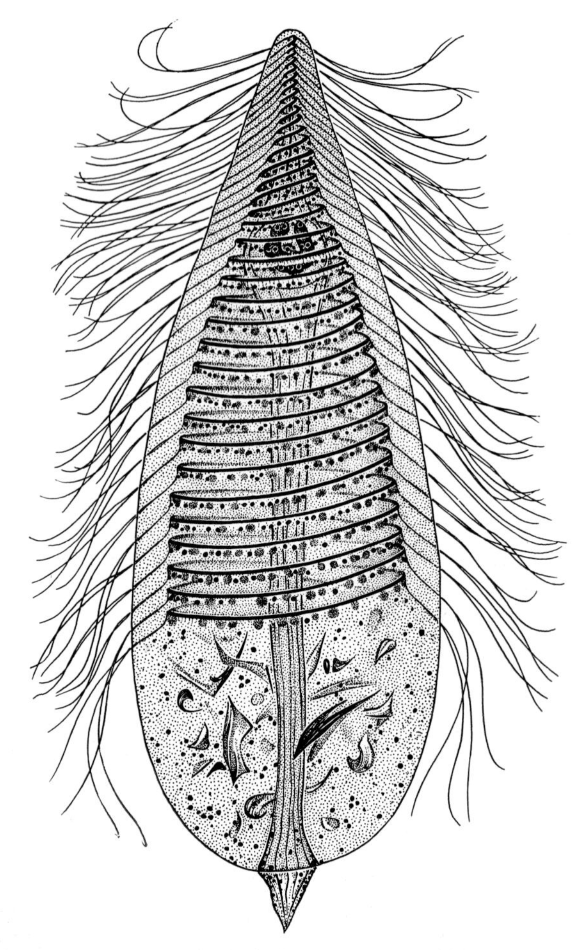

Fig. 341. *Spirotrichonympha bispira* CLEVELAND. × 940. After CLEVELAND

Trichonympha. The front end (rostrum) is set off from the remaining cell body by a circular cleft and is capped anteriorly by an operculum filled with liquid. Numerous species in termites and *Cryptocercus punctulatus* (Fig. 164).
T. acuta CLEVELAND (Fig. 342). In *Cryptocercus punctulatus.*

Leptospironympha. Flagella originate from two broad bands.
L. wachula CLEVELAND (Fig. 158 and 163). In *Cryptocercus punctulatus.*

Barbulanympha. Two flagellar groups at the front end. Numerous parabasal bodies and axostyles. Large centrioles visible in life.
B. ufalula CLEVELAND [Fig. 343). In *Cryptocercus punctulatus.*

Teratonympha. Flagella originate at the front end (rostrum) and from belt-like bands encircling the cell.
T. mirabilis KOIDZUMI (Fig. 344). In termites of the genus *Reticulotermes.*

▶

Fig. 342. *Trichonympha acuta* CLEVELAND. The top of the rostrum is covered by a double cap. Flagella originate only from the anterior cell half. The parabasal bodies surround the nucleus. In the posterior cell half: phagocytized pieces of wood. × 800. After CLEVELAND, 1962 [*307*]

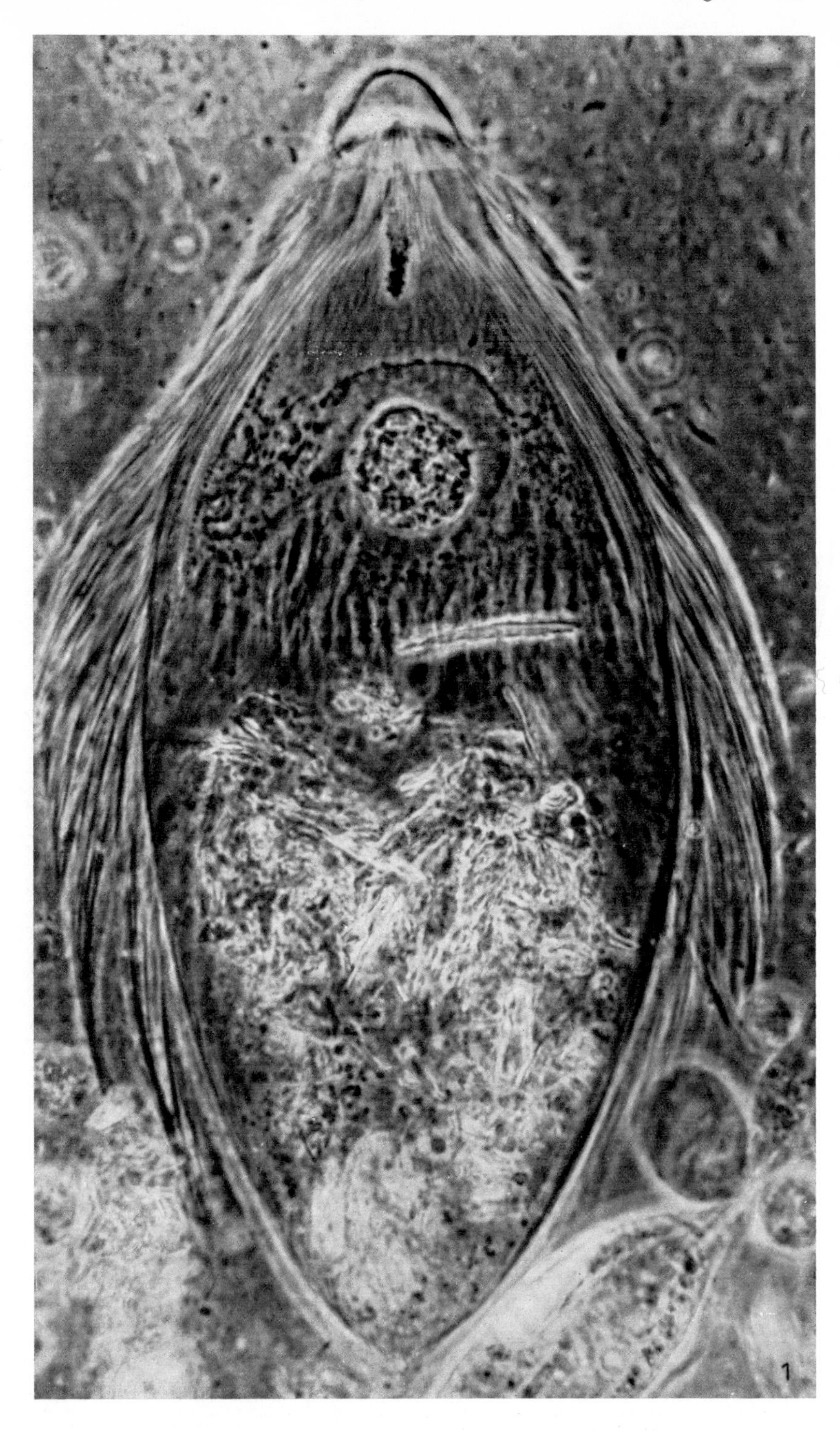

Fig. 342

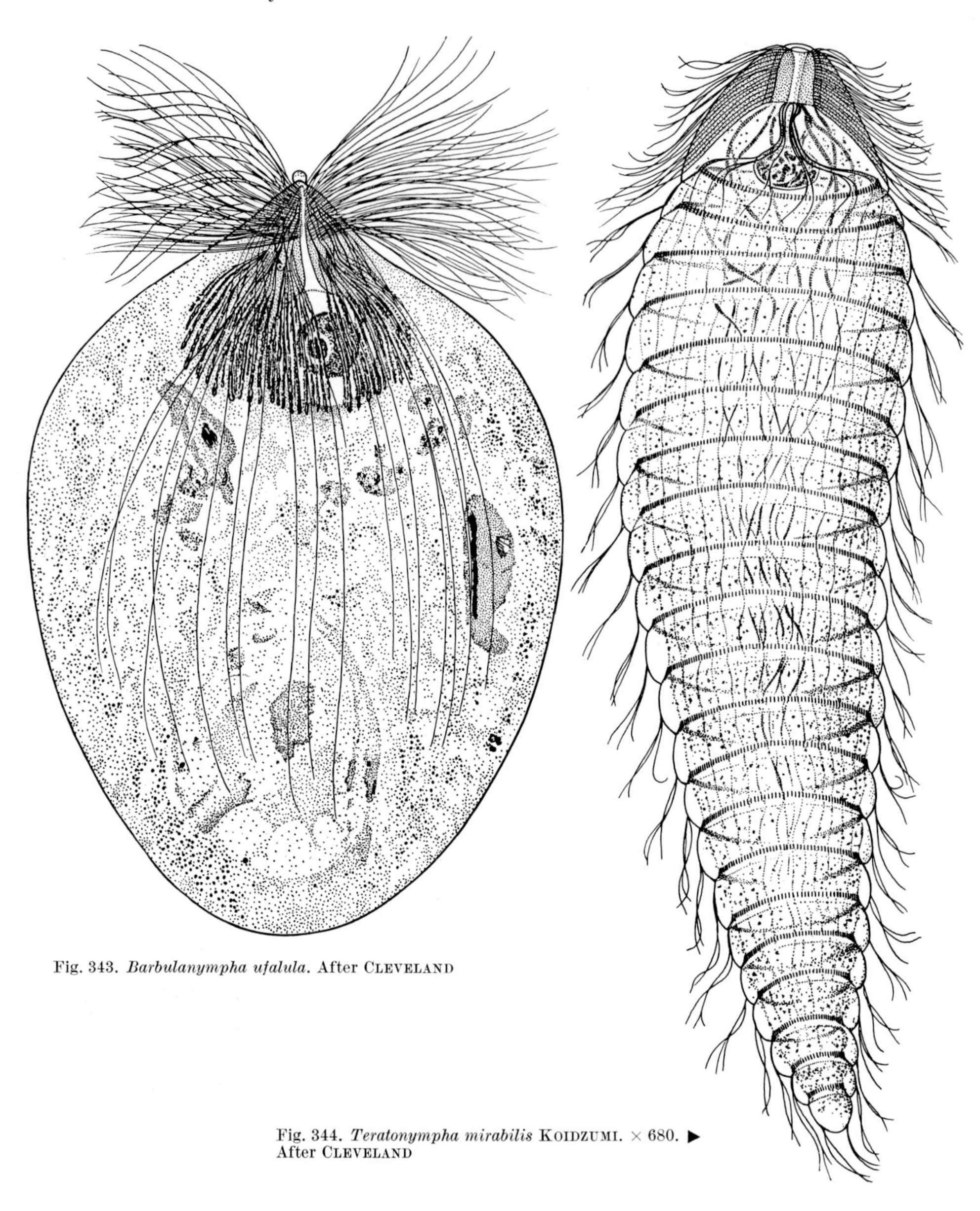

Fig. 343. *Barbulanympha ufalula*. After CLEVELAND

Fig. 344. *Teratonympha mirabilis* KOIDZUMI. × 680. ▶
After CLEVELAND

▶

Fig. 345. *Opalina*. Developmental cycle in the adult frog (*1*—*5*) and tadpole (*6*—*11*). After being ingested by a young frog a uninucleate agamont emerges from the cyst (*1*). After a growth phase connected with asynchronous nuclear divisions it becomes a multinucleate agamont (*2*). These agamonts which can be found in the recta of adult frogs throughout most of the year multiply by binary divisions being either longitudinal (*3*) or transverse. During the breeding season of the host, many agamonts undergo repeated, consecutive divisions without compensatory growth, thereby becoming successively smaller (*4*). These stages, having 2—5 nuclei, encyst and are evacuated with the feces of the host (*5*). Tadpoles ingest cysts. After hatching (*6*, *7*) the gamont undergoes meiosis. In this only one of the nuclei participates. It is not clear, whether the other nuclei degenerate (as supposed in the diagram) or undergo meiosis in subsequent divisions.The question-marks indicate that it is also uncertain whether meiosis is connected with two cellular divisions leading to gamete formation or gametogenesis occurs later on. In any case macro- and microgametes are formed (*8*, *9*) which join in pairs (*10*) to form zygotes. The zygotes encyst (*11*) and are voided with the tadpoles feces. Agamonts (*1*—*3*) are larger compared with the other stages drawn in the diagram. After investigations of NERESHEIMER (1907), WESSENBERG (1961) and KACZANOWSKI (1971)

Ninth Order: Opalinina

Corliss, J. O.: The opalinid infusorians: flagellates or ciliates? J. Protozool. **2**, 107—114 (1955).
Metcalf, M. M.: The opalinid ciliate infusorians. Smithonian Inst. U.S.N. Museum Bull. **120** (1923).
Wessenberg, H.: Studies on the life-cycle and morphogenesis of *Opalina*. Univ. Calif. Publ. Zool. *61*, 315—370 (1961).

The Opalinina* have many short uniform cilia arranged along the whole body in oblique longitudinal rows from the front end posteriorly. There are either two or numerous nuclei which divide by intranuclear mitosis. All species are parasites.

* Because of their ciliation, the Opalinina have frequently been classified as "Protociliata". However, they differ from "true" ciliates ("Euciliata") by the absence of nuclear differentiation and conjugation.

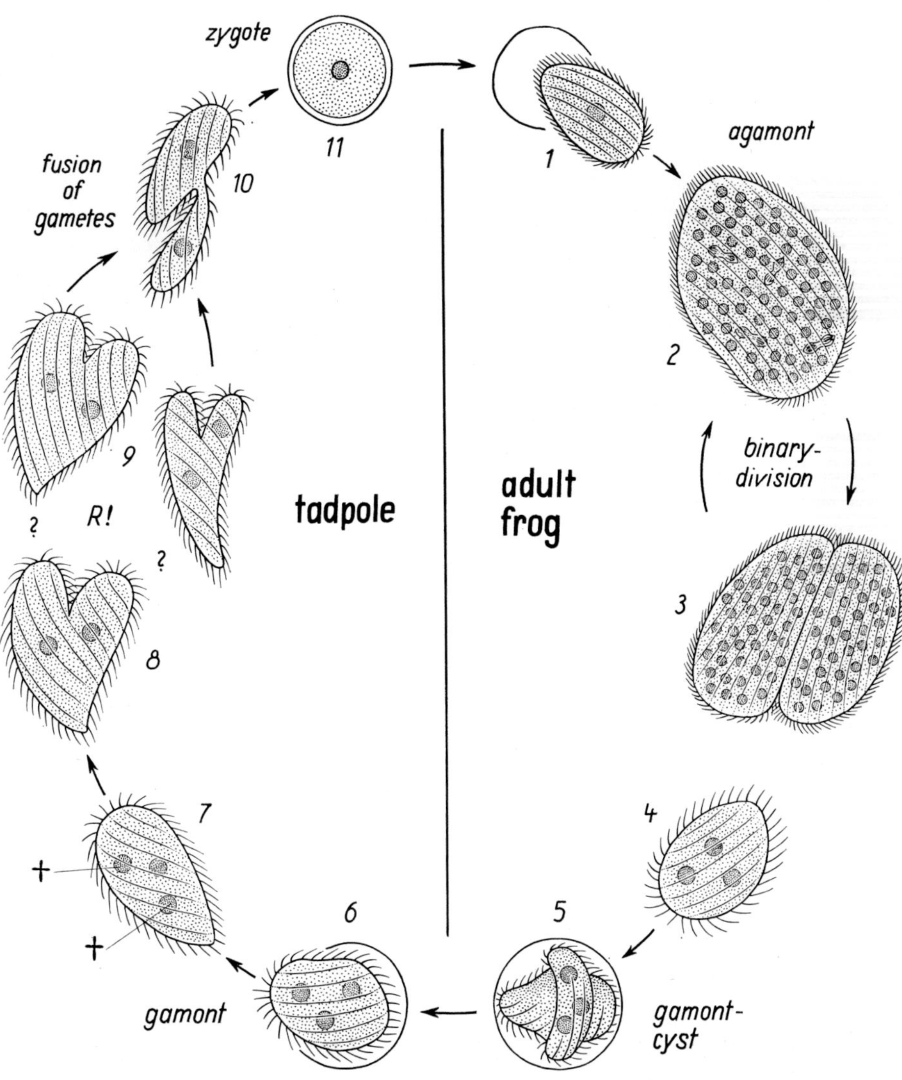

Fig. 345

Food is taken up by the whole body surface. Reproduction by binary fission which is sometimes longitudinal, sometimes transverse. Opalinids appear to be diploids with gametic meiosis. Fig. 345 shows the developmental cycle of *Opalina* which is, however, not yet clarified completely.

Opalinids live predominantly in the hind gut of amphibians, some species in fishes and reptiles.

Protopalina. Cell spindle-shaped, binucleated.
 P. saturnalis LEGER et DUBOSCQ. In the marine fish *Box boops*.

Zelleriella. Cell flat, binucleated.
 Z. elliptica CHEN. In toads.

Opalina. Cell flattened, many nuclei.
 O. ranarum EHRENBERG (Fig. 345). In the grass frog.

Second Class: Rhizopoda

CASH, J., WAILES, G. H.: The british fresh-water Rhizopoda and Heliozoa, Vol. 4. London: Ray Society 1919.

GRASSÉ, P. P.: Traité de Zoologie. **1**, Fasc. 2: Protozoaires (Rhizopodes, Actinopodes, Sporozoaires, Cnidosporidies). Paris: Masson et Cie. 1953.

GROSPIETSCH, TH.: Wechseltierchen (Rhizopoden). Stuttgart: Kosmos-Verlag 1958.

KÜHN, A.: Morphologie der Tiere in Bildern. Heft 2, Teil 2: Rhizopoden. Berlin: Bornträger 1926.

LEIDY, J.: Freshwater rhizopods of North America. Rep. S.U. Geol. Survey Terr. **12** (1879).

PENARD, E.: Faune rhizopodique du Bassin du Léman. Genève: Henry Kündig 1902.

Rhizopoda have no permanent organelles for locomotion. Instead, they move by the aid of *pseudopodia* which can also serve in food uptake. Their mode of nutrition is heterotrophic.

Rhizopoda reproduce by binary or multiple fission. Sexual processes are known only in Heliozoa and Foraminifera.

First Order: Amoebina

CHATTON, E.: Ordre des Amoebiens nus ou Amoebaea. In: GRASSÉ, P. P.: Traité de Zoologie. **1**, Fasc. 2. Paris: Masson et Cie. 1953.

DOBELL, C.: The Amoebas living in Man. London: J. Bale, Sons & Danielson, Ltd. 1919.

JEON, A. L. (Ed.): The Biology of *Amoeba*. New York-London: Academic Press 1972.

SCHAEFFER, A. A.: Taxonomy of the Amoebas. Papers from the Department of Marine Biology of the Carnegie Institution of Washington, Vol. **24**, 1926.

The Amoebina or "naked amoebae" have no fixed body shape, although a polarity is often clearly present. Their cytoplasm may be subdivided into a granular endoplasm filled with inclusions and a hyaline ectoplasm. The nucleus, usually with a central nucleolus, is located in the endoplasm. Many amoebae have several cell nuclei. Most fresh-water species can encyst under unfavorable environmental conditions.

The amoebae are a very heterogeneous group which is difficult to delimit. Although the pseudopodia are often lobular, they can also taper to a point or be thread-like throughout (Fig. 346e). Some species form multinucleated plasmodia with unlimited capacity for growth.

Amoebae living in moist soil often have a tendency to form aggregates prior to encystment. The distribution of such cysts or encysted aggregates may be facilitated by a stalk protruding into the air (Fig. 347).

Such species represent a transition to the so-called *collective amoebae (Acrasina)* which form a "pseudoplasmodium" when the food supply is exhausted. The latter differentiates into a sporophore (Fig. 4) consisting of the spores (= cysts) and stalk cells. Form and size of the sporophore is species-specific (Fig. 348).

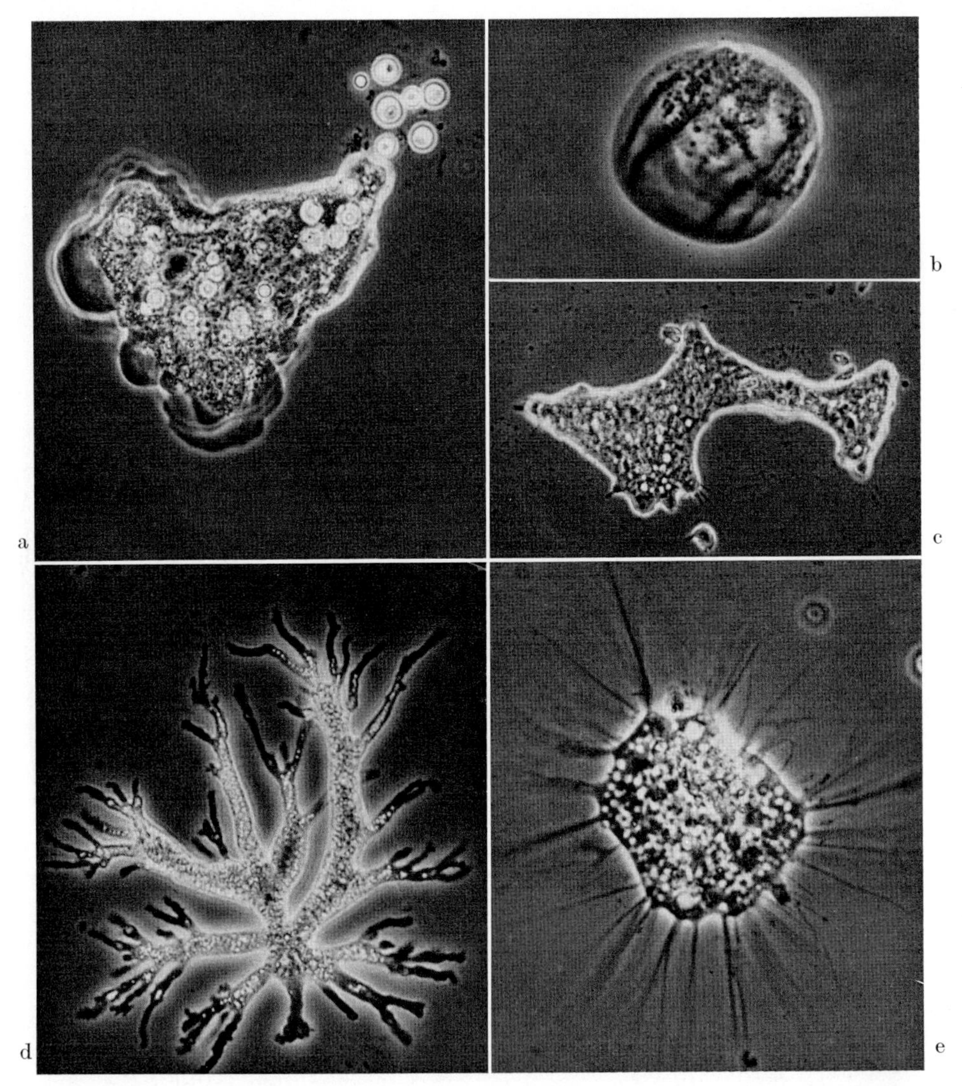

Fig. 346. Free-living amoebae. a *Thecamoeba verrucosa* EHRENBERG. × 390. b *Thecamoeba orbis* SCHAEFFER. × 1440. c *Pontifex maximus* SCHAEFFER. × 310. d *Stereomyxa angulosa* GRELL. × 500. e *Nuclearia simplex* CIENKOWSKI. × 570. From the film C 942 (GRELL)

a) Free-living amoebae

Amoeba. With numerous lobopodia radiating into different directions.
A. proteus PALLAS (Fig. 243a and 283), fresh water.

Thecamoeba. With a tough, usually wrinkled outer layer. Clear polarity.
T. verrucosa EHRENBERG (Fig. 346a), fresh water.
T. orbis SCHAEFFER (Fig. 346b), marine.

Hartmannella (Acanthamoeba). Small "Limax-amoebae", feed on bacteria and form cysts.
H. (A.) castellanii DOUGLAS (Fig. 282), fresh water.
H. (A.) astronyxis RAY et HAYES (Fig. 347), fresh water.

Naegleria. Amoeboid creeping form and flagellated swimming form ("amoeboflagellates"), forms cysts.

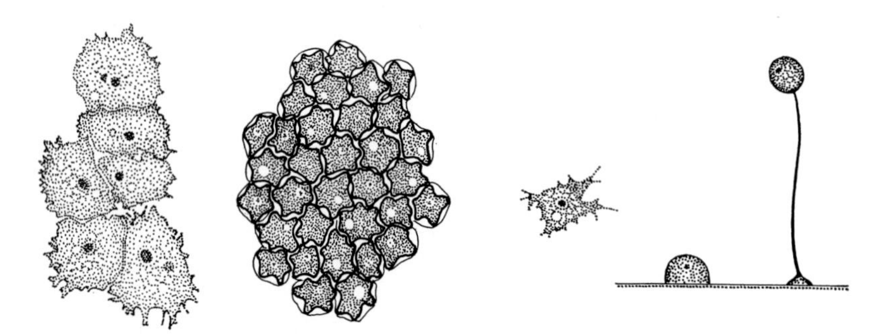

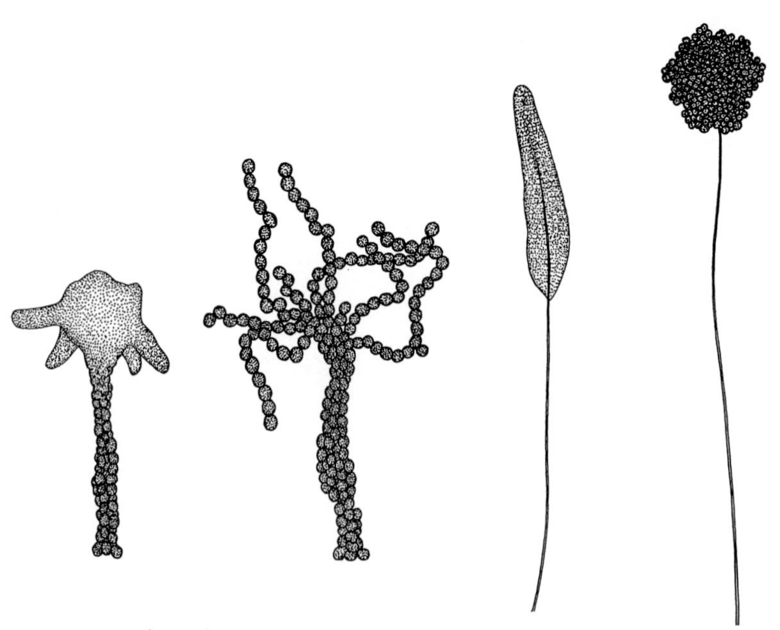

Fig. 347. Free-living amoebae which aggregate before encystment and form stalked cysts or cyst aggregates. Drawing by G. GERISCH after micrographs by K. B. RAPER, 1960 [*1415*]

N. gruberi SCHARDINGER (Fig. 238), fresh water.

"Limax-amoebae" and *Naegleria gruberi* have been suspected of being the causative agent of a type of meningoencephalitis in man [*209, 345, 837, 867*].

Pontifex. Multinucleated. No differentiation of ecto- and endoplasm. Attaches to the substrate by adhesive projections.

P. maximus SCHAEFFER (Fig. 346c), marine.

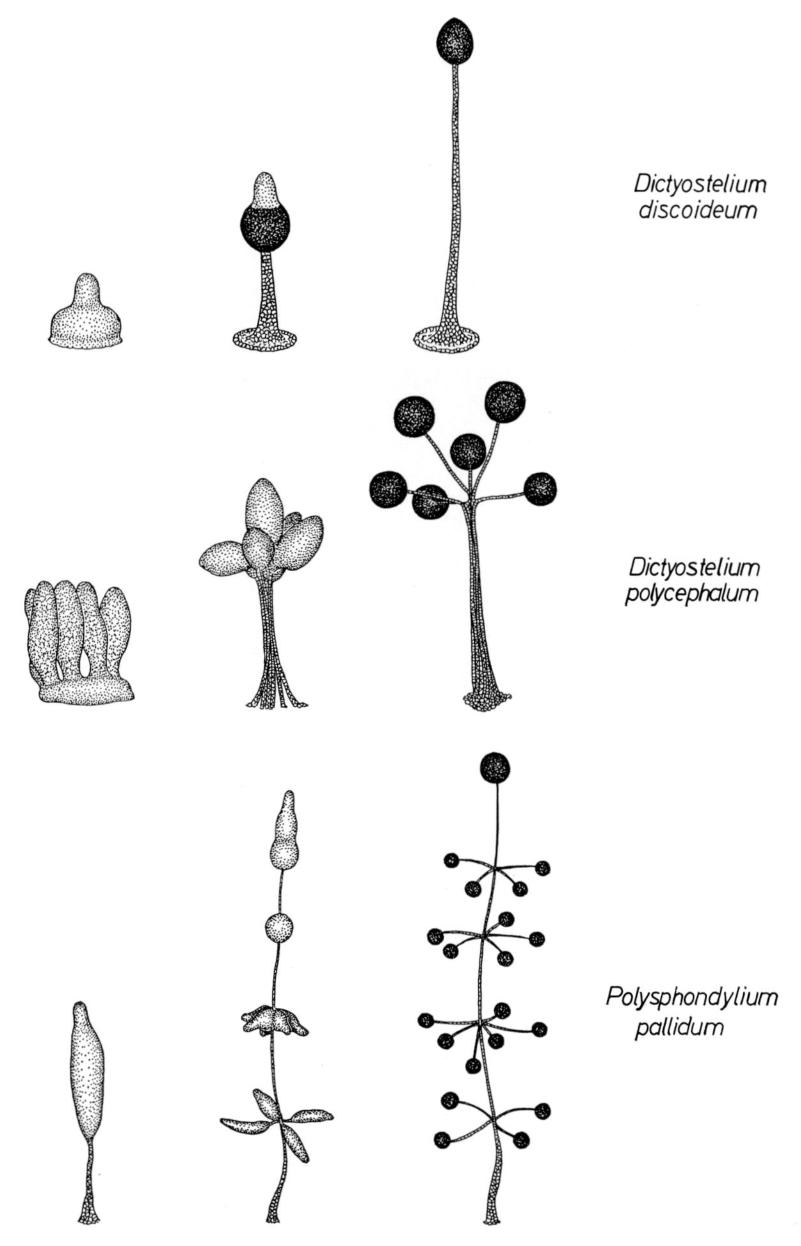

Fig. 348. Collective amoebae (Acrasina). Formation of the sporophore. Differentiation into stalk cells and spore cells (cyts). Drawing by G. GERISCH after micrographs by K. B. RAPER, 1960 [*1415*]

Pelomyxa. Multinucleated. No differentiation of ecto- and endoplasm. Rolling motion.
P. palustris GREEF, in the sapropel.
Paramoeba. One or several „Nebenkörper" (associated bodies) attached to the nucleus.
P. eilhardi SCHAUDINN (Fig. 304 and 307), marine.
Stereomyxa. Highly branched hyaline pseudopodia.
S. angulosa GRELL (Fig. 346d), marine.
S. ramosa GRELL (Fig. 244a), marine.
Corallomyxa. Multinucleated, forms "plasmodia" of unlimited growth. Budding.
C. mutabilis GRELL (Fig. 244b), marine.
Nuclearia. Filiform pseudopodia, resembling Heliozoa.
N. simplex CIENKOWSKI (Fig. 346e), fresh water.
Protostelium. Single amoebae can form stalked or unstalked cysts.
P. mycophaga OLIVE et STOIANOVITCH (Fig. 347).
Acrasis. Single amoebae aggregate before encystment, forming a pseudoplasmodium. The central cells of the pseudoplasmodium, which encyst first, form a "stalk", the peripheral cells which at first surround the stalk as a lobular mass form the "branches" of the sporophore.
A. rosea OLIVE et STOIANOVITCH (Fig. 347).
Acytostelium. After aggregation to a pseudoplasmodium, the amoebae form jointly a noncellular stalk of cellulose on top of which they encyst.
A. leptosomum RAPER et QUINLAN (Fig. 347).
The following two genera form sporophores of somatic stalk cells and generative spore cells (or cyst cells) (cell differentiation, p. 8).
Dictyostelium. The spherical spore clusters ("sori") are at the tip of the simple or branched stalk.
D. discoideum RAPER (Fig. 348).
D. polycephalum RAPER (Fig. 348).
Polysphondylium. The spherical spore clusters ("sori") are at the tip of the stalk or at the tip of side branches arranged in whorls.
P. pallidum OLIVE (Fig. 348).

b) Parasitic amoebae

Of the many species inhabiting the alimentary tracts of invertebrates and vertebrates only the *amoebae of man* are discussed in the following. With the exception of *Entamoeba histolytica,* they are all harmless commensals.

In the oral cavity:

Entamoeba gingivalis GROS. Frequently in the coating of the teeth.

In the gut (Fig. 349):

Entamoeba histolytica SCHAUDINN. Causative agent of amoebic dysentery.

Like the other entamoebae, *E. histolytica* has a nucleus with a fairly small nucleolus. Motility is due to broad pseudopodia of the type shown in Fig. 243b. The species appears in two modifications, only one being pathogenic. The nonpathogenic form is called the *minuta* or *intestinal lumen form* as it is relatively small (10—20 μm) and normally confined to the lumen of the intestine. It can multiply there and form cysts with four nuclei which are voided with the fecal matter and serve in spreading the parasite. It is only in the presence of definite environmental conditions that it changes into the larger *magna* or *tissue form* which ingests erythrocytes and penetrates into the gut wall. Since these conditions need not be realized in all people which are host to *E. histolytica,* there are many "carriers" which show no symptoms of amoebic dysentery.

Amoebic dysentery is a disease endemic in warm countries. It becomes manifest whenever the resistance of the gut is lowered in people infected with *E. histolytica.* In such cases, the *magna*-type is formed which secretes histolytic enzymes and penetrates through the gut wall into the tissues below, leading to the intestinal ulcers characteristic of amoebic dysentery. Abscesses can also be formed in other organs (e.g. the liver) if the amoebae spread.

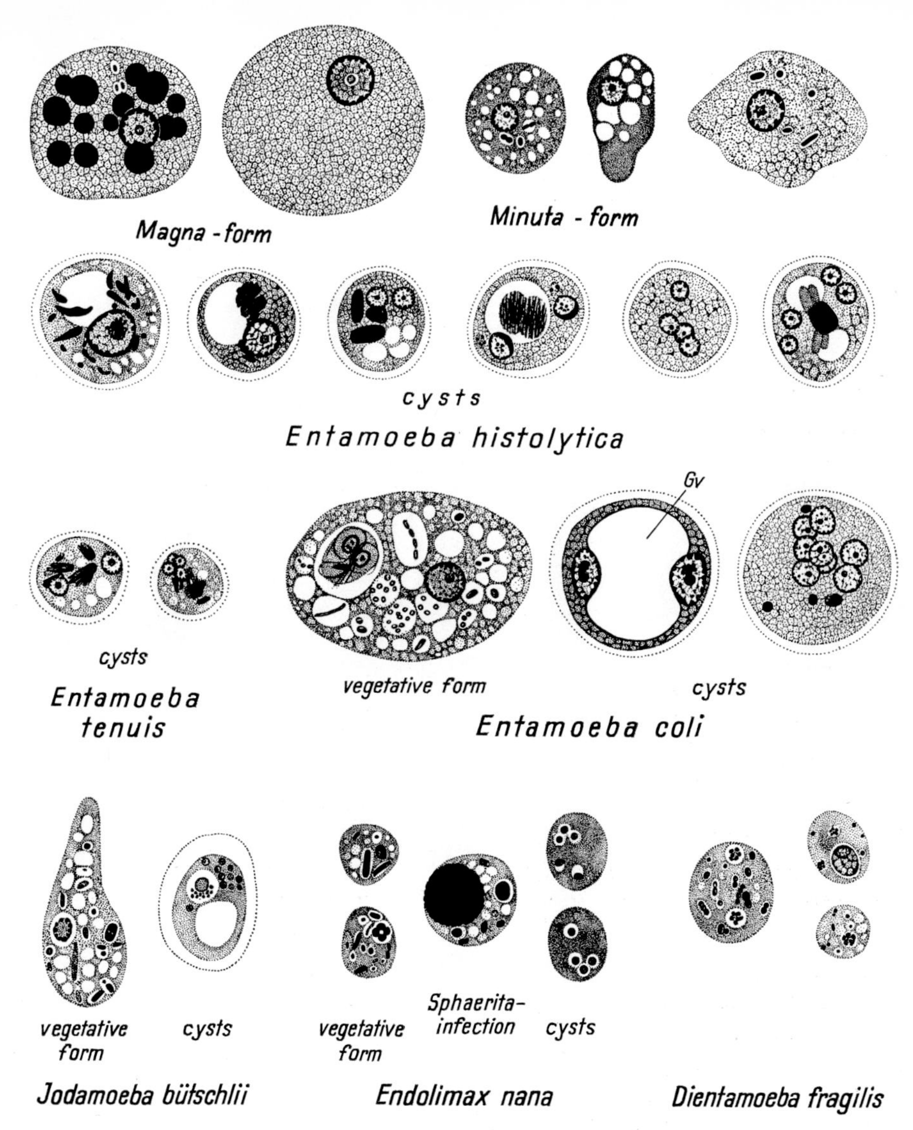

Fig. 349. Amoebae of the human gut. After REICHENOW

Entamoeba tenuis (syn. *hartmanni*) v. PROWAZEK, small species forming cysts with two or four nuclei.

Entamoeba coli LÖSCH is very similar to *E. histolytica*, but contains only bacteria, cysts of *Lamblia* etc., never red blood corpuscles in its food vacuoles. It forms cysts with eight nuclei. Still immature, binucleated cysts frequently contain a large glycogen vacuole (*Gv*).

Jodamoeba buetschlii v. PROWAZEK. Nucleus with a large nucleolus. The cyst contains one nucleus and a glycogen vacuole which stains reddish brown upon addition of iodine (Lugol's solution).

Endolimax nana WENYON et O'CONNOR. Vegetative stages fairly small (less than 10 μm), cyst usually with four nuclei.

Dientamoeba fragilis JEPPS et DOBELL. Small (4—10 μm), in stool smear preparations a very fragile species, usually with two nuclei.

Second Order: Testacea

DEFLANDRE, G.: Ordre des Thécamoebiens. In: GRASSÉ, P. P.: Traité de Zoologie. **1**, Fasc. 2. Paris: Masson et Cie. 1953.

The Testacea, or shelled amoebae, possess nonpartitioned tests or shells consisting of an organic coat often connected with inorganic constituents like platelets of silica (Fig. 350) or covered with foreign bodies, such as sand grains or diatom shells.

Fig. 350. *Euglypha rotunda* WAILES. Empty test. Stereoscan micrograph. Courtesy of H. NETZEL

The shell may have the shape of a bowl, an urn or an ampulla. The cytoplasm is frequently layered: the posterior portion of the cell contains the nucleus and stains intensely with basic dyes, while the anterior portion is rich in reserve substances and food vacuoles. The pseudopodia protrude from the shell opening, emerging frequently from a special cytoplasmic cone, and take the form of lobopodia or filopodia, rarely rhizopodia *(Lieberkuehnia)*.

Reproduction is by way of binary fission whose course may differ depending upon the type of shell (p. 123).

Testacea are found mainly in fresh water and are especially richly developed in swamps and turfs of moss.

a) With Lobopodia

Arcella. The shell has the shape of a watchglass, without attached foreign bodies. Its surface is subdivided into hexagonal fields by delicate ridges. With one or several nuclei.
A. vulgaris EHRENBERG (Fig. 351), fresh water.

Centropyxis. Shell cap-shaped, with or without spines.
C. aculeata STEIN, fresh water.

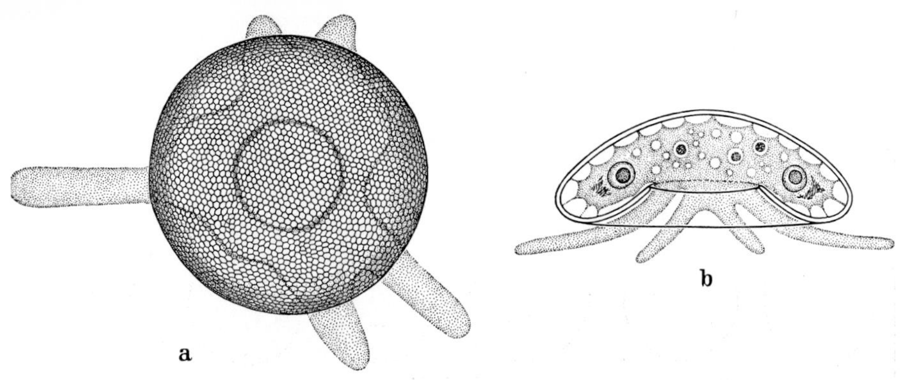

Fig. 351. *Arcella vulgaris* EHRENBERG. a From above. b Optical section and side view combined. × 250. a After VERWORN, b After KÜHN

Difflugia. Shell with foreign bodies.
D. pyriformis PERTY, fresh water.

b) With Filopodia

Euglypha. Shell of overlapping platelets of silica, like a shingled roof (Fig. 350).
E. alveolata DUJARDIN (Fig. 112), fresh water.
Chlamydophrys. Shell ampulla-shaped.
C. stercorea CIENKOWSKI. In manure infusions.
Pamphagus. Shell hyaline.
P. hylinus BELAR (Fig. 111), fresh water.

Third Order: Foraminifera

BRADY, H. B.: Foraminifera. In: Challenger Report Zool. **9** (1884).
CUSHMAN, J. A.: Foraminifera. Their classification and economic use. 4. Ed. Cambridge (Mass.): Harvard University Press 1948.
ELLIS, B. F., MESSINA, A. R.: Catalogue of Foraminifera. Published by Amer. Mus. Nat. Hist. New York.
GALLOWAY, J. J.: A manuel of Foraminifera. Bloomington (Indiana): Principia Press 1933.
LE CALVEZ, J.: Ordre des Foraminifères. In: GRASSÉ, P. P.: Traité de Zoologie. **1**, Fasc. 2. Paris: Masson et Cie. 1953.
LOEBLICH, A. R., TAPPAN, H.: Sarcodina (chiefly "Thecamoebians and Foraminiferida"). In: MOORE, R. C.: Treatise on Invertebrate Palaeontology, Part C, Protista 2 (2 Vols.). University of Kansas Press 1964.
PHLEGER, F. B.: Ecology and distribution of recent Foraminifera. Baltimore: John Hopkins Press 1960.
— PARKER, F. L., PEIRSON, J. F.: North Atlantic Foraminifera. Rep. Swedish Deep Sea Expedition **7** (1953).

The Foraminifera have shells which are perforated by pores. The pseudopodia are mostly rhizopodia. Heterophasic alternation of generations has been demonstrated for many species.

In contrast to the Testacea, the Foraminifera are exclusively inhabitants of the sea. Most species are found in sand, on stones or algae. Two families (Globigerinidae, Globorotalidae) have changed to a pelagic mode of life.

Since the *shells* can remain intact after reproduction or the death of their owners, they become part of the marine sand. Especially the shells of pelagic species con-

Fig. 352. Foraminiferan sand from the Bikini atoll. × 60. From the film C801 (GRELL)

tribute considerably to the *formation of sediment* in large oceanic basins (Fig. 352). The Foraminifera are thus of importance for geological research too. They are the major concern of the so-called micropaleontology.

The shell consists of an organic matrix which can be reinforced in various ways by the deposition of or incrustation with calcium carbonate or foreign bodies (usually grains of sand). Besides forms whose shells are either purely calcareous or reinforced only by sand, there are others which make use of both materials to strengthen their shells ("mixed shells"). The wall of the shell is perforated by numerous pore canals of varying width. Special openings are provided for the extension of pseudopodia.

In Foraminifera of a more simple organization, the shell consists of one chamber only *(Monothalamia)*, but in most species the shell is subdivided into many chambers which are interconnected by pores and added successively as the cell grows *(Polythalamia)*. Within the polythalamic groups, some forms have secondarily become single-chambered again.

The large diversity of foraminiferan shells is due mainly to differences in the arrangement of the chambers. These may be linearly arranged in a row (*Nodosaria*-type) or wound to a spiral (*Rotalia*-type). The spiral may be confined to a single plane (plano spiral) or form a helix (turbospiral). Occasionally, the chambers are arranged plait-like in two or three rows (*Textularia*-type), or they are added from the inside out as concentric circles (*Planorbulina*-type). Differences in chamber size, complicated systems of pore canals and surface structures such as spines, ridges, teeth etc. add further diversity to the appearance of foraminiferan shells. Examples are given in Fig. 353 and 354.

Shell size can vary between 20 micra and several centimeters. The fossil nummulites had the largest shells. One species of the miocene (Near East), related to the num-

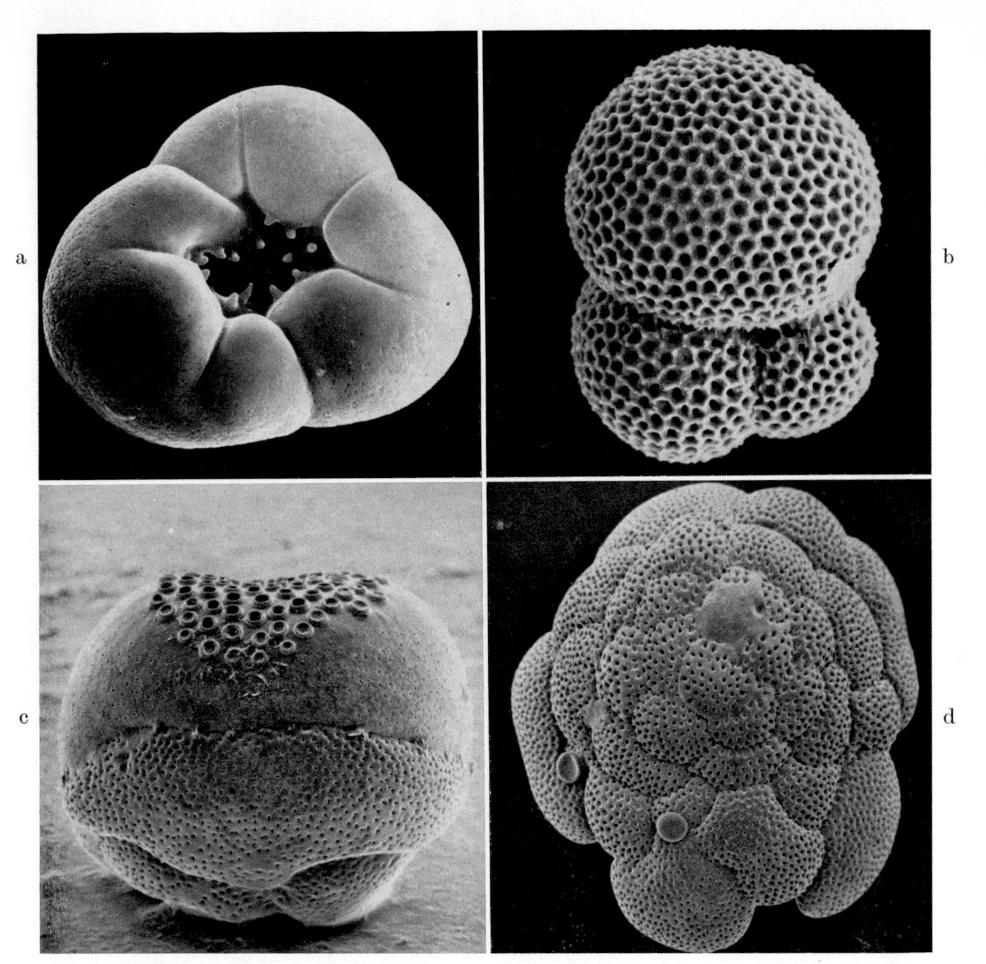

Fig. 353. Foraminiferan shells. a *Rotaliella heterocaryotica* GRELL, ventral surface. ×1400. b *Globigerinoides sacculifer* BRADY. ×105. c, d *Tretomphalus bulloides* D'ORBIGNY. c Gamont with floating chamber, ×120, d Agamont, dorsal surface, ×80. Stereoscan micrographs

mulites, *Lepidocyclina elephantina* LEMOINE et DOUVILLE, attained a diameter of 14.5 cm and a thickness of 1.5 cm.

The *rhizopodia*, which have been described in detail above (pp. 282), serve in locomotion as well as food capture. It is probable that they also play a role in the formation of new chambers.

The food includes diatoms and a variety of Protozoa, occasionally probably also small Metazoa. Many species feed exclusively on dead organic matter (detritus).

The mode of *reproduction* is known only from comparatively few species. Some Foraminifera (Milionidae, Textularidae) seem to reproduce only asexually (budding, multiple fission).

In many foraminiferan species, however, an *alternation of generations* has been shown to take place in which an asexually reproducing generation (agamont) alternates with a sexually reproducing generation (gamont). Both generations may

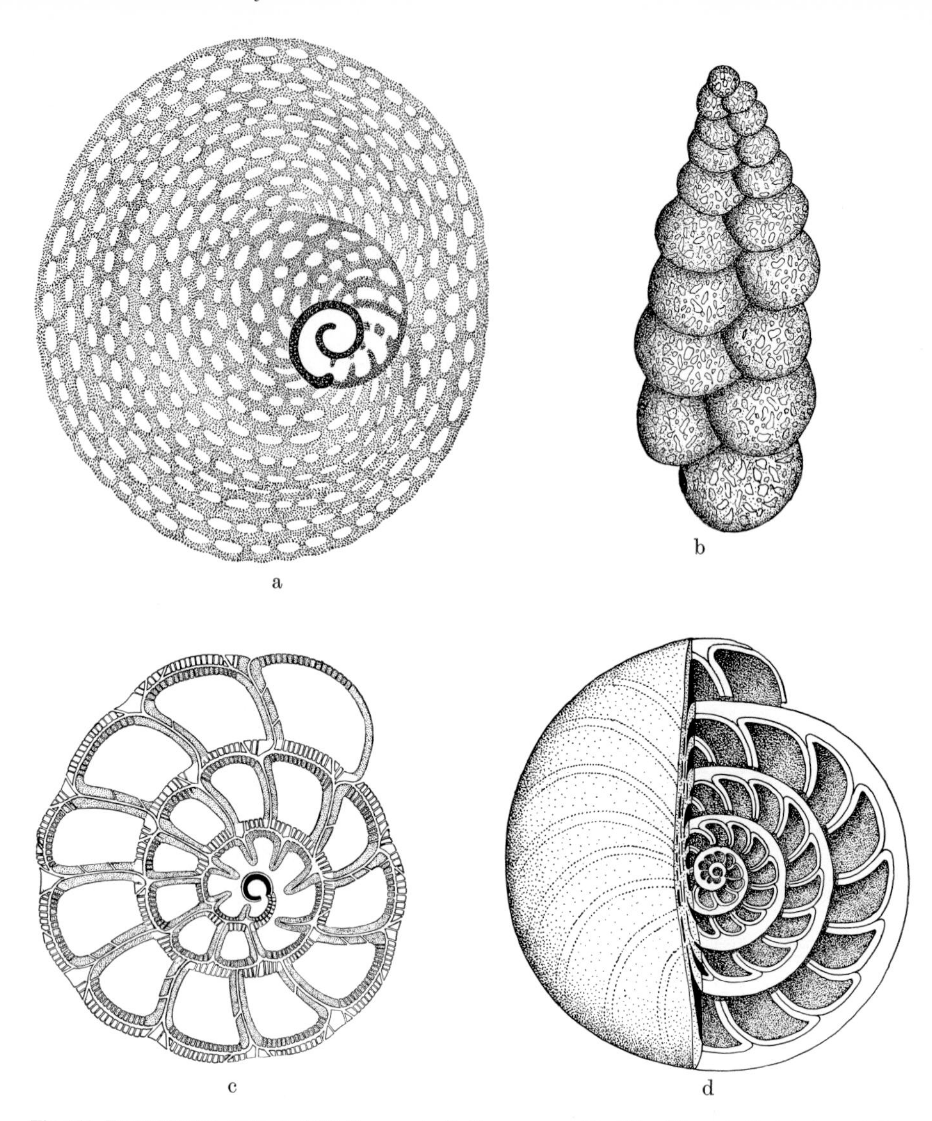

Fig. 354. Foraminiferan shells. a *Orbitolites marginalis* LAMARCK. × 70. After KÜHN. b *Textularia agglutinans* D'ORBIGNY. × 160. After RHUMBLER from KÜHN. c *Rotalia beccarii* LAMARCK. Diagrammatic section through the spirally arranged chambers. × 70. After KÜHN. d *Nummulites cummingii* CARPENTER. One half of the shell is cut longitudinally to show the chambers. After BRADY from KÜHN

be completely alike morphologically. In others, they are different to a greater or lesser degree. In monothalamic species (e.g. *Iridia*, Fig. 358; *Myxotheca*, Fig. 359; *Allogromia*) the two generations are either alike or differ only in size. The agamonts of *Patellina* (Fig. 360), *Rotaliella* (Fig. 362), *Metarotaliella* (Fig. 363), and *Rubratella* (Fig. 365) are on the average larger than the gamonts. In *Glabratella* (Fig. 367), with just the reversed size relationships, and in species of *Discorbis* (Fig. 356 and 357) the two generations differ also by the shapes of their shells so that they might be mistaken for different species.

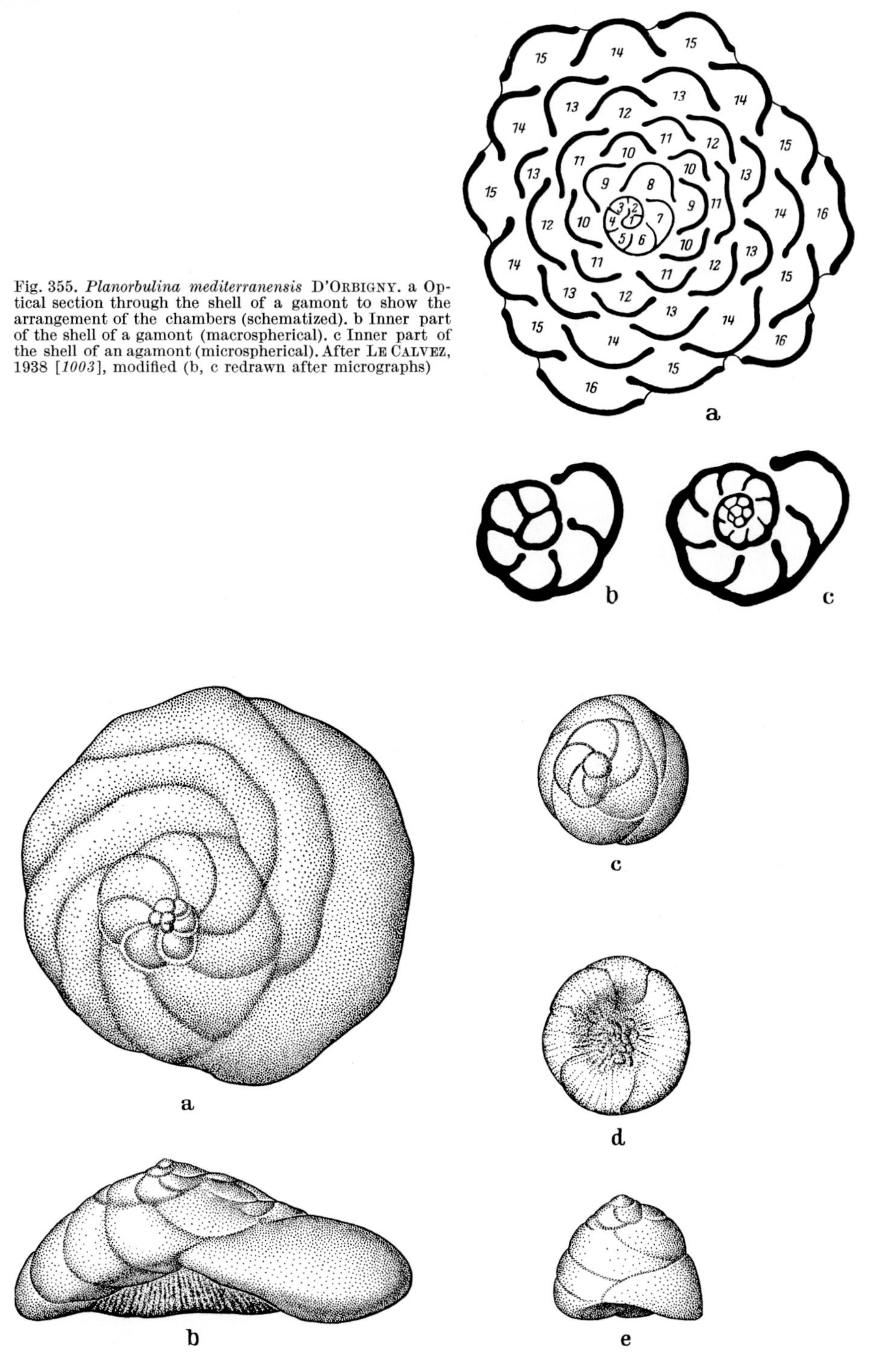

Fig. 355. *Planorbulina mediterranensis* D'ORBIGNY. a Optical section through the shell of a gamont to show the arrangement of the chambers (schematized). b Inner part of the shell of a gamont (macrospherical). c Inner part of the shell of an agamont (microspherical). After LE CALVEZ, 1938 [*1003*], modified (b, c redrawn after micrographs)

Fig. 356. *Discorbis mediterranensis*. a, b Adult agamont. c, d, e Adult gamont. × 80. After LE CALVEZ, 1950 [*1004*]

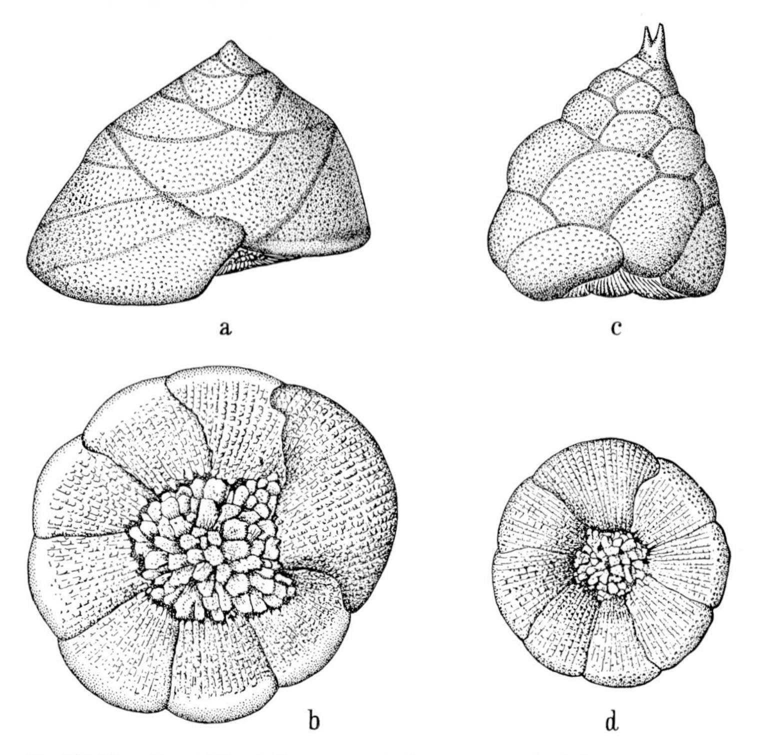

Fig. 357. *Discorbis patelliformis* BRADY — *erecta* SIDEBOTTOM. a, b adult agamont *("D. patelliformis")*. c, d Adult gamont *("D. erecta")*. × 180. After LE CALVEZ, 1952 [*1005*]

In many Foraminifera there is a difference in the size of the initial chamber (proloculus). The gamont which arises by multiple fission has an initial chamber which is larger than that of the agamont which arises from the zygote (Fig. 355). The gamont was therefore formerly referred to as the "macrospherical", the agamont as the "microspherical" generation.

In all closely studied cases it became evident that meiosis (p. 88) is intermediary. The gamont is therefore haploid, the agamont diploid.

With respect to sexual reproduction (see p. 175, 181, 185) three different possibilities must be distinguished, i.e. gametogamy (Fig. 358 and 359), autogamy (Fig. 85 and 362), and gamontogamy (Fig. 360, 363, 365, and 367).

Asexual reproduction usually includes two phases. The first of these takes place even before emergence of the young agamonts, or shortly thereafter and consists of a series of nuclear divisions. After this period of nuclear multiplication, the agamont grows to a given size. The second phase is associated with meiosis and leads to the formation of gamonts.

Whereas in *homokaryotic* Foraminifera (Fig. 358–360) the nuclei of the agamont are all alike, the first phase of agamogony in *heterokaryotic* species is followed by a simultaneous nuclear differentiation: one or several nuclei become somatic, the others remain generative (nuclear dualism, pp. 96).

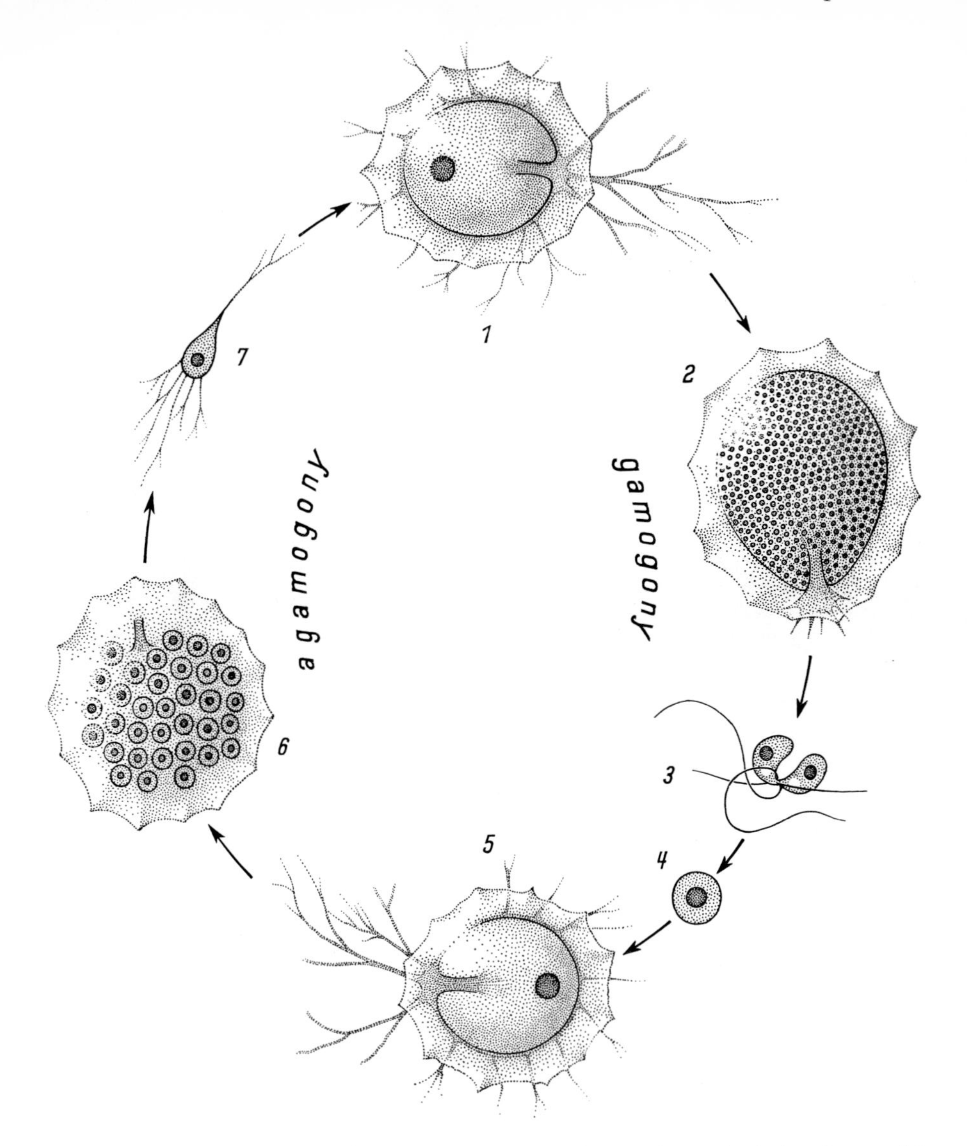

Fig. 358. *Iridia lucida* LE CALVEZ. Developmental cycle. Gamont and agamont are uninucleate and surrounded by a pseudochitinous envelope which covers them like a tent. *1* Gamont, *2* gamont after formation of the gamete nuclei, *3* fertilization, *4* zygote, *5* agamont, *6* formation of the agametes, *7* young creeping agamete (compare with Fig. 249). After investigations by LE CALVEZ, 1938 [*1003*]

Far-reaching *morphological transformations* may be associated with the reproductive processes. In some species the septa between successive chambers are dissolved at the onset of gametogenesis. Empty shells of the two generations can therefore clearly be distinguished even if they are otherwise largely identical (Fig. 361). In other species this dissolution takes place both at the onset of gamogony *and* meiosis (Fig. 85). Gamonts of *Tretomphalus bulloides* undergo an especially peculiar transformation (p. 175 and Fig. 353c).

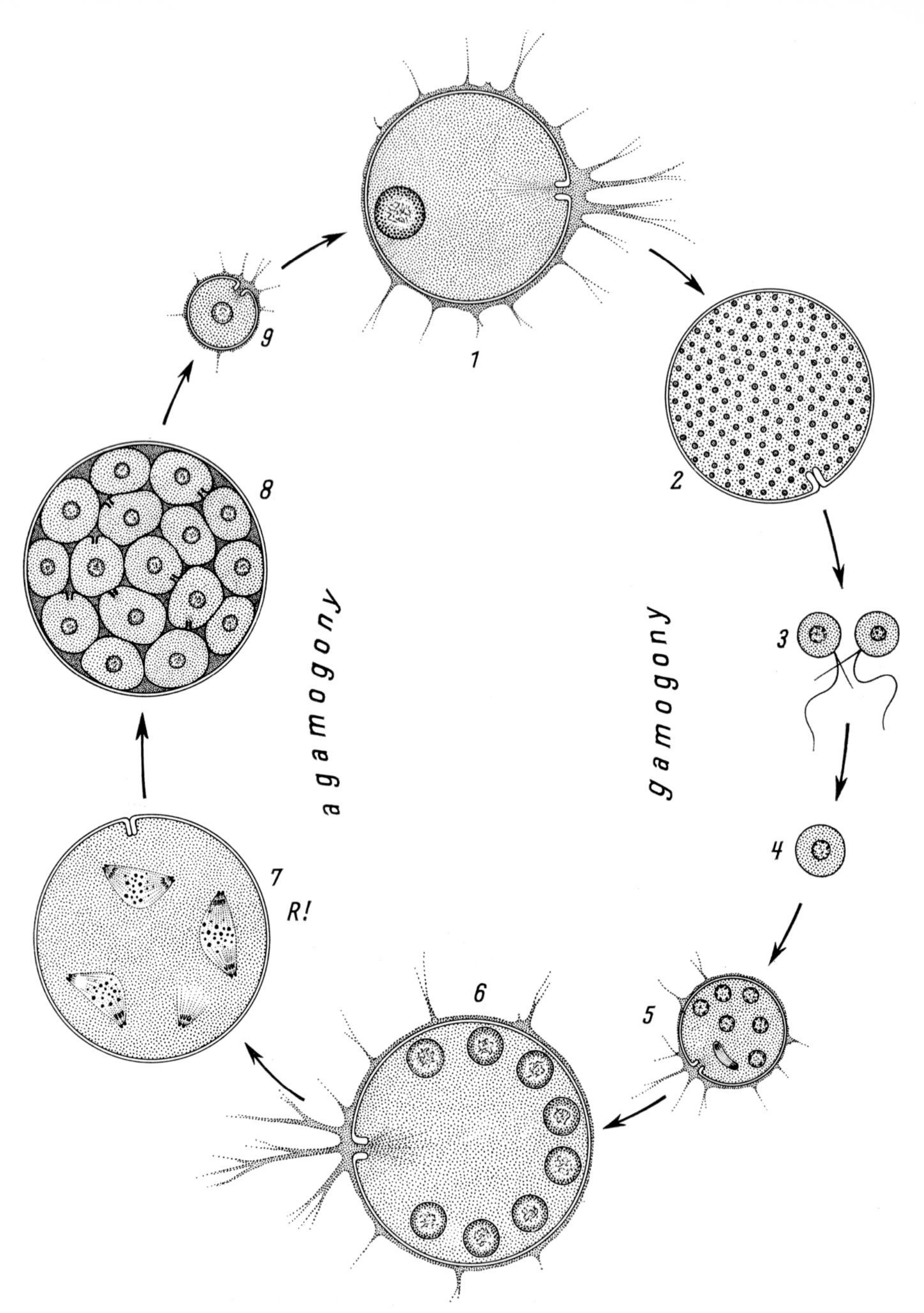

Fig. 359. *Myxotheca arenilega* SCHAUDINN. Developmental cycle. *1* Gamont (uninucleate), *2* gamont after formation of the gamete nuclei, *3* fertilization, *4* zygote, *5* young agamont (first phase of agamogony), *6* adult agamont, *7* meiosis, *8* formation of the agametes, *9* young agamete (= gamont). After investigations by the author

The *time* required for a full cycle of two alternating generations is highly different. While smaller species need only a few weeks, larger ones may take a whole year.

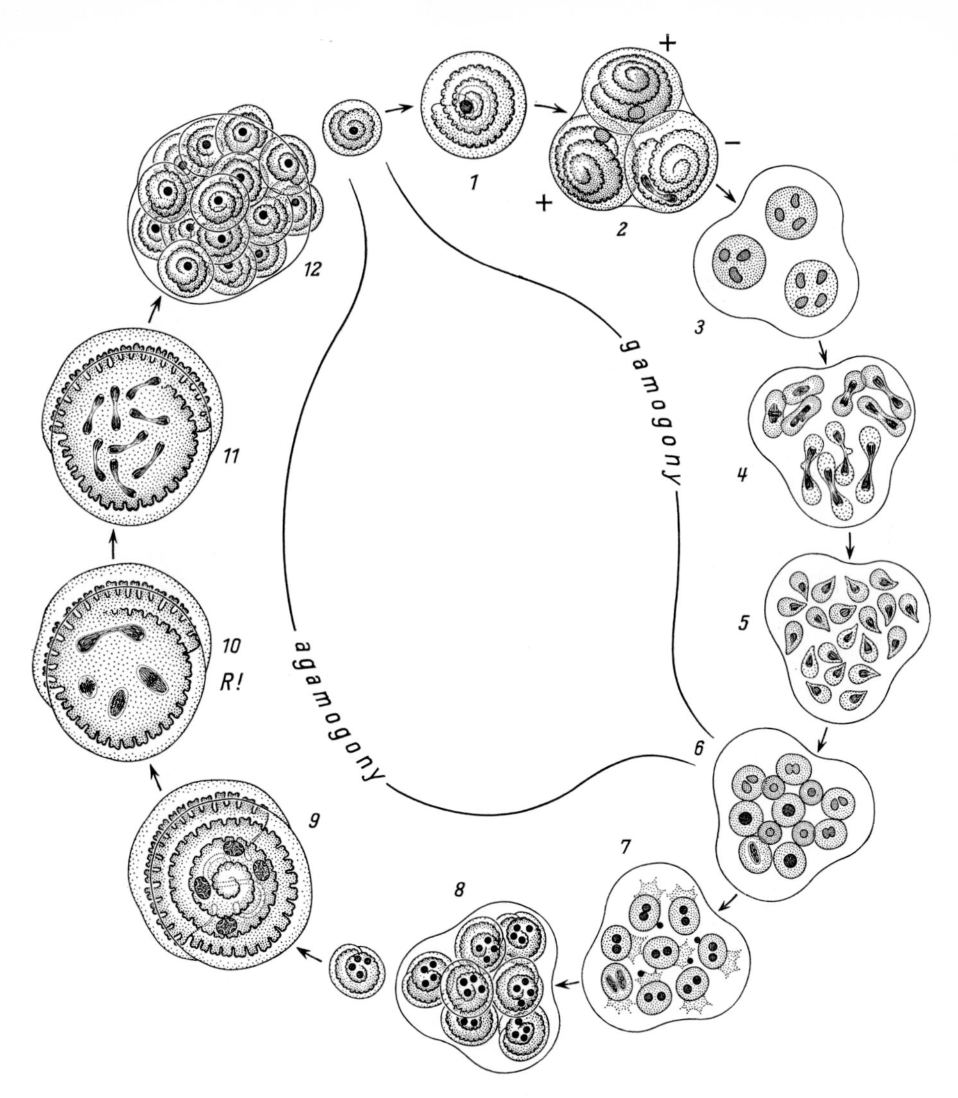

Fig. 360. *Patellina corrugata* WILLIAMSON. Developmental cycle. *1* Gamont. *2* Aggregate of three gamonts (two from the + sex, one from the — sex). *3* Protoplasts of the gamonts on the bottom of the space which is roofed over by the empty shells. *4* Last gamogony mitosis and gamete formation. *5* Gametes (twelve from the +, eight from the — sex). *6* Eight zygotes and four left-over gametes (from the + sex). *7* Binucleate agamonts (after the first mitosis). *8* Young agamonts (with four nuclei). *9* Adult agamont (with four nuclei). *10* Meiosis I. *11* Meiosis II. *12* Formation of the agametes. The gamonts and gametes of the + sex are more densely stippled than those of the — sex. After GRELL, 1959 [*667*]

Occasionally, it has been observed that one generation predominates in spring, the other in the fall.

The system of Foraminifera is based solely on the architecture of the shells. Since the few species investigated thus far exhibit large differences in their modes of reproduction, it seems possible that further investigations of reproduction modes may lead to a better understanding of taxonomic relationships than studies based on the morphology of shells. However, this goal lies far in the future.

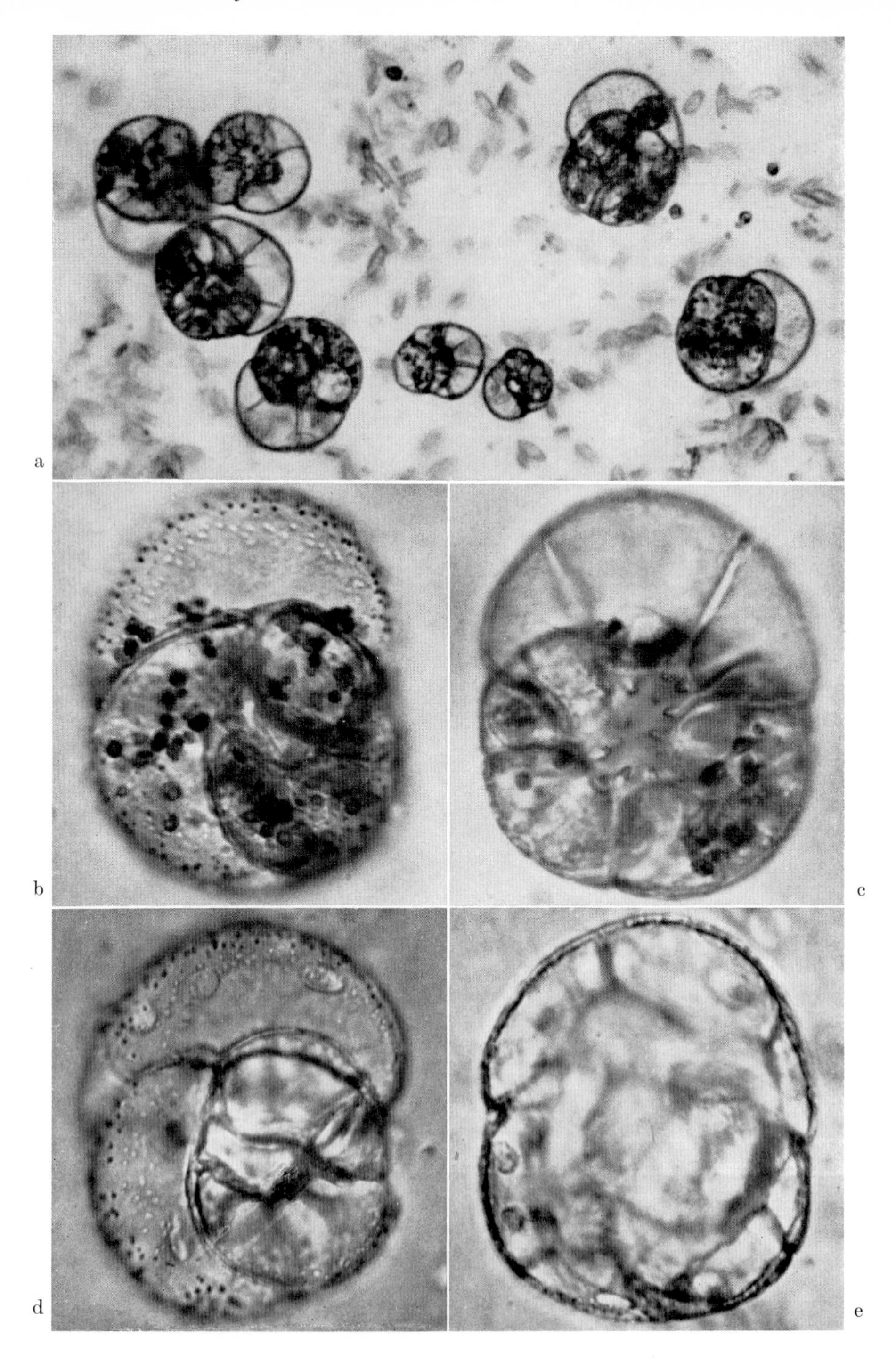

Fig. 361. *Rotaliella roscoffensis* GRELL. a Different growth stages in a culture dish. × 400, b, c Four-chambered agamont (b upper side, c lower side). × 960. d Empty shell of an agamont (walls between chambers retained). e Empty shell of a gamont (walls between chambers dissolved). × 960. After GRELL, 1957 [*663*]

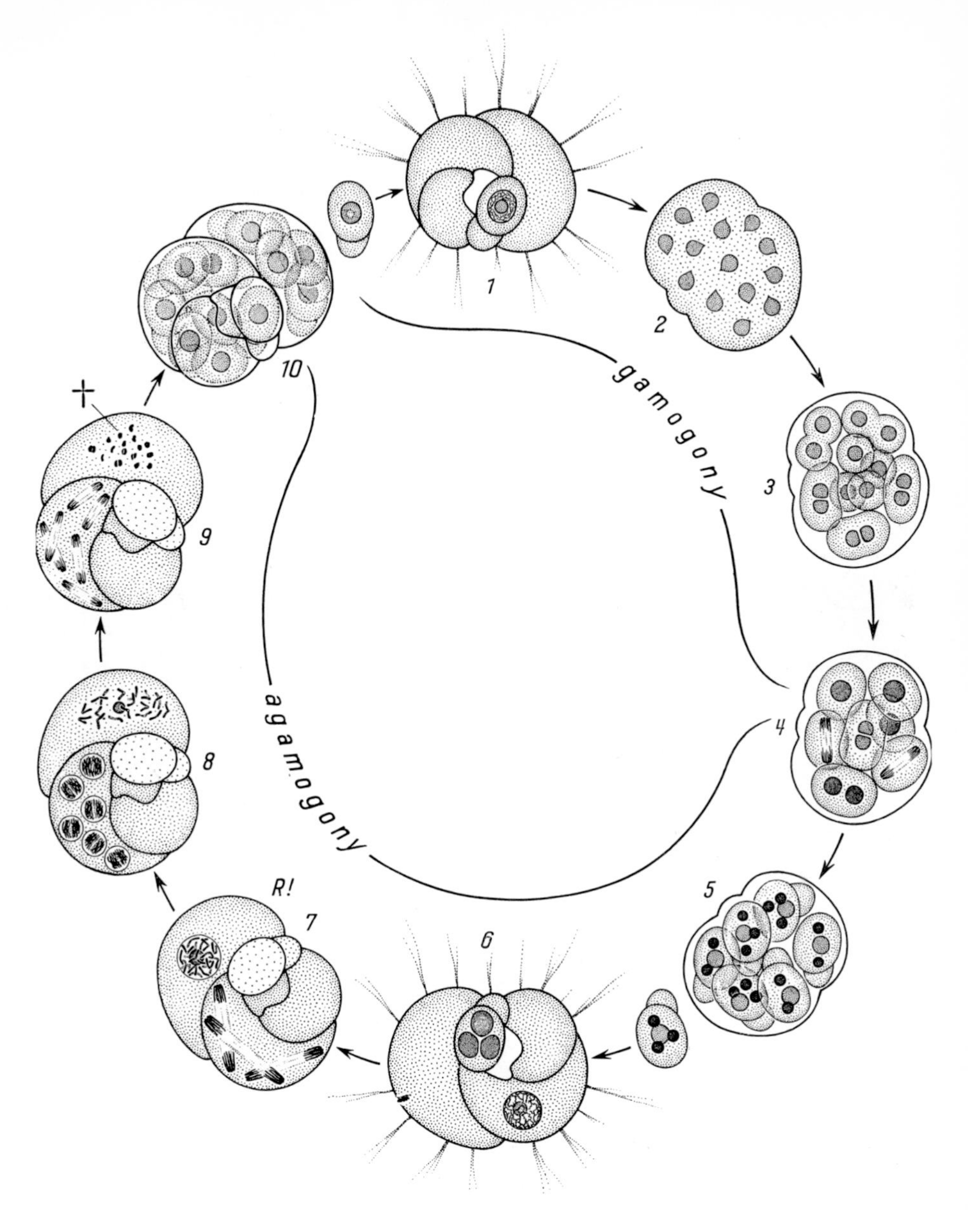

Fig. 362. *Rotaliella roscoffensis* GRELL. Developmental cycle. *1* Gamont. *2* Formation of the gamete nuclei. *3* Autogamic fertilization of the gametes. *4* Zygotes and the first metagamic division. *5* Young agamonts (some with three nuclei, some with four). *6* Adult agamont (generative nuclei swollen prior to meiosis). *7* Anaphase of the first meiotic division. *8* Metaphase. *9* Anaphase of the second meiotic division (dispersion and pycnosis of the chromosomes of the somatic nucleus). *10* Formation of the agametes. After GRELL, 1957 [*663*]

The subdivision of Foraminifera into single-chambered (Monothalamia) and multi-chambered ones (Polythalamia) is of formal importance only because it is probable that in many cases originally multi-chambered forms have secondarily become single-chambered.

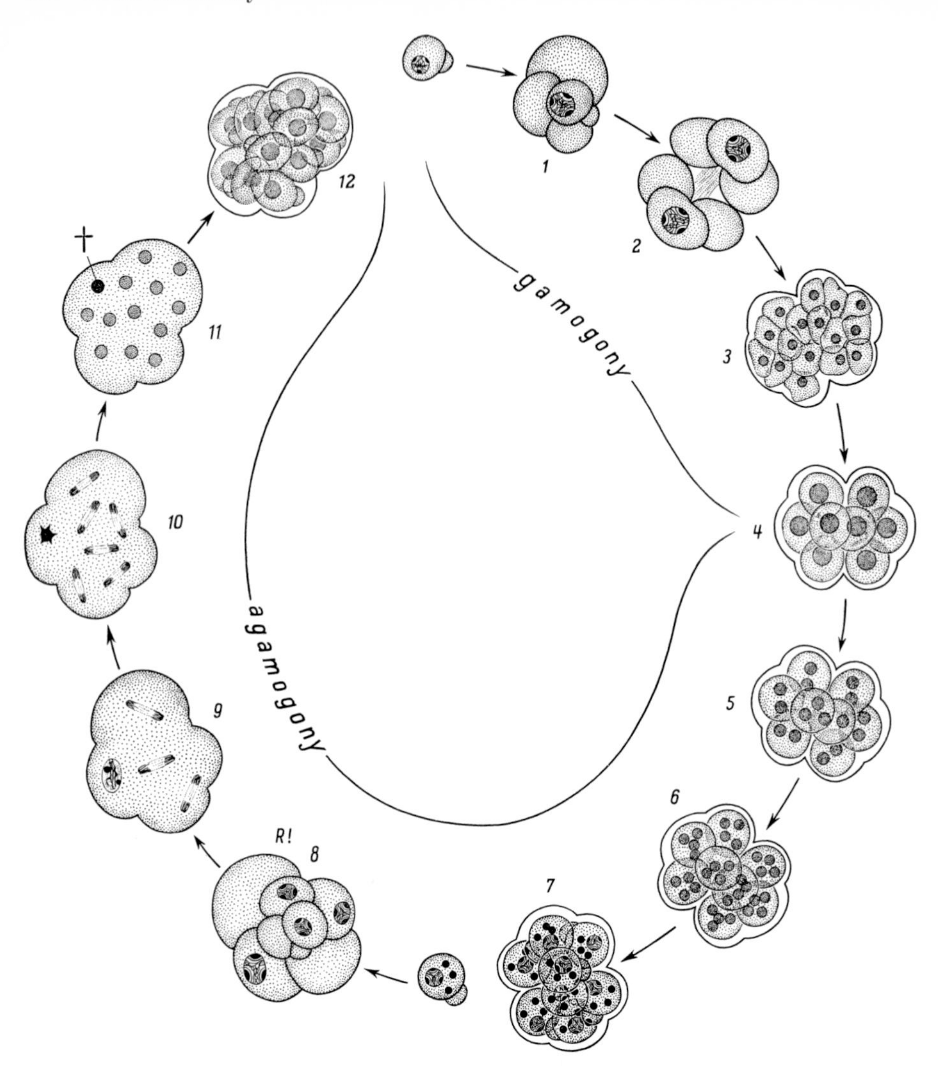

Fig. 363. *Metarotaliella parva* GRELL. Developmental cycle. *1* Gamont. *2* Mating of two gamonts. *3* Gametes. *4* Zygotes. *5* Binucleate agamonts (after the first metagamic nuclear division). *6* Agamonts with four nuclei (after the second metagamic nuclear division). *7* Agamonts with four nuclei (after nuclear differentiation: three generative nuclei, one somatic nucleus). *8* Adult agamont (the generative nuclei have also formed nucleoli). *9* First meiotic division. *10* Second meiotic division. *11* and *12* Formation of the agametes. After WEBER, 1965 [*1789*]

a) Single-chambered (monothalamic)

Iridia. Cell body flattened. Both generations uninucleate. Gametogamous.
I. lucida LE CALVEZ (Fig. 358).
Myxotheca. Cell body roundish, often with attached foreign bodies. Gametogamous.
M. arenilega SCHAUDINN (Fig. 359).
Allogromia. Cell body roundish. Autogamous.
A. laticollaris ARNOLD (Fig. 242).
Rhabdammina. Cell body rod-shaped, occasionally branched. Shell of foreign bodies.
R. abyssorum CARPENTER.
Astrorhiza. Cell body star-shaped, shell of foreign bodies.
A. limicola SANDAHL.

b) Multi-chambered (polythalamic)

Orbitolites. Spiral arrangement of chambers at first, later circular; subdivided by secondary walls. Calcareous.
O. marginalis LAMARCK (Fig. 354a).
Textularia. Chambers arranged to alternate in two or more rows. Shell sandy or calcareous.
T. agglutinans D'ORBIGNY (Fig. 354b).
Peneroplis. Chambers arranged at first in a spiral, later added on to a straight row. Calcareous. Gametogamous.
P. pertusus FORSCAL.
Spirillina. Shell planospiral, calcareous. Gamontogamous. Homokaryotic.
S. vivipara EHRENBERG.
Patellina. Shell cap-shaped, calcareous. Gamontogamous. Homokaryotic.
P. corrugata WILLIAMSON (Fig. 360).
Rotalia. Turbospiral arrangement of chambers, bi-convex, calcareous.
R. beccarii LAMARCK (Fig. 354c).
Rotaliella. Small, autogamous. Heterokaryotic.
R. heterocaryotica GRELL (Fig. 85 and 353a).
R. roscoffensis GRELL (Fig. 361 and 362).
Metarotaliella. Small, gamontogamous. Heterokaryotic.
M. parva GRELL (Fig. 363).
Rubratella. Small, red coloration. Every chamber is subdivided by a septum. Heterokaryotic.
R. intermedia GRELL (Fig. 364 and 365).
Discorbis. Chambers in turbospiral arrangement, plano-convex, calcareous. Usually gamontogamous.
D. mediterranensis D'ORBIGNY (Fig. 356).
D. patelliformis BRADY-SIDEBOTTOM (Fig. 357).
Glabratella. Similar to preceding genus. Gamontogamous. Heterokaryotic.
G. sulcata GRELL (Fig. 366 and 367).
Tretomphalus. Similar to preceding genus. Gamont forms "floating chamber". Gametogamous.
T. bulloides D'ORBIGNY (Fig. 165 and 353c, d).
Globigerina. Chambers considerably dilated, often with spines as an aid in floating, pelagic.
G. bulloides CARPENTER (Fig. 250).
Nummulites. Large extinct genus.
N. cummingii CARPENTER (Fig. 354d).

Fourth Order: Heliozoa

PENARD, E.: Les Héliozoaires d'eau douce. Genève: Henry Kündig 1904.
TRÉGOUBOFF, F.: Classe des Héliozoaires. In: GRASSÉ, P.P.: Traité de Zoologie. **1**, Fasc. 2. Paris: Masson et Cie. 1953.
VALKANOV, A.: Die Heliozoen und Proteomyxien. Arch. Protistenk. **93** (1940).

The Heliozoa or "sun animalcules" have a spherical cell body. The pseudopodia are axopodia which extend like rays in all directions. The cytoplasm is frequently subdivided into a more dense endoplasm and a coarsely vacuolated ectoplasm. The endoplasm contains one or several nuclei. Some species have a superficial layer of skeletal needles, others a gelatinous covering or a capsule with grid-like perforations.

Most Heliozoa float freely in the water, but some are attached to the substrate by a delicate stalk.

Reproduction is by binary fission. Flagellated stages have been described for some of the species. Sexual processes, probably always involving autogamy (p. 178), are so far known only from *Actinophrys sol* and the species of the genus *Actinosphaerium.*

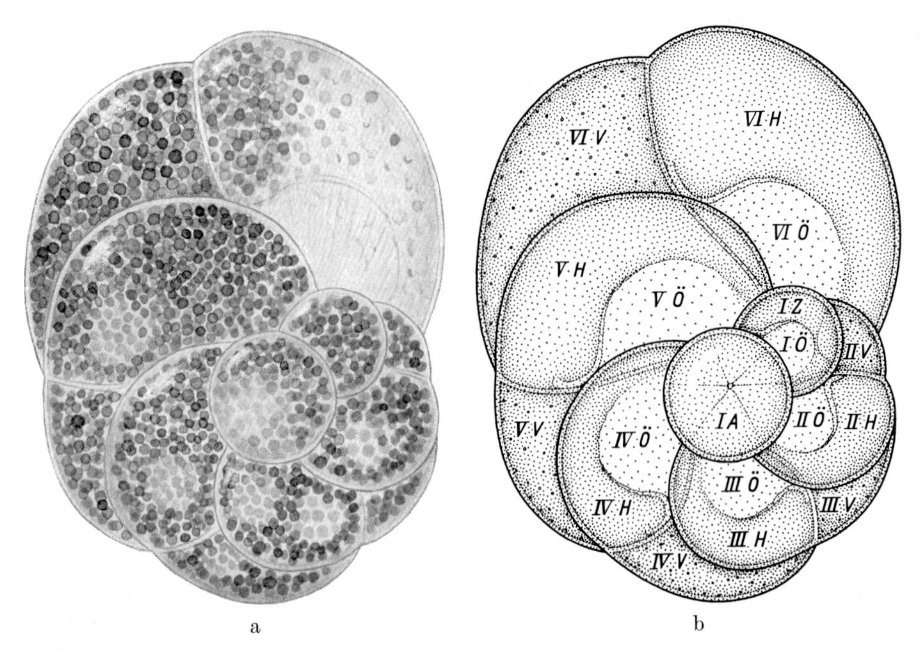

Fig. 364. *Rubratella intermedia* GRELL. Morphology of a six-chambered agamont. a Upper side of a living individual. b Empty shell, diagrammatic. × 1000. After GRELL, 1958 [*664*]

First Suborder: **Actinophrydia**

Axonemes of axopodia do not terminate at a central granule.

Actinophrys. Small, with a centrally located nucleus at which the axonemes terminate.
A. sol EHRENBERG (Fig. 172), fresh water.
Actinosphaerium. Large (1 mm), with numerous nuclei. Ectoplasm and endoplasm clearly separate.
A. eichhorni EHRENBERG (Fig. 368), fresh water.

Second Suborder: **Centrohelidia**

Axonemes of axopodia terminate at a central granule (centroplast).

Acanthocystis. Casing of radial siliceus spicules and tangential platelets.
A. aculeata HERTWIG (Fig. 369), fresh water.
Wagnerella. Sessile. The body consists of a small head with radiating axopodia, a stalk, and a basal plate containing the nucleus.
W. borealis MERESCHKOWSKI, marine.

Third Suborder: **Desmothoraca**

Mostly sedentary. Cell enclosed in a latticed organic capsule.

Clathrulina. Stalked capsule with large openings.
C. elegans CIENKOWSKI (Fig. 370). Asexual reproduction by biflagellated swarmers.

Fifth Order: Radiolaria

BORGERT, A.: Die Tripyleen-Radiolarien der Plankton-Expedition. Ergebn. der Plankton-Expedition, Humboldt-Stiftung **3** (1905—1913).

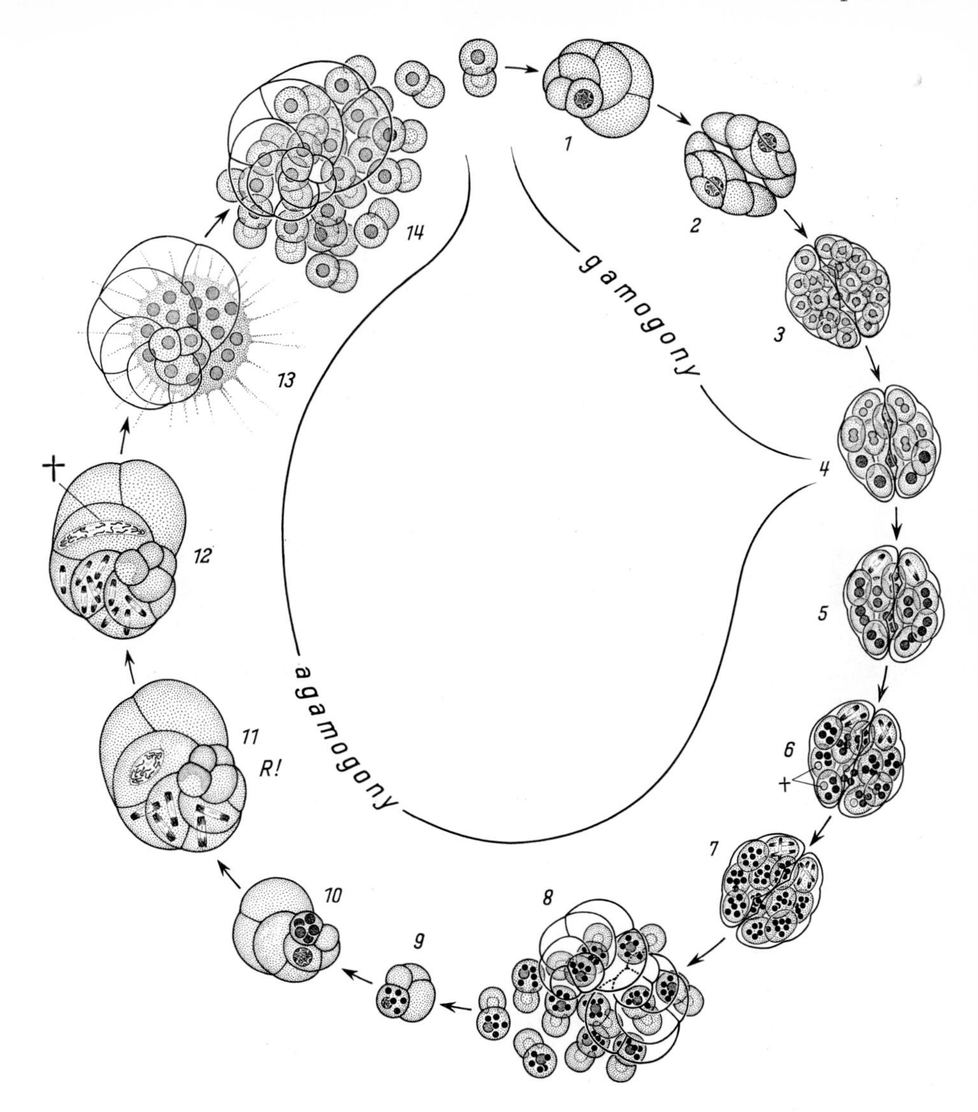

Fig. 365. *Rubratella intermedia* GRELL. Developmental cycle. *1* Gamont. *2* Mating of two gamonts. *3* Formation of the gametes. *4* Zygotes (some still in karyogamy). *5* Binucleate agamonts (partly still in the first metagamic division). *6* Agamonts with four nuclei (partly still in the second metagamic division, partly degeneration of one of the four sister nuclei). *7* Agamonts with six nuclei (partly still in the third metagamic division). *8* Hatching of the young agamonts (one somatic and five generative nuclei). *9*, *10* Growth stages of the agamonts. *11* First meiotic division. *12* Second meiotic division (somatic nucleus elongated). *13*, *14* Formation of the agametes. After GRELL, 1958 [*664*]

BRANDT, K.: Die koloniebildenden Radiolarien (Sphärozoen) des Golfs von Neapel und der angrenzenden Meeresteile. Fauna u. Flora d. Golfs von Neapel, Monogr. **13** (1885).

HAECKEL, E.: Die Radiolarien. Berlin: Reimer 1862.

— Report on the Radiolaria collected by H. S. M. Challenger during the years 1873—1876. Challenger Report Zool. **18** (1887).

HOLLANDE, A., ENJUMET, M.: Cytologie, évolution et systématique des Sphaeroidés (Radiolaires). Arch. Muséum Nat. Hist. Nat. **7**, sér. 7 (1964).

SCHEWIAKOFF, W.: Acantharia. Fauna u. Flora d. Golfs v. Neapel, Monogr. **37** (1926).

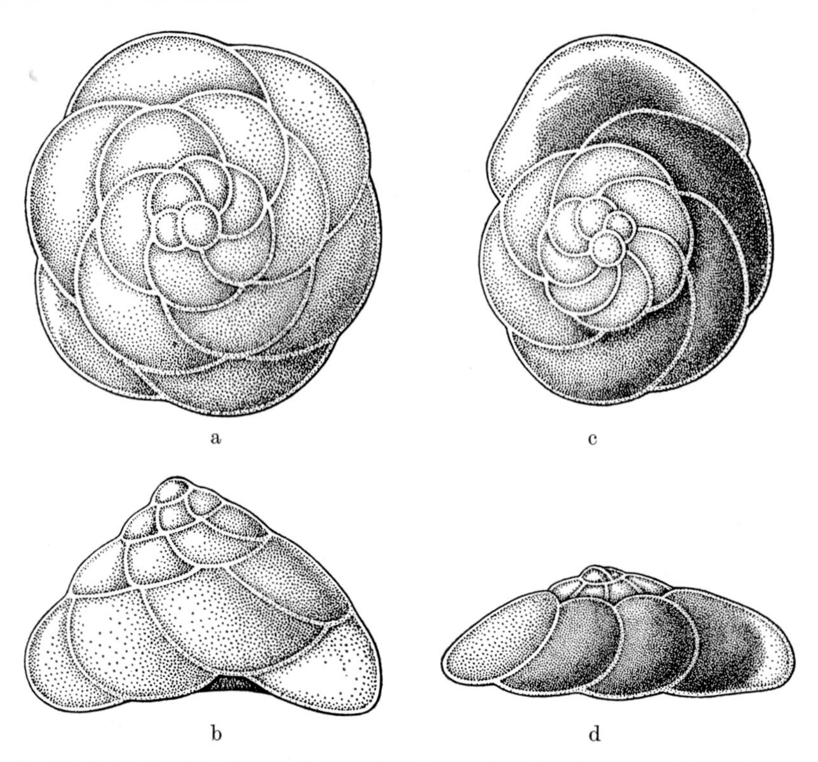

Fig. 366. *Glabratella sulcata* GRELL. a Gamont from above, b from the side. c Agamont from above, d from the side. The surface structure of the shell has not been depicted. × 270. After GRELL, 1956 *[661]*

TRÉGOUBOFF, G.: Classe des Acanthaires. Classe des Radiolaires. In: GRASSÉ, P.P.: Traité de Zoologie. **1**, Fasc. 2. Paris: Masson et Cie. 1953.

The Radiolaria resemble Heliozoa because their cell bodies are usually spherical with pseudopodia (axo- or filopodia, occasionally rhizopodia, p. 282) radiating towards all sides. However, they are clearly distinguished by the possession of a *central capsule*. Usually this is a central portion of the cell which is enclosed by a membrane and contains one or several nuclei. The interior of the central capsule is connected by pores with the exterior part, the so-called *extracapsularium*. If removed, the latter can in some species be regenerated by the central capsule. Inside the central capsule as well as outside, the frequently foamy cytoplasm may contain a variety of inclusions such as oil spheres, crystals and zooxanthellae (p. 359).

The possession of *skeletons* is especially characteristic for Radiolaria. They are usually composed of silicon (only Acantharia have skeletons of strontium sulphate) and appear in an immense variety of forms. In the simplest case, the skeleton may be composed of a system of either tangentially or radially arranged spicules or spines. The latter may be broadened at definite points and the broadened parts may fuse to form a spherical grid. Spines may also be branched or equipped with many kinds of extensions. In others, the spherical structure has been abandoned, the skeleton forming a flat disc or a helmet-, lantern- or cage-like framework, frequently even of bilateral symmetry. Because of the variety of their skeletal forms,

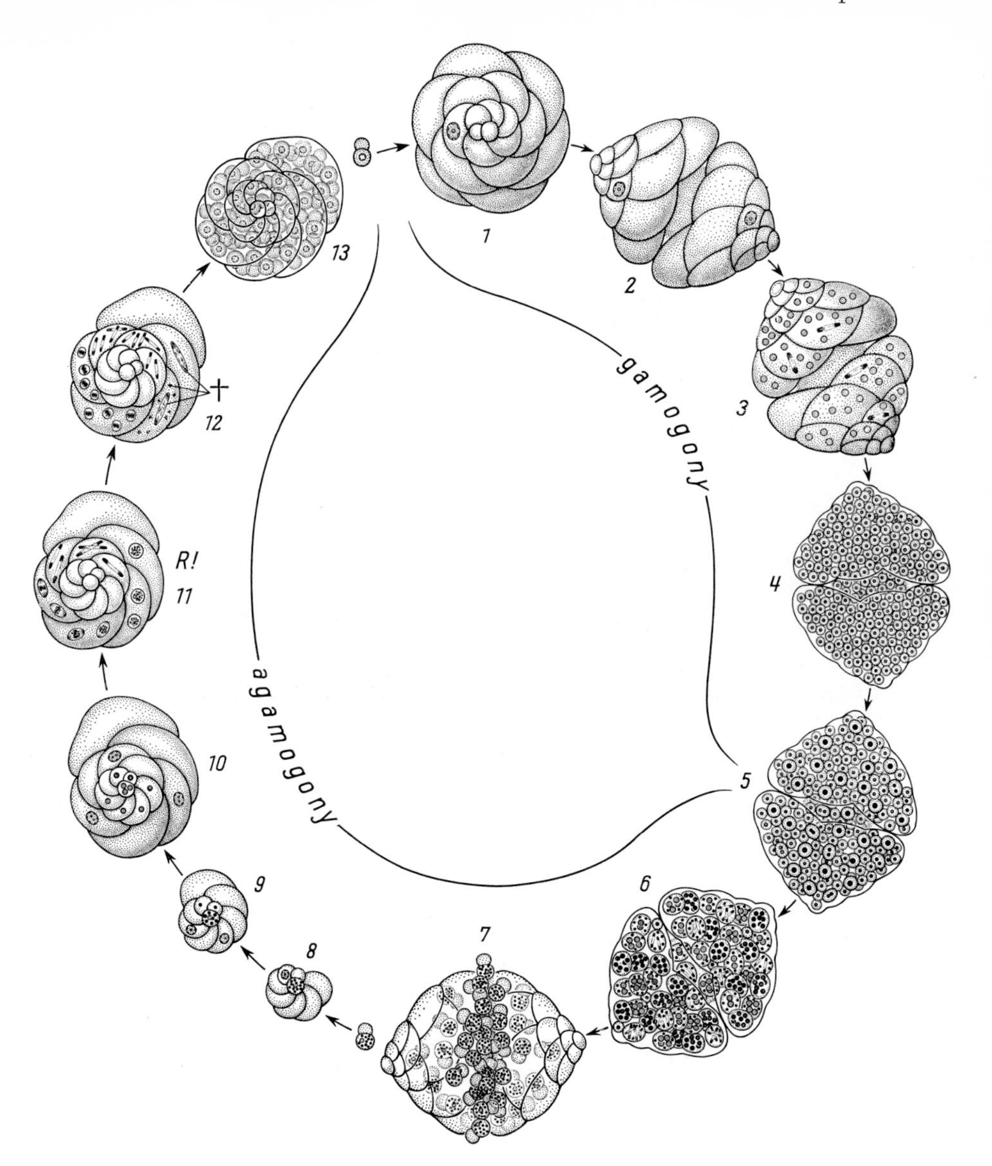

Fig. 367. *Glabratella sulcata* GRELL. Developmental cycle. *1* Gamont. *2* Mating of two gamonts. *3* Gamogony. *4* Gametes. *5* Fertilization (involving only a small fraction of the gametes) and zygotes. *6* Metagamic divisions. *7* Hatching of the young agamonts. *8*, *9* Growth stages of an agamont. *10*. Adult agamont (three somatic, nine generative nuclei). *11* First meiotic division. *12* Second meiotic division (elongation and disintegration of the somatic nuclei). 13 Formation of the agametes.
After GRELL, 1958 [*665*]

Radiolaria have been described as the most exquisite ,,Kunstformen der Natur" (HAECKEL).

Like the shells of Foraminifera, the skeletons of Radiolaria can also contribute to the formation of sediments on the ocean floor, although to a lesser degree. The tertiary marl of Barbados Island (Lesser Antilles) is especially rich in Radiolarian skeletons, which are still well preserved for the most part (Fig. 371).

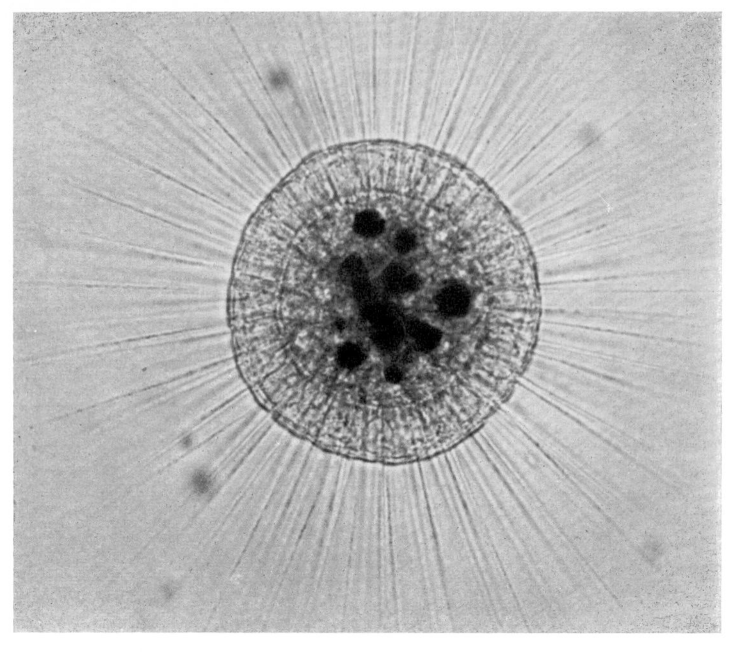

Fig. 368. *Actinosphaerium eichhorni* EHRENBERG. × 210. Photographed in life. Original by W. KUHL

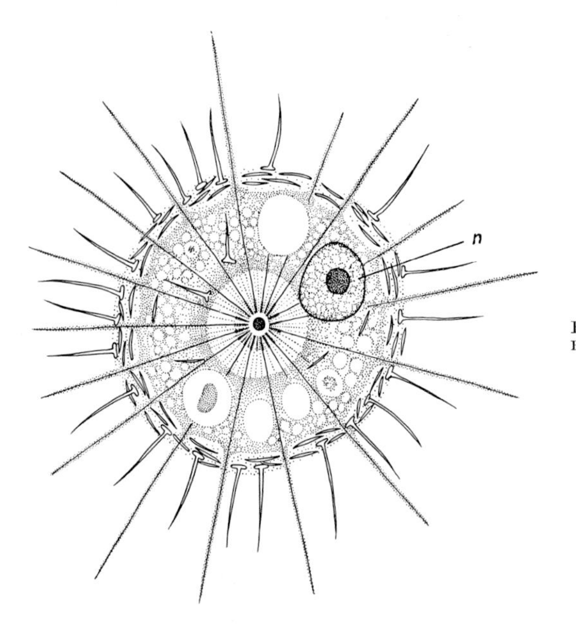

Fig. 369. *Acanthocystis aculeata* EHRENBERG. × 1200. *n* Nucleus. After STERN

Our knowledge of the mode of *reproduction* of Radiolaria is still very insufficient. Many appear to reproduce exclusively by multiple fission. The biflagellated swarmers thus formed bear crystalline inclusions and are therefore called crystal swarmers (Fig. 104 and 126). In other species binary fission takes place in addition

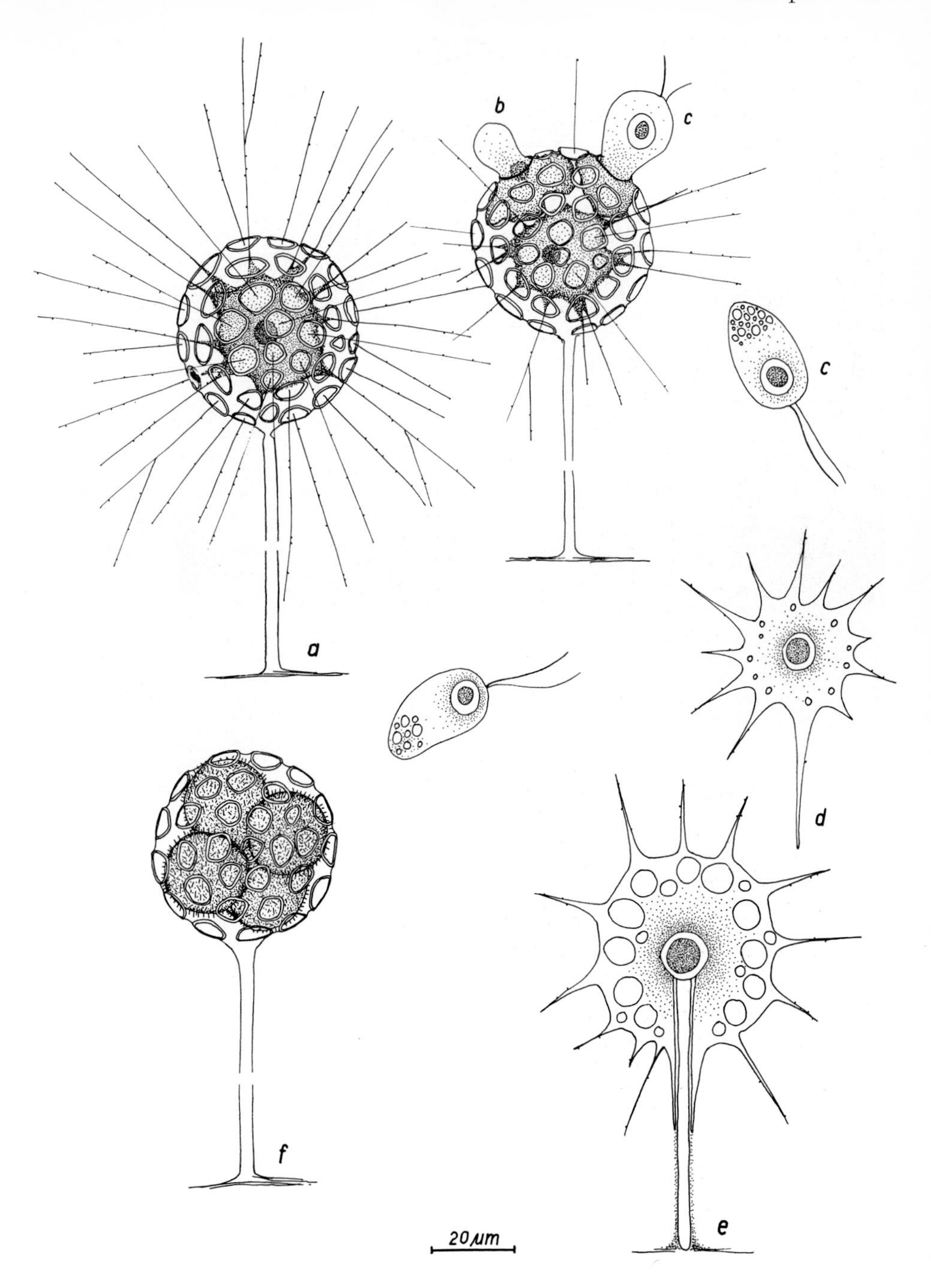

Fig. 370. *Clathrulina elegans* CIENKOWSKI. Developmental cycle. The encapsuled heliozoan-like stage (a) reproduces by the formation of amoeboid cells (b), resp. biflagellated swarmers (c), which leave the mother capsule. The amoeboid stage, resp. the flagellate stage transforms into a naked heliozoan (d) which differentiates a stalk-forming pseudopod (e) and later secretes its own capsule. Under unfavorable conditions all fission products encyst within the mother capsule (f). Upon excystation a swarmer emerges from each cyst and undergoes metamorphosis as depicted on the right side of the drawing. After BARDELE, 1972 [*89*]

Fig. 371. Radiolarian skeletons from the Upper Eocene of Barbados Island. × 450. Stereoscan micrograph (Dipl.-Phys. H. J. HUBER), Institute for Applied Microscopy, Photography and Cinematography of the Fraunhofer Society (Dir. Dr. H. REUMUTH)

to multiple fission (p. 116). It is probable that there are also Radiolaria which reproduce only by binary fission.

Radiolaria belong to the plankton of warmer seas. Many occur only at great depths. Smaller plankton organisms serve as food.

Within the system, a special position is held by the Acantharia. Whether it is justified to consider them as a separate order must be decided by further investigations.

First Suborder: **Peripylea (Spumellaria)**

The Peripylea have a round central capsule, penetrated all over by pores.

1. Family Group: Sphaerellaria

Skeleton consisting of one or more grid-like shells, often with radial spines. Group with very diverse forms.

Heliosphaera. With a grid-like shell, from which numerous spines of unequal length originate. *H. actinota* HAECKEL (Fig. 372).

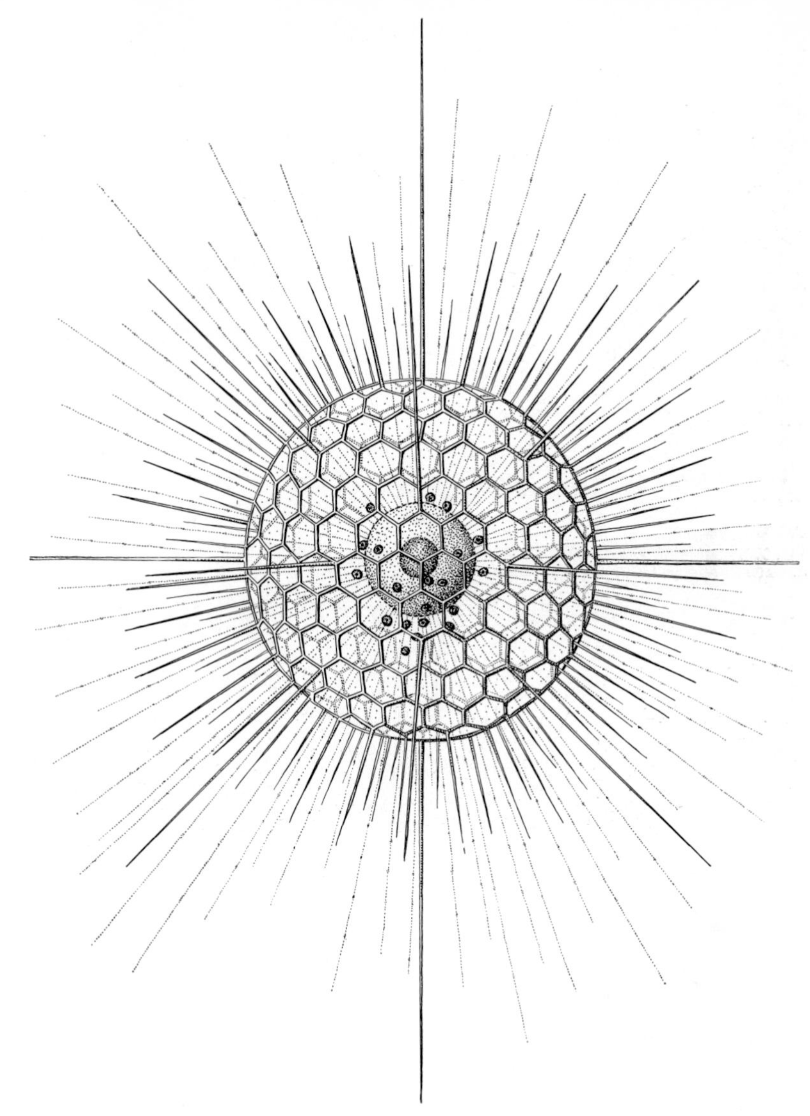

Fig. 372. *Heliosphaera actinota* HAECKEL. Within the grid-like sphere is the central capsule with the nucleus; outside of the central capsule are zooxanthellae. × 210. After HAECKEL from KÜHN

Actinosphaera. With two grid-like shells, connected by radial props. Spines equally long.
A. capillaceum HAECKEL (Fig. 373).
Hexacontium. With three grid-like shells and six radial major spines.
H. asteracanthion HAECKEL (Fig. 374).
Spongosphaera. With two small (intranuclear) grid-like shells and a spongy outer framework. Long spines.
S. polyacantha HAECKEL (Fig. 375).

2. Family Group: Collodaria

The Collodaria include spherical, usually larger forms, in which the skeleton consists

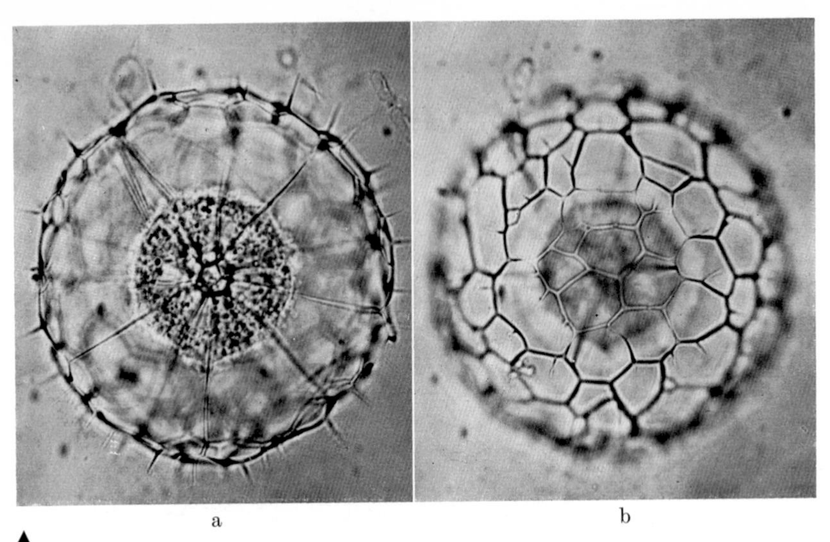

▲
Fig. 373. *Actinosphaera capillaceum* HAECKEL. The same individual in different focal planes. Photographed by J. CACHON and M. CACHON (Villefranche-sur-mer)

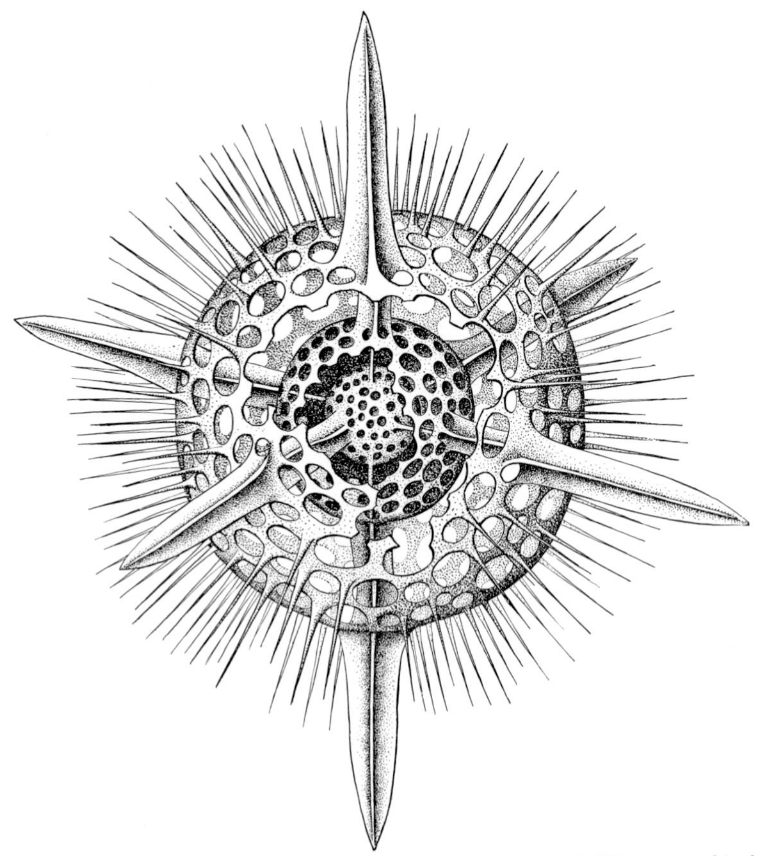

Fig. 374. *Hexacontium asteracanthion* HAECKEL. Skeleton (the two outer shell layers opened to show the inner one). × 480. After HAECKEL from KÜHN

Fig. 375. *Spongosphaera polyacantha* HAECKEL. Formation of pseudopodia. × 60

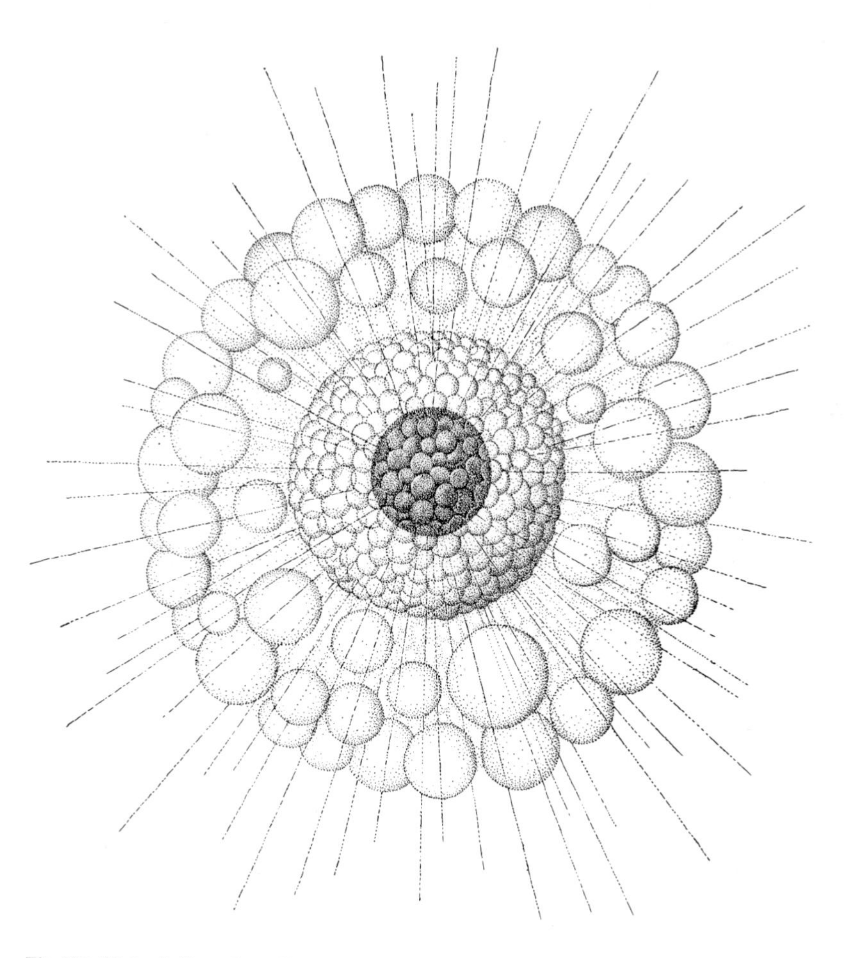

Fig. 376. *Thalassicolla nucleata*. HUXLEY. × 40. After HUTH from KÜHN

of single silicon needles only or is completely absent. Central capsule with a nucleus which is often very large.

Thalassicolla. With numerous extracapsulary vacuoles, arranged in layers. Without skeleton.
T. nucleata HUXLEY (Fig. 376).

3. Family Group: Polycyttaria

In the Polycyttaria or colony-forming Radiolaria numerous central capsules are found in a common gelatinous mass. The central capsules may be connected by anastomosing pseudopodia and usually contain many nuclei. Each central capsule frequently contains an oil sphere. In some species the central capsules are surrounded by latticed spheres, or single skeletal elements are found.

Collozoum. Without a skeleton.
C. inerme MÜLLER (Fig. 377).
C. pelagicum HAECKEL (Fig. 378c).
Sphaerozoum. With only single skeletal needles.
S. punctatum MÜLLER (Fig. 247 and 378a, b).
Collosphaera. Central capsules surrounded by grid-like shells.
C. huxleyi MÜLLER.

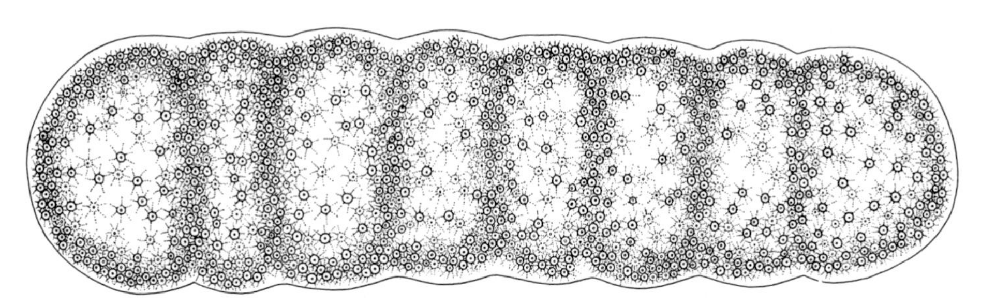

Fig. 377. *Collozoum inerme* MÜLLER. Colony. × 8. After BRANDT

Second Suborder: **Monopylea (Nassellaria)**

The central capsule of the Monopylea has only one axial opening which is not, however, uniform but represents a pore field. A highly refractile cone, furnished with canals ("pore canals") arises towards the inside above the pore field.

Cystidium. Without skeleton.
C. princeps HAECKEL (Fig. 379).
Cyrtocalpis. With urn-shaped, latticed shell.
C. urceolus HAECKEL (Fig. 380).
Theopilium. With helmet-like, latticed shell, divided in three parts.
T. cranoides HAECKEL (Fig. 381).

Third Suborder: **Tripylea (Phaeodoria)**

The central capsule of the Tripylea has three openings: a main opening (astropyle) and two secondary openings in the other half (parapyles). The main opening is closed by a radially striped cap with a tube arising from its middle. In the vicinity of the main opening, a yellow-brownish pigment mass (phaeodium) is found.

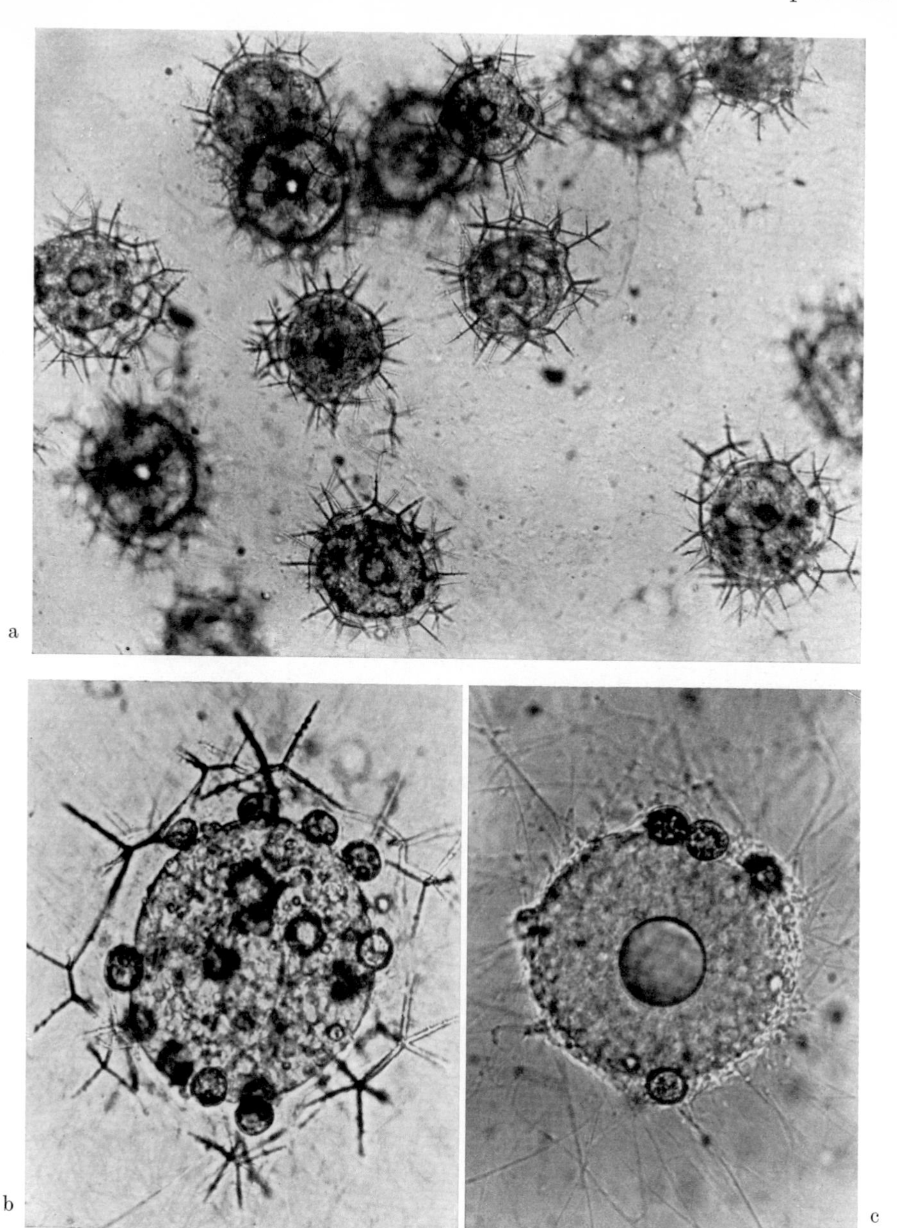

Fig. 378. Polycyttaria. a *Sphaerozoum punctatum* MÜLLER. Survey picture. Numerous central capsules in a common gelantinous matrix. b Single central capsule, surrounded by zooxanthellae and skeletal needles. c *Collozoum pelagicum* HAECKEL, single central capsule (without skeletal needles). Photographed in life

Aulacantha. Skeleton consisting of radially arranged spines and fine tangential needles.

A. scolymantha HAECKEL (Fig. 382). Reproduction, see pp. 118 (Fig. 104—108).

Caementella. With skeleton of foreign bodies.

C. stapedia HAECKEL.

Challengeron. With bilateral symmetrical shell.

C. wyvillei HAECKEL (Fig. 383).

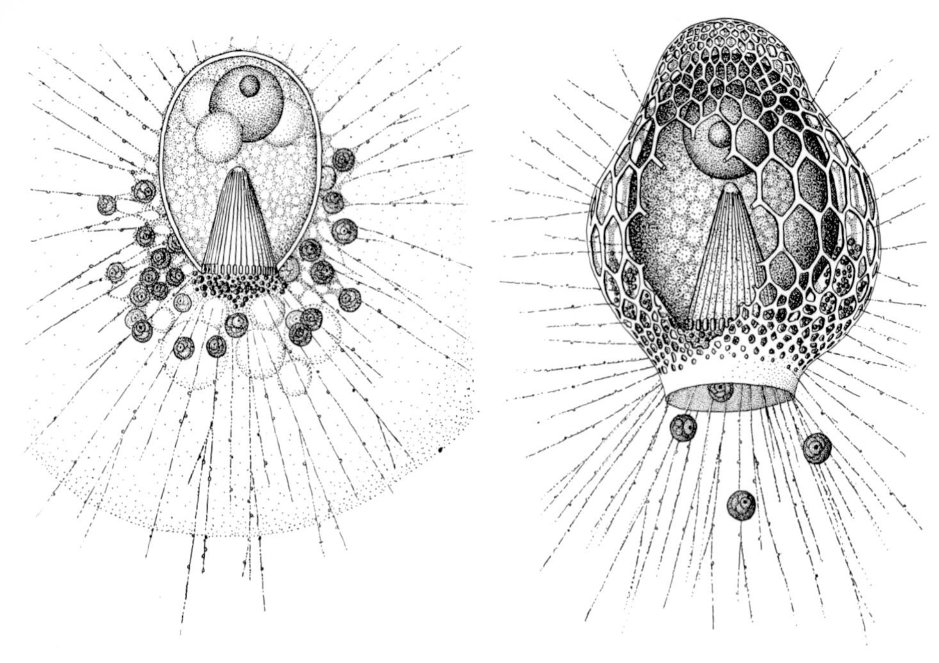

Fig. 379. *Cystidium princeps* HAECKEL. Part of a cell. × 300. After HAECKEL from KÜHN

Fig. 380. *Cyrtocalpis urceolus* HAECKEL. × 330. After HAECKEL from KÜHN

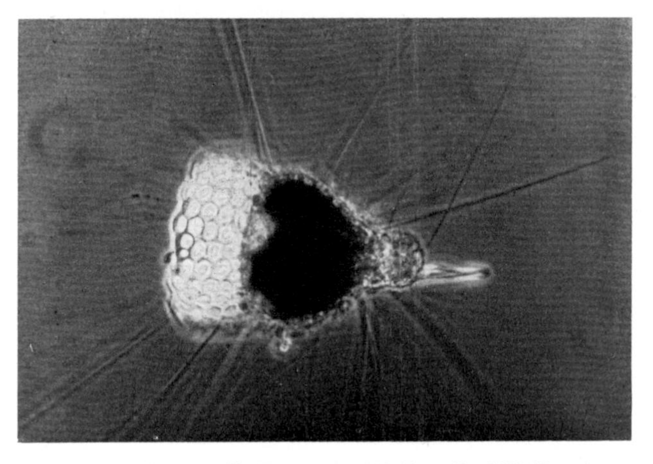

Fig. 381. *Theopilium cranoides* HAECKEL. × 380. From film C829 (GRELL)

Fourth Suborder: **Acantharia**

The Acantharia have a skeleton of strontium sulfate, consisting, as a rule, of 20 spines which converge in the middle of the cell. The spines are usually arranged in a definite geometrical pattern (Müller's law).

In many species bundles of fibers ("myophriscs") which are connected with the cell surface are fastened to the spines. Their function is still questionable. It has been

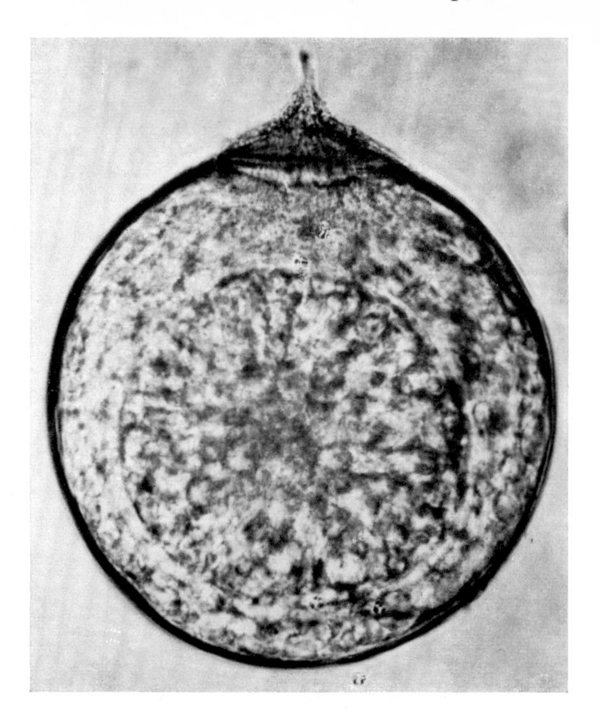

Fig. 382. *Aulacantha scolymantha* HAECKEL, isolated central capsule. × 380. Photographed in life

claimed that their contraction and expansion enables the cell to change its volume. Recent studies, however, did not confirm these observations [*511*]. A central capsule membrane is found in only some of the Acantharia (Arthracantha). It has been proved recently that many Acantharia can encyst [*778*]. This is accompanied by an extensive reduction of the skeleton.

Acanthometron. Regularly arranged, simple spines.
A. elasticum MÜLLER (Fig. 384).
Amphilonche. Two opposite spines are much longer and thicker than the remainder.
A. elongata MÜLLER (Fig. 385).
Lithoptera. Four spines arranged to form a cross, longer than the others and distally broadened to form grid-like plates. Cell body cross-shaped.
L. muelleri HAECKEL (Fig. 386).

Third Class: Sporozoa

GRASSÉ, P. P.: Sous-Embranchement des Sporozoaires. In: GRASSÉ, P. P.: Traité de Zoologie. 1, Fasc. 2. Paris: Masson et Cie. 1953.

LABBÉ, A.: Sporozoa. In: SCHULZE, F. E.: Das Tierreich. Teil 5. Berlin: Friedländer 1889.

NAVILLE, A.: Les Sporozoaires (cycle chromosomique et sexualité). Mém. Soc. Phys. Hist. Nat. Genève **41** (1931).

REICHENOW, E.: Sporozoa. In: GRIMPE-WAGLER: Die Tierwelt der Nord- und Ostsee. II. Leipzig: Akadem. Verlagsges. 1932.

All Sporozoa* are parasites. They have a *haplo-homophasic alternation of generations*, in which sexual (gamogony) and asexual (sporogony) multiple fissions alter-

* Only those forms which in other surveys have been designated as "Telosporidia" have been included under this heading.

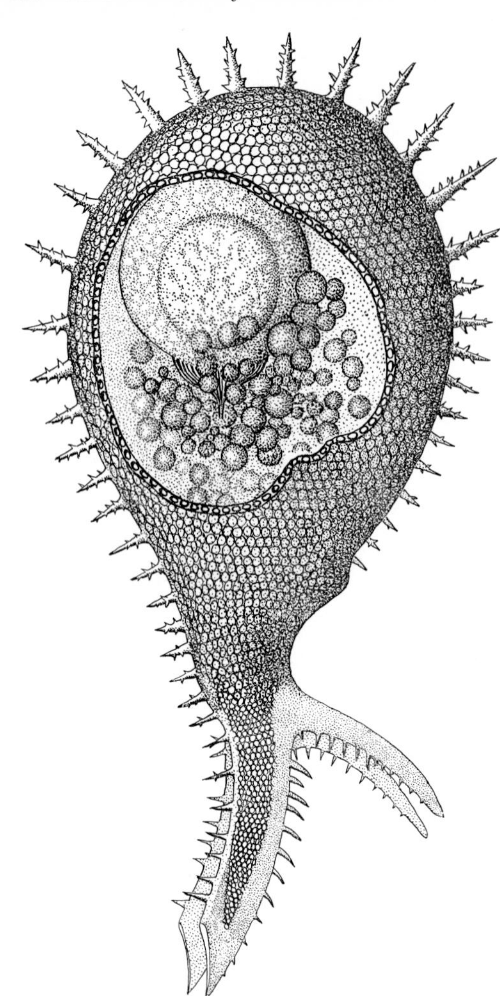

Fig. 383. *Challengeron wyvillei* HAECKEL. Side view; drawn with a part of the shell removed to show the central capsule and phaeodium. × 350. After HAECKEL from KÜHN

nate. In a number of Sporozoa a further asexual multiple fission (schizogony) takes place. The more pronounced this is, the more the other two modes of reproduction may lose in importance. The cysts, designated as spores, which serve in transmission to another host, arise by sporogony. Special organelles of locomotion appear only in microgametes in the form of flagella.

Two orders of Sporozoa can be distinguished, based upon the occurrence of multiple fission in the gamogony of either sex or the male sex alone.

First Order: Gregarinida

BATHIA, B. L.: Synopsis of the genera and classification of Haplocyte Gregarines. Parasitology **22** (1930).

GEUS, A.: Sporentierchen, Sporozoa. Die Gregarinida der land- und süßwasserbewohnenden Arthropoden Mitteleuropas. In: DAHL, F.: „Die Tierwelt Deutschlands", 57. Teil. Jena: VEB Gustav Fischer Verlag 1969.

GRASSÉ, P. P.: Classe des Grégarinomorphes. In: GRASSÉ, P. P.: Traité de Zoologie. **1**, Fasc. 2. Paris: Masson et Cie. 1953.

Fig. 384. *Acanthometron elasticum* Müller. Central capsule with numerous nuclei. "Myophriscs" (*My*) are located where the spines leave the cell body. After R. Hertwig from Kühn

Léger, L.: Les schizogrégarines des trachéates. I, II. Arch. Protistenk. **8**, **18** (1907/09).
Watson-Kamm, M. E.: Studies on Gregarines. I, II. Illinois Biol. Monogr. **2**, **7** (1916—1922).
— A list of new gregarines described from 1911 to 1920. Trans. Amer. microscop. Soc. **41** (1922).
Weisser, J.: A new classification of the Schizogregarina. J. Protozool. **2** (1955).

In the Gregarinida the *gamonts of both sexes* carry out *multiple fission* so that usually numerous male and female gametes are found. Sexual reproduction begins with the mating of gamonts *(gamontogamy)*. Mating can take place either during or at the end of the growth phase and leads as a rule to the formation of a common envelope. Within this gamont cyst the gametes then copulate. They may have the

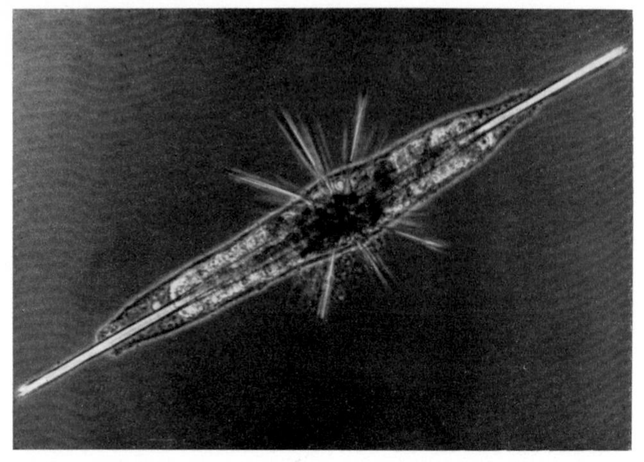

Fig. 385. *Amphilonche elongata* MÜLLER. × 440. From the film C829 (GRELL)

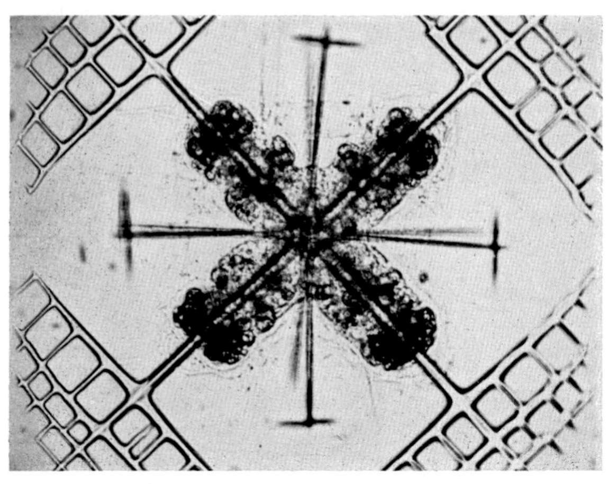

Fig. 386. *Lithoptera muelleri* HAECKEL. Photographed by J. CACHON and M. CACHON (Villefranche-sur-mer)

same shape (isogamety) or be of different shapes (anisogamety) (see p. 195). In this manner numerous zygotes originate which immediately change into spores. Sporogony, consisting only of *one* nuclear division sequence and leading to the formation of sporozoites, takes place in the spores. Most gregarines have spores containing eight sporozoites.

Gregarine development can proceed with or without schizogony.

First Suborder: **Eugregarinida**

Extracellular development prevailing, without schizogony.

The sporozoites, freed when the spores burst open, grow immediately to the sexually differentiated forms or gamonts. This growth phase, which usually takes place in body cavities (gut lumen, coelom), is the longest developmental period of the eugregarines. The mature gamonts and the cysts formed by them are thus comparatively large stages. The gamont cysts are often provided with special mechanisms to eject the spores.

The eugregarines are mainly parasites of annelids and arthropods. Two families can be distinguished, depending on whether the gamont is simple or subdivided into compartments.

1. Family: Monocystidae. Gamont simple

The Monocystidae are found chiefly in the seminal vesicles of oligochaetes, few species also in other invertebrates. In the seminal vesicles, the sporozoites often penetrate first into the sperm-forming cells (blastophores) of the host. After they

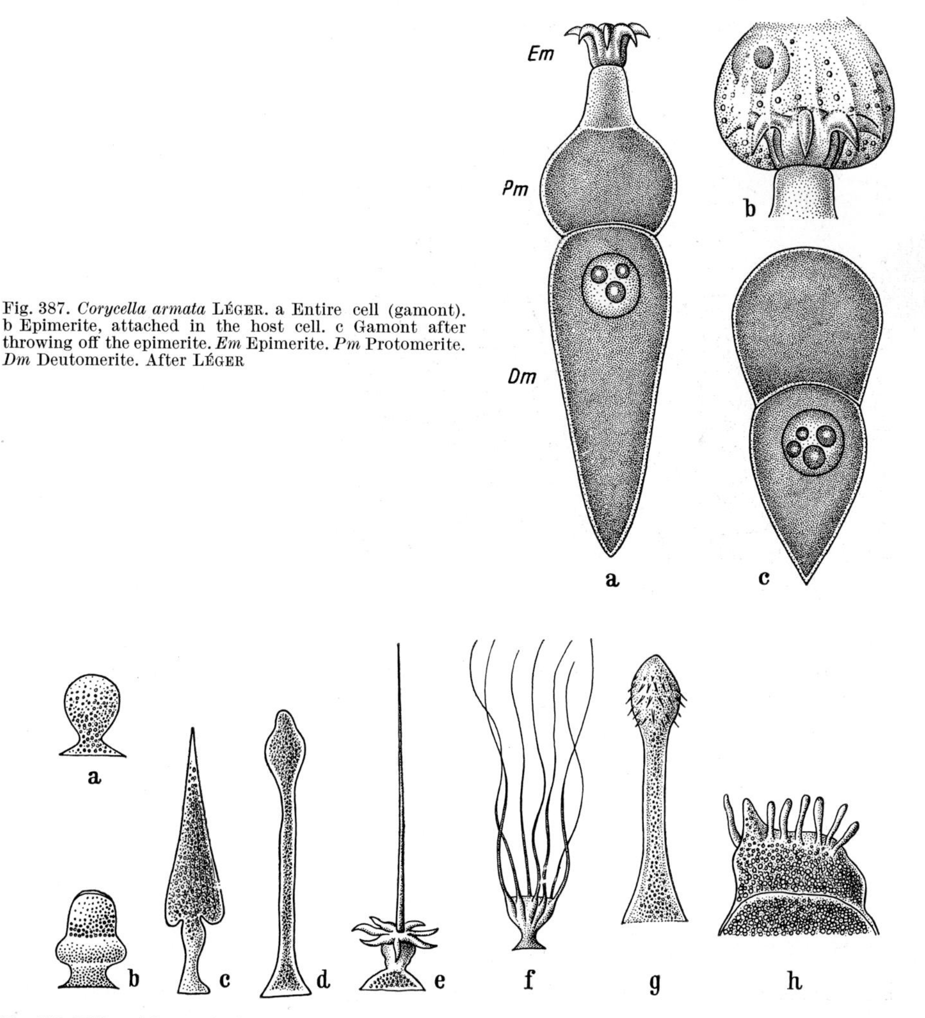

Fig. 387. *Corycella armata* LÉGER. a Entire cell (gamont). b Epimerite, attached in the host cell. c Gamont after throwing off the epimerite. *Em* Epimerite. *Pm* Protomerite. *Dm* Deutomerite. After LÉGER

Fig. 388. Different forms of epimerites in eugregarines. a *Gregarina longa*. b *Sycia inopinata*. c *Pileocephalus heeri*. d *Stylocephalus longicollis*. e *Beloides firmus*. f *Cometoides crinitus*. g *Geniorhynchus monnieri*. h *Echinomera hispida*. After LÉGER

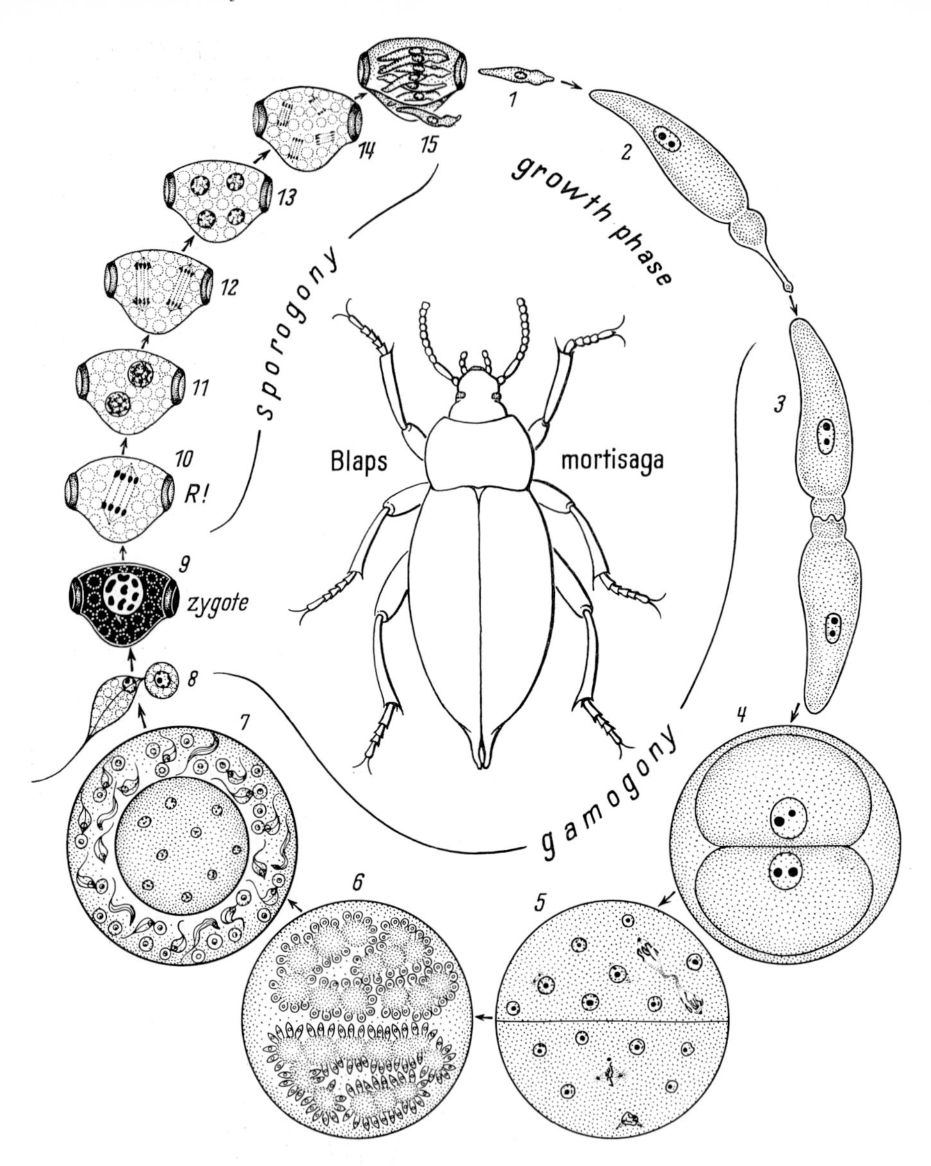

Fig. 389. *Stylocephalus longicollis* STEIN. Developmental cycle. Development takes place partly in the gut of the beetle *Blaps mortisaga* (*1—4*), partly outside the host (*5—14*). The spores reach the midgut of the beetle together with food. Here they burst due to the action of digestive juices, liberating the sporozoites (see Fig. 390d). The latter (*1*) grow in the gut lumen to gamonts (*2*). As soon as the gamonts have reached a certain size, they cast off their epimerites and unite pairwise with each other (*3*). The gamont pairs surround themselves with a common cyst envelope (*4*). Gamont cysts reach the outside with feces of the beetle. A period of nuclear multiplication takes place in each gamont (*5*). During the formation of gametes (*6*) not all of the cytoplasm is used up. The remaining cytoplasmic masses fuse to form a residual body. This is delimited from the liquid contents of the gamont cysts by a membrane. The gametes are morphologically distinct. In addition to the gametes which unite with one another (*7, 8*), so-called fusiform gametes appear which cannot fertilize. Their significance is unclear. After fertilization, the zygote, which is at first round, assumes the coin-purse shape of the spore (*9*). The first sporogony division (*10*) leads to reduction of the chromosome number. Eight sporozoites are formed ultimately (*11—15*). As in the following diagrams (Figs. 391, 393, 395, 397, 398, 403, 404), the various stages of development are shown at different magnifications. The diploid phase, in Sporozoa only the zygote, is drawn dark. Based on studies by LÉGER, 1904 [*1018*] and GRELL, 1940 [*648*]

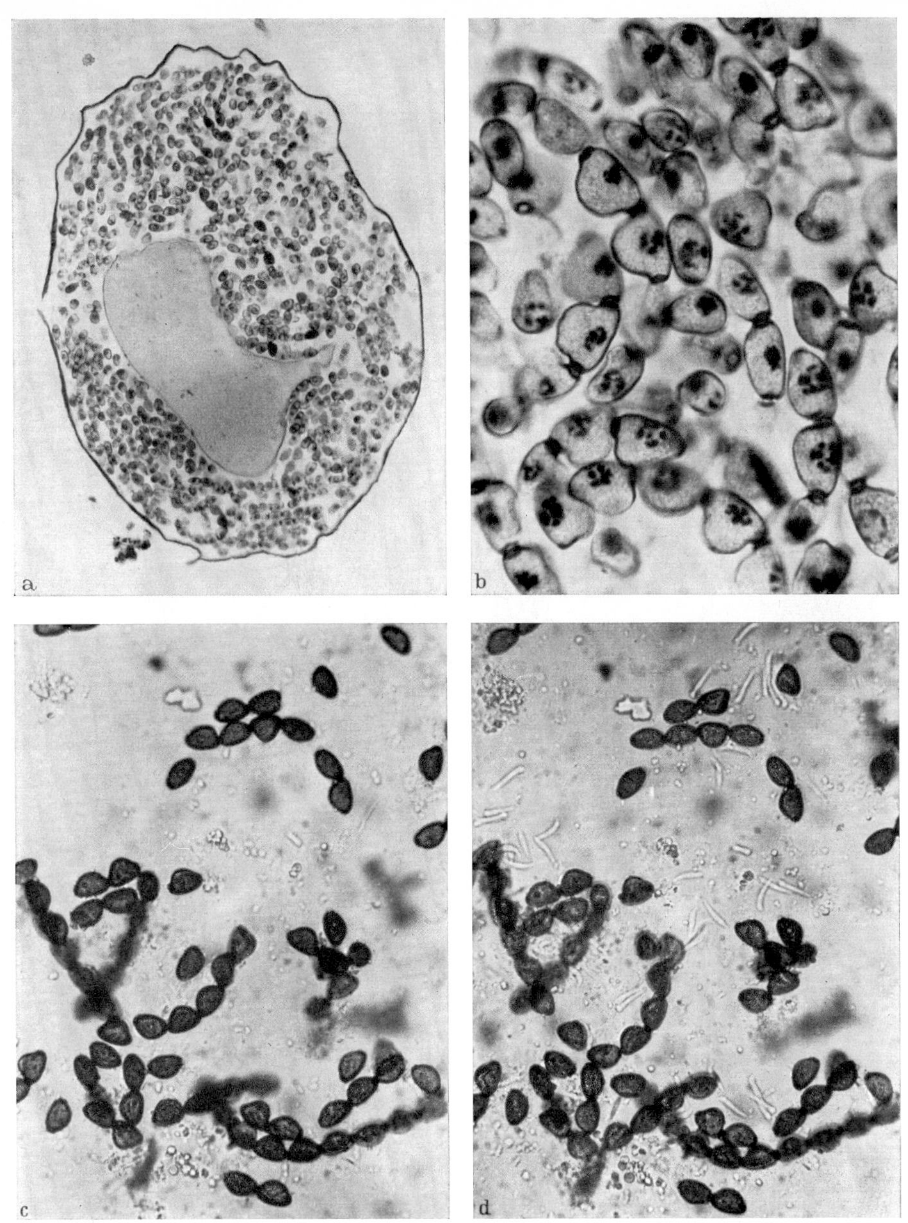

Fig. 390. *Stylocephalus longicollis* STEIN. a Gamont cysts after formation of the spores; in the middle, the residual body. b Section from such a gamont cyst. The nucleus with the chromosomes (late prophase of meiosis) is recognizable in the spores. c Spores before, d after action of the digestive juices of the beetle: sporozoites are set free. The spores of *Stylocephalus longicollis* are arranged in a row like a long string of pearls. The string has broken into pieces during transfer to the slide. a, b Sections. Carnoy, iron hematoxylin. a × 150. b × 800. c, d In life. After GRELL, 1940 [*648*]

have reached a certain size there, the sperm-forming cell bursts and the gamonts develop further floating in the lumen.

Monocystis agilis STEIN. In the seminal vesicle of the earthworm.
Lankesteria ascidiae LANKESTER. In the gut of the ascidian, *Ciona intestinalis*.

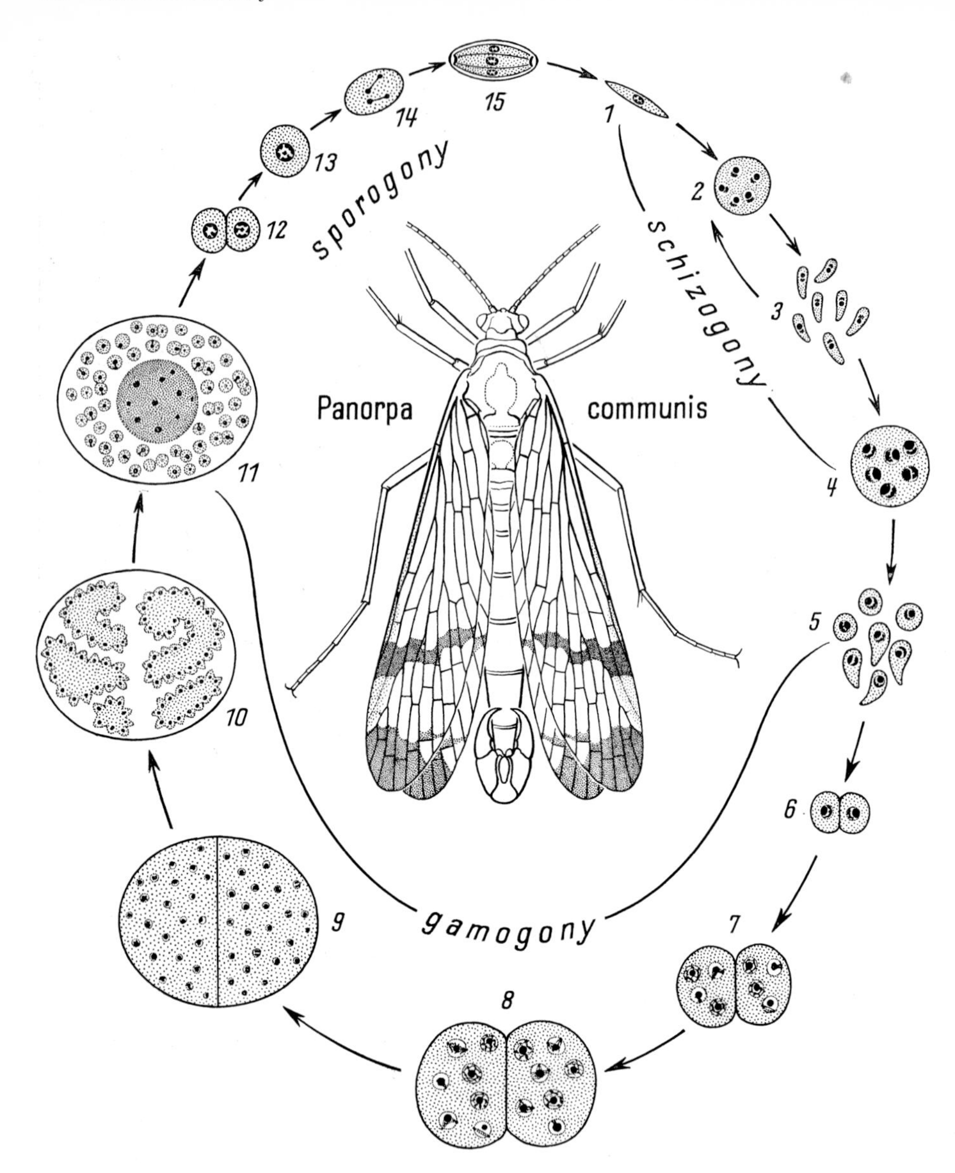

Fig. 391. *Lipocystis polyspora* GRELL. Developmental cycle. The whole course of development takes place within the fat body of the scorpion fly *Panorpa communis*. The spores burst in the gut of the host and the sporozoites (*1*) penetrate through the gut wall into the body cavity. In the cells of the fat tissue they grow at first into schizonts with small nuclei (*2*), which produce comparatively small merozoites (*3*). This schizogony can be repeated several times. When a certain degree of infection has been reached in the fat tissues, schizonts (*4*) with large nuclei are formed, which give rise to larger merozoites (*5*). The latter develop into gamonts and arrange themselves pairwise next to each other within the fat cells (*6*). The gamont pairs then grow at a rapid rate which is connected with an intense nuclear multiplication (*7—9*). During gamete formation (*10*) a portion of the cytoplasm remains as a residual body. Numerous isogametes arise which pair within the gamont cyst (*11, 12*). Accordingly, many spores are formed (*13—15*), which contain 8 sporozoites. After investigations by GRELL, 1938 [*647*]

2. Family: Polycystidae. Gamont subdivided into compartments

The gamont of the Polycystidae (Fig. 387) is usually elongated and consists of two compartments, the shorter anterior part designated as *protomerite* and the longer

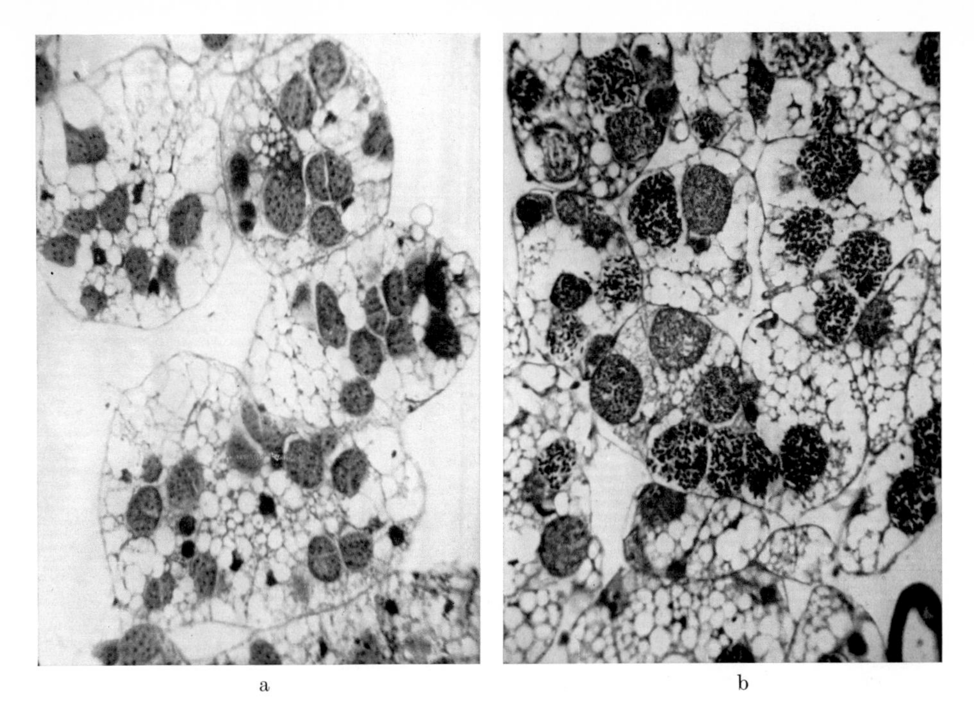

a b

Fig. 392. *Lipocystis polyspora* GRELL. a Stages of gamogony and sporogony (b) in the fat tissues of the scorpion fly *Panorpa communis*. Carnoy, Delafields hematoxylin. × 280

posterior as *deutomerite*. The two compartments are separated by an ectoplasmatic partitioning (septum). The nucleus, which sometimes contains several nucleoli, is found in the deutomerite. The protomerite can elongate anteriorly to form an extension or *epimerite* (Fig. 388) which varies considerably in shape in the individual species.

The sporozoite often penetrates at first into an epithelial cell, from which it later grows out for the most part. The gamont can, with the help of the epimerite, remain attached to the epithelial cell for quite a long time. Sooner or later the epimerite is cast off, and the gamont then develops further floating in the lumen.

The gamonts unite in some species even before the end of the growth phase, and, moreover, usually in such a manner that one gamont attaches by its protomerite to the deutomerite of another. The anterior one would then be designated as *primite*, the posterior as *satellite*. Final encystment takes place only after both gamonts have matured. In other cases the gamonts attach with their protomerites (Fig. 389). In some species the gamont cysts reach the outside prematurely, in others they remain within the host until the end of sporogony. The Polycystidae are predominantly parasites of the gut and body cavities.

Gregarina polymorpha HAMMERSCHMIDT, *G. steini* BERNDT, *G. cuneata* STEIN. In the gut of the larva of the flour-beetle, *Tenebrio molitor*.

Stylocephalus longicollis STEIN (Fig. 389 and 390). In the gut of the beetle, *Blaps mortisaga*.

Echinomera hispida SCHNEIDER. In the gut of the centipede, *Lithobius forficatus*.

Actinocephalus parvus WELMER. In the gut of the larva of the dog flea, *Ctenocephalus canis*.

Second Suborder: **Schizogregarinida**

Development primarily intracellular with schizogony.

As soon as they reach the tissue or organ in which they preferentially develop, the

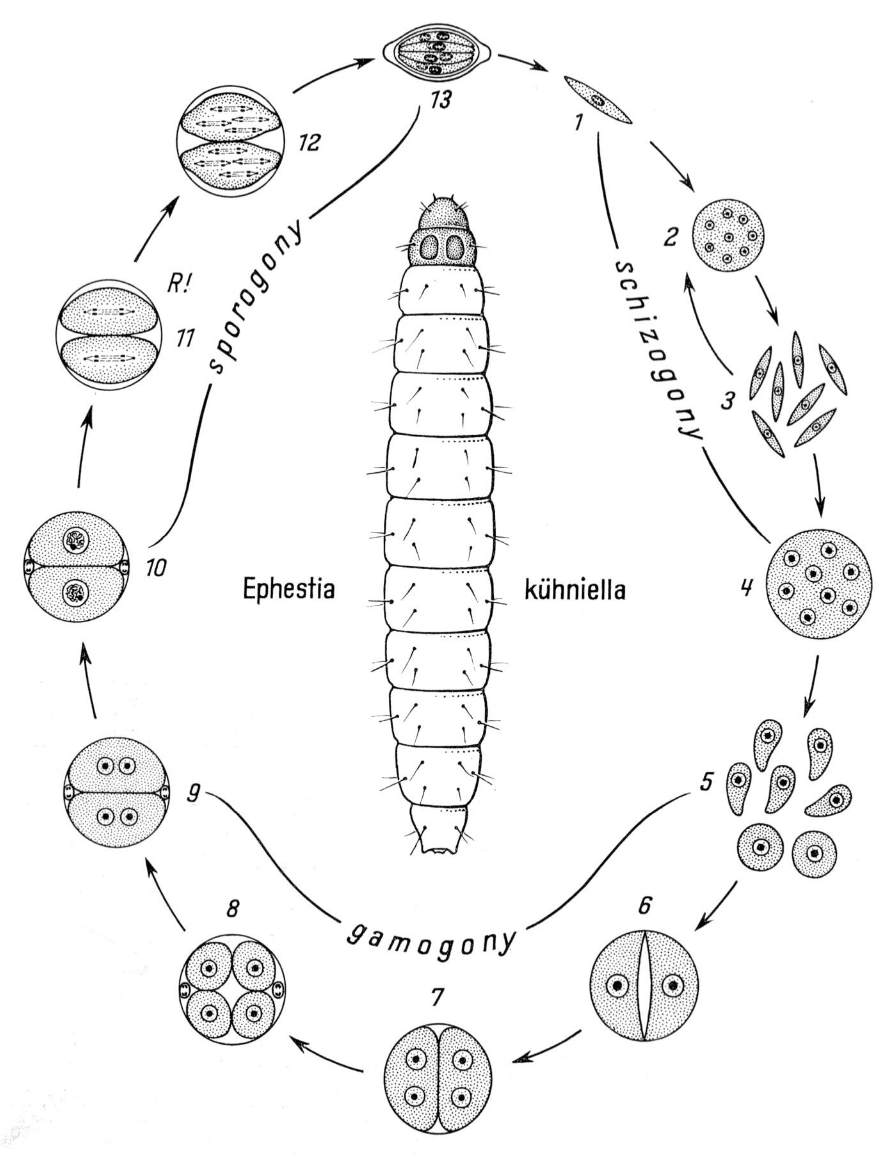

Fig. 393. *Mattesia dispora* NAVILLE. Developmental cycle. The entire course of development takes place in the fat tissue of the caterpillar of the flour moth *Ephestia kuehniella*. As in the case of *Lipocystis polyspora* (Fig. 391) schizonts with small nuclei are first formed, later with large nuclei (*1*—*5*). The gamonts, which unite pairwise and encyst (*6*), pass through a relatively insignificant growth phase. In each gamont four nuclei are formed. However, two degenerate within small residual bodies. Each gamont thus forms only two gametes (*7*—*8*). Occasionally it even happens that each gamont transforms into only a single gamete. As a rule, however, the gamont cyst gives rise to two zygotes (*9*, *10*) from which the spores with 8 sporozoites (*11*, *12*) originate. — Infection results not only from spores, but also due to direct transfer by ichneumon-wasps transmitting fluid from the body cavity, in which free merozoites as well as spores swim about regularly (see Fig. 394), to other caterpillars. After studies by NAVILLE, 1930 [*1222*]

sporozoites grow first to asexually reproducing cells, the so-called *schizonts*, rather than immediately to sexual forms. By multiple fission the schizonts issue numerous, for the most part spindle-shaped, daughter cells, the *merozoites*. Repetition of schizogony leads to a great increase of the parasites within the host organism which compensates to a certain extent for the fact that the growth phase is not as extensive as that of the eugregarines. In species living intracellularly, the period of increase stops only when the tissue concerned is entirely infected by parasites. Gamogony starts as soon as a certain degree of infection is reached. The further processes take a course similar to the eugregarines. In species with a very prolonged schizogony, multiplication by gamogony may be more or less restricted. Thus, a gamont from *Mattesia dispora* (Fig. 393) issues only two gametes, one from *Ophryocystis mesnili* even only a single gamete. The number of spores is reduced accordingly.

Damage to the host organism caused by the tissue parasitism of the schizogregarines is in general more extensive than that caused by the body cavity parasitism of the eugregarines. However, species of schizogregarines also exist which live exclusively in body cavities.

a) Gamont cysts with many spores

Schizocystis gregarinoides Léger. In the gut of larvae of *Ceratopogon*-gnats.
Lipocystis polyspora Grell (Fig. 391 and 392). In the fat body of the mecopteran, *Panorpa communis*.

b) Gamont cysts with one or two spores

Ophryocystis mesnili Léger. In the malpighian tubes of larvae of the flour-beetle, *Tenebrio molitor*.

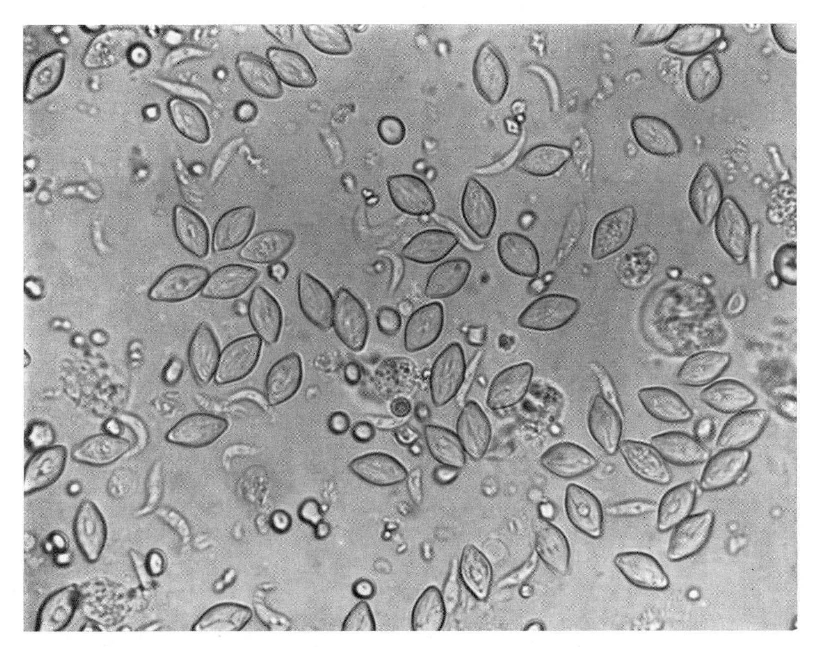

Fig. 394. *Mattesia dispora* Naville. Fluid from the body cavity of a caterpillar of *Ephestia kuehniella* with spores and merozoites. Photographed in life

Mattesia dispora NAVILLE (Fig. 393 and 394). In the fat body of the meal-moth, *Ephestia kuehniella*.

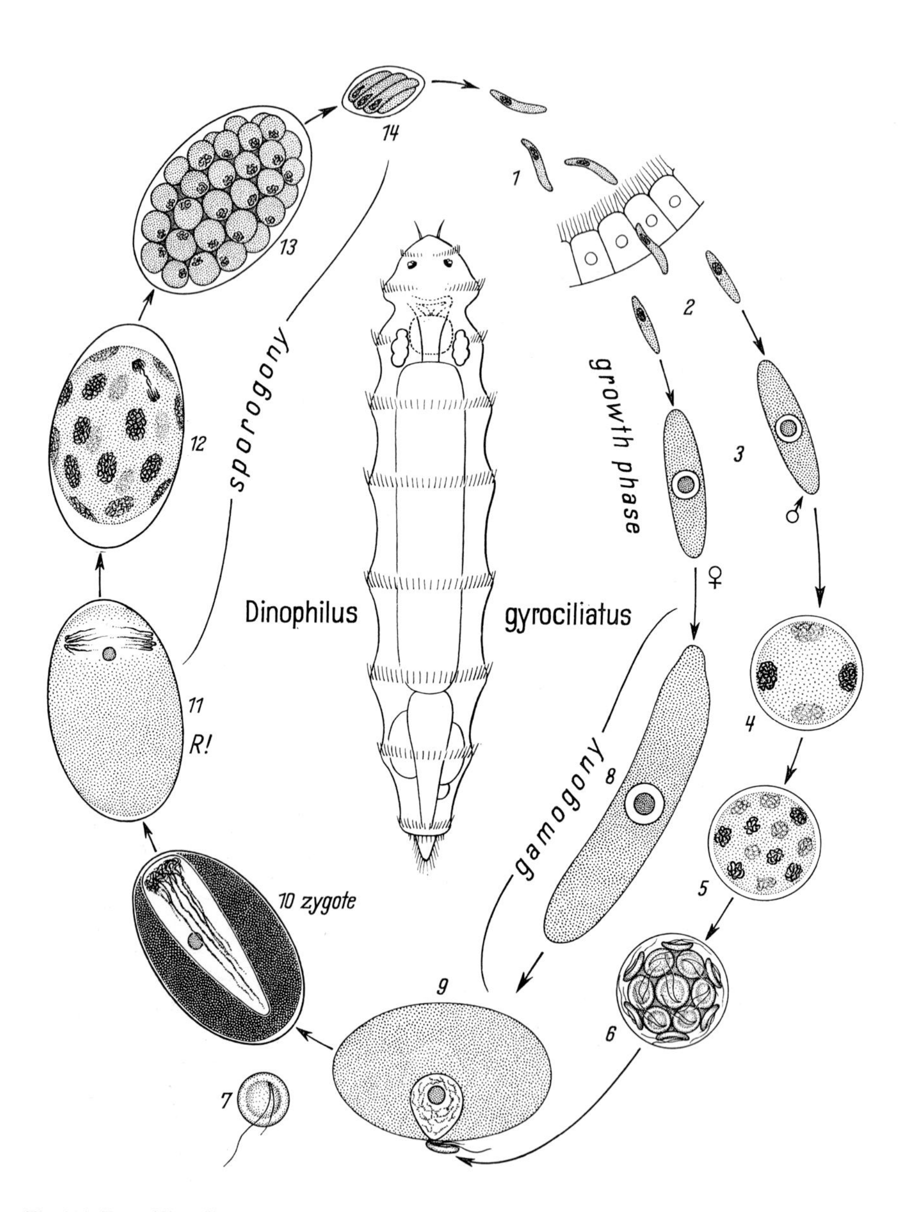

Fig. 395. *Eucoccidium dinophili* GRELL. Developmental cycle. The entire course of development is extracellular in the body cavity of the marine archiannelid *Dinophilus gyrociliatus*. The spores burst after being taken into the gut, and the sporozoites penetrate through the gut wall into the body cavity (*1*, *2*). There they immediately grow into gamonts (*3*). The macrogamonts are large and cigar-shaped (*8*), the microgamonts, small and round. By multiple fission the microgamonts give rise to approximately 12—32 microgametes (*4*—*6*). The latter are bowl-shaped and have two flagella of different lengths (*7*).The macrogamonts round off becoming macrogametes. After fertilization (*9*), the synkaryon elongates ("fertilization spindle", *10*). Sporogony begins with meiosis (*11*) and leads to formation of the spores (*12*, *13*). The oocysts then burst and the spores float about freely in the body cavity. Six sporozoites are formed in each spore (*14*). The spores reach the outside when the host dies and its integument bursts. After GRELL, 1953 [*657*]

Second Order: Coccidia

BECKER, E. R.: Coccidia and coccidiosis of domesticated and laboratory animals and of man. Ames (Iowa): Collegiate Press 1934.
GARNHAM, P. C. C.: Malaria parasites and other Haemosporidia. Oxford: Blackwell Sci. Publ. 1966.
GRASSÉ, P. P.: Classe des Coccidiomorphes. In: GRASSÉ, P. P.: Traité de Zoologie. **1**, Fasc. 2. Paris: Masson et Cie. 1953.
HAMMOND, D. M., LONG, P. L.: The Coccidia. *Eimeria*, *Isospora*, *Toxoplasma* and related genera, 450 pp. Baltimore: University Park Press 1972.
PELLÉRDY, L.: Catalogue of the genus *Eimeria* (Protozoa: Eimeriidae). Acta Veter. Acad. Sci. Hung. **6** (1956).
— Catalogue of the genus *Isospora* (Protozoa: Eimeriidae). Acta Veter. Acad. Sci. Hung. **7** (1956).
— Coccidia and Coccidiosis. Publishing house of the Hungarian Academy of Sciences, Budapest: Akadémiai Kiadó. 1965.
TYZZER, E. E.: Coccidiosis in gallinaceous birds. Amer. J. Hyg. **10** (1929).

In the Coccidia *only the male gamont* (microgamont) carries out *multiple fission*, leading to the formation of small, flagellated microgametes. The female gamont (macrogamont) transforms immediately into a large, immobile macrogamete. Thus, only one large zygote is formed, which is designated as an *oocyst*. Within the oocyst *sporogony* takes place which, in contrast to the gregarines, consists of *two phases*. The first phase leads to the formation of spores; the second takes place

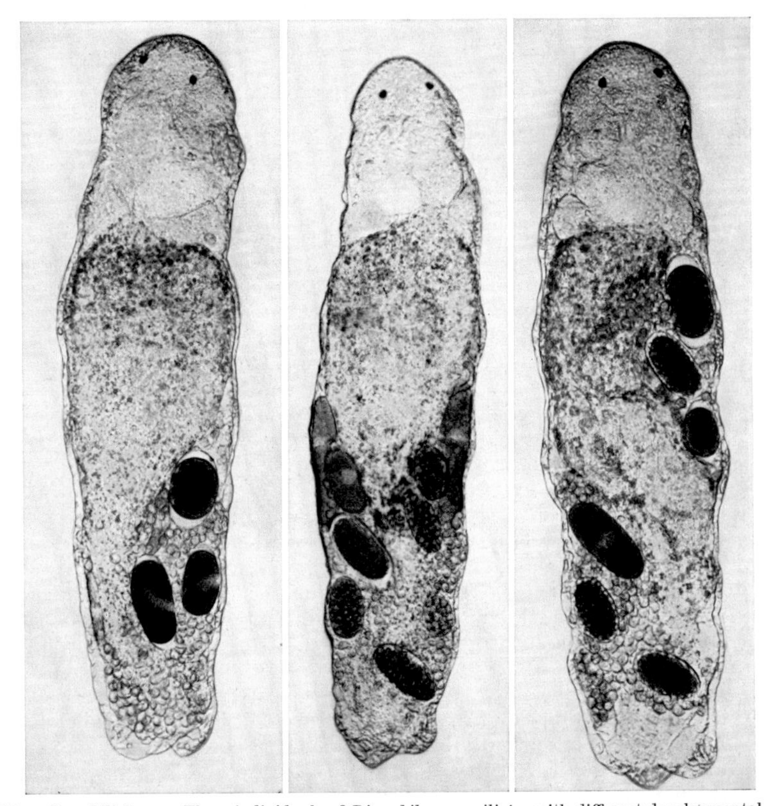

Fig. 396. *Eucoccidium dinophili* GRELL. Three individuals of *Dinophilus gyrociliatus* with different developmental stages in the body cavity, somewhat compressed under the coverslip. Photographed in life

within the spores and ends with the origin of the sporozoites. In some Coccidia the first phase can be very much reduced so that few spores are formed in the oocyst. Also the number of sporozoites within the spore can differ from species to species.

As in gregarines, schizogony may either be lacking or be present in Coccidia.

First Suborder: **Eucoccidia**

Development primarily extracellular, without schizogony

Eucoccidium dinophili GRELL (Fig. 396). In the body cavity of the archiannelid, *Dinophilus gyrociliatus*. Bisexual (Fig. 395) and parthenogenetic development (Fig. 170).

E. ophryotrochae GRELL (Fig. 300). In the body cavity of the polychaete, *Ophryotrocha puerilis*.

Coelotropha durchoni VIVIER et HENNERÉ. In the body cavity of the polychaete, *Nereis diversicolor*.

Second Suborder: **Schizococcidia**

Development primarily intracellular, with schizogony.

The Schizococcidia occur as parasites in vertebrates and invertebrates and infect above all epithelial tissues. Some species alternate between hosts, one part of development (schizogony) taking place in one host, the remaining part (gamogony and sporogony) in the other.

1. Family: Eimeridae

Macro- and microgamonts develop separately. Two phases of sporogony.

Eimeria. Oocyst with four spores. Each spore contains two sporozoites. One host only.

E. schubergi SCHAUDINN (Fig. 397). In the gut epithelium of the centipede, *Lithobius forficatus*.

E. stiedae LINDEMANN. Causative agent of rabbit coccidiosis.

The parasites attack the epithelium of the bile duct in the liver of domestic and wild rabbits. They bring about a proliferation of host tissue ("coccidian knots") which frequently lead to death. Other *E.*-species of rabbits (for example, *E. perforans* LEUCKART, *E. magna* PÉRARD) are parasites in the epithelium of the small intestine.

E. zürni RIVOLTA. Cause of the "bloody flux" in cattle.

E. tenella RAILLET et LUCET and other species. Cause of fowl coccidiosis.

Isospora. Oocyst with two spores. Each spore contains four sporozoites. Species of *I.* occur in dogs and cats. Infections in man with *I. belli* or *I. hominis* have been occasionally observed, but they do not lead to severe symptoms.

Toxoplasma gondii NICOLLE et MANCEAUX.

First discovered in an African rodent, *Toxoplasma gondii* turned out to be a widespread intracellular parasite infesting various birds and mammals including man. Preferential tissues are the reticuloendothelial and the central nervous systems (brain). Within the host cells the parasites multiply by a peculiar type of longitudinal binary fission (endodyogeny): Within a mother cell two daughter cells are formed which surround themselves with own membranes, leaving the mother cell after disruption of its envelope. Under certain conditions large cysts (about 150 μm in diameter) are produced comprising thousands of single cells.

Electron microscopic investigations suggest that *Toxoplasma gondii* is a coccidian. There is some — though not yet enough — evidence that it is related to *Isospora*.

Asymptomatic infections are common in men (30—50%). Clinical *Toxoplasmosis* caused by the parasite is rare, however. Most severe cases are seen in newborn babies which acquire the parasites through their mothers. Symptoms are hydrocephalus and chorioretinitis. Infants infected before birth usually die.

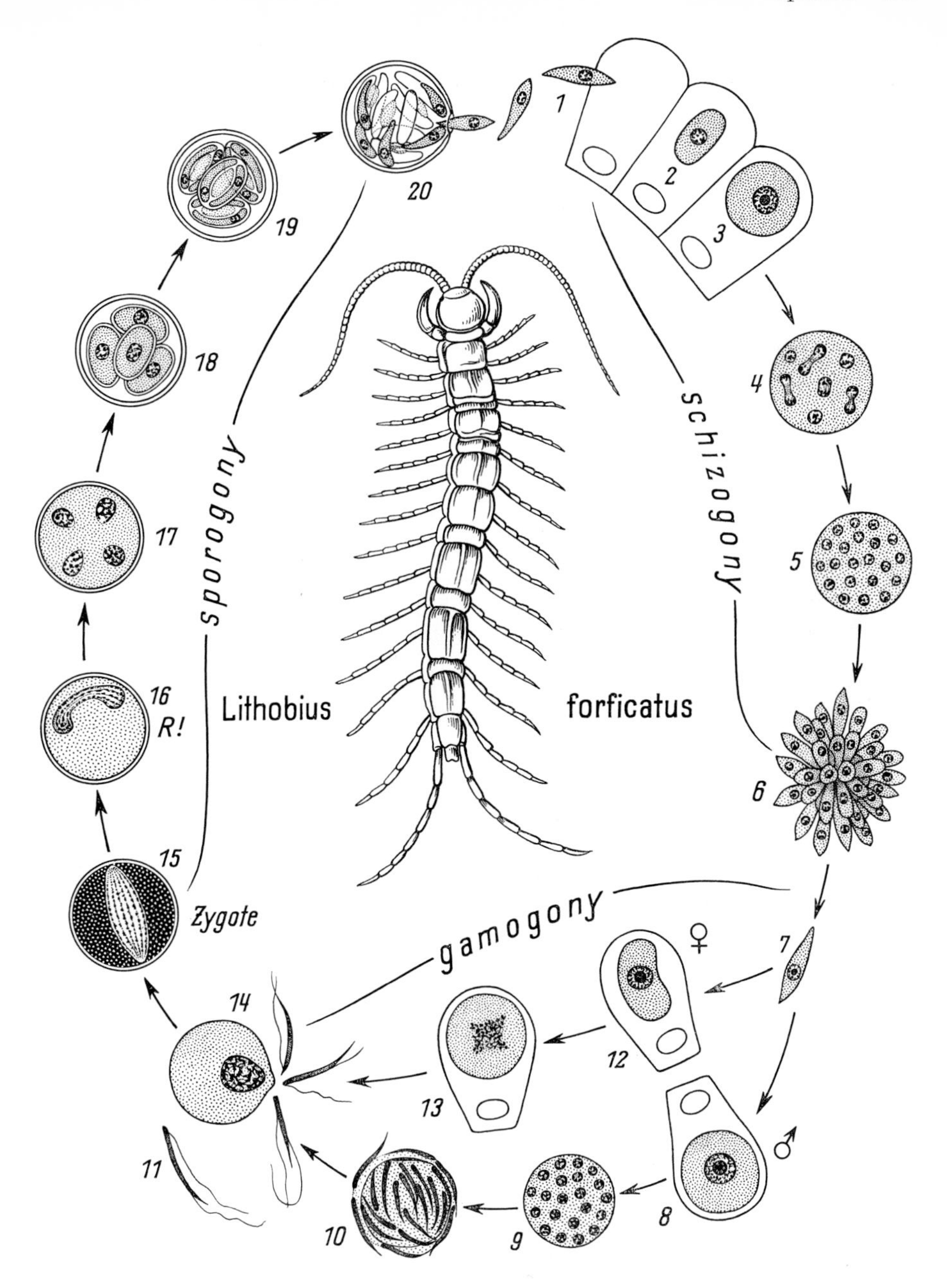

Fig. 397. *Eimeria schubergi* SCHAUDINN. Developmental cycle. The entire course of development takes place in the gut epithelium of the centipede *Lithobius forficatus*. The latter takes up the oocysts together with food. Within the oocysts the sporozoites at first hatch out of the spores and then reach the gut lumen through a special opening of the oocyst (micropyle). The sporozoites (*1*) penetrate into cells of the gut epithelium where they grow into schizonts (*2*, *3*). During formation of the merozoites (*4*—*6*), the schizonts generally fall out of the gut epithelium. Gamonts are formed after a certain degree of infection has been reached. The macrogamonts transform directly into macrogametes (*12*, *13*), while the microgamonts give rise to numerous flagellated microgametes by multiple fission (*8*—*11*). During fertilization (*14*) the macrogamete forms a "fertilization cone". As in the other Coccidia sporogony (*15*—*20*) consists of two phases, but leads only to the formation of four spores, each containing two sporozoites. In accordance with SCHAUDINN, 1900

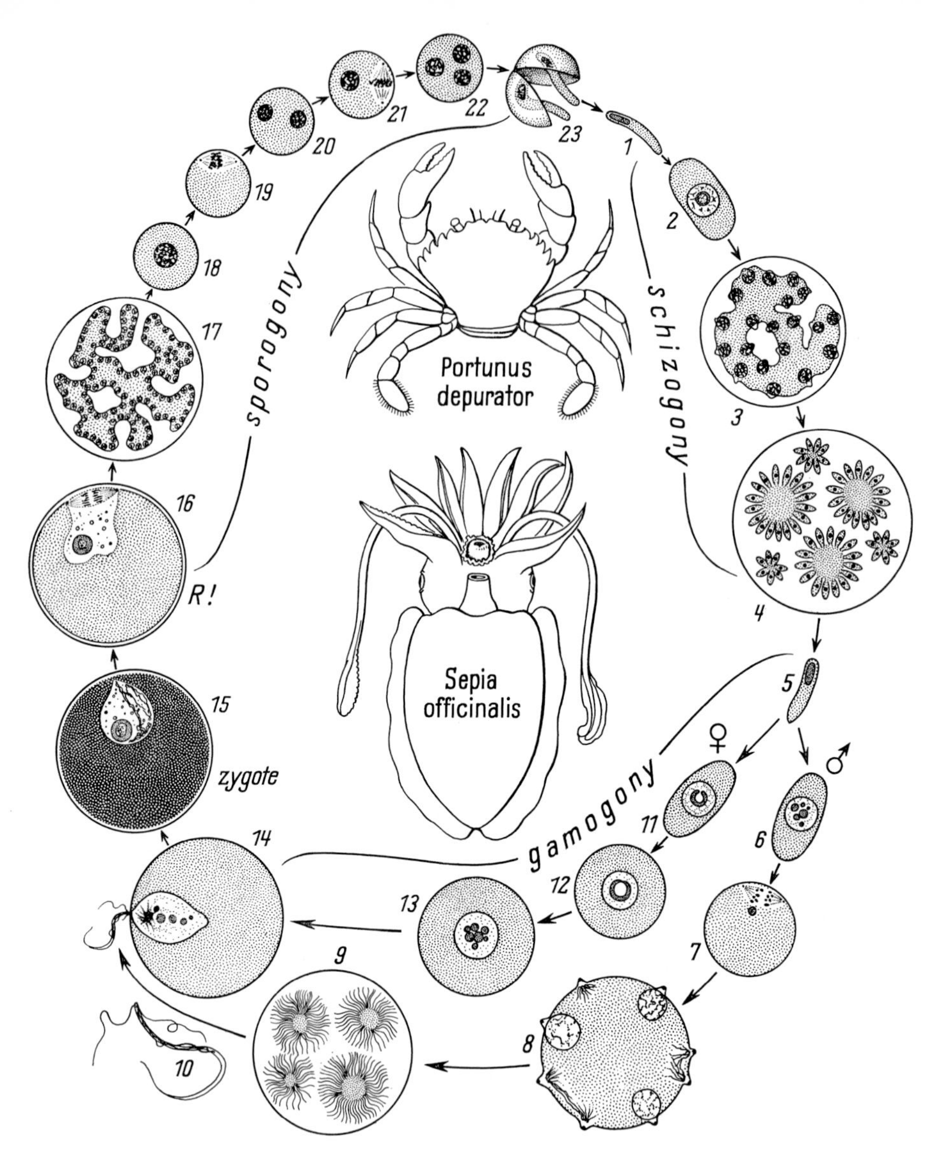

Fig. 398. *Aggregata eberthi* LABBÉ. Developmental cycle. Development is connected with a host alternation. Schizogony (*1*—*4*) takes place in a crab, for example, in the species *Portunus depurator*. The schizonts grow in the connective tissue cells surrounding the gut. As soon as the crab has been eaten by the cuttlefish *Sepia officinalis*, the merozoites penetrate through the gut wall into cells of the submucosa. There they grow into gamonts. The macrogamonts give rise to the macrogametes, which are especially large (*11*—*14*), while the microgamonts bring forth numerous biflagellated microgametes provided with an undulating membrane (*6*—*10*). During the first phase of sporogony (*15*—*17*) a great many globular spores are formed, in which three sporozoites originate (*18*—*23*). The infection ensues by way of spores rather than oocysts. Based on studies by DOBELL, 1925 [*410*]

Aggregata. Transmission to another host takes place by spores, which are formed in great numbers in each oocyst. Host alternation between crab (schizogony) and cuttlefish (gamogony and sporogony).

A. eberthi LABBÉ (Fig. 398). The cuttlefish *Sepia officinalis* and crabs of the genus *Portunus* (e.g. *P. depurator*) alternate as hosts.

2. Family: Haemosporidae

Macro- and microgamonts develop separately. Blood parasites with host alternation. Because of the special manner of transmission, sporogony exists only in *one* phase and is not combined with the development of spores.

The Haemosporidae alternate between a vertebrate and a bloodsucking dipteran as hosts. In the vertebrate, they infect above all the blood circulatory system. Their development is limited here for the most part to schizogony, while gamogony and sporogony take place in the insect. Transmission is not by spores. Instead, the stinging insect injects the sporozoites directly into the blood stream. The sporozoites penetrate probably at first always into endothelial cells lining blood vessels or into other tissue cells. After schizogony, which is either limited to the tissue cells or can continue in the red blood corpuscles, gamonts originate which are taken up when the insect sucks blood.

The formation of microgametes and fertilization take place in the insect's gut. The zygote is motile (ookinete) and develops to an oocyst on the wall of the gut. The sporozoites are formed in the oocyst which, when it bursts, releases them into the body cavity (Fig. 399) whence they penetrate into the salivary gland of the insect.

Haemoproteus. Schizogony exclusively in endothelial cells of the blood capillaries (mainly in the inner organs). After repeated cycles of schizogony, gamonts arise which penetrate into red blood corpuscles and are taken up by the insect when sucking blood. Gamogony and sporogony as in the following genus.

H. columbae Kruse. In the pigeon and its parasite, the louse-fly *Lynchia maura*.

H. oryzivorae Anschütz. In the Java sparrow.

Plasmodium. Schizogony at first in tissue cells (primary or pre-erythrocytic schizogony), later in red blood corpuscles (secondary or erythrocytic schizogony).

Species of the genus *Plasmodium* occur in reptiles, birds and mammals (bats, mice, primates). Transmission is due to mosquitos.

P. praecox Grassi et Feletti (Fig. 400). Vertebrate host: songbirds; invertebrate host: *Culex*-species. This species can be transmitted to canaries and thus played an important

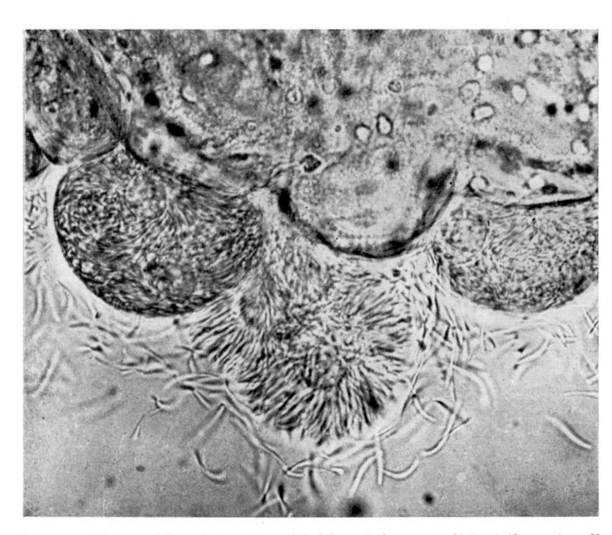

Fig. 399. *Plasmodium vivax* Grassi et Feletti. Ripe and burst oocysts with liberated sporozoites at the gut wall of *Anopheles maculipennis*. Photographed in life by Weyer and Plett from Ruge, Mühlens Zur Verth, 1942

role as a laboratory animal in the elucidation of the developmental stages [*1149*, *1423*] and in the investigation of remedies for malaria.

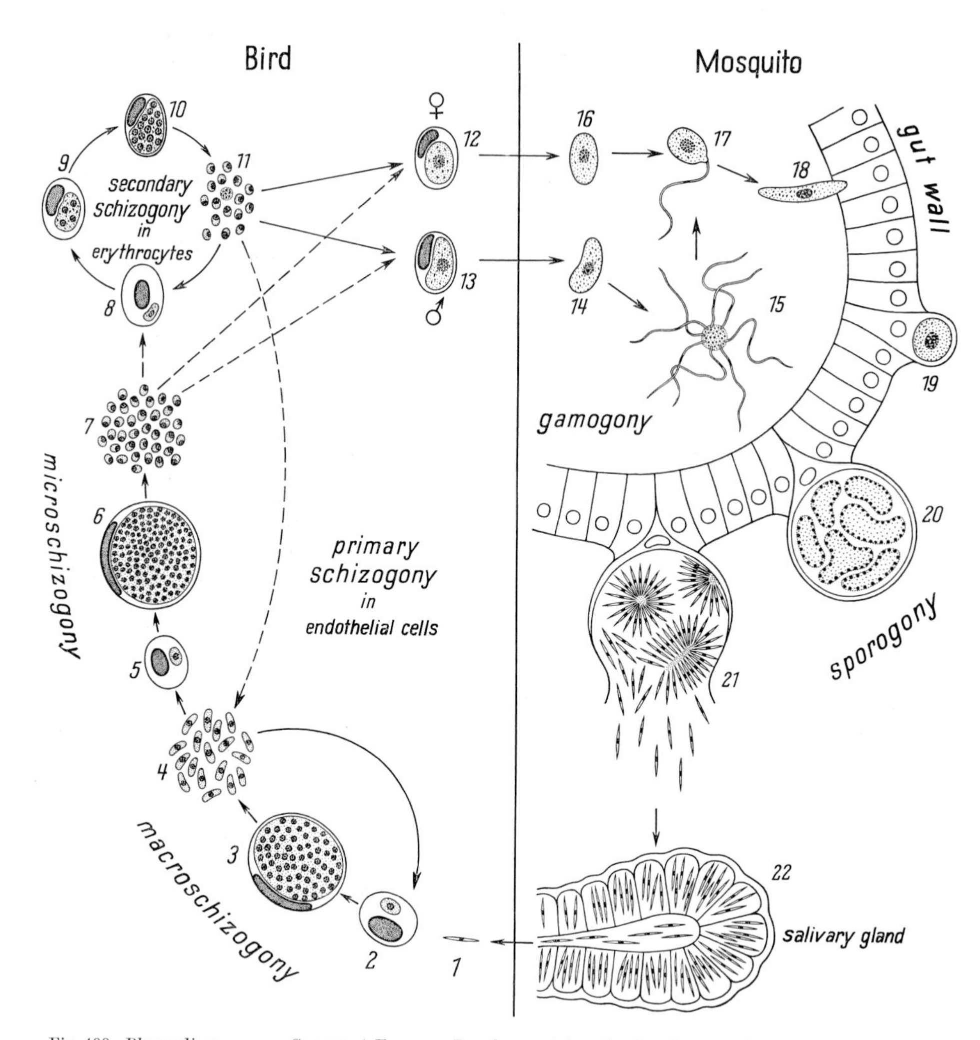

Fig. 400. *Plasmodium praecox* GRASSI et FELETTI. Developmental cycle. Development is connected with a host alternation. Schizogony (*1—13*) takes place in a songbird, gamogony (*14—18*) and sporogony (*19—22*) in a mosquito of the genus *Culex*. The sporozoites, which reach the blood stream of the bird together with saliva of the mosquito, develop at first into schizonts (*1—3*) in cells of the reticuloendothelial system or in phagocytes. The schizonts give rise to comparatively large merozoites containing much cytoplasm, and are therefore designated as macroschizonts. The macromerozoites (*4*) can infect other tissue cells. While macroschizogony is repeated, a portion of the macromerozoites grows into schizonts which give rise to smaller merozoites with less cytoplasm (*5—7*). They are therefore called microschizonts. Normally, the micromerozoites no longer penetrate into the tissue cells but infect the erythrocytes. As infection progresses, macroschizogony is replaced more and more by microschizogony. In this manner the infestation of the blood increases continuously. Further schizogony cycles take place in the red corpuscles (*8—11*). A portion of the merozoites derived from the microschizonts are transformed directly into sexual forms (gamonts). The gamonts can also arise from merozoites which are formed during erythrocytic schizogony. Occasionally it also appears that erythrocytic merozoites change into macromerozoites again, resulting in a secondary infection of the reticuloendothelial system. When, in the act of stinging, the mosquito sucks the bird's blood, the schizogony stages are digested in its gut. Only the gamonts develop further. The microgamont forms approximately 4—8 thread-like microgametes which wriggle about (*14—15*). After fertilization (*17*) the amoeboid zygote, which is motile (ookinete, *18*) penetrates into an epithelial cell of the gut where it grows into a large oocyst. The infected host cell moves out of the epithelial layer and finally degenerates. Within the oocysts, numerous sporozoites arise (*19—21*) without spore formation. As soon as the oocyst has reached maturity, it empties its contents into the body cavity. The sporozoites swim within the fluid of the body cavity to the salivary glands and penetrate into its cells (*22*). Finally, they accumulate in the lumen of the salivary gland. Based upon studies by different authors, especially MUDROW and REICHENOW, 1944 [*1149*]

P. berghei VINCKE et LIPS. Vertebrate host: African wild rats; invertebrate host: *Anopheles*-species. Transmissible to rats, mice and hamsters as laboratory animals.
In man, four species of P. occur which cause the so-called *malaria diseases*. All of them are transmitted by mosquitos of the genus *Anopheles*. In contrast to species of bird malaria (for example, *P. praecox*), at least part of the pre-erythrocytic development of species occurring in man takes place in parenchyma cells of the liver (Fig. 401).

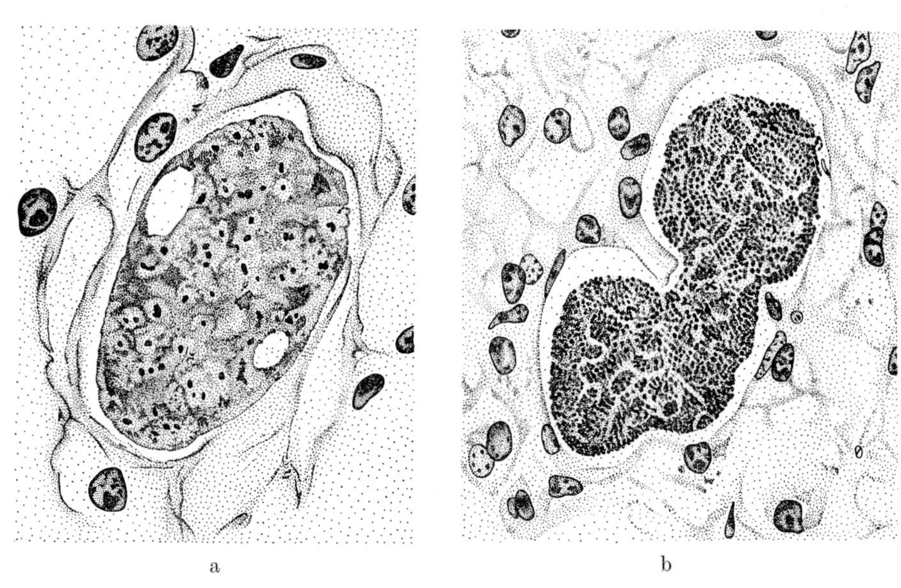

Fig. 401. Schizonts in liver parenchyma cells of man. a *Plasmodium vivax* (7th day after infestation). × 1200. b *Plasmodium falciparum* (6th day after infestation). × 800. Carnoy, Giemsa. a After SHORTT and GARNHAM, 1948; b after SHORTT, FAIRLEY, COVELL, SHUTE and GARNHAM, 1951

P. vivax GRASSI et FELETTI, *P. ovale* STEPHENS. Causative agents of malaria tertiana.
P. malariae LAVERAN. Causative agent of malaria quartana.
P. falciparum WELCH. Causative agent of malaria tropica.
In the patient's blood smear, *P. vivax*, *P. malariae*, and *P. falciparum* can be distinguished on the basis of the characteristics compiled in Fig. 402 (p. 446).
Of all the contagious diseases caused by Protozoa, *malaria diseases* are the most destructive for man. They are widespread above all in the tropics and subtropics, occur, however, also in certain areas of the temperate zones. Significant are the periodic attacks of fever which are repeated in the "tertiana" every second day and in the "quartana" every third day. In the "tropica", the attacks are more irregular and can occur every day. This periodicity is based upon the fact that erythrocytic schizogonies take place more or less synchronously. The attacks of fever take place immediately after the merozoites are liberated. During the pre-erythrocytic phase of parasitic development, which lasts about one or two weeks, no symptoms of disease can be observed (incubation period). In some cases, this phase can extend for months before the blood becomes infected (late manifestation). The relapses, characteristic above all for the "quartana", are due to renewed infestation of the blood by latent pre-erythrocytic stages. Pathogenicity is due primarily to erythrocytic schizogonies. They lead not only to destruction of red blood cells but also to a poisoning of the blood by breakdown products of erythrocytes and residual bodies from schizonts. The danger of malaria diseases lies, however, not so much in the severity of the individual cases but rather in the fact that, lacking appropriate combatting measures in the "malaria regions", large portions of the population are weakened by malaria every year. Mortality is comparatively low, even in "tropica" which generally takes the most serious course.

		Plasmodium vivax (tertiana)	Plasmodium malariae (quartana)	Plasmodium falciparum (tropica)
Schizogony	young stages	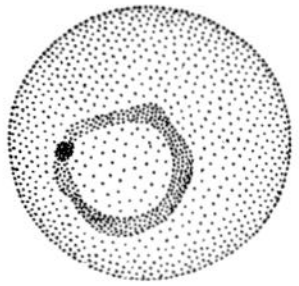 Mostly larger rings with broad cytoplasmic margin		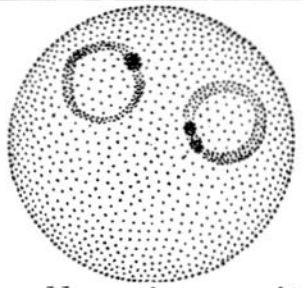 Mostly smaller rings with narrow cytoplasmic margin
	more advanced stages	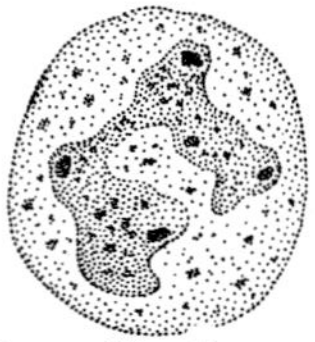 Amoeboid form. Erythrocyte enlarged and with "Schüffner's dots"	 Occasionally "tape-like" forms: Erythrocyte not enlarged and without dots	Schizonts with several nuclei and division stages do not normally appear in the peripheral blood stream
	division stages	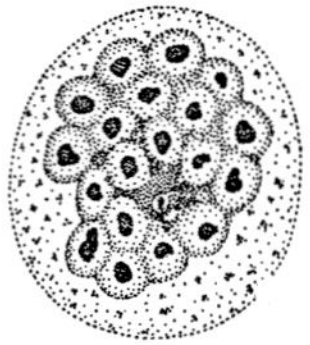 "morula"-stages (18—24 merozoites)	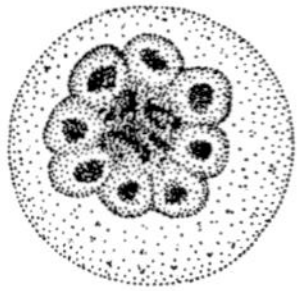 "daisy flower"-stages (8—12 merozoites)	
Gamonts		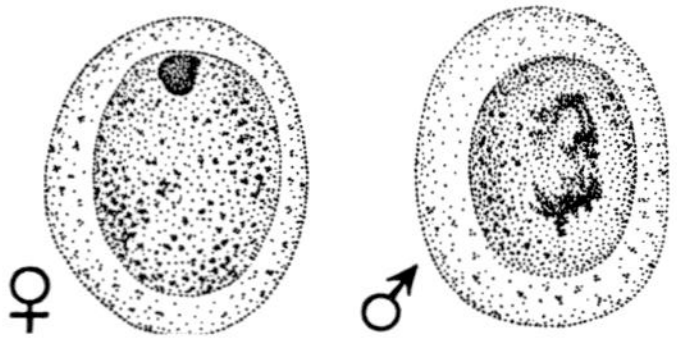 ♀ ♂ gamonts roundish	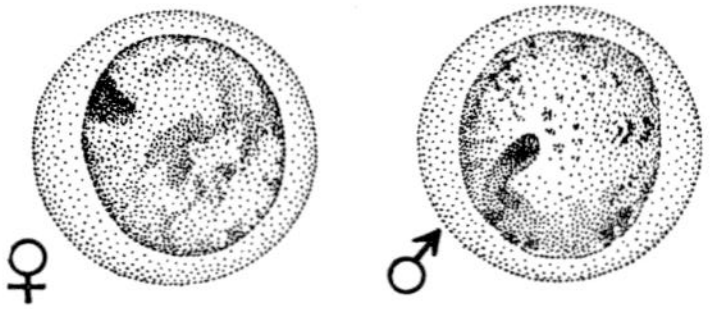 ♀ ♂	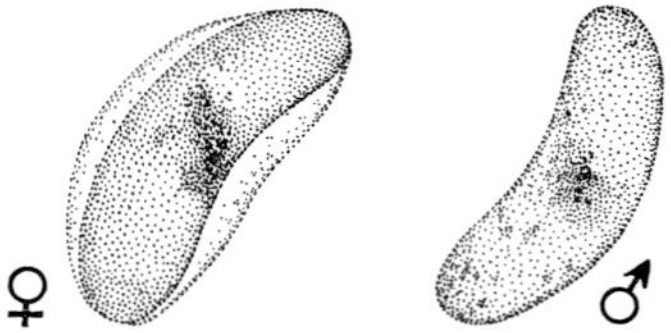 ♀ ♂ gamonts half-moon shaped ("tropica half moon")

Fig. 402. *Plasmodium* species of man. Tabulated survey of their distinguishing characteristics
◄

The *history of malaria research* [1150, 1420] begins with the discovery of the causative agent by LAVERAN in 1880. Soon thereafter, GOLGI described (1885) the course of erythrocytic schizogony. Several investigators (especially ROSS) recognized that part of the development takes place in the *Anopheles*-mosquitos. In 1898, MCCULLUM discovered the sexual forms and observed fertilization in the stomach of the mosquito. The development of oocysts in the gut wall of the mosquito and migration of the sporozoites into the salivary gland were described by GRASSI in 1898. SCHAUDINN, 1902, gave for the first time a comprehensive view of the development of *P. vivax*, relating his observation that the sporozoites can penetrate into red blood corpuscles. This observation led to the erroneous opinion that all developmental stages of the malaria parasites were known. In the malaria treatment of paralysis, the fact that a rather long period of incubation is always to be observed after inoculation of sporozoites, but is lacking after blood transfusions, caused JAMES in 1931 to conceive the idea that the sporozoites first go through a latent development in tissue cells. Between 1934—1938 various investigators found such "exoerythrocystic" stages in the *Plasmodium* species parasitizing birds. MUDROW and REICHENOW *(P. praecox, P. cathemerium)* as well as HUFF and COULSTON *(P. gallinaceum)* succeeded in 1944 in elucidating fully the development in birds. The "exoerythrocytic" stages were found in bird plasmodia only in cells of the reticuloendothelial system. For *Plasmodium* species of man, irrefutable proof of such stages could not be furnished. By study of the species of *Plasmodium* found in monkeys, the suspicion arose that in human malaria a part of the development takes place in the liver. As a matter of fact, schizogony stages (with very large schizonts) of *P. vivax* (1948), later also of *P. falciparum* (1951), were found by SHORTT and GARNHAM in liver parenchyma cells of man (Fig. 401).

3. Family: Adeleidae

Macro- and microgamonts unite with each other (gamontogamy). Two phases of sporogony.

The Adeleidae are similar to gregarines in that the gamonts unite in pairs. The gamonts always have a distinct size difference. While the macrogamont immediately changes into a macrogamete, the microgamont carries out one or two divisions so that only two or four microgametes are formed.

Adelina deronis HAUSCHKA et PENNYPACKER (Fig. 190). In the body cavity of the oligochaete *Dero limosa*.

Klossia. In the kidney epithelium of pulmonates (Fig. 301b and 403).
K. helicina SCHNEIDER. In species of *Helix*, *Cepaea*, and *Succinea*.
K. loossi NABIH. In slugs *(Arion, Limax)*.

Karyolysus. Host alternation between lizard (schizogony in endothelial cells) and mite (gamogony in the gut epithelium, sporogony partly in the gut epithelium, partly in the egg yolk).
K. lacertarum DANILEWSKY (Fig. 404).

Haemogregarina. Host alternation between turtle (schizogony in erythrocytes) and leech (gamogony and sporogony in the gut).
H. stepanowi DANILEWSKY. In the European terrapin *Emys orbicularis* and the leech *Placobdella catenigera*.

Fourth Class: Ciliata

CANELLA, M. F.: Studi e ricerche sui tentaculiferi nel quadro della biologia generale. Ann. Univ. Ferrara (N.S. Sect. III) **1** (1957).

COLLIN, B.: Études monographiques sur les Acinétiens. I, II. Arch. Zool. exp. gén. **48**, **51** (1911—1912).

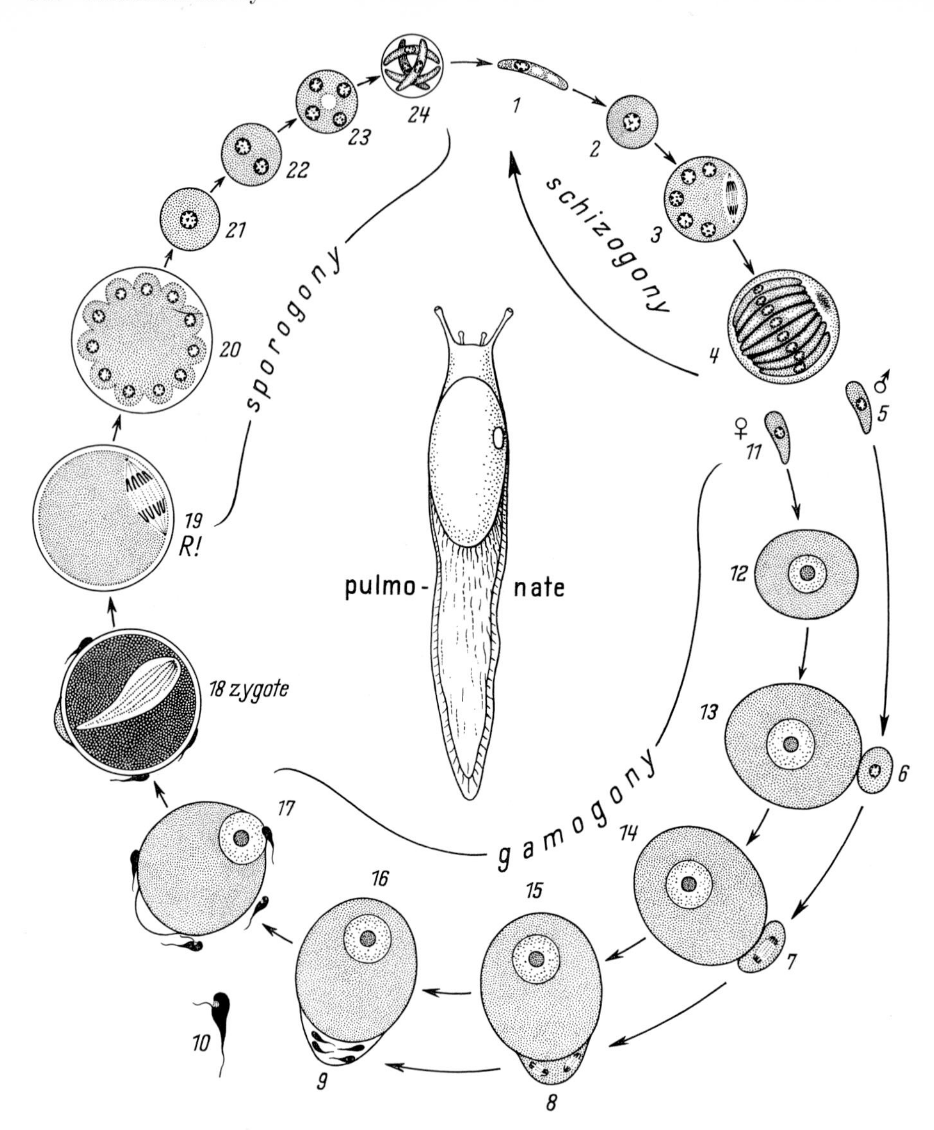

Fig. 403. *Klossia*. Developmental cycle. The entire course of development takes place in the kidney epithelium of a pulmonate. After schizogony (*1*—*4*), which takes place mostly in the spring months, has been repeated several times, the gamonts arise and arrange themselves in pairs. The macrogamonts develop into large, egg-like macrogametes (*11*—*16*), while the adhering lens-shaped microgamonts form four biflagellated microgametes (*5*—*10*). One of the microgametes carries out fertilization (*17*). Numerous round spores, containing four sporozoites, are formed in the oocyst (*18*—*24*). Based on investigations by Naville, 1927 [*1221*] and Nabih, 1938 [*1171*]

Corliss, J.C.: The ciliated Protozoa: Characterization, classification, and guide to the literature. New York: Pergamon Press 1961.

Dragesco, J.: Ciliés Mésopsammiques Littoraux. Travaux de la station biologique de Roscoff **12** (1960).

Kahl, A.: Ciliata libera et ectocommensalia. In: Grimpe-Wagler: Die Tierwelt der Nord- und Ostsee. II. Leipzig: Akadem. Verlagsges. 1933.

— Ciliata entocommensalia et parasitica. In: Grimpe-Waglre: Die Tierwelt der Nord- und Ostsee. II. Leipzig: Akadem. Verlagsges. 1934.

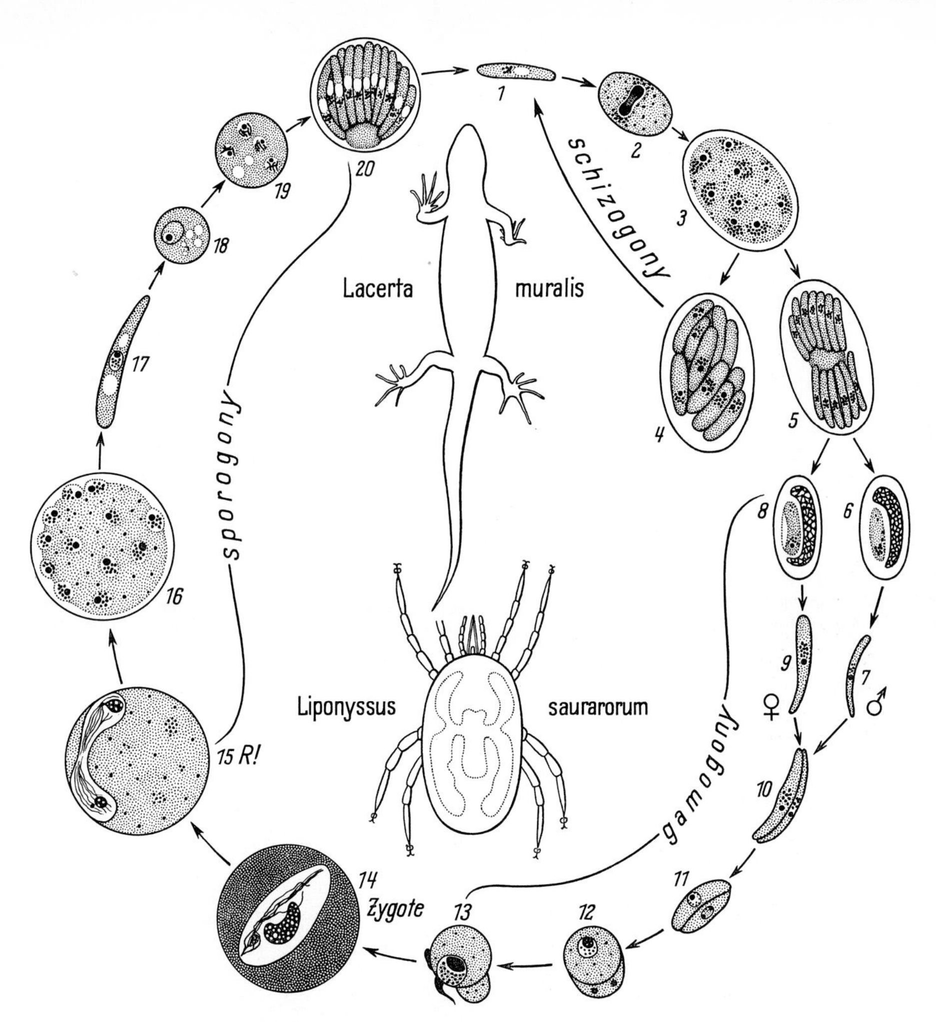

Fig. 404. *Karyolysus lacertarum* DANILEWSKY. Developmental cycle. Development is connected with an alternation of hosts. Schizogony takes place in the lizard *Lacerta muralis*, gamogony and sporogony in its parasitic mite *Liponyssus saurarorum*. The lizards become infected when they eat the mites which contain spores. The sporozoites penetrate through the gut epithelium into the blood stream of the lizard. There they infect the endothelial cells of the blood capillaries. After repeated schizogony, during which merozoites (macromerozoites) with much cytoplasm are formed (*1—4*), slimmer merozoites (micromerozoites, *5*, with less cytoplasm), finally arise which penetrate into the red corpuscles where they grow into gamonts (*6*, *8*). The latter are taken up by the mite when sucking blood. The gamonts are freed during digestion of the red corpuscles (*7*, *9*) and join lengthwise in pairs (*10*). The gamonts penetrate in this form into a gut epithelium cell. While the macrogamont grows faster and becomes a macrogamete, the microgamont carries out a nuclear division so that only two microgametes are formed (*11—13*). After fertilization the zygote grows to a larger size (*14*). At the end of the first phase of sporogony, spores do not form in the oocysts. Instead, large worm-like stages called sporokinetes arise (*15—17*). The latter are motile and penetrate into the eggs of the mite. There they carry out a brief growth phase, but finally round up and surround themselves with a membrane (*18—20*). From now on they can be designated as spores. Approximately 20—30 sporozoites are formed in the spore. After studies by REICHENOW, 1921 [*1419*]

KAHL, A.: Suctoria: In: GRIMPE-WAGLER: Die Tierwelt der Nord- und Ostsee. II. Leipzig: Akadem. Verlagsges. 1934.

— Urtiere oder Protozoa. I. Wimpertiere oder Ciliata (Infusoria). In: DAHL, F.: Die Tierwelt Deutschlands. Jena: Fischer 1935.

LEPSI, J.: Die Infusorien des Süßwassers und Meeres. Berlin-Lichterfelde: Bermühler 1926.

MATTHES, D., WENZEL, F.: Wimpertiere (Ciliata). (Einführung in die Kleinlebewelt.) Stuttgart: Kosmos-Verlag 1966.

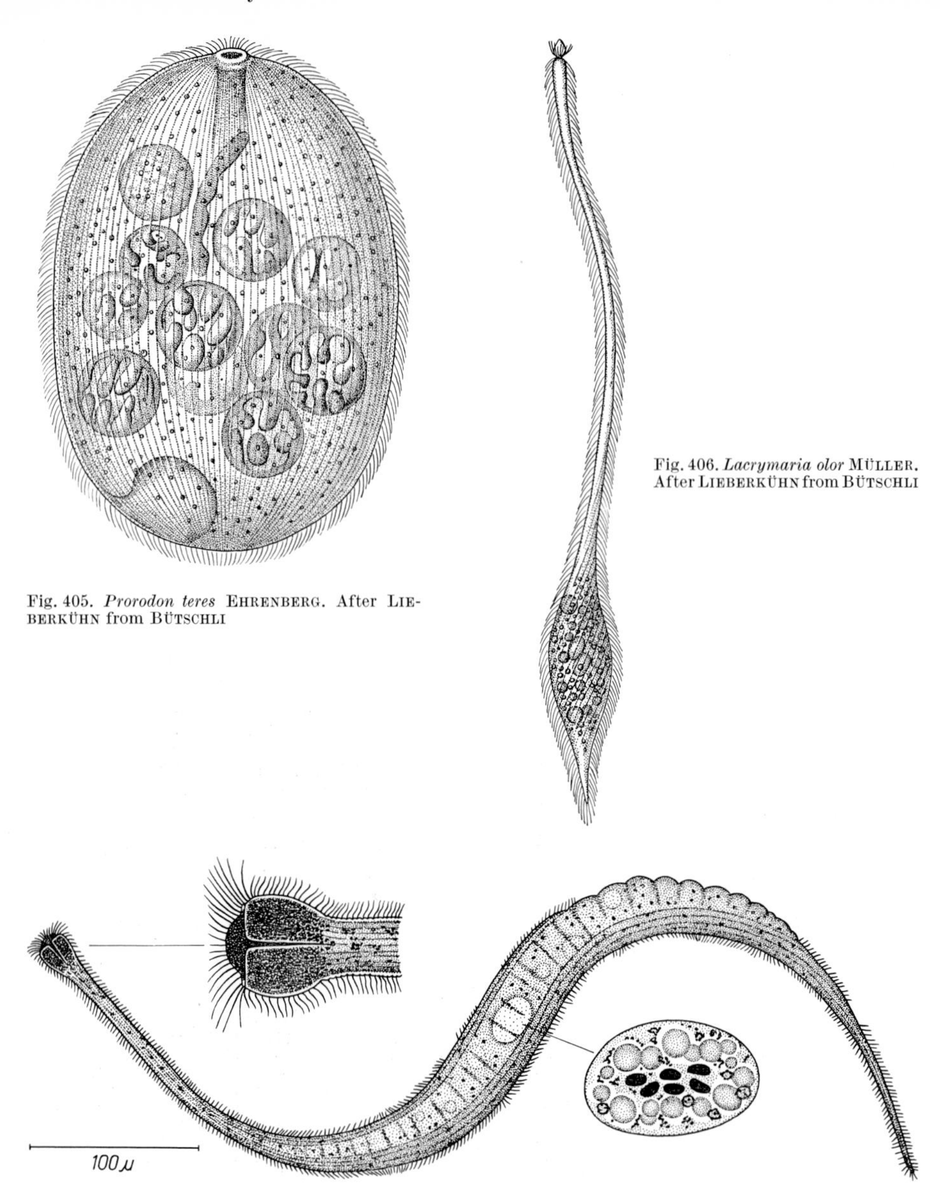

Fig. 405. *Prorodon teres* EHRENBERG. After LIEBERKÜHN from BÜTSCHLI

Fig. 406. *Lacrymaria olor* MÜLLER. After LIEBERKÜHN from BÜTSCHLI

Fig. 407. *Tracheloraphis phoenicopterus* COHN. The scale refers to the entire animal. Front end and cell nucleus (comp. Fig. 91) are shown magnified. After RAIKOV, 1962 [*1389*]

The Ciliata have two different types of cell nuclei: generative micronuclei and somatic macronuclei *(nuclear dimorphism)**. Locomotion due to the coordinated beat of *cilia*. Asexual reproduction is by binary or multiple fission, sexual reproduction by *conjugation*. Autogamy occurs in some species.

The first four orders are frequently grouped together under the heading *Euciliata* since they always have cilia, whereas the *Suctoria* are only ciliated as swarmers.

* An exception is the homokaryotic genus *Stephanopogon* (p. 100).

First Order: Holotricha

Body usually uniformly ciliated. Without adoral membranelle band.

First Suborder: **Gymnostomata**

Cytostome at the body surface, without special organelles for swirling food particles to the cytostome. Gulpers.

a) Rhabdophorina

Cell body round in cross section. Cytostome at the front end or at the side. Often with toxicysts. Carnivorous.

Prorodon. Cell body egg-shaped.
 P. teres EHRENBERG (Fig. 405). Fresh water.

Coleps. Cell body barrel-shaped. Pellicle strengthened by armor-like single plates. Carrion-eaters.
 C. hirtus NITSCH. Fresh water.

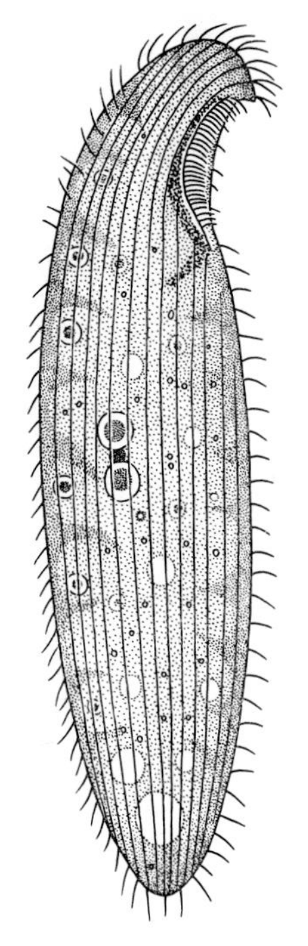

Fig. 408. *Loxodes rostrum* MÜLLER. One micronucleus (dark) and two (diploid) macronuclei (comp. Fig. 89). × 680. After DRAGESCO, 1965 [*444*]

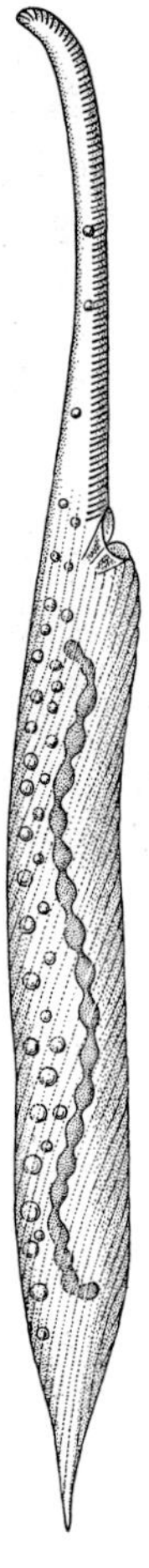

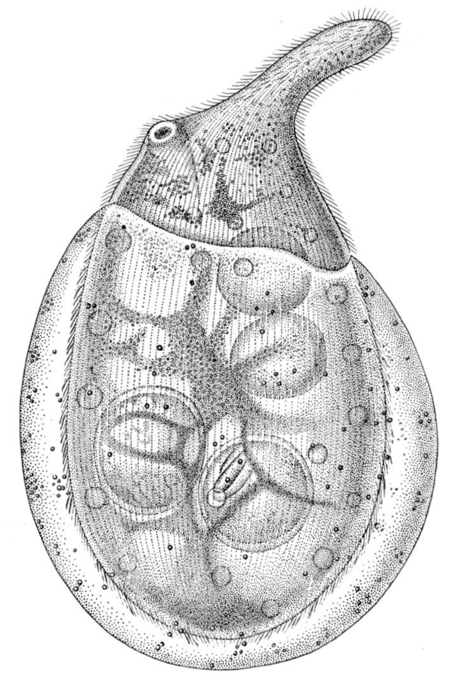

Fig. 410. *Trachelius ovum* EHRENBERG, hatching from the cyst. After LIEBERKÜHN from BÜTSCHLI

Fig. 409. *Dileptus anser* MÜLLER. After BÜTSCHLI

Lacrymaria. Cell body divided into a thin motile anterior end with "head part" and a broader posterior end. *L. olor* Müller (Fig. 406). Fresh water.
Tracheloraphis. Cell body worm-shaped. With composite cell nucleus (p. 102).
T. phoenicopterus Cohn (Fig. 407). Marine sands.
Didinium. Cell body barrel-shaped with "oral cone". Ciliation limited to two belt zones. Feeds on paramecia.
D. nasutum Müller (Fig. 287). Fresh water.
Loxodes. Anterior end bent like a beak. Cytostome at the concave side.
L. rostrum Müller (Fig. 408). Fresh water.
Dileptus. Cell body elongated, pointed posteriorly. Narrowed to form a trunk in front of the lateral cytostome.
D. anser Müller (Fig. 409). Fresh water.
Trachelius. Cell body globular, posteriorly rounded. Narrowed to form a finger in front of the lateral cytostome.
T. ovum Ehrenberg (Fig. 410). Fresh water.

b) Cyrtophorina

Cell body dorso-ventrally flattened. Cytostome ventral. Cytopharyngeal basket to pinch off algae (Fig. 285). Herbivorous.

Nassula. Cell body completely ciliated.
N. ornata Ehrenberg. Fresh water.
Chilodonella. Ventral side ciliated.
C. uncinata Ehrenberg. Fresh water.
Dysteria. Cytostome and cilia in a trough. With a special appendage (stylus) for attachment.
D. monostyla Ehrenberg. Fresh water.

Second Suborder: **Trichostomata**

Cytostome on the bottom of a depression, containing special rows of cilia used in swirling food particles. Swirlers.

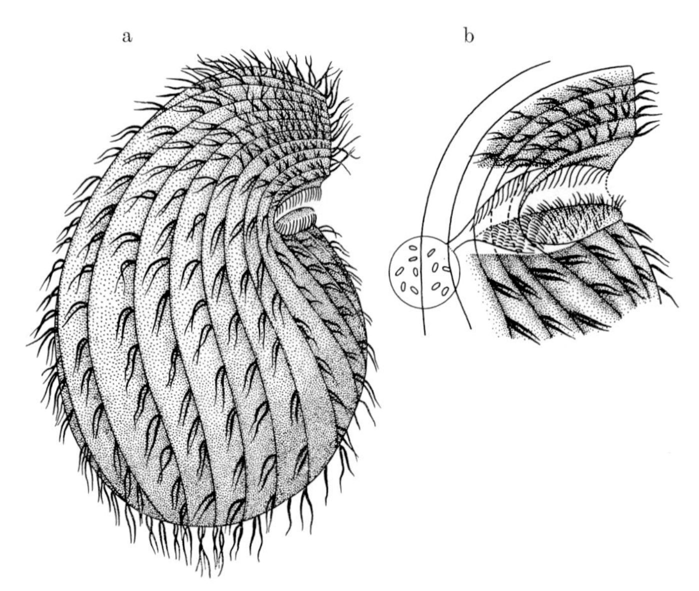

Fig. 411. *Colpoda cucullus* Müller. a The entire cell. × 600. b Region of the oral cavity. On the cytostome: a food vacuole. After Mackinnon and Hawes, 1961

Colpoda. Cell body kidney-shaped. Reproductive cysts.
C. cucullus MÜLLER (Fig. 411). Fresh water.
Isotricha. Cell body egg-shaped. Cytostome at posterior end.
I. prostoma STEIN (Fig. 20). In the ruminant paunch.

Third Suborder: **Hymenostomata**

Cytostome on the bottom of a depression in which membranes are formed by fusion of rows of cilia. Swirlers.

Tetrahymena. Cell body pear-shaped. Small buccal cavity behind the anterior end (with an undulating membrane and three membranelles).
T. pyriformis EHRENBERG (Fig. 412). Fresh water.
Colpidium. Cell body elongated and oval. Small buccal cavity at the side.
C. campylum STOKES. Fresh water.
Paramecium. Cell body elongated. The vestibulum forms the continuation of a deep lateral invagination (peristome). "Slipper animalcules".
P. caudatum EHRENBERG (Fig. 413).
P. multimicronucleatum POWERS et MITCHELL.
P. aurelia EHRENBERG.
P. trichium STOKES.
P. bursaria EHRENBERG.
Ichthyophthirius. Cell body oval. Parasite in the skin of fresh-water fish, in which it can grow to a size of about 800 μm. After a shorter or longer growth phase, encystment takes place. In the cyst, which falls out of the skin and sinks to the ground, a variable number of small swarmers develop which re-infect the fish.
I. multifiliis FOUQUET (Fig. 414). Fresh water.
Pleuronema. With a large sail-like membrane, projecting beyond the margin of the body.
P. marinum DUJARDIN (Fig. 415). Marine.

Fourth Suborder: **Astomata**

Without cytostome. Parasites or commensals which take up nutritive matter through the entire body surface. Reproduction is usually by terminal budding. The daughter individuals can remain connected to the mother cells for an extended period of time so that a chain of individuals results (Fig. 100c).

The Astomata occur for the most part in the gut and body cavity of oligochaetes.

Anoplophrya lumbrici SCHRANK and
Maupasella nova CÉPÈDE. In the gut of *Lumbricus*.
Intoshellina maupasi CÉPÈDE. In the gut of *Tubifex*.

The suborders **Apostomea** and **Thigmotricha** include marine forms which, because of their parasitic way of life, are extremely varied in structure and development.

Second Order: Peritricha

Anterior end of the cell body broadened to form a disc-shaped peristome from which two rows of cilia, coiled counterclockwise, lead to the cytostome. While the inner row can be doubled, the outer one, whose cilia fuse in front of the cytostome to form an undulating membrane, is always single. The food swirled in is collected in a vestibulum which can extend far into the cell body. The pulsating vacuole opens into the vestibulum. Myonemes may permit withdrawal of the peristome disc and other changes of the cell shape (Fig. 268).

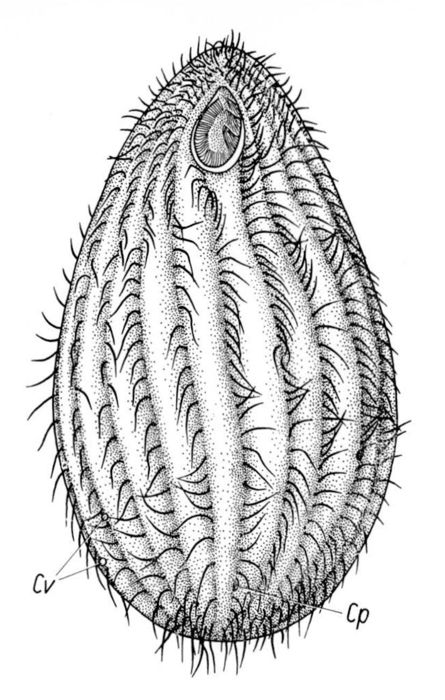

Fig. 412. *Tetrahymena pyriformis* EHRENBERG. *Cv* Pores of the pulsating vacuoles. *Cp* Anus. × 1500. After MACKINNON and HAWES, 1961

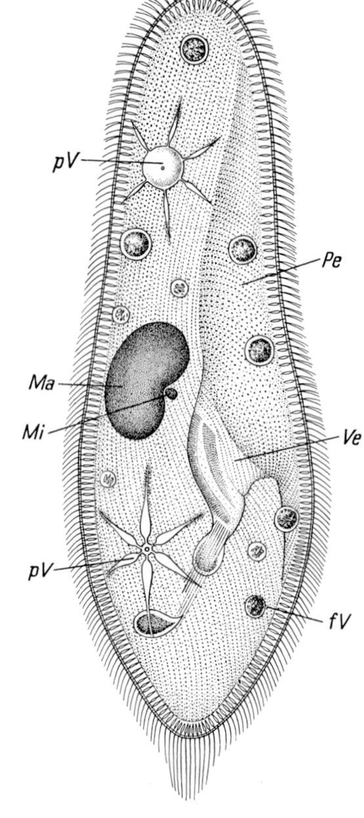

Fig. 413. *Paramecium caudatum*, diagrammatic. *Ma* Macronucleus, *Mi* micronucleus, *pV* pulsating vacuole, *Pe* peristome, *Ve* vestibulum, *fV* food vacuole. From KÜKENTHAL-MATTHES, somewhat modified

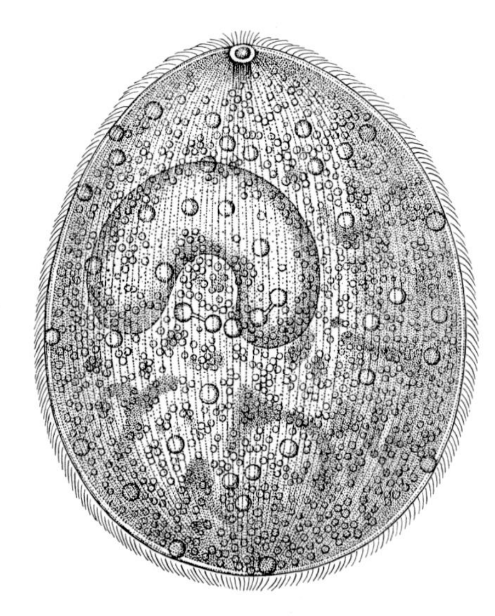

Fig. 414. *Ichthyophthirius multifiliis* FOUQUÉ. After BÜTSCHLI

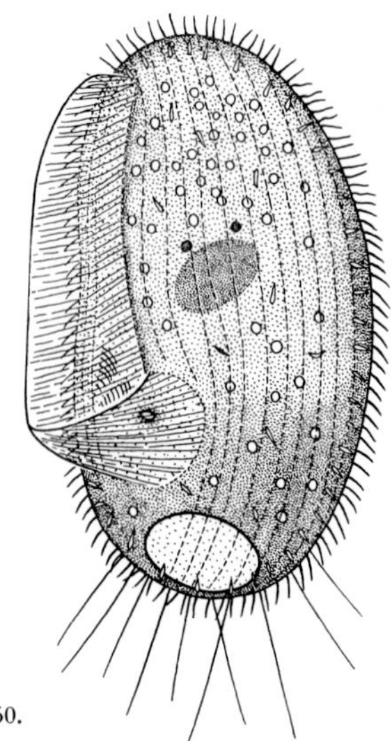

Fig. 415. *Pleuronema marinum* DUJARDIN. × 450. After DRAGESCO, 1960 [*439*]

Reproduction is by longitudinal fission. In the solitary Sessilia, one of the two daughter cells becomes a swarmer (telotroch) with an aboral ciliary wreath. Conjugation is modified to a unilateral fertilization (anisogamonty, p. 207).

The peritrichs are most likely derived from holotrichs (Hymenostomata). Because of their numerous special adaptations, connected for the most part with their sessile mode of life, it is, however, expedient to concede them an order of their own [*1377*].

Some species have secondarily changed back to a free-swimming way of life.

First Suborder: **Sessilia.** Sessile

a) Without test

Vorticella. Solitary, with contractile stalk.
V. nebulifera MÜLLER (Fig. 416). Fresh water.
Epistylis. Colonial, with a noncontractile, usually dichotomously branched stalk.
E. plicatilis EHRENBERG. Fresh water.

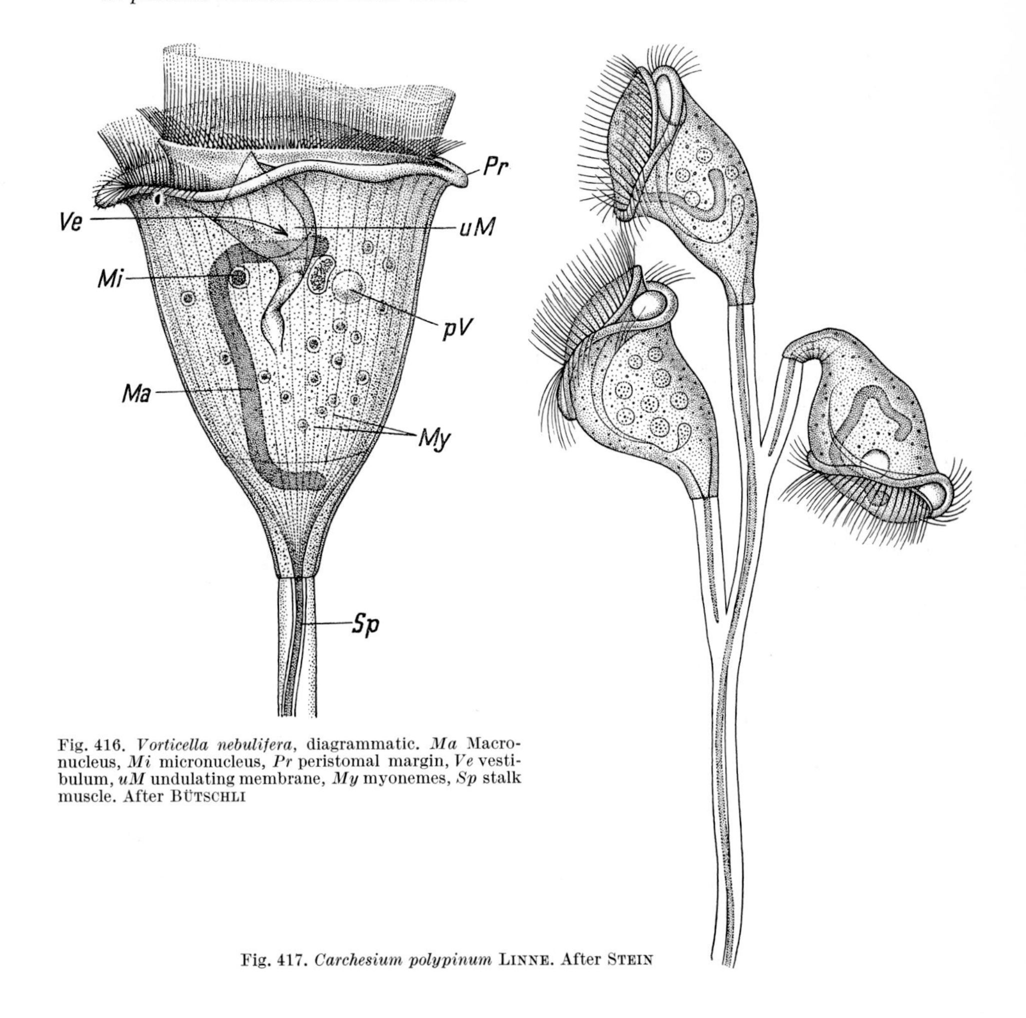

Fig. 416. *Vorticella nebulifera*, diagrammatic. *Ma* Macronucleus, *Mi* micronucleus, *Pr* peristomal margin, *Ve* vestibulum, *uM* undulating membrane, *My* myonemes, *Sp* stalk muscle. After BÜTSCHLI

Fig. 417. *Carchesium polypinum* LINNE. After STEIN

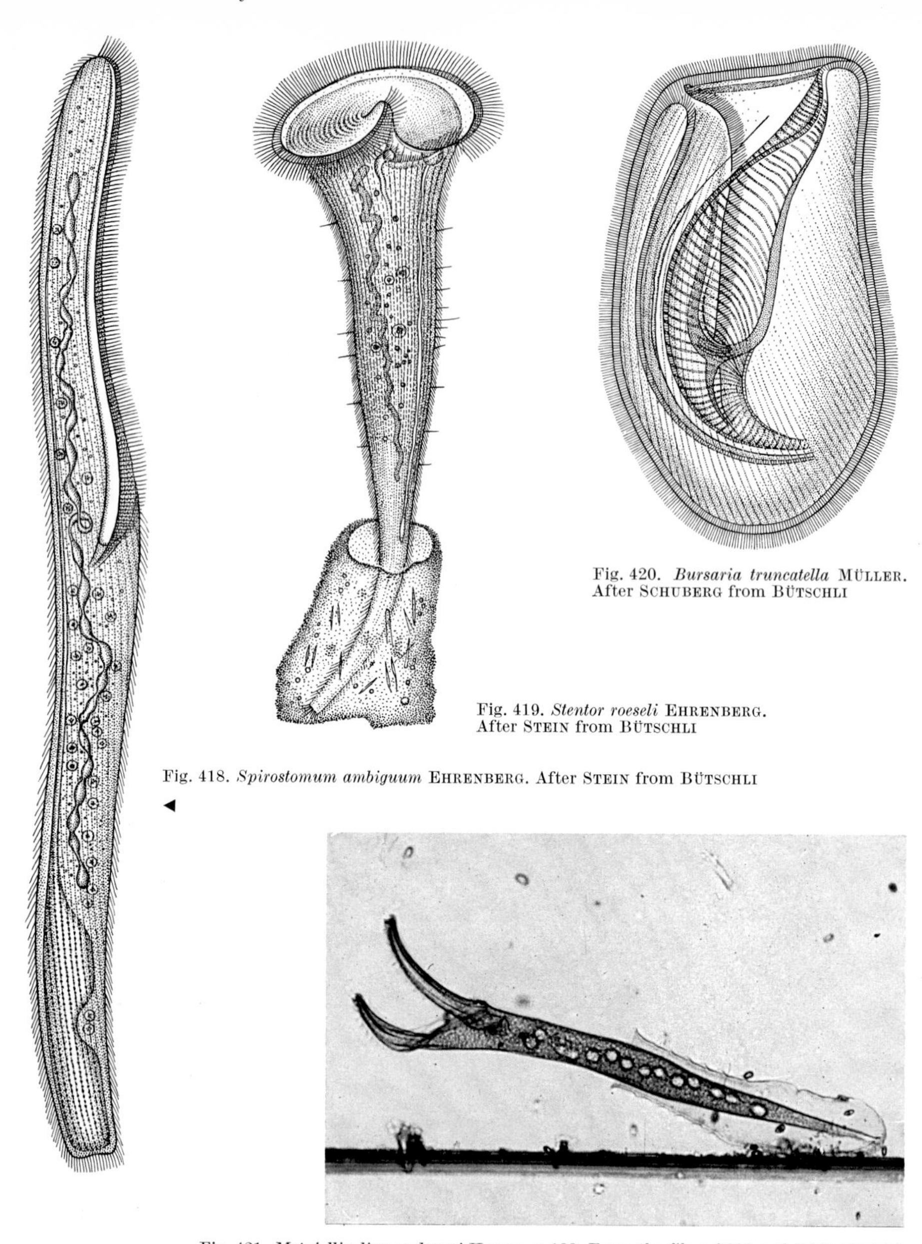

Fig. 420. *Bursaria truncatella* MÜLLER. After SCHUBERG from BÜTSCHLI

Fig. 419. *Stentor roeseli* EHRENBERG. After STEIN from BÜTSCHLI

Fig. 418. *Spirostomum ambiguum* EHRENBERG. After STEIN from BÜTSCHLI ◀

Fig. 421. *Metafolliculina andrewsi* HADZI. × 100. From the films C882 and E649 (UHLIG)

Carchesium. Colonial, with contractile stalk. The stalk muscles (spasmonemes) of the individuals are separate.

C. polypinum LINNÉ (Fig. 417). Fresh water.

Zoothamnium. Colonial, with contractile stalk, and common stalk muscle (spasmoneme).

Z. arbuscula EHRENBERG. Fresh water.

Z. alternans CLAPARÈDE et LACHMANN (Fig. 6). Marine.

Z. pelagicum DU PLESSIS. Marine (plankton).

b) With test

Cothurnia. Test with stalk.
C. astaci STEIN. On crayfish.
Vaginicola. Test without stalk.
V. terricola GREEF. In moist moss.

Second Suborder: **Mobilia.** Free-swimming

Trichodina pediculus EHRENBERG. On fresh water polyps ("polyp louse").

Third Order: Spirotricha

Characteristic for Spirotricha is the presence of an adoral band of membranelles which winds clockwise to the cytostome.

First Suborder: **Heterotricha**

Cell body round in cross section, covered all over with cilia.

Spirostomum. Cell body worm-shaped and contractile. Without undulating membrane.
S. ambiguum EHRENBERG (Fig. 418) Fresh water (2 mm long).
Blepharisma. Cell body pear-shaped. With undulating membrane at the right margin of the peristome.
B. undulans STEIN. Fresh water.
Stentor. Front end broadened to a funnel-shaped peristome. Provided with a holdfast for attachment.
S. polymorphus EHRENBERG. Colored green by zoochlorellae, with rosary-shaped macronucleus.
S. coeruleus EHRENBERG. Bluish, with rosary-shaped macronucleus.
S. roeseli EHRENBERG (Fig. 419). With rod-shaped macronucleus.
Bursaria. Cell body egg- or purse-shaped; the peristome is funnel-shaped and reaches almost to the posterior end.
B. truncatella MÜLLER (Fig. 420). Fresh water (larger than 1 mm).
Metafolliculina. Peristome widened into two wings. In a lorica resembling a hot-water bottle.
M. andrewsi HADZI (Fig. 421). Marine.

Second Suborder: **Hypotricha**

Cell body flattened dorso-ventrally. Membranelle band at the ventral side. Body cilia fused to form cirri at the ventral side and modified as short bristles at the upper surface.

Keronopsis. Two ventral rows of cirri.
K. gracilis KAHL (Fig. 422). Marine.
Urostyla. Numerous ventral rows of cirri.
U. grandis EHRENBERG. Fresh water.
Stylonychia. Cirri in groups, with bristle-like terminal cirri.
S. mytilus MÜLLER (Fig. 264). Fresh water.

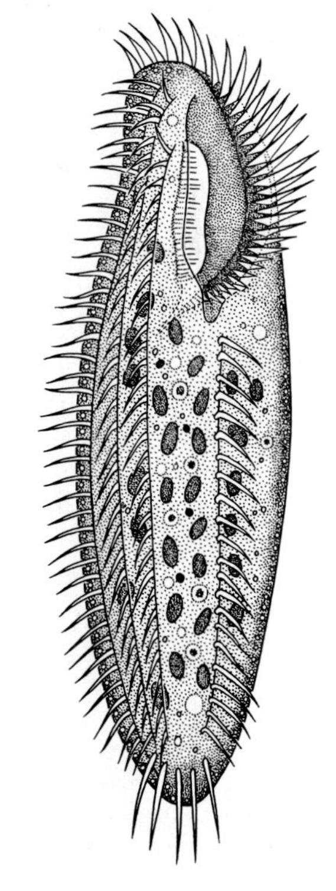

▶ Fig. 422. *Keronopsis gracilis* KAHL. Numerous micronuclei and macronuclei. × 800. After DRAGESCO, 1965 [*444*]

Euplotes. Cirri in groups, no marginal cirri.
E. patella MÜLLER. Fresh water.
E. vannus MÜLLER. Marine.

Uronychia. Several, very strong cirri at the rear end form a "saltatory apparatus".
U. transfuga MÜLLER. Marine.

Third Suborder: **Oligotricha**

Body ciliation reduced or absent. Adoral membranelles at the front end.

Strombidium. Without equatorial wreath of bristles.
S. arenicola DRAGESCO (Fig. 423). Marine.

Halteria. With equatorial wreath of bristles.
H. grandinella MÜLLER. Fresh water.

The following groups are now separated from the Oligotricha and listed as separate suborders of the Spirotricha. However, since they can only be distinguished from the other Spirotricha by the same negative criterion and because the relationships among these groups are unclarified, it seems more expedient to leave them in this suborder for the time being.

The *Tintinnoidea* are small ciliates within delicate loricae. They are found in large numbers of species in plankton, especially the marine plankton.

Tintinnopsis (Stenosemella) ventricosa CLAPARÈDE et LACHMANN (Fig. 424). Marine.

The *Odontostomata* have a laterally compressed cell body which is drawn out into thorn-like protrusions. The pellicle is armor-like. Body ciliation restricted to isolated groups of cilia. Only few membranelles, located in a groove. Saprozoic.

Saprodinium dentatum LAUTERBORN (Fig. 425).

The *Entodiniomorpha*, characterized by an armor-like pellicle and bizarre extensions at their posterior ends, live as commensals in the ruminant paunch.

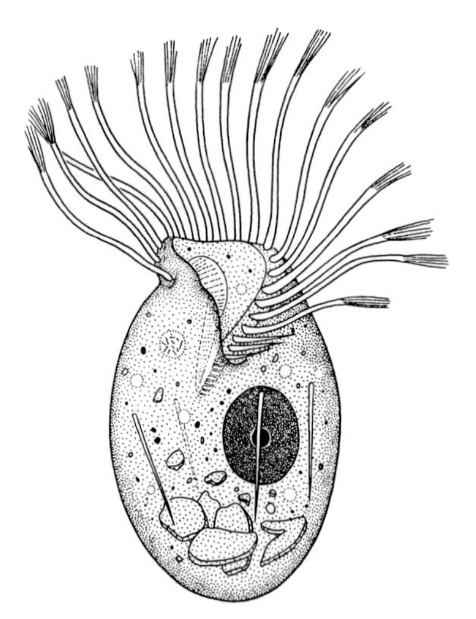

Fig. 423. *Strombidium arenicola* DRAGESCO. × 1000. After DRAGESCO, 1960 [*439*]

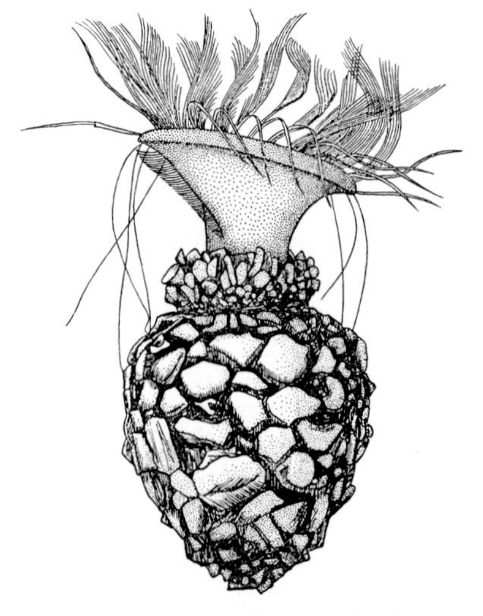

Fig. 424. *Tintinnopsis ventricosa* CLAPARÈDE et LACHMANN. After CORLISS, 1961 [*340*] originally redrawn from FAURÉ-FREMIET (1924)

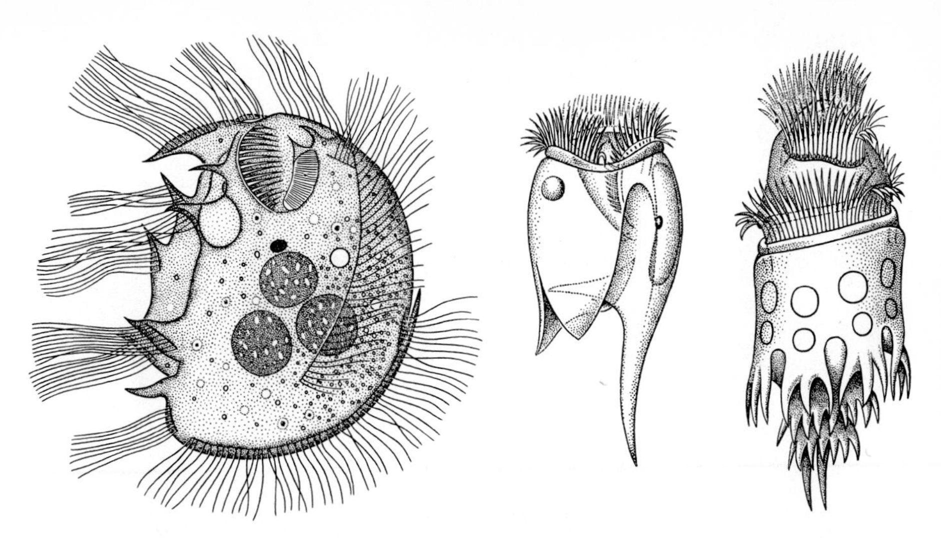

Fig. 425. *Saprodinium dentatum* LAUTERBORN. × 620. After DRAGESCO, 1960 [439]

Fig. 426. *Entodinium caudatum* STEIN. After SCHUBERG

Fig. 427. *Ophryoscolex purkinjei* STEIN. After BÜTSCHLI

Entodinium caudatum STEIN (Fig. 426).
Ophryoscolex purkinjei STEIN (Fig. 427).

Fourth Order: Chonotricha

The Chonotricha form a group of sessile ciliates comprising only few genera and species. The anterior end is transformed into a funnel-shaped swirling apparatus. The rows of cilia running along it transport food to the cytostome. The remaining cell body is not ciliated.

Reproduction by budding. The bud is a motile swarmer. Ecto-commensals on crustaceans.

Spirochona. Swirling apparatus shaped like a corkscrew staircase.
S. gemmipara STEIN (Fig. 428). On the gill lamellae of *Gammarus pulex*.

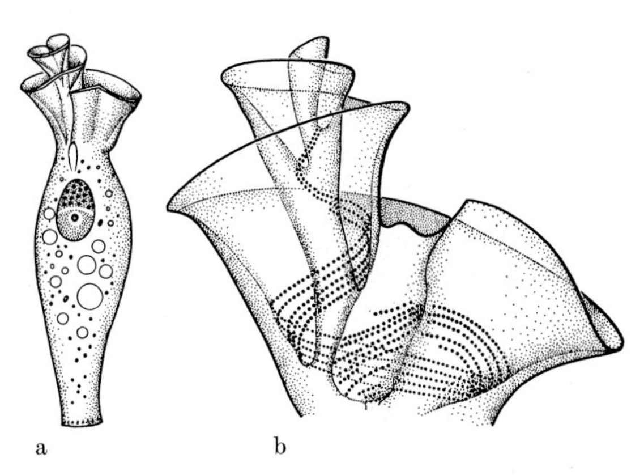

Fig. 428. *Spirochona gemmipara* STEIN. a Fully grown individual. × 375. After HERTWIG. b Collar with basal bodies (kinetosomes) of the cilia. × 1500. After GUILCHER, 1951 [701]

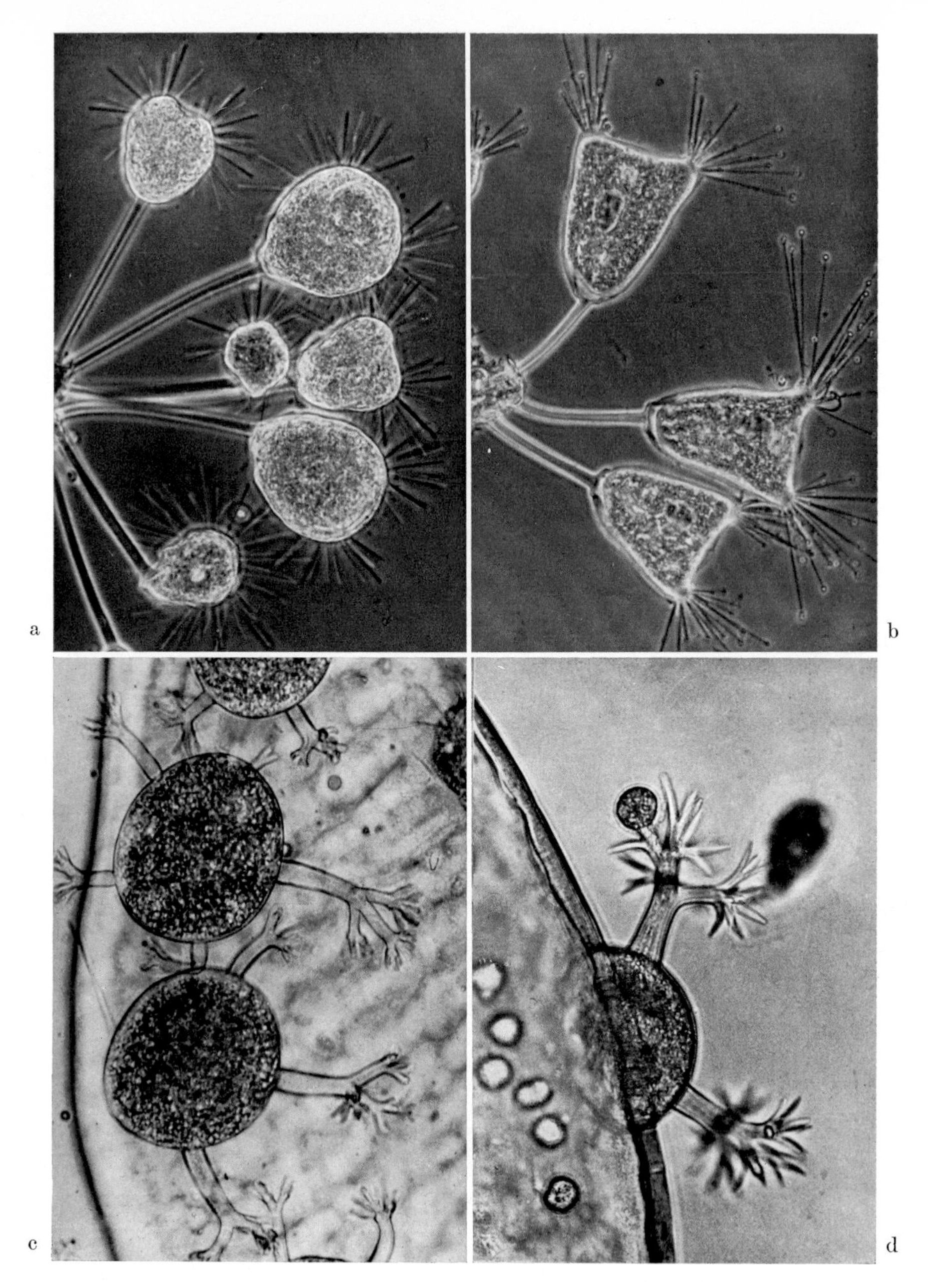

Fig. 429. Suctoria. a *Tokophrya lemnarum* STEIN. × 350. b *Acineta tuberosa* EHRENBERG. × 240. c, d *Dendrocometes paradoxus* STEIN on the gill lamellae of *Gammarus pulex* (c from the top, d from the side). × 280. From the film C 912 (GRELL)

Fifth Order: Suctoria

The Suctoria are sessile ciliates which take up food by tentacles rather than by a cytostome. Reproduction by simple or multiple budding. Only the swarmers are ciliated. During metamorphosis, the cilia are lost while tentacles appear as new structures. Many species form a stalk and/or a lorica.

a) With exogenous budding (p. 147)

Paracineta. In a stalked lorica. Tentacles uniformly distributed.
P. limbata MAUPAS (Fig. 136). Marine.
Ephelota. Feeding and prehensile tentacles. Multiple budding.
E. gemmipara HERTWIG (Fig. 430a). Marine.
Tachyblaston ephelotensis MARTIN (Fig. 430b, 431 and 432). Parasite of *Ephelota gemmipara*, with alternation of generations.

b) With endogenous budding (p. 149)

Tokophrya. Without lorica, but with stalk.
T. lemnarum STEIN (Fig. 429a). Fresh water.
Acineta. In a stalked lorica. Tentacles arranged in bundles.
A. tuberosa EHRENBERG (Fig. 429b). Marine.
Acinetopsis. In a stalked lorica. With one or more prehensile tentacles.
A. rara ROBIN (Fig. 290). Marine.

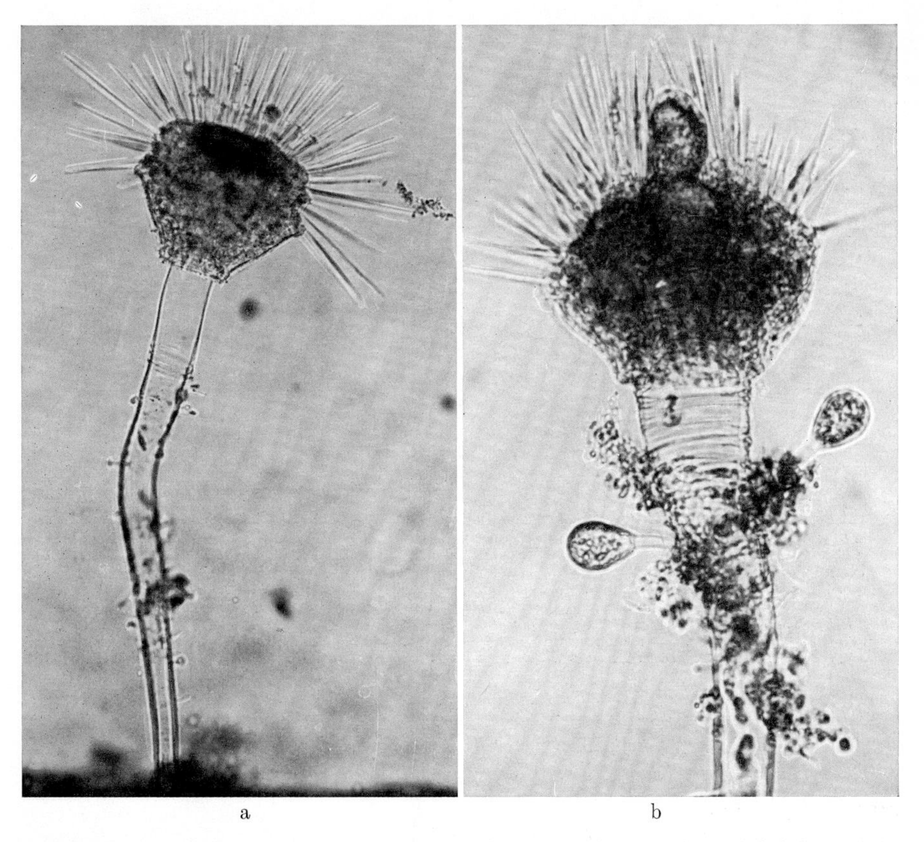

Fig. 430. *Ephelota gemmipara* HERTWIG, attached to the stolon of the hydrozoon polyp *Tubularia larynx*. b An *Ephelota* which is infested by the suctorian *Tachyblaston ephelotensis* MARTIN. The parasite, which has formed a swarmer, is recognizable within the cell body; attached to the stalk are two young *Dactylophrya* stages (comparable to stage 9 in Fig. 431). Photographed in life

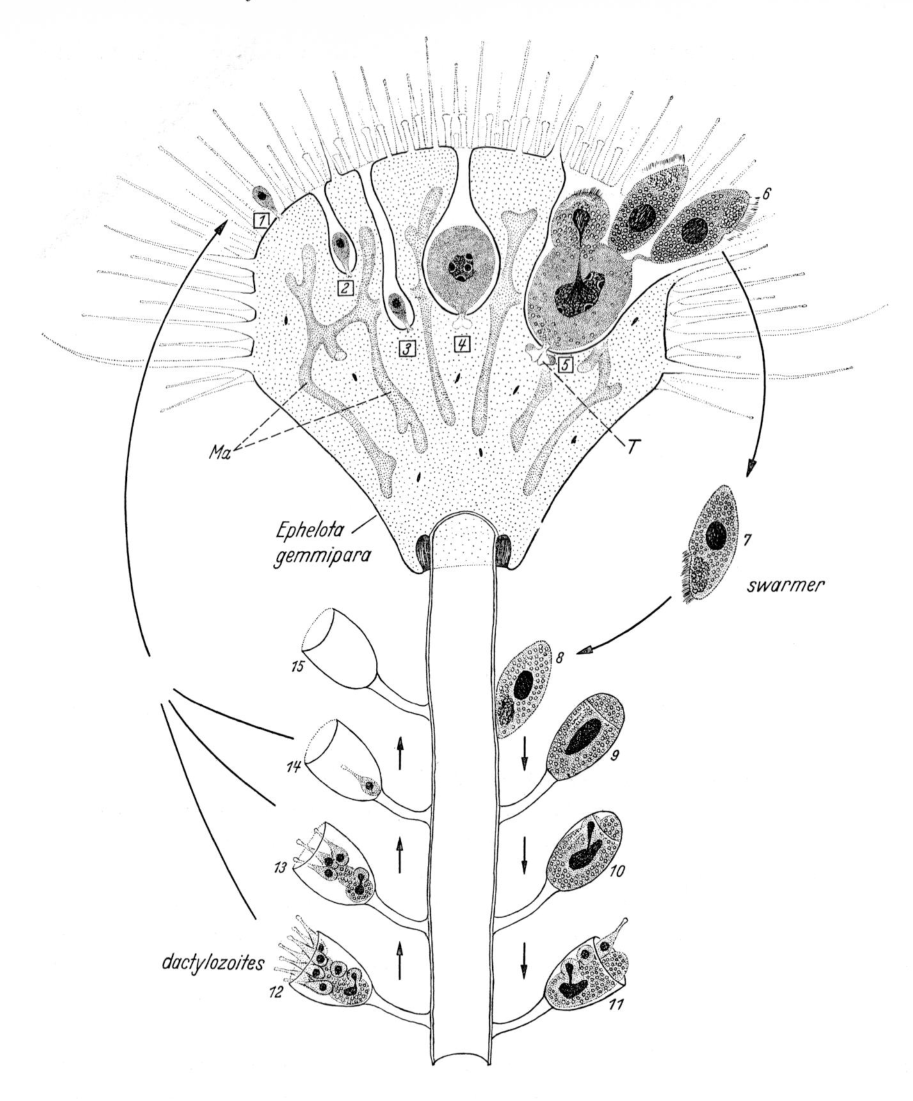

Fig. 431. *Tachyblaston ephelotensis* MARTIN. Diagram of the developmental cycle. Development involves an alternation of generations. The parasitic generation lives on the marine suctorian *Ephelota gemmipara*. Young infective stages (dactylozoites) pierce the pellicle of the host cell with their tentacles (*1*). Then they drive a pellicular canal into the host cell (*2*, *3*). At its end, they grow into large cells, with dense cytoplasm (*4*), which take up the cytoplasm of the host cell with their feeding tentacles (comp. with Fig. 303) and finally form one swarmer after another (*5*). The swarmers (*6*) detach themselves (*7*) and fasten to a suitable substrate, for example, the stalk of an *Ephelota* (*8*). Then they form a stalk of their own and a goblet-shaped test (*9*). This represents the free-living generation (*Dactylophrya* stages). After the anterior portion of the test has disintegrated, the cell divides by repeated budding into approximately 16 small buds which then infest other host cells (*10*—*15*). After GRELL, 1950 [*650*]

Dendrosoma. Cell body branched; at the ends of the branches are bundles of tentacles.
D. radians EHRENBERG. Fresh water.

Dendrocometes. Cell body flattened, with branched, noncontractile tentacles.
D. paradoxus STEIN (Fig. 429c, d). On the gill lamellae of *Gammarus pulex*.

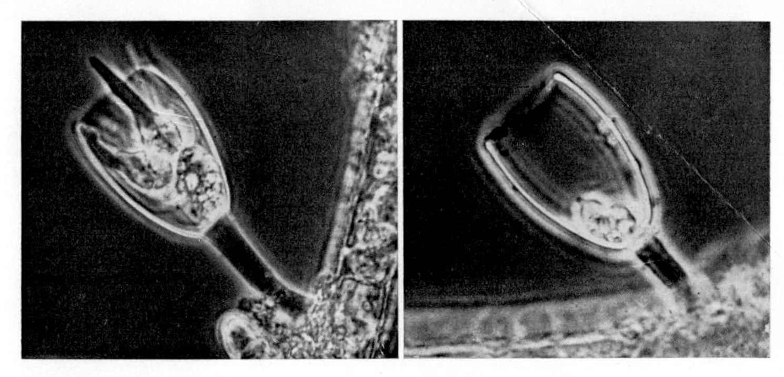

Fig. 432. *Tachyblaston ephelotensis* MARTIN. *Dactylophrya* stages. × 500. From the film C 913 (GRELL)

Fifth Class: "Cnidosporidia"

AUERBACH, M.: Die Cnidosporidien (Myxosporidien, Actinomyxidien, Mikrosporidien). Eine monographische Studie. Leipzig: Klinkhardt 1910.

GEORGÉVITCH, J.: Über Diplo- und Haplophase im Entwicklungskreis der Myxosporidien. Arch. Protistenk. **84** (1935).

KUDO, R.: A biologic and taxonomic study of the Microsporidia. Illinois Biological Monographs. Vol. **9**, Nos. 2 and 3 (1924).

NOBLE, E. R.: Life cycles in the Myxosporidia. Quart. Rev. Biol. **19** (1944).

WEISSER, J.: Die Mikrosporidien als Parasiten der Insekten. Monogr. angew. Ent. Beihefte z. Z. angew. Ent. **17** (1961).

Until now, the Myxosporidia, the Actinomyxidia, and the Microsporidia have been classified as "Cnidosporidia". Aside from the fact that they are without exception parasites, they have a common characteristic which distinguishes them from the "remaining" Protozoa: they form spores (cysts) which contain one or more "polar filaments". These are coiled up within the spore and, in transmission to another host, are extruded like an inverted glove finger. Instead of sporozoites the spores contain one or several amoeboid cells, the so-called *amoebulae*. A further, though negative, criterion is that the "Cnidosporidia" do not have flagellated or ciliated stages.

In the last few years it has been doubted whether these characteristics are sufficient to unite the groups named above in one class. Unfortunately, the developmental cycle is only inadequately known. Especially the nuclear phenomena are still awaiting clarification.

The **Myxosporidia** are primarily fish parasites. They can be present in body cavities (gall bladder, urinary bladder) as well as in tissues (muscular system, gills, spleen, kidney, liver). However, instead of penetrating into the cells, they inhabit the intercellular spaces ("diffuse infiltration"). In the opinion of most investigators, development takes place in the manner diagrammatically illustrated in Fig. 433. The uninucleated *amoebula* (*1*), which emerges from the spore, grows into a multinucleated plasmodium which in some species can attain a size of several millimeters. Even in early growth stages, generative and somatic nuclei are differentiated (*2*). A definite amount of cytoplasm segregates around each generative nucleus forming separate cells within the plasmodium (*3*). These cells give rise to

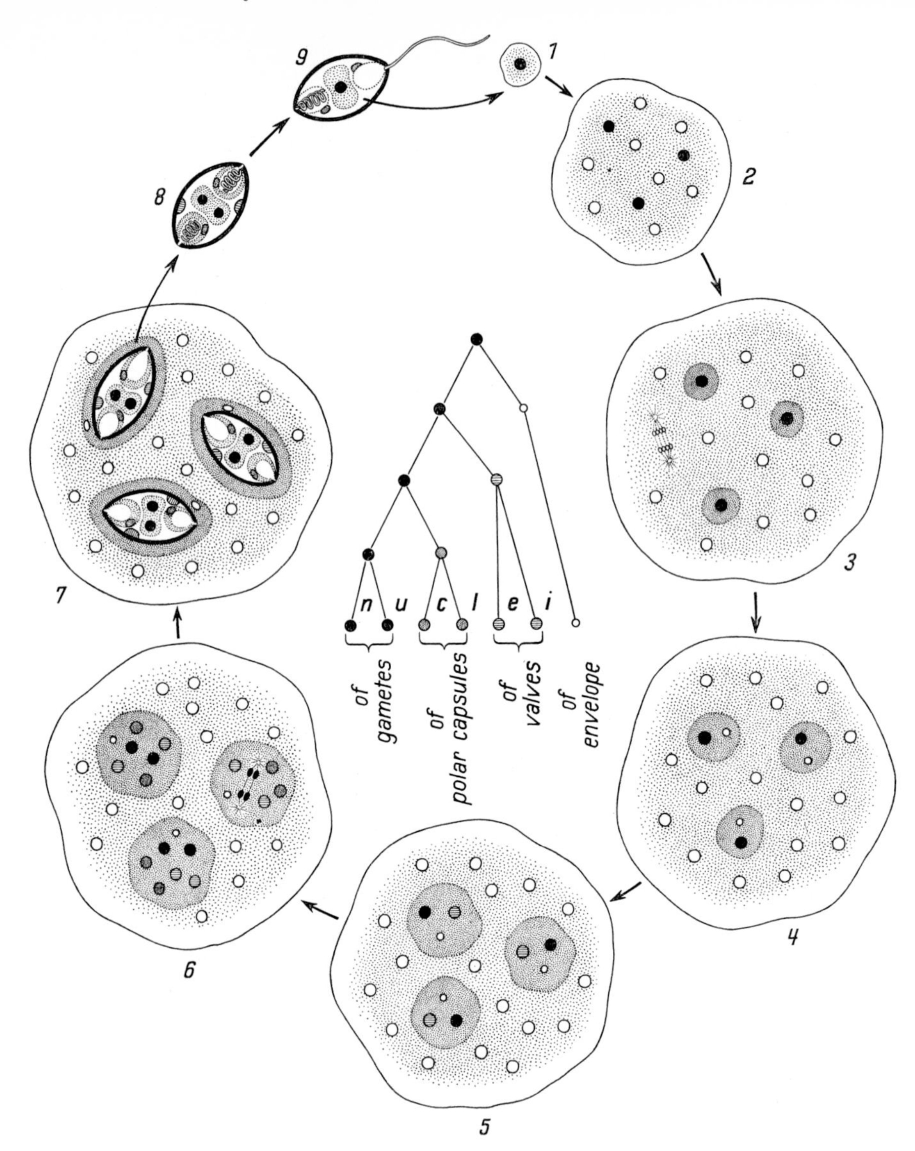

Fig. 433. Diagram of the development of a myxosporidian. *1* Uninucleate amoebula. *2* Multinucleate plasmodium: differentiation into generative (dark) and somatic (light) nuclei. *3* Segregation of the sporoblasts. *4—6* Different stages of nuclear multiplication in the sporoblasts, corresponding to the divisional sequence shown in the middle of the diagram (envelope nucleus, white; valve nuclei, cross-hatched; nuclei of the polar capsule, dotted; gamete nuclei (germ line!) black). *7* Plasmodium with spores. *8* Single spore with binucleate amoebula. *9* Single spore with uninucleate amoebula and a polar capsule with discharged polar filament.

the spores and are therefore termed *sporogonous cells* or *sporoblasts*. In most species the sporoblast forms several spores. It is then referred to as a *pansporoblast*. The sporoblast or pansporoblast grows within the plasmodium. Simultaneously, a period of nuclear divisions involving a definite number of differential mitoses takes place (*4—7*). In each division, one daughter nucleus remains generative while the other one becomes a somatic nucleus. It seems that successively formed

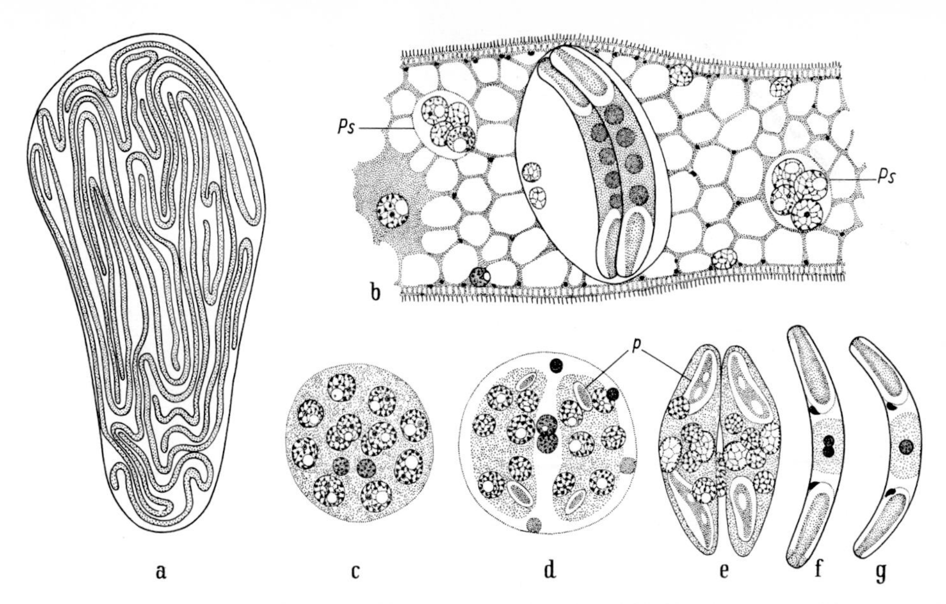

Fig. 434. *Sphaeromyxa sabrazesi* LAVERAN et MESNIL. a Section through a gall bladder of a sea horse with numerous plasmodia. b Section through a plasmodium with developing pansporoblasts (*Ps*) and a mature pansporoblast (center) which contains two spores. × 1200. c—g Development of a pansporoblast. *p* Polar capsules. × 1200 After SCHRÖDER, 1907 [*1543*]

somatic nuclei are specialized for different physiological tasks. The envelope nucleus is formed first and becomes located in a plasmatic envelope surrounding the completed spore. The next two are the nuclei located later inside the two valves of the spore, and finally the two nuclei are formed which are allocated to the characteristic *polar capsules* of the Myxosporidia. The generative nucleus which remains after the somatic nuclei have been formed also divides once more. Both these nuclei are within the amoebula occupying the center of the spore (*8*), but soon afterwards they fuse again. They are evidently to be regarded as gamete nuclei which fuse in autogamy. It seems likely that chromosome reduction takes place in the course of the divisions leading to the formation of gamete nuclei. When the spore, set free at disintegration of the plasmodium, gets into a new host, the polar capsules discharge their filaments. The amoebula is freed by the bursting spore and, after seeking out its final site of parasitism, it grows again to a large plasmodium.

Since spore formation within the plasmodium is not synchronous (as drawn in the diagram), a fixed and stained preparation shows a variety of stages side by side. It is therefore difficult to establish a time sequence of events, which has led to very diverging interpretations of the developmental cycle.

Sphaeromyxa sabrazesi LAVERAN et MESNIL (Fig. 434). In the gall bladder of the sea horse.
Myxidium lieberkuehni BÜTSCHLI. In the urinary bladder of the pike.
Myxobolus pfeifferi THÉLOHAN. In the muscular system of the barbel (cause of the "boil disease").
Myxosoma cerebrale HOFER. In the cartilage of trout (cause of the spin or twist disease).

The **Actinomyxidia** comprise only a few genera and species which are parasites in the body cavities or in the gut epithelium of oligochaetes of fresh and salt

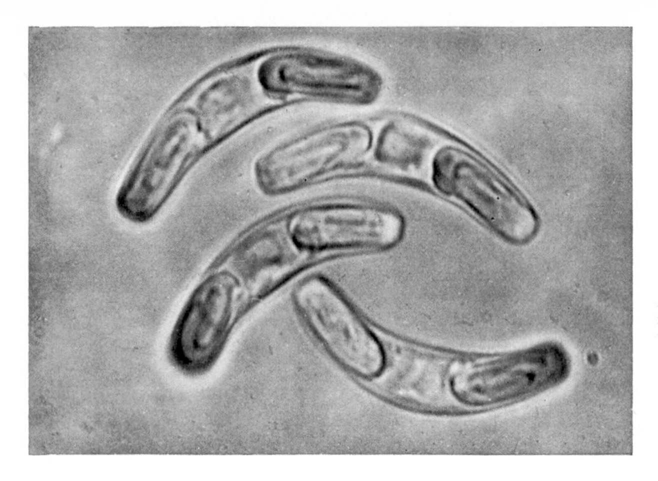

Fig. 435. Spores of *Sphaeromyxa*. Photographed in life

water (several in sipunculids). A fully-grown individual corresponds to a pansporoblast in which 8 spores are formed. The spores have radial three-point symmetry, contain three polar capsules and usually numerous amoebulae.

Sphaeractinomyxon stolci CAULLERY et MESNIL. In *Clitellio arenarius.*
Triactinomyxon ignotum STOLC. In *Tubifex tubifex.*

The **Microsporidia** are intracellular parasites of arthropods and fish and are much smaller than the Myxosporidia. The amoebula which penetrates into the host cell does not grow into a large plasmodium. It reproduces by repeated binary or multiple fission (Fig. 436) so that the host cell ultimately is entirely filled with parasites containing one or a few nuclei.

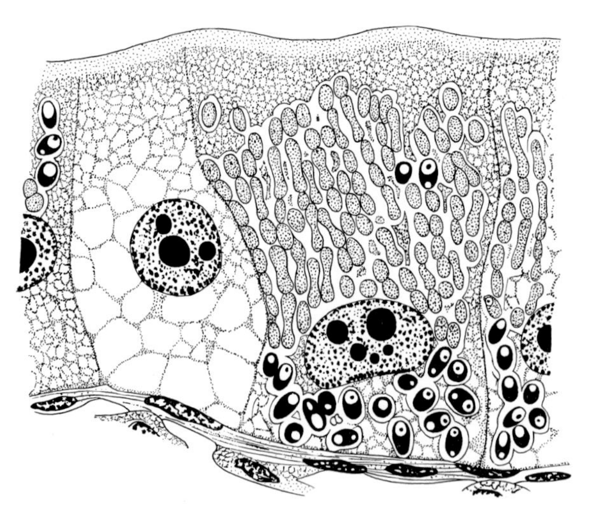

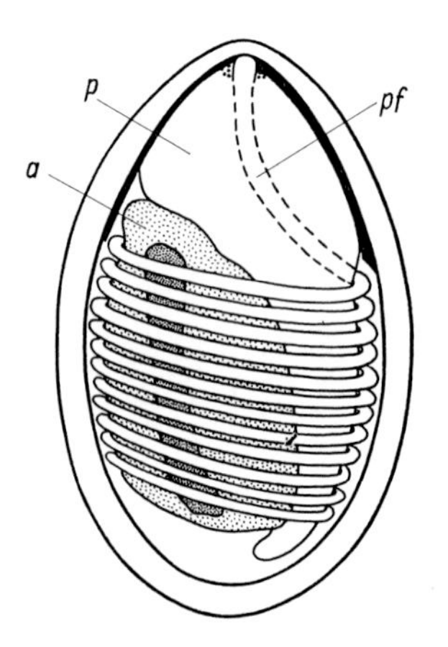

Fig. 436. *Nosema bombycis* NAEGELI. Developmental stages and spore formation in intestinal epithelial cells of silkworm. After STEMPELL, 1909 [*1671*]

Fig. 437 ▲

Fig. 437. *Thelohania californica* KELLEN et LIPA. Diagram of a spore (based on electron microscope studies). *a* amoebula with cell nucleus. *p* Polaroplast. *pf* Polar filament. After KUDO and DANIELS, 1963 [*973*]

In some cases, the host cell reacts to parasitic infection with a pronounced hypertrophy (p. 352). Salivary gland cells of Diptera when infected by Microsporidia show giant chromosomes of a higher degree of polyteny.

Little is known about spore formation. Whereas earlier investigators stated that the spores, like those of the Myxosporidia, are multinucleated, electron micrographs seem to indicate that they contain only one nucleus. The polar filament is not in a special capsule but is helically coiled below the spore envelope. A residual body capable of swelling, the so-called polaroplast, apparently plays a role in extrusion of the polar filament (Fig. 437).

Nosema apis ZANDER. In gut epithelial cells of the honey bee (cause of *Nosema* disease).
Nosema bombycis NÄGELI (Fig. 436). In larvae of the silkworm *Bombyx mori* (cause of the "spotting disease" or pébrine).
Glugea anomala MONIEZ. In the stickleback *Gasterosteus aculeatus*.

It is conceivable that the Microsporidia are descended from Myxosporidia and became so "simplified" by intracellular parasitism that only the amoebula and the polar filament remained in the spore. However, the polar filaments of Myxosporidia and Microsporidia might also represent convergent structures which are not due to any close relationship. In this case the Microsporidia would have to be regarded as a separate class of Protozoa.

N. Publications

I. Summary Presentations

BÜTSCHLI, O.: Protozoa. In: BRONN, H. G.: Klassen und Ordnungen des Tierreichs, Bd. 1. Heidelberg: Winter 1880—1889.

CALKINS, G. N., SUMMERS, F. M. (Ed.): Protozoa in biological research. New York: Columbia University Press 1941.

CHEN, T. T. (Ed.): Research in Protozoology. Vol. 1—4. New York: Pergamon Press 1967/68.

DOFLEIN, F., REICHENOW, E.: Lehrbuch der Protozoenkunde. 6. Aufl. Jena: Fischer 1949—1953.

DOGIEL, V. A., revised by POLJANSKIJ, J. I., CHEJSIN, E. M.: General Protozoology, 2. Ed. Oxford: Clarendon Press 1965.

GRASSÉ, P. P. (Ed.): Protozoaires. In: Traité de Zoologie, Vol. 1. Paris: Masson et Cie. 1952—1953.

HALL, R. P.: Protozoology. New York: Prentice-Hall Inc. 1953.

HUTNER, S. H. (Ed.): Biochemistry and physiology of Protozoa, Vol. 3. New York: Academic Press 1964.

KIDDER, G. W. (Ed.): Protozoa. In: FLORKIN, M., SCHEER, B. T.: Chemical Zoology, Vol. 1, 913 pp. New York: Academic Press 1967.

KUDO, R. R.: Protozoology, 5. Ed. Springfield (Illinois): Thomas 1966.

MACKINNON, D. L., HAWES, R. S. J.: An introduction to the study of Protozoa. Oxford: Clarendon Press 1961.

MANWELL, R. D.: Introduction to Protozoology. London: Edward Arnold 1961.

PITELKA, D. R.: Electron-microscopic structure of Protozoa. New York: Pergamon Press 1963.

WENYON, C. M.: Protozoology. London: Baillère, Tindall and Cox 1926.

II. Separate Works and Works from Associated Fields

The works listed are mainly recent publications. Reviews with numerous references are marked*.

1. ADOUTTE, A., BEISSON, J.: Hérédité cytoplasmique de mutations de résistance à l'érythromycine chez la Paramécie. J. Protozool. **17** (Suppl.), 31 (1970).

2. — — Cytoplasmic inheritance of erythromycin resistant mutations in *Paramecium aurelia*. Mol. gen. Genet. **108**, 70—77 (1970).

3. — — Evolution of mixed populations of genetically different mitochondria in *Paramecium aurelia*. Nature (Lond.) **235**, 393—396 (1972).

4. — BALMEFREZOL, M., BEISSON, J., ANDRE, J.: The effects of erythromycin and chloramphenicol on the ultrastructure of mitochondria in sensitive and resistant strains of *Paramecium*. J. Cell Biol. **54**, 8—19 (1972).

5. AFZELIUS, B. A.: The nucleus of *Noctiluca scintillans*. Aspects of nucleocytoplasmic exchanges and the formation of nuclear membrane. J. Cell Biol. **19**, 229—238 (1963).

6. AIKAWA, M.: The fine structure of the erythrocytic stages of three avian malarial parasites, *Plasmodium fallax*, *P. lophurae*, and *P. cathemerium*. Amer. J. trop. Med. **15**, 449—471 (1966).

7. — HEPLER, P. K., HUFF, C. G., SPRINZ, H.: The feeding mechanism of avian malarial parasites. J. Cell Biol. **28**, 355—373 (1966).

8. — HUFF, C. G., SPRINZ, H.: Comparative fine structure study of the gametocysts of avian, reptilian and mammalian malarial parasites. J. Ultrastruct. Res. **26**, 316—331 (1969).

*9. Allen, R. D.: Structure and function in ameboid movement. In: Goodwin, T. W., Lindberg, O.: Biological structure and function, Vol. 2, pp. 549—556. New York-London: Academic Press 1961.

*10. — Ameboid movement. In: Brachet, J., Mirsky, A. E.: The cell, Vol. 2, pp. 135—216. New York-London: Academic Press 1961.

11. — A new theory of ameboid movement and protoplasmic streaming. Exp. Cell Res. 8, 17—31 (1961).

12. — Fine structure, reconstruction and possible functions of components of the cortex of *Tetrahymena pyriformis*. J. Protozool. **14**, 553—565 (1967).

13. — The morphogenesis of basal bodies and accessory structures of the cortex of the ciliated protozoan *Tetrahymena pyriformis*. J. Cell Biol. **40**, 716—733 (1969).

*14. — Comparative aspects of amoeboid movement. Acta Protozool. **VII**, 291—299 (1970).

15. — Fine structure of membranous and microfibrillar systems in the cortex of *Paramecium caudatum*. J. Cell Biol. **49**, 1—20 (1971).

16. — Pattern of birefringence in the giant amoeba *Chaos carolinensis*. Exp. Cell Res. **72**, 34—45 (1972).

*17. — Kamiya, N.: Primitive motile systems in cell biology. New York: Academic Press 1964.

18. Allen, S. L.: Inherited variations in the esterases of *Tetrahymena*. Genetics **45**, 1051 to 1070 (1960).

19. — Genetic control of the esterases in the protozoan *Tetrahymena pyriformis*. Ann. N. Y. Acad. Sci. **94**, 753—773 (1961).

20. — Genomic exclusion in *Tetrahymena:* genetic basis. J. Protozool. **10**, 413—420 (1963).

21. — Linkage studies in variety 1 of *Tetrahymena pyriformis:* a first case of linkage in the ciliated Protozoa. Genetics **49**, 617—627 (1964).

22. — The esterase isozymes of *Tetrahymena:* their distribution in isolated cellular components and their behavior during the growth cycle. J. exp. Zool. **155**, 349—370 (1964).

23. — Genetic control of enzymes in *Tetrahymena*. Brookhaven Symp. Biol. **18**, 27—54 (1965).

*24. — The chemical genetics of the Protozoa. In: Florkin, M., Scheer, B. T.: Chemical Zoology, Vol. 1. New York-London: Academic Press 1967.

25. — Cytogenetics of genomic exclusion in *Tetrahymena*. Genetics **55**, 797—822 (1967).

26. — Genomic exclusion in *Tetrahymena*. Genetics **55**, 823—837 (1967).

27. — Genomic exclusion: a rapid means for inducing homozygous diploid lines in *Tetrahymena pyriformis*, Syngen 1. Science **155**, 575—577 (1967).

28. — A late-determined gene in *Tetrahymena* heterozygotes. Genetics **68**, 415—433 (1971).

29. — Byrne, B. C., Cronkite, D. L.: Intersyngenic variations in the esterases of bacterized *Paramecium aurelia*. Biochem. Genet. **5**, 135—150 (1971).

30. — Gibson, I.: Intersyngenic variations in the esterases of axenic stocks of *Paramecium aurelia*. Biochem. Genet. **5**, 161—181 (1971).

31. — — The purification of DNA from the genomes of *Paramecium aurelia* and *Tetrahymena pyriformis*. J. Protozool. **18**, 518—525 (1971).

32. — — Genome amplification and gene expression in the ciliate macronucleus. Biochem. Genet. **6**, 293—313 (1972).

33. — Nanney, D. L.: An analysis of nuclear differentiation in the selfers of *Tetrahymena*. Amer. Naturalist **92**, 139—160 (1958).

34. — Weremiuk, S. L.: Intersyngenic variations in the esterases and acid phosphatases of *Tetrahymena pyriformis*. Biochem. Genet. **5**, 119—133 (1971).

35. — — Patrick, C. A.: Is there selective mating in *Tetrahymena* during genomic exclusion. J. Protozool. **18**, 515—517 (1971).

36. — — Defective micronuclei and genomic exclusion in selected C* subclones of *Tetrahymena*. J. Protozool. **18**, 509—515 (1971).

37. Allera, A., Beck, R., Wohlfarth-Bottermann, K. E.: Weitreichende fibrilläre Protoplasmadifferenzierungen und ihre Bedeutung für die Protoplasmaströmung. VIII. Identifizierung der Plasmafilamente von *Physarum polycephalum* als F-Actin durch Anlagerung von heavy Meromyosin in situ. Cytobiol. **4**, 437—449 (1971).

38. ALLISON, A. C., HULANDS, G. H., NUNN, J. F., KITCHING, J. A., MACDONALD, A.: The effect of inhalational anaesthetics on the microtubular system in *Actinosphaerium nucleofilum*. J. Cell Sci. **7**, 483—499 (1970).

39. ALMEIDA, F. F. DE: Über den Einfluß des Parasiten *Eucoccidium dinophili* auf die freien Aminosäuren seines Wirtes *Dinophilus gyrociliatus*. Arch. Protistenk. **104**, 359—363 (1959).

40. ALVERDES, F.: Studien an Infusorien über Flimmerbewegung, Lokomotion und Reizbeantwortung. Arb. Geb. exp. Biol. **3**, 1—133 (1922).

41. AMMERMANN, D.: Cytologische und genetische Untersuchungen an dem Ciliaten *Stylonychia mytilus* EHRENBERG. Arch. Protistenk. **108**, 109—152 (1965).

42. — Das Paarungssystem der Ciliaten *Paramecium woodruffi* und *Paramecium trichium*. Arch. Protistenk. **109**, 139—146 (1965).

43. — Synthese und Abbau der Nucleinsäuren während der Entwicklung des Makronucleus von *Stylonychia mytilus* (Protozoa, Ciliata). Chromosoma (Berl.) **25**, 107—120 (1968).

44. — Release of DNA breakdown products into the culture medium of *Stylonychia mytilus* exconjugants (Protozoa, Ciliata) during the destruction of the polytene chromosomes. J. Cell Biol. **40**, 576—577 (1969).

45. — The micronucleus of the ciliate *Stylonychia mytilus*; its nucleic acid synthesis and its function. Exp. Cell Res. **61**, 6—12 (1970).

46. — Micronuclear DNA synthesis in *Stylonychia mytilus*. Naturwissenschaften **3**, 149 (1971).

47. — Morphology and development of the macronuclei of the ciliates *Stylonychia mytilus* and *Euplotes aediculatus*. Chromosoma (Berl.) **33**, 209—238 (1971).

48. AMON, J. P., PERKINS, F. O.: Structure of *Labyrinthula sp.* zoospores. J. Protozool. **15**, 543—546 (1968).

49. AMOS, W. B.: Structure and coiling of the stalk in the peritrich ciliates *Vorticella* and *Carchesium*. J. Cell Sci. **10**, 95—122 (1972).

50. ANDERSON, E.: A cytological study of *Chilomonas paramecium* with particular reference to the so-called trichocysts. J. Protozool. **9**, 380—395 (1962).

51. — Cytoplasmic organelles and inclusions of Protozoa. In: CHEN, T.-T. (Ed.): Research in Protozoology, Vol. 1, pp. 1—40. Oxford: Pergamon Press 1967.

52. — BEAMS, H. W.: The fine structure of the heliozoan *Actinosphaerium nucleofilum*. J. Protozool. **7**, 190—199 (1960).

53. — — The ultrastructure of *Tritrichomonas* with special reference to the blepharoplast complex. J. Protozool. **8**, 71—75 (1961).

54. — DUMONT, J. N.: A comparative study of the concrement vacuole of certain endocommensal ciliates — a so-called mechanoreceptor. J. Ultrastruct. Res. **15**, 414—450 (1966).

55. ANDERSON, T. F., PREER, JR., J. R., PREER, L. B., BRAY, M.: Studies on killing particles from *Paramecium:* The structure of refractile bodies from *kappa* particles. J. Microscop. **3**, 395—402 (1964).

56. ANDERSON, W. A., ELLIS, R. A.: Ultrastructure of *Trypanosoma lewisi*. Flagellum, microtubules and the kinetoplast. J. Protozool. **12**, 483—499 (1965).

57. — HILL, G. C.: Division and DNA synthesis in the kinetoplast of *Crithidia fasciculata*. J. Cell Sci. **4**, 611—620 (1969).

58. ANDRÉ, J.: Sur quelques détails nouvellement connus de l'ultrastructure des organites vibratiles. J. Ultrastruct. Res. **5**, 86—108 (1961).

59. — THIERY, J. P.: Mise en évidence d'une sousstructure fibrillaire dans les filaments axonématiques des flagelles. J. Microscop. **2**, 71—80 (1963).

60. ANDRESEN, N., CHAPMAN-ANDRESEN, C., NILSSON, J. R.: The fine structure of *Pelomyxa palustris*. C. R. Lab. Carlsberg **36**, 285—317 (1968).

61. ANGELOPOULOS, E.: Pellicular microtubules in the family Trypanosomatidae. J. Protozool. **17**, 39—51 (1970).

62. ANTIPA, G.: Structural differentiation in the somatic cortex of a ciliated protozoan *Conchophthirus curtus* ENGELMANN 1862. Protistologica **7**, 471—504 (1971).

63. Applewhite, P. B., Gardner, F. T., Lapan, E.: Physiology of habituation learning in a protozoan. Trans. N. Y. Acad. Sci. **31**, 842—849 (1969).

64. — — Theory of protozoan habituation. Nature (Lond.) **230**, 285—287 (1971).

65. Argetsinger, J.: The isolation of ciliary bodies (kinetosomes) from *Tetrahymena pyriformis*. J. Cell Biol. **24**, 154—157 (1965).

66. Armstrong, J. J., Surzycki, S. J., Moll, B., Levine, R. P.: Genetic transcription and translation specifying chloroplasts in *Chlamydomonas reinhardi*. Biochemistry **10**, 692—701 (1971).

67. Arnold, Z. M.: Life history and cytology of the foraminiferan *Allogromia laticollaris*. Univ. Calif. Publ. Zool. **61**, 167—252 (1955).

68. — Biological observations in the foraminifer *Spiroloculina hyalina* Schulze. Univ. Calif. Publ. Zool. **72**, 1—78 (1964).

69. — Biological observations on the foraminifer *Calcituba polymorpha* Roboz. Arch. Protistenk. **110**, 280—304 (1967).

70. Arnott, H. J., Brown, Jr., R. M.: Ultrastructure of the eyespot and its possible significance in phototaxis of *Tetracystis excentrica*. J. Protozool. **14**, 529—539 (1967).

71. Asato, Y., Fulsome, C. E.: Temporal genetic mapping of the blue-green alga *Anacystis nidulans*. Genetics **65**, 407—419 (1970).

72. Ax, P.: Personal communication

73. Babillot, C.: Étude de l'incorporation d'uridine-H^3 dans le noyau chez l'*Amphidinium carteri*, Dinoflagellé. C. R. Acad. Sci. (Paris) **271**, 828—831 (1970).

74. — Étude des effects de l'actinomycine D sur le noyau du dinoflagellé *Amphidinium carteri*. J. Microscop. **9**, 485—502 (1970).

75.* Balamuth, W.: Regeneration in Protozoa: A problem of morphogenesis. Quart. Rev. Biol. **15, 290—337 (1940).

76. Balbinder, E.: Two loci controlling the maintenance and stability of the cytoplasmic factor "*Kappa*" in stock 51, var. 4 killers of *Paramecium aurelia*. Genetics **41**, 634 (1956).

77. — The genotypic control of *kappa* in *Paramecium aurelia*, syngen 4, stock 51. Genetics **44**, 1227—1241 (1959).

**78.* Ball, G. H.: Organisms living on and in Protozoa. In: Chen, T.-T. (Ed.): Research in Protozoology, Vol. 3, pp. 565—718. New York: Pergamon Press 1969.

79. Balsley, M.: Dependence of the *kappa* particles of stock 7 of *Paramecium aurelia* on a single gene. Genetics **56**, 125—131 (1967).

80. Bannister, L. H.: The structure of trichocysts in *Paramecium caudatum*. J. Cell Sci. **11**, 899—929 (1972).

81. — Tatchell, E. C.: Contractility and the fibre systems of *Stentor coeruleus*. J. Cell Sci. **3**, 295—308 (1968).

82. Bardele, C. F.: Elektronenmikroskopische Untersuchungen an dem Sporozoon *Eucoccidium dinophili* Grell. Z. Zellforsch. **74**, 559—595 (1966).

83. — *Acineta tuberosa*. I. Der Feinbau des adulten Suktors. Arch. Protistenk. **110**, 403—421 (1968).

84. — *Acineta tuberosa*. II. Die Verteilung der Mikrotubuli im Makronucleus während der ungeschlechtlichen Fortpflanzung. Z. Zellforsch. **93**, 93—104 (1969).

85. — Ultrastruktur der „Körnchen" auf den Axopodien von *Raphidiophrys* (Centrohelida, Heliozoa). Z. Naturforsch. **24**b, 362—363 (1969).

86. — Budding and metamorphosis in *Acineta tuberosa*. An electron microscopic study on morphogenesis in Suctoria. J. Protozool. **17**, 51—70 (1970).

87. — Studies on the fine structure of *Eucoccidium dinophili* and *E. ophryotrochae*. J. Parasitol. **56**, 19—20 (1970).

88. — Microtubule model systems: Cytoplasmic transport in the suctorian tentacle and the centrohelidian axopod. In: Arceneaux, C. J. (Ed.): 29th Ann. Proc. Electron Microscopy Soc. Amer. Boston: Mass. 1971.

89. — Cell cycle, morphogenesis and ultrastructure in the pseudoheliozoan *Clathrulina elegans*. Z. Zellforsch. **130**, 219—142 (1972).

90. — A microtubule model for ingestion and transport in the suctorian tentacle. Z. Zellforsch. **126**, 116—134 (1972).

91. Bardele, C.F., Grell, K.G.: Elektronenmikroskopische Beobachtungen zur Nahrungsaufnahme bei dem Suktor *Acineta tuberosa* Ehrenberg. Z. Zellforsch. **80**, 108—123 (1967).

92. Barnett, A.: Cytology of conjugation in *Paramecium multimicronucleatum*, syngen 2, stock 11. J. Protozool. **11**, 147—153 (1964).

93. — A circadian rhythm of mating type reversals in *Paramecium multimicronucleatum*. In: Aschoff, I.: Circadian clocks, pp. 305—308. Amsterdam: North-Holland Publ. Co. 1964.

94. — A circadian rhythm of mating type reversals in *Paramecium multimicronucleatum*, syngen 2, and its genetic control. J. cell. Physiol. **67**, 239—270 (1966).

95. Batisse, A.: Les appendices préhenseurs d'*Ephelota gemmipara* Hertwig. C. R. Acad. Sci. (Paris) **261**, 5629—5632 (1965).

96. — L'ultrastructure des tentacules suceurs d'*Ephelota gemmipara* Hertwig, C. R. Acad. Sci. (Paris) **262**, 771—774 (1966).

97. — Le développement des phialocystes chez les Acinétiens. C. R. Acad. Sci. (Paris) **265**, 972—974 (1967).

98. — Données nouvelles sur la structure et le fonctionnement des ventouses tentaculaire des Acinétiens. C. R. Acad. Sci. (Paris) **265**, 1056—1058 (1967).

99. — Les ultrastructures squelettiques chez certains Thecacinetidae. Protistologica **4**, 447—492 (1968).

100. — Quelques aspects de l'ultrastructure de *Pseudogemma pachystyla* Collin. Protistologica **4**, 271—282 (1968).

101. — Les structures pédonculaires dans les genres *Tokophrya Bütschli* et *Choanophrya Hartog* (Ciliata, Suctorida). Protistologica **5**, 387—412 (1969).

102. Baudoin, A.: Contribution à l'étude des Grégarines des Trichoptères Limnophilidae et Sericostomidae. Protistologica **2** (2), 97—107 (1966).

103. — Étude comparée de quelques Grégarines Acanthosporinae. J. Protozool. **18**, 654—661 (1971).

104. Bazin, M.J.: Sexuality in a blue-green alga: Genetic recombination in *Anacystis nidulans*. Nature (Lond.) **218**, 282—283 (1968).

105. Beale, G.H.: The process of transformation of antigenic types in *Paramecium aurelia*, variety 4. Proc. Nat. Acad. Sci. **34**, 418—423 (1948).

106. — Antigen variation in *Paramecium aurelia*, variety 1. Genetics **37**, 62—74 (1952).

**107.* — The genetics of *Paramecium aurelia*. Monographs in experimental biology, Vol. 2, 178 pp. Cambridge: University Press 1954.

108. — The antigen system of *Paramecium aurelia*. Int. Rev. Cytol. **6**, 1—23 (1957).

109. — Genetic control of the cell surface. Proc. roy. Soc. Edinb. A **28**, 71—78 (1959).

110. — The role of cytoplasm in inheritance. Sci. Progr. (Lond.) **49**, 17—35 (1961).

111. — A note on the inheritance of erythromycin-resistance in *Paramecium aurelia*. Genet. Res. **14**, 341—342 (1969).

112. — Jurand, A.: Structure of the mate-killer (*mu*) particles in *Paramecium aurelia*, stock 540. J. gen. Microbiol. **23**, 234—252 (1960).

113. — — Three different types of mate-killer (*mu*) particles in *Paramecium aurelia* (Syngen 1). J. Cell Sci. **1**, 31—34 (1966).

114. — Jurand, A., Preer, J.R.: The classes of endosymbionts of *Paramecium aurelia*. J. Cell Sci. **5**, 65—91 (1969).

115. — Kacser, H.: Studies on the antigens of *Paramecium aurelia* with the aid of fluorescent antibodies. J. gen. Microbiol. **17**, 68—74 (1957).

116.* — Knowles, J.K.C., Tait, A.: Mitochondrial genetics in *Paramecium*. Nature (Lond.) **235, 396—397 (1972).

117. — McPhail, S.: Some additional results on the maintenance of *kappa* particles in *Paramecium aurelia* (stock 51) after loss of gene *K*. Genet. Res. **9**, 369—373 (1967).

118. — Mott, M.R.: Further studies on the antigens of *Paramecium aurelia* with the aid of fluorescent antibodies. J. gen. Microbiol. **28**, 617—623 (1962).

119. — Wilkinson, J.F.: Antigenic variation in unicellular organisms. Ann. Rev. Microbiol. **15**, 263—296 (1961).

*120. Beams, H. W., Anderson, E.: Fine structure of Protozoa. Ann. Rev. Microbiol. **15**, 47—68 (1961).

121. — King, R. L., Tahmisian, T. N., Devine, R.: Electron microscope studies on *Lophomonas striata* with special reference to the nature and position of the striations. J. Protozool. **7**, 91—101 (1960).

122. — Sekhon, S.: Further studies on the fine structure of *Lophomonas blattarum* with special reference to the so-called calyx, axial filament and parabasal body. J. Ultrastruct. Res. **26**, 296—316 (1969).

123. — Tahmisian, T. N., Anderson, E., Wright, W.: Studies on the fine structure of *Lophomonas blattarum* with special reference to the so-called parabasal apparatus. J. Ultrastruct. Res. **5**, 166—183 (1961).

124. — — Devine, R. L., Anderson, E.: The fine structure of the nuclear envelope of *Endamoeba blattae*. Exp. Cell Res. **18**, 366—369 (1959).

*125. Beermann, W.: Riesenchromosomen. In: Protoplasmatologia (Handbuch der Protoplasmaforschung) VI D. 161 S. Wien: Springer 1962.

*126. — Developmental studies on giant chromosomes. In: Results on problems of cell differentiation, Vol. 4, 227 p. Berlin-Heidelberg-New York: Springer 1972.

127. Beers, C. D.: Excystment in the ciliate *Bursaria truncatella*. Biol. Bull. **94**, 86—98 (1948).

128. Beisson, J.: Déterminants nucléaires et cytoplasmiques dans la biogenèse des structures chez les Protozoaires. Ann. Biol. **11**, 401—411 (1972).

129. — Rossignol, M.: The first case of linkage in *Paramecium aurelia*. Gent. Res. **13**, 85—90 (1969).

130. — Sonneborn, T. M.: Cytoplasmic inheritance of the organization of the cell cortex in *Paramecium aurelia*. Proc. nat. Acad. Sci. (Wash.) **53**, 275—282 (1965).

131. Belar, K.: Untersuchungen über Thecamöben der *Chlamydophrys*-Gruppe. Arch. Protistenk. **43**, 287—354 (1921).

132. — Untersuchungen an *Actinophrys sol* Ehrenberg. I. Die Morphologie des Formwechsels. Arch. Protistenk. **46**, 1—96 (1922).

133. — Untersuchungen an *Actinophrys sol*. II. Beiträge zur Physiologie des Formwechsels. Arch. Protistenk. **48**, 371—434 (1924).

*134. — Der Formwechsel der Protistenkerne. Ergebn. Fortschr. Zool. **6**, 420 (1926).

*135. — Die cytologischen Grundlagen der Vererbung. In: Handbuch der Vererbungswissenschaft, Bd. 1, 412 S. Berlin: Gebrüder Bornträger 1928.

136. Bennett, D. A., Francis, D.: Learning in *Stentor*. J. Protozool. **19**, 484—487 (1972).

137. Ben-Shaul, Y., Epstein, H. T., Schiff, J. A.: Studies of chloroplast development in *Euglena*. X. The return of the chloroplast to the proplastic condition during dark adaptation. Canad. J. Bot. **43**, 129—136 (1965).

138. — Schiff, J. A., Epstein, H. T.: Studies of chloroplast development in *Euglena*. VII. Fine structure of the developing plastid. Plant. Physiol. **39**, 231—240 (1964).

139. Bergström, S. R.: Lernen bei Einzellern. Bild der Wissenschaft **7**, 687—692 (1970).

140. Berthold, W.-U.: Untersuchungen über die sexuelle Differenzierung der Foraminifere *Patellina corrugata* Williamson mit einem Beitrag zum Entwicklungsgang und Schalenbau. Arch. Protistenk. **113**, 147—184 (1971).

141. Bhowmick, F. K.: Electronmicroscopy of *Trichamoeba villosa* and ameboid movement. Exp. Cell Res. **45**, 570—589 (1967).

142. Bingley, M., Bell, L. G. E., Jeon, K. W.: Pseudopod initiation and membrane depolarisation in *Amoeba proteus*. Exp. Cell Res. **28**, 208—209 (1962).

143. Bleyman, L.: The inhibition of mating reactivity in *Paramecium aurelia* by inhibitors of protein and RNA synthesis. Genetics **50**, 236 (1964).

*144. — Temporal patterns in the ciliated Protozoa. From: Cameron, I. L., Padilla, G. M., Zimmerman, A. M. (Eds.): Developmental aspects of the cell cycle. pp. 67—91. New York-London: Academic Press 1971.

145. — Simon, E. M., Brosi, R.: Sequential nuclear differentiation in *Tetrahymena*. Genetics **54**, 277—291 (1966).

146. Blum, J. J., Sommer, J. R., Kahn, V.: Some biochemical, cytological and morphogenetic comparisons between *Astasia longa* and a bleached *Euglena gracilis*. J. Protozool. **12**, 202—209 (1965).

147. Bohatier, J.: Structure et ultrastructure de *Lacrymaria olor* (O. F. M. 1786). Protistologica **6**, 331—342 (1970).

148. Bomford, R.: Infection of algae-free *Paramecium bursaria* with strains of *Chlorella, Scendesmus* and a yeast. J. Protozool. **12**, 221—224 (1965).

149. — The syngens of *Paramecium bursaria*: New mating types and intersyngenic mating reactions. J. Protozool. **13**, 497—501 (1966).

150. Borgert, A.: Untersuchungen über die Fortpflanzung der tripyleen Radiolarien, speziell von *Aulacantha scolymantha* H. (Teil I). Zool. Jb. Anat. **14**, 203—276 (1900).

151. — Untersuchungen über die Fortpflanzung der tripyleen Radiolarien, speziell von *Aulacantha scolymantha* H. (Teil II). Arch. Protistenk. **14**, 134—261 (1909).

152. Bouck, G. B.: The structure, origin, isolation and composition of the tubular mastigonemes of the *Ochromonas* flagellum. J. Cell Biol. **50**, 362—384 (1971).

153. — Sweeney, B. M.: The fine structure and ontogeny of trichocysts in marine dinoflagellates. Protoplasma **61**, 205—223 (1966).

154. Bouligand, Y., Puiseux-Dao, S., Soyer, M.-O.: Liaisons morphologiquements définies entre chromosomes et membrane nucléaire chez certains péridiniens. C. R. Acad. Sci. (Paris) **266**, 1287—1289 (1968).

155. — Soyer, M.-O., Puiseux Dao, S.: La structure fibrillaire et l'orientation des chromosomes chez les Dinoflagellés. Chromosoma (Berl.) **24**, 251—287 (1968).

156. Bourque, J. E., Boynton, J. E., Gillham, N. W.: Studies on the structure and cellular location of various ribosome and ribosomal RNA species in the green alga *Chlamydomonas reinhardi*. J. Cell Sci. **8**, 153—183 (1971).

157. Bowers, B., Korn, E. D.: The fine structure of *Acanthamoeba castellanii*. I. The trophozoite. J. Cell Biol. **39**, 95—111 (1968).

158. — — The fine structure of *Acanthamoeba castellanii* (Neff strain). II. Encystment. J. Cell Biol. **41**, 786—805 (1969).

159. Boynton, J. E., Gillham, N. W., Burkholder, B.: Mutations altering chloroplast ribosome phenotype in *Chlamydomonas*. II. A new mendelian mutation. Proc. nat. Acad. Sci. (Wash.) **67**, 1505—1512 (1970).

160. — — Chabot, J. F.: Chloroplast ribosome deficient mutants in the green alga *Chlamydomonas reinhardii* and the question of chloroplast ribosome function. J. Cell Sci. **10**, 267—305 (1972).

**161.* Brachet, J.: Nucleocytoplasmic interactions in unicellular organisms. In: Brachet, J., Mirsky, A. E.: The cell, Vol. 2, pp. 771—841. New York-London: Academic Press 1961.

162. Bradbury, P. C., Gallucci, B. B.: The fine structure of differentiating merozoites of *Haemoproteus columbae* Kruse. J. Protozool. **18**, 679—687 (1971).

163. Bradbury, P., Pitelka, D. R.: Observations on kinetosome formation in an apostome ciliate. J. Microscop. **4**, 805—810 (1965).

164. Bradley, D. E.: Observations on the flagella of two Chrysophyceae using negative staining. Quart. J. Microscop. Sci. **106**, 327—331 (1965).

165. — The ultrastructure of the flagella of three Chrysomonads with particular reference to the mastigonemes. Exp. Cell Res. **41**, 162—173 (1966).

166. Brandt, P. W.: A study of the mechanism of pinocytosis. Exp. Cell Res. **15**, 300—313 (1958).

167. — Pappas, G. D.: Mitochondria. II. The nuclear-mitochondrial relationship in *Pelomyxa carolinensis* Wilson (*Chaos chaos* L.). J. biophys. biochem. Cytol. **6**, 91—96 (1959).

168. — — An electron microscopic study of pinocytosis in ameba. I. The surface attachment phase. J. biophys. biochem. Cytol. **8**, 675—687 (1959).

169. — — An electron microscopic study of pinocytosis in ameba. II. The cytoplasmic uptake phase. J. Cell Biol. **15**, 55—57 (1962).

170. Brawerman, G., Chargaff, E.: A self-reproducing system concerned with the formation of chloroplasts in *Euglena gracilis*. Biochim. biophys. Acta (Amst.) **37**, 221—229 (1960).

171. Brawerman, G., Eisenstadt, J.: Deoxyribonucleic acid from the chloroplasts of *Euglena gracilis*. Biochim. biophys. Acta (Amst.) **91**, 477—485 (1960).

172. Brokaw, C. J.: Movement and nucleoside polyphosphatase activity of isolated flagella from *Polytoma uvella*. Exp. Cell Res. **22**, 151—162 (1961).

173. — Movement of flagella of *Polytoma uvella*. J. exp. Biol. **40**, 149—156 (1953).

174. Brooks, A. E.: The sexual cycle and intercrossing in the genus *Astrephomene*. J. Protozool. **13**, 368—375 (1963).

175. Brown, K. N., Armstrong, J. A., Valentine, R. C.: The ingestion of protein molecules by blood forms of *Trypanosoma rhodesiense*. Exp. Cell Res. **39**, 129—135 (1965).

176. Brugerolle, G.: Ultrastructure du genre *Trichomitus* Swezy 1915, Zooflagellata, Trichomonadina. Protistologica **7**, 171—176 (1971).

**177.* Buchner, P.: Endosymbiose der Tiere mit pflanzlichen Mikroorganismen. Basel: Birkhäuser 1953.

178. Bullington, W. E.: A study of spiral movement in the ciliate infusoria. Arch. Protistenk. **50**, 219—274 (1925).

**179.* Bullock, T. H.: Protozoa, Mesozoa and Porifera. In: Bullock, T. H., Horridge, G. A.: Structure and function in the nervous systems of invertebrates, Vol. 1, pp. 433—457. San Francisco: Freeman 1965.

180. Bünning, E., Schneiderhöhn, G.: Über das Aktionsspectrum der phototaktischen Reaktion von *Euglena*. Arch. Mikrobiol. **24**, 80—90 (1956).

181. — Tazawa, M.: Über die negativ-phototaktische Reaktion von *Euglena*. Arch. Mikrobiol. **27**, 306—310 (1956).

182. Burton, P. R., Dusanic, D. G.: Fine structure and replication of the kinetoplast of *Trypanosoma lewisi*. J. Cell Biol. **39**, 318—331 (1968).

183. Butzel, H. M.: Mating type mutations in variety 1 of *Paramecium aurelia*, and their bearing upon the problem of mating type determination. Genetics **40**, 321—330 (1955).

184. — Mating type determination in stock 51, syngen 4 of *Paramecium aurelia* grown in axenic culture. J. Protozool. **15**, 284—290 (1968).

185. Byrne, B. J.: *Kappa*, *mu* and the metagon hypothesis in *Paramecium aurelia*. Genet. Res. **13**, 197—211 (1969).

186. Cachon, J.: Contribution à l'étude des péridiniens parasites. Cytologie, cycles évolutifs. Ann. Sci. nat. Zool. **6**, 1—158 (1964).

187. Cachon-Enjumet, M.: Contribution à l'étude des Radiolaires Phaeodariés. Arch. Zool. exp. gen. **100**, 151—238 (1961).

188. — L'évolution sporogénétique des Phaeodariés (Radiolaires). C. R. Acad. Sci. (Paris) **259**, 2677—2679 (1964).

189. Cachon, J., Cachon-Enjumet, M.: Cytologie et ultrastructure de l'ergastoplasme et du système axopodial des Radiolaires Phaeodariés. Arch. Zool. exp. gen. **103**, 1—12 (1964).

190. — — Cycle évolutif et cytologie de *Neresheimeria catenata* Neresheimer, péridinien parasite d'appendiculaires. Rapports de l'hôte du parasite. Ann. Sci. nat. Zool. (Paris) **4**, 779—800 (1964).

191. — — *Pomatodinium impatients* nov. gen. nov. sp. péridinien noctilucidae Kent. Protistologica **2**, (1) 23—30 (1966).

192. — — Contribution à l'étude des Noctilucidae Saville Kent. I. Les Kofoidininae Cachon, J. et M. Evolution morphologique et systématique. Protistologica **3**, 427—444 (1967).

193. — — Cytologie et cycle évolutif des *Chytriodinium* Chatton. Protistologica **4**, 249—262 (1968).

194. — — *Filodinium hovassei* nov. gen. nov. sp., péridinien phorétique d'appendiculaires. Protistologica **4**, 15—18 (1968).

195. — — Contribution à l'étude des Noctilucidae Saville Kent. Evolution morphologique, cytologie, systématique. II. Les Leptodiscinae Cachon, J. et M. Protistologica **5**, 11—34 (1969).

196. — — Les processus sporogénétiques du Radiolaire *Sticholonche zanclea* Hertwig. Arch. Protistenk. **111**, 87—99 (1969).

197. Cachon, J., Cachon-Enjumet, M.: Révision systématique des Nassellaires Plectoidea à propos de la déscription d'un nouveau représentant, *Plectogonidium deflandrei* nov. gen. nov. sp. Arch. Protistenk. **111**, 236—251 (1969).

198. — — Ultrastructures des Amoebophryidae (Péridiniens Duboscquodinida). I. Manifestations des rapports entre l'hôte et le parasite. Protistologica **5**, 535—548 (1969).

199. — — Ultrastructure des Amoebophryidae (Péridiniens Duboscquodinida). II. Systèmes atractophoriens et microtubulaires; leur intervention dans la mitose. Protistologica **6**, 57—70 (1970).

200. — — Le système axopodial des Radiolaires Nassellaires. Origine, organisation et rapports avec les autres organites cellulaires. Arch. Protistenk. **113**, 80—97 (1971).

201. — — Ultrastructures du genre *Oodinium* Chatton. Différentiations cellulaires en rapport avec la vie parasitaire. Protistologica **7**, 153—169 (1971).

202. — — *Protoodinium chattoni* Hovasse. Manifestations ultrastructurales des rapports entre le péridinien et la méduse-hôte: fixation, phagocytose. Arch. Protistenk. **113**, 293—305 (1971).

203. — — Le système axopodial des Radiolaires sphaeroidés. I. Centroaxoplastidiés. Arch. Protistenk. **114**, 51—64 (1972).

204. — — Les modalités du dépôt de la silice chez les Radiolaires. Arch. Protistenk. **114**, 1—13 (1972).

205. — — Bouquaheux, F.: *Myxodinium pipiens* gen. nov., sp. nov., Péridinien parasite d'*Halosphaera*. Phycologia **8**, 157—163 (1969).

206. — — Ferru, G.: Rapports du squelette et du système axopodial chez les Radiolaires Nassellaires. C. R. Acad. Sci. (Paris) **267**, 1602—1604 (1968).

207. — — Greuet, C.: Le systéme pusulaire de quelques Péridiniens libres ou parasites. Protistologica **6**, 467—476 (1970).

208. — — Pyne, Ch.: Structure et ultrastructure de *Paradinium poucheti* Chatton 1910, et position systématique des Paradinides. Protistologica **4**, 303—311 (1968).

209. Callicott, J. H.: Amoebic meningo-encephalitis due to free-living amoebae of the *Hartmannella* (*Acanthamoeba*) *Naegleria* group. Amer. J. clin. Path. **49**, 84—91 (1968).

210. Canning, E. U., Anwar, M.: Studies on meiotic division in coccidial and malarial parasites. J. Protozool. **15**, 290—298 (1968).

211. Carasso, N., Fauré-Fremiet, E., Favard, P.: Ultrastructure de l'appareil excréteur chez quelques ciliés péritriches. J. Microscop. **1**, 455—468 (1962).

212. — Favard, P.: Microtubules fusoriaux dans les micro- et macronucleus de ciliés péritriches en division. J. Microscop. **4**, 395—402 (1965).

213. — — Mise en évidence du calcium dans les myonèmes pédonculaires de ciliés péritriches. J. Microscop. **5**, 759—770 (1966).

214. Carell, E. F.: Studies on chloroplast development and replication in *Euglena*. J. Cell Biol. **41**, 431—440 (1969).

215. Carter, R.: Enzyme variation in *Plasmodium berghei*. Trans. roy. Soc. trop. Med. Hyg. **64**, 401—406 (1970).

216. Casagrande, V. A., Harting, J. K., Hall, W. C., Diamond, I. T., Martin, G. F.: Contraction in *Stentor coeruleus:* A cinematic analysis. Science **177**, 447—449 (1972).

217. Chao, J., Ball, G. H.: In vitro culture of the vector phase of snake hemogregarines in mosquito cell lines. J. Parasit. **58**, 148—152 (1972).

218. Chao, P. K.: *Kappa* concentration per cell in relation to the life cycle, genotype and mating type in *Paramecium aurelia*, variety 4. Proc. nat. Acad. Sci. (Wash.) **39**, 103—113 (1953).

219.* Chapman-Andresen, C.: Studies on pinocytosis in amoebae. C. R. Lab. Carlsberg **33, 73—264 (1962).

220. — Measurement of material uptake by cells: Pinocytosis. In: Prescott, D. M.: Methods in cell physiology, Vol. 1, pp. 277—304. New York-London: Academic Press 1964.

221. — The induction of pinocytosis in amoebae. Arch. Biol. (Liège) **76**, 189—207 (1965).

222. — Studies on endocytosis in amoebae. The distribution of pinocytically ingested dyes in relation to food vacuoles in *Chaos chaos*. 1. Light microscopic observation. C. R. Lab. Carlsberg **36**, 9—10 (1967).

223. CHAPMAN-ANDRESEN, C., HOLTER, H.: Studies on the ingestion of C^{14} glucose by pinocytosis in the amoeba *Chaos chaos*. Exp. Cell Res., Suppl. **3**, 52—63 (1955).
224. — NILSSON, J. R.: Electronmicrographs of pinocytosis channels in *Amoeba proteus*. Exp. Cell Res. **19**, 631—633 (1960).
225. — — On vacuole formation in *Tetrahymena pyriformis* Gl. C. R. Lab. Carlsberg **36**, 405—432 (1968).
226. CHARRET, R.: Caractères cytologiques du thécamoebien *Arcella polypora*. Protistologica **3**, (1) 73—78 (1967).
227. CHATTON, E.: *Pleodorina californica* à Banyuls-sur-mer. Son cycle évolutif et sa signification phylogénique. Bull. Sci. Fr. Belg. **44**, 309—331 (1911).
228. — Les péridiniens parasites: Morphologie, reproduction, éthologie. Arch. Zool. exp. gén. **59**, 1—475 (1920).
229. — Essai d'un schéma de l'énergide d'après une image objective et synthétique: le dinoflagellé *Polykrikos schwartzi* BÜTSCHLI. Arch. zool. Ital. **16**, 169—187 (1931).
230. — L'origine péridinienne des radiolaires et l'interprétation parasitaire de l'anisosporogénèse. C. R. Acad. Sci. (Paris) **198**, 309 (1934).
231. — HOVASSE, R.: Sur les premiers stades de la cnidogénèse chez le péridinien *Polykrikos schwartzi*. C. R. Acad. Sci. (Paris) **218**, 60 (1944).
232. — LWOFF, A.: Les ciliés apostomes. I. Aperçu historique et général; Etude monographique des genres et des espèces. Arch. Zool. exp. gén. **77**, 1—453 (1935).
233. — — La constitution primitive de la strie ciliaire des infusoires. La desmodexie. C. R. Soc. Biol. (Paris) **118**, 1068—1072 (1935).
234. — SÉGUÉLA, J.: Sur la continuité génétique du cinétome chez quelques ciliés hypotriches. C. R. Acad. Sci. (Paris) **208**, 868 (1939).
235. CHEISSIN, E. M.: Ultrastructure of *Lamblia duodenalis*. I. Body surface, sucking disc and median bodies. J. Protozool. **11**, 91—98 (1964).
236. — Electron microscopic study of microgametogenesis in two species of coccidia from rabbit (*Eimeria magna* and *E. intestinalis*). Acta Protozool. **3**, 215—224 (1965).
237. — Ultrastructure of *Lamblia duodenalis*. II. The locomotory apparatus, axial rod and other organelles. Arch. Protistenk. **108**, 8—18 (1965).
238. — MOSSEVICH, T. N.: An electron microscope study of *Colpidium colpoda* (Ciliata, Holotricha). Arch. Protistenk. **106**, 181—200 (1962).
239. — OVCHINNIKOVA, L. P.: A photometric study of DNA content in macronuclei and micronuclei of different species of *Paramecium*. Acta Protozool. **2**, 225—236 (1964).
240. — — KUDRIAVTSEV, B. N.: A photometric study of DNA content in macronuclei and micronuclei of different strains of *Paramecium caudatum*. Acta Protozool. **2**, 237—245 (1964).
241.* — POLJANSKY, G. I.: On the taxonomic system of Protozoa. Acta Protozool. **1, 327—352 (1963).
242. — SNIGIREVSKAYA, E. S.: Some new data on the fine structure of the merozoites of *Eimeria intestinalis* (Sporozoa, Eimeriidea). Protistologica **1**, (1) 121—126 (1965).
243. CHEN, T. T.: Polyploidy and its origin in *Paramecium*. J. Hered. **31**, 175—184 (1940).
244. — A further study on polyploidy in *Paramecium*. (Chromosomes and mating types in *Paramecium bursaria*). J. Hered. **31**, 249—251 (1940).
245. — Conjugation of three animals in *Paramecium bursaria*. Proc. nat. Acad. Sci. (Wash.) **26**, 231—238 (1940).
246. — Polyploidy in *Paramecium bursaria*. Proc. nat. Acad. Sci. (Wash.) **26**, 239—240 (1940).
247. — Conjugation in *Paramecium bursaria* between animals with diverse nuclear constitutions. J. Hered. **31**, 185—196 (1940).
248. — Conjugation in *Paramecium bursaria*. I. Conjugation of three animals. J. Morph. **78**, 353—395 (1946).
249. — Conjugation in *Paramecium bursaria*. II. Nuclear phenomena in lethal conjugation between varieties. J. Morph. **79**, 125—262 (1946).
250. — Chromosomes in Opalinidae (Protozoa, Ciliata) with special reference to their behavior, morphology, individuality, diploidy, haploidy and association with nucleoli. J. Morph. **83**, 281—358 (1948).

251. CHEN, T. T.: Conjugation in *Paramecium bursaria*. III. Nuclear changes in conjugation between double monsters and single animals. J. Morph. **88**, 245—292 (1951).

252. — Conjugation in *Paramecium bursaria*. IV. Nuclear behavior in conjugation between old and young clones. J. Morph. **88**, 293—360 (1951).

253. — Paramecin 34, a killer substance produced by *Paramecium bursaria*. Proc. Soc. exp. Biol. (N.Y.) **88**, 541—543 (1955).

254. — Chromosomes and nucleoli in some opalinid Protozoa. In: CHEN, T.-T. (Ed.): Research in Protozoology, Vol. 4, pp. 351—392. New York: Pergamon Press 1972.

255. CHEN-SHAN, L.: Cortical morphogenesis in *Paramecium aurelia* following amputation of the posterior region. J. exp. Zool. **170**, 205—228 (1969).

256. CHILD, F. M.: The characterization of the cilia of *Tetrahymena pyriformis*. Exp. Cell Res. **18**, 258—267 (1959).

257. — Some aspects of the chemistry of cilia and flagella. Exp. Cell Res. **8** (Suppl.), 47—53 (1961).

258. CHO, P. L.: Cortical patterns in two syngens of *Glaucoma*. J. Protozool. **18**, 180—183 (1971).

259. — The genetics of mating type in a syngen of *Glaucoma*. Genetics **67**, 377—390 (1971).

260. CHRISTIANSEN, O.: Notes on the biology of Foraminifera. Vie et Milieu, C. Symp. Européen de Biol. Marine **22**, 465—478 (1971).

261. CHRISTIANSEN, R. G., MARSHALL, J. M.: A study of phagocytosis in the ameba *Chaos chaos*. J. Cell Biol. **25**, 443—457 (1965).

262. CHUNOSOFF, L., HIRSHFIELD, H. I.: Nuclear structure and mitosis in the dinoflagellate *Gonyaulax monilata*. J. Protozool. **14**, 157—163 (1967).

263. CLARK, A. M.: Some effects of removing the nucleus from amoeba. Aust. J. exp. Biol. med. Sci. **20**, 241—247 (1942).

264. — Some physiological functions of the nucleus in amoeba, investigated by micrurgical methods. Aust. J. exp. Biol. med. Sci. **21**, 251—220 (1943).

265. — Attempts to prolong the life of enucleated amoebae. Aust. J. exp. Biol. med. Sci. **22**, 179—183 (1944).

266. — The responses of enucleated amoebae to stimuli. Aust. J. exp. Biol. med. Sci. **22**, 185—196 (1944).

267. CLARK, G. M., ELLIOTT, A. M.: The induction of haploidy in *Tetrahymena pyriformis* following x-irradiation. J. Protozool. **3**, 181—188 (1956).

268. CLARK, T. B., WALLACE, F. G.: A comparative study of kinetoplast ultrastructure in the Trypanosomatidae. J. Protozool. **7**, 115—124 (1960).

269. CLAUS, C., GROBBEN, K., KÜHN, A.: Lehrbuch der Zoologie, 10. Aufl. Berlin: Julius Springer 1932.

270. CLEVELAND, L. R.: Symbiosis among animals with special reference to termites and their intestinal flagellates. Quart. Rev. Biol. **1**, 51—60 (1926).

271. — The centrioles of *Pseudotrichonympha* and their role in mitosis. Biol. Bull. **69**, 46—51 (1935).

272. — Longitudinal and transverse division in two closely related flagellates. Biol. Bull. **74**, 1—24 (1938).

273. — Origin and development of the achromatic figure. Biol. Bull. **74**, 41—55 (1938).

274. — Morphology and mitosis of *Teranympha*. Arch. Protistenk. **91**, 442—451 (1938).

275. — The whole life cycle of chromosomes and their coiling systems. Trans. Amer. Phil. Soc. **39**, 1—100 (1949).

276. — Hormone-induced sexual cycles of flagellates. I. Gametogenesis, fertilization, and meiosis in *Trichonympha*. J. Morph. **85**, 197—295 (1949).

277. — Hormone-induced sexual cycles of flagellates. II. Gametogenesis, fertilization, and one-division meiosis in *Oxymonas*. J. Morph. **86**, 185—214 (1950).

278. — Hormone-induced sexual cycles of flagellates. III. Gametogenesis, fertilization, and one-division meiosis in *Saccinobaculus*. J. Morph. **86**, 215—228 (1950).

279. — Hormone-induced sexual cycles of flagellates. IV. Meiosis after syngamy and before nuclear fusion in *Notila*. J. Morph. **87**, 317—348 (1950).

280. — Hormone-induced sexual cycles of flagellates. V. Fertilization in *Eucomonympha*. J. Morph. **87**, 349—368 (1950).

281. Cleveland, L. R.: Hormone-induced sexual cycles of flagellates. VI. Gametogenesis. fertilization, meiosis. oocysts, and gametocysts in *Leptospironympha*. J. Morph. **88**, 199—244 (1951).
282. — Hormone-induced sexual cycles of flagellates. VII. One-division meiosis and autogamy without cell division in *Urinympha*. J. Morph. **88**, 385—440 (1951).
283. — Hormone-induced sexual cycles of flagellates. VIII. Meiosis in *Rhynchonympha* in one cytoplasmic and two nuclear divisions followed by autogamy. J. Morph. **91**, 269—324 (1952).
284. — Hormone-induced sexual cycles of flagellates. IX. Haploid gametogenesis and fertilization in *Barbulanympha*. J. Morph. **93**, 381—403 (1953).
285. — Studies on chromosomes and nuclear division. I. Fusion of nucleoli independent of chromosomal homology. II. Spontaneous aberrations, homologous and non-homologous union of fragments. III. Pairing, segregation, and crossing-over. IV. Photomicrographs of living cells during meiotic divisions. Trans. Amer. Phil. Soc. **43**, 809—869 (1953).
286. — Hormone-induced sexual cycles of flagellates. X. Autogamy and endomitosis in *Barbulanympha* resulting from interruption of haploid gametogenesis. J. Morph. **95**, 189—212 (1954).
287. — Hormone-induced sexual cycles of flagellates. XI. Reorganization in the zygote of *Barbulanympha* without nuclear or cytoplasmic division. J. Morph. **95**, 213—236 (1954).
288, — Hormone-induced sexual cycles of flagellates. XII. Meiosis in *Barbulanympha* following fertilization, autogamy, and endomitosis. J. Morph. **95**, 557—620 (1954).
289. — Hormone-induced sexual cycles of flagellates. XIII. Unusual behavior of gametes and centrioles of *Barbulanympha*. J. Morph. **97**, 511—542 (1955).
290. — Hormone-induced sexual cycles of flagellates. XIV. Gametic meiosis and fertilization in *Macrospironympha*. Arch. Protistenk. **101**, 99—168 (1956).
291. — Cell division without chromatin in *Trichonympha* and *Barbulanympha*. J. Protozool. **3**, 78—83 (1956).
292.* — Brief accounts of the sexual cycles of the flagellates of *Cryptocercus*. J. Protozool. **3, 161—180 (1956).
293. — Additional observations on gametogenesis and fertilization in *Trichonympha*. J. Protozool. **4**, 164—168 (1957).
294. — Correlation between the molting period of *Cryptocercus* and sexuality in its Protozoa. J. Protozool. **4**, 168—175 (1957).
295. — Types and life cycles of centrioles of flagellates. J. Protozool. **4**, 230—241 (1957)
296. — Achromatic figure formation by multiple centrioles of *Barbulanympha*. J. Protozool. **4**, 241—248 (1957).
297. — A factual analysis of chromosomal movement in *Barbulanympha*. J. Protozool. **5**, 47—62 (1958).
298. — Movement of chromosomes in *Spirotrichonympha* to centrioles instead of the ends of central spindle. J. Protozool. **5**, 63—68 (1958).
299. — Photographs of fertilization in the smaller species of *Trichonympha*. J. Protozool. **5**, 105—115 (1958).
300. — Photographs of fertilization in *Trichonympha grandis*. J. Protozool. **5**, 115—122 (1958).
301. — Sex induced with ecdysone. Proc. nat. Acad. Sci. (Wash.) **45**, 747—753 (1959).
302. — The centrioles of *Trichonympha* from termites and their function in reproduction. J. Protozool. **7**, 326—341 (1960).
303. — Photographs of living centrioles in resting cells of *Trichonympha collaris*. Arch. Protistenk. **105**, 110—112 (1960).
304. — Photographs of fertilization in *Eucomonympha*. Arch. Protistenk. **105**, 137—148 (1961).
305. — The centrioles of *Trichomonas* and their functions in cell reproduction. Arch. Protistenk. **105**, 149—162 (1961).
306. — Pairing and segregation in haploids and diploids of *Holomastigotoides*. Arch. Protistenk. **105**, 163—172 (1961).

307. Cleveland, L. R.: Photographs of gametogenesis in living cells of *Trichonympha*. Arch. Protistenk. **105**, 497—508 (1962).
308. — Reproduction in *Deltotrichonympha*. Arch. Protistenk. **109**, 8—14 (1966).
309. — General features of the flagellate and amoeboid stages of *Deltotrichonympha operculata* and *D. nana, sp. nov.* Arch. Protistenk. **109**, 1—7 (1966).
310. — General features and reproduction in *Koruga bonita, gen. et sp. nov.* Arch. Protistenk. **109**, 18—23 (1966).
311. — General features and reproduction in *Mixotricha*. Arch. Protistenk. **109**, 26—36 (1966).
312. — Fertilization in *Deltotrichonympha*. Arch. Protistenk. **109**, 15—17 (1966).
313. — Fertilization in *Koruga*. Arch. Protistenk. **109**, 24—25 (1966).
314. — Fertilization in *Mixotricha*. Arch. Protistenk. **109**, 37—38 (1966).
315. — Nuclear division without cytokinesis followed by fusion of pronuclei in *Paranotila lata gen. et sp. nov.* J. Protozool. **13**, 132—136 (1966).
316. — Reproduction by binary and multiple fission in *Gigantomonas*. J. Protozool. **13**, 573—585 (1966).
317. — Burke, A. W.: Effects of temperature and tension on oxygen toxicity for the Protozoa of *Cryptocercus*. J. Protozool. **3**, 74—77 (1956).
318. — — Modifications induced in the sexual cycles of the Protozoa of *Cryptocercus* by change of host. J. Protozool. **7**, 240—245 (1960).
319, — — Karlson, P.: Ecdysone induced modifications in the sexual cycles of the Protozoa of *Cryptocercus*. J. Protozool. **7**, 229—239 (1960).
320. — Cleveland, B. T.: The locomotory waves of *Koruga, Deltotrichonympha* and *Mixotricha*. Arch. Protistenk. **109**, 39—63 (1966).
321. — Day, M.: Spirotrichonymphidae of *Stolotermes*. Arch. Protistenk. **103**, 1—53 (1958).
322. — Grimstone, A. V.: The fine structure of the flagellate *Mixotricha paradoxa* and its associated microorganisms. Proc. roy. Soc. Edinb. B, **159**, 668—686 (1964).
323. — Hall, D. R., Sanders, E. P., Collier, J.: The wood-feeding roach *Cryptocercus*, its Protozoa, and the symbiosis between Protozoa and roach. Mem. Amer. Acad. Arts Sci. **17**, I—X, 185—342 (1934).
324. — Nutting, W. L.: Suppression of sexual cycles and death of the Protozoa of *Cryptocercus* resulting from change of hosts during molting period. J. exp. Zool. **130**, 485—514 (1955).
325. Codreanu, R., Vavra, J.: The structure and ultrastructure of the microsporidian *Telomyxa glugeiformis* Leger and Hesse 1910, parasite of *Ephemera danica* Müll nymphs. J. Protozool. **17**, 374—384 (1970).
326. Cohen, L. W.: Diurnal intracellular differentiation in *Paramecium bursaria*. Exp. Cell Res. **36**, 398—406 (1964).
327. — The basis for the circadian rhythm of mating in *Paramecium bursaria*. Exp. Cell Res. **37**, 360—367 (1965).
328. — Siegel, R. W.: The mating-type substances of *Paramecium bursaria*. Genet. Res. **4**, 143—150 (1963).
329. Cole, J., Siegel, R. W.: A heterocaryon in *Paramecium bursaria*. Genetics **63**, 361—368 (1969).
330. Coleman, A. W.: Sexual isolation in *Pandorina morum*. J. Protozool. **6**, 249—264 (1959).
331. — Immobilization, agglutination and agar precipitin effects of antibodies to flagella of *Pandorina* mating types. J. Protozool. **10**, 141—148 (1963).
332. Colley, F. C.: Fine structure of microgametocytes and macrogametes of *Eimeria nieschulzi*. J. Protozool. **14**, 663—674 (1967).
333. — Fine structure of sporozoites of *Eimeria nieschulzi*. J. Protozool. **14**, 214—216 (1967).
334. — Fine structure of schizont and merozoites of *Eimeria nieschulzi*. J. Protozool. **15**, 374—382 (1968).
335. Collin, B.: Étude monographique sur les acinétiens. I. Recherches expérimentales sur l'étendue des variation et les facteurs tératogènes. Arch. Zool. exp. gén. 5e Sér. **8**, 421—497 (1911).
336. — Étude monographique sur les acinétiens. II. Morphologie, Physiologie, Systématique. Arch. Zool. exp. gén. **51**, 1—457 (1912).

337. COOPER, J. E.: A fast-swimming "mutant" in stock 51 of *Paramecium aurelia*, variety 4. J. Protozool. **12**, 381—384 (1965).
338. CORLISS, J. O.: Le cycle autogamique de *Tetrahymena rostrata*. C. R. Acad. Sci. (Paris) **235**, 399—402 (1952).
339.* — An illustrated key to the higher groups of the ciliated Protozoa with definition of terms. J. Protozool. **6, 265—281 (1959).
**340.* — The ciliated Protozoa. 310 pp. London: Pergamon Press 1961.
341. — L'autogamie et la sénescence du cilié hyménostome *Tetrahymena rostrata* (KAHL). Ann. Biol. **4**, 49—69 (1965).
342. COSGROVE, W. B.: Cytochemical demonstration of mitochondrial enzymes in kinetoplasts. J. Protozool. **13** (Suppl.), 16 (1966).
**343.* — The cell cycle and cell differentiation in trypanosomatids. In: CAMERON, I. L., PADILLA, G. M., ZIMMERMAN, A. M. (Eds.): Developmental aspects of the cell cycle, pp. 1—21. New York-London: Academic Press 1971.
344. CRAIG, N., GOLDSTEIN, L.: Studies on the origin of ribosomes in *Amoeba proteus*. J. Cell Biol. **40**, 622—632 (1969).
345. CULBERTSON, C. G.: Pathogenic free-living amoebas. Industr. Trop. Hth. **7**, 118—123 (1970).
346. CULBERTSON, J. R.: Physical and chemical properties of cilia isolated from *Tetrahymena pyriformis*. J. Protozool. **13**, 397—406 (1966).
347. CULLIS, C. A.: The basis of cell-to-cell transformation in *Paramecium bursaria*. I. Transfer of cytoplasmic material, J. Cell Sci. **11**, 601—609 (1972).
348. — The basis of cell-to-cell transformation in *Paramecium bursaria*. II. Investigation into the molecular nature of the transforming agent. J. Cell Sci. **11**, 611—619 (1972).
349. CZARSKA, L., GREBECKI, A.: Membrane folding and plasma-membrane ratio in the movement and shape transformation in *Amoeba proteus*. Acta Protozool. **4**, 201—239 (1966).
350. CZIHAK, G., GRELL, K. G.: Zur Determination der Zellkerne bei der Foraminifere *Rotaliella heterocaryotica*. Naturwissenschaften **47**, 211—212 (1960).
351. DAHLGREN, L.: On the ultrastructure of the gamontic nucleus and the adjacent cytoplasm of the monothalamous foraminifer *Ovammina opaca* DAHLGREN. Zool. Bidrag (Uppsala) **37**, 78—112 (1967).
352. — On the nuclear distribution of RNA and DNA and on the ultrastructure of nulcei and adjacent cytoplasm of the foraminifers *Hippocrepinella alba* HERON-ALLEN and EARLAND and *Globobulimina turgida* (BAILEY). Zool. Bidrag (Uppsala) **37**, 114—138 (1967).
353. DANFORTH, H. D., HAMMOND, D. M.: Stages of merogony in multinucleate merozoites of *Eimeria magna* PERARD 1925. J. Protozool. **19**, 454—457 (1972).
**354.* DANFORTH, W. F.: Respiratory metabolism. In: CHEN, T.-T. (Ed.): Research in Protozoology, Vol. 1, pp. 201—306. Oxford: Pergamon Press 1967.
355. DANIEL, W. A., MATTERN, C. F. T., HONIGBERG, B. M.: Fine structure of the mastigont system in *Tritrichomonas muris* GRASSI. J. Protozool. **18**, 575—592 (1971).
356. DANIELLI, J. F.: The cell-to-cell transfer of nuclei in amoebae and a comprehensive cell theory. Ann. N. Y. Acad. Sci. **78**, 675—687 (1959).
**357.* — Cellular inheritance as studied by nuclear transfer in amoebae. In: WALKER, P. M. B.: New approaches in cell biology, pp. 15—22. New York-London: Academic Press 1960.
358. DANIELS, E. W.: Electron microscopy of centrifuged *Amoeba proteus*. J. Protozool. **11**, 281—290 (1964).
359. — Origin of the Golgi system in amoebae. Z. Zellforsch. **64**, 38—51 (1964).
360. — BREYER, E.: Differences in mitochondrial fine structure during mitosis in amoebae. J. Protozool. **12**, 417—422 (1965).
361. — — Stratification within centrifuged amoeba nuclei. Z. Zellforsch. **70**, 449—460 (1966).
362. — — Ultrastructure of the giant amoeba *Pelomyxa palustris*. J. Protozool. **14**, 167—179 (1967).
363. — — KUDO, R. R.: *Pelomyxa palustris* GREEF. II. Its ultrastructure. Z. Zellforsch. **73**, 367—383 (1966).

364. Daniels, E. W., Roth, L. E.: Electronmicroscopy of mitosis in a radiosensitive giant amoeba. J. Cell Biol. **20**, 75—84 (1964).
365. Darden, Jr., W. H.: Sexual differentiation in *Volvox aureus*. J. Protozool. **13**, 239—255 (1966).
366. — Production of a male-inducing hormone by a parthenosporic *Volvox aureus*. J. Protozool. **15**, 412—414 (1968).
367. Dass, C. M. S.: Studies on the nuclear apparatus of peritrichous ciliates. I. The nuclear apparatus of *Epistylis articulata* (From). Proc. nat. Inst. Sci. India B **19**, 389—404 (1953).
368. — Studies on the nuclear apparatus of peritrichous ciliates. II. The nuclear apparatus of *Carchesium spectabile* Ehrenberg. Proc. nat. Inst. Sci. India B **20**, 174—186 (1954).
369. — Studies on the nuclear apparatus of peritrichous ciliates. III. The nuclear apparatus of *Epistylis sp.* Proc. nat. Inst. Sci. India B **20**, 703—715 (1954).
370. Deane, M. P., Kloetzel, J. K.: Differentiation and multiplication of dyskinetoplastic *Trypanosoma cruzi* in tissue culture and in the mammalian host. J. Protozool. **16**, 121—126 (1969).
371. Deason, T. R., Darden, Jr., W. H., Ely, S.: The development of sperm packets of the M5 strain of *Volvox aureus*. J. Ultrastruct. Res. **26**, 85—95 (1969).
372.* DeBruyn, P. P. H.: Theories of amoeboid movement. Quart. Rev. Biol. **22, 1—14 (1947).
373. Dedeken-Grenson, M.: The mass induction of white strains in *Euglena* as influenced by the physiological conditions. Exp. Cell Res. **18**, 185—186 (1959).
374. — Godts, A.: Descendance of *Euglena* cells isolated after various bleaching treatments. Exp. Cell Res. **19**, 376—382 (1960).
375. — Messin, S.: La continuité génétique des chloroplastes chez les euglènes. Biochim. biophys. Acta (Amst.) **27**, 145—155 (1958).
376. De Haller, G.: Morphogenèse expérimentale chez les ciliés: II. Effets d'une irradiation UV sur la différentiation des cils chez *Paramecium aurelia*. Rev. Suisse Zool. **75**, 583—588 (1968).
377.* — Interactions nucléocytoplasmiques dans la morphogenèse des ciliés. Ann. Biol. **8, 115—137 (1969).
378. Dellinger, O.: Locomotion of amoebae and allied forms. J. exp. Zool. **3**, 337—358 (1906).
379. Demar-Gervais, C.: Recherches sur le cycle biologique de *Fabrea salina* Henneguy. Faculté des Sciences de Paris. Thèse de Doctorat. (1971).
380. — Quelques précisions sur le déterminisme de la conjugaison chez *Fabrea salina* Henneguy. Protistologica **7**, 177—195 (1971).
381. Dembowski, J.: On conditioned reactions of *Paramecium caudatum* towards light. Acta Biol. exp. (Warszawa) **15**, 5—17 (1950).
382. De Terra, N.: A study of nucleo-cytoplasmic interactions during cell division in *Stentor coeruleus*. Exp. Cell Res. **21**, 41—48 (1960).
383. — Nucleocytoplasmic interactions during the differentiation of oral structures in *Stentor coeruleus*. Develop. Biol. **10**, 269—288 (1964).
384. — Differential growth in the cortical fibrillar system as the trigger for oral differentiation and cell division in *Stentor*. Exp. Cell Res. **56**, 142—153 (1969).
385.* — Cytoplasmic control over the nuclear events of cell reproduction. Int. Rev. Cytol. **25, 1—29 (1969).
386. — Evidence for cortical control of macronuclear behaviour in *Stentor*. J. Cell Physiol. **78**, 377—386 (1971).
387. Deutsch, K.: An electron microscope study of the *mu*-particles in a mate-killing strain of *Paramecium aurelia*. Dtsch. Acad. Wiss. Berlin **2**, 507—509 (1960).
388. Devauchelle, G.: Etude de l'ultrastructure de *Gregarina polymorpha* Hamm en syzygie. J. Protozool. **15**, 629—636 (1968).
389. Devi, R. V.: Autogamy in *Frontonia leucas* (Ehrbg.). J. Protozool. **8**, 277—283 (1961).
390. Devidé, Z., Geitler, L.: Die Chromosomen der Ciliaten. Chromosoma (Berl.) **3**, 110—136 (1947).
391. Dickinson, A. H.: Cytochemistry of the helices in the nucleus of *Amoeba proteus*. J. Cel Biol. **31**, 27 (1966).

392. Didier, P.: Sur les modalités d'élaboration des némadesmes chez certains ciliés hyménostomes. C. R. Soc. Biol. (Paris) **164**, 313—317 (1970).
393. — Sur l'ultrastructure et les modalités d'élaboration des némadesmes chez quelques ciliés hyménostomes péniculiens Frontoniidae. Protistologica **6**, 373—382 (1970).
394. — Contribution à l'étude comparée des ultrastructures corticales et buccales des ciliés hyménostomes péniculiens. Ann. Stn. biol. Besse-en-Chandesse **5**, 1—347 (1970).
395. Diller, W. F.: Nuclear reorganization processes in *Paramecium aurelia*, with descriptions of autogamy and "hemixis". J. Morph. **59**, 11—67 (1936).
396. — Nuclear variation in *Paramecium caudatum*. J. Morph. **66**, 605—633 (1940).
397. — Re-conjugation in *Paramecium caudatum*. J. Morph. **70**, 229—259 (1942).
398. — Nuclear behavior of *Paramecium trichium* during conjugation. J. Morph. **82**, 1—52 (1948).
399. — An abbreviated conjugation process in *Paramecium trichium*. Biol. Bull. **97**, 331—343 (1949).
400. — An extra postzygotic nuclear division in *Paramecium caudatum*. Trans. Amer. microscop. Soc. **69**, 309—316 (1950).
401. — Cytological evidence for pronuclear interchange in *Paramecium caudatum*. Trans. Amer. microscop. Soc. **69**, 317—323 (1950).
402. — Autogamy in *Paramecium polycaryum*. J. Protozool. **1**, 60—70 (1954).
403. — Studies on conjugation in *Paramecium polycaryum*. J. Protozool. **5**, 282—292 (1958).
404. — Correlation of ciliary and nuclear development in the life cycle of *Euplotes*. J. Protozool. **13**, 43—54 (1966).
405. — Kounaris, D.: Description of a Zoochlorella bearing form of *Euplotes*, *E. daidaleos* n. sp. (Ciliophora, Hypotrichida). Biol. Bull. **131**, 437—445 (1966).
406. Dingle, A. D.: Control of flagellum number in *Naegleria*. J. Cell Sci. **7**, 463—481 (1970).
407. — Fulton, C.: Development of the flagellar apparatus of *Naegleria*. J. Cell Biol. **31**, 43—54 (1966).
408. Dippell, R. V.: Reproduction of surface structure in *Paramecium*. 2nd Intern. Conf. Protozool. Excerpta Medica **91**, 65 (1965).
409. — The development of basal bodies in *Paramecium*. Proc. nat. Acad. Sci. (Wash.) **61**, 461—468 (1968).
410. Dobell, C. C.: The life history and chromosome cycle of *Aggregata eberthi*. Parasitology **17**, 1—136 (1925).
411. Dodge, J. D.: Chromosome structure in the Dinophyceae. I. The spiral chromonema. Arch. Mikrobiol. **45**, 46—57 (1963).
412. — The nucleus and nuclear division in the Dinophyceae. Arch. Protistenk. **106**, 442 to 452 (1963).
413. — Chromosome structure in the Dinophyceae. II. Cytochemical studies. Arch. Mikrobiol. **48**, 66—80 (1964).
414. — Nuclear division in the dinoflagellate *Gonyaulax tamarensis*. J. gen. Microbiol. **36**, 269—276 (1964).
415. — Fine structure of the dinoflagellate *Aureodinium pigmentosum* gen. et sp. nov. Brit. phycol. Bull. **3**, 327—336 (1967).
416. — The fine structure of chloroplasts and pyrenoids in some marine dinoflagellates. J. Cell Sci. **3**, 41—48 (1968).
417.* — A review of the fine structure of algal eyespots. Brit. phycol. J. **4, 199—210 (1969).
418. — The ultrastructure of *Chroomonas mesostigmatica* Butcher (Cryptophyceae). Arch. Mikrobiol. **69**, 266—280 (1969).
419. — A dinoflagellate with both a mesocaryotic and an eucaryotic nucleus. I. Fine structure of the nuclei. Protoplasma **73**, 145—157 (1971).
420.* — Fine structure of the Pyrrophyta. Botan. Rev. **37, 481—508 (1971).
421. — Crawford, R. M.: Observations on the fine structure of the eyespot and associated organelles in the dinoflagellate *Glenodinium foliaceum*. J. Cell Sci. **5**, 479—493 (1966).
422. — — Fine structure of the dinoflagellate *Amphidinium carteri* Hulbert. Protistologica **4**, 231—242 (1968).
423. — — Observations on the fine structure of the eyespot and associated organelles in the dinoflagellate *Glenodinium foliaceum*. J. Cell Sci. **5**, 479—493 (1969).

424. Dodge, J. D., Crawford, R. M.: A survey of thecal fine structure in the Dinophyceae. Bot. J. Linn. Soc. **63**, 53—67 (1970).

425. — — The morphology and fine structure of *Ceratium hirundinella* (Dinophyceae). J. Phycol. **6**, 137—149 (1970).

426. — — A fine structural survey of dinoflagellate pyrenoids and food-reserves. Botan. J. Linn. Soc. **64**, 105—115 (1971).

427. — — Fine structure of the dinoflagellate *Oxyrrhis marina*. I. The general structure of the cell. Protistologica **7**, 295—304 (1971).

428. — — Fine structure of the dinoflagellate *Oxyrrhis marina*. II. The flagellar system. Protistologica **7**, 399—409 (1971).

429. Dogiel, V. A.: Die Geschlechtsprozesse bei Infusorien (speziell bei den Ophryoscoleciden) neue Tatsachen und theoretische Erwägungen. Arch. Protistenk. **50**, 283—442 (1925).

430. — Die sog. „Konkrementvacuole" der Infusorien, als eine Statocyste betrachtet. Arch. Protistenk. **68**, 319—348 (1929).

431. Doroszewski, M.: The response of *Dileptus* and its fragments to the puncture. Acta Protozool. **1**, 313—319 (1963).

432. — The response of the ciliate *Dileptus* and its fragments to the water shake. Acta Biol. exp. (Warszawa) **23**, 3—10 (1963).

433. — The response of *Dileptus cygnus* to the bisection. Acta Protozool. **3**, 175—182 (1965).

434. — Responses of the ciliate *Dileptus* to mechanical stimuli. Acta Protozool. **7**, 353—362 (1970).

435. — The responses to bisections of dividing *Dileptus cygnus*. Acta Protozool. **10**, 109—113 (1972).

436. Downs, L. E.: Mating types and their determination in *Stylonychia putrina*. J. Protozool. **6**, 285—292 (1959).

437. Dragesco, J.: Sur la biologie du *Zoothamnium pelagicum* (de Plessis). Bull. Soc. Zool. Fr. **73**, 130—134 (1948).

438. — La capture des proies chez *Dileptus gigas* (cilié holotriche). Bull. Soc. Zool. Fr. **73**, 62—65 (1948).

439. — Les ciliés mésopsammiques littoraux (systématique, morphologie, écologie). Trav. St. Biol. Roscoff **12**, 356 (1960).

440. — Capture et ingestion des prois chez les infusoires ciliés. Bull. Biol. Fr. Belg. **96**, 123—167 (1962).

441. — Révision du genre *Dileptus*, Dujardin 1871 (Ciliata Holotricha) (Systématique, Cytologie, Biologie). Bull. Biol. Fr. Belg. **97**, 103—145 (1963).

442. — Compléments à la connaissance des ciliés mésopsammiques de Roscoff. I. — Holotriches. Cah. Biol. mar. **4**, 91—119 (1963).

443. — Compléments à la connaissance des ciliés mésopsammiques de Roscoff, II. — Hétérotriches, III. — Hypotriches. Cah. Biol. mar. **4**, 251—275 (1963).

444. — Ciliés mésopsammiques d'Afrique noire. Cah. Biol. mar. **6**, 357—399 (1965).

445. — Étude cytologique de quelques flagellés mésopsammiques. Cah. Biol. mar. **6**, 83—115 (1965).

446. — Observations sur quelques ciliés libres. Arch. Protistenk. **109**, 155—206 (1966).

447. — Armature fibrillaire interne chez *Harmannula acrobates* Entz (cilié holotriche gymnostome). Protistologica **3**, (1) 61—66 (1967).

448. — Auderset, G., Baumann, M.: Observations sur la structure et la genèse des trichocystes toxiques et des protrichocystes de *Dileptus* (ciliés holotriches). Protistologica **1**, (2) 81—95 (1965).

449. — Hollande, A.: Sur la présence de trichocystes fibreux chez les péridiniens; leur homologie avec les trichocystes fusiformes des ciliés. C. R. Acad. Sci. (Paris) **260**, 2073—2076 (1965).

450. — Raikov, I.: L'appareil nucléaire, la division et quelques stades de la conjugaison *de Tracheloraphis margaritatus* (Kahl) et *T. caudatus sp. nov.* (Ciliata Holotricha). Arch. Protistenk. **109**, 99—113 (1966).

*451. DROOP, M. R.: Algae and invertebrates in symbiosis, pp. 171—199. In: Symbiotic associations. Thirteenth Symposium of the Society f. gen. Microbiology, London. Cambridge: University Press 1963.

452. DRYL, S.: Chemotactic and toxic effects of lower alcohols on *Paramecium caudatum*. Acta Biol. exp. **19**, 95—104 (1959).

453. — Chemotaxis in *Paramecium caudatum* as adaptive response of organism to its environment. Acta Biol. exp. (Warszawa) **21**, 75—83 (1961).

454. — Contributions to mechanism of chemotactic response in *Paramecium caudatum*. Anim. Behav. **11**, 393—396 (1963).

455. — Oblique orientation of *Paramecium caudatum* in electric field. Acta Protozool. **1**, 193—199 (1963).

*456. — Response of ciliate Protozoa to external stimuli. Acta Protozool. **7**, 325—333 (1970).

*457. — GREBECKI, A.: Progress in the study of excitation and response in ciliates. Protoplasma **62**, 255—284 (1966).

458. — PREER, J. R.: The possible mechanism of resistance of *Paramecium aurelia* to *kappa* toxin from killer stock 7, syngen 2 during autogamy, conjugation and cell division. J. Protozool. **14** (Suppl.), 33—34 (1967).

459. DU BUY, H. G., MATTERN, C. F. T., RIDLEY, F. L.: Isolation and characterization of DNA from kinetoplasts of *Leishmania enrietti*. Science **147**, 754—756 (1965).

460. DUMONT, J. N.: Observations on the fine structure of the ciliate *Dileptus anser*. J. Protozool. **8**, 392—402 (1961).

461. DUNCAN, D., EADES, J., JULIAN, S. R., MICKS, D.: Electron microscope observations on malarial oocysts (*Plasmodium cathemerium*). J. Protozool. **7**, 18—26 (1960).

462. DUNNEBACKE, T. H., SCHUSTER, F. L.: Infectious agent from a free-living soil amoeba, *Naegleria gruberi*. Science **174**, 516—518 (1971).

463. DUPY-BLANC, J.: Etude par cytophotométrie des teneurs en ADN nucléaire chez trois espèces de Paramécies, chez différentes variété d'une même espèce et chez différents types sexuels d'une même variété. Protistologica **5**, 297—308 (1969).

464. — Etude cytophotométrique des teneurs en ADN des micronucleus de *Paramecium caudatum* au cours de la conjugaison et pendant la différentiation des „Anlage" en macronucleus. Protistologica **5**, 239—248 (1969).

465. DURCHON, M., VIVIER, E.: Influence des sécrétions endocrines sur le cycle des Grégarines chez les Néreidiens (Annelides polychètes). Ann. endocr. (Paris) **25**, 43—48 (1964).

466. EBERHARDT, R.: Untersuchungen zur Morphogenese von *Blepharisma* und *Spirostomum*. Arch. Protistenk. **106**, 241—341 (1962).

467. EBERSOLD, W. T.: Crossing over in *Chlamydomonas reinhardi*. Amer. J. Bot. **43**, 408—410 (1956).

*468. — Biochemical genetics. In: LEWIN, R. A.: Physiology and biochemistry of algae, pp. 731—739. New York-London: Academic Press 1962.

469. — Heterozygous diploid strains of *Chlamydomonas reinhardi*. Genetics **48**, 888 (1963).

470. — *Chlamydomonas reinhardi:* heterozygous diploid strains. Science **157**, 447—448 (1967).

471. — LEVINE, R. P.: A genetic analysis of linkage group I of *Chlamydomonas reinhardi*. Z. Vererbungsl. **90**, 74—82 (1959).

472. — — LEVINE, E. E., OLMSTED, M. A.: Linkage maps in *Chlamydomonas reinhardi*. Genetics **47**, 531—543 (1962).

473. ECKERT, R., NAITOH, Y.: Passive electrical properties of *Paramecium* and problems of ciliary coordination. J. gen. Physiol. **55**, 467—483 (1970).

474. — — Bioelectric control of locomotion in the ciliates. J. Protozool. **19**, 237—243 (1972).

475. EDELMAN, M., COWAN, C. A., EPSTEIN, H. T., SCHIFF, J. A.: Studies of chloroplast development in *Euglena*. VIII. Chloroplast-associated DNA. Proc. nat. Acad. Sci. (Wash.) **52**, 1214—1219 (1964).

476. — SCHIFF, J. A., EPSTEIN, H. T.: Studies of chloroplast development in *Euglena*. XII. Two types of satellite DNA. J. molec. Biol. **11**, 769—774 (1965).

477. EGELHAAF, A.: Cytologisch-entwicklungsphysiologische Untersuchungen zur Konjugation von *Paramecium bursaria* FOCKE. Arch. Protistenk. **100**, 447—514 (1955).

478. Ehret, C.F., DeHaller, G.: Origin, development, and maturation of organelles and organelle systems of the cell surface of *Paramecium*. J. Ultrastruct. Res. Suppl. **6**, 1—42 (1963).
479. — Powers, E.L.: The cell surface of *Paramecium*. Int. Rev. Cytol. 8, 97—133 (1959).
480. — Savage, N., Alblinger, J.: Patterns of segregation of structural elements during cell division. Z. Zellforsch. **64**, 129—139 (1964).
481. Eide, A.: Subunit composition of the macronucleus in the ciliate *Ichthyophthirius multifiliis*. Arbok Univ. Bergen Mat.-Naturv. Serie **7**, 1—16 (1968).
482.* Elliott, A.M.: Biology of *Tetrahymena*. Ann. Rev. Microbiol. **13, 79—95 (1959).
483.* — A quarter century exploring *Tetrahymena*. J. Protozool. **6, 1—7 (1959).
**484.* — (Ed.): Biology of *Tetrahymena*. New York: Appleton, Century Crofts, Inc. 1971.
485. — Bak, I.J.: The contractile vacuole and related structures in *Tetrahymena pyriformis*. J. Protozool. **11**, 250—261 (1964).
486. — Clark, G.M.: The induction of haploidy in *Tetrahymena pyriformis* following X-irradiation. J. Protozool. **3**, 181—188 (1956).
487. — — Further cytological studies on haploid and diploid strains of *Tetrahymena pyriformis* with special reference to univalent spindles. Cytologia **22**, 355—359 (1957).
488. — — Genetic studies of the pyridoxine mutant in variety two of *Tetrahymena pyriformis*. J. Protozool. **5**, 235—240 (1958).
489. — — Genetic studies of the serine mutant in variety nine of *Tetrahymena pyriformis*. J. Protozool. **5**, 240—246 (1958).
490. — Clemmons, G.L.: An ultrastructural study of ingestion and digestion in *Tetrahymena pyriformis*. J. Protozool. **13**, 311—323 (1966).
491. — Kennedy, J.R.: The morphology and breeding system of variety 9, *Tetrahymena pyriformis*. Trans. Amer. microscop. Soc. **81**, 300—308 (1962).
492. — Nanney, D.L.: Conjugation in *Tetrahymena*. Science **116**, 33—34 (1952).
**493.* Ephrussi, B.: Nucleo-cytoplasmic relations in microorganisms. Oxford: Clarendon Press 1953.
494. Epstein, H.T., Schiff, J.A.: Studies of chloroplast development in *Euglena*. IV. Electron and fluorescence microscopy of the proplastid and its development into a mature chloroplast. J. Protozool. 8, 427—437 (1961).
495. Ettl, H., Manton, I.: Die feinere Struktur von *Pedinomonas minor* Korschikoff. Nova Hedwigia 8, 421—451 (1964).
496. Evenson, D.P., Prescott, D.M.: RNA metabolism in the macronucleus of *Euplotes eurystomus* during the cell cycle. Exp. Cell Res. **61**, 71—78 (1970).
497. — — The disruption of DNA synthesis in *Euplotes* by heat shock. Exp. Cell Res. **63**, 245—252 (1970).
498. Eversole, R.A.: Biochemical mutants of *Chlamydomonas reinhardi*. Amer. J. Bot. **43**, 404—407 (1956).
499. — Tatum, E.L.: Chemical alteration of crossing over frequency in *Chlamydomonas*. Proc. nat. Acad. Sci. (Wash.) **42**, 68—72 (1956).
500. Falk, H., Wunderlich, F., Franke, W.W.: Microtubular structures in macronuclei of synchronously dividing *Tetrahymena pyriformis*. J. Protozool. **15**, 776—780 (1968).
501. Fauré-Fremiet, E.: Morphogénèse de bipartition chez *Urocentrum turbo* (cilié holotriche). J. Embryol. exp. Morphol. **2**, 227—238 (1954).
502.* — Les problèmes de la différenciation chez les protistes. Bull. Soc. zool. Fr. **79, 311—329 (1954).
503. — Le macronucleus hétéromère de quelques ciliés. J. Protozool. **4**, 7—17 (1957).
504.* — Cils vibratiles et flagelles. Biol. Rev. **36, 464—536 (1961).
505. — Microtubules et mécanismes morphopoiétiques. Ann. Biol. **9**, 1—61 (1970).
506. — Favard, P., Carasso, N.: Étude au microscope électronique des ultrastructures d'*Epistylis anastatica* (Cilié Péritriche). J. Microscopie **1**, 287—312 (1962).
507. — Rouiller, C.: Myonèmes et cinétodesmes chez les ciliés du genre *Stentor*. Bull. Micr. appl. 8, 117—119 (1958).
508. Favard, P., Carasso, N.: Mise en évidence d'un processus de micropinocytose interne au niveau des vacuoles digestives d'*Epistylis anastatica* (Cilié Péritriches). J. Microscopie **2**, 495—498 (1963).

509. Favard, P., Carasso, N.: Étude de la pinocytose au niveau des vacuoles digestives de ciliés Péritriches. J. Microscopie **3**, 671—696 (1964).
**510.* Fawcett, D. W.: Cilia and flagella. In: Brachet, J., Mirsky, A. E.: The cell, Vol. 2, pp. 217—298. New York-London: Academic Press 1961.
511. Febvre, J.: Le myonème d'Acanthaire: Essai d'interprétation ultrastructurale et cinétique. Protistologica **7**, 379—391 (1971).
512. Febvre-Chevalier, C.: Constitution ultrastructurale de *Globigerina bulloides* d' Orbigny, 1826 (Rhizopoda, Foraminifera). Protistologica **7**, 311—324 (1971).
513. Feldherr, C. M.: The nuclear annuli as pathways for nucleocytoplasmic exchanges. J. Cell Biol. **14**, 65—72 (1962).
514. — Marshall, J. M.: The use of colloidal gold for studies of intracellular exchanges in the amoeba *Chaos chaos*. J. Cell Biol. **12**, 640—645 (1962).
515. Finck, H., Holtzer, H.: Attempts to detect myosin and actin in cilia and flagella. Exp. Cell Res. **23**, 251—257 (1961).
516. Finley, H. E.: Sexual differentiation in *Vorticella microstoma*. J. exp. Zool. **81**, 209—229 (1939).
517. — The conjugation of *Vorticella microstoma*. Trans. Amer. microscop. Soc. **62**, 97—121 (1943).
518. — Sexual differentiation in peritrichous ciliates. J. Morph. **91**, 569—606 (1952).
519. — Brown, C. A., Daniel, W. A.: Electron microscopy of the ectoplasm and infraciliature of *Spirostomum ambiguum*. J. Protozool. **11**, 264—280 (1964).
520. Fjerdingstad, E. J.: Ultrastructure of the collar of the choanoflagellate *Codonosiga botrytis* Ehrenb. Z. Zellforsch. **54**, 499—510 (1961).
521. Flavell, R. A., Jones, I. G.: DNA from isolated pellicles of *Tetrahymena*. J. Cell Sci. **9**, 719—726 (1971).
522. Flickinger, C. J.: The fine structure of the nucleoli of normal and actinomycin D treated *Amoeba proteus*. J. Ultrastruct. Res. **23**, 260—272 (1968).
523. — The effects of enucleation on the cytoplasmic membranes of *Amoeba proteus*. J. Cell Biol. **37**, 300—315 (1968).
524. — The development of Golgi complexes and their dependence upon the nucleus in amoebae. J. Cell Biol. **43**, 250—262 (1969).
525. — The fine structure of the nuclear envelope in amoebae: alterations following nuclear transplantation. Exp. Cell Res. **60**, 225—236 (1970).
526. — Coss, R. A.: The role of the nucleus in the formation and maintenance of the contractile vacuole in *Amoeba proteus*. Exp. Cell Res. **62**, 326—330 (1970).
527. Förster, H., Wiese, L.: Untersuchungen zur Kopulationsfähigkeit von *Chlamydomonas eugametos*. Z. Naturforsch. **9**b, 470—471 (1954).
528. — — Gamonwirkungen bei *Chlamydomonas eugametos*. Z. Naturforsch. **9**b, 548—550 (1954).
529. — — Gamonwirkung bei *Chlamydomonas reinhardi*. Z. Naturforsch. **10**b, 91 (1955).
530. — — Braunitzer, G.: Über das agglutinierend wirkende Gynogamon von *Chlamydomonas eugametos*. Z. Naturforsch. **11**b, 315—317 (1956).
531. Föyn, B.: Über die Kernverhältnisse der Foraminifere *Myxotheca arenilega* Schaudinn. Arch. Protistenk. **87**, 272—295 (1936).
532. — Zur Kenntnis der asexuellen Fortpflanzung und der Entwicklung der Gamonten von *Discorbina vilardeboana* D'Orbigny. Bergens Museum Arbok, Naturvid. r. 5 (1937).
533.* Franceschi, T.: Ciclo vitale e sistemi ereditari di *Paramecium aurelia*. Riv. Sci. Nat. Natura **55, 1—48 (1964).
534. Frankel, J.: Morphogenesis in *Glaucoma chattoni*. J. Protozool. **7**, 362—376 (1960).
535. — Effects of localized damage on morphogenesis and cell division in a ciliate, *Glaucoma chattoni*. J. exp. Zool. **143**, 175—193 (1960).
536. — Cortical morphogenesis and synchronization in *Tetrahymena pyriformis* GL. Exp. Cell Res. **35**, 349—360 (1964).
537. — Morphogenesis and division in chains of *Tetrahymena pyriformis* GL. J. Protozool. **11**, 514—526 (1964).
538. — The effect of nucleic acid antagonists on cell division and oral organelle development in *Tetrahymena pyriformis*. J. exp. Zool. **159**, 113—148 (1965).

539. Frankel, J.: Studies on the maintenance of oral development in *Tetrahymena pyriformis* Gl-C. I. An analysis of mechanism of resorption of developing oral structures. J. exp. Zool. **164**, 435—460 (1967).

540. — Studies on the maintenance of oral development in *Tetrahymena pyriformis* Gl-C. II. The relationship of protein synthesis to cell division and oral organelle development. Cell Biol. **34**, 841—858 (1967).

541. — Participation of the undulating membrane in the formation of oral replacement primordia in *Tetrahymena pyriformis*. J. Protozool. **16**, 26—35 (1959).

542. — The relationship of protein synthesis to cell division and oral development in synchronized *Tetrahymena pyriformis* GL-C. An analysis employing cycloheximide. J. Cell Physiol. **74**, 135—148 (1969).

543. — An analysis of the recovery of *Tetrahymena* from effects of cycloheximide. J. Cell Physiol. **76**, 55—64 (1970).

544. — The synchronization of oral development without cell division in *Tetrahymena pyriformis* GL-C. J. exp. Zool. **173**, 79—100 (1970).

545. French, J. W.: Trial and error learning in *Paramecium*. J. exp. Physiol. **26**, 609—613 (1940).

546. Freudenthal, H. D.: *Symbiodinium gen. nov.* and *Symbiodinium microadriaticum sp. nov.*, a Zooxanthella: Taxonomy, life cycle, and morphology. J. Protozool. **9**, 45—62 (1962).

**547.* Frey-Wyssling, H.: Die submikroskopische Struktur des Cytoplasmas. In: Heilbrunn, L. V., Weber, F.: Protoplasmatologia. Wien. Springer 1955.

548. Friedberg, I., Goldberg, I., Ohad, I.: A prolamellar body-like structure in *Chlamydomonas reinhardi*. J. Cell Biol. **50**, 268—276 (1971).

549. — Colwin, A. L., Colwin, L. H.: Fine-structural aspects of fertilization in *Chlamydomonas reinhardi*. J. Cell Sci. **3**, 115—128 (1968).

550. Friend, D. S.: The fine structure of *Giardia muris*. J. Cell Biol. **29**, 317—332 (1966).

551. Friz, C. T.: Studies of the nucleic acids of the amoeba *Chaos chaos*. C. R. Trav. Lab. Carlsberg. **35**, 287—339 (1966).

552. Fuge, H.: Electron microscopic studies on the intra-flagellar structure of trypanosomes. J. Protozool. **16**, 460—466 (1969).

553. Fulton, C., Dingle, A. D.: Basal bodies, but not centrioles, in *Naegleria*. J. Cell Biol. **51**, 826—836 (1971).

**554.* Fulton, J. D.: Metabolism and pathogenic mechanisms of parasitic Protozoa. In: Chen, T.-T. (Ed.): Research on Protozoology, Vol. 3, pp. 389—504. New York: Pergamon Press 1969.

555. Furssenko, A.: Lebenscyclus und Morphologie von *Zoothamnium arbuscula* Ehrenberg (Infusoria Peritricha). Arch. Protistenk. **67**, 376—500 (1929).

556. Gall, J. G.: Macronuclear duplication in the ciliated protozoan *Euplotes*. J. biophys. biochem. Cytol. **5**, 295—308 (1959).

557. Gantt, E., Edwards, M. R., Provascoli, L.: Chloroplast structure of the Cryptophyceae. Evidence for phycobiliproteins within intrathylacoidal spaces. J. Cell Biol. **48**, 280—290 (1971).

558.* Garnham, P. C. C.: Locomotion in the parasitic Protozoa. Biol. Rev. **41, 561—586 (1966).

559. — Baker, J. R., Bird, R. G.: Fine structure of *Lankesterella garnhami*. J. Protozool. **9**, 107—114 (1962).

560. — Bird, R. G., Baker, J. R.: Electron microscope studies of motile stages of malaria parasites. I. The fine structure of the sporozoites of *Haemamoeba* (*Plasmodium*) *falcipara*. Trans. roy. Soc. trop. Med. Hyg. **54**, 274—278 (1960).

561. — — — Electron microscope studies of motile stages of malaria parasites. II. The fine structure of the sporozoites of *Laverania* (*Plasmodium*) *gallinacea*. Trans. roy. Soc. trop. Med. Hyg. **55**, 98—102 (1961).

562. — — — Electron microscope studies of motile stages of malaria parasites. III. The ookinete of *Haemamoeba* and *Plasmodium*. Trans. roy. Soc. trop. Med. Hyg. **56**, 116—120 (1962).

563. GARNHAM, P. C. C., BIRD, R. G., BAKER, J. R.: Electron microscope studies of motile stages of malaria parasites. IV. The fine structure of the sporozoites of four species of *Plasmodium*. Trans. roy. Soc. trop. Med. Hyg. **57**, 27—31 (1963).

564. — — — Electron microscope studies of motile stages of malaria parasites. V. Exflagellation in *Plasmodium, Hepatocystis* and *Leucocytozoon*. Trans. roy. Soc. Med. Hyg. **61**, 58—68 (1967).

**565.* GEITLER, L.: Endomitose und endomitotische Polyploidisierung. In: HEILBRUNN, L. V., WEBER, F.: Protoplasmatologia. Wien: Springer 1953.

566. GELEI, G. v.: Neue Beiträge zum Bau und zu der Funktion des Exkretionssystems von *Paramecium*. Arch. Protistenk. **92**, 384—400 (1939).

567. GÉNERMONT, J.: Le déterminisme génétique de la vitesse de multiplication chez *Paramecium aurelia* syng. 1. Protistologica **2**, (4), 45—51 (1966).

568. — Recherches sur les modifications durables et le déterminisme génétique de certains caractères quantitatifs chez *Paramecium aurelia*. Thèse à la Fac. Sci. Univ. Paris, 1966.

569. — Quelques charactéristiques des populations de *Paramecium aurelia* adaptée au chlorure de calcium. Protistologica **5**, 101—108 (1969).

570.* — Le rôle génétique des mitochondries et des chloroplastes chez les êtres unicellulaires. Ann. Biol. **10, 53—76 (1971).

571. — DUPY-BLANC, J.: Recherches biométriques sur la sexualité des Paramécies (*P. aurelia* et *P. caudatum*). Protistologica **7**, 197—212 (1971).

572. GERISCH, G.: Die Zelldifferenzierung bei *Pleodorina californica* SHAW und die Organisation der Phytomonadinenkolonien. Arch. Protistenk. **104**, 292—358 (1959).

573.* — Die zellulären Schleimpilze als Objekte der Entwicklungsphysiologie. Ber. dtsch. bot. Ges. **75, 82—89 (1962).

574. — *Dictyostelium discoideum* (Acrasina). Aggregation und Bildung des Sporophors. Begleitveröff. z. d. Film E631, 1964.

575. GIBBONS, I. R.: Studies on the protein components of cilia from *Tetrahymena pyriformis*. Proc. nat. Acad. Sci. (Wash.) **50**, 1002—1010 (1963).

**576.* — The organization of cilia and flagella. In: ALLEN, J. M. (Ed.): Molecular organization and biological function, pp. 211—237. New York-Evanston-London: Harper & Row Publ. 1967.

577. — GRIMSTONE, A. V.: On flagellar structure in certain flagellates. J. biophys. biochem. Cytol. **7**, 697—716 (1960).

578. GIBBS, S. P.: The fine structure of *Euglena gracilis* with special reference to the chloroplasts and pyrenoids. J. Ultrastruct. Res. **4**, 127—148 (1960).

579. — The ultrastructure of the pyrenoids of algae, exclusive of the green algae. J. Ultrastruct. Res. **7**, 247—261 (1962).

580. — The ultrastructure of the pyrenoids of green algae. J. Ultrastruct. Res. **7**, 262—272 (1962).

581. — The ultrastructure of the chloroplasts of algae. J. Ultrastruct. Res. **7**, 418—435 (1962).

582. — Nuclear envelope — chloroplast relationships in algae. J. Cell Biol. **14**, 433—444 (1962).

583. — Chloroplast development in *Ochromonas danica*. J. Cell Biol. **15**, 343—361 (1962).

584. GIBOR, A., GRANICK, S.: The plastid system of normal and bleached *Euglena gracilis*. J. Protozool. **9**, 327—334 (1962).

585.* — — Plastids and mitochondria: Inheritable systems. Science **145, 890—898 (1964).

586. GIBSON, I., BEALE, G. H.: Genic basis of the mate-killer trait in *Paramecium aurelia*, stock 540. Genet. Res. **2**, 82—91 (1961).

587. — — The mechanism whereby the genes M_1 and M_2 in *Paramecium aurelia*, stock 540, control growth of the mate-killer (*mu*) particles. Genet. Res. **3**, 24—50 (1962).

588. GIESBRECHT, P.: Vergleichende Untersuchungen an den Chromosomen des Dinoflagellaten *Amphidinium elegans* und denen der Bakterien. Zbl. Bakt. II. Abt. **187**, 452—498 (1962).

589. — Über das Ordnungsprinzip in den Chromosomen von Dinoflagellaten und Bakterien. Zbl. Bakt. II. Abt. **196**, 516—519 (1965).

590. Giese, A.: Mating types in *Paramecium multimicronucleatum*. J. Protozool. **4**, 120—124 (1957).

**591.* — Effects of radiation upon Protozoa. In: Chen, T.-T. (Ed.): Research in Protozoology, Vol. 2, pp. 267—356. Oxford: Pergamon Press 1967.

592. Gillham, N. W.: The nature of exceptions to the pattern of uniparental inheritance for high level streptomycin resistance in *Chlamydomonas reinhardi*. Genetics **48**, 431—440 (1963).

593. — Induction of chromosomal and nonchromosomal mutations in *Chlamydomonas reinhardi* with N-methyl-nitro-N-nitrosoguanidine. Genetics **52**, 529—537 (1965).

594. — Linkage and recombination between nonchromosomal mutations in *Chlamydomonas reinhardi*. Proc. nat. Acad. Sci. (Wash.) **54**, 1560—1567 (1965).

595.* — Uniparental inheritance in *Chlamydomonas reinhardi*. Amer. Naturalist **103, 355—367 (1969).

596. — Fifer, W.: Recombination of nonchromosomal mutations: a three-point cross in the green alga *Chlamydomonas reinhardi*. Science **162**, 683—684 (1968).

597. — Levine, R. P.: Studies on the origin of streptomycin resistant mutants in *Chlamydomonas reinhardi*. Genetics **47**, 1463—1474 (1962).

598. Gillies, C., Hanson, E. D.: A new species of *Leptomonas* parasitizing the macronucleus of *Paramecium trichium*. J. Protozool. **10**, 467—473 (1963).

599. — — Morphogenesis of *Paramecium trichium*. Acta Protozool. **6**, 13—32 (1968).

600. Gilman, L. C.: Mating types in diverse races of *Paramecium caudatum*. Biol. Bull. **80**, 384—402 (1941).

601. — Distribution of the varieties of *Paramecium caudatum*. J. Protozool. **3**, (Suppl.) 4 (1956).

602. — European varieties of *Paramecium caudatum*. J. Protozool. **5**, (Suppl.) 17 (1958).

603. Gliddon, R.: Ciliary organelles and associated fibre systems in *Euplotes eurystomus* (Ciliata, Hypotrichida). J. Cell Sci. **1**, 439—448 (1966).

**604.* Goldacre, R. J.: On the mechanism and control of ameboid movement. In: Allen, R. D., Kamiya, N.: Primitive motile systems in cell biology, pp. 237—255. New York-London: Academic Press 1964.

605. Goldberg, I., Ohad, I.: Biogenesis of chloroplast membranes. IV. Lipid and pigment changes during synthesis of chloroplast membranes in a mutant of *Chlamydomonas reinhardi* y-1. J. Cell Biol. **44**, 563—571 (1970).

606. — — Biogenesis of chloroplast membranes. V. A radioautographic study of membrane growth in a mutant of *Chlamydcmonas reinhardi* y-1. J. Cell Biol. **44**, 572—591 (1970)

607. Goldstein, M.: Speciation and mating behavior in *Eudorina*. J. Protozool. **11**, 317—344 (1964).

608. — Colony differentiation in *Eudorina*. Canad. J. Bot. **45**, 1591—1596 (1967).

609. Goldstein, L., and W. Plaut: Direct evidence for nuclear synthesis of cytoplasmic ribose nucleic acid. Proc. nat. Acad. Sci. (Wash.) **41**, 874—879 (1955).

610. Golikowa, M.N.: Der Aufbau des Kernapparates und die Verteilung der Nukleinsäuren und Proteine bei *Nyctotherus cordiformis* Stein. Arch. Protistenk. **108**, 191—216 (1965).

611. Goodenough, U. W., Levine, R. P.: Chloroplast structure and function in ac-20, a mutant strain of *Chlamydomonas reinhardi*. III. Chloroplast ribosomes and membrane organization. J. Cell Biol. **44**, 547—562 (1970).

612. — — The effects of inhibitors of RNA and protein synthesis on the recovery of chloroplast ribosomes, membrane organization, and photosynthetic electron transport in the ac-20 strain of *Chlamydomonas reinhardi*. J. Cell Biol. **50**, 50—62 (1971).

613. — Staehelin, L. A.: Structural differentiation of stacked and unstacked chloroplast membranes. Freeze-etch electron microscopy of wild-type and mutant strains of *Chlamydomonas*. J. Cell Biol. **48**, 594—619 (1971).

**614.* Goodwin, T. W.: The plastid pigments of flagellates. In: Hutner, S. H.: Biochemistry and physiology of Protozoa, Vol. 3, pp. 319—337. New York-London: Academic Press 1964.

615. Gorovsky, M. A.: Studies on nuclear structure and function in *Tetrahymena pyriformis*. II. Isolation of macro- and micronuclei. J. Cell Biol. **47**, 619—630 (1970).

616. GOROVSKY, M. A.: Studies on nuclear structure and function in *Tetrahymena pyriformis.* III. Comparison of the histones of macro- and micronuclei by quantitative polyacrylamide gel electrophoresis. J. Cell Biol. **47**, 631—636 (1970).

617. — WOODART, J.: Studies on nuclear structure and function in *Tetrahymena pyriformis.* I. RNA synthesis in macro- and micronuclei. J. Cell Biol. **42**, 673—682 (1969).

618. GÖSSEL, I.: Über das Aktionsspektrum der Phototaxis chlorophyllfreier Euglenen und über die Absorption des Augenflecks. Arch. Mikrobiol. **27**, 288—305 (1957).

619. GOWANS, C. S.: Some genetic investigations in *Chlamydomonas eugametos.* Z. Vererbungsl. **91**, 63—73 (1960).

620. GRAIN, J.: Étude cytologique de quelques cilés holotriches endocommensaux des ruminants et des équidés. Protistologica **2** (1), 5—52 (1966).

621. — Les systèmes fibrillaires chez *Stentor igneus* EHRENBERG et *Spirostomum ambiguum* EHRENBERG. Protistologica **4**, 27—36 (1968).

622.* — Le cinétosome et ses dérivés chez les Ciliés. Ann. Biol. **8, 53—97 (1969).

623. — GOLINSKA, K.: Structure et ultrastructure de *Dileptus cygnus* CLAPAREDE et LACHMANN 1859, cilié holotriche gymnostome. Protistologica **5**, 269—296 (1969).

624. GRASSÉ, P. P., DRAGESCO, J.: L'ultrastructure du chromosome des péridiniens et ses conséquences génétiques. C. R. Acad. Sci. (Paris) **245**, 2447—2452 (1957).

625. — HOLLANDE, A.: Cytologie et mitose des *Pseudotrichonympha* GRASSI et Foà 1911. Ann. Sci. nat. Zool. **13**, 237—246 (1951).

626. — — Recherches sur les symbiontes des termites Hodotermitidae nord-africains. I. Le cycle évolutif des flagellés du genre *Kirbynia.* II. Les Rhizonymphidae fam. nov. III. *Polymastigoides,* nouveau genre de Trichomonadidae. Ann. Sci. nat. Zool. **13**, 1—32 (1951).

627. — — Les flagellés des genres *Holomastigotoides* et *Rostronympha.* Structure et cycle de spiralisation des chromosomes chez *Holomastigotoides psammotermitidis.* Ann. Sci. nat. Zool. **5**, 749—792 (1963).

628. — — CACHON, J., CACHON-ENJUMET, M.: Nouvelle interprétation de l'ultrastructure du chromosome de certains péridiniens (*Prorocentrum, Gymnodinium, Amphidinium, Plectodinium* et Xanthelles d'Anémones). C. R. Acad. Sci. (Paris) **260**, 1743—1747 (1965).

629. — — — — Interprétation de quelques aspects infrastructuraux des chromosomes de péridiniens en division. C. R. Acad. Sci. (Paris) **260**, 6975—6978 (1965).

630. — MUGARD, H.: Les organites mucifères et la formation du kyste chez *Ophryoglena mucifera.* Progr. Protozool. (Prag) 417—418 (1963).

631. GREBECKI, A.: L'enregistrement microphotographique des courants d'eau autour d'un cilié. Experientia (Basel) **17**, 93—94 (1961).

632. — Phénomènes électrocinétiques dans le galvanotropisme de *Paramecium caudatum.* Bull. Biol. Fr. Belg. **96**, 723—754 (1962).

633. — Rebroussement ciliaire et galvanotaxie chez *Paramecium caudatum.* Acta Protozool. **1**, 99—112 (1963).

634. — Point isoélectrique superficiel et quelques réactions locomotrices chez *Paramecium caudatum.* Protoplasma **56**, 80—88 (1963).

635. — Electrobiologie d'ingestion des colorants par le cytostome de *Paramecium caudatum.* Protoplasma **56**, 89—98 (1963).

636. — Galvanotaxie transversale et oblique chez les ciliés. Acta Protozool. **1**, 91—98 (1963).

637. — Calcium substitution in staining the cilia. Acta Protozool. **2**, 375—377 (1964).

638.* — Modern lines in the study of amoeboid movement. Acta Protozool. **2, 379—402 (1964).

639. — Rôle des ions K^+ et Ca^{2+} dans l'excitabilité de la cellule protozoaire. I. Equilibrement des ions antagonistes. Acta Protozool. **2**, 69—79 (1964).

640. — Gradient stomato-caudal d'excitabilité des ciliés. Acta Protozool. **3**, 79—101 (1965).

641. — Rôle of Ca^{2+} ions in the excitability of protozoan cell. Decalcification, recalcification, and the ciliary reversal in *Paramecium caudatum.* Acta Protozool. **3**, 275—289 (1965).

642. — KUZNICKI, L.: Immobilization of *Paramecium caudatum* in the chloral hydrate solutions. Bull. Acad. pol. Sci. Cl. II. **9**, 459—462 (1961).

643. — — The influence of external pH on the toxicity of inorganic ions for *Paramecium caudatum.* Acta Protozool. **1**, 157—164 (1963).

644. GREBECKI, A., KUZNICKI, L., MIKOLAJCZYK, E.: Right spiralling induced in *Paramecium* by Ni-ions and the hydrodynamics of the spiral movement. Acta Protozool. **4**, 389—408 (1967).

645. — — — Some observations on the inversion of spiralling in *Paramecium caudatum*. Acta Protozool. **4**, 383—388 (1967).

646. — MIKOLAJCZYK, E.: Ciliary reversal and re-normalization in *Paramecium caudatum* immobilized by Ni-ions. Acta Protozool. **5**, 297—303 (1968).

647. GRELL, K. G.: Untersuchungen an Schizogregarinen. I. *Lipocystis polyspora n. g. n. sp.*, eine neue Schizogregarine aus dem Fettkörper von *Panorpa communis* L. Arch. Protistenk. **91**, 526—545 (1938).

648. — Der Kernphasenwechsel von *Stylocephalus* (*Stylorhynchus*) *longicollis* F. STEIN. (Ein Beitrag zur Frage der Chromosomenreduktion der Gregarinen). Arch. Protistenk. **94**, 161—200 (1940).

649. — Die Entwicklung der Makronucleusanlage im Exkonjuganten von *Ephelota gemmipara* R. HERTWIG. Biol. Zbl. **58**, 289—312 (1949).

650. — Der Generationswechsel des parasitischen Suktors *Tachyblaston ephelotensis* MARTIN. Z. Parasistenk. **14**, 499—534 (1950).

*651. — Der Kerndualismus der Ciliaten und Suktorien. Naturwissenschaften **37**, 347—356 (1950).

652. — Die Paarungsreaktion von *Stylonychia mytilus* MÜLLER. Z. Naturforsch. **6**b, 45—47 (1951).

653. — Die Konjugation von *Ephelota gemmipara* R. HERTWIG. Arch. Protistenk. **98**, 287 bis 326 (1953).

654. — Die Struktur des Makronucleus von *Tokophrya*. Arch. Protistenk. **98**, 466—468 (1953).

655. — Die Chromosomen von *Aulacantha scolymantha* HAECKEL. Arch. Protistenk. **99**, 1—54 (1953).

*656. — Der Stand unserer Kenntnisse über den Bau der Protistenkerne. Verh. dtsch. zool. Ges. Freiburg 1952, 212—215 (1953).

657. — Entwicklung und Geschlechtsbestimmung von *Eucoccidium dinophili*. Arch. Protistenk. **99**, 156—186 (1953).

658. — Zur Sexualität der Foraminiferen. Naturwissenschaften **41**, 44—45 (1954).

659. — Der Generationswechsel der polythalamen Foraminifere *Rotaliella heterocaryotica*. Arch. Protistenk. **100**, 268—286 (1954).

660. — Röntgeninduzierte Chromosomenmutationen bei *Eucoccidium dinophili*. Arch. Protistenk. **100**, 323—330 (1955).

661. — Der Kerndualismus der Foraminifere *Glabratella sulcata*. Z. Naturforsch. **11**b, 367—368 (1956).

662. — Über die Elimination somatischer Kerne bei heterokaryotischen Foraminiferen. Z. Naturforsch. **11**b, 759—761 (1956).

663. — Untersuchungen über die Fortpflanzung und Sexualität der Foraminiferen. I. *Rotaliella roscoffensis*. Arch. Protistenk. **102**, 147—164 (1957).

664. — Untersuchungen über die Fortpflanzung und Sexualität der Foraminiferen. II. *Rubratella intermedia*. Arch. Protistenk. **102**, 291—308 (1958).

665. — Untersuchungen über die Fortpflanzung und Sexualität der Foraminiferen. III. *Glabratella sulcata*. Arch. Protistenk. **102**, 449—472 (1958).

*666. — Studien zum Differenzierungsproblem an Foraminiferen. Naturwissenschaften **45**, 25—32 (1958).

667. — Untersuchungen über die Fortpflanzung und Sexualität der Foraminiferen. IV. *Patellina corrugata*. Arch. Protistenk. **104**, 211—235 (1959).

668. — Nachweis der sexuellen Differenzierung bei *Patellina corrugata* durch Teilbildanalyse eines Films. Z. Naturforsch. **15**b, 270 (1960).

669. — Reziproke Infektion mit Eucoccidien aus verschiedenen Wirten. Naturwissenschaften **47**, 47—48 (1960).

670. — Über den „Nebenkörper" von *Paramoeba eilhardi* SCHAUDINN. Arch. Protistenk. **105**, 303—312 (1961).

*671. GRELL, K. G.: Morphologie und Fortpflanzung der Protozoen (einschließlich Entwicklungsphysiologie und Genetik). Fortschr. Zool. **14**, 1—85 (1962).

672. — Entwicklung und Geschlechtsdifferenzierung einer neuen Foraminifere. Naturwissenschaften **49**, 241 (1962).

*673. — The protozoan nucleus. In: BRACHET, J., MIRSKY, A. E.: The cell, Vol. 6, pp. 1—79. New York-London: Academic Press 1964.

674. — Amöben der Familie Stereomyxidae. Arch. Protistenk. **109**, 147—154 (1966).

*675. — Sexual reproduction in Protozoa. In: CHEN, T.-T.: Research in Protozoology, Vol. 2, pp. 149—213. Oxford: Pergamon Press 1967.

676. — Eibildung und Furchung von *Trichoplax adhaerens* F. E. SCHULZE (Placozoa). Z. Morph. Tiere **73**, 297—314 (1972).

677. — BENWITZ, G.: Die Zellhülle von *Paramoeba eilhardi* SCHAUDINN. Z. Naturforsch. **21** b, 600—601 (1966).

678. — — Ultrastruktur mariner Amöben. I. *Paramoeba eilhardi* SCHAUDINN. Arch. Protistenk. **112**, 119—137 (1970).

679. — — Ultrastruktur mariner Amöben. II. *Stereomyxa ramosa*. Arch. Protistenk. **113**, 51—67 (1971).

680. — — Ultrastruktur mariner Amöben. III. *Stereomyxa angulosa*. Arch. Protistenk. **113**, 68—79 (1971).

681. — — Die Ultrastruktur von *Trichoplax adhaerens* F. E. SCHULZE. Cytobiol. **4**, 216—240 (1971).

682. — RUTHMANN, A.: Über die Karyologie des Radiolars *Aulacantha scolymantha* und die Feinstruktur seiner Chromosomen. Chromosoma (Berl.) **15**, 185—211 (1964).

683. — SCHWALBACH, G.: Elektronenmikroskopische Untersuchungen an den Chromosomen der Dinoflagellaten. Chromosoma (Berl.) **17**, 230—245 (1965).

684. — WOHLFARTH-BOTTERMANN, K. E.: Licht- und elektronenmikroskopische Untersuchungen an dem Dinoflagellaten *Amphidinium elegans*. Z. Zellforsch. **47**, 7—17 (1957).

685. GREUET, C.: Structure fine de l'ocelle d'*Erythroposis pavillardi* HERTWIG, péridinien Warnowiidae Lindemann. C. R. Acad. Sci. (Paris) **261**, 1904—1907 (1965).

686. — Organisation ultrastructurale du tentacule d'*Erythropsis pavillardi* KOFOID et SWEZY Péridinien Warnowiidae Lindemann. Protistologica **3**, 335—345 (1967).

687. — Organisation ultrastructurale de l'ocelle de deux péridiniens Warnoviidae, *Erythropsis pavillardi* KOFOID et SWEZY et *Warnowia pulchra* SCHILLER. Protistologica **4**, 209—230 (1968).

688. — Etude morphologique et ultrastructurale du trophonte d'*Erythropsis pavillardi* KOFOID et SWEZY. Protistologica **5**, 481—504 (1969).

689. — Etude ultrastructurale et évolution des cnidocystes de *Nematodinium*, péridinien Warnowiidae Lindemann. Protistologica **7**, 345—355 (1971).

690. GRIFFIN, J. L.: Movement, fine structure and fusion of pseudopods of an enclosed amoeba, *Difflugiella sp.* J. Cell Sci. **10**, 563—583 (1972).

691. — ALLEN, R. D.: The movement of particles attached to the surface of amoebae in relation to current theories of amoeboid movement. Exp. Cell Res. **20**, 619—622 (1960).

692. GRIM, J. N.: Fine structure of the surface and infraciliature of *Gastrostyla steinii*. J. Protozool. **19**, 113—126 (1972).

693. GRIMES, G. W.: Cortical structure in nondividing and cortical morphogenesis in dividing *Oxytricha fallax*. J. Protozool. **19**, 428—445 (1972).

694. — PREER, J. R.: Further observations on the correlation between *kappa* and phage-like particles in *Paramecium*. Genet. Res. **18**, 115—116 (1971).

695. GRIMSTONE, A. V.: Cytoplasmic membranes and the nuclear membrane in the flagellate *Trichonympha*. J. biophys. biochem. Cytol. **6**, 369—377 (1959).

*696. — Fine structure and morphogenesis in Protozoa. Biol. Rev. **36**, 97—150 (1961).

697. — The fine structure of some polymastigote flagellates. Proc. Linn. Soc. London **174**, 49—52 (1963).

698. — CLEVELAND, L. R.: The fine structure and function of the contractile axostyles of certain flagellates. J. Cell Biol. **24**, 387—400 (1965).

699. GRIMSTONE, A.V., GIBBONS, I.R.: The fine structure of the centriolar apparatus and associated structures in the complex flagellates *Trichonympha* and *Pseudotrichonympha*. Phil. Trans. B, **250**, 215—242 (1966).

700. GRUCHY, D.F.: The breeding system and distribution of *Tetrahymena pyriformis*. J. Protozool. **2**, 178—185 (1955).

701. GUILCHER, Y.: Contribution à l'étude des ciliés gemmipares, chonotriches et tentaculifères. Ann. Sci. nat. Zool. (sér. 11) **13**, 33—132 (1951).

702. GUILLARD, R.R.L.: A mutant of *Chlamydomonas moewusii* lacking contractile vacuoles. J. Protozool. **7**, 262—268 (1960).

703. HABEREY, M., WOHLFARTH-BOTTERMANN, K.E., STOCKEM, W.: Pinocytose und Bewegung von Amöben. VI. Kinematographische Untersuchungen über das Bewegungsverhalten der Zelloberfläche von *Amoeba proteus*. Cytobiol. **1**, 70—84 (1969).

*704. HALL, R.P.: Nutrition and growth of Protozoa. In: CHEN, T.-T. (Ed.): Research in Protozoology, Vol. 1, pp. 337—404. Oxford: Pergamon Press 1967.

*705. HALLDAL, P.: Phototaxis in Protozoa. In: HUTNER, S.H.: Biochemistry and physiology of Protozoa, Vol. 3, pp. 227—295. New York-London: Academic Press 1964.

706. HALLER, G. DE: Altération expérimentale de la stomatogenèse chez *Paramecium aurelia*. Rev. Suisse Zool. **71**, 592—600 (1964).

707. — Sur l'hérédité de characteristiques morphologiques du cortex chez *Paramecium aurelia*. Arch. Zool. exp. gén. **105**, 169—178 (1965).

708. — ROUILLER, C.: La structure fine de *Chlorogonium elongatum*. I. Etude systématique au microscope électronique. J. Protozool. **8**, 452—462 (1961).

709. HÄMMERLING, J.: Über die Geschlechtsverhältnisse von *Acetabularia mediterranea* und *Acetabularia wettsteinii*. Arch. Protistenk. **83**, 57—97 (1934).

*710. — Nucleo-cytoplasmic interactions in *Acetabularia* and other cells. Ann. Rev. Plant Phys. **14**, 67—92 (1963).

711. HAMMOND, D.M., ERNST, J.V., MINER, M.L.: The development of first generation schizonts of *Eimeria bovis*. J. Protozool. **13**, 559—564 (1966).

712. — SCHOLTYSECK, E., CHOBOTAR, B.: Fine structural study of the microgametogenesis of *Eimeria auburnensis*. Z. Parasitenk. **33**, 65—84 (1969).

713. — — MINER, M.L.: The fine structure of microgametocytes of *Eimeria perforans*, *E. bovis* and *E. auburnensis*. J. Parasit. **53**, 235—247 (1967).

714. HANSON, E.D.: Morphogenesis and regeneration of oral structures in *Paramecium aurelia*: An analysis of intracellular development. J. exp. Zool. **150**, 45—68 (1962).

715. — Evolution of the cell from primordial living systems. Quart. Rev. Biol. **41**, 1—12 (1966).

716. — GILLIES, C., KANEDA, M.: Oral structure development and nuclear behavior during conjugation in *Paramecium aurelia*. J. Protozool. **16**, 197—204 (1969).

717. — KANEDA, M.: Evidence for sequential gene action within the cell cycle of *Paramecium*. Genetics **60**, 793—805 (1968).

718. HARRIS, W.F.: The arrangement of the axonemal microtubules and links of *Echinosphaerium nucleofilum*. J. Cell Biol. **46**, 183—187 (1970).

719. HARTMANN, M.: Über experimentelle Unsterblichkeit von Protozoen-Individuen. Ersatz der Fortpflanzung von *Amoeba proteus* durch fortgesetzte Regeneration. Zool. Jb. Physiol. **45**, 973—987 (1928).

*720. — Allgemeine Biologie, 4. Aufl., 940 S. Stuttgart: Gustav Fischer 1953.

*721. — Sexualität, 2. Aufl., 463 S. Stuttgart: Gustav Fischer 1956.

722. HARTSHORNE, J.N.: The function of the eyespot in *Chlamydomonas*. New Phytologist **52**, 292—297 (1953).

723. HASCALL, G.K., RUDZINSKA, M.A.: Metamorphosis in *Tokophrya infusionum*; an electron-microscope study. J. Protozool. **17**, 311—323 (1970).

*724. HAUPT, W.: Die Orientierung der Pflanzen zum Licht. Naturw. Rundschau **18**, 261—267 (1965).

*725. — Phototaxis in plants. Int. Rev. Cytol. **19**, 267—299 (1966).

726. HAUSCHKA, T.: Life-history and chromosome cycle of the coccidian *Adelina deronis*. J. Morph. **73**, 529—581 (1943).

727. HAUSER, M.: Elektronenmikroskopische Untersuchung an dem Suktor *Paracineta limbata* MAUPAS. Z. Zellforsch. **106**, 584—614 (1970).

728. — The intranuclear mitosis of the ciliates *Paracineta limbata* and *Ichthyophtirius multifiliis*. I. Electron microscope observations on pre-metaphase stages. Chromosoma (Berl.) **36**, 158—175 (1972).

729. HAUSMANN, K.: Cytologische Studien an Trichocysten. III. Die Feinstruktur ausgeschiedener Mucocysten von *Tetrahymena pyriformis*. Cytobiol. **5**, 468—474 (1972).

730. — STOCKEM, W.: Pinocytose und Bewegung von Amöben. *VIII*. Mitteilung: Endocytose und intrazelluläre Verdauung bei *Hyalodiscus simplex*. Cytobiol. **5**, 281—300 (1972).

731. — — WOHLFARTH-BOTTERMANN, K. E.: Cytologische Studien an Trichocysten. I. Die Feinstruktur der gestreckten Spindeltrichocysten von *Paramecium caudatum*. Cytobiol. **5**, 208—227 (1972).

732. — — — Cytologische Studien an Trichocysten. II. Die Feinstruktur ruhender und gehemmter Spindeltrichocysten von *Paramecium caudatum*. Cytobiol. **5**, 228—246 (1972).

733. — — — Pinocytose und Bewegung von Amöben. VII. Mitteilung: Quantitative Untersuchungen zum Membran-Turnover bei *Hyalodiscus simplix*. Z. Zellforsch. **127**, 270—286 (1972).

*734. HAYES, W.: The genetics of Bacteria and their viruses. Oxford: Blackwell Scientific Publ. 2. ed. 1968.

735. HECKMANN, K.: Paarungssystem und genabhängige Paarungstypdifferenzierung bei dem hypotrichen Ciliaten *Euplotes vannus* O. F. MÜLLER. Arch. Protistenk. **106**, 393—421 (1963).

736. — Experimentelle Untersuchungen an *Euplotes crassus*. I. Paarungssystem, Konjugation und Determination der Paarungstypen. Z. Vererbungsl. **95**, 114—124 (1964).

737. — Totale Konjugation bei *Urostyla hologama* n. sp. Arch. Protistenk. **108**, 55—62 (1965).

738. — Age dependent intraclonal conjugation in *Euglotes crassus*. J. exp. Zool. **165**, 269 to 278 (1967).

739. — FRANKEL, J.: Genic control of cortical pattern in *Euplotes*. J. exp. Zool. **168**, 11—38 (1968).

740. — PREER, JR., J. R., STRAETLING, W. H.: Cytoplasmic particles in the killers of *Euplotes minuta* and their relationship to the killer substance. J. Protozool. **14**, 360—363 (1967).

741. — SIEGEL, R. W.: Evidence for the induction of mating-type substances by cell to cell contacts. Exp. Cell Res. **36**, 688—691 (1964).

*742. HEDLEY, R. H.: The biology of Foraminifera. Intern. Rev. gen. exp. Zool. **1**, 1—45 (1964).

743. — BERTAUD, W. S.: Electron-microscopic observations of *Gromia oviformis* (Sarcodina). J. Protozool. **9**, 79—87 (1962).

744. — PARRY, D. M., WAKEFIELD, J. St. J.: Fine structure of *Shepheardella taeniformis* (Foraminifera, Protozoa). J. roy. Microsc. Soc. **87**, 445—456 (1967).

745. — WAKEFIELD, J. St. J.: Fine structure of *Gromia oviformis* (Rhizopodea, Protozoa). Bull. Brit. Mus. nat. Hist. (Zool.) **18**, 1—89 (1969).

746. HELLER, G.: Elektronenmikroskopische Untersuchungen an *Aggregata eberthi* aus dem Spiraldarm von *Sepia officinalis* (Sporozoa, Coccidia). I. Die Fernstrukturen der Merozoiten, Mikrogameten und Sporen. Z. Parasitenk. **33**, 44—64 (1969).

747. — Elektronenmikroskopische Untersuchungen an *Aggregata eberthi* aus dem Spiraldarm von *Sepia officinalis* (Sporozoa, Coccidia). II. Die Entwicklung der Mikrogameten. Z. Parasitenk. **33**, 183—193 (1970).

748. — Die Feinstrukturen des peripheren Zellbereichs und ihre mögliche Bedeutung für die Nahrungsaufnahme bei den Makrogamonten von *Aggregata eberthi* (Sporozoa, Coccidia). Z. Parasitenk. **34**, 251—257 (1970).

749. — Elektronenmikroskopische Untersuchungen zur Schizogonie in den sog. kleinen Schizonten von *Eimeria stiedae* (Sporozoa, Coccidia). Protistologica **7**, 461—470 (1971).

750. HELLER, G.: Elektronenmikroskopische Untersuchungen zur Bildung und Struktur von Conoid, Rhoptrien und Mikronemen bei *Eimeria stiedae* (Sporozoa, Coccidia). Protistologica 8, 43—51 (1972).

751. — SCHOLTYSECK, E.: Feinstrukturuntersuchungen zur Merozoitenbildung bei *Eimeria stiedae* (Sporozoa, Coccidia). Protistologica 7, 451—460 (1971).

752. HEMLEBEN, CHR.: Ultrastrukturen bei kalkschaligen Foraminiferen. Naturwissenschaften **56**, 534-538 (1969).

752a — Zur Morphogenese planktischer Foraminiferen. Zitteliana **1**, 91—133 (1969).

753. HENNERÉ, E.: Etude cytologique des premiers stades du dévelopement d'une coccidie: *Myriosporides amphiglenae*. J. Protozool. **14**, 27—39 (1967).

754. HERTWIG, R.: Über Kernteilung, Richtungskörperbildung und Befruchtung von *Actinosphaerium eichhorni*. Abhandl. k. bayer. Akad. Wiss. II. Cl. **19**, 633—734 (1898).

*755. HILL, D. L.: The biochemistry and physiology of *Tetrahymena*. 230 pp. New York-London: Academic Press 1972

756. HILL, G. C., ANDERSON, W.: Effects of acriflavine on the mitochondria and kinetoplast of *Crithidia fasciculata*. J. Cell Biol. **41**, 547—561 (1969).

757. — — Electron transport systems and mitochondrial DNA in Trypanosomatidae: A review. Exp. Parasitol. **28**, 356—380 (1970).

758. — BROWN, C. A., CLARK, M. V.: Structure and function of mitochondria in *Crithidia fasciculata*. J. Protozool. **15**, 102—109 (1968).

759. — HUTNER, S. H.: Effect of trypanocidal drugs on terminal respiration of *Crithidia fasciculata*. Exp. Parasitol. **22**, 207—212 (1968).

760. HITCHCOCK, L.: Color sensitivity of the amoeba revisited. J. Protozool. 8, 322—324 (1961).

761. HIWATASHI, K.: Inheritance of mating types in variety 12 of *Paramecium caudatum*. Sci. Resp. Res. Inst. Tohoku Univ. Biol. **24**, 119—129 (1958).

762. — Induction of conjugation by ethylenediamine tetraacetic acid (EDTA) in *Paramecium caudatum*. Sci. Rep. Res. Inst. Tohoku Univ., Biol. **25**, 81—90 (1959).

763. — Analysis of the change of mating type during vegetative reproduction in *Paramecium caudatum*. Jap. J. Genet. **35**, 213—221 (1960).

764. — Inheritance of difference in the life feature of *Paramecium caudatum*, syngen 12. Bull. biol. Stn. Asamushi Tohoku Univ. **9**, 157—159 (1960).

765. — Locality of mating reactivity on the surface of *Paramecium caudatum*. Sci. Rep. Res. Inst. Tohoku Univ. Biol. **27**, 93—99 (1961).

766. — Mating type inheritance in *Paramecium caudatum*, syngen 3. Genetics **50**, 225—256 (1964).

767. — Serotype inheritance and serotypic alleles in *Paramecium caudatum*. Genetics **57**, 711—717 (1967).

768. — Determination and inheritance of mating type in *Paramecium caudatum*. Genetics **58**, 373—386 (1968).

769. — Genetic and epigenetic control of mating type in *Paramecium caudatum*. Jap. J. Genet. **44**, 383—387 (1969).

*770. — *Paramecium*. In: Fertilization, Vol. 2, pp. 255—293. New York-London: Academic Press 1969.

771. HOFFMAN, E. J.: The nucleic acids of basal bodies isolated from *Tetrahymena pyriformis*. J. Cell Biol. **25**, 217—228 (1965).

772. HOFFMANN-BERLING, H.: Geißelmodelle und Adenosintriphosphat (ATP). Biochem. biophys. Acta (Amst.) **16**, 146—154 (1955).

*773. — Physiologie der Bewegungen und Teilungsbewegungen tierischer Zellen. Fortschr. Zool. **11**, 142—207 (1958).

774. HOLBERTON, D. V., PRESTON, T. M.: Arrays of thick filaments in ATP-activated *Amoeba* model cells. Exp. Cell Res. **62**, 473—476 (1970).

775. HOLLANDE, A.: Infrastructure du complexe rostral et origine du fuseau chez *Staurojoenina caulleryi*. C. R. Acad. Sci. (Paris) **266**, 1283—1286 (1968).

*775a. HOLLANDE, A.: Le déroulement de la cryptomitose et les modalités de la ségrégation des chromatides dans quelques groupes de protozoaires I. Trichomonadinia HOLLANDE et CARRUETTE-VALENTIN, J. Dinoflagellida BÜTSCHLI. Radiolaria J. MÜLLER. Foraminiferida ZBORZEWSKI 1834. Telosporea SCHAUDINN. Ann. Biol. **9**, 427—466 (1972).

776. — CACHON, J., CACHON, M.: La dinomitose atractophorienne à fuseau endonucléaire chez les Radiolaires Thalassophysidae. Son homologie avec la mitose des Foraminifères et avec celle des Levures. C. R. Acad. Sci. (Paris) **269**, 179—182 (1969).

777. — — — La signification de la membrane capsulaire des Radiolaires et ses rapports avec le plasmalemme et les membranes du réticulum endoplasmique. Affinités entre Radiolaires, Heliozoaires et Péridiniens. Protistologica **6**, 311—318 (1970).

778. — — — Les modalités de l'enkystement présporogénétique chez les Acanthaires. Protistologica **1**, (2), 91—104 (1965).

779. — — — L'infrastructure des axopodes chez les Radiolaires Sphaerellaires Périaxoplastidiés. C. R. Acad. Sci. (Paris) **261**, 1388—1391 (1965).

780. — — — VALENTIN, J.: Infrastructure des axopodes et organisation générale de *Sticholonche zanclea* HERTWIG (Radiolaire, Sticholonchidae). Protistologica **3**, 155—164 (1967).

781. — CACHON-ENJUMET, M.: La polyploidie du noyau végétatif des Radiolaires. C. R. Acad. Sci. (Paris) **248**, 2641—2643 (1959).

782. — CARRUETTE-VALENTIN, J.: La lignée des Pyrsonymphines et les caractères infrastructuraux communs aux genres *Opisthomitus, Oxymonas, Saccinobacculus, Pyrsonympha* et *Streblomastix*. C. R. Acad. Sci. (Paris) **270**, 1587—1590 (1970).

783. — — Appariement chromosomique et complexes synaptonématiques dans le noyau en cours de dépolyploidisation chez *Pyrsonympha flagellata*: Le cycle évolutif des Pyrsonymphines symbiontes de *Reticulitermes lucifugus*. C. R. Acad. Sci. (Paris) **270**, 2550—2553 (1970).

784. — — Interprétation générale des structures rostrales des Hypermastigines et modalités de la pleuromitose chez les Flagellés du genre *Trichonympha*. C. R. Acad. Sci. (Paris) **270**, 1476—1479 (1970).

785. — — Les Atractophores, l'induction du fuseau et la division cellulaire chez les Hypermastigines. Etude infrastructurale et révision systématique des Trichonymphines et des Spirotrichonymphines. Protistologica **7**, 5—100 (1971).

786. — — Le problème du centrosome et la cryptopleuromitose atractophorienne chez *Lophomonas striata*. Protistologica 8, 267—278 (1972).

787. — ENJUMET, M.: Contribution à l'étude biologique des Sphaerocollides (Radiolaires collodaires et Radiolaires polycyttaires) et de leurs parasites. Ann. Sci. nat. Zool. **15**, 99—183 (1953).

788. — — Parasites et cycle évolutif des Radiolaires et des Acanthaires. Bull. Stn. Agric. Pêche Castiglione Nouv. sér. **7**, 153—176 (1955).

789. — — Cytologie, évolution et systématique des Sphaeroidés (Radiolaires). Arch. Mus. nat. Hist. nat. Paris **7**, (7e Sér.) 1—134 (1960).

790. — GARREAU DE LOUBRESSE, N.: Compléments à l'étude morphologique de *Deltotrichonympha turkestanica* BERNSTEIN. Ann. Sci. Nat. Zool. **12**, 815—820 (1963).

791. — GHARAGOZLOU, I.: Morphologie infrastructurale de *Pillotina calotermitidis* nov. gen. nov. sp., Spirochaetale de l'intestin de *Calotermes praecox*. C. R. Acad. Sci. (Paris) **265**, 1309—1312 (1967).

792. — VALENTIN, J.: Interprétation des structures dites centriolaires chez les Hypermastigines symbiontes des Termites et du *Cryptocercus*. C. R. Acad. Sci. (Paris) **264**, 1868—1871 (1967).

793. — — Morphologie et infrastructure du genre *Barbulanympha*, Hypermastigine symbiontique de *Cryptocercus punctulatus* SCUDDER. Protistologica **3**, 257—267 (1967).

794. — — Rélations entre cinétosomes, atractophores et complexe fibrillaire axostyloparabasal chez les Hypermastigines du genre *Barbulanympha*. C. R. Acad. Sci. (Paris) **264**, 3020—3022 (1967).

795. — — Données critiques sur la pleuromitose et affinités entre Trichomonadines et Joeniides. C. R. Acad. Sci. (Paris) **267**, 1383—1386 (1968).

796. Hollande, A., Valentin, J.: Infrastructure des centromères de la pleuromitose chez les Hypermastigines. C. R. Acad. Sci. (Paris) **266**, 367—370 (1968).
797. — — Infrastructure des centromères et déroulement de la pleuromitose chez les Hypermastigines. C. R. Acad. Sci. (Paris) **266**, 367—370 (1968).
798. — — Morphologie infrastructurale de *Trichomonas* (*Trichomitopsis* Kofoid et Swezy 1919), *termopsidis*, parasite intestinal de *Termopsis angusticollis* Walk. Critique de la notion de centrosome chez les Polymastigines. Protistologica **4**, 127—139 (1968).
799. — — Appareil de Golgi, pinocytose, lysosomes, mitochondries, bactéries symbiontiques, atractophores et pleuromitose chez les Hypermastigines du genre *Joenia*. Affinités entre Joeniides et Trichomonadines. Protistologica **5**, 39—86 (1969).
800. — — La cinétide et ses dépendances dans le genre *Macrotrichomonas*. Considérations générales sur la sous-famille des Macrotrichomonadinae. Protistologica **5**, 335—344 (1969).
801. Holt, P. A., Chapman, G. B.: The fine structure of the cyst wall of the ciliated protozoon *Didinium nasutum*. J. Protozool. **18**, 604—614 (1971).
802.* Holter, H.: Pinocytosis. Intern. Rev. Cytol. **8, 481—504 (1959).
803. — Membrane in correlation with pinocytosis. In: Seno, S., Cowdry, E. V.: Intracellular membranous structure, pp. 451—465. Okayama: Japan Soc. Cell Biol. 1965.
804. Holwill, M. E. J.: The motion of *Strigomonas oncopelti*. J. exp. Biol. **42**, 125—137 (1965).
805. — The motion of *Euglena viridis*: The role of flagella. J. exp. Biol. **44**, 579—588 (1966).
806.* — Physical aspects of flagellar movement. Physiol. Rev. **46, 696—785 (1966).
**807.* Holz, Jr., G. G.: Nutrition and metabolism of ciliates. In: Hutner, S. H.: Biochemistry and physiology of Protozoa, Vol. 3, pp. 199—233. New York-London: Academic Press 1964.
808. Honigberg, B. M.: Evolutionary and systematic relationships in the flagellate order Trichomonadina Kirby. J. Protozool. **10**, 6—10 (1963).
809. — Berett, C. J.: Lightmicroscopic observations on structure and division of *Histomonas meleagridis* Smith. J. Protozool. **18**, 687—697 (1971).
810.* — and Committee: A revised classification of the phylum Protozoa. J. Protozool. **11, 7—20 (1964).
811. — Daniel, W. A., Mattern, C. F. T.: Fine structure of *Trichomitus batrachorum* (Petry). J. Protozool. **19**, 446—453 (1972).
812. — Mattern, C. F. T., Daniel, W. A.: Structure of *Pentatrichomonas hominis* Davaine as revealed by electron microscopy. J. Protozool. **15**, 419—430 (1968).
813. — — — Fine structure of the mastigont system in *Tritrichomonas foetus* Riedmüller. J. Protozool. **18**, 183—198 (1971).
814. Hoober, J. K., Blobel, G.: Characterization of the chloroplastic and cytoplasmic ribosomes of *Chlamydomonas reinhardi*. J. molec. Biol. **41**, 121—138 (1969).
815. Hopkins, J. M.: Subsidiary components of the flagella of *Chlamydomonas reinhardi*. J. Cell Sci. **7**, 823—839 (1970).
816. — Watson, M. R.: The cilia of *Tetrahymena pyriformis*. Isolation of ciliary segments. Exp. Cell Res. **32**, 187—189 (1963).
817. Hovasse, R.: Quelques faits nouveaux concernant les trichocystes et nématocystes des *Polykrikos* (Dinoflagellés). Arch. Zool. exp. gén. **102**, 189 (1963).
**818.* — Trichocystes, Corps trichocystoides, Cnidocystes et Colloblastes. In: Protoplasmatologia (Handbuch der Protoplasmaforschung), Vol. 3, F, pp. 1—57. Wien-New York: Springer 1965.
819. — Ultrastructure comparée des axopodes chez les héliozoaires des genres *Actinosphaerium*, *Actinophrys* et *Raphidiophrys*. Protistologica **1** (1), 81—88 (1965).
820.* — Trichocystes ou corps trichocystoides et nematocystes chez les Protistes. Ann. Station Biol. Besse **4, 245—269 (1969).
821. — Mignot, J.-P., Joyon, L.: Nouvelles observations sur les trichocystes des Cryptomonadines et les "R bodies" des particules *kappa* de *Paramecium aurelia* killer. Protistologica **3**, 241—255 (1967).
822.* — — — Baudoin, J.: Etude comparée des dispositifs servant à la fixation chez les Protistes. Ann. Biol. **11, 1—61 (1972).

823. Hufnagel, L. A.: Structural and chemical observations on pellicles isolated from paramecia. J. Cell Biol. **27**, 46 A (1965).
824. — Properties of DNA associated with raffinose-isolated pellicles of *Paramecium aurelia.* J. Cell Sci. **5**, 561—573 (1969).
825. — Cortical ultrastructure of *Paramecium aurelia.* Studies on isolated pellicles. J. Cell Biol. **40**, 779—801 (1969).
826. Hull, R. W.: Studies on suctorian Protozoa: The mechanism of prey adherence. J. Protozool. **8**, 343—350 (1961).
827. — Studies on suctorian Protozoa: The mechanism of ingestion of prey cytoplasm. J. Protozool. **8**, 351—359 (1961).
828.* Hutner, S. H., Provasoli, L.: Nutrition of algae. Amer. Rev. Plant Physiol. **15, 37—56 (1964).
829.* — — Comparative physiology: Nutrition. Ann. Rev. Physiol. **27, 19—50 (1965).
830. Inaba, F., Kudo, N.: Electron microscopy of the nuclear events during binary fission in *Paramecium multimicronucleatum.* J. Protozool. **19**, 57—63 (1972).
831. — Sotokawa, Y.: Electron-microscopic observation on nuclear events during binary fission in *Blepharisma wardsi* (Ciliata, Heterotrichida). Jap. J. Genet. **43**, 335—348 (1968).
832. Ito, S., Chang, R. S., Pollard, T. D.: Cytoplasmic distribution of DNA in a strain of hartmannellid amoeba. J. Protozool. **16**, 638—645 (1969).
**833.* Jacherts, D., Jacherts, B.: Elemente der Bakterienphysiologie. 349 S. Frankfurt a. M.: Akadem. Verlagsgesellschaft 1964.
**834.* Jacob, F., Wollman, F. J.: Sexuality and the genetics of Bacteria. 374 pp. New York-London: Academic Press 1961.
835. Jacobs, M., Hopkins, J., Randall, J. Sir: Biochemistry of *Chlamydomonas* flagella. Proc. roy. Soc. B **173**, 61—62 (1969).
836. — McVittie, A.: Identification of the flagellar proteins of *Chlamydomonas reinhardi.* Exp. Cell Res. **63**, 53—61 (1970).
837. Jadin, J. B., Willaert, E.: Trois cas de méningo-encéphalite amibienne primitive à *Naegleria gruberi* observés a Anvers (Belgique). Protistologica **8**, 95—100 (1972).
838. Jahn, T. L.: The mechanism of ciliary movement. I. Ciliary reversal and activation by electric current; the Ludloff phenomenon in terms of core and volume conductors. J. Protozool. **8**, 369—380 (1961).
839. — The mechanism of ciliary movement. II. Ion antagonism and ciliary reversal. J. cell. comp. Physiol. **60**, 217—228 (1962).
840. — Relative motion in *Amoeba proteus.* In: Primitive motile systems in cell biology, pp. 279—302. New York-London: Academic Press 1964.
841. — The mechanism of ciliary movement. III. Theory of suppression of reversal by electrical potential of cilia reversed by Barium ions. J. Cell Physiol. **70**, 79—90 (1967).
**842.* — Bovée, E. C.: Protoplasmic movements and locomotion of Protozoa. In: Hutner, S. H.: Biochemistry and physiology of Protozoa, Vol. 3, pp. 62—119. New York-London: Academic Press 1964.
843.* — — Movement and locomotion of microorganisms. Ann. Rev. Microbiol. **19, 21—58 (1965).
**844.* — — Motile behavior of Protozoa. In: Chen, T.-T. (Ed.): Research in Protozoology, Vol. 1, pp. 41—200. Oxford: Pergamon Press 1967.
845. — — Protoplasmic movement within cells. Physiol. Rev. **49**, 793—862 (1969).
846. — Harmon, W. M., Landman, M.: Mechanisms of locomotion in flagellates. I. *Ceratium.* J. Protozool. **10**, 358—363 (1963).
847. — Votta, J. J.: Locomotion of Protozoa. Ann. Rev. Fluid Mechanics **4**, 93—116 (1972).
848. — Landman, M., Fonseca, J. R.: The mechanism of locomotion of flagellates. II. Function of the mastigonemes of *Ochromonas.* J. Protozool. **11**, 291—296 (1964).
849. — Rinaldi, R. A.: Protoplasmic movement in the foraminiferan, *Allogromia laticollaris*; and a theory of its mechanism. Biol. Bull. **117**, 100—118 (1959).
850. Jakus, M. A.: The structure and properties of the trichocysts of *Paramecium.* J. exp. Zool. **100**, 457—485 (1945).

851. Jakus, M. A., Hall, C. E.: Electron microscope observations of the trichocysts and cilia of *Paramecium*. Biol. Bull. mar. biol. Lab. Woods Hole **91**, 141—144 (1946).

852. Janet, Ch.: Le Volvox Mémoire I—III. Paris: Masson et Cie. 1912, 1922, 1923.

853. Janisch, R.: Regeneration of surface structures in *Paramecium caudatum*. Acta Protozool. **3**, 363—367 (1965).

854. Jankowski, A. W.: Cytogenetics of *Paramecium putrinum* C. et L., 1858. Acta Protozool. **10**, 285—394 (1972).

855. Jareno, A., Alonso, P., Perez-Silva, J.: Induced autogamy in two species of *Stylonychia*. J. Protozool. **17**, 384—388 (1970).

856. — — — Identification of some puffed regions in the polytene chromosomes of *Stylonychia mytilus*. Protistologica **8**, 237—244 (1972).

857. Jeffery, W. R., Stuart, K. D., Frankel, J.: The relationship between DNA replication and cell division in heat-synchronized *Tetrahymena*. J. Cell Biol. **46**, 533—543 (1970).

858. Jenkins, R. A., Sawyer, H. R.: Selective extirpation of oral ciliature of *Blepharisma* by laser microbeam. Exp. Cell Res. **63**, 192—194 (1970).

**859.* Jennings, H. S.: Behaviour of lower organisms. New York: Columbia Univ. Press 1906.

860. — Genetics of *Paramecium bursaria*. I. Mating types and groups, their interrelations and distribution: mating behavior and self sterility. Genetics **24**, 202—233 (1939).

**861.* — Inheritance in Protozoa. In: Calkins, G. N., Summers, F. M.: Protozoa in biological research, pp. 710—771. New York: Columbia University Press 1941.

862. Jensen, H. M., Wellings, S. R.: Development of the polar filament — polaroplast complex in a microsporidian parasite. J. Protozool. **19**, 297—305 (1972).

863. Jeon, J. W.: Nuclear-cytoplasmic relations in lethal Amoeba hybrids. Exp. Cell Res. **55**, 77—80 (1969).

864. — Bell, L. G.: Behaviour of cell membrane in relation to locomotion in *Amoeba proteus*. Exp. Cell Res. **33**, 531—539 (1964).

865. — Lorch, I. J.: Lethal effect of heterologous nuclei in Amoeba heterokaryons. Exp. Cell Res. **56**, 233—238 (1969).

866. Jerka-Dziadosz, M., Frankel, J.: An analysis of the formation of ciliary primordia in the hypotrich ciliate *Urostyla weissei*. J. Protozool. **16**, 612—637 (1969).

867. Jirovec, O.: Les amibes du type *Limax* comme agent vecteur des méningo-encéphalites chez l'homme. J. Méd. Lyon **50**, 1701—1710 (1969).

868. Johnson, U. G., Porter, K. R.: Fine structure of cell division in *Chlamydomonas reinhardi*. Basal bodies and microtubules. J. Cell Biol. **38**, 403—425 (1968).

869. Jollos, V.: Dauermodifikationen und Mutationen bei Protozoen. Arch. Protistenk. **83**, 197—219 (1934).

870. Jones, A. R., Jahn, T. L., Fonseca, J. R.: Contraction of protoplasm. IV. Cinematographic analysis of the contraction of some peritrichs. J. Cell Physiol. **75**, 9—20 (1970).

871. Jones, R. F., Lewin, R. A.: The chemical nature of the flagella of *Chlamydomonas moewusii*. Exp. Cell Res. **19**, 408—410 (1960).

872. Joyon, L.: Contribution à l'étude cytologique de quelques protozoaires flagellés. Ann. Fac. Sci. Univ. Clermont **22**, 1—96 (1963).

873. — Compléments à la connaissance ultrastructurale des genres *Haematococcus pluvialis* Flotow et *Stephanosphaera pluvialis* Cohn. Ann. Fac. Sci. Univ. Clermont **26**, 57—69 (1964).

874. — Sur la présence de glycogène dans l'axostyle de *Trichomonas lacertae* (Prowaczek). Arch. Zool. exp. gén. **105**, 285—288 (1965).

875. — Fott, B.: Quelques particularités infrastructurales du plaste des *Carteria* (Volvocales). J. Microscopie **3**, 159—166 (1964).

876. — Lom, J.: Sur l'ultrastructure de *Costia necatrix* Leclerq (Zooflagellé); place systématique de ce Protiste. C. R. Acad. Sci. (Paris) **262**, 660—663 (1966).

877. — — Etude cytologique, systématique et pathologique d'*Ichtyobodo necator* Henneguy 1883, Pinto 1928 (Zooflagellé). J. Protozool. **16**, 703—719 (1969).

878. — Mignot, J.-P., Kattar, M.-R., Brugerolle, G.: Compléments à l'étude des Trichomonadida et plus particulièrement de leur cinétide. Protistologica **5**, 309—326 (1969).

879. Joyon, L., Mignot, J.-P.: Données récentes sur la structure de la cinétide chez les Protozoaires flagellés. Ann. Biol. 8, 1—52 (1969).
880. Juergensmeyer, E. B.: Serotype expression and transformation in *Tetrahymena pyriformis*. J. Protozool. **16**, 344—352 (1969).
881. Jurand, A.: An electron microscope study of food vacuoles in *Paramecium aurelia*. J. Protozool. **8**, 125—130 (1961).
882. — Bomford, B.: The fine structure of the parasitic suctorian *Podophrya parameciorum*. J. Microscopie **4**, 509—522 (1965).
883. — Preer, L. B.: Ultrastructure of flagellated *lambda* symbionts in *Paramecium aurelia*. J. gen. Microbiol. **54**, 359—364 (1969).
884. — Rudman, B. M., Preer, J. R.: Prelethal effects of killing action by stock 7 of *Paramecium aurelia*. J. exp. Zool. **177**, 365—388 (1971).
**885.* — Selman, G. G.: The anatomy of *Paramecium aurelia*, pp. 1—218. Macmillan St. Martin's Press 1969.
886. — — Ultrastructure of the nuclei and intranuclear microtubules of *Paramecium aurelia*. J. gen. Microbiol. **60**, 357—364 (1970).
887. Kaczanowska, J.: Studies on topography of the cortical organelles of *Chilodonella cucullulus* (O. F. M.). III. Morphogenetic movements, regional multiplication of kinetosomes and cytokinesis in normal dividers and after phenethyl alcohol treatment. Acta Protozool. **9**, 83—103 (1971).
888. — Topography of cortical organelles in early dividers of *Chilodonella cucullulus* (O. F. M.). Acta Protozool. **8**, 231—250 (1971).
889. Kaczanowski, A.: Mitosis and polyploidy in nuclei of *Opalina ranarum*. Experientia (Basel) **24**, 846 (1968).
890. — *Opalina ranarum* Purkinje et Valentin: Meiosis and dimorphism of nuclear behaviour during meiosis. Acta Protozool. **9**, 105—106 (1971).
891. Kalley, J. P., Bisalputra, Th.: *Peridinium trochoideum*: The fine structure of the theca as shown by freeze-etching. J. Ultrastruct. Res. **31**, 95—109 (1970).
**892.* Kalmus, H.: *Paramecium*, das Pantoffeltierchen. 188 S. Jena: G. Fischer 1931.
893. Kamada, T.: Polar effect of electric current on the ciliary movements of *Paramecium*. J. Fac. Sci. Tokyo Imp. Univ. **2**, 285—298 (1931).
894. — Reversal of electric polar effect in *Paramecium* according to the change of current strength. J. Fac. Sci. Tokoy Imp. Univ. **2**, 299—307 (1931).
895. — Some observations on potential differences across the ectoplasm membrane of *Paramecium*. J. exp. Biol. **11**, 94—102 (1934).
896. — Intracellular calcium and ciliary reversal in *Paramecium*. Proc. imp. Acad. Japan Tokyo **14**, 260—262 (1938).
897. — Ciliary reversal of *Paramecium*. Proc. imp. Acad. Japan Tokyo **16**, 241—247 (1940).
898. — Kinosita, H.: Calcium-potassium factor in ciliary reversal of *Paramecium*. Proc. imp. Acad. Japan Tokyo **16**, 125—130 (1940).
899. Kaneda, M.: On the division of macronucleus in the living gymnostome ciliate, *Chlamydodon pedarius*, with special reference to the behaviors of chromonemata, nucleoli and endosome. Cytologia (Tokyo) **26**, 89—104 (1961).
900. — On the interrelation between the macronucleus and the cytoplasm in the gymnostome ciliate *Chlamydodon pedarius*. Cytologia (Tokyo) **26**, 408—418 (1961).
901. — Fine structure of macronucleus of the gymnostome ciliate, *Chlamydodon pedarius*. Jap. J. Genet. **36**, 223—234 (1961).
902. — Fine structure of the oral apparatus of the gymnostome ciliate *Chlamydodon pedarius*. J. Protozool. **9**, 188—195 (1962).
903. Karakashian, S. J.: Growth of *Paramecium bursaria* as influenced by the presence of algal symbionts. Physiol. Zool. **36**, 52—68 (1963).
904. — Karakashian, M. W.: Evolution and symbiosis in the genus *Chlorella* and related algae. Evolution **19**, 368—377 (1965).
905. — — Rudzinska, M. A.: Electron microscopic observations on the symbiosis of *Paramecium bursaria* and its intracellular Algae. J. Protozool. **15**, 113—128 (1968).
906. — Siegel, R. W.: A genetic approach to endocellular symbiosis. Exp. Parasit. **17**, 103—122 (1965).

907. KATASHIMA, R.: Mating types in *Euplotes eurystomus*. J. Protozool. **6**, 75—83 (1959).
908. — The intimacy of union between the two members of the conjugating pairs in *Euplotes eurystomus*. Jap. J. Zool. **12**, 329—343 (1959).
909. — Breeding system of *Euplotes patella* in Japan. Jap. J. Zool. **13**, 39—61 (1961).
910. — Mate-killing in *Euplotes patella*, syngen 1. Annotationes Zool. Japan **38**, 207—215 (1965).
911. KENNEDY, JR., J. R.: The morphology of *Blepharisma undulans* STEIN. J. Protozool. **12**, 542—561 (1965).
912. — BRITTINGHAM, E.: Fine structure changes during chloral hydrate deciliation of *Paramecium caudatum*. J. Ultrastruct. Res. **22**, 530—546 (1968).
913. KEVIN, M. J., HALL, W. T., McLAUGHLIN, J., ZAHL, P. A.: *Symbiodinium microadriaticum* FREUDENTHAL, a revised taxonomic description of the ultrastructure. J. Phycol. **5**, 341—350 (1969).
914. KILLBY, V. A. A., SILVERMAN, P. H.: Fine structural observations of the erythrocytic stages of *Plasmodium chabaudi* LANDAU 1965. J. Protozool. **16**, 354—370 (1969).
915. KIMBALL, R. F.: A delayed change of phenotype following a change of genotype in *Paramecium aurelia*. Genetics **24**, 49—58 (1939).
916. — The nature and inheritance of mating types in *Euplotes patella*. Genetics **27**, 269—285 (1942).
917.* — Mating types in the ciliate Protozoa. Quart. Rev. Biol. **18, 30—45 (1943).
**918.* — Physiological genetics of the ciliates. In: HUTNER, S. H.: Biochemistry and physiology of Protozoa, Vol. 3, pp. 244—275. New York-London: Academic Press 1964.
919. — GAITHER, N.: Behavior of nuclei at conjugation in *Paramecium aurelia*. I. Effect of incomplete chromosome sets and competition between complete and incomplete nuclei. Genetics **40**, 878—889 (1955).
920. — — Behavior of nuclei at conjugation in *Paramecium aurelia*. II. The effects of x-rays on diploid and haploid clones with a discussion of dominant lethals. Genetics **41**, 715—728 (1965).
921. — PRESCOTT, D. M.: Desoxyribonucleic acid synthesis and distribution during growth and amitosis of the macronucleus of *Eugplotes*. J. Protozool. **9**, 88—92 (1962).
922. — — RNA and protein synthesis in amacronucleate *Paramecium aurelia*. J. Cell Biol. **21**, 496—497 (1964).
923. — VOGT-KÖHNE, L., CASPERSSON, T. O.: Quantitative cytochemical studies on *Paramecium aurelia*. III. Dry weight and ultraviolet adsorption of isolated macronucle. during various stages of the interdivision interval. Exp. Cell Res. **20**, 368—377 (1960).
924. — — Quantitative cytochemical studies on *Paramecium aurelia*. IV. The effect of limited food and starvation on the macronucleus. Exp. Cell Res. **23**, 479—487 (1961).
925. — — Effects of radiation on cell and nuclear growth in *Paramecium aurelia*. Exp. Cell Res. **28**, 228—238 (1962).
926. KINASTOWSKI, W.: Der Einfluß der mechanischen Reize auf die Kontraktilität von *Spirostomum ambiguum* EHRBG. Acta Protozool. **1**, 201—222 (1963).
927. KINOSITA, H.: Electrical stimulation of *Paramecium* with two successive subliminal current pulses. J. cell. comp. Physiol. **12**, 103—117 (1938).
928. — Electrical stimulation of *Paramecium* with linearly increasing current. J. cell. comp. Physiol. **13**, 253—261 (1939).
929. — Electric potentials and ciliary response in *Opalina*. J. Fac. Sci. Tokyo Univ. (Sec. IV) **7**, 1—14 (1954).
930. — DRYL, S., NAITOH, Y.: Changes in the membrane potential and the responses to stimuli in *Paramecium*. J. Fac. Sci. Tokyo Univ. (Sec. IV) **10**, 291—301 (1964).
931. — — — Relation between the magnitude of membrane potential and ciliary activity in *Paramecium*. J. Fac. Sci. Tokyo Univ. (Sec. IV) **10**, 303—309 (1964).
932. — — — Spontaneous change in membrane potential of *Paramecium caudatum* induced by barium and calcium ions. Bull. Acad. pol. Sci. Cl. II **12**, 459—461 (1964).
933. — MURAKAMI, A., YASUDA, M.: Interval between membrane potential change and ciliary reversal in *Paramecium* immersed in Ba-Ca mixture. J. Fac. Sci. Tokyo Univ. (Sec. IV) **10**, 421—425 (1965).

*934. KIRBY, H.: Relationships between certain Protozoa and other animals. In: CALKINS, G.N., SUMMERS, F.M.: Protozoa in biological research, pp. 890—1008. New York: Columbia Univ. Press 1941.

*935. — Organisms living on and in Protozoa. In: CALKINS, G.N., SUMMERS, F.M.: Protozoa in biological research, pp. 1009—1113. New York: Columbia Univ. Press 1941.

936. — Some observations on cytology and morphogenesis in flagellate Protozoa. J. Morph. **75**, 361—421 (1944).

937. — Systematic differentiation and evolution of flagellates in termites. Rev. Soc. mex. Hist. nat. **10**, 57—79 (1949).

938. KITCHING, J.A.: Effects of high hydrostatic pressure on a feeding suctorian. Protoplasma **46**, 475—480 (1956).

*939. — Contractile vacuoles of Protozoa. Protoplasmatologia, Vol. 3, (D-3a), pp. 1—45. (Handbuch der Protoplasmaforschung). Wien-New York: Springer 1956.

*940. — Food vacuoles. Protoplasmatologia, Vol. 3 (D-3b), pp. 1—57. (Handbuch der Protoplasmaforschung). Wien-New York: Springer 1956.

*941. — The physiological basis of behavior in the Protozoa. In: RAMSAY, J.A., WIGGLESWORTH, V.B.: The cell and the organism, pp. 60—78. Cambridge: Cambridge Univ. Press 1961.

942. — The axopods of the sun animalcule *Actinophrys sol* (Heliozoa). In: Primitive motile systems in cell biology, pp. 445—456. New York-London: Academic Press 1964.

*943. — Contractile vacuoles, ionic regulation and excretion. In: CHEN, T.-T. (Ed.): Research in Protozoology, Vol. 1, pp. 307—336. Oxford: Pergamon Press 1967.

944. — CRAGGS, S.: The axopodial filaments of the heliozoon *Actinosphaerium nucleofilum*. Exp. Cell Res. **40**, 658—660 (1965).

945. KLOETZEL, J.A.: Compartmentalization of the developing macronucleus following conjugation in *Stylonychia* and *Euplotes*. J. Cell Biol. **47**, 395—407 (1970).

946. KLUG, S.H.: Cortical studies on *Glaucoma*. J. Protozool. **15**, 321—327 (1968).

947. KLUSS, B.C.: Electron microscopy of the macronucleus of *Euplotes eurystomus*. J. Cell Biol. **13**, 462—465 (1962).

*948. KNIGHT-JONES, E.W.: Relations between metachronism and the direction of ciliary beat in Metazoa. Quart. J. micr. Sci. **95**, 503—521 (1954).

949. KNOWLES, J.K.C.: Observations on two mitochondrial phenotypes in single *Paramecium* cells. Exp. Cell Res. **70**, 223—226 (1971).

950. — TAIT, A.: A new method for studying the genetic control of specific mitochondrial proteins in *Paramecium aurelia*. Molec. gen. Genet. **117**, 53—59 (1972).

951. KOCHERT, G.D.: Differentiation of reproductive cells in the NB-3 and NB-7 strains of *Volvox carteri*. Indiana University, Thesis (1967).

952. — Differentiation of reproductive cells in *Volvox carteri*. J. Protozool. **15**, 438—452 (1968).

953. KOEHLER, O.: Über die Geotaxis von *Paramecium*. Arch. Protistenk. **45**, 1—94 (1922).

*954. — Galvanotaxis. In: Handbuch der norm. und path. Physiol. Bd. 11, S. 1027—1049. 1925.

955. — Über die Geotaxis von *Paramecium*. II. Arch. Protistenk. **70**, 279—360 (1930).

956. KOFOID, C.A., SWEZY, C.: The free-living unarmored dinoflagellates. Mem. Univ. Calif. **5** (1921).

957. KOIZUMI, S.: Serotypes and immobilization antigens in *Paramecium caudatum*. J. Protozool. **13**, 73—76 (1966).

958. — PREER, J.R.: Transfer of cytoplasm by microinjection in *Paramecium aurelia*. J. Protozool. **13**, 27 (1966).

*959. KOMNICK, H., WOHLFARTH-BOTTERMANN, K.E.: Morphologie des Cytoplasmas. Fortschr. Zool. **17**, 1—154 (1964).

*960. — STOCKEM, W., WOHLFARTH-BOTTERMANN, K.E.: Ursachen, Begleitphänomene und Steuerung zellulärer Bewegungserscheinungen. Fortschr. Zool. **21**, 1—74 (1972).

961. KÖNIG, K.: Wirkung von Lithium- und Rhodanid-Ionen auf die polare Differenzierung und die Morphogenese von *Stentor coeruleus* EHRENBERG. Arch. Protistenk. **110**, 179—230 (1967).

962. Korfsmeier, K.: Strukturen des Stieles und Köpfchens peritricher sessiler Ciliaten. Zool. Jb. Anat. **70**, 199—224 (1949).

963. Kormos, J., Kormos, K.: Direkte Beobachtung der Kernveränderungen der Konjugation von *Cyclophrya katharinae* (Ciliata Protozoa). Acta. biol. hung. **10**, 373—394 (1960).

964. — — Experimentelle Untersuchung der Kernveränderungen der Konjugation von *Cyclophrya katharinae* (Ciliata Protozoa). Acta. biol. hung. **10**, 395—419 (1960).

965. Korohoda, W., Kurowska, A.: Quantitative estimation of the thresholds of electrotactic responses in *Amoeba proteus*. Acta Protozool. **7**, 375—382 (1970).

966. — Rakoczy, L., Walczak, T.: On the control mechanism of protoplasmic streamings in the plasmodia of Myxomycetes. Acta Protozool. **7**, 363—374 (1970).

967. Kovaleva, V. G.: Meiosis and some stages of conjugation in the holotrichous ciliate *Trachelonema sulcata*. Protistologica **8**, 83—90 (1972).

968. Kowallik, K.: The crystal lattice of the pyrenoid matrix of *Prorocentrum micans*. J. Cell Sci. **5**, 251—269 (1969).

969. Krassner, S. M.: Cytochromes, lactic dehydrogenase and transformation in *Leishmania*. J. Protozool. **13**, 286—290 (1966).

970. Krüger, F.: Untersuchungen über den Bau und die Funktion der Trichocysten von *Paramecium caudatum*. Arch. Protistenk. **72**, 91—134 (1930).

971. — Wohlfahrth-Bottermann, K. E.: Elektronenoptische Beobachtungen an Ciliatenorganellen. Mikroskopie **7**, 121—130 (1952).

972. Kubai, D. F., Ris, H.: Division in the dinoflagellate *Gyrodinium cohnii* Schiller. A new type of nuclear reproduction. J. Cell Biol. **40**, 508—528 (1969).

*972*a. Kubota, T., Tokoroyama, T., Tsukuda, Y., Koyama, H., Miyake, A.: Isolation and structure determination of Blepharismin, a conjugation initiating gamone in the ciliate *Blepharisma*. Science **179**, 400—402 (1973).

973. Kudo, R. R., Daniels, E. W.: An electron microscope study of the spore of a microsporidian, *Thelohania californica*. J. Protozool. **10**, 112—120 (1963).

974. Kudrjavtsev, B. N.: Changes of the DNA content in macro- and micronucleus of *Paramecium putrinum* in the interdivision phase. Acta Protozool. **4**, 51—57 (1966).

975. Kuhl, W.: Mikrodynamische Untersuchungen an der lebenden Zelle von *Actinosphaerium eichhorni* Ehrbg., unter Änderung des Zeitmomentes. Protoplasma **40**, 555—613 (1951).

*976. Kühn, A.: Die Orientierung der Tiere im Raum. Jena: Fischer 1919.

977. — Untersuchungen zur kausalen Analyse der Zellteilung. I. Zur Morphologie und Physiologie der Kernteilung von *Vahlkampfia bistadialis*. Wilhelm Roux Arch. Entwickl.-Mech. Org. **46**, 259—327 (1920).

*978. — Grundriß der Vererbungslehre. 4. Aufl. Heidelberg: Quelle und Meyer 1965.

979. Kümmel, G.: Die Gleitbewegung der Gregarinen. Elektronenmikroskopische und experimentelle Untersuchungen. Arch. Protistenk. **102**, 501—522 (1958).

980. Kung, C.: Genic mutants with altered system of excitation in *Paramecium aurelia*. I. Phenotypes of the behavioral mutants. Z. vergl. Physiol. **7**, 142—164 (1971).

*980*a. — Genic mutants with altered system of excitation in *Paramecium aurelia* II. Mutagenesis, screening, and genetic analysis of the mutants. Genetics **69**, 29—45 (1971).

981. — Eckert, R.: Genetic modification of electric properties of an excitable membrane. Proc. nat. Acad. Sci. (Wash.) **69**, 93—107 (1972).

*981*a. — Naitoh, Y.: Calcium-induced ciliary reversal in the extracted models of "Pawn", a behavioral mutant of *Paramecium*. Science **179**, 195—196 (1973).

982. Kusel, J. P., Moore, K. E., Weber, M. M.: The ultrastructure of *Crithidia fasciculata* and morphological changes induced by growth in acriflavine. J. Protozool. **14**, 283—296 (1967).

983. Kužnicki, L.: Recovery in *Paramecium caudatum* immobilized by chloral hydrate treatment. Acta Protozool. **1**, 177—185 (1963).

984. — Reversible immobilization of *Paramecium caudatum* by nickel ions. Acta Protozool. **1**, 301—312 (1963).

985. Kužnicki, L.: Role of Ca^{2+} ions in the excitability of protozoan cell. Calcium factor in the ciliary reversal induced by inorganic cations in *Paramecium caudatum*. Acta Protozool. **4**, 241—256 (1966).

986. — Ciliary reversal in *Paramecium caudatum* in relation to external pH. Acta Protozool. **4**, 257—261 (1966).

987. — Behavior of *Paramecium* in gravity fields. I. Sinking of immobilized specimens. Acta Protozool. **6**, 109—117 (1968).

988. — Mechanisms of the motor responses of *Paramecium*. Acta Protozool. **8**, 83—118 (1970).

989. — Jahn, T. L., Fonseca, J. R.: Helical nature of the ciliary beat of *Paramecium multimicronucleatum*. J. Protozool. **17**, 16—24 (1970).

990. — Sikora, J.: Inversion of spiralling of *Paramecium aurelia* after homologous antiserum treatment. Acta Protozool. **4**, 263—268 (1966).

991.* Lamy, L. H.: Protozoaires intracellulaires en culture cellulaire. Intérèt — possibilités — limites. Ann. Biol. **11, 145—183 (1972).

992. Lang, N. J.: An additional ultrastructural component of flagella. J. Cell Biol. **19**, 631—634 (1963).

993. — Electron-microscopic demonstration of plastids in *Polytoma*. J. Protozool. **10**, 333—339 (1963).

994. — Electron microscopy of the Volvocaceae and Astrephomenaceae. Amer. J. Bot. **50**, 279—300 (1963).

*994*a. Lanners, H. N.: Beobachtungen zur Konjugation von *Heliophrya* (*Cyclophrya*) *erhardi* (Rieder) Matthes, Ciliata, Suctoria. Arch. Protistenk. (in press).

*994*b.— Experimente zur Differenzierung des Makronucleus bei *Heliophrya* (*Cyclophrya*) *erhardi*. Arch. Protistenk. (in press).

995. Larison, L. L., Siegel, R. W.: Illegitimate mating in *Paramecium bursaria* and the basis for cell union. J. gen. Microbiol. **26**, 499—508 (1961).

996. Laurent, M., Steinert, M.: Electron microscopy of kinetoplastic DNA from *Trypanosoma mega*. Proc. nat. Acad. Sci. (Wash.) **66**, 419—424 (1970).

997. Laval, M.: Ultrastructure et mode de nutrition du Choanoflagellé *Salpingoeca pelagica*, sp. nov. Comparaison avec les choanocytes des Spongiaires. Protistologica **7**, 325—336 (1971).

998. Leadbeater, B. S.: The intracellular origin of flagellar hairs in the dinoflagellate *Woloszynskia micra* Leadbeater & Dodge. J. Cell Sci. **9**, 443—451 (1971).

999. — Dodge, J. D.: An electron microscope study of dinoflagellate flagella. J. gen. Microbiol. **46**, 305—314 (1967).

1000. — — An electronmicroscope study of nuclear and cell division in a dinoflagellate. Arch. Mikrobiol. **57**, 239—254 (1967).

1001. Le Calvez, J.: Flagellispores du radiolaire *Coelodendrum ramosissimum* (Haeckel). Arch. Zool. exp. gén. **77**, 99—102 (1935).

1002. — Observations sur le genre *Iridia*. Arch. Zool. exp. gén. **78**, 115—131 (1936).

1003. — Recherches sur les foraminifères. 1. Développement et reproduction. Arch. Zool. exp. gén. **80**, 163—333 (1938).

1004. — Recherches sur les foraminifères. 2. Place de la méiose et sexualité. Arch. Zool. exp. gén. **87**, 211—243 (1950).

1005. — *Discorbis patelliformis* (Brady), *erecta* (Sidebottom) et les *Discorbis* plastogamiques. Arch. Zool. exp. gén. **89** (1952).

1006. Lee, J. J., McEnery, M. E., Rubin, H.: Quantitative studies on the growth of *Allogromia laticollaris* (Foraminifera). J. Protozool. **16**, 377—395 (1969).

1007. — Zucker, W.: Algal-flagellate symbiosis in the foraminifer *Archaias*. J. Protozool. **16**, 71—81 (1969).

1008. Leedale, G. F.: Nuclear structure and mitosis in the Euglenineae. Arch. Mikrobiol. **32**, 32—64 (1958).

1009. — Pellicle structure in *Euglena*. Brit. Phycol. Bull. **2**, 291—306 (1964).

1010. — Endonuclear bacteria in species of *Euglena*, *Strombomonas* and *Trachelomonas*. Brit. phycol. Bull. **3**, 413 (1967).

*1011. Leedale, G. F.: Euglenida/Euglenophyta. Ann. Rev. Microbiol. **21**, 31—48 (1967).

1012. — The cytology and fine structure of *Trachelomonas oblonga* var. *punctata* Pringsheim, with special reference to envelope structure and formation. Brit. phycol. Bull. **3**, 602 (1968).

*1013. — Phylogenetic aspects of nuclear cytology in the Algae. Ann. N. Y. Acad. Sci. **175**, 429—453 (1970).

1014. — Buetow, D. E.: Observations on the mitochondrial reticulum in living *Euglena gracilis*. Cytobiol. **1**, 195—202 (1970).

1015. — Leadbeater, B. S. C., Massalski, A.: The intracellular origin of flagellar hairs in the Chrysophyceae and Xanthophyceae. J. Cell Sci. **6**, 701—719 (1970).

1016. — Meeuse, B. J. D., Pringsheim, E. G.: Structure and physiology of *Euglena spirogyra* I. and II. Arch. Mikrobiol. **50**, 68—102 (1965).

1017. — — — Structure and physiology of *Euglena spirogyra* III—VI. Arch. Mikrobiol. **50**, 133—155 (1965).

1018. Léger, L.: La reproduction sexuel chez les *Stylorhynchus*. Arch. Protistenk. **3**, 303—357 (1904).

1019. Legrand, B.: Recherches expérimentales sur le déterminisme de la contraction et les structures contractiles chez le Spirostome. Protistologica **6**, 283—300 (1970).

1020. Leibenguth, F.: Veränderungen der Haemolymphe ausgewachsener *Ephestia*-Raupen nach Infektion mit *Mattesia dispora*. Z. Parasitenk. **33**, 235—245 (1970).

1021. — Die Entwicklung von *Mattesia dispora* in *Habrobracon juglandis*. Z. Parasitenk. **38**, 162—173 (1972).

1022. — Mikrospektrophotometrische Untersuchungen zur Kernhypertrophie nach Infektion mit *Klossia helicina*. Z. Parasitenk. **39**, 211—220 (1972).

1023. Lengsfeld, A. M.: Nahrungsaufnahme und Verdauung bei der Foraminifere *Allogromia laticollaris*. Helgol. wiss. Meeresunters. **19**, 385—400 (1969).

1024. — Zum Feinbau der Foraminifere *Allogromia laticollaris*. I. Mitteilung: Zellen mit ausgestreckten und eingezogenen Rhizopodien. Helgol. wiss. Meeresunters. **19**, 230—261 (1969).

1025. — Zum Feinbau der Foraminifere *Allogromia laticollaris*. II. Mitteilung: Ausgestreckte und durch Abreißen isolierte Rhizopodien. Helgol. wiss. Meeresunters. **19**, 262—283 (1969).

1026. Lerche, W.: Untersuchungen über Entwicklung und Fortpflanzung in der Gattung *Dunaliella*. Arch. Protistenk. **88**, 236—268 (1937).

1027. Levine, M.: The diverse mate-killers of *Paramecium aurelia*, variety 8: their interrelations and genetic basis. Genetics **38**, 561—578 (1953).

1028. — The interaction of nucleus and cytoplasm in the isolation and evolution of species of *Paramecium*. Evolution **7**, 366—385 (1953).

1029. Levine, N. D.: Protozoology today. J. Protozool. **9**, 1—6 (1962).

*1030. — Relationship between certain Protozoa and other animals. In: Chen, T.-T. (Ed.): Research in Protozoology, Vol. 4, pp. 291—350. New York: Pergamon Press 1972.

1031. Levine, R. P.: Genetic control of photosynthesis in *Chlamydomonas reinhardi*. Proc. nat. Acad. Sci. (Wash.) **46**, 972—978 (1960).

1032. — Ebersold, W. T.: Gene recombination in *Chlamydomonas reinhardi*. Cold Spr. Harb. Symp. quant. Biol. **23**, 101—109 (1958).

*1033. — — The genetics and cytology of *Chlamydomonas*. Ann. Rev. Microbiol. **14**, 197—216 (1960).

1034. — Goodenough, U. W.: The genetics of photosynthesis and the chloroplast in *Chlamydomonas reinhardi*. A. Rev. Genet. **4**, 397—408 (1970).

1035. — Paszewski, A.: Chloroplast structure and function in ac-20, a mutant strain of *Chlamydomonas reinhardi*. II. Photosynthetic electron transport. J. Cell Biol. **44**, 540—546 (1970).

1036. — Volkman, D.: Mutants with impaired photosynthesis in *Chlamydomonas reinhardi*. Biochem. biophys. Res. Commun. **6**, 264—269 (1961).

1037. Lewin, R. A.: Gamete behaviour in *Chlamydomonas*. Nature (Lond.) **166**, 76 (1950).

1038. — Isolation of sexual strains of *Chlamydomonas*. J. gen. Microbiol. **5**, 926—929 (1951).

1039. LEWIN, R. A.: Ultraviolet induced mutations in *Chlamydomonas moewusii* GERLOFF. J. gen. Microbiol. **6**, 233—248 (1952).

1040. — The genetics of *Chlamydomonas moewusii* GERLOFF. J. Genetics **51**, 543—560 (1953).

**1041.* — Sex in unicellular algae. In: WENRICH, D. H.: Sex in microorganisms, pp. 100—133. Washington: Amer. Ass. Advanc. Sci. 1954.

1042. — Control of sexual activity in *Chlamydomonas* by light. J. gen. Microbiol. **15**, 170 to 185 (1956).

1043. LIESCHE, W.: Die Kern- und Fortpflanzungsverhältnisse von *Amoeba proteus* (Pall.). Arch. Protistenk. **91**, 135—186 (1938).

1044. LIU, T. P., DAVIES, D. M.: Fine structure of developing spores of *Thelohania bracteata* (STRICKLAND, 1913) (Microsporida, Nosematidae) emphasizing polar-filament formation. J. Protozool. **19**, 461—469 (1972).

1045. LOEFER, J. B., OWEN, R. D., CHRISTENSEN, E.: Serological types among 31 strains of the ciliated protozoan, *Tetrahymena pyriformis*. J. Protozool. **5**, 209—217 (1958).

1046. — SMALL, E. B., FURGASON, W. H.: Range of variation in the somatic infraciliature and contractile vacuole pores of *Tetrahymena pyriformis*. J. Protozool. **13**, 90—102 (1966).

1047. LOM, J.: The morphology and morphogenesis of the buccal ciliary organelles in some peritrichous ciliates. Arch. Protistenk. **107**, 131—162 (1964).

1048. — Notes on the extrusion and some other features of myxosporidian spores. Acta Protozool. **2**, 321—327 (1964).

1049. — CORLISS, J. O.: Ultrastructural observations on the development of the microsporidian protozoon *Plistophora hyphessobryconis* SCHÄPERCLAUS. J. Protozool. **14**, 141—152 (1967).

1050. — — Observations on the fine structure of two species of the peritrich ciliate genus *Scyphidia* and on their mode of attachment to their host. Trans. Amer. Microsc. Soc. **87**, 493—509 (1968).

1051. — — Attachment structures in ectoparasitic Protozoa of fishes and their possible relation to pathogenicity. J. Parasit. **56**, 212—213 (1970).

1052. — — Morphogenesis and cortical ultrastructure of *Brooklynella hostilis*, a dysteriid ciliate ectoparasitic on marine fishes. J. Protozool. **18**, 261—281 (1971).

1053. — — NOIROT-TIMOTHEE, C.: Observations on the ultrastructure of the buccal apparatus in thigmotrich ciliates and their bearing on thigmotrich-peritrich affinities. J. Protozool. **15**, 824—840 (1968).

1054. — KOZLOFF, E. N.: The ultrastructure of *Phalacrocleptes verruciformis*, an unciliated ciliate parasitizing the polychaete *Schizobranchia insignis*. J. Cell Biol. **33**, 355—364 (1967).

1055. — — Ultrastructure of the cortical regions of ancistrocomid ciliates. Protistologica **5**, 173—192 (1969).

1056. — VÁVRA, J.: The mode of sporoplasm extrusion in microsporidian spores. Acta Protozool. **1**, 81—90 (1963).

1057. — — Fine morphology of the spore in Microsporidia. Acta Protozool. **1**, 279—283 (1963).

1058. — — Notes on the morphogenesis of the polar filament in *Henneguya* (Protozoa, Cnidosporidia). Acta Protozool. **2**, 57—60 (1964).

1059. LORCH, I. J., DANIELLI, J. F.: Nuclear transplantation in amoebae. I. Some species characters of *Amoeba proteus* and *Amoeba discoides*. Quart. J. micr. Sci. **94**, 445—460 (1953).

1060. — — Nuclear transplantation in amoebae. II. The immediate results of transfer of nuclei between *Amoeba proteus* and *Amoeba discoides*. Quart. J. micr. Sci. **94**, 461—480 (1953).

1061. — JEON, K. W.: Character changes induced by heterologous nuclei in Amoeba heterokaryons. Exp. Cell Res. **57**, 223—229 (1969).

1062.* LOWNDES, A. G.: On flagellar movement in unicellular organism. Proc. Zool. Soc. Lond. **111 A, 111—134 (1941).

1063. Ludwig, F.-D.: Die Zooxanthellen bei *Cassiopea andromeda* Eschscholtz 1829 (Polypstadium) und ihre Bedeutung für die Strobilation. Zool. Jb. Anat. **86**, 238—277 (1969).

1064. Luporini, P., Nobili, R.: New mating types and the problem of syngens in *Euplotes minuta* Yocum (Ciliata, Hypotricha). Atti Ass. Genet. Ital. **12**, 345—360 (1967).

1065. Lwoff, A.: Le cycle nucléaire de *Stephanopogon mesnili* Lw. (cilié homocaryote). Arch. Zool. exp. gén. **78**, 117—132 (1936).

**1066.* — Problems of morphogenesis in ciliates. The kinetosomes in development, reproduction and evolution. 103 S. New York: John Wiley & Sons 1950.

**1067.* — Biochemistry and physiology of Protozoa. 434 S. New York-London: Academic Press 1951.

1068. Lyman, H., Epstein, H. T., Schiff, J. A.: Studies of chloroplast development in *Euglena*. I. Inactivation of green colony formation by ultraviolet light. Biochim. biophys. Acta (Amst.) **50**, 301—309 (1961).

1069. Machemer, H.: Abhängigkeit der Lebensdauer und Teilung bei *Stylonychia mytilus* von äußeren Faktoren. Zool. Jb. Physiol. **71**, 245—256 (1965).

1070. — Analyse langzeitlicher Bewegungserscheinungen des Ciliaten *Stylonychia mytilus* Ehrenberg. Arch. Protistenk. **108**, 91—107 (1965).

1071. — Analyse kurzzeitlicher Bewegungserscheinungen des Ciliaten *Stylonychia mytilus* Ehrenberg. Arch. Protistenk. **108**, 153—190 (1965).

1072. — Zur Koordination und Wirkungsweise der Membranellen von *Stylonychia mytilus*. Arch. Protistenk. **109**, 257—277 (1966).

1073. — Erschütterungsbedingte Sensibilisierung gegenüber rauhem Untergrund bei *Stylonychia mytilus*. Arch. Protistenk. **109**, 245—256 (1966).

1074. — Versuche zur Frage nach der Dressierbarkeit hypotricher Ciliaten unter Einsatz hoher Individuenzahlen. Z. Tierpsychologie **6**, 641—654 (1966).

1075. — Regulation der Cilienmetachronie bei der „Fluchtreaktion" von *Paramecium*. J. Protozool. **16**, 764—771 (1969).

1076. — Filmbildanalysen vier verschiedener Schlagmuster der Marginalcirren von *Stylonychia*. Z. vergl. Physiol. **62**, 183—196 (1969).

1077. — Eine 2-Gradienten-Hypothese für die Metachronieregulation bei Ciliaten. Arch. Protistenk. **111**, 100—128 (1969).

1078. — Primäre und induzierte Bewegungsstadien bei Osmiumsäurefixierung vorwärtsschwimmender Paramecien. Acta Protozool. **7**, 531—535 (1970).

1079. — Properties of polarized ciliary beat in *Paramecium*. Acta Protozool. **11**, 295—300 (1972).

1080. — Temperature influences on ciliary beat and metachronal coordination in *Paramecium*. J. Mechanochem. Cell Motility **1**, 57—66 (1972).

1081. — Ciliary activity and the origin of metachrony in *Paramecium*: Effects of increased viscosity. J. exp. Biol. **57**, 239—259 (1972).

**1082.* — Ciliary activity and metachronism in Protozoa. In: Sleigh, M. A. (Ed.): Cilia and Flagella. New York-London: Academic Press 1973.

1083. — Eckert, R.: Electrophysiological control of reversed ciliary beating in *Paramecium*. J. gen. Physiol. **61**, 572—587 (1973).

1084. Manton, I.: Electron microscopical observations on a very small flagellate: The problem of *Chromulina pusilla* Butcher. J. Mar. Biol. Ass. U. K. **38**, 319—333 (1959).

1085. — Further observations on the fine structure of the haptonema in *Prymnesium parvum*. Arch. Mikrobiol. **49**, 315—330 (1964).

1086. — Some possibly significant structural relations between chloroplasts and other cell components. In: Goodwin, T. W.: Biochemistry of chloroplasts, Vol. 1, pp. 23—47. New York: Academic Press 1966.

1087. — Further observations on the fine structure of *Chrysochromulina chiton*, with special reference to the pyrenoid. J. Cell Sci. **1**, 187—192 (1966).

1088. — Observations on scale production in *Prymnesium parvum*. J. Cell Sci. **1**, 375—380 (1966).

1089. — Observations on scale production in *Pyramimonas amylifera* Conrad. J. Cell Sci. **1**, 429—438 (1966).

1090. Manton, I.: Further observations on the fine structure of *Chrysochromulina chiton*, with special reference to the pyrenoid. J. Cell Sci. **1**, 187—192 (1966).

1091. — Further observations on the fine structure of *Chrysochromulina chiton* with special reference to the haptonema, "peculiar" Golgi structure and scale production. J. Cell Sci. **2**, 265—272 (1967).

1092. — Further observations on scale formation in *Chrysochromulina chiton*. J. Cell Sci. **2**, 411—418 (1967).

1093. — Ettl, H.: Observations on the fine structure of *Mesostigma viride* Lauterborn. J. Linn. Soc. Bot. **59**, 175—184 (1965).

1094. — Harris, K.: Observations on the microanatomy of the brown flagellate *Sphaleromantis tetragona* Skuja with special reference to the flagellar apparatus and scales. J. Linn. Soc. Bot. **59**, 397—403 (1966).

1095. — Leedale, G. F.: Observations on the fine structure of *Paraphysomonas vestita*, with special reference to the Golgi apparatus and the origin of scales. Phycologia **1**, 37—57 (1961).

1096. — — Further observations on the fine structure of *Chrysochromulina ericina* Parke and Manton. J. Mar. Biol. Ass. U. K. **41**, 145—155 (1961).

1097. — — Further observations on the fine structure of *Chrysochromulina minor* and *C. kappa* with special reference to the pyrenoids. J. Mar. Biol. Ass. U. K. **41**, 519—526 (1961).

1098. — — Observations on the microanatomy of *Coccolithus pelagicus* and *Cricosphaera carterae*, with special reference to the origin and nature of coccoliths and scales. J. Mar. Biol. Ass. U. K. **49**, 1—16 (1969).

1099. — Parke, M.: Further observations on small green flagellates with special reference to possible relatives of *Chromulina pusilla* Butcher. J. Mar. Biol. Ass. U. K. **39**, 275—298 (1960).

1100. — — Observations on the fine structure of two species of *Platymonas* with special reference to flagellar scales and the mode of origin of the theca. J. Mar. Biol. Ass. U. K. **45**, 743—754 (1965).

1101. — Rayns, D. G., Ettl, H., Parke, M.: Further observations on green flagellates with scaly flagella: The genus *Heteromastix* Korshikov. J. Mar. Biol. Ass. U. K. **45**, 241—255 (1965).

1102. Margolin, P.: The ciliary antigens of stock 172, *Paramecium aurelia*, variety 4. J. exp. Zool. **133**, 345—387 (1956).

1103. — Loefer, J. B., Owen, R. D.: Immobilizing antigens of *Tetrahymena pyriformis*. J. Protozool. **6**, 207—215 (1959).

1104. Marszalek, D. S.: Observations on *Iridia diaphana*, a marine foraminifer. J. Protozool. **16**, 599—911 (1969).

1105. Mattar, F. E., Byers, T. J.: Morphological changes and the requirements for macromolecule synthesis during excystment of *Acanthamoeba castellanii*. J. Cell Biol. **49**, 507—520 (1971).

1106. Mattern, C. F. T., Daniel, W. A., Honigberg, B. M.: Structure of *Hypotrichomonas acosta* Moskowitz (Monocercomonadidae, Trichomonadida) as revealed by electron microscopy. J. Protozool. **16**, 668—685 (1969).

1107. — Honigberg, B. M., Daniel, W. A.: The mastigont system of *Trichomonas gallinae* Rivolta as revealed by electron microscopy. J. Protozool. **14**, 320—339 (1967).

1108. Matthes, D., Gräf, W.: Ein Flavobacterium (*Flavobacterium buchneri n. sp.*) als Endosymbiont zweier Sauginfusorien. Z. Morph. Ökol. Tiere **58**, 381—395 (1967).

1109. Maupas, E.: Le rajeunissement karyogamique chez les ciliés. Arch. Zool. exp. gén. 2e Sér. **7**, 149—517 (1889).

1110. McCracken, M. D., Starr, R. C.: Induction and development of reproductive cells in the K-32 strains of *Volvox rousseletii*. Arch. Protistenk. **112**, 262—282 (1970).

1111. McDonald, B. B.: Synthesis of deoxyribonucleic acid by micro- and macronuclei of *Tetrahymena pyriformis*. J. Cell Biol. **13**, 193—203 (1962).

1112. — The exchange of RNA and protein during conjugation in *Tetrahymena*. J. Protozool. **13**, 277—285 (1966).

1113. McGee-Russel, S. M., Allen, R. D.: Reversible stabilization of labile microtubules in the reticulopodial network of *Allogromia*. Advanc. Cell molec. Biol. **1**, 153—184 (1971).

**1114.* McLaughlin, J. J. A.: Axenic culture. In: McGraw Hill Encyclopedia of Science and Technology. pp. 698—701. 1960.

1115. — Zahl, P. A.: Endozoic Algae. In: Henry, S. M. (Ed.): Symbiosis, Vol. 1, pp. 257 to 297. New York: Academic Press 1966.

1116. Mego, J. L., Buetow, D. E.: Influence of cell division on the degree of streptomycin bleaching of *Euglena gracilis*. J. Protozool. **13**, 20—23 (1966).

1117. Meister, H.: Über die Modifizierbarkeit der DNS-Verteilung im Macronucleus bei *Stentor coeruleus* Ehrenberg. Arch. Protistenk. **112**, 314—342 (1970).

1118. Messer, G., Ben-Shaul, Y.: Fine structure of *Peridinium westii* Lemm., a freshwater dinoflagellate. J. Protozool. **16**, 272—280 (1969).

1119. Metz, C. B.: The nature and mode of action of the mating type substances. Amer. Naturalist **82**, 85—95 (1948).

**1120.* — Mating substances and the physiology of fertilization in ciliates. In: Wenrich, D. H.: Sex in microorganisms, pp. 284—334. Washington: Amer. Ass. Advanc. Sci. 1954.

1121. — Foley, M. T.: Fertilization studies on *Paramecium aurelia:* An experimental analysis of a non-conjugating stock. J. exp. Zool. **112**, 505—528 (1949).

1122. Metzner, F.: Zur Mechanik der Geißelbewegung. Biol. Zbl. **40**, 78—83 (1920).

1123. Meyer, H.: The fine structure of the flagellum and kinetoplast — chondriome of *Trypanosoma* (*Schizotrypanum*) *cruzi* in tissue culture. J. Protozool. **15**, 614—621 (1968).

1124. Mignot, J. P.: Quelques particularités de l'*Entosiphon sulcatum*, (Duj.) Stein; Flagellé Euglénien. C. R. Acad. Sci. (Paris) **257**, 2530 (1963).

1125. — Etude ultrastructurale de *Cyathomonas truncata* From. (Flagellé Cryptomonadine). J. Microscopie **4**, 239—252 (1965).

1126. — Ultrastructure des Eugléniens. I. Etude de la cuticule chez différentes espèces. Protistologica **1** (1), 5—15 (1965).

1127. — Etude ultrastructurale des Eugléniens: II. A, dictyosomes et dictyocinèse chez *Distigma proteus* Ehrbg. B, mastigonèmes chez *Anisonema costatum* Christen. Protistologica **1**, (2), 17—22 (1965).

1128. — Structure et ultrastructure de quelques Euglénomonadines. Protistologica **2** (3), 51—117 (1966).

1129. — Affinités des euglénomonadines et des chloromonadines. Remarques sur la systématique des Euglenida. Protistologica **3** (1), 25—60 (1967).

1130. — Structure et ultrastructure de quelques chloromonadines. Protistologica **3** (1), 5—24 (1967).

1131. — Brugerolle, G., Metenier, G.: Compléments à l'étude des mastigonèmes des Protistes flagellés. Utilisation de la technique de Thiery pour la mise en évidence des polysaccharides sur coupes fines. J. Microscop. **14**, 327—342 (1972).

1132. — Hovasse, R., Joyon, L.: Nouvelles données sur le fonctionnement des trichocystes (= taeniobolocystes) des Cryptomonadines. J. Microscop. **9**, 127—132 (1970).

1133. — Joyon, L., Kattar, M. R.: Sur la structure de la cinétide et sur les affinités systématiques de *Devescovina striata* Foa, Protozoaire flagellé. C. R. Acad. Sci. (Paris) **268**, 1738—1741 (1969).

1134. — — Pringsheim, E. G.: Quelques particularités structurals de *Cyanophora paradoxa* Korsch., Protozoaire flagellé. J. Protozool. **16**, 138—145 (1969).

1135. Milder, R., Deane, M. P.: Ultrastructure of *Trypanosoma conorhini* in the crithidial phase. J. Protozool. **14**, 65—72 (1967).

1136. — — The cytostome of *Trypanosoma cruzi* and *T. conorhini*. J. Protozool. **16**, 730—737 (1969).

1137. Millecchia, L. L., Rudzinska, M. A.: Basal body replication and ciliogenesis in a Suctorian, *Tokophrya infusionum*. J. Cell Biol. **46**, 553—563 (1970).

1138. — — The ultrastructure of brood pouch formation in *Tokophrya infusionum*. J. Protozool. **17**, 574—583 (1970).

1139. — — The permanence of the infraciliature in Suctoria: An electronmicroscopic study of pattern formation in *Tokophrya infusionum*. J. Protozool. **19**, 473—483 (1972).

1140. MILLER, JR., O. L., STONE, G. E.: Fine structure of the oral area of *Tetrahymena patula*. Protozool. **10**, 280—288 (1963).

1141. MISHRA, N. C., THRELKELD, S. F. H.: Genetic studies in *Eudorina*. Genet. Res. **11**, 21—31 (1968).

1142. MIYAKE, A.: Artifical induction of conjugation by chemical agents in *Paramecium aurelia, P. multimicronucleatum, P. caudatum* and between them. J. Protozool. **7**, 15 (1960).

1143. — Artifical induction of conjugation by chemical agents in *Paramecium* of the "*aurelia* group" and some of its applications to genetic work. Amer. Zoologist **1**, 373—374 (1961).

1144. — Induction of conjugation by cell-free preparations in *Paramecium multimicronucleatum*. Science **146**, 1583—1585 (1964).

1145. — Induction of conjugation by chemical agents in *Paramecium*. J. exp. Zool. **167**, 359—380 (1968).

*1145*a.— BEYER, J.: Cell interaction by means of soluble factors (gamones) in conjugation of *Blepharisma intermedium*. Exp. Cell Res. **76**, 15—24 (1973).

1146. MOORE, J., CANTOR, M. H., SHEELER, P., KAHN, W.: The ultrastructure of *Polytomella agilis*. J. Protozool. **17**, 671—676 (1970).

*1146*a. MOOSEKER, M. S., TILNEY, L. G.: Isolation and reactivation of the axostyle. Evidence for a dynein-like ATPase in the axostyle. J. Cell Biol. **56**, 13—26 (1973).

1147. MORIBER, L. G., HERSHENOV, B., AARONSON, S., BENSKY, B.: Teratological chloroplast structures in *Euglena gracilis* permanently bleached by exogenous physical and chemical agents. J. Protozool. **10**, 80—86 (1963).

1148. MORNIN, L., FRANCIS, D.: The fine structure of *Nematodinium armatum*, a naked dinoflagellate. J. Microscop. **6**, 759—772 (1967).

1149. MUDROW, L., REICHENOW, E.: Endotheliale und erythrocytäre Entwicklung von *Plasmodium praecox*. Arch. Protistenk. **97**, 101—170 (1944).

*1150. MUDROW-REICHENOW, L.: Unser heutiges Wissen von der Plasmodienentwicklung im Wirbeltier. Z. Tropenmed. Parasit. **1**, 113—152 (1949).

1151. — REICHENOW, E.: Die Entwicklung von *Plasmodium cathemerium* im Endothel und im Blut des Kanarienvogels. Zool. Jb. Anat. **70**, 129—168 (1949).

1152. MÜGGE, E.: Die Konjugation von *Vorticella campanula* (EHRBG.). Arch. Protistenk. **102**, 165—208 (1957).

1153. MÜHL, D.: Beitrag zur Kenntnis der Morphologie und Physiologie der Mehlwurmgregarinen. Arch. Protistenk. **43**, 361—414 (1921).

*1154. MÜHLPFORDT, H.: Über den Kinetoplasten der Flagellaten. Z. Tropenmed. Parasit. **15**, 289—323 (1964).

1155. MÜLLER, H.: Zur Phototaxis von *Amoeba proteus*. Exp. Cell Res. **39**, 225—232 (1965).

1156. MUELLER, J. A.: Separation of *kappa* particles with infective activity from those with killing activity and identification of the infective particles in *Paramecium aurelia*. Exp. Cell Res. **30**, 492—508 (1963).

1157. — Cofactor for infection by *kappa* in *Paramecium aurelia*. Exp. Cell Res. **35**, 464—476 (1964).

1158. — Vitally stained *kappa* in *Paramecium aurelia*. J. exp. Zool. **160**, 369—372 (1965).

1159. — *Kappa*-affected paramecia develop immunity. J. Protozool. **12**, 278—281 (1965).

1160. — Resistance of *kappa*-bearing paramecia to *kappa* toxin. Exp. Cell Res. **41**, 131—137 (1966).

1161. — SONNEBORN, T. M.: Killer action on cells other than *Paramecium*. Anat. Rec. **134**, 613 (1959).

1162. MÜLLER, M.: Studies on feeding and digestion in Protozoa. V. Demonstration of some phosphatases and carboxylic esterases in *Paramecium multimicronucleatum* by biochemical methods. Acta biol. hung. **13**, 283—297 (1962).

1163. — RÖHLICH, P.: Studies on feeding and digestion in Protozoa. II. Food vacuole cycle in *Tetrahymena corlissi*. Acta morph. hung. **10**, 297—305 (1961).

1164. — — TÖRÖ, I.: Studies on feeding and digestion in Protozoa. VII. Ingestion of polystyrene latex particles and its early effect on acid phosphatase in *Paramecium multimicronucleatum* and *Tetrahymena pyriformis*. J. Protozool. **12**, 27—34 (1965).

1165. Müller, M., Röhlich, P., Toth, J., Törö, I.: Fine structure and enzymic activity of protozoan food vacuoles. In: De Reuch, A. V. S., Cameron, M. P.: Ciba foundation symposium on lysosomes, pp. 201—216. London: J. A. Churchill, Ltd. 1963.
1166. — Törö, I.: Studies on feeding and digestion in Protozoa. III. Acid phosphatase activity in food vacuoles of *Paramecium multimicronucleatum*. J. Protozool. **9**, 98—102 (1962).
1167. — — Polgár, M., Druga, A.: Studies on feeding and digestion in Protozoa. VI. The effect of ingestion of non-nutritive particles on acid phosphatase in *Paramecium multimicronucleatum*. Acta biol. hung. **14**, 209—213 (1963).
1168. Myers, E. H.: The life history of *Patellina corrugata* Wiliamson, a foraminifer. Bull. Scripps Inst. Oceanogr. tech. Ser. Univ. Calif. **3**, 355—392 (1935).
1169. — The life-cycle of *Spirillina vivipara* Ehrenberg, with notes on morphogenesis, systematics and distribution of the Foraminifera. J. roy. micr. Soc. **56**, 120—146 (1936).
1170. — Biology, ecology and morphogenesis of a pelagic foraminifer. Stanford Univ. Publs. Biol. Sci. **9**, 5—30 (1943).
1171. Nabih, A.: Studien über die Gattung *Klossia* und Beschreibung des Lebenscyclus von *Klossia loossi* (*nov. sp.*). Arch. Protistenk. **91**, 474—515 (1938).
1172. Nachmias, V. T.: Fibrillar structures in the cytoplasm of *Chaos chaos*. J. Cell Biol. **23**, 183—188 (1964).
1173. — Inhibition of streaming in *Amoeba* by pinocytosis inducers. Possible mechanism. Exp. Cell Res. **51**, 347—361 (1968).
1174. Naitoh, Y.: Direct current stimulation of *Opalina* with intracellular microelectrode. Annotationes zool. Jap. **31**, 59—73 (1958).
1175. — Local chemical stimulation of *Opalina*. I. The mode of action of K ions to induce ciliary reversal. Zool. Mag. **70**, 435—446 (1960).
1176. — Ciliary responses of *Paramecium* to external application of various chemicals under different ionic conditions. Zool. Mag. **73**, 207—212 (1964).
1177. — Effect of change in external Ca-concentration on the membrane potential of *Opalina*. Zool. Mag. **73**, 233—238 (1964).
1178. — Reversal response elicited in nonbeating cilia of *Paramecium* by membrane depolarization. Science **154**, 660—662 (1966).
1179. — Ionic control of the reversal response of cilia in *Paramecium caudatum:* A calcium hypothesis. J. gen. Physiol. **51**, 85—103 (1968).
1180. — Electrical properties of *Paramecium caudatum:* All-or-none electrogenesis. Z. vergl. Physiol. **61**, 453—472 (1968).
1181. — Ciliary orientation: Controlled by cell membrane or by intracellular fibrils? Science **166**, 1633—1635 (1969).
1182. — Ionic mechanism controlling behavioral responses of *Paramecium* to mechanical stimulation. Science **164**, 963—965 (1969).
1183. — Eckert, R.: Electrical properties of *Paramecium caudatum:* Modification by bound and free cations. Z. vergl. Physiol. **61**, 427—462 (1968).
1184. — Kaneko, H.: ATP-Mg-reactivated tritonextracted models of *Paramecium:* Modification of ciliary movement by calcium ions. Science **176**, 523—524 (1972).
1185. Nakata, A.: Mating types in *Glaucoma scintillans*. J. Protozool. **16**, 689—692 (1969).
1186. Nanney, D. L.: Caryonidal inheritance and nuclear differentiation. Amer. Naturalist **90**, 291—307 (1956).
1187. — Mating type inheritance at conjugation in variety 4 of *Paramecium aurelia*. J. Protozool. **4**, 89—95 (1957).
1188. — Inbreeding degeneration in *Tetrahymena*. Genetics **42**, 137—146 (1957).
1189. — Epigenetic control systems. Proc. nat. Acad. Sci. (Wash.) **44**, 712—717 (1958).
1190.* — Epigenetic factors affecting mating type expression in certain ciliates. Cold Spr. Harb. Symp. quant. Biol. **23, 327—335 (1958).
1191. — Genetic factors affecting mating type frequencies in variety 1 of *Tetrahymena pyriformis*. Genetics **44**, 1173—1184 (1959).
1192. — Serotype determination in *Tetrahymena pyriformis*, variety 1. Rec. Genet. Soc. Amer. **28**, 89 (1959).

1193. NANNEY, D. L.: Vegetative mutants and clonal senility in *Tetrahymena*. J. Protozool. **6**, 171—177 (1959).

1194. — Temperature effects on nuclear differentiation in variety 1 of *Tetrahymena pyriformis*. Physiol. Zool. **33**, 146—151 (1960).

1195. — The relationship between the mating type and the H serotype systems in *Tetrahymena*. Genetics **45**, 1351—1358 (1960).

1196. — Anomalous serotypes in *Tetrahymena*. J. Protozool. **9**, 485—487 (1962).

**1197.* — Cytoplasmic inheritance in Protozoa. In: BURDETTE, W. J.: Methodology in basic genetics, pp. 355—380. San Francisco: Holden-Day 1963.

1198. — The inheritance of H—L serotype differences at conjugation in *Tetrahymena*. J. Protozool. **10**, 152—155 (1963).

1199. — Irregular genetic transmission in *Tetrahymena* crosses. Genetics **48**, 737—744 (1963).

1200. — Macronuclear differentiation and subnuclear assortment in ciliates. In: LOCKE, M.: Role of chromosomes in development, pp. 253—273. New York-London: Academic Press 1964.

1201. — Corticotype transmission in *Tetrahymena*. Genetics **54**, 955—968 (1966).

1202. — Cortical integration in *Tetrahymena*: An exercise in Cytogeometry. J. exp. Zool. **161**, 307—318 (1966).

1203. — Corticotypic technics in *Tetrahymena* taxonomy. J. Protozool. **13**, 483—490 (1966).

1204. — Corticotypes in *Tetrahymena pyriformis*. Amer. Naturalist **100**, 303—318 (1966).

1205. — Comparative corticotype analysis in *Tetrahymena*. J. Protozool. **14**, 690—697 (1967).

1206. — Patterns of cortical stability in *Tetrahymena*. J. Protozool. **15**, 109—112 (1968).

1207.* — Cortical patterns in cellular morphogenesis. Science **160, 496—502 (1968).

1208. — Cortical characteristics of strains of syngens 10, 11 and 12 of of *Tetrahymena pyriformis*. J. Protozool. **18**, 33—36 (1971).

1209. — The constancy of cortical units in *Tetrahymena* with varying numbers of ciliary rows. J. exp. Zool. **178**, 177—182 (1971).

1210. — ALLEN, S. K.: Intranuclear co-ordination in *Tetrahymena*. Physiol. Zool. **32**, 221—229 (1959).

1211. — CAUGHEY, P. A.: Mating type determination in *Tetrahymena pyriformis*. Proc. nat. Acad. Sci. (Wash.) **39**, 1057—1063 (1953).

1212. — — An unstable nuclear condition in *Tetrahymena pyriformis*. Genetics **40**, 388—398 (1955).

1213. — — TEFANKJIAN, A.: The genetic control of mating type potentialities in *Tetrahymena pyriformis*. Genetics **40**, 668—680 (1955).

1214. — DUBERT, J. M.: The genetics of the H serotype system in variety 1 of *Tetrahymena pyriformis*. Genetics **45**, 1335—1358 (1960).

1215. — NAGEL, M. J.: Nuclear misbehavior in an aberrant inbred *Tetrahymena*. J. Protozool. **11**, 465—473 (1964).

1216. — — TOUCHBERRY, R. W.: The timing of H antigenic differentiation in *Tetrahymena*. J. exp. Zool. **155**, 25—42 (1964).

1217. — REEVE, S. J., NAGEL, J., DE PINTO, S.: H serotype differentiation in *Tetrahymena*. Genetics **48**, 803—813 (1963).

**1218.* — RUDZINSKA, M. A.: Protozoa. In: BRACHET, J., MIRSKY, A. E.: The cell, Vol. 4, pp. 109—149. New York-London: Academic Press 1960.

1219. NARASIMHA, R., AMMERMANN, D.: Polytene chromosomes and nucleic acid metabolism during macronuclear development in *Euplotes*. Chromosoma (Berl.) **29**, 246—254 (1970).

1220. NAVILLE, A.: Recherches sur le cycle sporogonique des *Aggregata*. Revue suisse Zool. **32**, 12—179 (1925).

1221. — Recherches sur le cycle évolutif et chromosomique de *Klossia helicina* (A. SCHNEIDER). Arch. Protistenk. **57**, 427—474 (1927).

1222. — Recherches cytologiques sur les Schizogrégarines. I. Le cycle évolutif de *Mattesia dispora* n. g. n. sp. Z. Zellforsch. **11**, 375—396 (1930).

1223. NELSEN, E. M.: Division delays and abnormal oral development produced by colchicine in *Tetrahymena*. J. exp. Zool. **175**, 69—84 (1970).

1224. NENNINGER, U.: Die Peritrichen der Umgebung von Erlangen mit besonderer Berücksichtigung ihrer Wirtsspezifität. Zool. Jb. (Systematik) **77**, 169—266 (1948).

1225. NETZEL, H.: Die Schalenbildung bei der Thekamöben-Gattung *Arcella* (Rhizopoda, Testacea). Cytobiol. **3**, 89—92 (1971).

1226. — Die Schalenbildung bei *Difflugia oviformis* (Rhizopoda, Testacea). Z. Zellforsch. **135**, 55—61 (1972).

1227. — Die Bildung der Gehäusewand bei der Thekamöbe *Centropyxis discoides* (Rhizopoda, Testacea). Z. Zellforsch. **135**, 45—54 (1972).

1228. — Morphogenese des Gehäuses von *Euglypha rotunda* (Rhizopoda, Testacea). Z. Zellforsch. **135**, 63—69 (1972).

1229. — HEUNERT, H. H.: Die Zellteilung bei *Arcella vulgaris* var. *multinucleata* (Rhizopoda, Testacea). Arch. Protistenk. **112**, 285—292 (1971).

1230. NEWTON, B. A.: Biochemical peculiarities of trypanosomatid flagellates. Ann. Rev. Microbiol. **22**, 109—130 (1968).

1231. NIELSEN, M. H., LUDWIG, J., NIELSEN, R.: On the ultrastructure of *Trichomonas vaginalis* DONNÉ. J. Microscopie **5**, 229—250 (1966).

1232. NILSSON, J. R.: Suggestive structural evidence for macronuclear "subnuclei" in *Tetrahymena pyriformis* GL. J. Protozool. **17**, 539—548 (1970).

1233. NOBILI, R.: On conjugation between *Euplotes vannus* O. F. MÜLLER and *Euplots minuta* YOCUM. Crayologia **17**, 393—397 (1964).

1234. — Mating types and mating type inheritance in *Eugplotes minuta* YOCUM (Ciliata, Hypotrichida). J. Protozool. **13**, 38—41 (1966).

1235. — LUPORINI, P.: Maintenance of heterozygosity at the *mt* locus after autogamy in *Euplotes minuta* (Ciliata, Hypotrichida). Genet. Res. **10**, 35—43 (1967).

1236. NOIROT-TIMOTHÉE, C.: Recherches sur l'ultrastructure d'*Opalina ranarum*. Ann. Sci. nat. Zool. 12e sér., 265—281 (1959).

1237. — Etude d'une Famille des ciliés: les "Ophryoscolecidae". Structures et ultrastructures (Thèse, Paris). Ann. Sci. nat. Zool. 12e sér, 527—718 (1960).

1238. — Présence simultanée de deux types de vésicules de micropinocytose chez *Cepedea dimidiata* (Protozoa Opalinina). C. R. Acad. Sci. (Paris) **263**, 1230—1233 (1966).

1239. — LOM, J.: L'ultrastructure de l'haplocinétie des cilié péritriches. Comparaison avec la membrane ondulante des hyménostomes. Protistologica **1** (1), 33—40 (1965).

1240. NOLAND, L. E.: Conjugation in the ciliate *Metopus sigmoides* C. and L. J. Morph. **44**, 341—361 (1927).

1241.* — Protoplasmic streaming: a perennial puzzle. J. Protozool. **4, 1—6 (1957).

1242. — GOJDICS, M.: Ecology of free-living Protozoa. In: CHEN, T.-T. (Ed.): Research in Protozoology, Vol. 2, pp. 215—266. Oxford: Pergamon Press 1967.

1243. NUTTING, W. L., CLEVELAND, L. R.: Effects of glandular extirpations on *Cryptocercus* and the sexual cycles of its Protozoa. J. exp. Zool. **137**, 13—37 (1958).

1244. O'DONNELL, E. H. J.: Nucleolus and chromosomes in *Euglena gracilis*. Cytologia (Tokyo) **30**, 118—154 (1965).

1245. OKAJIMA, A.: Studies on the metachronal wave in *Opalina*. I. Electrical stimulation with the micro-electrode. Jap. J. Zool. **11**, 87—100 (1953).

1246. — Studies on the metachronal wave in *Opalina*. II. The regulating mechanism of ciliary metachronism and of ciliary reversal. Annotationes Zool. japon. **27**, 40—45 (1954).

1247. — Studies on the metachronal wave in *Opalina*. III. Time-change of effectiveness of chemical and electrical stimuli during adaption in various media. Annotationes Zool. japon. **27**, 46—51 (1954).

1248. — KINOSITA, H.: Ciliary activity and coordination in *Euplotes eurystomus*. I. Effect of microdissection of neuromotor fibers. Comp. Biochem. Physiol. **19**, 115—131 (1966).

1249. O'NEILL, C. H.: Isolation and properties of the cell surface membrane of *Amoeba proteus*. Exp. Cell Res. **35**, 477—496 (1964).

1250. — WOLPERT, L.: Isolation of the cell membrane of *Amoeba proteus*. Exp. Cell Res. **24**, 592—595 (1961).

1251. ORD, M. J.: The synthesis of DNA through the cell cycle of *Amoeba proteus*. J. Cell Sci. **3**, 483—491 (1968).

1252. Ord, M. J.: The viability of the anucleate cytoplasm of *Amoeba proteus*. J. Cell Sci. **3**, 81—88 (1968).
1253. — Immediate and delayed effects of N-methyl-N-nitrosourethane on *Amoeba proteus*. Exp. Cell Res. **53**, 73—84 (1968).
1254. — Mutations induced in *Amoeba proteus* by the carcinogen N-methyl-N-nitrosourethane. J. Cell Sci. **7**, 531—548 (1970).
1255. Orias, E.: Mating-type determination in variety 8, *Tetrahymena pyriformis*. Genetics **48**, 1509—1518 (1963).
1256. Ossipov, D. V.: Methods of obtaining homozygous *Paramecium caudatum* clones. Genetika **2**, 41—48 (1966) (in Russian).
1257. — Analysis of hereditary mechanisms determining thermostability of *Paramecium caudatum*. Genetika **1**, 119—131 (1966) (in Russian).
1258. — On macronuclear regeneration in *Paramecium caudatum*. Cytologia **8**, 108—110 (1966) (in Russian).
1259. — Skoblo, I. I.: The autogamy during conjugation in *Paramecium caudatum* Ehrbg. II. The ex-autogamont stages of nuclear reorganization. Acta Protozool. **6**, 33—48 (1968).
1260. Outka, D. E.: Conditions for mating and inheritance of mating type in variety seven of *Tetrahymena pyriformis*. J. Protozool. **8**, 179—184 (1961).
1261. — The amoeba-to-flagellate transformation in *Tetramitus rostratus*. I. Population dynamics. J. Protozool. **12**, 85—93 (1965).
1262. — Williams, D. C.: Sequential coccolith morphogenesis in *Hymenomonas carterae*. J. Protozool. **18**, 285—298 (1971).
1263. Ovchinnikova, L. P.: Variability of DNA content in micronuclei of *Paramecium bursaria*. Acta Protozool. **7**, 211—220 (1970).
1264. — Selivanova, G. V., Cheissin, E. M.: Photometric study of the DNA content in the nuclei of *Spirostomum ambiguum* (Ciliata, Heterotricha). Acta Protozool. **3**, 69—78 (1965).
1265. Pantin, C.: On the physiology of amoeboid movement. I. J. Mar. Biol. Ass. U. K. **13**, 24—69 (1923).
1266. Pappas, G. D.: The fine structure of the nuclear envelope of *Amoeba proteus*. J. biophys. biochem. Cytol. **2**, 431—434 (1956).
1267. — Helical structures in the nucleus of *Amoeba proteus*. J. biophys. biochem. Cytol. **2**, 221—222 (1956).
1268. — Electron microscope studies on Amoebae. Ann. N. Y. Acad. Sci. **78**, Art. 2, 448—473 (1959).
1269. — Brandt, P. W.: The fine structure of the contractile vacuole in ameba. J. biophys. biochem. Cytol. **4**, 485—488 (1958).
1270. — — Mitochondria. I. Fine structure of the complex patterns in the mitochondria of *Pelomyxa carolinensis* Wilson (*Chaos chaos* L.). J. biophys. biochem. Cytol. **6**, 85—90 (1959).
1271. Parducz, B.: Die Fixation als Reizwirkung in der Tätigkeit der Zellorganellen. Acta biol. hung. **3**, 1—17 (1952).
1272. — Reizphysiologische Untersuchungen an Ziliaten. I. Über das Aktionssystem von *Paramecium*. Acta microbiol. hung. **1**, 175—221 (1954).
1273. — Reizphysiologische Untersuchungen an Ziliaten. II. Neuere Beiträge zum Bewegungs- und Koordinationsmechanismus der Ziliatur. Acta biol. hung. **5**, 169—212 (1954).
1274. — Reizphysiologische Untersuchungen an Ziliaten. III. Über die Tätigkeit der Peristomalcilien von *Paramecium*. Ann. hist.-nat. Mus. nat. hung. (Ser. Nov.) **6**, 189—195 (1955).
1275. — Reizphysiologische Untersuchungen an Ziliaten. IV. Über das Empfindungs- bzw. Reaktionsvermögen von *Paramecium*. Acta biol. hung. **6**, 289—316 (1956).
1276. — Reizphysiologische Untersuchungen an Ziliaten. V. Zum physiologischen Mechanismus der sog. Fluchtreaktion und der Raumorientierung. Acta biol. hung. **7**, 73—99 (1956).

1277. Parducz, B.: Reizphysiologische Untersuchungen an Ziliaten. VI. Eine interessante Variante der Fluchtreaktion bei *Paramecium*. Ann. hist.-nat. Mus. nat. hung. (Ser. Nov.) **7**, 363—370 (1956).
1278. — Reizphysiologische Untersuchungen an Ziliaten. VII. Das Problem der vorbestimmten Leitungsbahnen. Acta biol. hung. 8, 219—251 (1958).
1279. — Reizphysiologische Untersuchungen an Ziliaten. VIII. Ablauf der Fluchtreaktion bei allseitiger und anhaltender Reizung. Ann. hist.-nat. Mus. nat. hung. **51**, 227—246 (1959).
1280. — Bewegungsbilder über Didinien. Ann. hist.-nat. Mus. nat. hung. (Pars Zool.) **53**, 267—280 (1961).
1281. — Die ektoplasmatischen Fibrillensysteme der Ciliaten im Lichte der neueren elektronenmikroskopischen Befunde. Biol. Közlem. **9**, 41—54 (1961).
1282. — Studies on reactions to stimuli in ciliates. IX. Ciliary coordination of right spiraling paramecia. Ann. hist.-nat. Mus. nat. hung. (Pars Zool.) **54**, 221—230 (1962).
1283. — On a new concept of cortical organization in *Paramecium*. Acta biol. hung. **13**, 299—322 (1962).
1284. — Reizphysiologische Untersuchungen an Ciliaten. X. „Momentbilder" über galvanotaktisch frei schwimmende Paramecien. Acta biol. hung. **13**, 421—429 (1963).
1285. — On the nature of metachronal ciliary control in *Paramecium*. J. Protozool. **9** (Suppl.), 27 (1962).
1286. — Swimming and its ciliary mechanism in *Ophryoglena sp*. Acta Protozool. **2**, 267—374 (1964).
1287.* — Ciliary movement and coordination in ciliates. Int. Rev. Cytol. **21, 91—128 (1967).
1288. Parker, J. W., Giese, A. C.: Nuclear activity during regeneration in *Blepharisma intermedium* Bhandary. J. Protozool. **13**, 617—622 (1966).
1289. Parke, M., Manton, I.: Preliminary observations on the fine structure of *Prasinocladus marinus*. J. Mar. Biol. Ass. U. K. **45** (2), 525—536 (1965).
1290. — — The specific identity of the algal symbiont in *Convoluta roscoffensis*. J. Mar. Ass. U. K. **47**, 445—464 (1967).
1291. Parsons, J. A.: Mitochondrial incorporation of tritiated thymidine in *Tetrahymena pyriformis*. J. Cell Biol. **25**, 641—646 (1965).
1291a. Pasternak, J.: Differential genic activity in *Paramecium aurelia*. J. exp. Zool. **165**, 395—418 (1967).
1292. Paulin, J. J.: The fine structure of *Nyctotherus cordiformis* (Ehrenberg). J. Protozool. **14**, 183—196 (1967).
1293. — Bussey, J.: Oral regeneration in the ciliate *Stentor coeruleus:* A scanning and transmission electron optical study. J. Protozool. **18**, 201—213 (1971).
1294. — Corliss, J. O.: Ultrastructural and other observations which suggest suctorian affinities for the taxonomically enigmatic ciliate *Cyathodinium*. J. Protozool. **16**, 216—223 (1969).
1295. Pease, D.: The ultrastructure of flagellar fibrils. J. Cell Biol. **18**, 313—326 (1963).
1296. Pérez-Silva, J., Alonso, P.: Demonstration of polytene chromosomes in the macronuclear anlage of oxytrichous ciliates. Arch. Protistenk. **109**, 65—70 (1966).
1297. Perkins, F. O.: Formation of centriole and centriole-like structures during meiosis and mitosis in *Labyrinthula* sp. (Rhizopodea. Labyrinthulida). J. Cell Sci. **6**, 629—653 (1970).
1298. — Amon, J. P.: Zoosporulation in *Labyrinthula* sp., an electron microscope study. J. Protozool. **16**, 235—257 (1969).
1299. Peters, U.: Orts- und Geißelbewegung bei marinen Dinoflagellaten. Arch. Protistenk. **67**, 291—321 (1929).
1300. Phillips, R. B.: Inheritance of T serotypes in *Tetrahymena*. Genetics **56**, 667—681 (1967).
1301. — T serotype differentiation in *Tetrahymena*. Genetics **56**, 683—692 (1967).
1302. — Mating-type inheritance in syngen 7 of *Tetrahymena pyriformis:* intra- and interallelic interactions. Genetics **63**, 349—359 (1969).
1303. — Induction of competence for mating in *Tetrahymena* by cell-free fluids. J. Protozool. **18**, 163—165 (1971).

1304. Phillips, R. B., Abraham, I.: Mating type inheritance in *Glaucoma*. J. Protozool. **16**, 375—377 (1969).

1305. Pickett-Heaps, J.: The autonomy of the centriole: Fact or fallacy? Cytobios. **3**, 205—214 (1971).

1306. Piekarski, G.: Endomitose beim Großkern der Ciliaten? Versuch einer Synthese. Biol. Zbl. **61**, 416—426 (1941).

1307. Pietrowicz-Kosmynka, D.: The influence of definite ionic medium on the negative chemotaxis in *Stentor coeruleus*. Acta Protozool. **9**, 305—321 (1971).

1308. — Chemotactic effects of cations and of pH on *Stentor coeruleus*. Acta Protozool. **9**, 235—245 (1971).

1309. — The potassium-calcium equilibrum and chemotactic sensitivity in *Stentor coeruleus*. Acta Protozool. **9**, 349—363 (1972).

**1310.* Pitelka, D. R.: Electron microscopic structure of Protozoa. 269 S. New York-Oxford: Pergamon Press 1963.

1311. — New observations on cortical ultrastructure in *Paramecium*. J. Microscopie **4**, 373—394 (1965).

**1312.* — Fibrillar systems in Protozoa. In: Chen, T.-T. (Ed.): Research in Protozoology, Vol. 3, pp. 279—388. New York: Pergamon Press 1969.

1313. — Fibrillar structures of the ciliate cortex: The organization of kinetosomal territories. Progr. Protozool. Leningrad **1969**, 44—46.

1314.* — Ciliate ultrastructure: Some problems in cell biology. J. Protozool. **17, 1—10 (1970).

**1315.* — Child, F. M.: The locomotor apparatus of ciliates and flagellates: Relations between structure and function. In: Hutner, S. H.: Biochemistry and physiology of protozoa, Vol. 3, pp. 131—198. New York-London: Academic Press 1964.

1316. Pokorny, K. L.: *Labyrinthula*. J. Protozool. **14**, 697—708 (1967).

1317. Poljansky, G.: Geschlechtsprozesse bei *Bursaria truncatella* O. F. Müll. Arch. Protistenk. **81**, 420—546 (1934).

1318.* — Raikov, I. B.: The role of polyploidy in the evolution of Protozoa. Cytologia (Moskau) **2, 509—518 (1960) (in Russian).

1319.* — — Nature et origine du dualisme nucléaire chez les infusoires ciliés. Bull. Soc. zool. Fr. **86, 402—411 (1961).

1320. Pollard, T. D., Ito, S.: Cytoplasmic filaments of *Amoeba proteus*. I. The role of filaments in consistency change and movement. J. Cell Biol. **46**, 267—289 (1970).

1321. — Korn, E. D.: Filaments of *Amoeba proteus*. II. Binding of heavy meromyosin by thin filaments in motile cytoplasmic extracts. J. Cell Biol. **48**, 216—218 (1971).

1322. — Shelton, E., Weihing, R. R., Korn, E. D.: Ultrastructural characterization of F-actin isolated from *Acanthamoeba castellanii* and identification of cytoplasmic filaments as F-actin by reaction with rabbit heavy meromysin. J. molec. Biol. **50**, 91—97 (1970).

1323. Porchet-Henneré, E.: Etude des premiers stades de développement de la Coccidie *Coelotropha durchoni*. Z. Zellforsch. **80**, 556—569 (1967).

1324. — Corrélations entre le cycle biologique d'une Coccidie: *Coelotropha durchoni* Vivier, et celui de son hôte: *Nereis diversicolor* O. F. Müller (Annélide Polychaete). Z. Parasitenk. **31**, 299—314 (1969).

1325. — La microgamétogenèse chez la Coccidie *Coelotropha durchoni* Vivier-Henneré; étude au microscope électronique. Arch. Protistenk. **112**, 21—29 (1970).

1326. — La fécondation et la sporogenèse chez la Coccidie *Coelotropha durchoni*. Etude en microscopie photonique et électronique. Z. Parasitenk. **37**, 94—125 (1971).

1327. — Richard, A.: Structure fine du sporoblaste immature uninuclée d'*Aggregata eberthi* Labbe (Sporozoaire coccidiomorphe). C. R. Acad. Sci. (Paris) **269**, 1681—1683 (1969).

1328. — — Ultrastructure des stades végétatifs d'*Aggregata eberthi* Labbe: le trophozoite et le schizonte. Z. Zellforsch. **103**, 179—191 (1970).

1329. — — Structure fine des microgamètes d'*Aggregata eberthi* Labbe. Protistologica **6**, 71—81 (1970).

1330. — — La schizogonie chez *Aggregata eberthi*. Etude en microscope électronique. Protistologica **7**, 227—260 (1971).

1331. Porchet-Henneré, E., Richard, A.: La sporogenèse chez la Coccidie *Aggregata eberthi*. Etude en microscope électronique. J. Protozool. **18**, 614—629 (1971).

1332.* — Vivier, E.: Ultrastructure comparée des germes infectieux (sporozoites, mérozoites, schizontes, endozoites etc.) chez les Sporozoaires. Ann. Biol. **10, 77—113 (1971).

1333. Porter, E. D.: The buccal organelles in *Paramecium aurelia* during fission and conjugation with special reference to the kinetosomes. J. Protozool. **7**, 211—217 (1960).

1334. Preer, J. R.: Microscopically visible bodies in the cytoplasm of the "killer" strains of *Paramecium aurelia*. Genetics **35**, 344—362 (1950).

**1335.* — Genetics of the Protozoa. In: Chen, T.-T. (Ed.): Research in Protozoology, Vol. 3, pp. 129—278. New York: Pergamon Press 1969.

1336.* — Extrachromosomal inheritance: Hereditary symbionts, mitochondria, chloroplasts. Ann. Rev. Genet. **5, 361—406 (1971).

1337. — Hufnagel, L. A., Preer, L. B.: Structure and behavior of "R" bodies from killer paramecia. J. Ultrastruct. Res. **15**, 131—143 (1966).

1338. — Jurand, A.: The relation between virus-like particles and R-bodies of *Paramecium aurelia*. Genet. Res. Camb. **12**, 331—340 (1968).

1339. — Preer, L. B., Rudman, B., Jurand, A.: Isolation and composition of bacteriophage-like particles from *kappa* of killer Paramecia. Molec. Gen. Genet. **111**, 202—208 (1971).

1340. Preer, L. B.: Alpha, an infectious macronuclear symbiont of *Paramecium aurelia*. J. Protozool. **16**, 570—578 (1969).

1341. — A study of killer stock 562 of *Paramecium aurelia:* killing effects of *kappa* and ultrastructure of the R-body. Progr. Protozool. Abstr. 3rd Int. Congr. Protozool. **118**.

1342. — Jurand, A., Preer, J. R., Rudman, B. M.: The classes of *kappa* in *Paramecium aurelia*. J. Cell Sci. **11**, 581—599 (1972).

1343. — Preer, J. R.: Killing activity from lysed *kappa* particles of *Paramecium*. Genet. Res. **5**, 230—239 (1964).

1344. Preer, J. R., Preer, L. B.: Virus-like bodies in killer paramecia. Proc. nat. Acad. Sci. (Wash.) **58**, 1774—1781 (1967).

1345. Prescott, D. M.: Nuclear synthesis of cytoplasmic ribonucleic acid in *Amoeba proteus*. J. biophys. biochem. Cytol. **6**, 203—206 (1959).

1346. — The nuclear dependence of RNA synthesis in *Acanthamoeba sp.* Exp. Cell Res. **19**, 29—34 (1960).

1347. — Nucleic acid and protein metabolism in the macronuclei of two ciliated Protozoa. J. Histochem. Cytochem. **10**, 145—153 (1962).

1348. — The syntheses of total macronuclear protein, histone, and DNA during the cell cycle in *Euplotes eurystomus*. J. Cell Biol. **31**, 1—9 (1966).

1349. — Kimball, R. F.: Relation between RNA, DNA, and protein syntheses in the replicating nucleus of *Euplotes*. Proc. nat. Acad. Sci. **47**, 686—693 (1961).

1350. — — Carrier, R. F.: Comparison between the timing of micronuclear and macronuclear DNA synthesis in *Euplotes eurystomus*. J. Cell Biol. **13**, 175—176 (1962).

**1351.* — Stone, G. E.: Replication and function of the protozoan nucleus. In: Chen, T.-T. (Ed.): Research in Protozoology, Vol. 2, pp. 117—146. Oxford: Pergamon Press 1967.

1352. Preston, T. M.: The form and function of the cytostome-cytopharynx of the culture forms of the elasmobranch haemoflagellate *Trypanosoma raiea* Laveran & Mesnil. J. Protozool. **16**, 320—333 (1969).

**1353.* Pringsheim, E.: Algenreinkulturen. 109 S. Jena: Fischer 1954.

**1354.* — Farblose Algen. Ein Beitrag zur Evolutionsforschung. 471 S. Stuttgart: Fischer 1963.

1355.* — Entwicklung und biologische Bedeutung der Sexualität. Biol. Zbl. **83, 739—756 (1964).

1356. — Ondracek, K.: Untersuchungen über die Geschlechtsvorgänge bei *Polytoma*. Beih. Bot. Zbl. **59**, 118—172 (1939).

1357.* Provasoli, L.: Nutrition and ecology of Protozoa and Algae. Ann. Rev. Microbiol. **12, 279—308 (1958).

1358. Prusch, R. D., Dunham, P. B.: Contraction of isolated contractile vacuoles from *Amoeba proteus*. J. Cell Biol. **46**, 431—434 (1970).

1359. Puytorac, P. de: Observations sur l'ultrastructure de la microsporidie *Mrazekia lumbriculi*, Jirovec. J. Microscopie **1**, 39—46 (1962).

1360. — Contribution à l'étude des ciliés astomes Haptophryidae Cépède, 1903 (Cytologie. Ultrastructure, Taxinomie). Ann. Sci. nat. Zool. Biol. Animale (12e sér.) **5**, 173—190 (1963).

1361. — Observations sur l'ultrastructure du cilié astome. *Mesnilella trispiculata* K. J. Microscopie **2**, 189—196 (1963).

1362. — L'ultrastructure des cnidocystes de l'actinomyxidie: *Sphaeractinomyxon amanieui sp. now.* C. R. Acad. Sci. (Paris) **256**, 1594—1596 (1963).

1363. — Sur le cytosquelette de quelques ciliés Hysterocinetidae. Arch. Zool. exp. gén. **102**, 213—224 (1963).

1364. — Quelques aspects de l'ultrastructure du cilié: *Prorodon viridis* Ehrbg. Kahl. Acta Protozool. **2**, 147—151 (1964).

1365. — Sur l'ultrastructure des trichocystes mucifères chez le cilié *Holophrya vesiculosa* Kahl. C. R. Séanc. Soc. Biol. **158**, 526—528 (1964).

1366. — Kattar, M. R.: Observations sur l'ultrastructure du Cilié *Helicoporodon multinucleatum* Dragesco 1960. Protistologica **5**, 549—560 (1969).

1367. — Njine, T.: Sur l'ultrastructure des *Loxodes* (Ciliés Holotriches Trichostomes). Protistologica **6**, 427—444 (1970).

1368. Pyne, C. K.: L'ultrastructure de l'appareil basal des flagelles chez *Cryptobia helicis* (Flagellé Bodonidae). C. R. Acad. Sci. (Paris) **250**, 1912 (1960).

1369. — Etudes sur la structure inframicroscopique du cinétoplaste chez *Leishmania tropica*. C. R. Acad. Sci. (Paris) **251**, 2776—2778 (1960).

1370. — Etudes préliminaires sur l'organisation de la chromatine pendant l'interphase chez *Euglena gracilis*. Protistologica **3**, 291—294 (1967).

1371. — L'ultrastructure des flagellés des familles Trypanosomidae et Bodonidae. Ann. Sci. nat. Zool. **9**, 23—424 (1967).

1372. — Sur l'absence d'incorporation de la thymidine tritiée dans les cinétosomes de *Tetrahymena pyriformis*. C. R. Acad. Sci. (Paris) **267**, 755—757 (1968).

1373. — Tuffrau, M.: Structure et ultrastructure de l'appareil cytopharyngien et des tubules complexes en relation avec celui-ci chez le Cilié Gymnostome *Chilodonella uncinata* Ehrbg. J. Microscop. **9**, 503—516 (1970).

1374. Raabe, H.: L'appareil nucléaire d'*Urostyla grandis* Ehrbg. I. Appareil micronucléaire. Ann. Univ. M. Curie-Sklodowska, Sect. C. **1**, 1—34 (1946).

1375. — L'appareil nucléaire d'*Urostyla grandis* Ehrbg. II. Appareil macronucléaire. Ann. Univ. M. Curie-Sklodowska, Sect. C. **1**, 133—170 (1947).

1376.* Raabe, Z.: Remarks on the principles and outline of the system of Protozoa. Acta Protozool. **2, 1—18 (1964).

1377. — The taxonomic position and rank of Peritricha. Acta Protozool. **2**, 19—32 (1964).

1378. Rabinovitch, M., Plaut, W.: Cytoplasmic DNA synthesis in *Amoeba proteus*. II. On the behavior and possible nature of the DNA-containing elements. J. Cell Biol. **15**, 535—540 (1962).

1379. Radzikowski, S.: Changes in the heteromeric macronucleus in division of *Chilodonella cucullulus* (Müller) Acta Protozool. **3**, 233—238 (1965).

1380. — Study on morphology, division and postconjugation morphogenesis in *Chilodonella cuccullulus* (Müller). Acta Protozool. **4**, 89—95 (1966).

1381. — Die Entwicklung des Kernapparates und die Nucleinsäuresynthese während der Konjugation von *Chilodonella cucullulus* O. F. M. Arch. Protistenk. (in press).

1382. Rae, P. M. M.: The nature and processing of ribosomal ribonucleic acid in a dinoflagellate. J. Cell Biol. **46**, 106—113 (1970).

1383. Raff, R. A., Mahler, H. R.: The non symbiotic origin of mitochondria. The question of the origin of the eucaryotic cell and its organelles is reexamined. Science **177**, 575—582 (1972).

1384. Rahat, M., Parnas, I., Nevo, A.C.: Extensibility and tensile strength of the stalk "muscle" of *Carchesium*. Exp. Cell Res. **54**, 58—68 (1969).

1385. Raikov, I.B.: Der Formwechsel des Kernapparates einiger niederer Ciliaten. I. Die Gattung *Trachelocerca*. Arch. Protistenk. **103**, 129—192 (1958).

1386. — Der Formwechsel des Kernapparates einiger niederer Ciliaten. II. Die Gattung *Loxodes*. Arch. Protistenk. **104**, 1—42 (1959).

1387. — Cytological and cytochemical peculiarities of the nuclear apparatus and division in the holotrichous ciliate *Geleia nigriceps* Kahl. Cytologia (Moskau) **1**, 566—579 (1959) (in Russian).

1388. — Der Kernapparat von *Nassula ornata* Ehrbg. (Ciliata, Holotricha). Zur Frage über den Chromosomenaufbau des Makronucleus. Arch. Protistenk. **105**, 463—488 (1962).

1389. — Les ciliés mésopsammiques du littoral de la Mer Blanche (U.R.S.S.) avec une description de quelques espèces nouvelles ou peu connues. Cah. Biol. mar. **3**, 325—361 (1962).

1390. — The nuclear apparatus of the holotrichous ciliates *Geleia orbis* Fauré-Fremiet and *G. murmanica* Raikov. Acta Protozool. **1**, 21—30 (1963).

1391. — The nuclear apparatus of *Remanella multinucleata* Kahl (Ciliata, Holotricha). Acta biol. hung. **14**, 221—229 (1963).

1392. — Some stages of conjugation of the holotrichous ciliate *Trachelocerca coluber* Kahl. Cytologia (Moskau) **5**, 685—689 (1963) (in Russian).

1393. — DNA content of the nuclei and nature of macronuclear chromatin strands of the ciliate *Nassulopsis elegans* (Ehrbg.). Acta Protozool. **2**, 339—355 (1964).

1394. — Elektronenmikroskopische Untersuchung des Kernapparates von *Nassula ornata* Ehrbg. (Ciliata, Holotricha). Arch. Protistenk. **109**, 71—98 (1966).

1395. — The nuclear apparatus and some cytoplasmic structures of *Helicoprorodon gigas* (Holotricha, Gymnostomatida). Acta Protozool. **5**, 49—58 (1967) (in Russian).

**1396.* — The macronucleus of ciliates. In: Chen, T.-T. (Ed.): Research in Protozoology, Vol. 3, pp. 1—128. New York: Pergamon Press 1969.

1397. — Bactéries épizoiques et mode de nutrition du Cilié psammophile *Kentrophoros fistulosum* Faure-Fremiet (étude au microscope électronique). Protistologica **7**, 365—378 (1971).

**1398.* — Nuclear phenomena during conjugation and autogamy in ciliates. In: Chen, T.-T. (Ed.): Research in Protozoology, Vol. 4, pp. 147—290. New York: Pergamon Press 1972.

1399. — Cheissin, E.M., Buze, E.G.: A photometric study of DNA content of macro- and micronuclei in *Paramecium caudatum*, *Nassula ornata* and *Loxodes magnus*. Acta Protozool. **1**, 285—300 (1963).

1400. — Dragesco, J.: Ultrastructure des noyaux et de quelques organites cytoplasmiques du Cilié *Tracheloraphis caudatus* Dragesco et Raikov (Holotricha, Gymnostomatida). Protistologica **5**, 193—208 (1969).

1401. Randall, J.T.: The fine structure of the protozoan *Spirostomum ambiguum*. Symp. Soc. exp. Biol. **10**, 185—198 (1957).

1402. — The flagellar apparatus as a model organelle for the study of growth and morphopoiesis. Proc. Roy. Soc. Lond. **173**, 31—58 (1969).

1403. — Cavalier-Smith, T., McVittie, A., Warr, J.R., Hopkins, J.M.: Development and control processes in the basal bodies and flagella of *Chlamydomonas reinhardi*. Develop. Biol. **1**, 43—83 (1967).

1404. — Disbrey, C.: Evidence for the presence of DNA at basal body sites in *Tetrahymena pyriformis*. Proc. roy. Soc. B **162**, 473—491 (1965).

1405. — Hopkins, J.M.: On the stalks of certain peritrichs. Phil. Trans. roy. Soc. **245**, 59—79 (1962).

1406. — Jackson, F.S.: Fine structure and function in *Stentor polymorphus*. J. biophys. biochem. Cytol. **4**, 807—830 (1958).

1407. — Warr, J.R., Hopkins, J.M., McVittie, A.: A single-gene mutation of *Chlamydomonas reinhardi* affecting motility: A genetic and electron microscope study. Nature (Lond.) **203**, 912—914 (1964).

1408. Rannestad, J., Williams, N. E.: The synthesis of microtubule and other proteins of the oral apparatus in *Tetrahymena pyriformis*. J. Cell Biol. **50**, 709—720 (1971).

1409. Rao, M. V. N.: Nuclear behavior of *Euplotes woodruffi* during conjugation. J. Protozool. **11**, 296—304 (1964).

1410. — Conjugation in *Kahlia sp.* with special reference to meiosis and endomitosis. J. Protozool. **13**, 565—573 (1966).

1411. — Macronuclear development in *Euplotes woodruffi* following conjugation. J. Cell Biol. **31**, 90 (1966).

1412. — Nuclear behavior of *Spirostomum ambiguum* during conjugation with special reference to macronuclear development. J. Protozool. **15**, 748—752 (1968).

1413. — Ammermann, D.: Polytene chromosomes and nucleic acid metabolism during macronuclear development in *Euplotes*. Chromosoma (Berl.) **29**, 246—254 (1970).

1414. — Prescott, D. M.: Micronuclear RNA synthesis in *Paramecium caudatum*. J. Cell Biol. **33**, 281—285 (1967).

1415.* Raper, K. B.: Levels of cellular interaction in amoeboid populations. Proc. Amer. phil. Soc. **104, 579—604 (1960).

1416. Ray, C.: Meiosis and nuclear behavior in *Tetrahymena pyriformis*. J. Protozool. **3**, 88—96 (1965).

1417. Ray, D. S., Hanawalt, P. C.: Properties of the satellite DNA associated with the chloroplasts of *Euglena gracilis*. J. molec. Biol. **9**, 812—824 (1964).

1418. Reger, J. F.: The fine structure of the gregarine *Pyxinoides balani*, parasitic in the barnacle *Balanus tintinnnabulum*. J. Protozool. **14**, 488—497 (1967).

1419. Reichenow, E.: Die Hämococcidien der Eidechsen. Vorbemerkungen und erster Teil: Die Entwicklungsgeschichte von *Karyolysus*. Arch. Protistenk. **42**, 179—291 (1921).

**1420.* — Die endothelialen Entwicklungsformen der Malariaparasiten im Lichte der Phylogenie der Hämosporidien. 3. Intern. Congr. Microbiol. New York, pp. 139—151, 1940.

1421. — Zur Frage der Bedeutung des Blepharoplast der Trypanosomen. Archos Inst. biol. S. Paulo **11**, 433—436 (1940).

1422. — Die Entwicklung der Malariaplasmodien im Vogelkörper. Zbl. Bakt., II. Abt. **152**, 272—284 (1947).

1423. — Mudrow, L.: Der Entwicklungsgang von *Plasmodium praecox* im Vogelkörper. Dtsch. tropenmed. Z. **47**, 289—299 (1943).

1424.* Reichenow, L.: Die Fortschritte der Malaria-Parasitologie. Zbl. Bakt., II. Abt. **157, 3—13 (1951).

1425. Reiff, I.: Die genetische Determination multipler Paarungstypen bei dem Ciliaten *Uronychia transfuga* (Hypotricha, Euplotidae). Arch. Protistenk. **110**, 372—397 (1968).

**1426.* Reinert, J., Urspruing, H. (Eds.): Origin and continuity of cell organelles. In: Results and problems of cell differentiation, Vol. 2, 342 pp. Berlin-Heidelberg-New York: Springer 1971.

1427. Renaud, F. J., Rowe, A. J., Gibbons, I. R.: Some properties of the protein forming the outer fibers of cilia. J. Cell Biol. **31**, 92 A (1966).

1428. Renger, H. C., Wolstenholme, D. R.: Kinetoplast desoxyribonucleic acid of the haemoflagellate *Trypanosoma lewisi*. J. Cell Biol. **47**, 689—702 (1970).

1429. — — Kinetoplast and other satellite DNAs of kinetoplastic and dyskinetoplastic strains of *Trypanosoma*. J. Cell Biol. **50**, 533—539 (1971).

1430. Richter, I.-E.: Bewegungsphysiologische Untersuchungen an polycystiden Gregarinen unter Anwendung des Mikrozeitrafferfilmes. Protoplasma **51**, 197—241 (1959).

1431. Rinaldi, R. A., Jahn, T. L.: On the mechanism of ameboid movement. J. Protozool. **10**, 344—357 (1963).

1432. Ringertz, N. R., Bolund, L., Debault, L. E.: Isolation and chemical composition of macronuclei from *Tetrahymena*. Exp. Cell Res. **45**, 519—532 (1967).

1433. — Hoskins, G. C.: Cytochemistry of macronuclear reorganisation. Exp. Cell Res. **38**, 160—179 (1965).

1434. Ringo, D. L.: Electron microscopy of *Astasia longa*. J. Protozool. **10**, 167—173 (1963).

1435. Ringo, D. L.: The arrangement of subunits in flagellar fibers. J. Ultrastruct. Res. **17**, 266—277 (1967).

1436. Riou, G., Pautrizel, R.: Nuclear and kinetoplastic DNA from trypanosomes. J. Protozool. **16**, 509—513 (1969).

1437. Ris, H.: Interpretation of ultrastructure in the cell nucleus. In: Harris, R. J. C.: The interpretation of ultrastructure, pp. 69—88. New York-London: Academic Press 1962.

1438. — Plaut, W.: Ultrastructure of DNA-containing areas in the chloroplast of *Chlamydomonas*. J. Cell Biol. **13**, 383—391 (1962).

1439. Roberts, W. L., Hammond, D. M., Anderson, L. C., Speer, C. A.: Ultrastructural study of schizogony in *Eimeria callospermophili*. J. Protozool. **17**, 584—592 (1970).

1440. — — Ultrastructural and cytologic studies of the sporozoites of four *Eimeria* species. J. Protozool. **17**, 76—86 (1970).

**1441.* Robertson, J. D.: A molecular theory of cell membrane structure. Verhandl. 4. intern. Kongr. Elektronenmikroskopie, Berlin, Bd. 2, S. 159—171, 1960.

1442. Röhlich, P., Toth, J., Törö, I.: Fine structure and enzymic activity of protozoan food vacuoles. In: De Reuck, A. V. S., Cameron, M. P.: Lysosomes, pp. 201—225. London: Churchill 1963.

1443. Roque, M.: Recherches sur les infusoires ciliés: Les hymenostomes peniculiens. Bull. biol. Fr. Belg. **95**, 432—519 (1961).

1444. — De Puytorac, P., Lom, J.: L'architecture buccale et la stomatogénèse d'*Ichthyophthirius multifiliis* Fouquet 1876. Protistologica **3** (1), 79—90 (1967).

1445. Rosenbaum, J. L., Moulder, J. E., Ringo, D. L.: Flagellar elongation and shortening in *Chlamydomonas*. The use of cycloheximide and colchicine to study the synthesis and assembly of flagellar proteins. J. Cell Biol. **41**, 600—619 (1969).

1446. Roth, L. E.: Electron microscopy of pinocytosis and food vacuoles in *Pelomyxa*. J. Protozool. **7**, 176—185 (1960).

1447. — Daniels, E. W.: Infective organisms in the cytoplasm of *Amoeba proteus*. J. biophys. biochem. Cytol. **9**, 317—323 (1961).

1448. — — Electron microscopic studies of mitosis in amebae. II. The giant ameba *Pelomyxa carolinensis*. J. Cell Biol. **12**, 57—78 (1962).

1449. — Minick, O. T.: Electron microscopy of nuclear and cytoplasmic events during division in *Tetrahymena pyriformis*, strains W and HAM 3. J. Protozool. 8, 12—21 (1961).

1450. — Obetz, S. W., Daniels, E. W.: Electron microscopic studies of mitosis in amebae. I. *Amoeba proteus*. J. biophys. biochem. Cytol. 8, 207—220 (1960).

1451. — Pihlaja, D. J., Shigenaka, Y.: Microtubules in the heliozoan axopodium. I. The gradion hypothesis of allosterism in structural proteins. J. Ultrastruct. Res. **30**, 7—38 (1970).

1452. — Shigenaka, Y.: The structure and formation of cilia and filaments in rumen Protozoa. J. Cell Biol. **20**, 249—270 (1964).

1453. — — Microtubules in the heliozoan axopodium. II. Rapid degradation by cupric and nickelous ions. J. Ultrastruct. Res. **31**, 356—375 (1970).

1454. Rouiller, C., Fauré-Fremiet, E.: L'ultrastructure des trichocystes fusiformes de *Frontonia atra*. Bull. Micr. appl. **7**, 135—139 (1957).

1455. — — Structure fine d'un flagellé chrysomonadien: *Chromulina psammobia*. Exp. Cell Res. **14**, 47—67 (1958).

1456. — — Gauchery, M.: Fibres scléroprotéiques d'origine ciliaire chez les infusoires péritriches. C. R. Acad. Sci. (Paris) **242**, 180—182 (1956).

1457. Rudzinska, M. A.: An electron microscope study of the contractile vacuole in *Tokophrya infusionum*. J. biophys. biochem. Cytol. **4**, 195—202 (1958).

1458. — The fine structure and function of the tentacle in *Tokophrya infusionum*. J. Cell Biol. **25**, 459—477 (1965).

1459. — Ultrastructures involved in the feeding mechanism of Suctoria. Trans. N. Y. Acad. Sci. **II/29**, 512—525 (1967).

1460.* — The fine structure of malaria parasites. Int. Rev. Cytol. **25, 161—199 (1969).

1461. RUDZINSKA, M. A.: The mechanism of food intake in *Tokophrya infusionum* and ultrastructural changes in food vacuoles during digestion. J. Protozool. **17**, 626—641 (1970).
1462. — D'ALESANDRO, P. A., TRAGER, W.: The fine structure of *Leishmania donovani* and the role of the kinetoplast in the Leishmania-leptomonad transformation. J. Protozool. **11**, 166—191 (1964).
1463. — BRAY, R. S., TRAGER, W.: Intracellular phagotrophy in *Plasmodium falciparum* and *Plasmodium gonderi*. J. Protozool. **7**, 24—25 (1960).
1464. — JACKSON, G. J., TUFFRAU, M.: The fine structure of *Colpoda maupasi* with special emphasis on food vacuoles. J. Protozool. **13**, 440—459 (1966).
1465. — TRAGER, W.: The role of the cytoplasm during reproduction in a malarial parasite (*Plasmodium lophurae*) as revealed by electron microscopy. J. Protozool. 8, 307—322 (1961).
1466. — — The fine structure of trophozoites and gametocystes in *Plasmodium coatneyi*. J. Protozool. **15**, 73—88 (1968).
1467. — — BRAY, R. S.: Pinocytotic uptake and the digestion of hemoglobin in malaria parasites. J. Protozool. **12**, 563—576 (1965).
1468. RUTHMANN, A.: Die Struktur des Chromatins und die Verteilung der Ribonucleinsäure im Makronucleus von *Loxophyllum meleagris*. Arch. Protistenk. **106**, 422—436 (1963).
1469. — Autoradiographische und mikrophotometrische Untersuchungen zur DNS-Synthese im Makronucleus von *Bursaria truncatella*. Arch. Protistenk. **107**, 117—130 (1964).
1470. — GRELL, K. G.: Die Feinstruktur des intracapsulären Cytoplasmas bei dem Radiolar *Aulacantha scolymantha*. Z. Zellforsch. **63**, 97—119 (1964).
1471. — HECKMANN, K.: Formwechsel und Struktur des Makronucleus von *Bursaria truncatella*. Arch. Protistenk. **105**, 313—340 (1961).
1472. SAGER, R.: Mendelian and non-Mendelian inheritance of streptomycin resistance in *Chlamydomonas reinhardi*. Proc. nat. Acad. Sci. (Wash.) **40**, 356—363 (1954).
1473. — Inheritance in the green alga *Chlamydomonas reinhardi*. Genetics **40**, 476—489 (1955).
1474. — The architecture of the chloroplast in relation to its photosynthetic activities. In: The photochemical apparatus, its structure and function. Brookhaven Symp. Biol. **11**, 101—117 (1958).
1475. — Genetic systems in *Chlamydomonas*. Science **132**, 1459—1465 (1960).
1476. — Streptomycin as a mutagen for nonchromosomal genes. Proc. nat. Acad. Sci. (Wash.) **48**, 2018—2026 (1962).
**1477.* — Studies of cell heredity with *Chlamydomonas*. In: HUTNER, S. H.: Biochemistry and physiology of protozoa, Vol. 3, pp. 297—318. New York-London: Academic Press 1964.
1478.* — Genes outside the chromosomes. Sci. Amer. **212, 70—79 (1965).
**1479.* — On the evolution of genetic systems. In: BRYSON, V., VOGEL, H. J.: Evolving genes and proteins, pp. 591—597. New York-London: Academic Press 1965.
**1480.* — On non-chromosomal heredity in microorganisms. In: 15th Symp. Soc. Genl. Microbiol., pp. 324—342. Cambridge: University Press 1965.
1481.* — Mendelian and non-Mendelian heredity: A reappraisal. Proc. roy. Soc. (London) **164, 290—297 (1966).
**1482.* — Cytoplasmic genes and organelle formation. From: WARREN, K. G. (Ed.): Formation and fate of cell organelles. Ann. Symp. Int. Soc. Cell Biol., pp. 317—334. New York-London: Academic Press 1968.
**1483.* — Cytoplasmic genes and organelles, 405 p. New York-London: Academic Press 1972.
1484. — GRANICK, S.: Nutritional studies with *Chlamydomonas reinhardi*. Ann. N. Y. Acad. Sci. **56**, 831—838 (1953).
1485. — — Nutritional control of sexuality in *Chlamydomonas reinhardi*. J. gen. Physiol. **37**, 729—742 (1954).
1486. — HAMILTON, M. G.: Cytoplasmic and chloroplast ribosomes of *Chlamydomonas:* Ultracentrifugal characterization. Science **157**, 709—711 (1967).

1487. SAGER, R., ISHIDA, M. R.: Chloroplast DNA in *Chlamydomonas*. Proc. nat. Acad. Sci. (Wash.) **50**, 725—730 (1963).
1488. — PALADE, G. E.: Chloroplast structure in green and yellow strains of *Chlamydomonas*. Exp. Cell Res. **7**, 584—588 (1954).
1489. — RAMANIS, Z.: The particulate nature of nonchromosomal genes in *Chlamydomonas*. Proc. nat. Acad. Sci. (Wash.) **50**, 260—268 (1963).
1490. — — Recombination of nonchromosomal genes in *Chlamydomonas*. Proc. nat. Acad. Sci. (Wash.) **53**, 1053—1061 (1965).
1491. — — Biparental inheritance of nonchromosomal genes induced by ultraviolet irradiation. Proc. nat. Acad. Sci. (Wash.) **58**, 931—937 (1967).
1492. — — The pattern of segregation of cytoplasmic genes in *Chlamydomonas*. Proc. nat. Acad. Sci. (Wash.) **61**, 324—331 (1968).
1493. — — A genetic map of non-mendelian genes in *Chlamydomonas*. Proc. nat. Acad. Sci. (Wash.) **65**, 593—600 (1970).
1494.* — — Formal genetic analysis of organelle genetic systems. Stadler Symposia **1, **2**, 65—78 (1971).
1495. — TSUBO, Y.: Genetic analysis of streptomycin-resistance and -dependence in *Chlamydomonas*. Z. Vererbungsl. **92**, 430—438 (1961).
1496. — — Mutagenic effects of Streptomycin in *Chlamydomonas*. Arch. Microbiol. **42**, 159—175 (1962).
1497. SAITO, M.: Studies in the mitosis of *Euglena*. I. On the chromosome cycle of *Euglena viridis* EHRBG. J. Protozool. 8, 307—322 (1961).
1498. — A note on the duplication process of macronuclear chromosomes in a peritrichous ciliate, *Vorticella campanula*. Jap. J. Genet. **36**, 184—186 (1961).
1499. SANDERS, E. J.: Pinocytosis in centrifuged and bisected amoebae. Exp. Cell Res. **61**, 461—464 (1970).
1500. SAPRA, G. R., DASS, C. M. S.: Organization and development of the macronuclear anlage in *Stylonychia notophora* STOKES. J. Cell Sci. **6**, 351—363 (1970).
1501. — — The cortical anatomy of *Stylonychia notophora* and morphogenetic changes during binary fission. Acta Protozool. **7**, 193—204 (1970).
**1502.* SATIR, P.: Structure and function in cilia and flagella. In: Protoplasmatologia (Handbuch der Protoplasmaforschung), Vol. 3, pp. 1—52. Wien-New York: Springer 1965.
1503. — Morphological aspects of ciliary motility. J. gen. Physiol. **50**, 241—258 (1967).
1504. SCHÄFER-DANNEEL, S.: Strukturelle und funktionelle Voraussetzungen für die Bewegung von *Amoeba proteus*. Z. Zellforsch. **78**, 441—462 (1967).
1505. SCHENSTED, I. v.: Model of subnuclear segregation in the macronucleus of ciliates. Amer. Naturalist **92**, 161—170 (1958).
**1506.* SCHERBAUM, O. H., LOEFFER, J. B.: Environmentally induced growth oscillations in Protozoa. In: HUTNER, S. H.: Biochemistry and physiology of protozoa, Vol. 3, pp. 10—55. New York-London: Academic Press 1964.
1507. SCHIMMER, O., ARNOLD, C. G.: Untersuchungen zur Lokalisation eines außerkaryotischen Gens bei *Chlamydomonas reinhardi*. Arch. Microbiol. **66**, 199—202 (1969).
1508. — — Über die Zahl der Kopien eines außerkaryotischen Gens bei *Chlamydomonas reinhardi*. Mol. gen. Genet. **107**, 366—371 (1970).
1509. — — Untersuchungen über Reversions- und Segregationsverhalten eines außerkaryotischen Gens von *Chlamydomonas reinhardi* zur Bestimmung des Erbträgers. Mol. gen. Genet. **107**, 281—290 (1970).
1510. — — Hin- und Rücksegregation eines außerkaryotischen Gens bei *Chlamydomonas reinhardi*. Mol. gen. Genet. **108**, 33—40 (1970).
1511. SCHMIDT, W. J.: Polarisationsmikroskopische Beobachtungen an *Actinosphaerium eichhorni*. Biol. Zbl. **64**, 314 (1954).
1512. SCHMITTER, R. E.: The fine structure of *Gonyaulax polyedra*, a bioluminescent marine dinoflagellate. J. Cell Sci. **9**, 147—173 (1971).
1513. SCHMOLLER, H.: Die Labyrinthulen und ihre Beziehung zu den Amöben. Naturwissenschaften **58**, 142—146 (1971).
1514. SCHNEIDER, L.: Elektronenmikroskopische Untersuchungen über das Nephridialsystem von *Paramecium*. J. Protozool. **7**, 75—90 (1960).

1515. Schneider, L.: Elektronenmikroskopische Untersuchungen der Konjugation von *Paramecium.* I. Die Auflösung und Neubildung der Zellmembran bei den Konjuganten. Protoplasma **56**, 109—140 (1963).

1516. — Elektronenmikroskopische Untersuchungen an den Ernährungsorganellen von *Paramecium.* II. Die Nahrungsvakuolen und die Cytopyge. Z. Zellforsch. **62**, 225—245 (1964).

1517. — Wohlfarth-Bottermann, K. E.: Protistenstudien. IX. Elektronenmikroskopische Untersuchungen an Amöben unter besonderer Berücksichtigung der Feinstruktur des Cytoplasmas. Protoplasma **51**, 377—389 (1959).

1518. Schneller, M. V.: Genic interrelationships between two particle-bearing stocks of syngen 4, *Paramecium aurelia.* Amer. Zoologist **1**, 386—387 (1961).

1519. — Some notes on the rapid lysis type of killing found in *Paramecium aurelia.* Amer. Zoologist **2**, 446 (1962).

1520. — Sonneborn, T. M., Mueller, J. A.: The genetic control of *kappa*-like particles in *Paramecium aurelia.* Genetics **44**, 533—534 (1959).

1521. Scholtyseck, E.: Electron microscope studies on *Eimeria perforans* (Sporozoa). J. Protozool. **9**, 407—414 (1962).

1522. — Elektronenmikroskopische Untersuchungen über die Wechselwirkung zwischen dem Zellparasiten *Eimeria perforans* und seiner Wirtszelle. Z. Zellforsch. **61**, 220—230 (1963).

1523. — Vergleichende Untersuchungen über die Kernverhältnisse und das Wachstum bei Coccidiomorphen unter besonderer Berücksichtigung von *Eimeria maxima.* Z. Parasitenk. **22**, 428—474 (1963).

1524. — Die Mikrogametenentwicklung von *Eimeria perforans.* Z. Zellforsch. **66**, 625—642 (1965).

1525. — Elektronenmikroskopische Untersuchungen über die Schizogonie bei Coccidien (*Eimeria perforans* und *E. stiedae*). Z. Parasitenk. **26**, 50—62 (1965).

1526. — Die Feinstruktur der Makrogameten des Hühnercoccids *Eimeria tenella.* Z. Parasitenk. **33**, 31—43 (1969).

1527. — Friedhoff, K., Piekarski, G.: Über morphologische Übereinstimmungen bei Entwicklungsstadien von Coccidien, Toxoplasmen und Piroplasmen. Z. Parasitenk. **35**, 119—129 (1970).

1528. — Hammond, D. M.: Electron microscope studies of macrogametes and fertilization in *Eimeria bovis.* Z. Parasitenk. **34**, 310—318 (1970).

1529. — — Ernst, V.: Fine structure of the macrogametes of *Eimeria perforans, E. stiedae, E. bovis* and *E. auburnensis.* J. Parasitology **52**, 975—987 (1966).

1530. — Mehlhorn, H., Friedhoff, K.: The fine structure of the conoid of Sporozoa and related organisms. Z. Parasitenk. **34**, 68—94 (1970).

1531. — — Ultrastructural study of characteristic organelles (paired organelles, micronemes, micropores) of Sporozoa and related organisms. Z. Parasitenk. **34**, 97—127 (1970).

1532. — — Hammond, D. M.: Electron microscope studies of microgametogenesis in Coccidia and related groups. Z. Parasitenk. **38**, 95—131 (1972).

1533. — Schäfer, D.: Über schlauchförmige Ausstülpungen an der Zellmembran der Makrogametocyten von *Eimeria perforans.* Z. Zellforsch. **61**, 214—219 (1963).

1534. — Voigt, W. H.: Die Bildung der Oocystenhülle bei *Eimeria perforans* (Sporozoa). Z. Zellforsch. **62**, 279—292 (1964).

1535. — Volkmann, B., Hammond, D.: Spezifische Feinstrukturen bei Parasit und Wirt als Ausdruck ihrer Wechselwirkungen am Beispiel von Coccidien. Z. Parasitenk. **28**, 78—94 (1966).

1536. Schreiber, E.: Zur Kenntnis der Physiologie und Sexualität höherer Volvocales. Z. Bot. **17**, 337—376 (1925).

1537. Schrevel, J.: Recherches sur le cycle des Lecudinidae, Grégarines parasites d'Annélides Polychètes. Protistologica **5**, 561—588 (1969).

1538. — Contribution à l'étude des Selenidiidae, parasites d'Annélides Polychètes. I. Cycles biologiques. Protistologica **6**, 389—426 (1970).

1539. Schrevel, J.: Contribution à l'étude des Selenidiidae, parasites d'Annélides Polychètes. II. Ultrastructure de quelques trophozoites. Protistologica **7**, 101—130 (1971).
1540.* — Les polysaccharides de réserve chez les Sporozoaires. Ann. Biol. **10, 31—51 (1971).
1541. — Facteurs contrôlant la réproduction des Grégarines. Etude de l'enkystement de la Grégarine intestinale, *Lecudina tuzetae* Schr., parasite de *Nereis diversicolor* O. F. Müller (Annélide Polychète). Protistologica **7**, 439—450 (1971).
1542. — Vivier, E.: Etude de l'ultrastructure et du rôle de la région antérieure (mucron et épimerite) de Grégarines parasites d'Annélides Polychètes. Protistologica **2** (3), 17—28 (1966).
1543. Schröder, O.: Beiträge zur Entwicklungsgeschichte der Myxosporidien, *Sphaeromyxa sabrazesi* (Laveran et Mesnil). Arch. Protistenk. **9**, 359—429 (1907).
1544. Schuster, F.: An electron microscope study of the amoeboflagellate, *Naegleria gruberi* (Schardinger). I. The amoeboid and flagellate stages. J. Protozool. **10**, 297—313 (1963).
1545. — An electron microscope study of the amoeboflagellate, *Naegleria gruberi* (Schardinger). II. The cyst stage. J. Protozool. **10**, 313—320 (1963).
1546. — Trichocysts of the cryptomonad *Cryptomonas truncata*. J. Cell Biol. **31**, 102 (1966).
1547. — The gullet and trichocysts of *Cyathomonas truncata*. Exp. Cell Res. **49**, 277—284 (1968).
1548. — Intranuclear virus-like bodies in the amoebo-flagellate *Naegleria gruberi*. J. Protozool. **16**, 724—727 (1969).
1549. — The trichocysts of *Chilomonas paramecium*. J. Protozool. **17**, 521—526 (1970).
1550. — Dunnebacke, T. H.: Formation of bodies associated with virus-like particles in the amoebo-flagellate *Naegleria gruberi*. J. Ultrastruct. Res. **36**, 659—668 (1971).
1551. Schwab, D.: Elektronenmikroskopische Untersuchung an der Foraminifere *Myxotheca arenilega* Schaudinn. Z. Zellforsch. **96**, 295—324 (1969).
1552. — Elektronenmikroskopische Untersuchungen an der Foraminifere *Allogromia laticollaris* Arnold. Der heranwachsende Agamont. Z. Zellforsch. **108**, 35—45 (1970).
1553. — Elektronenmikroskopische Untersuchungen an intrazellulär lebenden Einzellern in Foraminiferen. Z. Naturforsch. **26**, 12 (1971).
1554. — Electron microscopic studies on the foraminifer *Allogromia laticollaris* Arnold. Mitosis in agamonts. Protoplasma **75**, 79—89 (1972).
1555. — Schwab-Stey, H.: Fibrilläre und tubuläre Strukturen im Cytoplasma der Foraminifere *Allogromia laticollaris* Arnold. Cytobiol. **6**, 234—242 (1972).
1556. Schwartz, V.: Versuche über Regeneration und Kerndimorphismus bei *Stentor coeruleus* Ehrbg. Arch. Protistenk. **85**, 100—139 (1935).
1557. — Konjugation micronucleusloser Paramecien. Naturwissenschaften **27**, 724 (1939).
1558. — Der Formwechsel des Makronukleus in der Konjugation mikronukleusloser Paramecien. Biol. Zbl. **65**, 89—94 (1946).
1559. — Über die Physiologie des Kerndimorphismus bei *Paramecium bursaria*. Z. Naturforsch. **2**b, 369—381 (1947).
1560.* — Die Sexualität der Infusorien. Fortsch. Zool., N. F. **9, 605—619 (1952).
1561. — Nukeolenformwechsel und Zyklen der Ribosenukleinsäure in der vegetativen Entwicklung von *Paramecium bursaria*. Biol. Zbl. **75**, 1—16 (1956).
1562. — Über den Formwechsel achromatischer Substanz in der Teilung des Makronucleus von *Paramecium bursaria*. Biol. Zbl. **76**, 1—23 (1957).
1563. — Chromosomen im Makronucleus von *Paramecium bursaria*. Biol. Zbl. **77**, 347—364 (1958).
1564.* — Die Sicherung der arttypischen Zellformen bei Ciliaten. Naturwissenschaften **50, 631—640 (1963).
1565. — Die Teilungsspindel des Mikronucleus von *Paramecium bursaria*. Verh. dtsch. zool. Ges. Kiel, S. 123—131, 1964.
1566. — Einleitende Beobachtungen am Beutefang von *Didinium nasutum*. Z. Naturforsch. **20**b, 383—391 (1965).
1567. — Modifizierbarkeit des polaren Differenzierungsmusters der Heterotrichen. Verh. dtsch. zool. Ges. Göttingen (30. Suppl.) 490—494 (1966).

1568. SCHWARTZ,V.: Reaktionen der polaren Differenzierung von *Spirostomum ambiguum* auf Lithium- and Rhodanidionen. Roux' Arch. Entwicklungsmech. **158**, 89—102 (1967).

1569. SCHWEIKHARDT,F.: Zytochemisch-entwicklungsphysiologische Untersuchungen an *Stentor coeruleus* EHRBG. Roux' Arch. Entwickl.-Mech. Org. **157**, 21—74 (1966).

1570. SEAMAN,G.R.: Large-scale isolation of kinetosomes from the ciliated protozoan *Tetrahymena pyriformis*. Exp. Cell Res. **21**, 292—302 (1960).

1571. — Protein synthesis by kinetosomes isolated from the protozoan *Tetrahymena*. Biochim. biophys. Acta (Amst.) **55**, 889 (1962).

1572. SELMAN,G.G., JURAND,A.: Trichocyst development during the fission cycle of *Paramecium*. J. gen. Microbiol. **60**, 365—372 (1970).

*1572*a. SEPSENWOL,S.: Leucoplast of the cryptomonad *Chilomonas paramecium*. Evidence for presence of true plastid in a colorless flagellate. Exp. Cell Res. **76**, 395—409 (1973).

1573. SERAVIN,L.N.: A critical survey of the modern state of the problem of ameboid movement. Citologija (Moskau) **6**, 653—667 (1964) (in Russian).

1574. — The role of mechanical and chemical stimulators on the induction of phagocytic reactions in *Amoeba proteus* and *A. dubia*. Acta Protozool. **6**, 97—108 (1968) (in Russian).

1575. — Left and right spiralling round the long body axis in ciliate Protozoa. Acta Protozool. **7**, 313—323 (1970).

1576. SESHACHAR,B.R.: Observations on the fine structure of the nuclear apparatus of *Blepharisma intermedium* BHANDARY (Ciliata: Spirotricha). J. Protozool. **11**, 402—409 (1964).

1577. — The fine structure of the nuclear apparatus and the chromosomes of *Spirostomum ambiguum* EHRBG. Acta Protozool. **3**, 337—343 (1965).

1578. — SAXENA,D.M.: DNA-dependent RNA synthesis in isolated macronuclei of *Blepharisma intermedium* (Ciliata). J. Protozool. **15**, 697—700 (1968).

1579. SHEFFIELD,H.G., GARNHAM,P.C.C., SHIROISHI,T.: The fine structure of the sporozoite of *Lankesteria culicis*. J. Protozool. **18**, 98—105 (1971).

1580. SHIGENAKA,Y., ROTH,L.E., PIHLAJA,D.J.: Microtubules in the heliozoan axopodium. III. Degradation and reformation after dilute urea treatment. J. Cell Sci. **8**, 127—151 (1971).

1581. SIDDIQUI,W.A., RUDZINSKA,M.: The fine structure of axenically-grown trophozoites of *Entamoeba invadens* with special reference to the nucleus and helical ribonucleoprotein bodies. J. Protozool. **12**, 448—459 (1965).

1582. SIEGEL,R.W.: The genetic analysis of mate-killing in *Paramecium aurelia*. Genetics **37**, 625—626 (1952).

1583. — A genetic analysis of the mate-killer trait in *Paramecium aurelia*, variety 8. Genetics **38**, 550—560 (1953).

1584. — Mate-killing in *Paramecium aurelia*, variety 8. Physiol. Zool. **27**, 89—100 (1954).

1585. — Mating types in *Oxytricha* and the significance of mating type systems in ciliates. Biol. Bull. **110**, 352—357 (1956).

1586. — An analysis of the transformation from immaturity to maturity in *Paramecium aurelia*. Genetics **42**, 394—395 (1957).

1587. — Hereditary endosymbiosis in *Paramecium bursaria*. Exp. Cell Res. **19**, 239—252 (1960).

1588. — The genic control of complementary sex substances in *Paramecium bursaria*. "Progress in Protozoology". Proc. 1st. Intern. Conf. Protozool. Prague, pp. 115 to 119, 1961.

1589. — Nuclear differentiation and transitional cellular phenotypes in the life cycle of *Paramecium*. Exp. Cell Res. **24**, 6—20 (1961).

1590. — New results on the genetics of mating types in *Paramecium bursaria*. Genet. Res. **4**, 132—142 (1963).

1591. — Hereditary factors controlling development in *Paramecium*. In: Genetic control of differentiation. Brookhaven Symp. Biol. **18**, 55—65 (1965).

1592. Siegel, R. W.: Genetics of ageing and the life cycle in ciliates. Symp. Soc. exp. Biol. **21**, 127—148 (1967).
1593. — Organellar damage and revision as a possible basis for intraclonal variation in *Paramecium*. Genetics **66**, 305—314 (1970).
1594. — Cohen, L. W.: A temporal sequence for genic expression: cell differentiation in *Paramecium*. Amer. Zoologist **3**, 127—134 (1963).
1595. — Heckmann, K.: Inheritance of autogamy and the killer trait in *Euplotes minuta*. J. Protozool. **13**, 34—38 (1966).
1596. — Karakashian, S. J.: Dissociation and restoration of endocellular symbiosis in *Paramecium bursaria*. Anat. Rec. **134**, 639 (1959).
1597. — Larison, L. L.: The genic control of mating types in *Paramecium bursaria*. Proc. nat. Acad. Sci. (Wash.) **46**, 344—349 (1960).
1598. Simard-Duquesne, N., Couillard, P.: Ameboid movement. I. Reactivation of glycerinated models of *Amoeba proteus* with adenosine triphosphate. Exp. Cell Res. **28**, 85—91 (1962).
1599. — — Ameboid movement. II. Research of contractile proteins in *Amoeba proteus*. Exp. Cell Res. **28**, 92—98 (1962).
1600. Simpson, L.: The leishmania-leptomonad transformation of *Leishmania donovani:* Nutritional requirements, respiration changes and antigenic changes. J. Protozool. **15**, 201—208 (1968).
1601. — Effect of acriflavine on the kinetoplast of *Leishmania tarentolae*. J. Cell Biol. **37**, 660—682 (1968).
1602.* — The kinetoplast of the haemoflagellates. Int. Rev. Cytol. **32, 139—207 (1972).
1603. Sleigh, M. A.: Metachronism and frequency of beat in the peristomial cilia of *Stentor*. J. exp. Biol. **33**, 15—28 (1956).
1604. — Further observations on co-ordination and the determination of frequency in the peristomial cilia of *Stentor*. J. exp. Biol. **34**, 106—115 (1957).
1605. — The form of beat in cilia of *Stentor* and *Opalina*. J. exp. Biol. **37**, 1—10 (1960).
**1606.* — The biology of cilia and flagella. 242 S. Oxford: Pergamon Press 1962.
1607. — Flagellar movement of the sessile flagellates *Actinomonas*, *Codonosiga*, *Monas*, and *Poteriodendron*. Quant. J. micr. Sci. **105**, 405—414 (1964).
1608.* — The co-ordination and control of cilia. Symp. Soc. exp. Biol. **20, 11—31 (1965).
1609.* — Coordination of the rhythm of beat in some ciliary systems. Int. Rev. Cytol. **25, 31—54 (1969).
1610. — Some factors affecting the excitation of contraction in *Spirostomum*. Acta Protozool. **7**, 335—352 (1970).
1611.* — Cilia. Endeavour **30, 11—17 (1971).
1612. Small, E. B., Antipa, G. A., Marszalek, D.: The cytological sequence of events in the binary fission of *Didinium nasutum*. Acta Protozool. **9**, 275—282 (1971).
1613. — Marszalek, D. S.: Scanning electron microscopy of fixed, frozen and dried Protozoa. Science **163**, 1064—1065 (1969).
1614. — —Antipa, G. A.: A survey of ciliate surface patterns and organelles as revealed with scanning electron microscopy. Trans. Amer. Microscop. Soc. **90**, 283—294 (1971).
1615. Smith-Johannsen, H., Gibbs, S. P.: Effects of chloramphenicol on chloroplast and mitochondrial ultrastructure in *Ochromonas danica*. J. Cell Biol. **52**, 598—614 (1972).
1616. Smith-Sonneborn, J. E., Green, L., Marmur, J.: Desoxyribonucleic acid base composition of *kappa* and *Paramecium aurelia* stock 51. Nature (Lond.) **197**, 385 (1963).
1617. — Plaut, W.: Evidence for the presence of DNA in the pellicle of *Paramecium*. J. Cell Sci. **2**, 225—234 (1967).
1618. — — Studies on the autonomy of pellicular DNA in *Paramecium*. J. Cell Sci. **5**, 365—372 (1969).
1619. — Van Wagtendonk, W.: Purification and chemical characterization of *kappa* of stock 51, *Paramecium aurelia*. Exp. Cell Res. **33**, 50—59 (1964).
1620. Soldo, A. T., Godoy, G. A., van Wagtendonk, W. J.: Growth of particle-bearing and particle-free *Paramecium aurelia* in axenic culture. J. Protozool. **13**, 492—497 (1966).

1621. Soldo, A.T., Musil, G., Godoy, G.A.: Action of penicillin G on endosymbiote *lambda* particles of *Paramecium aurelia*. J. Bacteriol. **104**, 966—980 (1970).

1622. — van Wagtendonk, W.J.: A method for the mass collection of axenically cultivated *Paramecium aurelia*. J. Protozool. **14**, 497—498 (1967).

1623. — — Godoy, G.A.: Nucleic acid and protein content of purified endosymbiote particles of *Paramecium aurelia*. Biochem. biophys. Acta **204**, 325—333 (1970).

1624. Soltynska, M.S.: Morphology and fine structure of *Chilodonella cucullulus* O. F. M. Cortex and cytopharyngeal apparatus. Acta Protozool. **9**, 49—82 (1971).

1625. Sommer, J.R.: The ultrastructure of the pellicle complex of *Euglena gracilis*. J. Cell Biol. **24**, 253—257 (1965).

1626. — Blum, J.J.: Pellicular changes during division in *Astasia longa*. Exp. Cell Res. **35**, 423—425 (1964).

1627. Sommerville, J.: Immobilization antigen synthesis in *Paramecium aurelia:* Labelling antigen in vivo. Exp. Cell Res. **50**, 660—663 (1968).

1628. Sonneborn, T.M.: Sex, sex inheritance and sex determination in *Paramecium aurelia*. Proc. nat. Acad. Sci. (Wash.) **23**, 378—385 (1937).

1629. — Genetic evidence of autogamy in *Paramecium aurelia*. Anat. Rec. **75**, 85 (1939).

1630. — *Paramecium aurelia:* Mating types and groups; lethal interactions; determination and inheritance. Amer. Naturalist **73**, 390—413 (1939).

1631. — Gene and cytoplasm. I. The determination and inheritance of the killer character in variety 4 of *Paramecium aurelia*. II. The bearing of the determination and inheritance of characters in *Paramecium aurelia* on problems of cytoplasmic inheritance, pneumococcus transformations, mutations and development. Proc. nat. Acad. Sci. (Wash.) **29**, 329—343 (1943).

1632.* — Recent advances in the genetics of *Paramecium* and *Euplotes*. Advanc. Genet. **1, 263—358 (1947).

1633.* — Methods in the general biology and genetics of *Paramecium aurelia*. J. exp. Zool. **113, 87—148 (1950).

1634. — The cytoplasm in heredity. Heredity **4**, 11—36 (1950).

1635. — Patterns of nucleocytoplasmic integration in *Paramecium*. Proc. IXth Intern. Congr. Genet. Bellagio/Italy, 1953 (1954).

1636. — The relation of autogamy to senescence and rejuvenescence in *Paramecium aurelia*. J. Protozool. **1**, 38—53 (1954).

1637.* — Breeding systems, reproductive methods, and species problems in Protozoa. In: Mayr, E.: The species problem, pp. 155—324. Washington: Amer. Ass. Advanc. Sci. Publ. **50 (1957).

1638.* — *Kappa* and related particles in *Paramecium*. Advanc. Virus Res. **6, 229—356 (1959).

1639. — The gene and cell differentiation. Proc. nat. Acad. Sci. **46**, 149—165 (1960).

1640. — *Kappa* particles and their bearing on host-parasite relations. In: Pollard, M.: Perspectives in virology. Vol. 2, pp. 5—12. Minneapolis: Burgess Publ. (1961).

1641. — Does preformed cell structure play an essential role in cell heredity? In: Allen, J.M.: The nature of biological diversity, pp. 165—221. New York-London: McGraw-Hill 1963.

1642. — The differentiation of cells. Proc. nat. Acad. Sci. (Wash.) **51**, 915—929 (1964).

1643. — Gene action in development. Proc. roy. Soc. (Lond.) **176**, 347—366 (1970).

1644. — (Personal communication).

1645. — Dippell, R.V.: Sexual isolation, mating types, and sexual responses to diverse conditions in variety 4, *Paramecium aurelia*. Biol. Bull. **85**, 36—43 (1943).

1646. — — Mating reactions and conjugation between varieties of *Paramecium aurelia* in relation to conceptions of mating type and variety. Physiol. Zool. **19**, 1—18 (1946).

1647. — — Self-reproducing differences in the cortical organization in *Paramecium aurelia*, syngen 4. Genetics **46**, 900 (1961).

1648. — — The modes of replication of cortical organization in *Paramecium aurelia*, syngen 4. Genetics **46**, 899—900 (1961).

1649. — LeSuer, A.: Antigenic characters in *Paramecium aurelia* (Variety 4): Determination, inheritance and induced mutations. Amer. Naturalist **82**, 69—78 (1948).

1650. SONNEBORN, T. M., SCHNELLER, M. V., CRAIG, M. F.: The basis of variation in phenotype of gene controlled traits in heterozygotes of *Paramecium aurelia*. J. Protozool. 3 (Suppl.) 8 (1956).

1651. SOYER, M. O.: Rapports existant entre chromosomes et membrane nucléaire chez un Dinoflagellé parasite du genre *Blastodinium* CHATTON. C. R. Acad. Sci. (Paris) **268**, 2082—2084 (1969).

1652. — Etude ultrastructurale des inclusions vacuolaires chez *Noctiluca miliaris* SURIRAY, Dinoflagellé Noctilucidae et observations concernant leur rôle dans la genèse des trichocystes fibreux et muqueux. Protistologica **5**, 327—334 (1969).

1653. — Structure du noyau des *Blastodinium* (Dinoflagellés parasites). Division et condensation chromatique. Chromosoma (Berl.) **33**, 70—114 (1971).

1654. — Les ultrastructures nucléaires de la Noctiluque (Dinoflagellé libre) au cours de la sporogenèse. Chromosoma (Berl.) **39**, 419—441 (1972).

1655. STARR, R. C.: Sexuality in *Gonium sociale* (DUJARDIN) Warming. J. Tenn. Acad. Sci. **30**, 90—93 (1955).

1655*a. — Cellular differentiation in *Volvox*. Proc. nat. Acad. Sci. (Wash.) **59, 1082—1088 (1968).

1656. — Structure, reproduction and differentiation in *Volvox carteri* f. *nagariensis* IYENGAR, strains HK 9 & 10. Arch. Protistenk. **111**, 204—222 (1969).

1657.* — Control of Differentiation in *Volvox*. Develop. Biol. Suppl. **4, 59—100 (1970).

1658. STEHBENS, W. E.: The ultrastructure of *Lankesterella hylae*. J. Protozool. **13**, 63—73 (1966).

1659. STEIN, J. R.: A morphological and genetic study of *Gonium pectorale*. Amer. J. Bot. **45**, 664—672 (1958).

1660. — The four-celled species of *Gonium*. Amer. J. Bot. **46**, 366—371 (1959).

1661. — Sexual populations of *Gonium pectorale* (Volvocales). Amer. J. Bot. **52**, 379—388 (1965).

1662. — Growth and mating of *Gonium pectorale* (Volvocales) in defined media. J. Phycol. **2**, 23—28 (1966).

1663. — Effect of temperature on sexual populations of *Gonium pectorale* (Volvocales). Amer. J. Bot. **53**, 941—944 (1966).

1664. STEINERT, G., FIRKET, H., STEINERT, M.: Synthèse de l'acide désoxyribonucléique dans le corps parabasal de *Trypanosoma mega*. Exp. Cell Res. **15**, 632—635 (1958).

1665. STEINERT, M.: Etudes sur le déterminisme de la morphogénèse d'un trypanosome. Exp. Cell Res. **15**, 560—569 (1958).

1666. — Mitochondria associated with the kinetonucleus of *Trypanosoma mega*. J. biophys. biochem. Cytol. 8, 542—546 (1960).

1667. — LAURENT, M.: Structure moléculaire du DNA de *Trypanosoma mega*. J. Protozool. **17**, 38 (1970).

1668. — NOVIKOFF, A. B.: The existence of a cytostome and the occurrence of pinocytosis in the trypanosome, *Trypanosoma mega*. J. biophys. biochem. Cytol. **8**, 563—570 (1960).

1669. — STEINERT, G.: Synthèse de l'acide désoxyribonucléique au cours du cycle de division de *Trypanosoma mega*. J. Protozool. **9**, 203—211 (1962).

1670. — VAN ASSEL, S.: The loss of kinetoplastic DNA in two species of Trypanosomatidae treated with acriflavine. J. Cell Biol. **34**, 489—503 (1967).

1671. STEMPELL, W.: Über *Nosema bombycis* NÄG. nebst Bemerkungen über Mikrophotographie mit gewöhnlichem und ultraviolettem Licht. Arch. Protistenk. **16**, 281—358 (1909).

1672. STEVENS, A. R.: Machinery for exchange across the nuclear envelope. In: GOLDSTEIN, L.: The control of nuclear activity, pp. 189—271. Englewood Cliffs, N. J.: Prentice-Hall, Inc. 1967.

1673. STEWART, J. M., MUIR, A. R.: The fine structure of the cortical layers in *Paramecium aurelia*. Quart. J. micr. Sci. **104**, 129—134 (1963).

1674. STEY, H.: Electronenmikroskopische Untersuchung an *Labyrinthula coenocystis* SCHMOLLER. Z. Zellforsch. **102**, 387—418 (1969).

1675. STOCKEM, W.: Pinocytose und Bewegung von Amöben. I. Die Reaktion von *Amoeba proteus* auf verschiedene Markierungssubstanzen. Z. Zellforsch. **74**, 372—400 (1966).

1676. — Pinocytose und Bewegung von Amöben. III. Die Funktion des Golgiapparates von *Amoeba proteus* und *Chaos chaos*. Histochem. **18**, 217—240 (1969).

1677. — WOHLFARTH-BOTTERMANN, K.E., HABEREY, M.: Pinocytose und Bewegung von Amöben. V. Konturveränderungen und Faltungsgrad der Zelloberfläche von *Amoeba proteus*. Cytobiol. **1**, 37—57 (1969).

**1677a.* — — Pinocytosis (Endocytosis). In: LIMA DE FARIA, A. (Ed.): Handbook of Molecular Cytology. Amsterdam: North-Holland Publ. Comp. 1969.

1678. — — Zur Feinstruktur der Trichocysten von *Paramecium*. Cytobiol. **1**, 420—436 (1970).

1679. STONE, G.E., MILLER, O.L.: A stable mitochondrial DNA in *Tetrahymena pyriformis*. J. exp. Zool. **159**, 33—37 (1965).

1680. — — PRESCOTT, D.M.: H^3-thymidine derivate pools in relation to macronuclear DNA synthesis in *Tetrahymena pyriformis*. J. Cell Biol. **25**, 171—177 (1965).

1681. — PRESCOTT, P.M.: Cell division and DNA synthesis in *Tetrahymena pyriformis* deprived of essential amino acids. J. Cell Biol. **21**, 275—281 (1964).

1682. STÖSSEL-MÜGGE, E.: Über die Wirkung einer Röntgenbestrahlung während der Wachstumsphase auf die Entwicklung von *Eucoccidium dinophili*. Z. Naturforsch. **16**b, 598—604 (1961).

1683. STUART, K.D.: Evidence for the retention of kinetoplast DNA in an acriflavine-induced dyskinetoplastic strain of *Trypanosoma brucei* which replicates the altered central element of the kinetoplast. J. Cell Biol. **49**, 189—195 (1971).

1684. SUMMERS, F.M.: The division and reorganization of the macronuclei of *Aspidisca lynceus* MÜLLER, *Diophrys appendiculata* STEIN, and *Stylonychia pustulata* EHRBG. Arch. Protistenk. **85**, 173—208 (1935).

1685. — Some aspects of normal development in the colonial ciliate *Zoothamnium alternans* Biol. Bull. **74**, 41—55 (1938).

1686. SUMMERS, K.E., GIBBONS, I.R.: Adenosine triphosphate-induced sliding of tubules in trypsin-treated flagella of sea urchin sperm. Proc. nat. Acad. Sci. (Wash.) **68**, 3092—3096 (1971).

1687. SURZYCKI, S.J.: Genetic functions of the chloroplast of *Chlamydomonas reinhardi*. Effects of rifampin on chloroplast DNA-dependant RNA polymerase. Proc. nat. Acad. Sci. (Wash.) **63**, 1327—1334 (1969).

1688. — GILLHAM, N.W.: Organelle mutations and their expression in *Chlamydomonas reinhardi*. Proc. nat. Acad. Sci. (Wash.) **68**, 1301—1306 (1971).

1689. — GOODENOUGH, U.W., LEVINE, R.P., ARMSTRONG, J.J.: Nuclear and chloroplast control of chloroplast structure and function in *Chlamydomonas reinhardi*. Symp. Soc. exp. Biol. **24**, 13—37 (1970).

1690. SUYAMA, Y., PREER, J.R.: Mitochondrial DNA from Protozoa. Genetics **52**, 1051—1058 (1965).

1691. SUZUKI, S.: Conjugation in *Blepharisma undulans japonicus* SUZUKI, with special reference to the nuclear phenomena. Bull. Yamagata Univ. Nat. Sci. **4**, 43—84 (1957).

1692. — Morphogenesis in the regeneration of *Blepharisma undulans japonicus* SUZUKI. Bull. Yamagata Univ. Nat. Sci. **4**, 85—192 (1957).

1693. TAIT, A.: Altered mitochondrial ribosomes in an erythromycin resistant mutant of *Paramecium*. Febs Letters **24**, 117—120 (1972).

1694. TAKAGI, Y.: Sequental expression of sex-traits in the clonal development of *Paramecium multimicronucleatum*. Jap. J. Genet. **46**, 83—91 (1971).

1695. TAKAHASHI, M., HIWATASHI, K.: Disappearance of mating reactivity in *Paramecium caudatum* upon repeated washing. J. Protozool. **17**, 667—670 (1970).

**1696.* TALIAFERRO, W.H., STAUBER, L.A.: Immunology of protozoan infections. In: CHEN, T.-T. (Ed.): Research in Protozoology, Vol. 3, pp. 505—564. New York: Pergamon Press 1969.

1697. TAMM, S.L.: The effect of enucleation on flagellar regeneration in the protozoon *Peranema trichophorum*. J. Cell Sci. **4**, 171—178 (1969).

1698. TAMM, S. L.: Free kinetosomes in australian flagellates. I. Types and spatial arrangement. J. Cell Biol. **54**, 39—55 (1972).
1699. — Ciliary motion in *Paramecium*. A scanning electron microscope study. J. Cell Biol. **55**, 250—255 (1972).
1700. — HORRIDGE, G. A.: The relation between the orientation of the central fibrils and the direction of beat in the cilia of *Opalina*. Proc. roy. Soc. (Lond.) **175**, 219—233 (1970).
1701. TAMURA, S., TSURUHARA, R., WATANABE, Y.: Function of nuclear microtubules in macronuclear division of *Tetrahymena pyriformis*. Exp. Cell Res. **55**, 351—358 (1969).
**1702.* TARTAR, V.: The biology of Stentor. 413 S. New York-London: Pergamon Press 1961.
1703. — Induced division and division regression by cell fusion in *Stentor*. J. exp. Zool. **163**, 297—310 (1963).
1704. — Extreme alteration of the nucleocytoplasmic ratio in *Stentor coeruleus*. J. Protozool. **10**, 445—461 (1963).
1705. — Morphogenesis in homopolar tandem grafted *Stentor coeruleus*. J. exp. Zool. **156**, 243—252 (1964).
1706. — Fission and morphogenesis in a marine ciliate under osmotic stress. J. Protozool. **12**, 444—447 (1965).
1707. — Fission after division and primordium removal in the ciliate *Stentor coeruleus* and comparable experiments on reorganizers. Exp. Cell Res. **42**, 357—370 (1966).
1708. — Synchronization of oral primordia in *Stentor coeruleus*. J. exp. Zool. **161**, 53—62 (1966).
**1709.* — Morphogenesis in Protozoa. In: CHEN, T.-T. (Ed.): Research in Protozoology, Vol. 2, pp. 1—116. Oxford: Pergamon Press 1967.
1710. — Regeneration in situ of membranellar cilia in *Stentor coeruleus*. Trans. Amer. Microscop. Soc. **87**, 297—306 (1968).
1711. — Fission in heteropolar tandem grafts of the ciliate *Stentor coeruleus*. J. Protozool. **17**, 624—625 (1970).
1712. — Caffeine bleaching of *Stentor coeruleus*. J. exp. Zool. **181**, 245—252 (1972).
1713. — Anucleate *Stentors:* Morphogenetic and behavioral capabilities. Biology and radiobiology of anucleate systems I. Bacteria and animal cells, pp. 125—144. New York-London: Academic Press 1972.
1714. — PITELKA, D. R.: Reversible effects of antimitotic agents on cortical morphogenesis in the marine ciliate *Condylostoma magnum*. J. exp. Zool. **172**, 201—218 (1969).
1715. TAUB, S. R.: The genetics of mating type determination in syngen 7 of *Paramecium aurelia*. Genetics **44**, 541—542 (1959).
1716. — The effect of nuclear genes on nuclear differentiation in syngen 7. *Paramecium aurelia*. Genetics **47**, 990—991 (1962).
1718. — The genetic control of mating type differentiation in *Paramecium aurelia*. Genetics **48**, 815—834 (1963).
1719. — Unidirectional mating type changes in individual cells from selfing cultures of *Paramecium aurelia*. J. exp. Zool. **163**, 141—150 (1966).
1720. — Regular changes in mating type composition in selfing cultures and in mating type potentiality in selfing caryonides of *Paramecium aurelia*. Genetics **54**, 173—189 (1966).
1721. TAYLOR, C. V.: Demonstration of the function of the neuromotor apparatus in *Euplotes* by the method of microdissection. Univ. Calif Publ. Zool. **19**, 403—470 (1920).
1722. TAYLOR, D. L.: The nutritional relationship of *Anemonia sulcata* PENNANT and its dinoflagellate symbiont. J. Cell Sci. **4**, 751—762 (1969).
1723. — Idendity of zooxanthellae isolated from some pacific Tridacnidae. J. Phycol. **5**, 336—340 (1969).
1724.* — Chloroplasts as symbiontic organelles. Int. Rev. Cytol. **27, 29—64 (1970).
1725. TILNEY, L. G.: Studies on the microtubules in Helizoa. IV. The effect of colchicine on the formation and maintenance of the axopodia and the redevelopment of pattern in *Actinosphaerium nucleofilum* BARRETT. J. Cell Sci. **3**, 649—562 (1968).

1726. Tilney, L. G.: How microtubule patterns are generated. The relative importance of nucleation and bridging of microtubules in the formation of the axoneme of *Raphidiophrys*. J. Cell Biol. **51**, 837—854 (1971).
1727. — Byers, B.: Studies on the microtubules in Heliozoa. V. Factors controlling the organization of microtubules in the axonemal pattern in *Echinosphaerium* (*Actinosphaerium*) *nucleofilum*. J. Cell Biol. **43**, 148—165 (1969).
1728. — Porter, K. R.: Studies on the microtubules in Heliozoa. I. The fine structure of *Actinosphaerium nucleofilum* (Barrett), with particular reference to the axial rod structure. Protoplasma **60**, 317—344 (1965).
1729. — — Studies on the microtubules in Heliozoa. II. The effect of low temperature on these structures in the formation and maintenance of the axopodia. J. Cell Biol. **34**, 327—343 (1967).
1730. Togasaki, R. K., Levine, R. P.: Chloroplast structure and function in ac-20, a mutant strain of *Chlamydomonas reinhardi*. I. CO_2 fixation and ribolose-1,5-diphosphate carboxylase synthesis. J. Cell Biol. **44**, 531—539 (1970).
1731. Tokuyasu, K., Scherbaum, O. H.: Ultrastructure of mucocysts and pellicle of *Tetrahymena pyriformis*. J. Cell Biol. **27**, 67—81 (1965).
1732. Torch, R.: The nuclear apparatus of a new species of *Tracheloraphis* (Protozoa, Ciliata). Biol. Bull. **121**, 410—411 (1961).
1733. — Autoradiographic studies of nucleic acid synthesis in a gymnostome ciliate, *Tracheloraphis* sp. J. Cell Biol. **23**, 98 A (1964).
1734. Townes, M. M., Brown, D. E. S.: The involvement of pH, adenosine triphosphate, calcium, and magnesium in the contraction of the glycerinated stalks of *Vorticella*. J. cell. comp. Physiol. **65**, 261—270 (1965).
**1735.* Trager, W.: Intracellular parasitism and symbiosis. In: Brachet, J., Mirsky, A. E.: The Cell, Vol. 4, pp. 151—214. New York-London: Academic Press 1960.
**1736.* — The cytoplasm of Protozoa. In: Brachet, J., Mirsky, A. E.: The Cell, Vol. 6, pp. 81—137. New York-London: Academic Press 1964.
1737.* — Differentiation in Protozoa. J. Protozool. **10, 1—6 (1963).
1738.* — The kinetoplast and differentiation in certain parasitic Protozoa. Amer. Naturalist **99, 255—266 (1965).
**1739.* — Krassner, S. M.: Growth of parasitic Protozoa in tissue cultures. In: Chen, T.-T. (Ed.): Research in Protozoology, Vol. 2, pp. 357—382. Oxford: Pergamon Press 1967.
1740. — Rudzinska, M. A.: The riboflavin requirement and the effects of acriflavin on the fine structure of the kinetoplast of *Leishmania tarentolae*. J. Protozool. **11**, 133—145 (1964).
1741. Tschermak-Woess, E.: Extreme Anisogamie und ein bemerkenswerter Fall der Geschlechtsbestimmung bei einer neuen *Chlamydomonas*-Art. Planta **52**, 606—622 (1959).
1742. — Zur Kenntnis von *Chlamydomonas suboogama*. Planta **59**, 68—76 (1962).
1743. — Das eigenartige Kopulationsverhalten von *Chloromonas saprophila*, einer neuen Chlamydomonadacee. Öst. bot. Z. **110**, 294—307 (1963).
1744. Tucker, J. B.: Changes in nuclear structure during binary fission in the ciliate *Nassula*. J. Cell Sci. **2**, 481—498 (1967).
1745. — Fine structure and function of the cytopharyngeal basket in the ciliate *Nassula*. J. Cell Sci. **3**, 493—514 (1968).
1746. — Morphogenesis of a large microtubular organelle and its association with basal bodies in the ciliate *Nassula*. J. Cell Sci. **6**, 385—429 (1970).
1747. — Initiation and differentiation of microtubule patterns in the ciliate *Nassula*. J. Cell Sci. **7**, 793—821 (1970).
1748. — Spatial discrimination in the cytoplasm during microtubule morphogenesis. Nature (Lond.) **232**, 387—389 (1971).
1749. — Microtubules and a contracting ring of microfilaments associated with a cleavage furrow. J. Cell Sci. **8**, 557—571 (1971).
1750. — Development and deployment of cilia basal bodies and other microtubular organelles in the cortex of the ciliate *Nassula*. J. Cell Sci. **9**, 539—567 (1971).

1751. Tucker, J. B.: Microtubule-arms and propulsion of food particles inside a large feeding organelle in the ciliate *Phascolodon vorticella*. J. Cell Sci. **10**, 883—903 (1972).

1752. Tuffrau, M.: Les processus régulateurs de la "caryophtisis" du macronucleus de *Nassulopsis lagenula* Fauré-Fremiet, 1959. Arch. Protistenk. **106**, 201—210 (1962).

1753. — Les différenciations fibrillaires d'origine cinétosomienne chez les ciliés hypotriches. Arch. Zool. exp. gén. **105**, 83—96 (1965).

1754.* — Le phototropisme chez les protozoaires. Revue des données essentielles résultant des principaux travaux. Ann. Biol. **3, 267—281 (1965).

1755. Uhlig, G.: Entwicklungsphysiologische Untersuchungen zur Morphogenese von *Stentor coeruleus* Ehrbg. Arch. Protistenk. **105**, 1—109 (1960).

1756. — Der Gehäusebau bei *Metafolliculina andrewsi* (Ciliata, Heterotricha). Verh. dtsch. zool. Ges. München, S. 498—507 (1963).

1757. — Komnick, H., Wohlfarth-Bottermann, K. E.: Intrazelluläre Zellzotten in Nahrungsvakuolen von Ciliaten. Helgol. wiss. Meeresunters. **12**, 61—77 (1965).

1758. Uspenskaja, A. V.: On the mode of nutrition of vegetative stages of *Myxidium lieberkühni* (Bütschli). Acta Protozool. **4**, 81—88 (1966).

1759. — Ovchinnikova, L. P.: Quantitative changes of DNA and RNA during the life cycle of *Ichthyophthirius multifiliis*. Acta Protozool. **4**, 127—141 (1966).

1760. Vanderberg, J., Rhodin, J., Yoeli, M.: Electron microscopic and histochemical studies of sporozoite formation in *Plasmodium berghei*. J. Protozool. **14**, 82—103 (1967).

1761. Vandeberg, W. J., Starr, R. C.: Structure, reproduction and differentiation in *Volvox gigas* and *Volvox powersii*. Arch. Protistenk. **113**, 195—219 (1971).

1762. van Wagtendonk, W. J., Clark, J. A. D., Godoy, G. A.: The biological status of *lambda* and related particles in *Paramecium aurelia*. Proc. nat. Acad. Sci. (Wash.) **50**, 835—838 (1963).

1763. — Goldman, P. H., Smith, W. L.: The axenic culture of strains of the various syngens of *Paramecium aurelia*. J. Protozool. **17**, 389—391 (1970).

1764. Vavra, J., Joyon, L., de Puytorac, P.: Observation sur l'ultrastructure du filament polaire des microsporidies. Protistologica **2** (2), 109—112 (1966).

1765. — Small, E. B.: Scanning electron microscopy of gregarines (Protozoa, Sporozoa) and its contribution to the theory of gregarine movement. J. Protozool. **16**, 745—757 (1969).

1766. Vickerman, K.: The fine structure of *Trypanosoma congolense* in its bloodstream phase. J. Protozool. **16**, 54—69 (1969).

1767. — On the surface coat and flagellar adhesion in trypanosomes. J. Cell Sci. **5**, 163—193 (1969).

1768. — Preston, T. M.: Spindle microtubules in the dividing nuclei of trypanosomes. J. Cell Sci. **6**, 365—383 (1970).

1769. Vinckier, D.: Organisation ultrastructurale corticale de quelques Monocystidées, parasites du ver oligochète *Lumbricus terrestris* L. Protistologica **5**, 505—518 (1969).

1770. — Devauchelle, G., Prensier, G.: Etude ultrastructurale du développement de la microsporidie *Nosema vivieri* (V. D. et P., 1970). Protistologica **7**, 273—288 (1971).

1771. — Vivier, E.: Organisation ultrastructurale corticale de la grégarine *Monocystis herculea*. C. R. Acad. Sci. (Paris) **266**, 1737—1739 (1968).

1772. Vivier, E.: L'Organisation ultrastructurale corticale de la grégarine *Lecudina pellucida*; ses rapports avec l'alimentation et la locomotion. J. Protozool. **15**, 230—246 (1968).

1773. — Devauchelle, G., Petitprez, A., Porchet-Henneré, E., Prensier, G., Schrevel, J., Vinckier, D.: Observations de cytologie comparée chez les Sporozoaires. I. Les structures superficielles chez les formes végétatives. Protistologica **6**, 127—150 (1970).

1774. — Henneré, E.: Cytologie, cycle et affinités de la coccidie *Coelotropha durchoni*, nomen novum (= *Eucoccidium durchoni* Vivier), parasite de *Nereis diversicolor* O. F. Müller (Annélide Polychète). Bull. Biol. Fr. Belg. **98**, 153—206 (1964).

1775. — — Ultrastructure des stades végétatifs de la coccidie *Coelotropha durchoni*. Protistologica **1** (1), 89—104 (1965).

1776. VIVIER, E., PETITPREZ, A., CHIVÉ, A. F.: Observations ultrastructurales sur les Chlorelles symbiotes de *Paramecium bursaria*. Protistologica **3**, 325—334 (1967).

1777. — SCHREVEL, J., HENNERÉ, E.: Corrélations entre le cycle de quelques sporozoaires et le cycle de leurs hôtes (Annélides Polychètes). Arch. Zool. exp. gén. **102**, 231—238 (1963).

1778. — — Etude, au microscope électronique, d'une grégarine du genre *Selenidium* parasite de *Sabellaria alveolata* L. J. Mircoscopie **3**, 651—670 (1964).

1779. — — Les ultrastructures cytoplasmique de *Selenidium hollandei*, n. sp. Grégarine parasite de *Sabellaria alveolata* L. J. Microscopie **5**, 213—228 (1966).

1780. — — HENNERÉ, E.: L'ultastructure de l'enveloppe nucléaire et de ses pores chez des sporozaires. J. Microscopie **5**, 84a—85a (1966).

1781.* WALKER, P. J.: Reproduction and heredity in trypanosomes. Int. Rev. Cytol. **17, 51—98 (1964).

1782. WALSH, JR., R. D., CALLAWAY, C. S.: The fine structure of the gregarine *Lankesteria culicis*, parasitic in the yellow fever mosquito *Aedes aegypti*. J. Protozool. **16**, 536—545 (1969).

1783. WARNER, F. D.: The fine structure of *Rhynchocystis pilosa* (Sporozoa, Eugregarinida) J. Protozool. **15**, 59—73 (1968).

1784. WARR, J. R., MCVITTIE, A., RANDALL, J., HOPKINS, J. M.: Genetic control of flagellar structure in *Chlamydomonas reinhardi*. Genet. Res. **7**, 335—351 (1966).

1785. WASIELEWSKI, T. V., KÜHN, A.: Untersuchungen über Bau und Teilung des Amöbenkerns. Zool. Jb. Anat. **38**, 253—326 (1914).

1786. WATSON, M. R., ALEXANDER, J. B., SILVESTER, N. R.: The cilia of *Tetrahymena pyriformis*. Fractionation of isolated cilia. Exp. Cell Res. **33**, 112—129 (1964).

1787. WATTERS, C.: Studies on the motility of the Heliozoa. II. The locomotion of *Actinosphaerium eichhorni* and *Actinophrys* sp. J. Cell Sci. **3**, 231—244 (1968).

1788. WEBB, T. L., FRANCIS, D.: Mating types in *Stentor coeruleus*. J. Protozool. **16**, 758—763 (1969).

1789. WEBER, H.: Über die Paarung der Gamonten und den Kerndualismus der Foraminifere *Metarotaliella parva* GRELL. Arch. Protistenk. **108**, 217—270 (1965).

1790.* WEISZ, P. B.: Morphogenesis in Protozoa. Quart. Rev. Biol. **29, 207—229 (1954).

1791. WESSENBERG, H.: Studies of the life cycle and morphogenesis of *Opalina*. Univ. Calif. Publ. Zool. **61**, 315—370 (1961).

1792. — Observations on cortical ultrastructure in *Opalina*. J. Microscopie **5**, 471—492 (1966).

1793. — ANTIPA, G.: Studies on *Didinium nasutum*. I. Structure and ultrastructure. Protistologica **4**, 427—448 (1968).

1794. — — Capture and ingestion of *Paramecium* by *Didinium nasutum*. J. Protozool. **17**, 250—270 (1970).

1795. WICHTERMAN, R.: Cytogamy: a sexual process occurring in living joined pairs of *Paramecium caudatum* and its relations to other sexual phenomena. J. Morph. **66**, 423—451 (1940).

**1796.* — The biology of *Paramecium*. 527 S. New York: Blakiston 1953.

1797. — Survival and reproduction of *Paramecium* after X-irradiation. J. Protozool. 8, 158—162 (1961).

1798. — Studies on *Euplotes*. I. Structure and life cycle of a new species of marine *Euplotes*. Biol. Bull. **123**, 516 (1962).

1799. — Studies on *Euplotes*. II. Mating types and conjugation in a marine species of *Euplotes*. Biol. Bull. **123**, 516—517 (1962).

1800. — Mating types, breeding system, conjugation and nuclear phenomena in the marine ciliate *Euplotes cristatus* KAHL from the Gulf of Naples. J. Protozool. **14**, 49—58 (1967).

1801. WIDMAYER, D. J.: A nonkiller resistant *kappa* and its bearing on the interpretation of *kappa* in *Paramecium aurelia*. Genetics **51**, 613—623 (1965).

1802. WIESE, L.: On sexual agglutination and mating type substances (gamones) in isogamous heterothallic *Chlamydomonas*. I. Evidence of the identity of the gamones with the surface components responsible for sexual flagellar contact. J. Phycol. **1**, 46—54 (1965).

*1802a. WIESE, L.: Algae. In: Ferilization, Vol. 12, New York-London: Academic Press 1969.
1803. — JONES, R. F.: Studies on gamete copulation in heterothallic Chlamydomonads. J. cell. comp. Physiol. **61**, 265—274 (1963).
1804. WILKIE, D.: Reproduction of mitochondria and chloroplasts. Symp. Soc. Gen. Microbiol. **20**, 381—400 (1970).
1805. WILLE, J. J.: Abnormal morphogenesis and altered cellular localization of DNA-like RNA in *Paramecium aurelia*. Genetics **50**, 294—295 (1964).
1806. — Induction of altered patterns of cortical morphogenesis and inheritance in *Paramecium aurelia*. J. exp. Zool. **163**, 191—214 (1967).
1807. WILLMER, E. N.: Amoeba-flagellate transformation. Exp. Cell Res. **8** (Suppl.) 32—46 (1961).
1808. WISE, B. N.: The morphogenetic cycle in *Euplotes eurystomus* and its bearing on problems of ciliate morphogenesis. J. Protozool. **12**, 626—648 (1965).
1809. — Effects of ultraviolet microbeam irradiation on morphogenesis in *Euplotes*. J. exp. Zool. **159**, 241—268 (1965).
1810. WISE, G. E., FLICKINGER, C. J.: Relation of the Golgi apparatus to the cell coat in amoebae. Exp. Cell Res. **61**, 13—23 (1970).
1811. — — Cytochemical staining of the Golgi apparatus in *Amoeba proteus*. J. Cell Biol. **46**, 620—626 (1970).
1811a. — STEVENS, A. R., PRESCOTT, D. M.: Evidence of RNA in the helices of *Amoeba proteus*. Exp. Cell Res. **75**, 347—352 (1972).
1812. WITMAN, G. B., CARLSON, K., BERLINER, J., ROSENBAUM, J. L.: *Chlamydomonas* flagella. I. Isolation and electrophoretic analysis of microtubules, matrix, membranes, and mastigonemes. J. Cell Biol. **54**, 507—539 (1972).
1813. — — ROSENBAUM, J. L.: *Chlamydomonas* flagella. II. The distribution of Tubulins 1 and 2 in the outer doublet microtubules. J. Cell Biol. **54**, 540—555 (1972).
1814. WITTMANN, H.: Untersuchungen zur Dynamik einiger Lebensvorgänge von *Amoeba sphaeronucleolosus* (GREE) bei natürlichem „Zeitmoment" und unter Zeitraffung. Protoplasma **40**, 23—47 (1951).
1815. WOHLFARTH-BOTTERMANN, K. E.: Experimentelle und elektronenoptische Untersuchungen zur Funktion der Trichocysten von *Paramecium caudatum*. Arch. Protistenk. **98**, 169—226 (1953).
1816. — Protistenstudien. X. Licht- und elektronenmikroskopische Untersuchungen an der Amöbe. *Hyalodiscus simplex n. sp.* Protoplasma **52**, 58—107 (1960).
*1817. — Cell structures and their significance for ameboid movement. Int. Rev. Cytol. **16**, 61—131 (1964).
1818. — Weitreichende, fibrilläre Protoplasmadifferenzierungen und ihre Bedeutung für die Protoplasmaströmung. III. Entstehung und experimentell induzierbare Musterbildungen. Wilhelm Roux' Arch. Entwickl.-Mech. Org. **156**, 371—403 (1965).
1819. — STOCKEM, W.: Pinocytose und Bewegung von Amöben. II. Permanente und induzierte Pinocytose bei *Amoeba proteus*. Z. Zellforsch. **73**, 444—474 (1966).
1820. WOHLMAN, A., ALLEN, R. D.: Structural organization associated with pseudopod extension and contraction during cell locomotion in *Difflugia*. J. Cell Sci. **3**, 105—114 (1968).
1821. WOLFE, J.: Structural analysis of basal bodies of the isolated oral apparatus of *Tetrahymena pyriformis*. J. Cell Sci. **6**, 679—700 (1970).
1822. WOLPERT, L.: Cytoplasmic streaming and amoeboid movement. Symp. Soc. gen. Microbiol. **15**, 270—293 (1965).
1823. — O'NEILL, C. H.: Dynamics of the membrane of *Amoeba proteus* studied with labelled specific antibody. Nature (Lond.) **196**, 1261—1266 (1962).
1824. WOLSTENHOLME, D. R.: Electronmicroscopic identification of the interphase chromosomes of *Amoeba proteus* and *Amoeba discoides* using autoradiography; with some notes on helices and other nuclear components. Chromosoma (Berl.) **19**, 449—468 (1966).
1825. WOODARD, J., GELBER, B., SWIFT, H.: Nucleoprotein changes during the mitotic cycle in *Paramecium aurelia*. Exp. Cell Res. **23**, 258—264 (1961).

1826. Woodard, J., Woodard, M., Gelber, B., Swift, H.: Cytochemical studies of conjugation in *Paramecium aurelia*. Exp. Cell Res. **41**, 55—63 (1966).

1827. Wunderlich, F., Speth, V.: The macronuclear envelope of *Tetrahymena pyriformis* GL in different physiological states. IV. Structural and functional aspect of nuclear pore complexes. J. Microsc. **13**, 361—382 (1972).

1828. Wurmbach, H.: Über die Beeinflussung des Wirtsgewebes durch *Aggregata octopiana* und *Klossia helicina*. Arch. Protistenk. **84**, 257—284 (1935).

1829. Yagiu, R., Shigenaka, Y.: Electron microscopy of the longitudinal fibrillar bundle and the contractile fibrillar system in *Spirostomum ambiguum*. J. Protozool. **10**, 364—369 (1963).

1830. — — Electron microscopy of the ectoplasm and the proboscis in *Didinium nasutum*. J. Protozool. **12**, 363—381 (1965).

1831. Yamaguchi, T.: Studies on the modes of ionic behavior across the ectoplasmic membrane of *Paramecium*. I. Electric potential differences measured by the intracellular microelectrode. J. Fac. Sci. Univ. Tokyo, Sec. IV **8**, 573—591 (1960).

1832. — Studies on the modes of ionic behavior across the ectoplasmic membrane of *Paramecium*. II. In- and outfluxes of radioactive calcium. J. Fac. Sci. Univ. Tokyo, Sec. IV **8**, 593—601 (1960).

1833. Younger, K. B., Banerjee, S., Kelleher, J. K., Winston, M., Margulis, L.: Evidence that the synchronized production of new basal bodies is not associated with DNA synthesis in *Stentor coeruleus*. J. Cell Sci. **11**, 621—637 (1972).

1834. Yudin, A. L., Sopina, V. A.: On the role of nucleus and cytoplasm in the inheritance of some characters in amoebae (nuclear transfer experiments). Acta Protozool. 8, 1—40 (1970).

1835. Yusa, A.: An electron microscope study on regeneration of trichocysts in *Paramecium caudatum*. J. Protozool. **10**, 253—262 (1963).

1836. — Fine structure of developing and mature trichocysts in *Frontonia vesiculosa*. J. Protozool. **12**, 51—60 (1965).

**1837.* Yuyama, S.: Cell transformation studies on the amoebo-flagellate *Naegleria gruberi*. From: Cameron, I. L., Padilla, G. M., Zimmerman, A. M. (Eds.): Developmental aspects of the cell cycle. 41—66 pp. New York-London: Academic Press 1971.

1838. Zahalsky, A. C., Hutner, S. H., Keane, M., Burger, R. M.: Bleaching *Euglena gracilis* with antihistamines and streptomycin-type antibiotics. Arch. Microbiol. **42**, 46—55 (1962).

1839. Zebrun, W., Corliss, J. O., Lom, J.: Electronmicroscopical observations on the mucocysts of the ciliate *Tetrahymena rostrata*. Trans. Amer. Microsc. Soc. **86**, 28—36 (1967).

1840. Zech, L.: Zytochemische Messungen an den Zellkernen der Foraminiferen *Patellina corrugata* und *Rotaliella heterocaryotica*. Arch. Protistenk. **107**, 295—330 (1964).

1841. — Cytochemical studies on the distribution of DNA in the macronucleus of *Stentor coeruleus*. J. Protozool. **13**, 532—534 (1966).

**1842.* Zeuthen, E., Rasmussen, L.: Synchronized cell division in Protozoa. In: Chen, T.-T. (Ed.): Research in Protozoology, Vol. 4, pp. 9—146. New York: Pergamon Press 1972.

16a. Allen, R. D., Francis, D., Zeh, R.: Direct test of the positive pressure gradient theory of pseudopad extension and retraction in amoebae. Science, **174**, 1237—1240 (1971).

173a. Brokaw, C. J.: Flaggelar movement: A sliding filament model. Science, **178**, 455—462 (1972).

1138a. Millecchia, L. L., Rudzinska, M. A.: The ultrastructure of nuclear division in a suctorian, *Tokophrya infusionum*. Z. Zellforsch. **115**, 149—164 (1971).

III. Films*

1. Protozoa in General

C 836. Mittelmeerplankton-Protozoen. (K. G. GRELL, 1961) $11^1/_2$ min.
[Pelagic Protozoa of the Mediterranean Sea].

2. Flagellata

C 883. Morphologie und Fortpflanzung der Phytomonadinen. (K. G. GRELL, 1964) $13^1/_2$ min.
[Morphology and reproduction of Phytomonadina].

E 1318. *Chlamydomonas reinhardii* (Volvocales)-Asexuelle Fortpflanzung (U. SCHLÖSSER, 1966/1967) 10 min.

E 656. *Gonium pectorale* (Phytomonadina). Ungeschlechtliche Fortpflanzung. (K. G. GRELL, 1963) $4^1/_2$ min.
[Asexual reproduction of *Gonium pectorale*].

E 657. *Pleodorina californica* (Phytomonadina). Ungeschlechtliche Fortpflanzung. (K. G. GRELL, 1963) $4^1/_2$ min.
[Asexual reproduction of *Pleodorina californica*].

E 566. *Hertwigella volvocicola* (Rotatoria). Parasitismus bei *Volvox aureus*. (K. G. GRELL, 1963) $8^1/_2$ min.
[Parasitism in *Volvox aureus* of *Hertwigella volvocicola* (Rotatoria)].

E 1634. *Dissodinium lunula* (Dinophyceae). — Vegetative Vermehrung (G. DREBES, 1969) $11^1/_2$ min.
[Vegetative reproduction of *Dissodinium lunula* (Dinophyceae)].

C 1069. Entwicklung des Ektoparasiten *Dissodinium pseudocalani* (Dinophyceae). (G. DREBES, 1970) 10 min.
[Development of the ectoparasite *Dissodinium pseudocalani* (Dinophyceae)].

C 897. Entwicklung von *Noctiluca miliaris*. (G. UHLIG, 1965) 15 min.
[Development of *Noctiluca miliaris*].

3. Rhizopoda

C 942. Form und Bewegung freilebender Amöben. (K. G. GRELL, 1967) 11 min.
[Morphology and movement of free-living amoebae].

C 943. Nahrungsaufnahme und Fortpflanzung freilebender Amöben. (K. G. GRELL, 1967) 11 min.
[Ingestion of food and reproduction of free-living amoebae].

E 1171. *Amoeba proteus* (Amoebina). Nahrungsaufnahme und Fortpflanzung. (K. G. GRELL, 1967) 5 min.
[Ingestion of food and reproduction of *Amoeba proteus* (Amoebina)].

E 1169. *Hartmannella castellanii* (Amoebina). Nahrungsaufnahme und Fortpflanzung. (K. G. GRELL, 1967) 6 min.
[Ingestion of food and reproduction of *Hartmannella castellanii* (Amoebina)].

E 1170. *Naegleria gruberi* (Amoebina). Nahrungsaufnahme. (K. G. GRELL, 1967) 10 min.
[Ingestion of food of *Naegleria gruberi* (Amoebina)].

---

* All films marked E and C can be hired or bought from: Institut für den wissenschaftlichen Film *D-34 Göttingen*, Nonnenstieg 72, Federal Republic of Germany

All films marked E (Encyclopaedia cinematographica) can be hired from: The Pennsylvania State University, Audio-Visual Services, 6 Willard Building, *University Park*, Pa. 16802, USA

or: ECJA-EC Japan Archives, Shimonaka Memorial Foundation, Heibonsha Building, 4 Yonbancho, Chiyodaku, *Tokyo*, Japan

or: Stichting Film en Wetenschap, Hengeveldstraat 29, *Utrecht*, Netherlands

or: Bundesstaatliche Hauptstelle für wissenschaftliche Kinematographie, A-*1050 Wien*, Schönbrunner Straße 56 Austria

Some films are availabe for hire in France, Great Britain, Portugal, Switzerland, Brasil, Canada, Turkey.

E 407. *Paramoeba eilhardi* (Amoebina). Fortbewegung. (K. G. GRELL, 1960) 4 min. [Locomotion of *Paramoeba eilhardi* (Amoebina)].

E 1174. *Paramoeba eilhardi* (Amoebina). Bakterieninfektion des Zellkerns. (K. G. GRELL, 1967) $4^1/_2$ min. [Parasitic bacteria in the cell nucleus of *Paramoeba eilhardi* (Amoebina)].

W 130. Amibes ingérant des Algues oscillaires. (J. COMMANDON et P. DE FONBRUNE, 1935) $9^1/_2$ min. [*Thecamoeba verrucosa* ingesting blue algae].

W 129. Observation sur une Amibe (Acanthamoeba). (J. COMMANDON et P. DE FONBRUNE, 1935) 12 min. [Observation of *Acanthamoeba*].

E 1173. *Corallomyxa mutabilis*. Formwechsel. (K. G. GRELL, 1967) 4 min. [Morphology of *Corallomyxa mutabilis*].

C 876. Entwicklung von *Dictyostelium* (Acrasina). (G. GERISCH, 1963) $14^1/_2$ min. [Development of *Dictyostelium* (Acrasina)].

E 1172. *Labyrinthula coenocystis* (Protomyxoidea). (K. G. GRELL, 1967) 7 min.

C 1059. Morphogenese und Fortpflanzung beschalter Amöben (Testacea). (H. NETZEL, 1969) 11 min. [Morphogenesis and reproduction of testaceous amoebae (Testacea)].

C 1060. Form und Bewegung beschalter Amöben (Testacea). (H. NETZEL, 1969) 11 min. [Morphology and movement of testaceous amoebae (Testacea)].

E 1640. *Chlamydophrys minor* (Testacea). — Bewegung und Fortpflanzung. (H. NETZEL, 1969) 5 min. [Movement and reproduction of *Chlamydophrys minor* (Testacea)].

E 1641. *Difflugia oviformis* (Testacea). — Bewegung und Fortpflanzung. (H. NETZEL, 1969) 5 min. [Movement and reproduction of *Difflugia oviformis* (Testacea)].

E 1642. *Euglypha rotunda* (Testacea). — Bewegung und Fortpflanzung. (H. NETZEL, 1969) 6 min. [Movement and reproduction of *Euglypha rotunda* (Testacea)].

E 1643. *Arcella vulgaris var. multinucleata* (Testacea). — Bewegung und Fortpflanzung. (H. NETZEL, 1969) 9 min. [Movement and reproduction of *Arcella vulgaris var. multinucleata* (Testacea)].

E 1644. *Arcella dentata* (Testacea). — Bewegung und Fortpflanzung. (H. NETZEL, 1969) 10 min. [Movement and reproduction of *Arcella dentata* (Testacea)].

E 1645. *Centropyxis aculeata* (Testacea). — Bewegung und Fortpflanzung (H. NETZEL, 1969) 5 min. [Movement and reproduction of *Centropyxis aculeata* (Testacea)].

E 1646. *Lieberkühnia wagneri* (Testacea). — Bewegung und Fortpflanzung. (H. NETZEL, 1969) 6 min. [Movement and reproduction of *Lieberkühnia wagneri* (Testacea)].

C 801. Morphologie der Foraminiferen (K. G. GRELL, 1959) $4^1/_2$ min. [Morphology of Foraminifera].

C 802. Fortpflanzung der Foraminiferen. (K. G. GRELL, 1959) $14^1/_2$ min. [Reproduction of Foraminifera].

E 259. *Allogromia laticollaris* (Foraminifera). Nahrungsaufnahme. (K. G. GRELL, 1958) 3 min. [Ingestion of food of *Allogromia laticollaris* (Foraminifera)].

E 258. *Patellina corrugata* (Foraminifera). Fortpflanzung. (K. G. GRELL, 1958) 11 min. [Reproduction of *Patellina corrugata* (Foraminifera)].

C 627. *Actinosphaerium eichhorni* EHRBG. Bewegung, Defäkation, Plasmogamie, Verhalten nach Zentrifugieren und Pressen (W. u. G. KUHL, 1952) 18 min. [Movement, defecation, plasmogamy, behavior after centrifugation and compression of *Actinosphaerium eichhorni* EHRBG.].

E 648. *Actinosphaerium arachnoideum* (Heliozoa). Fortpflanzung. (K. G. GRELL, 1963) 8 min. [Reproduction of *Actinosphaerium arachnoideum* (Heliozoa)].

C 829. Morphologie der Radiolarien. (K. G. GRELL, 1960) 13½ min. [Morphology of Radiolaria].

4. Sporozoa

C 683. Die Entwicklung von *Eucoccidium dinophili*. (K. G. GRELL, 1954) 12 min. [Development of *Eucoccidium dinophili*].

E 1752. *Eimeria stiedai* (Sporozoa). — Exogene Phase. (U. DÜRR, 1970) 10½ min. [Exogenous phase of *Eimeria stiedai* (Sporozoa)].

E 485. *Isospora sylvianthina* (Sporozoa). Exogene Entwicklungsphase (Sporulation). (G. SCHWALBACH, K. G. LICKFELD, 1962) 13 min. [Exogenous developmental phase of *Isospora sylvianthina* (Sporozoa)].

E 1325. *Toxoplasma gondii*. — Entwicklung proliferativer Formen in Zellkulturen. (W. BOMMER, 1966/67) 13½ min. [Development of proliferative forms in cell cultures of *Toxoplama gondii*].

5. Ciliata

C 881. Morphologie der Ciliaten I. Holotricha. (K. G. GRELL, 1963) 9½ min. [Morphology of ciliates. I. Holotricha].

C 882. Morphologie der Ciliaten II. Spirotricha. Peritricha, Chonotricha, Suctoria. (K. G. GRELL, 1963) 11½ min. [Morphology of ciliates. II. Spirotricha, Peritricha, Chonotricha, Suctoria].

C 878. Fortpflanzung der Ciliaten. (K. G. GRELL, 1963) 10½ min. [Reproduction of ciliates].

W 437. Alimentation des Infusoires Ciliés. I. Nutrition des Ciliés Végétivores. (J. DRAGESCO, 1948—1958) 14½ min. [Ingestion of food by ciliated Infusoria. I. Herbivorous ciliates].

W 438. Alimentation des Infusoires Ciliés. II. Nutrition des Ciliés Histophages. (J. DRAGESCO, 1948—1958) 10 min. [Ingestion of food by ciliated Infusoria. II. Histophagous ciliates].

W 439. Alimentation des Infusoires Ciliés. III. Nutrition des Ciliés Gymnostomes Prédateus. (J. DRAGESCO, 1948—1958) 25½ min. [Ingestion of food by ciliated Infusoria. III. Predatory gymnostomatous ciliates].

W 440. Alimentation des Infusoirex Ciliés. IV. Nutrition des Tentaculifères (Acinétiens). (J. DRAGESCO, 1948—1958) 13½ min. [Ingestion of food by ciliated Infusoria. IV. Suctoria].

E 649. *Metafolliculina andrewsi* (Ciliata). Fortpflanzung. (G. UHLIG, 1963) 12 min. [Reproduction of *Metafolliculina andrewsi* (Ciliata)].

C 903. Morphogenese der Folliculiniden (Ciliata). I. Morphologie und Zellteilung. (G. UHLIG, 1965) 8 min. [Morphogenesis of folliculinids (Ciliata). I. Morphology and cell division].

C 904. Morphogenese der Folliculiniden (Ciliata). II. Gehäusebau und Reorganisation. (G. UHLIG, 1965) 12 min. [Morphogenesis of folliculinids (Ciliata). II. Formation of lorica and reorganization].

C 912. Morphologie der Suktorien. (K. G. GRELL, 1964/65) 9½ min. [Morphology of Suctoria].

C 913. Fortpflanzung der Suktorien. (K. G. GRELL, 1964/65) 14 min. [Reproduction of Suctoria].

C 907. Parasiten und Räuber von *Ephelota gemmipara* (Suctoria). (K. G. GRELL, 1965) $9^1/_2$ min.
[Parasites and predators of *Ephelota gemmipara* (Suctoria)].

E 1017. *Ephelota gemmipara* (Suctoria). Nahrungsaufnahme und Fortpflanzung. (K. G. GRELL, 1965) 13 min.
[Ingestion of food and reproduction of *Ephelota gemmipara* (Suctoria)].

E 914. *Acineta tuberosa* (Suctoria). Nahrungsaufnahme und Schwärmerbildung. (K. G. GRELL, 1965) 10 min.
[Ingestion of food and swarmer formation of *Acineta tuberosa* (Suctoria)].

E 913. *Tokophrya lemnarum* (Suctoria). Nahrungsaufnahme und Schwärmerbildung. (K. HECKMANN, 1964/65) 11 min.
[Ingestion of food and swarmer formation of *Tokophyra lemnarum* (Suctoria)].

O. Subject Index

P. Genera and Species

W. B. Vernberg and F. J. Vernberg

Environmental Physiology of Marine Animals

With 116 figures
X, 346 pages. 1972
Cloth DM 62,40
US $ 23.10
ISBN 3-540-05721-8

Prices are subject
to change without notice

Springer-Verlag
Berlin
Heidelberg
New York
München · London · Paris
Sydney · Tokyo · Wien

The sea represents the last unknown frontier on Earth. It is an area vast in size with a rich diversity in biota and habitats, and undeniably is one of man's greatest natural resources. Competition for the resource of the marine environment is already keen in coastal and estuarine areas, and is destined to increase in the future. Furthermore, the attendant problems of competition for this finite resource are international in scope.

Within the past few years there has been an increasing awareness of man's environmental impact on the marine environment. This awareness has sharpened the realization that if man is to be able to intelligently understand and manage this great resource, then it is imperative to have an understanding of the physiological capabilities of marine animals to adapt to both normal environmental fluctuations and man-induced factors. This book emphasizes the environmental physiology of the marine fauna representing the intertidal zone, estuaries, coastal and oceanic waters, and the benthic regions. The responses of the whole organism as well as its component organ systems to various ecological factors are examined Toleration physiology, as influenced by such factors as temperature, salinity, oxygen and pollutants, is discussed. Capacity adaptations of marine animals living in different regions of the oceans are discussed in terms of how they perceive their environment, find and use available food, respond metabolically to fluctuations in their environment, utilize their circulatory system and blood system to meet environmental stress, how osmoregulatory functions correlate with habitat preference, and how they find a mate and sucessfully reproduce.

Alfred Kühn Lectures on Developmental Physiology

Translated from the second revised and expanded German edition and edited by **Roger Milkman,**

620 illustrations
XVI, 535 pages. 1971
Cloth DM 68,–

Prices are subject to change without notice

Springer-Verlag
Berlin
Heidelberg
New York
München · London · Paris
Tokyo · Sydney

From the reviews of the second German Edition:

"In the ten years since the first edition a whole generation of new problems has arisen. This book covers both generations. Each of the thirty six lectures is an attempt at a simple exposition of an area in developmental biology. The breadth of the coverage is unusual. It includes animal and plant development and extends from genes to whole organisms." *The Quarterly Review of Biology*

"When the first edition of Kühn's book appeared in 1955 it was widely hailed as a great scientific event. It was and still is the only book in the world literature which presents the facts and theories of experimental embryology in a coherent extensive and critical fashion." *Biologisches Zentralblatt*

". . . Thanks to his decades of experience as a research worker and teacher, Alfred Kühn has succeeded in making a selection from the wealth of material available that not only covers all the significant lines of the physiology of development but at the same time reveals the nature of the problems of this branch of research. It provides experienced research workers and teachers with a valuable supplementation of their own work and young biologists with a guide to the immense wealth of knowledge that has been accumulated." *Mundus*